MEASURE AND INTEGRAL

This is a volume in
PROBABILITY AND MATHEMATICAL STATISTICS
A Series of Monographs and Textbooks

Editors: Z. W. Birnbaum and E. Lukacs

A complete list of titles in this series appears at the end of this volume.

MEASURE AND INTEGRAL

KONRAD JACOBS

Mathematisches Institut
Universität Erlangen–Nürnberg
Erlangen, Federal Republic of Germany

With an Appendix by
Jaroslav Kurzweil
Prague, Akademie Ved CSSR

ACADEMIC PRESS New York San Francisco London 1978
A Subsidiary of Harcourt Brace Jovanovich, Publishers

ACADEMIC PRESS, INC.
111 Fifth Avenue, New York, New York 10003

United Kingdom Edition published by
ACADEMIC PRESS, INC. (LONDON) LTD.
24/28 Oval Road, London NW1 7DX

Library of Congress Cataloging in Publication Data

Jacobs, Konrad, Date
Measure and integral.

(Probability and mathematical statistics series ;)
1. Measure theory, 2. Integrals, Generalized.
I. Title.
QA312.J23 515'.42 78-210
ISBN 0-12-378550-2

AMS (MOS) 1970 Subject Classifications: 28-02, 28 A 30, 52 A 05, 60 G 05

PRINTED IN THE UNITED STATES OF AMERICA

CONTENTS

Chapter III

EXTENSION OF POSITIVE σ- AND τ-MEASURES, AFTER DANIELL

Chapter IV

TRANSFORM OF σ-CONTENTS

Chapter V

CONTENTS AND MEASURES IN TOPOLOGICAL SPACES. PART I: REGULARITY

Chapter VI

CONTENTS AND MEASURES IN PRODUCT SPACES

Chapter VII

SET FUNCTIONS IN GENERAL

Chapter VIII

THE VECTOR LATTICE OF SIGNED CONTENTS

Chapter IX

THE VECTOR LATTICE OF SIGNED MEASURES

Chapter X

THE SPACES L^p

Chapter XI

CONTENTS AND MEASURES IN TOPOLOGICAL SPACES. PART II: THE WEAK TOPOLOGY

Chapter XII

THE HAAR MEASURE ON LOCALLY COMPACT GROUPS

Chapter XIII

SOUSLIN SETS, ANALYTIC SETS, AND CAPACITIES

Chapter XIV

ATOMS, CONDITIONAL ATOMS, AND ENTROPY

Chapter XV

CONVEX COMPACT SETS AND THEIR EXTREMAL POINTS

Chapter XVI

LIFTING

Appendix A

THE PERRON–WARD INTEGRAL AND RELATED CONCEPTS Jaroslav Kurzweil

Appendix B

CONTENTS WITH GIVEN MARGINALS

PREFACE

Although there is an abundance of excellent textbooks on measure theory and an output of more than four books on that subject from Erlangen alone (HAUPT–AUMANN–PAUC [1], BAUER [4, 5], ANGER–BAUER [1], KÖLZOW [4]), there seems to be no graduate text and reference book that covers all the material needed by a senior student of the field or a research worker in probability, ergodic theory, or functional analysis. The purpose of the present book is to fill this gap at least to a certain extent. I dare say this although a number of topics have deliberately not been included in the present volume.

Anyone who considers the immense wealth of present-day measure theory and its rapid adaptation to more and more purposes will realize that it is nearly impossible, if not undesirable, to present the theory in a form from which one can draw without thinking. Only a thorough training in the basic ideas and their applications can provide a reader with the capability he needs. I have therefore resolved to display every basic idea once or twice in the full proof of some fundamental theorem and to put many important applications, special cases, and extensions in the form of additional remarks and, in particular, of exercises. The reader is therefore invited to pay special attention to the exercises. Experience will show whether this resolution was wise in itself and appropriately realized in detail.

In spite of this basic decision I have tried to prepare the text in such a fashion that a solid nucleus of the story can be simply read from the definitions, lemmas, propositions, and theorems.

The great variety of definitions in measure theory has forced me to make some decisions concerning notation and assumptions. I have always hated to be forced to scan hundreds of pages preceding a theorem in order to find the definitions of all the relevant notation and assumptions. The reader will find that I have been very patient in stating assumptions and notations over and over again. Moreover, I have made up my mind to use a somewhat lengthy notation so as to remind the reader of the key words. When I had to decide between a very succinct notation like $\mathcal{M}^{\mathrm{b}}$ and a compound like loc $\mathrm{mble}^{\mathrm{b}}(\Omega, \mathcal{B}^{00}, \mathbb{R})$, after some deliberation I chose the latter since it quickly tells the reader more about "where he is." Also, it proved to be necessary to choose among types of set systems and set functions. I found it feasible to avoid the values $\pm\infty$ of set functions when possible. Thus the notion of a local σ-ring (also called δ-ring) came into focus. This brought me into a desirable neighborhood of one of the most well-developed branches of the theory: measure theory in locally compact spaces. Another decision was to abstain entirely from the term *measure* as meaning a σ-additive set function. I have rigorously used the word *content* (resp. σ-*content*) for such functions since extensions to various types of set systems, however fundamental they may be, are routine nowadays. The term *measure*, in conformity with Bourbaki, is strictly reserved for linear forms throughout this book. Throughout, I have tried to emphasize that σ-measures and σ-contents are merely two aspects of the same matter.

It may be useful to acquaint the reader with some details of the notation employed in this book. The "local" viewpoint is signaled mostly by an upper index ^{00}. Thus $\mathcal{B}^{00}$ is the symbol for a local σ-ring ($=\delta$-ring), and $\mathcal{C}^{00}(\Omega, \mathbb{R})$ denotes the continuous real functions vanishing outside compact sets for a topological space Ω. Wherever an upper index ^{00} seems unfitting, a prefix loc is used. Shortened words frequently figure as symbols; thus cont denotes the space of contents (and cont^{σ} the space of σ-contents), meas the space of measures, and mble the space of measurable functions. It was, of course, undesirable to replace all well-established symbols by such agglomerations. Still, $\mathcal{C}$ refers to continuous mappings and $\mathcal{L}$ integrable functions. The indicator ($=$characteristic) function of a set E is denoted by 1_E. I indulge frequently in a certain abuse of notation, which can be exemplified by saying that "$\mathcal{E} \ni f_1 \geq f_2 \geq \cdots \searrow 0$" means as much as "$f_1, f_2, \ldots \in \mathcal{E}, f_1 \geq f_2 \geq \cdots, \lim_{n\to\infty} f_n = 0$ pointwise." The arrow $\Rightarrow$ means "implies," and the double arrow $\Leftrightarrow$ means "iff" (= "if and only if"). I distinguish strictly between sets and families; a set of sets is often called a set system.

σ stands for countable operations and countable continuity properties, and τ stands for the corresponding uncountable items. By tradition, σ also stands for countable unions, and δ for countable intersections; F_σ sets are countable unions of closed sets, and G_δ sets are countable intersections of open sets (in a topological space, of course). $\mathbb{R}$ denotes the set of all real numbers, $\mathbb{Q}$ the set of all rational numbers, $\mathbb{Z}$ the set of all integers, and $\mathbb{N}$ the set $\{1, 2, \ldots\}$ of all natural numbers.

The book falls into two parts. Chapters I–III are meant to have the character of a graduate textbook, and students are invited to work through these carefully. Afterward they should know by heart, e.g., how the machinery of σ-extension works and how a σ-content is derived from a σ-measure. From Chapter IV onward a more summary and advanced style is employed, and various branches of measure and content theory are presented in a way that tries to lead quickly to basic information but trusts more than in Chapters I–III in the ability of the reader to fill in obvious details. I would be very much satisfied if there were some mathematicians who found some of Chapters IV–XVI to be a concise introduction to the respective topic.

The contents of the book may be sketched as follows. After the preliminary Chapter 0 on general mathematical notions and notations, the text proper begins with Chapter I, where are introduced the notions, on the one hand, of rings, fields, σ-rings, σ-fields, and, above all, local σ-rings, and, on the other hand, of positive contents and σ-contents defined on these objects. The extension problem is touched upon in a section on Eudoxos extension, and then fully formulated; the treatment includes the proof of uniqueness of an extension (after Dynkin). Next are considered positive measures and σ-measures on elementary domains of real functions (= vector lattices satisfying Stone's axiom). The extension problem is tackled once again, in the form of Riemann integration, and then more clearly formulated. The classical Hahn–Banach theorem is proved and Banach limits are introduced in the form of a major exercise. Chapter II takes up Carathéodory's extension theory for σ-contents, along with some routine techniques for further extensions, such as passing from a local σ-ring to the generated σ-ring or σ-field, and completion.

In Chapter III is presented the Daniell extension theory for positive σ-measures (= positive σ-continuous linear forms on elementary domains). The basic theorems of integration theory are proved. It is shown how to pass forth and back between σ-contents and σ-measures. The extension theory for τ-measure is sketched. The Hilbert space L^2 is introduced. Measurable functions and their integration are studied. Integration of complex-valued functions is treated.

Chapter IV deals with the transform of σ-contents and σ-measures by measurable mappings and kernels. The rudiments of ergodic theory—in particular, ergodic theorems—are given.

In Chapter V are studied contents and measures in topological spaces from the viewpoint of regularity. An abstract regularity theory is given and applied to the standard cases of compact, locally compact, and Polish spaces.

Chapter VI deals with contents and measures in product spaces. It turns out that employment of local σ-rings makes all assumptions about σ-finiteness obsolete so long as finite product spaces are considered. With the theory of infinite product spaces probability theory enters. The existence of infinite product probabilities is proved, as well as Kolmogorov's theorem about the extension of compatible marginal contents. The probabilistic theory of independence is given. The last section deals with Markovian semigroups and path spaces.

Chapter VII deals with rather general set functions. The values $\pm\infty$ are occasionally allowed for these (as in part of Chapter II). It is shown how to split a superadditive function into an additive and a purely superadditive part, and an additive one into a σ-additive and a purely additive part. The Vitali–Hahn–Saks theorem is proved, and total variation is examined.

Chapter VIII is devoted to a thorough study of the vector lattice of signed contents. How to carry out the vector lattice operations is shown, as is how to obtain Hahn and Lebesgue decompositions. The Radon–Nikodym theorem is proved for a σ-content of finite total variation that is absolutely continuous with respect to an arbitrary positive σ-content, and an integrable density is obtained. With this as a basis, probability theory is considered next; conditional expectations are introduced and the rudiments of martingale theory, including the submartingale convergence theorem, are presented. In the last problem is discussed the question of proving a Radon–Nikodym theorem without the assumption of finite total variation; disjoint bases (which always exist in the locally compact case) and locally integrable functions are introduced, and the Radon–Nikodym theorem is proved under the assumption that a disjoint base exists.

In Chapter IX Riesz band decompositions in general vector lattices are studied and applied to vector lattices of signed measures. Absolute continuity for signed measures is introduced in a purely vector lattice theoretical fashion: Lebesgue decompositions are obtained. In the last section these results are linked with the corresponding results about signed contents (Chapter VIII).

Chapter X presents, with the Radon–Nikodym theorem as a basic tool, the theory of the spaces L^p and their duality.

Chapter XI is concerned with the weak topology for measures and contents in topological spaces. A general theory is followed by applications to Polish spaces, which are significant in probability theory.

In Chapter XII Haar measure in locally compact groups is studied. The compact case is treated separately, as a prelude, so to speak; it is shown that continuous functions on a compact group are almost periodic, and a combinatorial theorem is used in order to establish a mean value for almost periodic functions, and thus a two-sided invariant Haar measure, on compact groups. The idea of using a generalized marriage theorem also in the locally compact case goes back to Kakutani and von Neumann and is carried out in Halmos' handwritten notes of von Neumann's lectures on measure theory delivered in 1940–1941. I have found this approach too unwieldy and instead followed HEWITT–ROSS [1], for the locally compact case, who in turn follow CARTAN [1]. In this way, Zorn's lemma is completely avoided. Examples of Haar measures are given. We do not penetrate more deeply into Haar measure theory, but the interested reader is referred to HALMOS [4], BERBERIAN [1], LOOMIS [2], HEWITT–ROSS [1], and NACHBIN [1].

Souslin sets, analytic sets, and the extension of capacities are the theme of Chapter XIII. Kuratowski's theorem on the measurability of "forward" images of measurable sets, the continuum hypothesis for Borel sets in Polish spaces, Choquet's capacitability theorem, and the measurable choice theorem are the main results. Following MEYER [2], we present a capacity approach to σ-content extension.

Chapter XIV is devoted to the Hanen–Neveu theory of atoms and conditional atoms and an entropy theory that is free of regularity assumptions. Most of the applications lie in ergodic theory but are not treated here.

Chapter XV starts with an introduction to the theory of topological vector spaces and proves the Krein–Milman theorem and Choquet's barycentric representation theorem, including simplexes and uniqueness. I have tried to present the quickest approaches to these matters, thus discarding a good portion of the beautiful machinery to be found in CHOQUET [5]. As an application of the Krein–Milman theorem, Lyapunov's convexity theorem is given. Choquet's theorem is applied to ergodic decompositions. Bochner's theorem on positive definite functions would provide a beautiful application of Choquet's theory, but is left out here for technical reasons (see CHOQUET [5]). The rest of the chapter is devoted to Choquet's theory of barycentric representation in convex compact sets.

The Krein–Milman theorem is one of the tools of lifting theory as treated here. Its rudiments, along with some of their applications, are presented in Chapter XVI. The existence of liftings in the presence of a disjoint base is proved, and the existence of strong lifting under separability assumptions follows. Kernel representations of linear mappings, conditional probabilities, disintegration of measures, and Strassen's theorem are among the applications set forth. The reader is referred to IONESCU–TULCEA [1] for further studies.

The book has two appendixes. Appendix A, an outline of the theory of the Perron–Ward integral and related topics, was kindly contributed by J. Kurzweil (Prague), the outstanding specialist in the field. Appendix B, by myself, treats the topic of contents and σ-contents with given marginals along the lines of a recent paper by HANSEL–TROALLIC [1], who introduce the max-flow–min-cut theorem of Ford and Fulkerson in order to obtain most of the basic theorems in an easy way.

General topology and functional analysis have been confined to a subsidiary role throughout this book. However, while the reader is assumed to have a solid background (the specifics are given in Section 0.2), functional analysis is treated in some detail when it is needed. Thus the Hahn–Banach theorem in its classical form is proved as Theorem I.6.1, and Section XV.1 serves as a rapid course in topological vector spaces, including the Krein–Milman theorem. I have tried, however, to restrict the use of general topology in Chapters I–III to a few points, and preferred substituting $\mathbb{R}^n$ for a general locally compact space in many of the examples.

It can be seen from this outline that some rudiments of topological vector space theory are contained in this book. They are proved where needed, but not too many of their consequences are followed up since they play only a subsidiary role here.

There is a considerable list of topics in measure theory that have deliberately not been included in this volume: measure and category (see OXTOBY [2, 3]), measure algebras (see CARATHÉODORY [2], KAPPOS [1, 2]), partitions in Lebesgue spaces (ROKHLIN [1]), vector-valued measures (see DINCULEANU [1]) and integration of vector-valued functions (see DINCULEANU [1] and IONESCU–TULCEA [1]), contents and measures in topological vector spaces, and other themes. They were left out for lack of time and space. Moreover, I think that this book enables a reader to handle any one of the monographs and papers mentioned above. I beg the reader to test his knowledge by reading classics like HAUSDORFF [1], CARATHÉODORY [1], and SAKS [3, 4], reputed textbooks like HALMOS [4], BERBERIAN [1], HAUPT–AUMANN [1], BAUER [4, 5], and ANGER–BAUER [1], historically important papers listed in the bibliography, and monographs or treatises like IONESCU–TULCEA [1], KÖLZOW [4], DUNFORD–SCHWARTZ [1], CHOQUET [5], PARRY [1], FREMLIN [1], and BICHTELER [1].

Note added in proof: (1) The most exciting result of 1978 is: V. Losert (Wien), An example of a compact Hausdorff space with no strong lifting, *Math. Annalen* (to appear). (2) A fine little textbook has appeared: H. Michel, "Mass- und Integrationstheorie I." VEB Deutscher Verlag der Wissenschaften, Berlin, 1978. (3) A reader interested in vector–valued measures should consult the book: J. Diestel and J. J. Uhl, Jr., "Vector Measures." American Math. Soc., Providence, Rhode Island, 1977.

Any comments and corrections will be highly appreciated.

Reference numbers and headings follow the following scheme: The book is divided into a chapter entitled "Basic Notions and Notations," which bears the number 0, and sixteen subsequent chapters, labeled with the Roman numbers I–XVI. Each chapter falls into at least two sections—1, 2, etc.; in some cases these split into further subsections. The definitions, remarks, exercises, lemmas, propositions, theorems, etc. that make up a section are numbered in the order of their appearance;

each of their numbers is preceded by the number of the section, and followed by the word *definition*, etc. Thus Section 2 consists, e.g., of **2.1. Definition, 2.2. Remark, 2.3. Lemma, 2.4. Theorem, ...** A definition, theorem, etc., may itself have several sections. These are numbered by the double number of the definition followed by simple or double "local" numbers in ordinary print. Often the numbers denoting the sections of a proof of a theorem parallel these ordinary numbers in the theorem. Thus a **2.4. Theorem** may have sections **2.4.1.1, 2.4.1.2, 2.4.2.1, . . .** , and the proof is then subdivided into 1.1, 1.2, 2.1, ...

When citing, we let the word definition, etc., precede all numbers. The numbers refer to the present chapter if no roman number appears. Otherwise the roman numeral is included. Thus it should be clear where to find Theorem IX.2.4.1.1. It is the same as Theorem 2.4.1.1 if we are in Chapter IX.

I thank J. Kurzweil (Prague) for his contribution of an appendix to this book. I thank D. Kölzow and J. Lembcke (both Erlangen) for the information that V. Losert (Vienna) has constructed recently an example of a compact measure space that has no strong lifting. My thanks go to E. Lukacs and Academic Press for accepting this book in spite of its size. Academic and the printer are to be congratulated for their careful editorial and typsetting work. My special thanks go to Frau H. Zech who carefully and swiftly typed the manuscript from my handwritten notes.

CHAPTER 0

BASIC NOTIONS AND NOTATION

In this preliminary chapter we present the basic facts about sets, families, orderings, etc. which will be used throughout the book without particular citation. No proofs are given and no completeness is aimed at. The purpose of this chapter is only to fix certain standards of wording in such a way that it will be easy for us and the reader to adapt the wording to all situations which might occur. The reader is advised to read this chapter and to check his own understanding of it carefully before he sets out to go ahead with any of the subsequent chapters.

1. SETS, RELATIONS, AND MAPPINGS

Throughout this book we rely on what is nowadays called naïve set theory. For an account of it, see KAMKE [1], HALMOS [2, 3]. (For nonnaïve (namely, axiomatic) set theory, see, e.g., TAKEUTI and ZARING [1, 2]; JECH [1].) The purpose of the present section is mainly to recall and standardize notation. We also rely on standard logic and use, e.g., the implication arrow $\Rightarrow$ in the usual way: If P, Q are statements, then $P \Rightarrow Q$ means "if P, then Q" or "Q whenever P." For this we write sometimes $P(Q)$.

A *set* is, naïvely, an entity that is a certain totality of other entities called its *elements* (sometimes, *points*), and the set is said to consist of its elements. To specify a set means to exactly specify which entities constitute its elements. This can be done by listing (possibly multiply) the elements or writing a precise description of them between braces. Thus $\{1\}$ is the set whose only element is the natural number 1, and so is, by the way, $\{1, 1\} = \{1, 1, 1\}$; $\{1, 2, 3, 4\}$ is the set whose elements are exactly the natural numbers 1, 2, 3 and 4; $\{1, 2, \ldots, 100\} = \{1, \ldots, 100\}$ is the set of all natural numbers not exceeding 100, and $\mathbb{N} = \{1, 2, \ldots\}$ is the set of all natural numbers. A set whose elements are sets again is also called a *set system.* A set that contains exactly one element is also called a *one-element set* or a *singleton.* It has turned out to be feasible to introduce a set that has no elements, the so-called *empty set* $\varnothing$. If M is a set with $M \neq \varnothing$, M is called *nonempty.*

If a is an element of the set M, we write $a \in M$ or $M \ni a$. The negation of this statement is expressed by $a \notin M$ or $M \not\ni a$. Sets are usually denoted by capital letters. If E, F are sets and every element of E is also an element of F, then E is also called a *subset* of F and we write $E \subseteq F$ or $F \supseteq E$. $E \subseteq F$ is also stated as: E is *contained* in F, or: E is *smaller* than F. Clearly $\varnothing \subseteq E$ holds for every set E; and $E \subseteq F$, $F \subseteq E$ is equivalent to $E = F$. The set of all subsets of F is also called the *power set* of E and denoted by $\mathscr{P}(E)$. If $E \subseteq F$ and there is an element of F that is not an element of E, then E is also called a *proper subset* of F, and we write $E \subset F$ or $E \subsetneqq F$ or $E \subseteq F$, $E \neq F$, the latter expression being nothing but a restatement of the definition of the notion of a proper subset.

A set E is often specified as a subset of another set F given in advance, e.g., as the set of all those elements of F having a certain property P which is defined for elements of F. A standard way of writing this is to write a "variable symbol" like x between an opening brace and a vertical bar (or a colon) and to write "$x \in F$, x has the property P" between the bar (resp. colon) and a closing brace. It is clear how to slightly modify or abbreviate this notational technique according to stylistic requirements. Thus

$$\begin{aligned}\{2, 4, 6, \ldots\} &= \{x \mid x \in \mathbb{N},\ x \text{ even}\} \\ &= \{n \mid n \in \mathbb{N},\ n \text{ is divisible by } 2\} \\ &= \{n \in N : n \text{ even}\}\end{aligned}$$

form a variety of usual descriptions of the set of all even natural numbers. If $\mathbb{Q}$ denotes the set of all rational numbers, we have, e.g.,

$$\mathbb{N} = \{x \mid x \in \mathbb{Q},\ x = a/b \text{ with } b = 1\},$$

and

$$\{x \mid x \in \mathbb{Q},\ 0 \leqq x \leqq 1\}$$

describes the set of all rationals not smaller than 0 and not larger than 1. The reader sees here how to bundle several properties into one. If $\mathbb{R}$ denotes the set of all real numbers, then

$$\{x \mid x \in \mathbb{R},\ x \geqq 0\}$$

describes the set of all nonnegative reals, which is also denoted by $\mathbb{R}_+$.

Similarly

$$\mathbb{R}_- = \{x \mid \mathbb{R} \ni x \leqq 0\}$$

denotes the set of all nonpositive reals; and, for any $a, b \in \mathbb{R}$,

$$[a, b] = \{x \mid x \in \mathbb{R}, a \leqq x \leqq b\}$$

denotes the set of all reals not smaller than a and not larger than b. In particular, using the "logical arrow" $\Rightarrow$ as an abbreviation for "implies," we have $a, b \in \mathbb{R}, a > b \Leftrightarrow [a, b] = \varnothing$. If $a \leqq b$, then $[a, b] \neq \varnothing$ and we call $[a, b]$ the *closed interval* from a to b. $[0, 1]$ is also called the (*closed*) *unit interval* (on $\mathbb{R}$). If $a < b$, then

$$]a, b[= \{x \mid x \in \mathbb{R}, a < x < b\}$$

is called the *open interval* from a to b, and

$$]a, b] = \{x \mid x \in \mathbb{R}, a < x \leqq b\}$$
$$[a, b[= \{x \mid x \in \mathbb{R}, a \leqq x < b\}$$

are called the *half-open intervals* from a to b ($]a, b]$ is left open and right closed, $[a, b[$ is left closed and right open). $\mathbb{C}$ is used as a notation for the set of all complex numbers, $\mathbb{Z}$ for the set of all integers, $\mathbb{Q}$ for the set of all rational numbers. The members of the set $\overline{\mathbb{R}} = \mathbb{R} \cup \{-\infty, \infty\}$ are called *extended real numbers*.

There are some obvious ways how to get some new sets from given ones.

Let E, F be two sets. Then the set $\{x \mid x \in E, x \in F\}$ is called the *intersection* of E and F and is denoted by $E \cap F$. Using the "logical arrows" $\Rightarrow$, $\Leftarrow$ in an obvious way and the double arrow $\Leftrightarrow$ as an abbreviation for "both $\Rightarrow$ and $\Leftarrow$" (verbally: "if and only if" = "iff"), we get the obvious statement $E \cap F = E \Leftrightarrow E \subseteq F$. Clearly $E \cap F \subseteq E$, $\subseteq F$ and $M \subseteq E$, $M \subseteq F \Rightarrow M \subseteq E \cap F$. In short: $E \cap F$ is the biggest set contained in both E and F. If $E \cap F \neq \varnothing$, we say that E and F have a nonempty intersection, or more concisely, they *intersect* or they *meet*. If $\mathscr{S}$ is a set of sets, i.e., a set whose elements are sets, then the set $\{x \mid x \in F$ for every $F \in \mathscr{S}\}$ is called the *intersection* of (all sets in) $\mathscr{S}$ and is denoted by $\bigcap_{F \in \mathscr{S}} F$ or $\bigcap \mathscr{S}$. If $E \cap F = \varnothing$, we say that E and F are *disjoint* or E is disjoint from F or vice versa. Every set E is disjoint from $\varnothing$.

Let E, F be two sets. Then the set $\{x \mid x \in E$ or $x \in F$ (or both)$\}$ is called the *union* of E and F and is denoted by $E \cup F$. We often write $E + F$ instead of $E \cup F$ if $E \cap F = \varnothing$ and extend this notion to similar situations in an obvious way. If $F = F_1 + \cdots + F_n$, we say that $F_1, \ldots, F_n$ form a *disjoint decomposition* of F. Obviously $E \subseteq F \Leftrightarrow E \cup F = F$. Clearly $E \subseteq E \cup F$, $F \subseteq E \cup F$ and $E \subseteq M$, $F \subseteq M \Rightarrow E \cup F \subseteq M$. In short: $E \cup F$

is the smallest set containing both E and F. If $\mathscr{S}$ is a set of sets, then the set $\{x \mid x \in F$ for at least one $F \in \mathscr{S}\}$ is called the *union* of (all sets in) $\mathscr{S}$ and is denoted by $\bigcup_{E \in \mathscr{S}} E$ or $\bigcup \mathscr{S}$. If Ω is a set and $\Omega \subseteq \bigcup_{E \in \mathscr{S}} E$, then $\mathscr{S}$ is said to *cover* Ω.

If $E, F \in \mathscr{S}$, $E \neq F \Rightarrow E \cap F = \varnothing$, and $\Omega = \bigcup \mathscr{S}$, then we say that $\mathscr{S}$ is a *disjoint decomposition* or a *partition* of Ω, and the sets $\varnothing \neq E \in \mathscr{S}$ are also called the *parts* or *atoms* of that decomposition. Let ξ, ζ be two partitions of Ω (lowercase greek letters are common symbols for partitions). We say that ξ is *finer* than ζ (or equivalently ζ is *coarser* than ξ) if every atom of ξ is contained in some atom of ζ (or equivalently every atom of ζ is a union of some atoms of ξ). For any two partitions α, β of Ω there is obviously a coarsest partition that is both finer than α and β. It is called the *common refinement* of α and β, and is denoted by $\alpha \vee \beta$; its atoms are all nonempty sets obtained by intersecting some atom of α and some atom of β. This carries easily over to arbitrary sets of partitions of Ω. There is also a finest partition that is coarser than α and β, and is denoted by $\alpha \wedge \beta$; we leave the identification of its atoms as an exercise for the reader as well as its generalization to arbitrary sets of partitions of Ω.

Let E, F be two sets. Then the set $\{x \mid x \in E, x \notin F\}$ is called the *difference* of E and F or the *complement* of F in E and is denoted by $E \backslash F$. If $E \supseteq F$, we speak of a *proper difference*. Clearly $E \backslash F \subseteq E$, $(E \backslash F) \cap F = \varnothing$ and $M \subseteq E$, $M \cap F = \varnothing \Rightarrow M \subseteq E \backslash F$. In short, $E \backslash F$ is the largest subset of E disjoint from F.

Fix any set Ω. Then for every $E \subseteq \Omega$, we also write $\complement E$ instead of $\Omega \backslash E$ and call it simply the *complement* of E. For every $\mathscr{S} \subseteq \mathscr{P}(\Omega)$, we denote the subset $\{\complement E \mid E \in \mathscr{S}\}$ of $\mathscr{P}(\Omega)$ by $\mathscr{S}_{\complement}$. We have the following (easily verifiable) duality relations between union and intersections

$$\complement \bigcup \mathscr{S} = \bigcap \mathscr{S}_{\complement}.$$

If $\mathscr{S} = \{E, F\}$, this boils down to $\complement(E \cup F) = \complement E \cap \complement F$, i.e.,

$$\Omega \backslash (E \cup F) = (\Omega \backslash E) \cap (\Omega \backslash F).$$

Let M, N be two nonempty sets. We shall abbreviate this statement also by "let $M, N \neq \varnothing$," implying tacitly that we are speaking of sets. Assume that to every $x \in M$ a unique element $f(x) \in N$ is attached. We then say that a *mapping* f or $f(\cdot)$ of M into N or an N-valued function f on M is given and write $f\colon M \to N$, $x \to f(x)$ or a suitable modification (e.g., $x \to fx$) thereof. M is also called the *domain of definition* of f and $f(M) = \{f(x) \mid x \in M\}$ the *image* of M under (the action of) f. $f(x)$ is called the *image* of x $(x \in M)$. If for some $A \subseteq M$ the set $\{f(x) \mid x \in A\}$ is a singleton, then f is said to be *constant* on A. If $f(M) = N$, then $f\colon M \to N$ is said to be *surjective* or "onto" or a *surjection*. If $x, y \in M$, $x \neq y \Rightarrow f(x) \neq f(y)$, then f is said to be *one-to-one* (in symbols, 1-1), or *injective* or an *injection*, or to *separate the points* of M. If $f\colon M \to N$ is injective and surjective, it is called *bijective*, a *bijection*, or *1-1 onto*, and a *permutation* or *bijection* of M if $N = M$. If $M \subseteq N$, then the mapping $f\colon M \to N$ defined by $f(x) = x$ $(x \in M)$ is called the *identity* on M, or the *identical mapping* of M into N, or the *natural embedding* of M into N; in case $M = N$, it is sometimes denoted by id_M or by 1.

If $f\colon M \to N$ is a mapping and $\varnothing \neq M_0 \subseteq M$, then the mapping $f_0\colon M_0 \to N$ defined by $f_0(x) = f(x)$ $(x \in M_0)$ is called the *restriction* of f to M_0. Some authors write $f_0 = \mathrm{rest}_{M_0} f$ in this case. Mappings $f\colon M \to \mathbb{R}$ are called *real* or *real-valued functions* on M. For any real function f on M, we call the set $\{x \mid x \in M,\ f(x) \neq 0\}$ the *support* of f and denote it by $\mathrm{supp}(f)$. This notation is used analogously in many similar situations. The set of all mappings $f\colon M \to N$ is also denoted by N^M (see also below), and we shall use the notation $f \in N^M$ besides $f\colon M \to N$ freely.

The concept of a mapping is as easy and as difficult to understand as the expression "attached." This difficulty can be reduced to the level where M consists of two different elements, say 0 and 1, or 1 and 2, by introducing the cartesian product of two sets and the notion of a relation. This goes as follows.

A mapping of $\{1, 2\}$ into a set N can be written as an "ordered couple" (x, y), where the left component x denotes the element attached to 1, and the right component denotes the element attached to 2, under the given mapping. If M, N are sets, then we consider all mappings (x_1, x_2) of $\{1, 2\}$ into $M \cup N$ and define $\{(x_1, x_2) \mid x_1 \in M,\ x_2 \in N\}$ as the *cartesian product* of M and N, which we denote by $M \times N$. It is obvious that:

$$M \times N \neq \varnothing \quad \Leftrightarrow \quad M \neq \varnothing \neq N;$$

$$M \times N \subseteq M' \times N \quad \Leftrightarrow \quad M \subseteq M' \qquad \text{for any set } M' \text{ if } N \neq \varnothing;$$

$$(M' \backslash M) \times N = (M' \times N) \backslash (M \times N) \qquad \text{for any set } M';$$

$$M \times N = N \times M \quad \Leftrightarrow \quad M = N.$$

Let M, N be sets. A subset R of $M \times N$ is also called a (*binary*) *relation* between (the elements of) M and (the elements of) N, and a binary relation in M if $M = N$. For every $x \in M$, the set $\{y \mid (x, y) \in R\}$ is called the *x-section* of R and is denoted by ${}_xR$. Similarly, $R_y = \{x \mid (x, y) \in R\}$ defines the *y-section* of R, for every $y \in N$. By ${}_RD$ we denote the set $\{x \mid {}_xR \neq \varnothing\}$, which we call the *M-domain* of R, and similarly $\{y \mid R_y \neq \varnothing\}$ defines the *N-domain* D_R of R. A relation $R \subseteq M \times N$ is called *maplike* if ${}_RD = M$ and ${}_xR$ is a singleton for every $x \in M$. It is now clear that a mapping $f\colon M \to N$ and a maplike binary relation R between M and N are merely two aspects of the same thing: If a mapping $f\colon M \to N$ is given, $R = \{(x, f(x)) \mid x \in M\}$ defines a maplike relation R between M and N. It is also called the *graph* of f. Different f define different maplike relations in this way, and every maplike relation arises from exactly one mapping. The reader now has the choice between an introduction of mappings through maplike relations, and the general definition of mappings given before.

Let M and I be arbitrary nonempty sets. A mapping $I \to M$ is also called a *family* in (or of elements of) M with *index set* I, and is also denoted by the "I-tuple" $(x_\iota)_{\iota \in I}$. The x_ι $(\iota \in I)$ are called the *members* of the family. Clearly f: $M \to N$ and $(f(x))_{x \in M}$ denote the same thing. $x_\bullet$ is a possible shorthand for $(x_\iota)_{\iota \in I}$. In case I is comparatively simple, we indulge ourselves in simpler writings, such as:

If $I = \{1, 2\}$, then we write $(x_\iota)_{\iota \in I} = (x_1, x_2)$ or simply x_1, x_2; and call this a *pair, couple*, or *2-tuple.*

If $I = \{1, \ldots, 100\}$, then we write $(x_\iota)_{\iota \in I} = (x_1, \ldots, x_{100})$ or simply $x_1, \ldots, x_{100}$.

If $I = \{1, \ldots, n\}$, then we write $(x_\iota)_{\iota \in I} = (x_1, \ldots, x_n)$, or simply $x_1, \ldots, x_n$ and call this an *n-tuple.*

If $I = \mathbb{N}$, then we write $(x_\iota)_{\iota \in I} = (x_1, x_2, \ldots)$ or simply $x_1, x_2, \ldots$ and call this a sequence, etc. Further simplifications are introduced wherever we find them feasible. The reader should carefully distinguish between a family $(x_\iota)_{\iota \in I}$ in a set M and the subset $\{x_\iota | \iota \in I\}$ of M determined by it; it may happen that all x_ι are the same element of M and hence the subset in question is a singleton only.

If the x_ι $(\iota \in I)$ are sets, we speak of a *family of sets.* If $(M_\iota)_{\iota \in I}$ is a family of sets, we define $\bigcap_{\iota \in I} M_\iota = \{x | x \in M_\iota \ (\iota \in I)\}$ to be its *intersection* and $\bigcup_{\iota \in I} M = \{x | x \in M_\iota$ for some $\iota \in I\}$ to be its *union.* We speak of finite resp. countable resp. uncountable unions or intersections if I is finite resp. countable resp. uncountably infinite. We speak of *disjoint unions* or decompositions or of partitions, and replace $\bigcup$ by $\sum$ if $\iota, \kappa \in I$, $\iota \neq \kappa \Rightarrow M_\iota \cap M_\kappa = \varnothing$; the M_ι $(\iota \in I)$ are in this case also called the *parts* or *atoms* of the partition. This way of speaking is appropriately adapted to particular situations.

Every nonempty set M gives rise to a particular family, by the identity mapping: $(x)_{x \in M}$. We call this "indexing M by itself." It is tacitly assumed whenever we write in terms of sets notions originally defined for families only.

Let $(M_\iota)_{\iota \in I}$ be a family of sets. Then the set $\{(x_\iota)_{\iota \in I} | x_\iota \in M_\iota \ (\iota \in I)\}$ (we usually write $(\iota \in I)$ instead of "for all $\iota \in I$," and carry this notation over also to other situations; it thus stands simply for $\iota \in I \Rightarrow$ written left of the statement to whose right $(\iota \in I)$ is written) is called the *cartesian product* of the given family of sets, and is also denoted by $\prod_{\iota \in I} M_\iota$. It is nonempty only if $M_\iota \neq \varnothing$ $(\iota \in I)$. The corresponding "if" statement if nothing but a version of the famous *axiom of choice*. We indulge ourselves in even stronger simplifications than in the case of families. The cartesian product of the family $(M_1, \ldots, M_n)$ is denoted by $M_1 \times \cdots \times M_n$, also written as $\{(x_1, \ldots, x_n) | x_1 \in M_1, \ldots, x_n \in M_n\}$ or $\prod_{k=1}^{n} M_k$, and even simply called the cartesian product of $M_1, \ldots, M_n$ or of the M_k $(k = 1, \ldots, n)$. We proceed in a similar fashion in other cases. If $M_\iota = M$ $(\iota \in I)$, then we also write M^I for $\prod_{\iota \in I} M_\iota$; this is the set of all mappings of I into M; in particular, $\mathbb{R}^I$ is the set of all real functions on I. If $I = \{1, \ldots, n\}$, we even write M^n instead of $M^{\{1, \ldots, n\}}$. The classical notation $\mathbb{R}^n$ now fits into our scheme.

It is an obvious exercise to formulate and prove an associativity law for cartesian multiplication. In particular, $\mathbb{N}^n$ denotes the set of all n-tuples of natural numbers $(n = 1, 2, \ldots)$ and $\mathbb{N}^{\mathbb{N}}$ is the set of all infinite sequences of natural numbers. $\mathbb{N}^*$ is a customary notation for the set $\bigcup_{n=1}^{\infty} \mathbb{N}^n$ of all n-tuples, of any length, of natural numbers.

If $G \neq \varnothing$, then a mapping $G \times G \to G$ is often called a *composition*. A subset M of G is said to be *stable* under the composition if $M \times M$ is mapped into M. The same terminology is adapted for mappings of arbitrary cartesian products of sets into one of these sets. Thus for any set X the set $\mathscr{P}(X)$ is stable, e.g., under the mapping $\mathscr{P}(X)^I \to \mathscr{P}(X)$, $(F_\iota)_{\iota \in I} \to \bigcap_{\iota \in I} F_\iota$ $(F_\iota \in \mathscr{P}(X)$ $(\iota \in I))$, i.e., under arbitrary intersections as we shall say, and $\{0\} \subseteq \mathbb{R}^n$ is stable under the mapping $\mathbb{R} \times \mathbb{R}^n \to \mathbb{R}^n$, $x \to \alpha x$ $(\alpha \in \mathbb{R}, x \in \mathbb{R}^n)$, etc.

If $\mathscr{C} \subseteq \mathscr{P}(X)$, then we define $\mathscr{C}_\sigma = \{C_1 \cup C_2 \cup \cdots | C_1, C_2, \ldots \in \mathscr{C}\}$, $\mathscr{C}_\delta = \{C_1 \cap C_2 \cap \cdots | C_1, C_2, \ldots \in \mathscr{C}\}$, $\mathscr{C}_{\sigma\delta} = (\mathscr{C}_\sigma)_\delta$, $\mathscr{C}_{\delta\sigma} = (\mathscr{C}_\delta)_\sigma$. Clearly $\mathscr{C}_\sigma$ is stable under the mapping $\mathscr{P}(X)^{\mathbb{N}} \to \mathscr{P}(X)$, $(F_n)_{n \in \mathbb{N}} \to F_1 \cup F_2 \cup \cdots$, i.e., under countable unions; and likewise $\mathscr{C}_\delta$ is stable under countable intersections (the Greek letter σ is a reminder to *sum*, and the Greek letter δ is a reminder to the German word *Durchschnitt* (= intersection) here).

Let $E, F, G \neq \varnothing$ and $f\colon E \to F$, $g\colon F \to G$ be mappings. Then $[g(f)](x) = g(f(x))$ $(x \in X)$ uniquely defines a mapping $g(f)\colon E \to G$, which we also denote by $g \circ f$. $\circ$ is also called *composition* or *multiplication* of mappings. It is clear how to carry this over to longer "chains" of mappings and to prove, e.g., an associativity law for the "product" $\circ$. If $f\colon E \to F$ is bijective, then there is a unique bijection $f^{-1}\colon F \to E$ such that $f^{-1} \circ f = \mathrm{id}_E$. It also satisfies $f \circ f^{-1} = \mathrm{id}_F$ and is called the *inverse (mapping)* of f.

Composition is one way to get a new mapping from given ones. Another, and an even simpler one, is *restriction*. If $f\colon E \to F$ is a mapping and $\varnothing \neq E_0 \subseteq E$, then $f_0(x) = f(x)$ $(x \in E_0)$ uniquely defines a mapping $f_0\colon E_0 \to F$, the *restriction* of f to the subset $E_0 \neq \varnothing$ of E. f is then also called an extension of f_0 (from E_0) to E. If f is written as a family

$(f(x))_{x \in E}$, we denote its restriction simply by $(f(x))_{x \in E_0}$. Let us apply this to cartesian products. If $M_\iota \neq \varnothing$ $(\iota \in I)$, then $\prod_{\iota \in I} M_\iota = \{(x_\iota)_{\iota \in I} | x_\iota \in M_\iota\}$ consists of mappings defined on I. If we choose some $\varnothing \neq J \subseteq I$ and attach to every $(x_\iota)_{\iota \in I}$ from $\prod_{\iota \in I} M_\iota$ its restriction $(x_\iota)_{\iota \in J}$ to J, we define a mapping φ_J^I: $\prod_{\iota \in I} M_\iota \to \prod_{\iota \in J} M_\iota$ which we call the *natural projection* of $\prod_{\iota \in I} M_\iota$ onto $\prod_{\iota \in J} M_\iota$ (the "onto" again follows from the axiom of choice), sometimes denoted by φ_J, and by φ_ι if $J = \{\iota\}$; in the latter case it is also called the *natural component projection* (or *component mapping*) onto M_ι $(\iota \in I)$. It is easy to verify $I \supseteq J \supseteq K \neq \varnothing \Rightarrow \varphi_K^I = \varphi_K^J \circ \varphi_J^I$.

Finally, we define a third way of getting new mappings from given ones: *cartesian multiplication*. Let $E, I \neq \varnothing$ and $(M_\iota)_{\iota \in I}$ a family of nonempty sets. For every $\iota \in I$, let ψ_ι: $E \to M_\iota$ be a mapping. Then $(\prod_{\iota \in I} \psi_\iota)(x) = (\psi_\iota(x))_{\iota \in I}$ uniquely defines a mapping $\prod_{\iota \in I} \psi_\iota$: $E \to \prod_{\iota \in I} M_\iota$. It is called the *cartesian product* of the family $(\psi_\iota)_{\iota \in I}$. We say that $(\psi_\iota)_{\iota \in I}$ *separates the points* of E if $\prod_{\iota \in I} \psi_\iota$ is injective. This means simply that for any $x, y \in E$ with $x \neq y$, there is an $\iota \in I$ with $\psi_\iota(x) \neq \psi_\iota(y)$. We indulge ourselves in obvious simplifications of writing in special situations. It is an easy exercise to show that, for $E = \prod_{\iota \in I} M_\iota$ and $\psi_\iota = \varphi_\iota =$ the ιth component mapping of $\prod_{\iota \in I} M_\iota$ onto M_ι $(\iota \in I)$, the product mapping $\prod_{\iota \in I} \varphi_\iota$ is nothing but the identity mapping of $\prod_{\iota \in I} M_\iota$ onto itself.

Let $\Omega \neq \varnothing$. There is a unique bijection of $\mathscr{P}(\Omega)$ onto $\{0, 1\}^\Omega$ which attaches to every $E \subseteq \Omega$ the $\{0, 1\}$-valued function 1_E defined on Ω by $1_E(\omega) = 0$ $(\omega \notin E)$, $1_E(\omega) = 1$ $(\omega \in E)$. 1_E is called the *indicator function* of E (as a subset of Ω). We have $f = 1_E$ with $E = \{\omega | \omega \in \Omega, f(\omega) = 1\}$ for every $f \in \{0, 1\}^\Omega$. The mentioned bijection allows us to express statements about sets and elements by statements about their indicator functions. We give a small list of examples:

$$\omega \in E \quad \Leftrightarrow \quad 1_E(\omega) = 1;$$
$$E \subseteq F \quad \Leftrightarrow \quad 1_E \leqq 1_F \qquad \text{(i.e., } 1_E(\omega) \leqq 1_F(\omega) \quad (\omega \in \Omega));$$
$$1_{E \cup F}(\omega) = \max[1_E(\omega), 1_F(\omega)] \quad (\omega \in \Omega);$$
$$1_{E \cap F}(\omega) = \min[1_E(\omega), 1_F(\omega)] \quad (\omega \in \Omega).$$

A property of elements of a given set Ω can be identified with the subset of all those elements of Ω having that property. Logical statements about such properties then find a natural set-theoretical expression in terms of the corresponding subsets of Ω. To pass to the negation of a property means passage from a set to its complement. Intersection corresponds to "and," union to (the nonexclusive) "or," etc.

Let $X \neq \varnothing$ and $R \subseteq X \times X$ be a binary relation in X. In this situation we often write xRy instead of $(x, y) \in R$ and often replace R by another and more suggestive symbol, such as $\sim$. A binary relation $\sim$ in X is called:

reflexive if $x \sim x$ $(x \in X)$, i.e., $R = \sim$ contains the "diagonal" $\{(x, x) | x \in X\}$;
symmetric if $x \sim y \Rightarrow y \sim x$ $(x, y \in X)$, i.e., $(x, y) \in R \Rightarrow (y, x) \in R$;
transitive if $x \sim y$, $y \sim z \Rightarrow x \sim z$ $(x, y, z \in X)$.

If $\sim$ is reflexive, symmetric, and transitive, then $\sim$ is called an *equivalence relation.* In this case we define, for every $x \in X$, its so-called *equivalence class* $[x] = \{y | y \in X, x \sim y\}$. This defines a family $([x])_{x \in X}$ of subsets of X. It is easily seen that $x \in [x]$, and hence $[x] \neq \varnothing$ $(x \in X)$, $X = \bigcup_{x \in X} [x]$. Furthermore, $[x] \neq [y] \Rightarrow [x] \cap [y] = \varnothing$. The set $\{[x] | x \in X\}$ consists therefore of pairwise disjoint nonempty subsets of X covering X; i.e., it is a partition of X whose atoms are the equivalence classes $[x]$ $(x \in X)$. It is called the *partition of X into equivalence classes* for $\sim$, or the *factor space* of X for $\sim$, and also denoted by $X/\sim$. By $x \to [x]$, a surjection of X onto $X/\sim$ is defined. We call it the *natural* (*standard, canonical*) *mapping* of X onto $X/\sim$, for $\sim$. Clearly $X/=$ is the partition of X into all singletons $\subseteq X$. Every partition of X is an $X/\sim$ for a unique equivalence relation $\sim$: define $x \sim y$ iff x and y are in the same atom of the partition. We call $\sim$ the equivalence relation corresponding to the given partition. Let ξ and ζ be partitions of X and $\sim_\xi$, $\sim_\zeta$ the corresponding equivalence relations. Then the equivalence relation $\sim_{\xi \vee \zeta}$ is nothing but the simultaneous conjunction of $\sim_\xi$ and $\sim_\zeta$: $x \sim_{\xi \vee \zeta} y \Leftrightarrow x \sim_\xi y$ and $x \sim_\zeta y$. The identification of $\sim_{\xi \wedge \zeta}$ as well as the generalization of these results to arbitrary sets of equivalence relations is left as an exercise to the reader.

A reflexive and transitive relation $\leqq$ in X is called a *partial ordering.* In this case, $(X, \leqq)$ is called a *partially ordered set.* $\leqq$ is often omitted in this notation. It is clear what the *majorant, minorant, upper bound, lower bound* of an (order) bounded subset of X mean. A partial ordering $\leqq$ in X is called *sharp* if $x \leqq y$, $y \leqq x \Rightarrow x = y$ $(x, y \in X)$. Let $\leqq$ be any partial ordering of X. Define $x \sim y$ iff $x \leqq y$ and $y \leqq x$. Then $\sim$ is an equivalence relation. It is called the "equivalence for $\leqq$" relation. There is a unique partial ordering $\leqq$ on $X/\sim$ such that $x \leqq y \Leftrightarrow [x] \leqq [y]$ $(x, y \in X)$, where $[x]$ denotes the equivalence class containing x. This partial ordering on $X/\sim$ is sharp. A partial ordering $\leqq$ on X is called a *total ordering* if for any $x, y \in X$ either $x \leqq y$ or $y \leqq x$ (or both). It is clear what we mean by a *totally ordered subset* of a partially ordered set $(X, \leqq)$. A partial ordering $\leqq$ in X is called (*upper*) *inductive* if every totally ordered subset of X has a majorant in X. The *lemma of* (Hausdorff–Kuratowski–) Zorn states that for every (upper) inductive partial ordering of a set $X \neq \varnothing$, there is at least one *maximal element* in X, i.e., an element $m \in X$ such that $x \in X$, $m \leqq x \Rightarrow x \leqq m$ (and thus $x = m$ in the case of a sharp partial ordering). It is well known that this lemma is equivalent to the axiom of choice. Replacing $\leqq$ by its inverse $\geqq$ (defined by $x \geqq y \Leftrightarrow y \leqq x$), we arrive at an equivalent ("downward") version of the (Hausdorff–Kuratowski–) Zorn lemma. Observe that maximal elements as considered here need not be majorants.

Let $(X, \leqq)$, $(Y, \leqq)$ be partially ordered sets. A mapping $\varphi\colon X \to Y$ is said to be *isotone, monotone, increasing,* or *order preserving* if $x, x' \in X$, $x \leqq x' \Rightarrow \varphi(x) \leqq \varphi(x')$, and *antitone, monotone decreasing,* or *order inverting,* if $x, x' \in X$, $x \leqq x' \Rightarrow \varphi(x') \leqq \varphi(x)$. An isotone bijection $\varphi\colon X \to Y$ is called an isomorphism of $(X, \leqq)$ and $(Y, \leqq)$ if its inverse $\varphi^{-1}\colon Y \to X$ is also isotone, i.e., if $x, x' \in X \Rightarrow [x \leqq x' \Leftrightarrow \varphi(x) \leqq \varphi(x')]$.

Let $(X, \leqq)$ be a partially ordered set. Let $\varnothing \neq M \subseteq X$. If the set $\overline{M} = \{x \mid x \in X, x \geqq y \ (y \in M)\}$ of all majorants of M contains a minimal element b (i.e., $\overline{M} \ni x \leqq b \Rightarrow x \geqq b$), this element b is called a *minimal upper bound* of M. It may happen that a set M has several different minimal upper bounds which are not equivalent for $\leqq$. If any two minimal upper bounds are equivalent for $\leqq$, we call any of them a *least upper bound (l.u.b.)* for M and denote it by $\sup M$, $\sup_{x \in M} x$, $\sup\{x \mid x \in M\}$, $\bigvee_{x \in M} x$ or the like. In particular it should be clear what $\sup_n a_n$ means. In a similar fashion we define the *greatest lower bound (g.l.b.)* of a set $\varnothing \neq M \subseteq X$, it it exists, and denote it by $\inf M$, $\bigwedge_{x \in M} x$ etc. $\bigvee_{x \in M} x$ is uniquely determined up to equivalence for $\leqq$ if it exists, and so is $\bigwedge_{x \in M} x$. In the case of a sharp partial ordering $\leqq$ we get uniqueness.

A partially ordered set $(X, \leqq)$ is called a *lattice* if for any two (and hence for any finitely many) elements of X a l.u.b. and a g.l.b. exist. We write $x \vee y$ for $\sup_{a \in \{x, y\}} a$, $x_1 \vee \cdots \vee x_n$ for $\sup_{x \in \{x_1, \ldots, x_n\}} x$, etc; notations like $x \wedge y$, $x_1 \wedge \cdots \wedge x_n$, etc., are equally obvious. In a lattice a set $\varnothing \neq M \subseteq X$ and the set $M' = \{x_1 \vee \cdots \vee x_n \mid n \geqq 1, x_1, \ldots, x_n \in M\}$ have the same majorants, and an analogous statement holds about minorants. Also, if $\varnothing \neq M \subseteq X$ has a l.u.b. $\overline{x}$ in X, then for any $y \in X$, $M \vee y = \{x \vee y \mid x \in M\}$ has l.u.b. $\overline{x} \vee y$. A symmetric statement holds for minorants.

A lattice $(X, \leqq)$ is called a *conditionally complete lattice* if every set $\varnothing \neq M \subseteq X$ having a majorant in X has a l.u.b. in X, and every set $\varnothing \neq M \subseteq X$ having a minorant in X, has a g.l.b. in X.

Let $(X, \leqq)$ be a partially ordered set. A family $(x_\iota)_{\iota \in I}$ in X is said to be *increasingly filtered* or *upward directed* if for any $\iota, \kappa \in I$, there is a $\lambda \in I$ such that $x_\iota, x_\kappa \leqq x_\lambda$. A subset M' of X is said to be *increasingly filtered* or *upward directed* if the family obtained by indexing M by itself is increasingly filtered. If $(X, \leqq)$ is a lattice and $M \subseteq X$, then

$$M = \{x_1 \vee \cdots \vee x_n \mid n \geqq 1, x_1, \ldots, x_n \in M\}$$

is increasingly filtered (actually, M' is stable under (the formation of) finite suprema). We speak of *decreasingly filtered = downward directed families and subsets* of X in an obvious analogous fashion.

Let $I \neq \varnothing$ be increasingly filtered by a partial ordering $\leqq$. Then every family $(x_\iota)_{\iota \in I}$ is also called a *generalized sequence* or a *Moore–Smith sequence* or a *net*, with the directed *index set* I. Let $(X, \leqq)$ be a partially ordered set and $(x_\iota)_{\iota \in I}$ a net in X. Assume that for every $\iota \in I$, $\sup_{\kappa \geqq \iota} x_\kappa = y_\iota$ exists and moreover $\inf_{\iota \in I} y_\iota = \bar{x}$ exists in X. Then we say that $(x_\iota)_{\iota \in I}$ has a lim sup in X and write $\bar{x} = \limsup_{\iota \in I} x$. Similarly we get $\underline{x} = \liminf_{\iota \in I} x$ (all needed existences provided). The net $(x_\iota)_{\iota \in I}$ is said to be order convergent with the (order) limit x if $x = \limsup_{\iota \in I} x_\iota = \liminf_{\iota \in I} x_\iota$. If $(x_\iota)_{\iota \in I}$ is order convergent with limits x, x', then $x \leqq x' \leqq x$; hence the order limit is uniquely determined in case the partial ordering $\leqq$ in X is sharp.

If $X \neq \varnothing$, then $(\mathscr{P}(X), \subseteq)$ is a conditionally complete lattice and a lattice. This example is in a way universal: If $(\Omega, \leqq)$ is a sharply partially ordered set, then $\Omega' = \{\Omega_a = \{x \mid x \in \Omega, x \leqq a\} \mid a \in \Omega\}$ is a set system whose partial ordering $\subseteq$ is in an obvious correspondence to $\leqq$. In fact the mapping $a \to \Omega_a$ constitutes an isomorphism.

If $(X, \leqq)$, $(Y, \leqq)$ are partially ordered sets and φ: $X \to Y$ is isotone, then φ sends every upper bound of a set $M \subseteq X$ into an upper bound of $\varphi(M) \subseteq Y$. If φ is an isomorphism, $(X, \leqq)$ is a lattice iff $(Y, \leqq)$ is; and similar statements for the other lattice theoretical concepts developed above are easy to deduce.

A set M is called (nonempty) and *finite* (of power $|M| = n \in \mathbb{N}$) if there is a bijection of M onto the set $\{1, \ldots, n\}$ of the first n natural numbers. It can be proved that this uniquely defines the power of the set. A set $\neq \varnothing$ is finite iff it allows of no bijection onto a proper subset of itself (this is Dedekind's definition of a finite set). All proper subsets of a finite set are empty or finite with a smaller power than the original set. $\varnothing$ is given the power 0. A finite disjoint union $M_1 + \cdots + M_n$ of finite sets is finite again, and $|M_1 + \cdots + M_n| = |M_1| + \cdots + |M_n|$. A finite cartesian product $M_1 \times \cdots \times M_n$ of finite sets $\neq \varnothing$ is finite, and $|M_1 \times \cdots \times M_n| = |M_1| \cdots |M_n|$. A set M that is not finite is called infinite, and we then write $|M| = \infty$. Accordingly, $|M| < \infty$ stands for finiteness of M.

A set M is called *countable* or of power $|M| = \aleph_0$ (aleph 0) if there is a bijection of the set onto the set $\mathbb{N}$ of all natural numbers. A set that is empty, finite, or countable is also called *at most countable.* Subsets of countable set are at most countable. If $(M_\iota)_{\iota \in I}$ is a countable family of countable sets, i.e., if I is countable and M_ι is countable $(\iota \in I)$, then $\bigcup_{\iota \in I} M$ is countable. In short, a countable union of countable sets is countable. A cartesian product of finitely many countable sets is countable.

If an infinite set is not countable, it is called *uncountable* or *uncountably infinite.* A countable cartesian product of nonempty at most countable sets is uncountable. In particular, $\{0, 1\}^{\mathbb{N}}$ is uncountably infinite.

Two sets $M, N \neq \emptyset$ are said to be *of equal power* if there is a bijection of M onto N. This definition is conformal with the definitions of power that we have introduced for at most countable sets. It is easy to see that all countable cartesian products of at most countable sets among which infinitely many have at least two different elements, have equal powers, and equal power with $\mathbb{R}$, the set of all reals (= the continuum). *F. Bernstein's equivalence theorem* says that two sets are of equal power if each of them is of equal power with a subset of the other. If M is of equal power with a subset of N, but not vice versa, then we say that M has smaller power than N, and write $|M| < |N|$. It can be shown that $|M| < |\mathscr{P}(M)|$ for every set M.

2. VECTOR SPACES AND VECTOR LATTICES. BANACH SPACES

Let $G \neq \emptyset$. A mapping $\cdot$: $G \times G \to G$, $(x, y) \to x \cdot y$ (the dot is usually omitted here) is called a *multiplication* in G if it is *associative*, i.e., if $x \cdot (y \cdot z) = (x \cdot y) \cdot z$ $(x, y, z \in G)$. The couple $(G, \cdot)$ is then called a *semigroup* or G is called a semigroup under $\cdot$. Two elements x, y of G are said to *commute* if $x \cdot y = y \cdot x$ holds. Clearly every element of G commutes with itself. The semigroup $(G, \cdot)$ is said to be *commutative* or *abelian* if the *commutative law* $x \cdot y = y \cdot x$ $(x, y \in G)$ holds, i.e., if any two elements of G commute. $e \in G$ is said to be a left (right) *neutral element* for the multiplication if $e \cdot x = x$ $(x \cdot e = x)$ for every $x \in G$. For any given $a \in G$, the mapping L_a: $G \to G$, $x \to ax$ is called the *left translation* by a, and the mapping R_a: $x \to xa$ the *right translation* by a; clearly $L_a \circ L_b = L_{a \cdot b}$, $R_a \circ R_b = R_{b \cdot a}$. If $a \in G \supseteq M$, we write $a \cdot M$ for the set $\{a \cdot x | x \in M\}$ and $M \cdot a$ for the set $\{x \cdot a | x \in M\}$. If $M, N \in G$, we write $M \cdot N$ for the set $\{x \cdot y | x \in M, y \in N\}$. A set $U \subseteq G$ is called a *left (right) ideal* in G if $G \cdot U \subseteq U$ $(U \cdot G \subseteq U)$, and a *subsemigroup* of G if $U \cdot U \subseteq U$.

Let $(G, \cdot)$, $(H, \cdot)$ be two semigroups (we denote both multiplications by the same dot, for the sake of brevity). A mapping φ: $G \to H$ is called a (semigroup) *homomorphism* if $\varphi(x \cdot y) = \varphi(x) \cdot \varphi(y)$ $(x, y \in G)$ holds; a homomorphism φ: $G \to H$ is called an *isomorphism* if it is bijective; in this case, φ^{-1}: $H \to G$ is an isomorphism, too, and we say that G and H are *isomorphic*. The mapping log: $\{x | 0 < x \in \mathbb{R}\} \to \mathbb{R}$ is an isomorphism if we employ ordinary multiplication in the first, and addition in the second set. It is clear that the composition of two homomorphisms resp. isomorphisms yields a homomorphism resp. isomorphism again. The identity mapping id G: $G \to G$ is always an isomorphism. If $H = G$, we also say *endomorphism* for homomorphism, and *automorphism* for isomorphism.

A semigroup $(G, \cdot)$ is called a *group* if (1) it has a left neutral element e and (2) for every $x \in G$ there is a $x^{-1} \in G$ such that $x^{-1} \cdot x = e$ (existence of a left inverse). One then proves that e is (a) unique and (b) a right neutral element at the same time, that (c) x^{-1} is uniquely determined by x, for every $x \in G$, and that (d) the mapping $x \to x^{-1}$ is a bijection $G \to G$. e is called the *neutral element* of G, and x^{-1} is called the *inverse* of x $(x \in G)$; one proves $x \cdot x^{-1} = e$ $(x \in G)$. Homomorphisms of groups send the neutral elements into the neutral element and the inverse of x into the inverse of the image of x.

$\mathbb{R}$ is an abelian group under addition, $\mathbb{R}\backslash\{0\}$ is an abelian group under multiplication, the set S_n of all permutations (= bijections) of $\{1, \ldots, n\}$ is a group under composition $\circ$, nonabelian for $n \geqq 3$.

A subsemigroup U of the semigroup G is called a *subgroup* of G if it is a group under the restriction of $\cdot$ to $U \times U$. If U is a subgroup of the group G, then $x \sim y \Leftrightarrow x \cdot U = y \cdot U$ $(x, y \in G)$ defines an equivalence relation in G as well as $x \sim y \Leftrightarrow U \cdot x = U \cdot y$ $(x, y \in G)$. If $x \cdot U = U \cdot x$ $(x \in G)$, then U is called a *normal subgroup* of G. In this case both equivalence relations coincide, we write G/U instead of $G/\sim$, and $(x \cdot U) \cdot (y \cdot U) = (x \cdot y) \cdot U$ shows how to make G/U into a group (called the *factor* or *quotient group* of G for U) such that the natural surjection φ: $G \to G/U$ becomes a homomorphism. If $(G, \cdot)$, $(H, \cdot)$ are groups and φ: $G \to H$ is a surjective homomorphism, then the so-called *kernel* $\ker \varphi = \{x \mid \varphi(x) = e\}$ (where e denotes the neutral element in H) is a normal subgroup of G, φ is constant on every equivalence class in $G/\ker \varphi$, and the mapping φ: $G/\ker \varphi \to H$ thus defined (and again denoted by φ) is an isomorphism. This is the so-called *homomorphism theorem* for groups.

Let $(H, +)$ be an abelian group and $\cdot$: $\mathbb{R} \times H \to H$, $(\alpha, x) \to \alpha \cdot x$ $(\alpha \in \mathbb{R}, x \in H)$ be a mapping such that (1) for every $0 \neq \alpha \in \mathbb{R}$ the mapping $x \to \alpha x$ of $H \to H$ is an automorphism of $(H, +)$, (2) $1 \cdot x = x$ $(x \in H)$, (3) $\alpha \cdot (\beta \cdot x) = (\alpha\beta) \cdot x$ and $(\alpha + \beta) \cdot x = (\alpha \cdot x) + (\beta \cdot x)$ (= $\alpha x + \beta x$ in the usual notation) $(\alpha, \beta \in \mathbb{R}, x \in H)$. Then $(H, +, \cdot)$ is called a *real vector space* or a *real linear space* under $+$ and $\cdot$ (which are however often omitted in the notation). The elements of H are also called *vectors*. $\varnothing \neq G \subseteq H$ is called a linear subspace of H if G, with the restriction of $+$ to $G \times G$, and of $\cdot$ to $\mathbb{R} \times G$, is a linear space. For any set $\varnothing \neq M \in H$, the intersection of all linear subspaces of H containing M is a linear subspace of H, the smallest linear subspace of H containing M. It is called the *linear hull* or the *linear span* of M, and is denoted by $\operatorname{lin}(M)$. We have $\operatorname{lin}(M) = \{\alpha_1 x_1 + \cdots + \alpha_n x_n \mid n \geqq 1, \alpha_1, \ldots, \alpha_n \in \mathbb{R}, x_1, \ldots, x_n \in M\}$. If G is a linear subspace of H, then the factor group H/G (for $+$) is a linear space in a natural way. Its elements are called the *linear manifolds* parallel to G in H. Let H, L be linear spaces. A mapping φ: $H \to L$ is called *linear* (or a vector space homomorphism) if $\varphi(x + y) = \varphi(x) + \varphi(y)$, $\varphi(\alpha x) = \alpha\varphi(x)$ ($\alpha \in \mathbb{R}$, $x, y \in H$). A linear mapping $\varphi\colon H \to L$ is called a *vector space isomorphism* if it is bijective; its inverse is then also an isomorphism. An obvious vector space analogue to the homomorphism theorem for groups is valid. A subset $M \neq \varnothing$ of a vector space H is said to be *linearly independent* if $\varnothing \neq N \subseteq M$, $\text{lin}(N) = \text{lim}(M) \Rightarrow N = M$. A linearly independent subset $B \neq \varnothing$ of H is called a *basis* of H if $\text{lin}(B) = H$. It can be shown (e.g., by an infinite version of the marriage theorem (XII.1.6)) that any two bases of H are of the same power. This power is called the *dimension* $\dim H$ of H. H is called *finite* or *infinite dimensional* according to whether $\dim H < \infty$ or $\dim H = \infty$.

The same theory can be built with $\mathbb{C}$ in place of $\mathbb{R}$. One then speaks of *complex vector spaces*. Clearly every complex vector space can also be considered as a real vector space.

A linear mapping $\varphi\colon H \to \mathbb{R}$ of a real vector space into the reals is called a *linear form* on H. A subset H_0 of H is called a *hyperplane* if there is a linear form φ on H that is different from the constant 0, and a real α such that $H_0 = \{x \mid x \in H,\ \varphi(x) = \alpha\}$. The sets $\{x \mid x \in H,\ \varphi(x) \leqq \alpha\}$ and $\{x \mid x \in H, \varphi(x) \geqq \alpha\}$ are then called the two *closed half-spaces* associated with (the mentioned representation of) the hyperplane H_0.

The set H^{alg} of all linear forms on a real vector space H is again a real vector space in a natural fashion. It is called the (*algebraic*) *dual* (*space*) of H.

A subset K of a real vector space H is called *convex* if

$$\alpha x + \beta y \in K \qquad (x, y \in K,\ \ 0 \leqq \alpha, \beta \in \mathbb{R},\ \ \alpha + \beta = 1).$$

It is called a *cone* with *tip* $a \in H$ if $a + \alpha(x - a) \in K$ ($x \in K$, $0 \leqq \alpha \in \mathbb{R}$). If $K \subseteq H$ is a convex cone with tip $0 \in H$, then $x \leqq y \Leftrightarrow y - x \in K$ defines a partial ordering $\leqq$ of H such that $x \leqq y \Rightarrow a + x \leqq a + y$ ($a \in H$) and $x \leqq y \Rightarrow \alpha x \leqq \alpha y$ ($0 \leqq \alpha \in \mathbb{R}$) for any $x, y \in H$. Conversely, if $\leqq$ is a partial ordering in H that has these two properties of compatibility with the linear structure of H, then $H_+ = \{x \mid 0 \leqq x \in H\}$ is a convex cone such that $x \leqq y \Leftrightarrow y - x \in H_+$ ($x, y \in H$). The union G of all linear subspaces contained in K is the largest such subspace and the elements of H/G are exactly the equivalence classes for $\leqq$. The cone K is called *sharp* if $\leqq$ is sharp. If $K \subseteq H$ is a sharp convex cone with tip 0, and if $\leqq$ denotes the attached sharp partial ordering of H, then $(H, \leqq)$ is called a *partially ordered vector space* and $H_+ = \{f \mid 0 \leqq f \in H\}$ its *positive cone*. Let $(H, \leqq)$ and $(L, \leqq)$ be two partially ordered vector spaces. A linear mapping $\varphi\colon H \to L$ is called *positive* if it is isotone. This is the case iff $0 \leqq f \in H \Rightarrow \varphi(f) \geqq 0$. If $H_+ - H_+ = H$, then a positive linear $\varphi\colon H \to L$ can be well defined given a $\varphi_+\colon H_+ \to L_+$ with $x, y \in H_+ \Rightarrow \varphi_+(x + y) = \varphi_+(x) + \varphi_+(y)$ and $0 \leqq \alpha \in \mathbb{R}$,

$x \in H_+ \Rightarrow \varphi_+(\alpha x) = \alpha\varphi_+(x)$. We need only put $\varphi(x - y) = \varphi_+(x) - \varphi_+(y)$ $(x, y \in H_+)$.

It is easy to give examples of sharp convex cones K in real vector spaces H such that $(H, \leqq)$ is not a lattice (take, e.g., $H = \mathbb{R}^3$, $K = \{(x, y, z) | z \geqq 0, x^2 + y^2 \leqq z^2\}$). If $(H, \leqq)$ is however a lattice, then K is called a *lattice cone* and $(H, +, \cdot, \leqq)$ is called a *vector lattice*. We shall often omit $+, \cdot, \leqq$ from the notation if there is no danger of confusion. It is obvious what to understand by a *vector lattice isomorphism*. In any partially ordered vector space $(H, \leqq)$, the existence of $\sup M$ for some $\varnothing \neq M \subseteq H$ implies the existence of $\sup(x + M)$ (where $x + M = \{x + y | y \in M\}$, of course) and its equality to $x + \sup M$. This often provides a tool for a proof that $(H, \leqq)$ is a vector lattice. It is sometimes useful to have the following lemma in mind.

Lemma 2.1. *Let $(H, +, \cdot)$ be a real space and $K \subseteq H$ a sharp convex cone with tip $0 \in H$, and $\leqq$ the corresponding partial ordering. Assume that for any $x \in H$ there are $y, z \in K$ such that*

(*a*) $x = y - z$;
(*b*) $y \wedge z = \min\{y, z\}$ *exists and is* $= 0 \in H$.

Then K is a lattice cone and $(H, +, \cdot, \leqq)$ is a vector lattice. (Actually, $y = x \vee 0$, $z = (-x) \vee 0$ in a.)

Proof. We have to prove that for any $x, x' \in H$ $x \wedge x' = \min\{x, x'\}$ and $x \vee x' = \max\{x, x'\}$ exist. For this, it is sufficient to show that for any $x \in H$, $x \vee 0$ exists; it is in fact easy to see (from the compatability of $\leqq$ with the linear structure of H) that the existence of $(x - x') \vee 0$ implies the existence of $x \vee x' = (x - x' + x') \vee (0 + x')$ and that this is $= [(x - x') \vee 0] + x'$; furthermore, the existence of $x \vee x'$ for arbitrary $x, x' \in H$ implies the existence of $x \wedge x'$, and that this is $= -[(-x) \vee (-x')]$. Let us now choose any $x \in H$, represent it in the form $x = y - z$ with $y, z \in K$, $y \wedge z = 0$, and prove that $x \vee 0$ exists and is $= y$. But $y \wedge z = 0 \Rightarrow (-y) \vee (-z) = 0 \Rightarrow (y - y) \vee (y - z) = y$ $\Rightarrow 0 \vee x = y$.

In a vector lattice, one usually writes x_+ for $x \vee 0$, x_- for $(-x) \vee 0$, and $|x|$ for $x \vee (-x)$ and calls this the *positive* resp. *negative part* of x, resp. the *modulus* of x. The following relations are easy to prove:

$$x = x_+ - x_-, \qquad x_+ \wedge x_- = 0,$$

$$|x| = x_+ + x_-, \qquad |x + y| \leqq |x| + |y|,$$

$$x_+ = \frac{x + |x|}{2}, \qquad x_- = \frac{|x| - x}{2},$$

$$x \vee y = x + (y - x)_+ = y + (x - y)_+,$$

$$x \wedge y = x - (x - y)_+ = y - (y - x)_+,$$

$$x + y = (x \vee y) + (x \wedge y).$$

They show that, for a linear subspace G of H, stability of G under one of the operations $x \to x_+$, $x \to x_-$, $x \to |x|$ implies stability under all of them, hence G is, with the restrictions of $+$, $\cdot$, $\leqq$, a vector lattice again; and the finite lattice operations $\wedge$, $\vee$ in G are restrictions of the same operations in H. If this is the case, G is called a *vector sublattice* of the vector lattice H.

Looking at the above identities with an eye for existence questions, we see that a partially ordered vector space $(H, \leqq)$ is a vector lattice if, e.g., $x \vee 0$ exists in H for every $x \in H$.

If $(H, +, \cdot, \leqq)$ is a vector lattice, $x, y \in H$ are said to be *orthogonal* if $|x| \wedge |y| = 0$. In this case we write also $x \perp y$. For example x_+ and x_- are always orthogonal. If $S \subseteq H$, then $S^\perp = \{y | x \in S \Rightarrow x \perp y\}$ is called the *orthogonal complement* of S. Clearly $S^\perp$ is a vector sublattice of H, and $(S^\perp)^\perp \supseteq S$.

A convex subset C of a real vector space H is called a *simplex* if there is a real vector space H', a hyperplane H_0' in H, a sharp convex lattice cone K' with tip 0 in H', and an $x' \in H'$ such that H is a subspace of H' and $C + x' = H_0' \cap K'$. A function $\|\cdot\|: H \to \mathbb{R}_+$ on a real vector space H is called a *seminorm* if it satisfies (a) $\|\alpha \cdot X\| = |\alpha| \cdot \|x\|$ $(\alpha \in \mathbb{R}, x \in H)$ and (b) the triangle inequality $\|x + y\| \leqq \|x\| + \|y\|$ $(x, y \in H)$. For any seminorm $\|\cdot\|$ on H, $\{x | \|x\| = 0\}$ is a linear subspace of H; we call it the *zero space* $H_{\|\cdot\|}$ for $\|\cdot\|$. If the zero space for the seminorm $\|\cdot\|$ in H is $= \{0\}$, then $\|\cdot\|$ is called a *norm.* Every seminorm defines a norm on the factor space $H/H_{\|\cdot\|}$ in a natural fashion since it is constant on every equivalence class that is an element of that space. If $\|\cdot\|$ is a norm in H, then $(H, \|\cdot\|)$ is called a *normed vector space* or a *normed linear space.* It is clear what norm convergence of sequences, norm-closedness of sets, etc. mean. We also speak of $\|\cdot\|$-convergence etc. in this context. A normed vector space $(H, \|\cdot\|)$ is called *complete* or a *Banach space* if the Cauchy criterion holds, i.e., if $x_1, x_2, \ldots \in H$, $\lim_{j,k \to \infty} \|x_j - x_k\| = 0$ implies the existence of an $x \in H$ such that $x_k \to x$ in norm, i.e., $\lim_{k \to \infty} \|x_k - x\| = 0$—in words, every $\|\cdot\|$-fundamental sequence in H is $\|\cdot\|$-convergent. A *ball* $\{x | \|x - x_0\| \leqq r\}$ with center $x_0 \in H$ and radius $r > 0$ in a Banach space is compact iff H is finite dimensional. Let $(H, \|\cdot\|)$, $(L, \|\cdot\|)$ be normed linear spaces. A linear mapping $\varphi: H \to L$ is called a *homomorphism* of the normed linear spaces $(H, \|\cdot\|)$, $(L, \|\cdot\|)$ if there is a real $A > 0$ such that $\|\varphi(x)\| \leqq A\|x\|$ $(x \in H)$. This is equivalent to norm continuity at $0 \in H$, and this in turn is equivalent to norm continuity everywhere on H. If such a homomorphism is a bijection and its inverse $\varphi^{-1}: L \to H$ is also a homomorphism, then φ is called an isomorphism of the normed linear spaces $(H, \|\cdot\|)$ and $(L, \|\cdot\|)$. If $\|\cdot\|_1$ and $\|\cdot\|_2$ are two norms on the real vector space H and id_H is an isomorphism of $(H, \|\cdot\|_1)$ and $(H, \|\cdot\|_2)$, then the norms $\|\cdot\|_1$ and $\|\cdot\|_2$ are said to be *equivalent*; this is the case iff there is an $A > 0$ such that $A^{-1}\|x\|_1 \leqq \|x\|_2 \leqq A\|x\|_1$ $(x \in H)$. The passage from $\|\cdot\|_1$ to an equivalent $\|\cdot\|_2$ is also called *renorming.*

The usual modulus is a norm on $\mathbb{R}$. A norm continuous linear mapping of a normed vector space H into $\mathbb{R}$ is also called a *continuous linear form* on H. The continuous linear forms on H form a vector space H^* in a natural fashion, and $\|L\| = \sup\{|Lx| \mid x \in H,\ \|x\| \leqq 1\}$ $(L \in H^*)$ defines a norm in H^*. In fact, the Hahn–Banach theorem (I.6.1) will prove the existence, for any given $0 \neq x \in H$, of an $L \in H^*$ such that $\|L\| \leqq 1$ and $Lx = \|x\|$ (see Example I.6.2; define L on $\mathrm{lin}\{x\} = \{\alpha x \mid \alpha \in \mathbb{R}\}$ by $L(\alpha x) = \alpha\|x\|$ and extend it to all of H, with the majorant function $\|\cdot\|$ on H). Norm convergence in H^* means uniform convergence on every norm-bounded subset of H, and it is practically obvious from this that $(H^*, \|\cdot\|)$ is a Banach space, the *dual Banach space* of the normed vector space $(H, \|\cdot\|)$.

Let $(H, +, \cdot, \leqq)$ be a vector lattice. Let $\|\cdot\|$ be a norm on H such that $\|x_+\| \leqq A\|x\|$ $(x \in H)$ for some real $A > 0$. Then $(H, +, \cdot, \leqq)$ is called a *normed vector lattice*; $+, \cdot, \leqq, \|\cdot\|$ are often omitted in the notation. If in addition $(H, \|\cdot\|)$ is a Banach space, $(H, +, \cdot, \|\cdot\|)$ is called a *Banach lattice.* In a normed vector lattice H, there are real constants $B, C > 0$ such that

$$\|(x \vee y) - (x' \vee y')\| \leqq B(\|x - x'\| + \|y - y'\|),$$
$$\|(x \wedge y) - (x' \wedge y')\| \leqq C(\|x - x'\| + \|y - y'\|) \qquad (x, x', y, y' \in H)$$

and the minorants of a fixed vector form a norm-closed set as well as its majorants. The same holds for majorants. All these statements are equivalent to the norm-closedness of H_+. It is clear what an isomorphism of normed vector lattices is. A Banach lattice $(H, +, \cdot, \leqq, \|\cdot\|)$ is called an *L-space* if $x, y \in H_+ \Rightarrow \|x + y\| = \|x\| + \|y\|$, and an *M-space* if $x, y \in H_+$, $x \wedge y = 0 \Rightarrow \|x + y\| = \max\{\|x\|, \|y\|\}$. *L*-spaces have an important property which might be called the *monotone convergence theorem.* We concentrate it, together with a corollary, into

2.2. Theorem. *Let $(H, +, \cdot, \leqq, \|\cdot\|)$ be an L-space.*

2.2.1. *Let $x_1, x_2, \ldots \in H$ be such that $x_1 \leqq x_2 \leqq \cdots$ and $\sup_n \|x_n\| < \infty$. Then there is an $x \in H$ such that $\lim_n \|x_n - x\| = 0$.*

2.2.2. *$(H, \leqq)$ is a conditionally complete lattice.*

Proof. 1. Let $y_n = x_n - x_1$ $(n = 1, 2, \ldots)$. Then $0 = y_1 \leqq y_2 \leqq \cdots$, and $\|y_n\| \leqq \|x_n\| + \|x_1\|$ implies $\sup_n \|y_n\| < \infty$. Thus we may assume

$0 \leqq x_1 \leqq x_2 \leqq \cdots$. Let $\sup_n \|x_n\| = \alpha$. Choose $\varepsilon > 0$ arbitrarily and determine n such that $\|x_n\| > \alpha - \varepsilon$. Let now $k \geqq n$. Then $\|x_k\| = \|x_k - x_n + x_n\| = \|x_k - x_n\| + \|x_n\| > \alpha - \varepsilon$. Here we have used $x_k - x_n, x_n \geqq 0$ plus the fact that H is an L-space. It is now clear that $\lim_{n, k \to \infty} \|x_k - x_n\| = 0$, whence the claimed convergence follows because H is a Banach space, i.e., complete.

2. Let $\varnothing \neq M \subseteq H$ and $\bar{x} \in H$ such that $x \leqq \bar{x}$ $(x \in M)$. The set $\{x_1 \vee \cdots \vee x_n | n \geqq 1, x_1, \ldots, x_n \in M\}$ is increasingly filtered and has the same majorants as M. Hence we may assume that M is stable under finite suprema and thus in particular increasingly filtered. Choose any $x_0 \in M$. Then $M_0 = \{x | x_0 \leqq x \in M\}$ has the same majorants as M. Thus we may assume that M contains a minorant x_0 of M. Let now $M' = \{x - x_0 | x \in M\}$. Then $0 \in M' \subseteq H_+$, M' is increasingly filtered, and $\bar{x}$ is a majorant of M iff $\bar{x} - x_0$ is a majorant of M. Thus we may assume $M \subseteq H_+$, $\bar{x} \in H_+$. Clearly $x, y \in H_+, x \leqq y \Rightarrow \|y\| = \|x + (y - x)\| = \|x\| + \|y - x\| \geqq \|x\|$. In particular, $x \in M \Rightarrow \|\bar{x}\| \geqq \|x\|$, and thus $\alpha = \sup_{x \in M} \|x\|$ is a finite real. Choose $x_1, x_2 \in M$ such that $\|x_n\| \to \alpha$. Since M is increasingly filtered, we may assume $x_1 \leqq x_2 \leqq \cdots$. We now deduce from 1 the existence of some $\tilde{x} \in H$ with $\lim_n \|\tilde{x} - x_n\| = 0$. By $|\|x - y\|| \leqq \|x - y\|$ $(x, y \in H)$ the norm is norm continuous and $\|\tilde{x}\| = \alpha$ follows. We claim that $\tilde{x} \geqq x$ $(x \in M)$. Assume there is some $x' \in M$ such that $\tilde{x} \geqq x'$ is not true. Since $\tilde{x} \geqq x'$ is tantamount to $\tilde{x} \vee x' = \tilde{x}$, we find $\tilde{x} \vee x' \neq x$. On the other hand, H is a normed vector lattice, which implies $\lim[x_n \vee x'] = \tilde{x} \vee x'$. But $x_n \vee x' \in M$, hence $\|\tilde{x} \vee x'\| \leqq \alpha$. On the other hand,

$$\begin{aligned} \|\tilde{x} \vee x'\| &= \|[(\tilde{x} \vee x') - \tilde{x}] + \tilde{x}\| \\ &= \|(\tilde{x} \vee x') - \tilde{x}\| + \|x'\| = \|(\tilde{x} \vee x') - \tilde{x}\| + \alpha. \end{aligned}$$

Thus $\|(\tilde{x} \vee x') - \tilde{x}\| = 0$, hence $\tilde{x} \vee x' = \tilde{x}$, hence $x' \leqq \tilde{x}$, a contradiction. Thus $\tilde{x}$ is an upper bound for M. Every upper bound $\bar{x}$ for M satisfies $\bar{x} \geqq x_n$, i.e., $\bar{x} \vee x_n = \bar{x}$ $(n = 1, 2, \ldots)$, which implies $\bar{x} \vee \tilde{x} = \bar{x}$, hence $\bar{x} \geqq \tilde{x}$. It follows that $\tilde{x}$ is the l.u.b. of M. Multiplication by -1 yields the existence of a g.l.b. for every set bounded from below.

Let $(H, +, \cdot)$ be a real vector space. A mapping $(\cdot, \cdot)$: $H \times H \to \mathbb{R}$ is called *bilinear* if for every $x \in H$ the mappings $(\cdot, x)$: $y \to (y, x)$ and $(x, \cdot)$: $y \to (x, y)$ of $H \to \mathbb{R}$ are linear. $(\cdot, \cdot)$ is called *positive semidefinite* if $x \in H \Rightarrow (x, x) \geqq 0$. A positive semidefinite $(\cdot, \cdot)$ is called *positive definite* or a *scalar product* in H if $(x, x) = 0 \Rightarrow x = 0$ for every $x \in H$. If $(\cdot, \cdot)$ is positive definite, then $\|x\| = \sqrt{(x, x)}$ $(x \in H)$ defines a seminorm on H. It is a norm iff $(\cdot, \cdot)$ is positive definite. $\|\cdot\|$ is called the (square) seminorm resp. norm *attached* to $(\cdot, \cdot)$. If it is a norm, $(H, (\cdot, \cdot))$ is called a *pre-Hilbert space*. If $(H, (\cdot, \cdot))$ is a pre-Hilbert space such that the attached normed linear space $(H, \|\cdot\|)$ is a Banach space, i.e., norm complete, then $(H, (\cdot, \cdot))$ is called a (*real*) *Hilbert space*. Let $(H, (\cdot, \cdot))$ be a Hilbert space. A linear isomorphism φ of H onto itself is called *unitary* if $(\varphi(x), \varphi(y)) = (x, y)$ $(x, y \in H)$.

Let H be a complex linear space. A mapping $(\cdot, \cdot)$: $H \times H \to \mathbb{C}$ is called a *scalar product* on H if (1) for every $x \in H$ the mapping $(\cdot, x)$: $y \to (y, x)$ of $H \to \mathbb{C}$ is (complex) linear, (2) $(x, y) = \overline{(y, x)}$ = the complex conjugate of (x, y) $(x, y \in H)$, and (3) $(\cdot, \cdot)$ is positive definite, i.e., $(x, x) \geqq 0$ $(x \in H)$ with equality iff $x = 0$. If $(\cdot, \cdot)$ is a scalar product on H, then $(H, (\cdot, \cdot))$ is called a *complex pre-Hilbert space*, and $\|\cdot\|$: $x \to \sqrt{(x, x)}$ is a norm on H which we call the (square) norm on H attached to $(\cdot, \cdot)$. The concept of a complex Hilbert space and of a unitary transformation are now defined literally as in the real case.

For reference, see SCHAEFER [1], HALMOS [1].

3. TOPOLOGICAL SPACES

Basic notions of point set topology are presented in this section. KELLEY [1] and STEEN and SEEBACH [1] are standard references. Let $\Omega \neq \varnothing$. A system $\mathscr{T}$ of subsets of Ω is called a *topology* in Ω if $\varnothing, \Omega \in \mathscr{T}$, and $\mathscr{T}$ is stable under finite intersections (i.e., $n \geqq 1$, $U_1, \ldots, U_n \in \mathscr{T} \Rightarrow U_1 \cap \cdots \cap U_n \in \mathscr{T}$) and stable under arbitrary unions (i.e., if $\bigcup_{\iota \in I} U \in \mathscr{T}$ for any family $(U_\iota)_{\iota \in J}$ in $\mathscr{T}$). $\mathscr{P}(\Omega)$ is called the *discrete topology* of Ω. If $\mathscr{T}$ is a topology in Ω, then $(\Omega, \mathscr{T})$ is called a *topological space*; $\mathscr{T}$ is often omitted from this notation. If $\Omega \ni \omega \in U \in \mathscr{T}$, then U is called an *open neighborhood* of ω. Every set containing an open neighborhood of ω is called a *neighborhood* of ω. Let $\Omega \neq \varnothing$ and $\mathscr{T}$, $\mathscr{T}'$ be two topologies in Ω. If $T \subseteq T'$, we say that $\mathscr{T}$ is *coarser* than $\mathscr{T}'$, and that $\mathscr{T}'$ is *finer* than $\mathscr{T}$. The discrete topology clearly is always finest; $\{\varnothing, \Omega)$ coarsest. If $\mathscr{A} \subseteq \mathscr{P}(\Omega)$, then the intersection of all topologies containing $\mathscr{A}$ is a topology, the coarsest of all those, or the *topology generated by* $\mathscr{A}$; we denote it by $\mathscr{T}(\mathscr{A})$; it consists of $\varnothing$, Ω, and all unions of finite intersections of set from $\mathscr{A}$. $\mathscr{A}$ is then also called a subbasis of $\mathscr{T}(\mathscr{A})$; if $\mathscr{A}$ is already stable under finite intersections, then $\mathscr{T}(\mathscr{A})$ consists of all unions of sets from $\mathscr{A}$, and $\mathscr{A}$ is called a basis of $\mathscr{T}(\mathscr{A})$. $\mathscr{T}$ resp. $(\Omega, \mathscr{T})$ is said to satisfy the *second countability axiom* or to be *second countable* if $\mathscr{T} = \mathscr{T}(\mathscr{A})$ for some countable $\mathscr{A}$.

If $(\Omega, \mathscr{T})$ is a topological space and $\Omega_0 \subseteq \Omega$, then $\mathscr{T}_0 = \mathscr{T} \cap \Omega_0 = \{U \cap \Omega_0 \mid U \in \mathscr{T}\}$ is a topology in Ω_0. It is called the *restriction* of $\mathscr{T}$ to Ω_0, or the *relative topology* in Ω_0 for $\mathscr{T}$.

Let $(\Omega, \mathscr{T})$ be a topological space. A set $G \subseteq \Omega$ is called *open* (for $\mathscr{T}$) if $G \in \mathscr{T}$. Clearly this is the case iff G contains a neighborhood of every

$\omega \in G$. A set $F \subseteq \Omega$ is called *closed* (for $\mathscr{T}$) if $\Omega \backslash F$ is open. A set that is both closed and open (for $\mathscr{T}$) is called *clopen* ($\varnothing$ and Ω are trivial examples). If $\mathscr{T}$ has a basis consisting of clopen sets, $(\Omega, \mathscr{T})$ is called *totally disconnected.* $(\Omega, \mathscr{P}(\Omega))$ is always totally disconnected. If $(\Omega, \mathscr{T})$ is totally disconnected, then every $\varnothing \neq \Omega_0 \subseteq \Omega$, endowed with the relative topology $\mathscr{T}_0 = \mathscr{T} \cap \Omega_0$ yields a totally disconnected topological space $(\Omega_0, \mathscr{T}_0)$. A set K is called *compact* if it has the finite covering property: If $(U_\iota)_{\iota \in I}$ is a family of open sets covering K, then there is a finite subset J of I such that $K \subseteq \bigcup_{\iota \in J} U_\iota$. Every closed subset of a compact set is compact. Every set that is contained in a compact set is called *conditionally compact.* If $\mathscr{F}$ is a decreasingly filtered system of nonempty compact sets in Ω, then $\bigcap_{K \in \mathscr{F}} K \neq \varnothing$. A set $M \subseteq \Omega$ is called a G_δ (set, for $\mathscr{T}$) if there is a sequence $G_1, G_2, \ldots$ of open sets with $M = G_1 \cap G_2 \cap \cdots$, i.e., if $M \in \mathscr{T}_\delta$ (the letter δ is a reminder for the german word *Durchschnitt* for "intersection," in the countable case). A set $M \subseteq \Omega$ is called an F_σ (set, for $\mathscr{T}$) if there is a sequence $F_1, F_2, \ldots$ of closed sets such that $M = F_1 \cup F_2 \cup \cdots$ (the letter is a reminder of "sigma" symbols for countable summation). Countable unions of compact sets are called K_σ sets.

Let $(\Omega, \mathscr{T})$ be a topological space and $M \subseteq \Omega$. A point ω of M is called an *inner* or *interior point* of M if some neighborhood of ω is contained in M (i.e., if M is a neighborhood of ω). The set of all inner points of M in the largest open set contained in M (namely, the union of all such sets); it is called the *interior* or the *open kernel* of M and denoted by M° or by int M. A point $\omega \in \Omega$ is called a *cluster point* or *accumulation point* of M if every neighborhood of ω has a nonempty intersection with M (i.e., intersects M). The set of all cluster points of M is the smallest closed set containing M (namely, the intersection of all such sets). It is called the *closure* of M and denoted by $\overline{M}$ or cl M. Obviously $\Omega \backslash \overline{M} = \operatorname{int}(\Omega \backslash M)$. $\overline{M} \backslash \operatorname{int} M$ is called the *boundary* of M and is denoted by ∂M. A point ω is called a *density point* of M if every neighborhood of ω contains an infinity of points of M. Points of M that are not density points of M are called *discrete points* of M. If M is contained in the closure of N, then N is said to be *dense* in M. In particular, it is clear what a dense set in Ω is. $(\Omega, \mathscr{T})$ is said to be *separable* or *first countable* or to satisfy the *first countability axiom* if there is a countable subset that is dense in Ω.

A topological space $(\Omega, \mathscr{T})$ is called *locally compact* if every point in Ω has a compact neighborhood, and *Hausdorff* (or a T_2 space) if the so-called *Hausdorff separation axiom* holds: Given $\omega, \eta \in \Omega$, $\omega \neq \eta$, there exist open sets U, V such that $U \cap V = \varnothing$, $\omega \in U$, $\eta \in V$. This axiom is also called the T_2 axiom. For other separation axioms see KELLEY [1], STEEN and SEEBACH [1]. In a Hausdorff space, every compact set is closed. Let $(\Omega, \mathscr{T})$ be a locally compact space and ∞ an entity $\notin \Omega$. Then

$$\mathscr{T}' = \mathscr{T} \cup \{(\Omega \backslash K) \cup \{\infty\} | K \subseteq \Omega \text{ compact}\}$$

defines a compact topology in $\Omega' = \Omega + \{\infty\}$. $(\Omega', \mathscr{T}')$ is called the *Alexandroff* or *one-point compactification* of $(\Omega, \mathscr{T})$; it is Hausdorff iff $(\Omega, \mathscr{T})$ is.

Let $\|\cdot\|$ be a seminorm in a real vector space $(H, +, \cdot)$. For every $x \in H$ and every $\varepsilon > 0$, denote by $U_\varepsilon(x)$ the so-called *open ε-neighborhood* $\{y | y \in H, \|y - x\| < \varepsilon\}$ of x. Call a set $G \subseteq H$ open (for $\|\cdot\|$) if for every $x \in G$, there is an $\varepsilon > 0$ such that $U_\varepsilon(x) \subseteq G$. Then $\mathscr{T}_{\|\cdot\|} = \{G | G \subseteq H$ open for $\|\cdot\|\}$ is a topology in H. It is Hausdorff iff $\|\cdot\|$ is a norm. If $\|\cdot\|$ is a norm, $(H, \mathscr{T})$ is locally compact iff H is of finite dimension.

With a few exceptions, all topological spaces encountered in this book are Hausdorff, and whenever we meet such an exception, we shall immediately try to modify it such that it becomes a Hausdorff space.

Let $(\Omega, \mathscr{T})$ be a topological space. A net $(\omega_\iota)_{\iota \in I}$ in Ω is said to *converge* to $\omega \in \Omega$ or to have the *limit* x, in symbols $\lim_{\iota \in I} \omega_\iota = \omega$ if for every neighborhood U of ω there is an $\iota \in I$ such that $\iota \leqq \kappa \in I \Rightarrow \omega_\kappa \in U$. $(\Omega, \mathscr{T})$ is Hausdorff iff every net in Ω has at most one limit in Ω. A point $\omega \in \Omega$ is called a *limit point* of $(\omega_\iota)_{\iota \in I}$ if for every $\iota \in I$ and every neighborhood U of ω, there is an $\iota \leqq \kappa \in I$ with $\omega_\kappa \in U$. If $(\Omega, \mathscr{T})$ is compact, then every net $(\omega_\iota)_{\iota \in I}$ has at least one limit point in Ω. If it has exactly one limit point ω in Ω, then $\lim_{\iota \in I} \omega_\iota = \omega$.

A system $\mathscr{F}$ of subsets of a set $\Omega \neq \varnothing$ is called a *filter base* if it consists of nonempty sets only and is decreasingly filtered for the partial ordering $\subseteq$; the latter means that for any $E, F \in \mathscr{F}$ there is a $G \in \mathscr{F}$ such that $G \subseteq E \cap F$. A filter base $\mathscr{F}$ is called a *filter* if $E \in \mathscr{F}$, $E \subseteq F \subseteq \Omega \Rightarrow F \in \mathscr{F}$. Every net $(\omega_\iota)_{\iota \in I}$ in Ω gives rise to a filter base $\{\{\omega_\kappa | \iota \leqq \kappa\} | \iota \in I\}$, which is called the *filter base associated* to $(\omega_\iota)_{\iota \in I}$, and every filter base $\mathscr{F}$ in Ω gives rise to a filter $\overline{\mathscr{F}} = \bigcup_{E \in \mathscr{F}} \{F | E \subseteq F \subseteq \Omega\}$. A filter base $\mathscr{F}$ in Ω is said to be *finer* than the filter base $\mathscr{G}$ in Ω if $\overline{\mathscr{F}} \supseteq \overline{\mathscr{G}}$. If $\overline{\mathscr{F}} \supseteq \overline{\mathscr{G}} \Rightarrow \overline{\mathscr{F}} = \overline{\mathscr{G}}$, then $\mathscr{G}$ is called an *ultrafilter basis* in Ω. By the lemma of Kuratowski and Zorn, there exists, for any given filter base $\mathscr{F}$ in Ω, an ultrafilter base $\mathscr{G}$ in Ω that is finer than $\mathscr{F}$. The system of all sets containing a fixed point of ω is always an ultrafilter in Ω; there are no other ultrafilters in Ω iff Ω is finite.

Let $(\Omega, \mathscr{T})$ be a topological space. A point $\omega \in \Omega$ is called a *limit point* of the filter base $\mathscr{F}$ in Ω if every neighborhood of ω meets every member of $\mathscr{F}$. If $\mathscr{F}$ is the filter base associated to the net $(\omega_\iota)_{\iota \in I}$, then $\mathscr{F}$ and $(\omega_\iota)_{\iota \in I}$ have exactly the same limit points. For every $\omega \in \Omega$, the set system $\mathscr{U}(\omega) = \{U | \omega \in U \in \mathscr{T}\}$ is a filter base in Ω; it is called the *neighborhood filter base* in ω. ω is a limit point of the filter base $\mathscr{F}$ in Ω iff there is a filter base finer both than $\mathscr{U}(\omega)$ and $\mathscr{F}$. $\mathscr{F}$ is said to converge to ω if $\mathscr{F}$ is finer than $\mathscr{U}(\omega)$; we write $\lim \mathscr{F} = \omega$ or $\lim_{\mathscr{F}} \eta = \omega$ in this case. Let $X \neq \varnothing$ and φ be a mapping of X into the topological space Ω. Let $\mathscr{F}$ be a filter base in X. φ is said to *converge* to ω along $\mathscr{F}$ if the filter base $\varphi F = \{\varphi F \mid F \in \mathscr{F}\}$ converges to ω. In this case we write $\lim_{\mathscr{F}} \varphi = \omega$.

The reader is urgently advised to work through these definitions and statements carefully such that he becomes able to recognize them easily if slightly varied or replaced by equivalent ones. He should be familiar with symbols like $\lim_n$, $\lim_{n \to \infty}$, $\lim_{t \to a+0}$, etc.

If $\varnothing \neq M \subseteq \Omega$, a point $\omega \in \Omega$ is a cluster point of M iff there is a filter basis in M converging to ω. If $(\Omega, \mathscr{T})$ is Hausdorff, a set $M \subseteq \Omega$ is compact iff every ultrafilter base in M has a limit in M, or, equivalently, if every filter base in M has a limit point in M.

Let $(X, \mathscr{S})$, $(\Omega, \mathscr{T})$ be topological spaces. A mapping $\varphi\colon X \to \Omega$ is called *continuous* (for $\mathscr{S}$, $\mathscr{T}$) if one of the following equivalent conditions holds: (1) The inverse image $\varphi^{-1}V = \{x \mid x \in X, \varphi(x) \in V\}$ of every open $V \subseteq \Omega$ is open (in short: $\varphi^{-1}\mathscr{T} \subseteq \mathscr{S}$). (2) The inverse image of every closed subset of Ω is closed (in short, $\varphi^{-1}\mathscr{T}_{\mathbf{c}} \subseteq \mathscr{S}_{\mathbf{c}}$). (3) For every $\omega \in \Omega$ and every neighborhood V of $\varphi(\omega)$, there is a neighborhood U of ω such that $\varphi U \subseteq V$. The set of all continuous mappings from X to Ω is denoted (omitting $\mathscr{S}$, $\mathscr{T}$ for shortness) by $\mathscr{C}(X, \Omega)$. It is obvious that the composition of continuous mappings yields a continuous mapping again. In short,

$$\mathscr{C}(Y, \Omega) \circ \mathscr{C}(X, Y) \subseteq \mathscr{C}(X, \Omega).$$

It is less obvious but not hard to prove that in the case of a compact space $(X, \mathscr{S})$ and a continuous $\varphi\colon X \to \Omega$ the set $\varphi X \subseteq \Omega$ is compact; and in the case of a compact Hausdorff space $(X, \mathscr{S})$ and a continuous bijection $\varphi\colon X \to \Omega$ the inverse mapping $\varphi^{-1}\colon \Omega \to X$ is continuous again. If $\varphi\colon X \to \Omega$ is a bijection such that φ and φ^{-1} are continuous, φ is called a *homeomorphism* (of X onto Ω).

If $X \neq \varnothing$ and $((\Omega_\iota, \mathscr{T}_\iota))_{\iota \in I}$ is a family of topological spaces, $\varphi_\iota\colon X \to \Omega_\iota$ a mapping from X into Ω_ι, then $\mathscr{T}(\bigcup_{\iota \in I} \varphi_\iota^{-1}\mathscr{T}_\iota)$ is the coarsest topology such that all φ_ι are continuous; it is also called the *topology generated* by $(\varphi_\iota)_{\iota \in I}$ (and $((\Omega_\iota, \mathscr{T}_\iota))_{\iota \in I}$, of course).

The space $\mathscr{C}(\Omega, \mathbb{R})$ of all continuous real functions on a topological space Ω is always a vector lattice (for the pointwise ordering $f \leqq g \Leftrightarrow f(\omega) \leqq g(\omega)$ $(\omega \in \Omega)$, of course) and an *algebra*, i.e., a vector space stable under (pointwise) multiplication. The subspace of all bounded continuous real functions on Ω is denoted by $\mathscr{C}^{\mathrm{b}}(\Omega, \mathbb{R})$; it is a Banach lattice (for the pointwise ordering and) for the so-called sup norm or uniform (convergence) norm $\|f\| = \sup_{\omega \in \Omega} |f(\omega)|$. If Ω is compact, then $\mathscr{C}(\Omega, \mathbb{R}) =$

$\mathscr{C}^b(\Omega, \mathbb{R})$. A function $f: \Omega \to \mathbb{R}$ is said to *vanish at infinity* if for every $\varepsilon > 0$, there is a compact K such that $\omega \in \Omega \backslash K \Rightarrow |f(\omega)| < \varepsilon$. The space $\mathscr{C}^0(\Omega, \mathbb{R})$ of all continuous real functions on Ω vanishing at infinity is contained in $\mathscr{C}^b(\Omega, \mathbb{R})$ and a Banach lattice with the sup norm. For every real function f on a set Ω, the set $\{\omega | f(\omega) \neq \varnothing\}$ is called the *carrier* or *support* of f and is denoted by supp(f). If Ω is a topological space, the closure $\overline{\text{supp}(f)}$ is called the *topological carrier* or *support* of f; its complement is the largest open set on which f is constant $= 0$ (one also says that f vanishes identically on it). For any topological space $(\Omega, \mathscr{T})$ the space $\mathscr{C}^{00}(\Omega, \mathbb{R}) = \{f | f \in \mathscr{C}(\Omega, \mathbb{R}), \overline{\text{supp}(f)} \text{ is compact}\}$ is contained in $\mathscr{C}^0(\Omega, \mathbb{R})$ and a vector lattice, but in general not a Banach lattice for the sup norm. If Ω is compact, then $\mathscr{C}^{00}(\Omega, \mathbb{R}) = \mathscr{C}^0(\Omega, \mathbb{R}) = \mathscr{C}^b(\Omega, \mathbb{R}) = \mathscr{C}(\Omega, \mathbb{R})$.

Dini's theorem says: If $(\Omega, \mathscr{T})$ is a topological space and $F \subseteq \mathscr{C}(\Omega, \mathbb{R})_+$ is decreasingly filtered with pointwise infimum 0, then for every compact $K \subseteq \Omega$ and every $\varepsilon > 0$, there is an $f \in F$ with $f \leqq \varepsilon$ uniformly on K. In particular, if Ω is compact, then $\mathscr{C}(\Omega, \mathbb{R}) \ni f_1 \geqq f_2 \geqq \cdots \searrow 0$ pointwise implies $f_n \to 0$ uniformly on Ω.

The fundamental *Weierstrass–Stone theorem* says: Let $(\Omega, \mathscr{T})$ be a compact Hausdorff space and $\mathscr{A}$ a sup norm closed linear subspace of $\mathscr{C}(\Omega, \mathbb{R})$ that is either a vector sublattice of $\mathscr{C}(\Omega, \mathbb{R})$ or an algebra, and separates the points of Ω ($\omega, \eta \in \Omega, \omega \neq \eta \Rightarrow f(\omega) \neq f(\eta)$ for some $f \in \mathscr{A}$); then either $\mathscr{A} = \mathscr{C}(\Omega, \mathbb{R})$ or there is a point $\omega_0 \in \Omega$ such that $\mathscr{A} = \{f | f \in \mathscr{C}(\Omega, \mathbb{R}), f(\omega_0) = 0\}$. It is hoped that the reader will be able to deduce variants of this theorem according to the needs of future situations.

A topological space $(\Omega, \mathscr{T})$ is said to be *completely regular* if for any $\omega \in \Omega$ and any open $\omega \in U \subseteq \Omega$, there is an $f \in \mathscr{C}(\Omega, \mathbb{R})$ such that $f(\omega) = 1$ and $0 \leqq f \leqq 1_U$, i.e., f takes on values between 0 and 1 only and vanishes outside U. Every locally compact Hausdorff space is, after a theorem of *Urysohn*, completely regular. In a completely regular space one can, again by a theorem named after *Urysohn*, for every compact set K and any closed set F disjoint from K (i.e., $K \cap F = \varnothing$) find a continuous real function f with $0 \leqq f \leqq 1, f(\omega) = 1$ $(\omega \in K), f(\omega) = 0$ $(\omega \in F)$.

Recall $\overline{\mathbb{R}} = \mathbb{R} \cup \{-\infty, \infty\}$ with its usual total ordering and its obvious topology. Let $(\Omega, \mathbb{R})$ be a topological space. A function $f: \Omega \to \overline{\mathbb{R}}$ is called *lower semicontinuous* if for every and every $\mathbb{R} \ni a < f(\omega)$, there is a neighborhood U of ω such that $f(\eta) > a$ $(\eta \in U)$; or, equivalently, if for every $a \in \mathbb{R}$, the set $\{\omega | f(\omega) > a\}$ is open. *Upper semicontinuity* is defined analogously. A function $f: \Omega \to \mathbb{R}$ is *continuous* iff it is upper and lower semicontinuous. The lower continuous $\overline{\mathbb{R}}$-valued functions on Ω form a lattice under pointwise ordering; they are even stable under arbitrary suprema; in particular, any supremum of functions from $\mathscr{C}(\Omega, \mathbb{R})$ is a lower semicontinuous function. $(\Omega, \mathscr{T})$ is completely regular iff every lower semicontinuous function bounded from below is a supremum of some subset of $\mathscr{C}^b(\Omega, \mathbb{R})$.

Analogous statements hold for upper semicontinuity.

Let $((\Omega_\iota, \mathscr{T}_\iota))_{\iota \in I}$ be a family of topological spaces, $\varphi_\iota: \prod_{\kappa \in I} \Omega_\kappa \to \Omega_\iota$ the natural projection ($\iota \in I$). Then the topology generated in $\prod_{\iota \in I} \Omega_\iota$ by $(\varphi_\iota)_{\iota \in I}$ is called the *product topology* of the topologies $\mathscr{T}_\iota$ ($\iota \in I$) or the *topology of componentwise convergence* in $\prod_{\iota \in I} \Omega_\iota$, and is also denoted by $\prod_{\iota \in I} \mathscr{T}_\iota$. It is Hausdorff iff all $\mathscr{T}_\iota$ ($\iota \in I$) are Hausdorff. We also write $\prod_{\iota \in I} (\Omega_\iota, \mathscr{T}_\iota)$ in place of $(\prod_{\iota \in I} \Omega_\iota, \prod_{\iota \in I} \mathscr{T}_\iota)$.

With the help of ultrafilters one proves *Tychonov's theorem*: If $(\Omega_\iota, \mathscr{T}_\iota)$ is compact ($\iota \in I$), then $\prod_{\iota \in I} (\Omega_\iota, \mathscr{T}_\iota)$ is compact.

If $(\Omega_\iota, \mathscr{T}_\iota)$ is totally disconnected ($\iota \in I$), then $\prod_{\iota \in I} (\Omega_\iota, \mathscr{T}_\iota)$ is totally disconnected.

Let $\Omega \neq \emptyset$. A mapping $|\cdot, \cdot|: \Omega \to \mathbb{R}_+$ is called a *pseudo-metric* (or *quasi-metric*) in Ω if the *triangle inequality* $|\omega, \eta| \leqq |\omega, \zeta| + |\zeta, \eta|$ ($\omega, \zeta, \eta \in \Omega$) holds. For every $\omega \in \Omega$, the set $U_\varepsilon(\omega) = \{\eta | \eta \in \Omega, |\omega, \eta| < \varepsilon\}$ is then called the (pseudo-metric) *open ε-neighborhood* of ω. A pseudo-metric $|\cdot, \cdot|$ in Ω is called a *metric* if $|\omega, \eta| = 0 \Rightarrow \omega = \eta$.

Let $\Omega \neq \emptyset$ and $f_1, f_2, \ldots \in \mathbb{R}^\Omega$, $0 \leqq f_n \leqq 1$ ($n = 1, 2, \ldots$). Then $|\omega, \eta| = \sum_{n=1}^\infty 2^{-n} |f_n(\omega) - f_n(\eta)|$ defines a pseudo-metric on Ω. It is a metric iff the $f_1, f_2, \ldots$ separate the points of Ω. This device is the general tool for the *metrization* of topological spaces, the $f_1, f_2, \ldots$ mostly coming from applications of a Urysohn theorem. Every locally compact space with a countable base is metrizable. Every metrizable compact space is second countable. There are compact separable Hausdorff spaces that are not metrizable. A countable product of metrizable spaces is metrizable again. If $(\Omega, \mathscr{T})$ is compact and metrizable, then $(\mathscr{C}(\Omega, \mathbb{R}), \|\cdot\|)$ is separable.

A metric $|\cdot,\cdot|$ in $\Omega \neq \emptyset$ is called *complete* if *Cauchy's convergence criterion* ($\omega_1, \omega_2, \ldots \in \Omega$, $\lim_{j,k \to \infty} |\omega_j, \omega_k| = 0 \Rightarrow \lim_{k \to \infty} |\omega_k, \omega| = 0$ for some $\omega \in \Omega$) holds. A separable topological space that is metrizable with a complete metric is called a *Polish space*. Clearly $\mathbb{R}$ is a Polish space and every closed subset of a Polish space is a Polish space again when endowed with the relative topology, and so are all G_δ subsets $\neq \emptyset$. In a Polish space every open set is an F_σ set. The structure of Polish spaces is well known: By a celebrated theorem of *Urysohn–Alexandroff* they are exactly those spaces that are homeomorphic to G_δ subsets of the compact metric space $\{0, 1\}^{\mathbb{N}}$.

The metric $|x, y| = |\arctan x - \arctan y|$ ($x, y \in \mathbb{R}$) in $\mathbb{R}$ shows that in a Polish space $(\Omega, \mathscr{T})$ not every metric $|\cdot, \cdot|$ such that $\mathscr{T} = \mathscr{T}_{|\cdot,\cdot|}$ must necessarily be complete.

If $(\Omega, |\cdot, \cdot|)$ is a complete metric space, then every countable intersection of open dense subsets of Ω is dense in Ω (*Baire's category theorem*).

Let $(\Omega, |\cdot, \cdot|)$ and $(\Omega', |\cdot, \cdot|)$ be two pseudo-metric spaces. A mapping $\varphi: \Omega \to \Omega'$ is called an *isometry* if $|\varphi(\omega), \varphi(\eta)| = |\omega, \eta|$ $(\omega, \eta \in \Omega)$. If both spaces are metric spaces, then every isometry is an injection. If it is, in addition, bijective, then $(\Omega, |\cdot, \cdot|)$ and $(\Omega', |\cdot, \cdot|)$ have the same metric properties: Both of them are complete or not, etc.

CHAPTER I

POSITIVE CONTENTS AND MEASURES

The *raison d'être* of content and measure theory is the desire to measure the content (say, of volume, area, or mass) of a great variety of subsets of a given basic set Ω, and to "integrate" a great variety of, say, real functions on Ω. This desire is channeled by a very simple idea: It seems to be obvious how to measure the content of certain comparatively simple subsets of Ω (such as rectangles if $\Omega = \mathbb{R}^2$ and we want to measure areas), or how to integrate certain comparatively elementary functions (such as step functions, which are constant on "simple" sets); the problem is how to *extend* this measurement resp. integration to more complicated sets resp. functions.

Thus we are to define real functions $\geqq 0$ on a certain domain consisting of subsets of Ω which we consider as "simple," and this leads to the notion of a (positive) *content* on a ring of subsets of Ω. Similarly, we are led to consider functionals on a certain set of real functions on Ω that we consider as "elementary" and this leads to the notion of a (positive) *measure* on an elementary domain of functions on Ω.

The purpose of this chapter is a careful definition and discussion of these two notions and their simplest properties. For each of them we offer an extension procedure which we name after Eudoxos (ca. 408–355 B.C.) resp. Riemann (1826–1866) (Jordan (1838–1922) could compete as well). The discussion will show why these extensions are too simple in order to satisfy certain "higher" desires, and thus will motivate the more subtle extension procedures that are to be presented in Chapters II and III.

1. CONTENTS ON SET RINGS

1.1. Definition. *Let $\Omega = \{\omega, \eta, \ldots\}$ be an arbitrary set $\neq \varnothing$. A system (i.e., a set) $\mathscr{R}$ of subsets of Ω is called a (set)* **ring** *on (or in (the basic set)) Ω if:*

1.1.1. *$\mathscr{R}$ is stable under finite unions, i.e.,*

$$E_1, \ldots, E_n \in \mathscr{R} \quad \Rightarrow \quad E_1 \cup \cdots \cup E_n \in \mathscr{R} \qquad (n = 0, 1, \ldots).$$

According to the usual convention that a union of 0, *i.e., no sets, also called an empty union, is* $\varnothing$; *our requirement implies* $\varnothing \in \mathscr{R}$. *In particular,* $\mathscr{R}$ *is nonempty.*

1.1.2. *$\mathscr{R}$ is stable under differences, i.e.*

$$E, F \in \mathscr{R} \quad \Rightarrow \quad E \backslash F \in \mathscr{R}.$$

A ring containing the basic set Ω *is also called a (set)* **field.** *If* $\mathscr{R}$ *is a ring in* Ω, *then the couple* $(\Omega, \mathscr{R})$ *is also called a measurable* **prespace**.

1.2. Example. The system $\mathscr{P}(\Omega)$ of all subsets of a set $\Omega \neq \varnothing$ is a set field in Ω. The system of all finite subsets of $\Omega \neq \varnothing$ is a ring in Ω, and a field iff Ω itself is finite. The system of all those subsets of Ω that are either finite or have a finite complement is a field in Ω.

1.3. Simple observations on set rings and fields.

1.3.1. Every ring in an $\Omega \neq \varnothing$ is stable under finite intersections: Let $F_1, \ldots, F_n \in \mathscr{R}$, put $F = F_1 \cup \cdots \cup F_n$, then $F \in \mathscr{R}$ and $F_1 \cap \cdots \cap F_n = F \backslash [(F \backslash F_1) \cup \cdots \cup (F \backslash F_n)] \in \mathscr{R}$.

1.3.2. Every at most countable union of sets from a ring $\mathscr{R}$ can be written as an increasing union of sets from $\mathscr{R}$: If $F_1, F_2, \ldots \in \mathscr{R}$, $E = F_1 \cup F_2 \cup \cdots$, we put $E_n = F_1 \cup \cdots \cup F_n$ $(n = 1, 2, \ldots)$ and get $\mathscr{R} \ni E_1 \subseteq E_2 \subseteq \ldots$, $E_1 \cup E_2 \cup \cdots = E$.

1.3.3. *First entrance decomposition:* This is a simple device which transforms every countable union of members (= elements) of a ring $\mathscr{R}$ in Ω into a countable disjoint union of smaller members of $\mathscr{R}$. Let $F_1, F_2, \ldots \in \mathscr{R}$ and put $E_0 = \varnothing$, $E_n = F_1 \cup \cdots \cup F_n$, $F'_n = F_n \backslash E_{n-1}$ $(n = 1, 2, \ldots)$. Then clearly $E_0, E_1, \ldots \in \mathscr{R}$ and $F'_1, F'_2, \ldots \in \mathscr{R}$, $F'_1 \subseteq F_1$, $F'_2 \subseteq F_2, \ldots$, $F'_j \cap F'_k = \varnothing$ $(j \neq k)$ and (recall that $+$ means disjoint union)

$$E_n = F'_1 + \cdots + F'_n \qquad (n = 1, 2, \ldots)$$

$$F_1 \cup F_2 \cup \cdots = F'_1 + F'_2 + \cdots.$$

The proof (by induction) is left to the reader.

1.3.4. Every field $\mathscr{F}$ in a set $\Omega \neq \varnothing$ is stable under the passage to complements:

$$F \in \mathscr{F} \quad \Rightarrow \quad \complement F = \Omega \backslash F \in \mathscr{F}.$$

1.4. Example. Let $\Omega = \mathbb{R}$ be the real axis. A half-open interval of the form

$$\left[\frac{k}{2^r}, \frac{k+1}{2^r}\right[= \left\{x \,\middle|\, x \in \mathbb{R}, \frac{k}{2^r} \leqq x < \frac{k+1}{2^r}\right\}$$

with integers r, k and $r \geqq 0$ is called a (half-open) *dyadic interval of order* r in $\mathbb{R}$.

1.4.1. For every $r = 0, 1, \ldots$, the system $\mathscr{D}_r$ of all finite disjoint unions of dyadic intervals of order r is a ring in $\mathbb{R}$.

1.4.2. $\mathscr{D}_0 \subseteq \mathscr{D}_1 \subseteq \cdots$, and $\mathscr{D} = \mathscr{D}_0 \cup \mathscr{D}_1 \cup \cdots$ is the system of all finite unions of dyadic intervals of any order in $\mathbb{R}$. It is a ring in $\mathbb{R}$. We call $(\mathbb{R}, \mathscr{D})$ the *dyadic measurable prespace* on $\mathbb{R}$.

The proof for 1 and 2 (Hint:

$$[k/2^r, (k+1)/2^r[= [2k/2^{r+1}, (2k+1)/2^{r+1}[+ [(2k+1)/2^{r+1}, (2k+2)/2^{r+1}[)$$

is left to the reader.

1.5. Exercises. Let $\Omega \neq \varnothing$ be a set. Recall that a system (= set) ξ of nonempty subsets of Ω is called a *partition* of Ω if it consists of pairwise disjoint sets with union Ω. The elements (= members) of ξ are then also called the *atoms* of the partition ξ. A partition η is said to be *finer* than the partition ξ of Ω if every atom of η is contained in some atom of ξ. In this case we write $\xi \leqq \eta$ or $\eta \geqq \xi$. Prove:

1.5.1. If ξ, η are two partitions of Ω such that $\xi \leqq \eta$, then there is a unique partition τ of η and a natural bijection between ξ and τ such that every atom of ξ is the union of those atoms of η that are elements of the corresponding atom of τ. η is called *finite* over ξ if every atom of τ is finite, i.e., if every atom of ξ is a finite union of atoms of η.

1.5.2. If ξ, η are any two partitions of Ω, then there is a unique partition ρ, also denoted $\rho = \xi \vee \eta$ and called the *common refinement* of ξ and η, of Ω such that $\xi \leqq \rho$, $\eta \leqq \rho$, and $\rho \leqq \sigma$ for any partition σ of Ω such that $\xi \leqq \sigma$, $\eta \leqq \sigma$. The atoms of ρ are exactly those $A \cap B$ with $A \in \xi$, $B \in \eta$ that are nonempty.

1.5.3. Let $\Omega = \mathbb{R}$. The half-open intervals of order r form a partition δ_r of $\mathbb{R}$ $(r = 0, 1, \ldots)$ such that $\delta_0 \leqq \delta_1 \leqq \cdots$ and δ_{r+1} is finite over δ_r $(r = 0, 1, \ldots)$.

1.5.4. Let $\Omega \neq \varnothing$ be arbitrary again. If ξ is a partition of a nonempty subset Ω_0 of Ω, then the system of all finite unions of atoms of ξ is a ring $\mathscr{R}(\xi)$ in Ω. If $\mathscr{R}'$ is a ring in ξ, then $\mathscr{R} = \{\bigcup_{A \in E} A \,|\, E \in \mathscr{R}'\}$ is a ring in Ω. If $\Omega_0 = \Omega$ and $\mathscr{R}'$ is a field, then so is $\mathscr{R}$.

1.5.5. Let Ω, Ω' and f: $\Omega \to \Omega'$ a mapping. Then the sets of the form

$$f^{-1}(\omega') = \{\omega \mid \omega \in \Omega,\ f(\omega) = \omega'\} \qquad (\omega' \in f(\Omega))$$

form a partition of Ω, the partition *derived* from f. If $\mathscr{R}'$ is a ring in Ω', then $f^{-1}(\mathscr{R}') = \{f^{-1}(F') = \{\omega \mid f(\omega) \in F'\} \mid F' \in \mathscr{R}'\}$ is a ring in Ω. If $\mathscr{R}'$ is a field, so is $\mathscr{R}$.

1.5.6. Let $\mathscr{R}$ be a ring in Ω and $\Omega_0 = \bigcup_{A \in \mathscr{R}} A$. Call $\omega, \omega' \in \Omega_0$ equivalent for $\mathscr{R}$ if for every $A \in \mathscr{R}$ either $\omega \in A \ni \omega'$ or $\omega \notin A \not\ni \omega'$. Show that this in fact defines an equivalence relation in Ω_0, and consequently the set of all equivalence classes is a partition of Ω_0. Prove that every $F \in \mathscr{R}$ is a union of some atoms of ξ. Prove that for a finite $\mathscr{R}$, $\xi \subseteq \mathscr{R}$ and $\mathscr{R} = \mathscr{R}(\xi)$.

1.5.7. Let $\xi_0 \leqq \xi_1 \leqq \cdots$ be partitions of $\Omega \neq \varnothing$ such that ξ_{r+1} is finite over ξ_r $(r = 0, 1, \ldots)$. Then $\mathscr{R}(\xi_0) \subseteq \mathscr{R}(\xi_1) \subseteq \cdots$ and $\mathscr{R} = \mathscr{R}(\xi_0) \cup \mathscr{R}(\xi_1) \cup \cdots$ is a ring in Ω.

1.5.8. Let $n \geqq 1$ be an integer. For any

$$x = (x_1, \ldots, x_n),\ y = (y_1, \ldots, y_n) \in \mathbb{R}^n$$

define

$$x \leqq y \quad \text{iff} \quad x_1 \leqq y_1, \ldots, x_n \leqq y_n,$$
$$x < y \quad \text{iff} \quad x_1 < y_1, \ldots, x_n < y_n.$$

For any $a = (a_1, \ldots, a_n)$, $b = (b_1, \ldots, b_n) \in \mathbb{R}^n$, define

$$[a, b[= \{x \mid x \in \mathbb{R}^n,\ a \leqq x < b\}$$

and call this the (possibly empty) *half-open interval* (in $\mathbb{R}^n$) from a to b. It can clearly be written as a cartesian product $[a_1, b_1[\times \cdots \times [a_n, b_n[$ of half-open intervals in $\mathbb{R}$, and it is nonempty iff $a < b$. It is called *dyadic of order* r if $[a_1, b_1[, \ldots, [a_n, b_n[$ are dyadic intervals of order r in $\mathbb{R}$. Let δ_r^n be the system of all half-open dyadic intervals of order r in $\mathbb{R}^n$. Prove that $\delta_0^n \leqq \delta_1^n \leqq \cdots$ and δ_{r+1}^n is finite over δ_r^n $(r = 0, 1, \ldots)$. Put $D_r^n = R(\delta_r^n)$ $(r = 0, 1, \ldots)$. From 1.5.7 we have $\mathscr{D}_0^n \subseteq \mathscr{D}_1^n \subseteq \cdots$ and $\mathscr{D}^n = \mathscr{D}_0^n \cup \mathscr{D}_1^n \cup \cdots$ is a ring in $\mathbb{R}^n$. In particular, $\mathscr{D}^1$ is the $\mathscr{D}$ of Example 1.4. We call $(\mathbb{R}^n, \mathscr{D}^n)$ the *dyadic* (*elementary, geometric*) *measurable prespace* in $\mathbb{R}^n$.

1.5.9. Let Δ be a system of rings in Ω. Assume that Δ is increasingly filtered, i.e., that for any $\mathscr{R}$, $\mathscr{S} \in \Delta$ there is a $\mathscr{T} \in \Delta$ such that $\mathscr{R} \cup \mathscr{S} \subseteq \mathscr{T}$. Prove that $\bigcup_{\mathscr{R} \in \Delta} \mathscr{R}$ is a ring in Ω again.

1.5.10. Let $\Omega = \mathbb{R}^n$ for some $n = 1, 2, \ldots$ and $\mathscr{R}$ the system of all disjoint finite unions of half-open intervals in $\mathbb{R}^n$ (see 1.5.8). Use 1.5.9 in order to prove that $\mathscr{R}$ is a ring in $\mathbb{R}^n$ and equals the system of all (not necessarily disjoint) unions of half-open intervals in $\mathbb{R}^n$. We call this the *elementary or geometric measurable prespace in* $\mathbb{R}^n$.

1.6. Example. Let $A \neq \varnothing$ be a finite set. We shall call it the alphabet in the present context. The set

$$A \times A \times \cdots = \{\omega = \omega_0 \omega_1 \cdots | \omega_0, \omega_1, \ldots \in A\}$$

of all one-sided infinite sequences (we deviate here from the usual way of writing sequences in some way, omitting some brackets and commas, for the sake of simplicity) of symbols from A is called the (one-sided) *shift space* or *Bernoulli space* over A. Clearly it has only one element if A has, and a continuum of elements if A has more than one element. For every $r = 0, 1, \ldots$, the sets of the form

$$[k_0, \ldots, k_r] = \{\omega = \omega_0 \omega_1 \cdots | \omega_0 = k_0, \omega_1 = k_1, \ldots, \omega_r = k_r\}$$

are called the (*special*) *cylinders* of order r in Ω. They form a finite partition γ_r of Ω (with $|A|^{r+1}$ atoms, where $|A|$ denotes the power of A, as usual). From

$$[k_0, \ldots, k_r] = \bigcup_{k \in A} [k_0, \ldots, k_r, k] \qquad (r = 0, 1, \ldots)$$

we see that $\gamma_0 \leqq \gamma_1 \leqq \cdots$ and γ_{r+1} is finite over γ_r $(r = 0, 1, \ldots)$. If $\mathscr{C}_r$ denotes the system of all finite unions of special cylinders of order r, then $\mathscr{C}_r$ is a field in Ω, $\mathscr{C}_0 \subseteq \mathscr{C}_1, \ldots$, and $\mathscr{C} = \mathscr{C}_0 \cup \mathscr{C}_1 \cdots$ is also a field in Ω. It consists of all finite unions of special cylinders of any order (1.5.7). $(A \times A \times \cdots, \mathscr{C})$ is also called the *Bernoulli* or *shift measurable prespace* for the alphabet A.

1.7. Definition. *Let $(\Omega, \mathscr{R})$ be a measurable prespace.*

1.7.1. *A real function m on $\mathscr{R}$ is called* **additive** *if*

$$m(F_1 + \cdots + F_n) = m(F_1) + \cdots + m(F_n)$$

holds for any pairwise disjoint $F_1, \ldots, F_n \in \mathscr{R}$.

1.7.2. *An additive real function m on $\mathscr{R}$ is called a (positive)* **content** *if it is nonnegative. The triple $(\Omega, \mathscr{R}, m)$ is then called a* **content prespace** *(over Ω).*

1.7.3. *If m is a positive content on the ring $\mathscr{R}$, then*

$$\|m\| = \sup_{F \in \mathscr{R}} m(F)$$

is called its **total mass** *or* **total variation.**

1.7.4. *A content m on a ring $\mathscr{R}$ is called* **bounded** *if $\|m\| < \infty$, and* **normalized** *if $\|m\| = 1$. If $\|m\| = \infty$, m is called* **unbounded.**

1.7.5. *A content prespace $(\Omega, \mathscr{R}, m)$ is called a* **probability prespace** *if $\Omega \in \mathscr{R}$ (i.e., $\mathscr{R}$ is a field) and $m(\Omega) = 1$. In this case m is also called a* **probability (content).**

1.7.6. *Let* $(\Omega, \mathscr{R}, m)$, $(\Omega, \mathscr{R}', m')$ *be two content prespaces with the same basic set* $\Omega \neq \emptyset$. *If* $\mathscr{R} \subseteq \mathscr{R}'$ *and* m' *coincides with* m *on* $\mathscr{R}$, *then* $(\Omega, \mathscr{R}', m')$ *is called an* **extension** *of* $(\Omega, \mathscr{R}, m)$ *(and* m' *an extension of* m *from* $\mathscr{R}$ *to* $\mathscr{R}'$, *of course). It is called a* **proper extension** *if* $\mathscr{R} \neq \mathscr{R}'$.

1.7.7. *A function* $m: \mathscr{R} \to \bar{\mathbb{R}}_+ = \mathbb{R} \cup \{\infty\}$ *or a function* $m: \mathscr{R} \to \bar{\mathbb{R}}_- = \mathbb{R} \cup \{-\infty\}$ *is called* **additive** *if*

$$m(F_1) + \cdots + m(F_n) = m(F_1) + \cdots + m(F_n) \qquad (F_1, \ldots, F_n \in \mathscr{R},\ F_j \cap F_k = \emptyset (j \neq k)).$$

1.8. Example (point mass distribution). Let $\mathscr{R}$ be a ring in $\Omega \neq \emptyset$. Choose any $\omega_1, \ldots, \omega_n \in \Omega$, $\alpha = (\alpha_1, \ldots, \alpha_n) \in \mathbb{R}^n$. Then

$$m(F) = \sum_{\omega_k \in F} \alpha_k \qquad (F \in \mathscr{R})$$

defines an additive real function m on $\mathscr{R}$. If $\alpha \geqq 0$, then m is a content with total mass $\alpha_1 + \cdots + \alpha_n$. Assume that $\mathscr{R}$ separates the points $\omega_1, \ldots, \omega_n$, i.e., that for any $1 \leqq j,\ k \leqq n,\ j \neq k$, there is either an $E \in \mathscr{R}$ with $\omega_j \in E \not\ni \omega_k$ or an $F \in \mathscr{R}$ with $\omega_k \in F \not\ni \omega_j$. Then different $\alpha \in \mathbb{R}^n$ lead to different m, and m is a content iff $\alpha \geqq 0$. It is easy to generalize this to countably many points $\omega_1, \omega_2, \ldots \in \Omega$ and reals $\alpha_1, \alpha_2, \ldots$ with $|\alpha_1| + |\alpha_2| + \cdots < \infty$. It is equally easy to generalize this to an arbitrary family $(\omega_\iota)_{\iota \in I}$ of points $\omega_\iota \in \Omega$ and an arbitrary family $(\alpha_\iota)_{\iota \in I}$ of reals, provided $|\{\iota | \omega_\iota \in E\}| < \infty$ $(E \in \mathscr{R})$.

1.9. Simple observations about additive set functions:

1.9.1. Since $\emptyset \in \mathscr{R}$ for every ring $\mathscr{R}$, additivity of $m: \mathscr{R} \to \mathbb{R}$ implies

$$m(\emptyset) = m(\emptyset + \emptyset) = m(\emptyset) + m(\emptyset) = 0.$$

It also implies

$$\begin{aligned} m(E \cup F) + m(E \cap F) &= m(E + (F \backslash E)) + m(E \cap F)) \\ &= m(E) + m(F \backslash E) + m(E \cap F) = m(E) + m(F). \end{aligned}$$

1.9.2. Every additive function m on a ring $\mathscr{R}$ is *isotone*, i.e.,

$$E, F \in R,\ E \subseteq F \quad \Rightarrow \quad m(E) \leqq m(F).$$

In fact, $m(F) = m(E + (F \backslash E)) = m(E) + m(F \backslash E) \geqq m(E)$.

1.9.3. Every content m on a ring $\mathscr{R}$ is *subtractive*, i.e.,

$$m(E \backslash F) = m(E) - m(F) \qquad (E, F \in \mathscr{R}, \quad E \supseteq F).$$

In fact, $m(E) = m(F + (E \backslash F)) = m(F) + m(E \backslash F)$.

1.9.4. Every content m on a ring $\mathscr{R}$ is *subadditive*, i.e.,

$$m(F_1 \cup \cdots \cup F_n) \leqq m(F_1) + \cdots + m(F_n) \qquad (F_1, \ldots, F_n \in \mathscr{R}).$$

In fact, if we construct, according to the *first entrance decomposition* 1.3.3, pairwise disjoint members $F'_1 \subseteq F_1, \ldots, F'_n \subseteq F_n$ of $\mathscr{R}$ such that $F'_1 + \cdots + F'_n = F_1 \cup \cdots \cup F_n$, we see

$$m(F_1 \cup \cdots \cup F_n) = m(F'_1 + \cdots + F'_n) = m(F'_1) + \cdots + m(F'_n) \leqq m(F_1) + \cdots + m(F_n).$$

1.9.5. Any finite linear combination of additive real functions on a ring $\mathscr{R}$ in some $\Omega \neq \varnothing$ is additive again. Any finite linear combination with nonnegative coefficients of contents on $\mathscr{R}$ is a content on $\mathscr{R}$ again, and

$$\|\alpha_1 m_1 + \cdots + \alpha_n m_n\| = \alpha_1 \|m_1\| + \cdots + \alpha_n \|m_n\|$$

in this case. If $\mathscr{R}$ is a field, the probabilities on $\mathscr{R}$ form a convex set, i.e., a set that is stable under finite linear combinations with nonnegative coefficients summing to 1.

1.9.6. Observe that 1.7.7 makes sense since the possibility of a pathological addition $\infty + (-\infty)$ is excluded. If we modify Example 1.8 such that $\alpha_k = \infty$ for some k, we get a simple example of additivity with value ∞ admitted. $m \equiv \infty$ would be another example. The only cases of real interest are those for which $m(\varnothing) = 0$.

1.10. Example. Let $\Omega = \mathbb{R}$. For every $r = 0, 1, \ldots, F \in \mathscr{D}_r$ (Example 1.4) put

$$m_r(F) = c/2^r$$

if F is a disjoint union of c dyadic intervals of order r (clearly F uniquely determines c). Then m is an unbounded content on $\mathscr{D}_r$. Since every dyadic interval of order r can be represented as a disjoint union of two dyadic intervals of order $r + 1$, we see

$$m_r(F) = \frac{c}{2^r} = \frac{2c}{2^{r+1}} = m_{r+1}(F) \qquad (F \in \mathscr{D}_r),$$

and this implies that m_{r+1} is an extension of m_r from $\mathscr{D}_r$ to $\mathscr{D}_{r+1}$ $(r = 0, 1, \ldots)$. Consequently, there is a unique real function m on $\mathscr{D} = \mathscr{D}_0 \cup \mathscr{D}_1 \cup \cdots$ such that the restriction of m to $\mathscr{D}_r$ is m_r $(r = 0, 1, \ldots)$. m is an unbounded content on $\mathscr{D}$. It is called the *elementary, geometric, Riemann, Jordan, or Lebesgue content* on $\mathscr{D}$, and $(\mathbb{R}, \mathscr{D}, m)$ is called the *dyadic elementary etc. content prespace* over $\mathbb{R}$.

1.11. Exercises

1.11.1. Let ξ be a finite partition of a nonempty subset of $\Omega \neq \varnothing$, and α a real function on ξ. For every $F \in \mathscr{R}(\xi)$ (Exercise 1.5.4) let $m(F) = \sum_{A \subseteq F} \alpha(A)$. Prove that m is an additive function on $\mathscr{R}(\xi)$ and that it is a content iff $\alpha \geqq 0$, its total mass then being $\sum_{A \in \xi} \alpha(A)$.

1.11.2. Use Exercises 1.5.7, 8 and carry Example 1.10 over to the dyadic elementary measurable prespace $(\mathbb{R}^n, D^n)$ in such a fashion that $m([a, b)) = 1/2^{nr}$ for any dyadic half-open interval $[a, b[$ of order r in $\mathbb{R}^n$ $(n = 1, 2, \ldots)$. The resulting content prespace $(\mathbb{R}^n, D^n, m)$ is called the *dyadic elementary etc. content prespace* on $\mathbb{R}^n$.

1.11.3. Let Δ be an increasingly filtered system of rings in $\Omega \neq \varnothing$. For every $\mathbb{R} \in \Delta$, let $m_{\mathscr{R}}$ be a content on $\mathscr{R}$. Assume that for $\mathscr{R}, \mathscr{S} \in \Delta$, $\mathscr{R} \subseteq \mathscr{S}$ the restriction of $m_{\mathscr{S}}$ to $\mathscr{R}$ coincides with $m_{\mathscr{R}}$. Show the unique existence of a content m on the ring $\bigcup_{\mathscr{R} \in \Delta} \mathscr{R}$ (Exercise 1.5.9) such that for every $\mathscr{R} \in \Delta$, the restriction of m to $\mathscr{R}$ is $m_{\mathscr{R}}$.

1.11.4. Use 3 and carry Exercise 2 over to the situation envisaged in Exercise 1.5.10, providing $m([a_1, b_1[\times \cdots \times [a_n, b_n[) = (b_1 - a_1) \cdots (b_n - a_n)$ for every half-open interval $[a_1, b_1[\times \cdots \times [a_n, b_n[\subseteq \mathbb{R}^n$. The resulting content prespace is called the *elementary etc. content prespace* on $\mathbb{R}^n$.

1.12. Example. Let $A \neq \varnothing$ be a finite set. A real function p on A such that

$$p(k) \geqq 0 \qquad (k \in A)$$

$$\sum_{k \in A} p(k) = 1$$

is also called a *probability vector* (over, or indexed by A). Let $\Omega = A \times A \times \cdots$ be the (one-sided) shift space over A and $\mathscr{C}_0 \subseteq \mathscr{C}_1 \subseteq \cdots \subseteq \mathscr{C}$ as in Example 1.6. For every $r = 0, 1, \ldots$, there is a unique probability content m_r on $\mathscr{C}_r$ such that

$$m_r([k_0, \ldots, k_r]) = p(k_0) \cdots p(k_r) \qquad (k_0, \ldots, k_r \in A).$$

(Use Exercise 1.11.1) or direct reasoning.) From

$$\begin{aligned} m_{r+1}([k_0, \ldots, k_r]) &= \sum_{k \in A} m_{r+1}([k_0, \ldots, k_r, k]) \\ &= \sum_{k \in A} p(k_1) \cdots p(k_r) p(k) \\ &= p(k_1) \cdots p(k_r) \sum_{k \in A} p(k) \\ &= p(k_1) \cdots p(k_r) \\ &= m_r([k_0, \ldots, k_r]) \end{aligned}$$

we infer that m_{r+1} is an extension of m_r from $\mathscr{C}_r$ to $\mathscr{C}_{r+1}$. Thus (Exercise 1.11.3) there is a unique probability content m on $\mathscr{C} = \mathscr{C}_0 \cup \mathscr{C}_1 \cup \cdots$ such that the restriction of m to $\mathscr{C}_r$ is m_r $(r = 0, 1, \ldots)$, m is called the *Bernoulli content* for (or associated with the probability vector) p and $(\Omega, \mathscr{C}, m)$ the corresponding ("one-sided") *Bernoulli probability* prespace.

1.13. Exercises

1.13.1. Carry out an obvious "two-sided" analogue of Examples 1.6 and 1.12.

1.13.2. Carry out an analogue of Example 1.12 where the probability vector p may "vary from component to component." Carry this over to the situation of Exercise 1.

1.14. Example. Let A, Ω, p as in Example 1.12. A real function P on $A \times A$ such that

$$\text{(a)}\quad P \geqq 0, \text{ i.e., } P(j, k) \geqq 0 \qquad (j, k \in A)$$

$$\text{(b)}\quad \sum_{k \in A} P(j, k) = 1 \qquad (j \in A).$$

is also called a *stochastic matrix* (over, or indexed by A). For every $r = 1, 2, \ldots$, there is a unique probability content m_r on $\mathscr{C}_r$ such that

$$m_r([k_0, \ldots, k_r]) = p(k_0)P(k_0, k_1) \cdots P(k_{r-1}, k_r) \qquad (k_0, \ldots, k_r \in A).$$

From

$$\begin{aligned} m_{r+1}([k_0, \ldots, k_r]) &= \sum_{k \in A} m_{r+1}([k_0, \ldots, k_r, k]) \\ &= p(k_0)P(k_0, k_1) \cdots P(k_{r-1}, k_r) \sum_{k \in A} P(k_r, k) \\ &= p(k_0)P(k_0, k_1) \cdots P(k_{r-1}, k_r) = m_r([k_0, \ldots, k_r]) \end{aligned}$$

we infer that m_{r+1} is an extension of m_r from $\mathscr{C}_r$ to $\mathscr{C}_{r+1}$. The unique content m on $\mathscr{C} = C_0 \cup \mathscr{C}_1 \cup \cdots$ whose restriction to $\mathscr{C}_r$ is m_r is called the *Markov content* for (or associated with) the *initial distribution* p and the *transition matrix* P, and $(A \times A \times \cdots, \mathscr{C}, m)$ is then also called the associated *Markov probability prespace.*

1.15. Exercise. Carry out an analogue of Example 1.14 where the transition matrix P "may vary from transition to transition."

2. σ-CONTENTS

In many cases a content has a property that is sharper than additivity: σ-additivity. This will turn out to be the basis of the sophisticated extension procedures to be discussed in Chapters II and III. In this section we set up its definition and discuss its occurrence and most direct implications.

2.1. Definition

2.1.1. *Let* $(\Omega, \mathscr{R}, m)$ *be a content prespace. The positive content* m *is called* **σ-additive** *or a* **σ-content** *if*

$$m(F_1 + F_2 + \cdots) = m(F_1) + m(F_2) + \cdots$$
$$(F_1, F_2, \ldots \in \mathscr{R}, \quad F_j \cap F_k = \varnothing \ (j \neq k), \quad F_1 + F_2 + \cdots \in \mathscr{R})$$

(recall that $+$ *also denotes disjoint union). In this case* $(\Omega, \mathscr{R}, m)$ *is also called a* **σ-content prespace**. *If a content* m *is a probability, we call it also a* **σ-probability** *and speak of the* **σ-probability prespace** $(\Omega, \mathscr{R}, m)$.

2.1.2. *A σ-content prespace* $(\Omega, \mathscr{R}, m)$ *is called* **full** *if* $F_1, F_2, \ldots \in \mathscr{R}$, $m(F_1) + m(F_2) + \cdots < \infty$ *implies* $F_1 \cup F_2 \cup \cdots \in \mathscr{R}$.

2.1.3. *Let* $\mathscr{S}$ *be a nonempty system of subsets of* $\Omega \neq \varnothing$. *A function* $m: \mathscr{S} \to \overline{\mathbb{R}}_+ = \mathbb{R} \cup \{\infty\}$ *or* $m: \mathscr{S} \to \overline{\mathbb{R}}_- = \mathbb{R}_- \cup \{-\infty\}$ *is called* **σ-additive** *if for any* $F_1, F_2, \ldots \in \mathscr{S}$ *such that* $F_j \cap F_k = \varnothing \ (j \neq k)$ *and* $F_1 + F_2 + \cdots \in \mathscr{S}$, *the series* $m(F_1) + m(F_2) + \cdots$ *(whose members may attain one of the values* $\pm\infty$*), converges to the (possibly infinite) value* $m(F_1 + F_2 + \cdots)$.

An easy criterion for σ-additivity is given in

2.2. Proposition. *Let* $(\Omega, \mathscr{R}, m)$ *be a content prespace. Then* m *is σ-additive iff* m *is "σ-continuous at* $\varnothing$*", i.e., iff*

$$(1) \qquad m(E_k) \to 0 \qquad (\mathscr{R} \ni E_1 \supseteq E_2 \supseteq \ldots, \quad E_1 \cap E_2 \cap \cdots = \varnothing).$$

Proof. I. Let m be σ-additive and $E_1, E_2, \ldots \in \mathscr{R}$, $E_1 \supseteq E_2 \supseteq \cdots$, $E_1 \cap E_2 \cap \cdots = \varnothing$. Then $E_1 = (E_1 \backslash E_2) + (E_2 \backslash E_3) + \cdots$ represents $E_1 \in \mathscr{R}$ as a disjoint countable union of members of $\mathscr{R}$, hence

$$m(E_1) = m(E_1 \backslash E_2) + m(E_2 \backslash E_3) + \cdots.$$

Now $E_k = E_{k+1} + (E_k \backslash E_{k+1})$ implies $m(E_k) = m(E_{k+1}) + m(E_k \backslash E_{k+1})$, and we get

$$\begin{aligned} m(E_1) &= [m(E_1) - m(E_2)] + [m(E_2) - m(E_3)] + \cdots \\ &= \lim_{n \to \infty} [m(E_1) - m(E_n)] \\ &= m(E_1) - \lim_{n \to \infty} m(E_n); \end{aligned}$$

and hence $m(E_k) \to 0$, the convergence being a decreasing one, by the way.

II. Let m satisfy (1) and let $F_1, F_2, \ldots \in \mathscr{R}$ be pairwise disjoint with $F_1 + F_2 + \cdots \in \mathscr{R}$. Then we put $E_k = F_k + F_{k+1} + \cdots$ and see that

$$E_k = (F_1 + F_2 + \cdots) \backslash (F_1 + \cdots + F_{k-1}) \in \mathscr{R}, \qquad E_1 \supseteq E_2 \supseteq \cdots,$$

and $E_1 \cap E_2 \cap \cdots = \varnothing$. Thus by (1) we have $m(E_k) \to 0$ and therefore

$$\begin{aligned} m(F_1 + F_2 + \cdots) &= m(F_1 + \cdots + F_{k-1} + E_k) \\ &= m(F_1) + \cdots + m(F_{k-1}) + m(E_k) \\ &\to m(F_1) + m(F_2) + \cdots, \end{aligned}$$

i.e., m is σ-additive.

2.3. Exercises

2.3.1. Prove that the contents defined in Example 1.8 are σ-additive.

2.3.2. Let $(\Omega, \mathscr{R}, m)$ be a content prespace. Prove that m is σ-additive iff m is "*σ-continuous from above*" (i.e.,

$$F, F_1, F_2, \ldots \in \mathscr{R}, F_1 \supseteq F_2 \supseteq \cdots, F_1 \cap F_2 \cap \cdots = F \quad \Rightarrow \quad m(F_n) \to m(F)$$

iff m is "*σ-continuous from below*" (i.e.,

$$F, F_1, F_2, \ldots, \in \mathscr{R}, F_1 \subseteq F_2 \subseteq \cdots, F_1 \cup F_2 \cup \cdots = F \quad \Rightarrow \quad m(F_n) \to m(F)).$$

2.3.3. Let $\mathscr{R}$ be a ring in $\Omega \neq \varnothing$ and $m\colon \mathscr{R} \to \overline{R}_+$ additive. Prove that m is σ-additive iff it is "*σ-continuous from below*" (which is defined as in 2 but with values ∞ allowed), and that σ-additivity implies "restricted σ-continuity from above," i.e., $\mathscr{R} \ni F_1 \supseteq F_2 \supseteq \cdots$, $F_1 \cap F_2 \cap \cdots = F \in \mathscr{R}$, $\lim m(F_n) < \infty \Rightarrow \lim m(F_n) = m(F)$. Prove that the latter property is equivalent to "restricted σ-continuity at $\varnothing$," i.e., its own special case with $F = \varnothing$.

2.3.4. Let $\mathscr{B}$ be a σ-field in $\Omega \neq \varnothing$ and $m\colon \mathscr{B} \to \overline{\mathbb{R}}_+$ σ-additive. Let $\mathscr{B}^{00} = \{F \mid F \in B, m(F) < \infty\}$ and let m also denote the restriction of m to $\mathscr{B}^{00}$. Prove that $(\Omega, \mathscr{B}^{00}, m)$ is a full σ-content prespace.

2.3.5. Let $\Omega = \mathbb{N} = \{1, 2, \ldots\}$, $\mathscr{R} = \mathscr{P}(\Omega)$, and $m(F) = \infty$ if $|F| = \infty$, $m(F) = \sum_{n \in F} 1/2^n$ if $|F| < \infty$. Prove that m is additive but not σ-additive.

2.3.6. Let Ω, $\mathscr{R}$ be as in 5. Prove that $m\colon F \to |F|$ is σ-additive, but not "σ-continuous at $\varnothing$." (*Hint*: Consider $F_n = \{n, n+1, \ldots\}$.)

2.3.7. Let Ω, $\mathscr{R}$ be as in 5. Put $\alpha_n = 1/2^{n/2}$ for $n \in \Omega$, n even, $\alpha_n = 1$ for $n \in \Omega$, n odd. Prove that $m(F) = \sum_{n \in F} \alpha_n$ $(F \in \mathscr{R})$ attains every value in $\overline{\mathbb{R}}_+$ and is σ-additive on $\mathscr{R}$.

2.4. Example. Let $(\mathbb{R}, \mathscr{D}, m)$ be the dyadic geometric content prespace on $\mathbb{R}$ (Example 1.10). Then m is σ-additive. We prove this by a compactness argument, employing Proposition 2.2. Let

$$\mathscr{D} \ni E_1 \supseteq E_2 \supseteq \cdots, E_1 \cap E_2 \cap \cdots = \varnothing.$$

Clearly the $m(E_k)$ form a decreasing sequence of reals. Assume their limit is some $\alpha > 0$. For every $k = 1, 2, \ldots$, we construct some $E_k' \subseteq E_k$ such that:

(a) $m(E_k') > m(E_k) - \alpha/2^{k+1}$;
(b) the closure $\overline{E}_k'$ of E_k' is contained in E_k.

This is an easy thing if E_k is a dyadic interval $[k/2^n, (k+1)/2^n[$ of some order n: Choose $r > n$ and observe that

$$E_k = \left[\frac{k2^{r-n}}{2^r}, \frac{k2^{r-n} + 2^{r-n}}{2^r}\right[$$

$$= \left[\frac{k2^{r-n}}{2^r}, \frac{k2^{r-n} + 1}{2^r}\right[+ \cdots + \left[\frac{k2^{r-n} + 2^{r-n} - 1}{2^r}, \frac{k2^{r-n} + 2^{r-n}}{2^r}\right[$$

represents E_k as a disjoint union of 2^{r-n} dyadic intervals of order r. Leaving off the rightmost of them, we define

$$E_k' = \sum_{j=0}^{2^{r-n}-2} \left[\frac{k2^{r-n} + j}{2^r}, \frac{k2^{r-n} + j + 1}{2^r}\right[$$

and get $E_k' \in \mathscr{D}_r \subseteq \mathscr{D}$,

$$\bar{E}_k' = \left[\frac{k}{2^n}, \frac{k+1}{2^n} - \frac{1}{2^r}\right] \subseteq E_k, \qquad m(E_k') = m(E_k) - \frac{1}{2^r}.$$

If r is large enough, we get $1/2^r < \alpha/2^{k+1}$, and we are through for this special case. If E_k is, say, in $\mathscr{D}_n$, i.e., a disjoint union of finitely many (say c) dyadic intervals of order n, we apply the above "curtailing" procedure to each of the latter and get $E_k' \in \mathscr{D}_r$, $E_k' \subseteq E_k$, and $m(E_k') = m(E_k) - (c/2^r)$ which is $> m(E_k) - \alpha/2^{k+1}$ for r sufficiently large. Put now $E_k'' = E_1' \cap \cdots \cap E_k'$. Then $\mathscr{D} \ni E_k'' \subseteq E_k'$, hence $\bar{E}_k'' \subseteq E_k$. We have

$$E_k \backslash E_k'' \subseteq (E_1 \backslash E_1') \cup \cdots \cup (E_k \backslash E_k'),$$

hence

$$\begin{aligned} m(E_k \backslash E_k'') &\leqq m((E_1 \backslash E_1') \cup \cdots \cup (E_k \backslash E_k')) \\ &\leqq m(E_1 \backslash E_1') + \cdots + m(E_n \backslash E_n') \\ &< \frac{\alpha}{2^2} + \cdots + \frac{\alpha}{2^{k+1}} < \frac{\alpha}{2}. \end{aligned}$$

Consequently

$$m(E_k'') = m(E_k) - m(E_k \backslash E_k'') > \alpha - \frac{\alpha}{2} = \frac{\alpha}{2} > 0,$$

which implies that $E_1'', E_2'', \ldots$ are nonempty. Thus $\bar{E}_1'', \bar{E}_2'', \ldots$ is a decreasing sequence of nonempty compact sets. A well-known topological result says that $\bar{E}_1'' \cap \bar{E}_2'' \cap \cdots \neq \varnothing$ (prove it as an exercise). The still larger set $E_1 \cap E_2 \cap \cdots$ is therefore also nonempty, in contradiction to our assumption. By Proposition 2.2 m is σ-additive.

2.5. Example. Let $(A \times A \times \cdots, \mathscr{C}, m)$ be the Bernoulli probability prespace for a given probability vector p over the finite set $A \neq \emptyset$ (Example 1.12). Then m is σ-additive. For this, choose any $\mathscr{C} \ni E_1 \supseteq E_2 \supseteq \cdots$ with $\lim_{k\to\infty} m(E_k) = \alpha > 0$. If we can prove $E_1 \cap E_2 \cap \cdots \neq \emptyset$, we are through by Proposition 2.2. We shall in fact prove it by a compactness argument which turns out to be simpler than that used in Example 2.4 because we can exploit the finiteness of A. Choose $\omega^{(k)} = \omega_0^{(k)}\omega_1^{(k)} \cdots \in E_k$ $(k = 1, 2, \ldots)$ arbitrarily. For every $t = 0, 1, \ldots$, the component sequence $\omega_t^{(k)}$ $(k = 1, 2, \ldots)$ runs in A, hence contains a finally constant subsequence. By an obvious diagonal procedure we may pass to a subsequence $\omega^{(k_1)}, \omega^{(k_2)}, \ldots$ of $\omega^{(1)}, \omega^{(2)}, \ldots$ such that for each $t = 0, 1, \ldots$, the sequence $\omega_t^{(k_n)}$ $(n = 1, 2, \ldots)$ is finally constant, say $= \omega_t$. Put $\omega = \omega_0\,\omega_1 \cdots$. We show $\omega \in E_1 \cap E_2 \cap \cdots$, i.e., $\omega \in E_k$ $(k = 1, 2, \ldots)$. For any k, we represent E_k as a member of some $\mathscr{C}_r$, namely, as the disjoint union of cylinders of order r. Every $\omega^{(k)}, \omega^{(k+1)}, \ldots$ belongs to some of these cylinders, thus every $\omega^{(k_n)}$ with n sufficiently large does so. If we choose n so large that $\omega_0^{(k_n)} = \omega_0, \ldots, \omega_r^{(k_n)} = \omega_r$, it follows that $[\omega_0, \ldots, \omega_r]$ is one of the cylinders constituting E_k, and thus $\omega \in E_k$ follows. This does it. The reader sees that we have not used m at all. Thus our proof applies to the effect that any content on $\mathscr{C}$ is σ-additive. The topologist sees, of course, what the reason is: All members of $\mathscr{C}$ are *compact* subsets of the *compact metric space* $\Omega = A \times A \times \cdots$, hence $\mathscr{C} \ni E_1 \supseteq E_2 \supseteq \cdots$, $E_1 \cap E_2 \cap \cdots = \emptyset$ implies $E_k = E_{k+1} = \cdots = \emptyset$ for some k, by a well-known theorem from topology.

2.6. Exercises

2.6.1. Carry over the result of Example 2.5 to the "two-sided infinite" situation envisaged in Exercise 1.13.1.

2.6.2. Carry over the result of Example 2.4 to $\Omega = \mathbb{R}^n$ (Exercise 1.5.8, Exercise 1.11.2).

2.6.3. Carry over the result of 2 to the situation envisaged in Exercises 1.5.10 and 1.11.4.

3. *EUDOXOS* EXTENSION OF CONTENTS

In many cases where a content prespace $(\Omega, \mathscr{R}, m)$ is given it seems desirable to "measure by m" some sets that are not yet in $\mathscr{R}$, as we have said in the introduction to this chapter.

Let, e.g., $(\mathbb{R}^2, \mathscr{D}^2, m)$ be the dyadic geometric content prespace on $\mathbb{R}^2$ (Exercise 1.11.2). The unit disk $D = \{(x, y) | x^2 + y^2 \leqq 1\}$ is not in $\mathscr{D}^2$, and even a set as simple as $\{(x, y) | 0 \leqq x, y < \frac{1}{3}\}$ is not, to say nothing of a set like $\{(x, y) | x, y \text{ rational}\}$. The ancient problem of "measuring" the unit disk D (and hopefully to find $m(D) = 3.14\cdots$) is an example of an *extension problem:* We have to find an extension $(\mathbb{R}^2, \mathscr{R}, m)$ of $(\mathbb{R}^2, \mathscr{D}^2, m)$ such that $D \in \mathscr{R}$.

Let us consider another example that stems from probability theory. Let $A \neq \varnothing$ be finite and p a probability vector over A. Let $(A \times A \times \cdots, \mathscr{C}, m)$ be the Bernoulli σ-content prespace for p (Examples 1.12, 2.5), specify any $k \in A$ and put $E = \{\omega = \omega_0 \omega_1 \cdots | \omega_t = k$ for at least one $t = 0, 1, \ldots\}$. Then E is not in $\mathscr{C}$ if $|A| \geqq 2$ (exercise); but probabilists would, of course, like to ascribe a "probability" to it, that is, to find an *extension* $(A \times A \times \cdots, \mathscr{R}, m)$ of $(A \times A \times \cdots, \mathscr{C}, m)$ such that $E \in \mathscr{R}$.

We shall now present a rather simple extension procedure which goes back more or less to Eudoxos (ca. 408–455 B.C.) and Archimedes (287–212 B.C.) and can be described by the following:

3.1. Definition. *Let $(\Omega, \mathscr{R}, m)$ be a content prespace. A set $F \subseteq \Omega$ is called* **Eudoxos** (*or* **Jordan**) **measurable** *for $(\Omega, \mathscr{R}, m)$ or, briefly, for m if for every $\varepsilon > 0$ there are $E, G \in \mathscr{R}$ such that*

$$E \subseteq F \subseteq G, \qquad m(G \backslash E) < \varepsilon,$$

or, equivalently (obey the rule $\inf \varnothing = \infty$, $\sup \varnothing = -\infty$)

$$-\infty < \sup_{\mathscr{R} \ni E \subseteq F} m(E) = \inf_{F \subseteq G \in \mathscr{R}} m(G) < \infty.$$

Let $\mathscr{R}_{(m)}$ denote the system of all Eudoxos measurable sets (for $(\Omega, \mathscr{R}, m)$).

3.2. Theorem. *Let $(\Omega, \mathscr{R}, m)$ be a content space and $\mathscr{R}_{(m)}$ the system of the corresponding Eudoxos measurable sets. Then:*

3.2.1. *$\mathscr{R}_{(m)} \supseteq \mathscr{R}$ and*

$$m(F) = \sup_{\mathscr{R} \ni E \subseteq F} m(E) \qquad (F \in \mathscr{R}_{(m)})$$

defines an extension (again denoted by m) of m from $\mathscr{R}$ to $\mathscr{R}_{(m)}$.

3.2.2. *$(\Omega, \mathscr{R}_{(m)}, m)$ is a content prespace again. It is a σ-content prespace iff $(\Omega, \mathscr{R}, m)$ is.*

3.2.3. *Every nonnegative additive extension of m from $\mathscr{R}$ to $\mathscr{R}_{(m)}$ coincides with the m given in 1, and $(\mathscr{R}_{(m)})_{(m)} = \mathscr{R}_{(m)}$.*

Proof. 1. If $F \in \mathscr{R}$, then clearly

$$m(F) = \sup_{\mathscr{R} \ni E \subseteq F} m(E) = \inf_{F \subseteq G \in \mathscr{R}} m(G).$$

2. We start by proving that $\mathscr{R}_{(m)}$ is a ring. Let $F, F' \in \mathscr{R}_m$; we show that $F \cup F' \in \mathscr{R}_{(m)}$. Let $\varepsilon > 0$ be arbitrary and $E, G, E', G' \in \mathscr{R}$ be such that

$E \subseteq F \subseteq G$, $E' \subseteq F' \subseteq G'$ and $m(G \backslash E) < \varepsilon/2 > m(G' \backslash E')$. Clearly $E \cup E'$, $G \cup G' \in \mathscr{R}$, $E \cup E' \subseteq F \cup F' \subseteq G \cup G'$,

$$m((G \cup G') \backslash (E \cup E')) \leqq m((G \backslash E) \cup (G' \backslash E'))$$
$$\leqq m(G \backslash E) + m(G' \backslash E') < \varepsilon/2 + \varepsilon/2 = \varepsilon.$$

The proof for $F \cap F' \in \mathscr{R}_{(m)}$ goes similarly and is left as an exercise to the reader.

Next we show that the extended (according to 1) m is additive again (its nonnegativity is obvious). Let F, $F' \in \mathscr{R}_{(m)}$ be disjoint. Then $\mathscr{R} \ni E \subseteq F$, $\mathscr{R} \ni E' \subseteq F'$ implies $E \cap E' = \varnothing$. Thus we obtain

$$\begin{aligned} m(F + F') &= \sup_{\mathscr{R} \ni H \subseteq F + F'} m(H) \\ &\geqq \sup_{\mathscr{R} \ni E \subseteq F,\, \mathscr{R} \ni E' \subseteq F'} m(E + E') \\ &\geqq \sup_{\mathscr{R} \ni E \subseteq F,\, \mathscr{R} \ni E' \subseteq F'} [m(E) + m(E')] \\ &= \sup_{\mathscr{R} \ni E \subseteq F} m(E) + \sup_{\mathscr{R} \ni E' \subseteq F'} m(E') \\ &= m(F) + m(F'). \end{aligned}$$

But we also have

$$\begin{aligned} m(F + F') &= \inf_{F + F' \subseteq H \in \mathscr{R}} m(H) \\ &\leqq \inf_{F \subseteq G \in \mathscr{R},\, F' \subseteq G' \in \mathscr{R}} m(G \cup G') \\ &\leqq \inf_{F \subseteq G \in \mathscr{R},\, F' \subseteq G' \in \mathscr{R}} [m(G) + m(G')] \\ &= \inf_{F \subseteq G \in \mathscr{R}} m(G) + \inf_{F' \subseteq G' \in \mathscr{R}} m(G') \\ &= m(F) + m(F') \end{aligned}$$

and thus $m(F + F') = m(F) + m(F')$ follows.

Let finally m be σ-additive on $\mathscr{R}$; we prove that its extension m to $\mathscr{R}_{(m)}$ is σ-additive there. For this we use the criterion given in Proposition 2.2. Let $\mathscr{R}_{(m)} \ni F_1 \supseteq F_2 \supseteq \cdots$, $F_1 \cap F_2 \cap \cdots = \varnothing$. For a given $\varepsilon > 0$, choose $E_1', E_2', \ldots \in \mathscr{R}$ such that $E_k' \subseteq F_k$, $m(E_k') > m(F_k) - \varepsilon/2^k$ $(k = 1, 2, \ldots)$. Put $E_k = E_1' \cap \cdots \cap E_k'$ $(k = 1, 2, \ldots)$. Clearly

$$E_k' \subseteq E_k \cup (F_{k-1} \backslash E_{k-1}') \cup \cdots \cup (F_1 \backslash E_1')$$

and thus

$$m(E_k') \leqq m(E_k) + \varepsilon/2^{k-1} + \cdots + \varepsilon/2 < m(E_k) + \varepsilon,$$

hence $m(E_k) > m(F_k) - 2\varepsilon$ $(k = 1, 2, \ldots)$. But we have $\mathscr{R} \ni E_1 \supseteq E_2 \supseteq \cdots$, $E_1 \subseteq F_1$, $E_2 \subseteq F_2$, $\ldots$, hence $E_1 \cap E_2 \cap \cdots = \varnothing$, which implies $\lim_{k\to\infty} m(E_k) = 0$ by the σ-additivity of m and Proposition 2.2. We conclude $\lim_{k\to\infty} m(F_k) \leqq 2\varepsilon$. Since $\varepsilon > 0$ was arbitrary, $\lim_{k\to\infty} m(F_k) = 0$ follows, and the σ-additivity of m follows from Proposition 2.2. The rest is obvious. 3 is an easy exercise.

3.3. Exercises

3.3.1. Let $(\mathbb{R}^2, \mathscr{D}^2, m)$ be the dyadic geometric content prespace on $\mathbb{R}^2$ (Exercise 1.11.2). Prove that every subset of $\mathbb{R}^2$ whose boundary is a finite union of smooth curves (such as the unit disk $\{(x, y) | x^2 + y^2 \leqq 1\}$ is Eudoxos measurable for m.

3.3.2. Try to find weaker sufficient conditions for geometric Eudoxos measurability of subsets of $\mathbb{R}^2$ than that given in 1.

3.3.3. Carry over 1 (and possibly 2) over to arbitrary dimensions $n = 1, 2, \ldots$.

3.3.4. Let $(A \times A \times \cdots, C, m)$ be the Bernoulli σ-content prespace for a probability vector p over the finite set $A \neq \varnothing$ (Examples 1.12, 2.5). Assume $|A| \geqq 2$ and $p(j) > 0$ $(j \in A)$. Prove that for every fixed $k \in A$, the set $E = \{\omega = \omega_0 \omega_1 \cdots | \omega_t = k \text{ for some } t = 0, 1, \ldots\}$ is Eudoxos measurable for m with $m(E) = 1$. (*Hint:* Start with the case $|A| = 2$ and look at the complement of E.)

3.3.5. Consider $(\mathbb{R}^2, D^2, m)$ as in 1 and prove that its Eudoxos extension $(\mathbb{R}^2, D^2_{(m)}, m)$ is translation and rotation invariant in the following sense: Two congruent subsets of $\mathbb{R}^2$ (i.e., two subsets that coincide after suitable translation and rotation) are either both in $D^2_{(m)}$ or both not in $D^2_{(m)}$ and yield the same value for m in the first case. (*Hint:* Begin by considering intervals and prove translation invariance; observe that a rotation about the origin carries the unit disk into itself.)

3.3.6. Carry 5 over to arbitrary dimensions $n = 1, 2, \ldots$.

3.4. Remark. The above exercises show that Eudoxos extension settles our problems in many cases. It does not settle, e.g., the following problem: Let $(\mathbb{R}, \mathscr{D}, m)$ be the dyadic geometric content. We would like to "measure" the set $F = \{x | 0 \leqq x < 1, x \text{ rational}\}$ (namely, with $m(F) = 0$ since it is a countable union of one-point sets which are clearly Eudoxos measurable with content value 0). But F is not Eudoxos measurable: Since F is dense in $[0, 1[$, any finite union of half-open dyadic intervals covering F must also cover $[0, 1[$ (exercise), but any half-open interval contained in F must be empty. Thus we find ourselves induced to search for an extension method that is more efficient than Eudoxos' device.

4. σ-RINGS, LOCAL σ-RINGS, AND σ-FIELDS

In this section we present a variety of types of systems of subsets of a given basic set Ω which seems to be more feasible for a sound formulation of an extension problem for σ-contents, and for an attempt to its solution than the concept of a ring or field.

The leading point of view for the ideas staged in the next definition may already be guessed from Definition 2.1 of σ-additivity of a content: It is awkward to always have to check whether a certain countable union of sets from the ring $\mathscr{R}$ is again a member of $\mathscr{R}$. This was also the main point in Remark 3.4.

4.1. Definition. *A set ring $\mathscr{R}$ in $\Omega \neq \varnothing$ is called:*

4.1.1. *a (set)* **σ-ring** *or* **σ-algebra** *(in Ω) if it is stable under countable unions:*

$$F_1, F_2, \ldots \in \mathscr{R} \quad \Rightarrow \quad F_1 \cup F_2 \cup \cdots \in \mathscr{R};$$

4.1.2. *a (set)* **σ-field** *(in Ω) if it is a σ-ring and a field (i.e., $\Omega \in \mathscr{R}$);*

4.1.3. *a local* **σ-ring** *or a* **δ-ring** *if for every $\Omega_0 \in \mathscr{R}$, the restriction $\{F \mid \Omega_0 \supseteq F \in \mathscr{R}\}$ of $\mathscr{R}$ to Ω_0 is a σ-field in Ω_0, i.e., if $\mathscr{R}$ is stable under bounded countable unions, i.e., if for every $\Omega_0 \in \mathscr{R}$*

$$\Omega_0 \supseteq F_1, F_2, \ldots \in \mathscr{R} \quad \Rightarrow \quad F_1 \cup F_2 \cup \cdots \in \mathscr{R}.$$

4.1.4. *If $\mathscr{B}^{00}$ is a local σ-ring in Ω, then the couple $(\Omega, \mathscr{B}^{00})$ is called a* **local measurable space**. *If $\mathscr{B}^{0}$ is a σ-ring in Ω, then $(\Omega, \mathscr{B}^{0})$ is called a* **measurable space**. *If $(\Omega, \mathscr{B}^{00}, m)$ is a σ-content prespace such that $(\Omega, \mathscr{B}^{00})$ is a local measurable space, i.e., $\mathscr{B}^{00}$ is a local σ-ring, then $(\Omega, \mathscr{B}^{00}, m)$ is also called a* **σ-content space**. *If m is a σ-probability here (hence in particular $\mathscr{B}^{00}$ is a σ-field), $(\Omega, \mathscr{B}^{00}, m)$ is also called a σ-probability space.*

4.2. Remarks

4.2.1. Note that we speak of measurable spaces instead of prespaces as soon as stability under bounded countable unions is secured. Clearly a full σ-content prespace is automatically a (full) σ-content space. See Exercise 2.3.4.

4.2.2. σ is, as already in Definition 2.1, by tradition a symbol for the occurrence of countable unions. Similarly, δ is the traditional symbol for the occurrence of countable intersections. Our synonymity of "local σ-ring" with "δ-ring" is underlined by the following fact:

A ring is a local σ-ring iff it is stable under countable intersections. In particular, every δ-ring is stable under countable intersections.

In fact, if $\mathscr{R}$ is a local σ-ring in $\Omega \neq \varnothing$ and $E_1, E_2, \ldots \in \mathscr{R}$, put $\Omega_0 = E_1$, $F_k = \Omega_0 \backslash E_k$ $(k = 1, 2, \ldots)$; then

$$E_1 \cap E_2 \cap \cdots = \Omega_0 \backslash (F_1 \cup F_2 \cup \cdots) \in \mathscr{R}.$$

Likewise, if $\mathscr{R}$ is a ring that is stable under countable intersections, $\Omega_0, F_1, F_2, \ldots \in \mathscr{R}$, $\Omega_0 \supseteq F_1, F_2, \ldots$, put $E_k = \Omega_0 \backslash F_k$ $(k = 1, 2, \ldots)$; then $F_1 \cup F_2 \cup \cdots = \Omega_0 \backslash (E_1 \cap E_2 \cap \cdots) \in \mathscr{R}$.

4.2.3. Every σ-field is a field and a σ-ring. Every σ-ring is a local σ-ring. Every local σ-ring is a ring.

4.3. Example. Let Ω be an infinite set. Then:

4.3.1. The system of all (at most) countable subsets of Ω is a σ-ring in Ω, and a σ-field iff Ω is countable.

4.3.2. The system of all subsets of Ω that are either (at most) countable or have an (at most) countable complement is a σ-field.

4.4. Example. Let $\Omega = \mathbb{R}$. The system of all (at most) countable conditionally compact subsets of $\mathbb{R}$ is a local σ-ring but not a σ-ring in $\mathbb{R}$.

Wherever in mathematics a certain type of object is defined by stability under a certain family of compositions, the idea of *generation* shines through. It always follows the same pattern, which we display in the next proposition for those special cases that we have under consideration right now.

4.5. Proposition. *Let $\mathscr{S}$ be a (possibly empty) system of subsets of $\Omega \neq \varnothing$. Then:*

4.5.1.1. *The system of all local σ-rings in Ω which contain $\mathscr{S}$ is nonempty: it contains $\mathscr{P}(\Omega) = \{F \mid F \subseteq \Omega\}$.*

4.5.1.2. *The intersection of an arbitrary system of local σ-rings in Ω is again a local σ-ring in Ω.*

4.5.1.3. *In particular, the intersection of all local σ-rings in Ω containing $\mathscr{S}$ is a local σ-ring in Ω: the smallest local σ-ring (in Ω) containing $\mathscr{S}$. It is denoted by $\mathscr{B}^{00}(\mathscr{S})$ and called the local σ-ring* **generated** *by $\mathscr{S}$. It equals $\mathscr{S}$ if $\mathscr{S}$ is already a local σ-ring.*

4.5.2. *Analogous statements hold for rings, σ-rings, and σ-fields.*

4.5.2.1. *$\mathscr{R}(\mathscr{S})$ denotes the ring generated by $\mathscr{S}$.*

4.5.2.2. *$\mathscr{F}(\mathscr{S})$ denotes the field in Ω generated by $\mathscr{S}$.*

4.5.2.3. *$\mathscr{B}^0(\mathscr{S})$ denotes the σ-ring generated by $\mathscr{S}$.*

4.5.2.4. *$\mathscr{B}(\mathscr{S})$ denotes the σ-field in Ω generated by $\mathscr{S}$ (it is important to mention Ω here since $\Omega \in \mathscr{B}(\mathscr{S})$).*

4.5.3. *If* $\mathscr{S} \subseteq \mathscr{T} \subseteq \mathscr{P}(\Omega)$, *then:*

4.5.3.1. $\mathscr{R}(\mathscr{S}) \subseteq \mathscr{R}(\mathscr{T})$.

4.5.3.2. $\mathscr{B}^{00}(\mathscr{S}) \subseteq \mathscr{B}^{00}(\mathscr{T}) = \mathscr{B}^{00}(\mathscr{R}(\mathscr{T}))$.

4.5.3.3. $\mathscr{B}^{0}(\mathscr{S}) \subseteq \mathscr{B}^{0}(\mathscr{T}) = \mathscr{B}^{0}(\mathscr{B}^{00}(\mathscr{T})) = \mathscr{B}^{0}(\mathscr{R}(\mathscr{T}))$.

4.5.3.4.
$$\mathscr{B}(\mathscr{S}) \subseteq \mathscr{B}(\mathscr{T}) = \mathscr{B}(\mathscr{B}^{0}(\mathscr{T})) = \mathscr{B}(\mathscr{B}^{00}(\mathscr{T})) = \mathscr{B}(\mathscr{F}(\mathscr{T})) = \mathscr{B}(\mathscr{R}(\mathscr{T})).$$

The proof is an easy exercise and left to the reader (use, e.g., Remark 4.2.3.).

4.6. Remarks

4.6.1. We emphasize once more that the generated system, no matter of what type, increases if the generating system $\mathscr{S}$ increases: $\mathscr{S} \subseteq \mathscr{S}'$ implies $\mathscr{R}(\mathscr{S}) \subseteq \mathscr{R}(\mathscr{S}')$, etc. We generally say "countably generated" instead of "generated by a countable $\mathscr{S}$."

4.6.2. Whenever in mathematics we have a generation definition like 4.5, the next question is whether the elements of the generated item can be *constructed* in a more or less explicit manner from the generating system. In our case the question is natural whether, e.g., every member of $\mathscr{B}(\mathscr{S})$ can be expressed by a finite chain of countable unions and intersections, and possibly some differences, through countably many members of $\mathscr{S}$. We state here without proof that this is generally not the case. We shall deal at length with this problem in Chapter XIII. Here we offer a few special results in the form of the next proposition and the exercises following it.

4.7. Proposition. *Let* $\Omega \neq \varnothing$.

4.7.1. *Let* $\mathscr{R}$ *be a ring in* Ω. *For every* $\Omega_0 \in \mathscr{R}$, *let* $\mathscr{R} \cap \Omega_0$ *denote the ring* $\{F \mid \mathscr{R} \ni F \subseteq \Omega_0\}$ *and* $\mathscr{B}_{\Omega_0}$ *the σ-ring in* Ω_0 *generated by it. Then*

$$\mathscr{B}^{00}(\mathscr{R}) = \bigcup_{\Omega_0 \in \mathscr{R}} \mathscr{B}_{\Omega_0}. \tag{1}$$

In particular, every $E \in \mathscr{B}^{00}(\mathscr{R})$ *is contained in some* $F \in \mathscr{R}$.

4.7.2. *Let* $\mathscr{B}^{00}$ *be a local σ-ring in* Ω. *Then*

4.7.2.1.
$$\begin{aligned}\mathscr{B}^{0}(\mathscr{B}^{00}) &= \{F_1 \cup F_2 \cup \cdots \mid F_1, F_2, \ldots \in \mathscr{B}^{00}\} \\ &= \{E_1 + E_2 + \cdots \mid E_1, E_2, \ldots \in \mathscr{B}^{00},\ E_j \cap E_k = \varnothing\ (j \neq k)\}.\end{aligned}$$

4.7.2.2. $E \in \mathscr{B}(\mathscr{B}^{00})$, $F \in \mathscr{B}^{00} \Rightarrow E \cap F \in \mathscr{B}^{00}$. *In particular* $E \in \mathscr{B}(\mathscr{B}^{00})$, $E \subseteq F$ *for some* $F \in \mathscr{B}^{00} \Rightarrow E \in \mathscr{B}^{00}$.

4.7.3. *Let* $\mathscr{B}^{0}$ *be a σ-ring in* Ω. *Then:*

4.7.3.1. $\mathscr{B}(\mathscr{B}^0) = \{F \mid F \in \mathscr{B}^0 \text{ or } \Omega\backslash F \in \mathscr{B}^0\}$.

4.7.3.2. $E \in \mathscr{B}(\mathscr{B}^0),\ F \in \mathscr{B}^0 \Rightarrow E \cap F \in \mathscr{B}^0$.

4.7.4. *For every* $\mathscr{S} \subseteq \mathscr{P}(\Omega)$ *we have:*

4.7.4.1. $\mathscr{R}(\mathscr{S}) \subseteq \mathscr{F}(\mathscr{S}) \subseteq \mathscr{B}(\mathscr{S})$.

4.7.4.2. $\mathscr{R}(\mathscr{S}) \subseteq \mathscr{B}^{00}(\mathscr{S}) \subseteq \mathscr{B}^0(\mathscr{S}) \subseteq \mathscr{B}(\mathscr{S})$.

4.7.5.1. *Let* Σ *be an increasing σ-filtered system of σ-rings in* Ω: $\mathscr{B}_1^0, \mathscr{B}_2^0, \ldots \in \Sigma \Rightarrow \mathscr{B}_1^0 \cup \mathscr{B}_2^0 \cup \cdots \subseteq \mathscr{B}^0$ *for some* $\mathscr{B}^0 \in \Sigma$. *Then* $\bigcup_{\mathscr{B}^0 \in \Sigma} \mathscr{B}^0$ *is a σ-ring.*

4.7.5.2. *Let* $\mathscr{S} \subseteq \mathscr{P}(\Omega)$ *and* $\Sigma = \{\mathscr{M} \mid \mathscr{M} \subseteq \mathscr{S},\ \mathscr{M}$ *countable*$\}$. *Then*

$$\mathscr{B}^{00}(\mathscr{S}) = \bigcup_{\mathscr{M} \in \Sigma} \mathscr{B}^{00}(\mathscr{M}), \qquad \mathscr{B}^0(\mathscr{S}) = \bigcup_{\mathscr{M} \in \Sigma} \mathscr{B}^0(\mathscr{M}), \qquad \mathscr{B}(\mathscr{S}) = \bigcup_{\mathscr{M} \in \Sigma} \mathscr{B}(\mathscr{M}).$$

Proof. 1. Since $\Omega_0 \in \mathscr{R} \cap \Omega_0$, we have $\mathscr{B}_{\Omega_0} = \mathscr{B}^{00}(\mathscr{R}_{\Omega_0}) \subseteq \mathscr{B}^{00}(\mathscr{R})$ $(\Omega_0 \in \mathscr{R})$ and thus $\supseteq$ holds in (1). But the right-hand member in (1) is a local σ-ring containing $\mathscr{R}$: If $F, F_1, F_2, \ldots \in \bigcup_{\Omega_0 \in \mathscr{R}} \mathscr{B}_{\Omega_0}$, $F_1, F_2, \ldots \subseteq F$, choose some $\Omega_0 \in \mathscr{R}$ with $F \subseteq \Omega_0$, and $F_1 \cup F_2 \cup \cdots \in \mathscr{B}_{\Omega_0}$ follows; thus we get stability under bounded countable union; the other properties of a local σ-ring are still easier to check; thus equality holds in (1).

2. Put $\mathscr{B}^0 = \{F_1 \cup F_2 \cup \cdots \mid F_1, F_2, \ldots \in \mathscr{B}^{00}\}$. Clearly $\mathscr{B}^0$ is stable under countable unions, in particular finite unions. But it is also stable under differences:

$$F_1, G_1, F_2, G_2, \ldots \in \mathscr{B}^{00} \Rightarrow \quad \left(\bigcup_n F_n\right)\Big\backslash\left(\bigcup_k G_k\right) = \bigcup_n \left[F_n \Big\backslash \bigcup_k G_k\right]$$
$$= \bigcup_n \left[\bigcap_k (F_n \backslash G_k)\right]$$

which belongs to $\mathscr{B}^0$ since $\bigcap_k (F_n\backslash G_k) \in \mathscr{B}^{00}$ $(n = 1, 2, \ldots)$ (we know by Remark 4.2.2 that a local σ-ring is stable under countable intersections). Thus $\mathscr{B}^0$ is a σ-ring. Since it contains $\mathscr{B}^{00}$ and clearly is contained in $\mathscr{B}^0(\mathscr{B}^{00})$, we have $\mathscr{B}^0(\mathscr{B}^{00}) = \mathscr{B}^0$. The second equality in 7.2.1 now follows by the first entrance decomposition 1.3.3.

2.2 is now an easy exercise in case $E \in \mathscr{B}^0(\mathscr{B}^{00})$. The general case will follow from 3.1.

3.1. Let $\mathscr{B} = \{F \mid F \in \mathscr{B}^0 \text{ or } \Omega\backslash F \in \mathscr{B}^0\}$. Clearly $\mathscr{B}^0 \subseteq \mathscr{B} \subseteq \mathscr{B}(\mathscr{B}^0)$. For 1, it suffices to prove that $\mathscr{B}$ is a σ-field. $\Omega \in \mathscr{B}$ is obvious.

Let us now prove that $\mathscr{B}$ is stable under differences. For $E, F \in \mathscr{B}^0$, we have $E\backslash F \in \mathscr{B}^0 \subseteq B$, $(\Omega\backslash E)\backslash F = \Omega\backslash(E \cup F) \in \mathscr{B}$ since $E \cup F \in \mathscr{B}^0$, $E\backslash(\Omega\backslash F) = E \cap F \in \mathscr{B}^0 \subseteq \mathscr{B}$, $(\Omega\backslash E)\backslash(\Omega\backslash F) = F\backslash E \in \mathscr{B}^0 \subseteq \mathscr{B}$. Observe that here $E\backslash F$, $E\backslash(\Omega\backslash F)$ are even in $\mathscr{B}^0$; i.e., subtraction of members of $\mathscr{B}$ does not lead out of $\mathscr{B}^0$. Finally, let us prove that $\mathscr{B}$ is stable under countable unions:

choose $F_1, F_2, \ldots \in \mathscr{B}$; if $F_1, F_2, \ldots \in \mathscr{B}^0$, then $F_1 \cup F_2 \cup \cdots \in \mathscr{B}^0 \subseteq \mathscr{B}$; if, say $\Omega \backslash F_1 \in \mathscr{B}^0$, then

$$(\Omega \backslash F_1) \cap (\Omega \backslash F_2) \cap \cdots = \bigcap_{n=1}^{\infty} [\cdots((\Omega \backslash F_1) \backslash F_2) \backslash \cdots) \backslash F_n];$$

by our above observation, $[\cdots((\Omega \backslash F_1) \backslash F_2) \backslash \cdots) \backslash F_n] \in \mathscr{B}^0$; since $\mathscr{B}^0$ is stable under countable intersections,

$$\Omega \backslash (F_1 \cup F_2 \cup \cdots) = (\Omega \backslash F_1) \cap (\Omega \backslash F_2) \cap \cdots \in \mathscr{B}^0$$

follows, hence $F_1 \cup F_2 \cup \cdots \in \mathscr{B}$.

3.2 follows from 3.1.

4 follows because every local σ-ring is a ring, every σ-ring is a local σ-ring, every σ-field is a σ-ring, every field is a ring, and every σ-field is a field.

5.1. Let us prove that $\bigcup_{\mathscr{B}^0 \in \Sigma} \mathscr{B}^0$ is stable under countable unions. Choose $F_1, F_2, \ldots \in \bigcup_{\mathscr{B}^0 \in \Sigma} \mathscr{B}^0$ and assume $F_1 \in \mathscr{B}_1^0 \in \Sigma$, $F_2 \in \mathscr{B}_2^0 \in \Sigma, \ldots$. Find a $\mathscr{B}_\infty^0 \in \Sigma$ containing $\mathscr{B}_1^0, \mathscr{B}_2^0, \ldots$. Then $F_1, F_2, \ldots \in \mathscr{B}_\infty^0$, hence $F_1 \cup F_2 \cup \cdots \in \mathscr{B}_\infty^0 \subseteq \bigcup_{\mathscr{B}^0 \in \Sigma} \mathscr{B}^0$. The stability of $\bigcup_{\mathscr{B}^0 \in \Sigma} \mathscr{B}^0$ under differences is still easier to prove (exercise).

5.2 follows from 1 and 5.1 (exercise).

4.8. Exercises

4.8.1. Let $\Omega \neq \varnothing$ be arbitrary and $\mathscr{S} = \{\{\omega\} | \omega \in \Omega\}$ the system of all one-point sets in Ω. Prove:

4.8.1.1. $\mathscr{R}(\mathscr{S}) = \mathscr{B}^{00}(\mathscr{S})$ is the system of all finite subsets of Ω.

4.8.1.2. $\mathscr{F}(\mathscr{S})$ is the system of all sets $\subseteq \Omega$ that are either finite or have a finite complement.

4.8.1.3. $\mathscr{B}^0(\mathscr{S})$ is the system of all (at most) countable subsets of Ω.

4.8.1.4. $\mathscr{B}(\mathscr{S})$ is the system of all sets $\subseteq \Omega$ that are either countable or have a countable complement.

4.8.2. Let $\mathscr{R}$ be a ring in $\Omega \neq \varnothing$. Show that $\mathscr{F}(\mathscr{R}) = \{F | F \in \mathscr{R}$ or $\Omega \backslash F \in \mathscr{R}\}$.

4.8.3. Let $\Omega \neq \varnothing$ and $\mathscr{S} \subseteq \mathscr{P}(\Omega)$ be finite, say, $\mathscr{S} = \{F_1, \ldots, F_n\}$. Let $\Omega_0 = F_1 \cup \cdots \cup F_n$. For every $k = 1, \ldots, n$, let $F_k^0 = F_k$, $F_k^1 = \Omega_0 \backslash F_k$. Prove that the nonempty sets among the sets of the form $F_1^{\varepsilon_1} \cap \cdots \cap F_n^{\varepsilon_n}$ $(\varepsilon_1, \ldots, \varepsilon_n = 0$ or $1)$ form a partition of Ω_0 and that $\mathscr{R}(\mathscr{S})$ consists of all finite unions of atoms of this partition. Prove that $\mathscr{F}(\mathscr{S})$ is obtained by putting $\Omega_0 = \Omega$.

4.8.4. Let $\Omega \neq \varnothing$ and $\varnothing \neq \mathscr{T} \subseteq \mathscr{S} \subseteq \mathscr{P}(\Omega)$ be such that every $F \in \mathscr{S}$ is contained in some $\Omega_0 \in \mathscr{T}$, and $\mathscr{T}$ is increasingly filtered:

$$\Omega_1, \ldots, \Omega_n \in \mathscr{T} \quad \Rightarrow \quad \Omega_1, \ldots, \Omega_n \subseteq \Omega_0$$

for some $\Omega_0 \in \mathscr{T}$. For every $\Omega_0 \in \mathscr{T}$, let:

4.8.4.1. $\mathscr{R}_{\Omega_0}$ be the ring generated by $\{F \mid \mathscr{S} \ni F \subseteq \Omega_0\}$. Prove that $\mathscr{R}(\mathscr{S}) = \bigcup_{\Omega_0 \in \mathscr{T}} \mathscr{R}_{\Omega_0}$.

4.8.4.2. $\mathscr{B}_{\Omega_0}$ be the σ-field in Ω_0 generated by $\{F \mid \mathscr{S} \ni F \subseteq \Omega_0\}$. Prove that $\mathscr{B}^{00}(\mathscr{S}) = \bigcup_{\Omega_0 \in \mathscr{T}} \mathscr{B}_{\Omega_0}$.

Assume now in addition that $\mathscr{T}$ is increasingly σ-filtered:

$$\Omega_1, \Omega_2, \ldots \in \mathscr{T} \quad \Rightarrow \quad \Omega_1, \Omega_2, \ldots \subseteq \Omega_0$$

for some $\Omega_0 \in \mathscr{T}$.

4.8.4.3. Prove that $\mathscr{B}^0(\mathscr{S}) = \bigcup_{\Omega_0 \in \mathscr{T}} \mathscr{B}_{\Omega_0}$.

4.8.4.4. Let $\mathscr{B}'_{\Omega_0}$ be the σ-field in Ω generated by $\{F \mid \mathscr{S} \ni F \subseteq \Omega_0\}$. Prove that $\mathscr{B}(\mathscr{S}) = \bigcup_{\Omega_0 \in \mathscr{T}} \mathscr{B}'_{\Omega_0}$.

4.8.5. Let $\Omega \neq \varnothing$ and $(F_\iota)_{\iota \in I}$ a family (indexed by I) of subsets of Ω. For every $\iota \in I$ let $F_\iota^0 = F_\iota$, $F_\iota^1 = \Omega \backslash F_\iota$. Prove that the nonempty sets among the sets of the form $\bigcup_{\iota \in I} F_\iota^{\varepsilon_\iota}$ ($\varepsilon_\iota = 0$ or 1 ($\iota \in I$)) form a partition ξ of Ω. Put $\mathscr{S} = \{F_\iota \mid \iota \in I\}$ and show that all atoms of ξ belong to $\mathscr{B}(\mathscr{S})$ if I is at most countable. Prove that every $F \in \mathscr{B}(\mathscr{S})$ is a union of atoms of ξ.

4.8.6. Let $\Omega = \mathbb{R}$ and

$$\begin{aligned}
\mathscr{S}_1 &= \mathscr{D} \qquad \text{(Example 1.4)},\\
\mathscr{S}_2 &= \{[a, b[\mid -\infty < a < b < \infty\},\\
\mathscr{S}_3 &= \{[a, b] \mid -\infty < a < b < \infty\},\\
\mathscr{S}_4 &= \{]a, b[\mid -\infty < a < b < \infty\},\\
\mathscr{S}_5 &= \{]a, b] \mid -\infty < a < b < \infty\},\\
\mathscr{S}_6 &= \{]-\infty, b[\mid b \in \mathbb{R}\},\\
\mathscr{S}_7 &= \{]-\infty, b] \mid b \in \mathbb{R}\},\\
\mathscr{S}_8 &= \{]a, \infty[\mid a \in \mathbb{R}\},\\
\mathscr{S}_9 &= \{[a, \infty[\mid a \in \mathbb{R}\},\\
\mathscr{S}_{10} &= \{F \mid F \subseteq \mathbb{R},\ F \text{ compact}\},\\
\mathscr{S}_{11} &= \{G \mid G \subseteq \mathbb{R},\ G \text{ bounded and open}\},\\
\mathscr{S}_{12} &= \{F \mid F \subseteq \mathbb{R},\ F \text{ closed}\},\\
\mathscr{S}_{13} &= \{G \mid G \subseteq \mathbb{R},\ G \text{ open}\}.
\end{aligned}$$

Prove:

4.8.6.1.

$$\begin{aligned}
\mathscr{B}^{00}(\mathscr{S}_1) = \mathscr{B}^{00}(\mathscr{S}_2) = \mathscr{B}^{00}(\mathscr{S}_3) = \mathscr{B}^{00}(\mathscr{S}_4) = \mathscr{B}^{00}(\mathscr{S}_5) = \mathscr{B}^{00}(\mathscr{S}_{10})\\
= \mathscr{B}^{00}(\mathscr{S}_{11}).
\end{aligned}$$

The members of the local σ-ring described here in various fashions are called the *bounded Borel sets* of $\mathbb{R}$.

4.8.6.2. $\mathscr{S}_1, \ldots, \mathscr{S}_{13}$ all generate the same σ-ring, and this σ-ring is a σ-field $\mathscr{B}(\mathbb{R})$. The members of this σ-field are called the *Borel subsets* of the real line $\mathbb{R}$.

4.8.7. Carry over 4.8.6:

4.8.7.1. to $\overline{\mathbb{R}}$. The result is the σ-field $\mathscr{B}(\overline{\mathbb{R}})$ of all *Borel subsets* of $\overline{\mathbb{R}}$. Prove that $\mathscr{B}(\overline{\mathbb{R}}) = \{E \mid E \subseteq \overline{\mathbb{R}}, E \cap \mathbb{R} \text{ is a Borel set in } \mathbb{R}\}$.

4.8.7.2. to $\mathbb{R}^n$. The result is the σ-field $\mathscr{B}(\mathbb{R}^n)$ of all *Borel subsets* of $\mathbb{R}^n$ $(n = 1, 2, \ldots)$.

4.8.7.3. to $\mathbb{C}$ by an obvious identification of $\mathbb{C}$ with $\mathbb{R}^2$. The result is the σ-field $\mathscr{B}(\mathbb{C})$ of all *Borel subsets* of $\mathbb{C}$.

4.8.7.4. to $\mathbb{C}^n$ by an obvious identification of $\mathbb{C}^n$ with $\mathbb{R}^{2n}$. The result is the σ-field of all *Borel subsets* of $\mathbb{C}^n$.

4.8.8. Let $n \geqq 1$ and $\mathscr{S}_1, \ldots, \mathscr{S}_n$ be systems of subsets of $\mathbb{R}$ such that $\mathscr{B}(\mathscr{S}_k) = \mathscr{B}(\mathbb{R})$ $(k = 1, \ldots, n)$. Prove that $\mathscr{B}(\mathbb{R}^n)$ is generated by the system

$$\mathscr{S} = \bigcup_{k=1}^{n} \{\mathbb{R}^{k-1} \times F \times \mathbb{R}^{n-k-1} \mid F \in \mathscr{S}_k\}.$$

4.9. Proposition. *Let $\mathscr{S}$ be a system of subsets of $\Omega \neq \varnothing$ that is stable under finite intersections:*

$$F_1, \ldots, F_n \in \mathscr{S} \quad \Rightarrow \quad F_1 \cap \cdots \cap F_n \in \mathscr{S}.$$

Let $\mathscr{B}^{00} \subseteq \mathscr{B}^{00}(\mathscr{S})$ be such that

(a) $\mathscr{S} \subseteq \mathscr{B}^{00}$
(b) $\mathscr{B}^{00}$ *is stable under proper differences,* i.e.,

$$E, F \in \mathscr{B}^{00}, \ E \subseteq F \quad \Rightarrow \quad F \setminus E \in \mathscr{B}^{00},$$

(c) $\mathscr{B}^{00}$ *is stable under (possibly empty) finite disjoint unions (thus in particular $\varnothing \in \mathscr{B}^{00}$), and under bounded countable disjoint unions. The latter means*

$$\mathscr{B}^{00} \ni F_1, F_2, \ldots \subseteq F \in \mathscr{B}^{00}, \quad F_j \cap F_k = \varnothing \quad (j \neq k)$$
$$\Rightarrow \quad F_1 + F_2 + \cdots \in \mathscr{B}^{00}.$$

Then $\mathscr{B}^{00} = \mathscr{B}^{00}(\mathscr{S})$.

Proof. Let $\mathscr{B}^{00}_{\min}$ be the intersection of all set systems with the properties (a)–(c). Clearly $\mathscr{B}^{00}_{\min}$ has properties (a)–(c) again, and it suffices to prove $\mathscr{B}^{00}_{\min} = \mathscr{B}^{00}(\mathscr{S})$. For this, it is sufficient to prove that $\mathscr{B}^{00}_{\min}$ is stable against arbitrary finite intersections. Indeed, this allows us to write every difference $F \setminus E$ of two sets $E, F \in \mathscr{B}^{00}_{\min}$ as a proper difference $F \setminus (E \cap F)$ of two sets from $\mathscr{B}^{00}_{\min}$ and every countable union of members of $\mathscr{B}^{00}_{\min}$ as a countable disjoint union of other (namely smaller) members of $\mathscr{B}^{00}_{\min}$, by the first entrance decomposition (1.3.3). From properties (a)–(c) of $\mathscr{B}^{00}_{\min}$ it will then follow that $\mathscr{B}^{00}_{\min}$ is a local σ-ring containing $\mathscr{S}$, and thus $\mathscr{B}^{00}_{\min} = \mathscr{B}^{00}(\mathscr{S})$ as desired. The proof that $\mathscr{B}^{00}_{\min}$ is finite intersection stable is achieved in two steps:

(I) Choose any $E \in \mathscr{S}$ and consider $\mathscr{B}^{00}_E = \{F \mid F \in \mathscr{B}^{00}_{\min}, E \cap F \in \mathscr{B}^{00}_{\min}\}$. Since $\mathscr{S}$ is finite intersection stable, we see $S \subseteq \mathscr{B}^{00}_E$, i.e., $\mathscr{B}^{00}_E$ satisfies (a). But it also satisfies (b). Indeed, if F, $G \in \mathscr{B}^{00}_E$, $F \subseteq G$, then $E \cap (G \backslash F) = (E \cap G) \backslash (E \cap F)$, being a proper difference of members of $\mathscr{B}^{00}_{\min}$, belongs to $\mathscr{B}^{00}_{\min}$, hence $G \backslash F \in \mathscr{B}^{00}_E$. And $\mathscr{B}^{00}_E$ satisfies (c). Indeed $\varnothing \in \mathscr{B}^{00}_{\min}$ is obvious. If $F_1, F_2, \ldots \in \mathscr{B}^{00}_E$, $F_j \cap F_k = \varnothing$ $(j \neq k)$, then $E \cap F_1, E \cap F_2, \ldots \in \mathscr{B}^{00}_{\min}$ are pairwise disjoint and hence their union $E \cap (F_1 + F_2 + \cdots)$ belongs to $\mathscr{B}^{00}_{\min}$ if only finitely many of the F_k are $\neq \varnothing$ or else all F_k are contained in some $F \in \mathscr{B}^{00}$. We conclude $F_1 + F_2 + \cdots \in \mathscr{B}^{00}_E$ in either case. Now $\mathscr{B}^{00}_E \supseteq \mathscr{B}^{00}_{\min}$ follows. Since $E \in \mathscr{S}$ was arbitrary, we see that $E \in \mathscr{S}$, $F \in \mathscr{B}^{00}_{\min} \Rightarrow E \cap F \in \mathscr{B}^{00}_{\min}$.

(II) Choose any $E \in \mathscr{B}^{00}_{\min}$ and define $\mathscr{B}^{00}_E$ literally as above. Using the result of (I) we prove in the same fashion as above $\mathscr{B}^{00}_E \supseteq \mathscr{B}^{00}_{\min}$, and thus finally $E, F \in \mathscr{B}^{00}_{\min} \Rightarrow E \cap F \in \mathscr{B}^{00}_{\min}$, as desired.

5. UNIQUENESS OF EXTENSION OF σ-CONTENTS

We now formulate the

5.1. Extension problem for contents. Let $(\Omega, \mathscr{R}, m)$ *be a content prespace. Does there exist a σ-content space* $(\Omega, \mathscr{B}^{00}, m^{00})$ *that is an extension of* $(\Omega, \mathscr{R}, m)$*? If yes, how many extensions exist?*

5.2. Discussion

5.2.1. Apparently, the σ-additivity of m is a necessary condition for a positive answer to the first question.

5.2.2. It seems reasonable to envisage not an arbitrary local σ-ring $\mathscr{B}^{00} \supseteq \mathscr{R}$ but $\mathscr{B}^{00}(\mathscr{R})$ in the first line. This will in fact enable us to prove a *uniqueness statement* immediately.

5.3. Proposition. *Let m, m' be σ-contents on the local σ-ring $\mathscr{B}^{00}$ in $\Omega \neq \varnothing$. Then*

$$\mathscr{B}^{00}_0 = \{F \mid F \in \mathscr{B}^{00}, m(F) = m'(F)\}$$

has the following properties:

(a) $\varnothing \in \mathscr{B}^{00}_0$;

(b) $\mathscr{B}_0^{00}$ *is stable under proper differences;*
(c) $\mathscr{B}_0^{00}$ *is stable under finite disjoint unions and under bounded countable disjoint unions.*

If m and m′ coincide on a set system $\mathscr{S} \subseteq \mathscr{B}^{00}$ *that is stable under finite intersections, then m and m′ coincide on* $\mathscr{B}^{00}(\mathscr{S})$*, in particular on* $\mathscr{B}^{00}$ *if* $\mathscr{S}$ *generates* $\mathscr{B}^{00}$.

Proof. (a) $m(\varnothing) = 0 = m'(\varnothing)$.
(b) $E, F \in \mathscr{B}_0^{00}$, $E \subseteq F$, $m(E) = m'(E)$, $m(F) = m'(F)$ imply

$$\begin{aligned} m(F \setminus E) &= m(F) - m(E) \\ &= m'(F) - m'(E) = m'(F \setminus E) \end{aligned}$$

(c) $\mathscr{B}_0^{00} \ni F_1, F_2, \ldots \subseteq F \in \mathscr{B}_0^{00}$, $F_j \cap F_k = \varnothing$ $(j \neq k)$, $m(F_1) = m'(F_1)$, $m(F_2) = m'(F_2)$ imply

$$F_1 + F_2 + \cdots \in \mathscr{B}_0^{00},$$

and

$$\begin{aligned} m(F_1 + F_2 + \cdots) &= m(F_1) + m(F_2) + \cdots \\ &= m'(F_1) + m'(F_2) + \cdots \\ &= m'(F_1 + F_2 + \cdots) \end{aligned}$$

by σ-additivity of m and m'. The proof that $\mathscr{B}_0^{00}$ is stable against finite disjoint unions is still easier.

The rest of the proof follows from Proposition 4.9.

The reader should observe that this proposition implies its own special case for σ-fields: If m and m' coincide on a *finite intersection stable set system* $\mathscr{S} \ni \Omega$, then they coincide on $\mathscr{B}(\mathscr{S})$.

As a consequence of Proposition 5.3, we get the following result concerning *uniqueness of extensions*:

5.4. Proposition. *If* $(\Omega, \mathscr{R}, m)$ *is a σ-content space, then there is at most one σ-content on* $\mathscr{B}^{00}(\mathscr{R})$ *extending m.*

The corresponding *existence statement* is also valid and will be proved in Chapter II (Theorem II.2.3 resp. Corollary II.2.4), and once more in Chapter III (Exercise 6.5.7).

5.5. Exercises

5.5.1. Let $\Omega \neq \varnothing$. A system $\mathscr{M}$ of subsets of Ω is called an *isotone class* if it is stable under countable monotone unions, i.e., if $E_1, E_2, \ldots \in \mathscr{M}$ implies

$$E_1 \subseteq E_2 \subseteq \cdots \quad \Rightarrow \quad E_1 \cup E_2 \cup \cdots \in \mathscr{M}$$

and an *antitone class* if it is stable under countable monotone intersections, i.e., if

$$E_1 \supseteq E_2 \supseteq \cdots \quad \Rightarrow \quad E_1 \cap E_2 \cap \cdots \in \mathscr{M}.$$

An isotone and antitone class is also called a **monotone class**. It is clear what we understand by the monotone class $\mathscr{M}(\mathscr{S})$ *generated* by a $\mathscr{S} \subseteq \mathscr{P}(\Omega)$, etc. Prove that:

5.5.1.1. A set ring is a σ-ring iff it is a monotone class.

5.5.1.2. If $\mathscr{R}$ is a ring in Ω, then $\mathscr{B}^{00}(\mathscr{R}) =$ the antitone class generated by $\mathscr{R}$.

5.5.1.3. If $\mathscr{S} \subseteq \mathscr{P}(\Omega)$ is finite intersection stable, then so is $\mathscr{M}(\mathscr{S})$.

5.5.2. Prove Proposition 5.4 anew, using monotone classes.

6. THE HAHN–BANACH THEOREM

Having pushed forward the theory of contents to a certain degree, we shall now leave it for a while and study what we will call *measures*. Measures will be nothing but certain *linear forms* on certain real vector spaces, and measure theory will turn out to be an extension theory for these linear forms. Thus it seems to be feasible to prelude that theory with one of the most powerful extension theorems for linear forms, the Hahn–Banach theorem in its classical form. We present it together with some standard applications, one of which (Banach limits) will later on yield an important example of a measure. The proof of the Hahn–Banach theorem will also be given in its now classical form. It should be noted that the Hahn–Banach theorem can be incorporated into a theory of convexity in topological vector spaces, and proved by a different method in that context (see BOURBAKI [1], KÖTHE [1, 2], SCHAEFER [1]).

6.1. Theorem (HAHN–BANACH). *Let H be an arbitrary real vector space and $p\colon H \to \mathbb{R}$ a so-called* **majorant function** *on H, i.e., a function such that*

$$p(f+g) \leqq p(f) + p(g) \qquad (f, g \in H).$$
$$p(\alpha f) = \alpha p(f) \qquad (0 \leqq \alpha \in \mathscr{R}, \quad f \in H).$$

Let H_0 be a linear subspace of H and m_0 a linear form on H_0 such that

$$m_0(f) \leqq p(f) \qquad (f \in H_0).$$

Then there is at least one linear form m on H extending m_0, i.e., coincides with m_0 on H_0, such that

$$m(f) \leqq p(f) \qquad (f \in H).$$

Proof. A pair (H_1, m_1) where $H_1 \supseteq H_0$ is a linear subspace of H and m_1 is a linear form on H_1 that coincides with m_0 on H_0 and satisfies $m_0(f) \leqq p(f)\,(f \in H_1)$ is called an admissible extension. If (H_1, m_1), (H_2, m_2) are admissible extensions, $H_1 \subseteq H_2$ and m_2 coincides with m_1 on H_1, we call (H_2, m_2) an extension of (H_1, m_1) and we write $(H_1, m_1) \preccurlyeq (H_2, m_2)$, and $\prec$ if $H_1 \neq H_2$, in which case we speak of a proper extension. It is easy to see (exercise) that the set Σ of all admissible extensions, endowed with the partial ordering $\preccurlyeq$, satisfies the hypothesis of Zorn's lemma, hence there exists an admissible extension (H_1, m_1) that is maximal in the sense that there is no proper extension of it in Σ. We shall be through if we can deduce $H_1 = H$ from this. Assume $H_1 \neq H$ and choose some $h_2 \in H \backslash H_1$. For any $f_1, g_1 \in H_1$, we have

$$\begin{aligned} m_1(g_1) - m_1(f_1) &= m_1(g_1 - f_1) \\ &\leqq p(g_1 - f_1) = p((g_1 + h_2) - (f_1 + h_2)) \\ &\leqq p(g_1 + h_2) + p(-(f_1 + h_2)) \end{aligned}$$

hence

$$-p(-(f_1 + h_2)) - m_1(f_1) \leqq p(g_1 + h_2) - m_1(g_1).$$

If we put

$$\alpha = \sup_{f_1 \in H_1} [-p(-(f_1 + h_2)) - m_1(f_1)], \qquad \beta = \inf_{g_1 \in H_1} [p(g_1 + h_2) - m_1(g_1)],$$

we have $-\infty < \alpha \leqq \beta < \infty$. Choose any real γ with $\alpha \leqq \gamma \leqq \beta$ and define

$$m_2(f_1 + \lambda h_2) = m_1(f_1) + \lambda\gamma \qquad (f_1 \in H_1, \lambda \in \mathbb{R}).$$

Observe that because $h_2 \notin H_1$, every f from the linear span H_2 of H_1 and h_2 has a unique representation $f = f_1 + \lambda h_2$ with $f_1 \in H_1$, $\lambda \in \mathbb{R}$, f_1 and λ depending linearly on f. Thus we have well defined a linear form m_2 on $H_2 \supseteq H_1$ with $H_1 \neq H_2$ which clearly coincides with m_1 on H_1. If we can show that (H_2, m_2) is admissible, we are through. But this follows for $\lambda > 0$ from

$$\begin{aligned} m_2(f_1 + \lambda h_2) &= m_1(f_1) + \lambda\gamma \leqq m_1(f_1) + \lambda\beta \\ &\leqq m_1(f_1) + \lambda\left[p\left(\frac{f_1}{\lambda} + h_2\right) - m_1\left(\frac{f_1}{\lambda}\right)\right] \\ &= p(f_1 + \lambda h_2), \end{aligned}$$

and by a similar argument for $\lambda < 0$.

6.2. Example. Let $\|\cdot\|$ be a *seminorm* on the real vector space H, i.e., a nonnegative real function on H satisfying

$$\|\alpha f\| = |\alpha|\cdot\|f\| \quad (\alpha \in \mathbb{R},\quad f \in H), \qquad \|f + g\| \leqq \|f\| + \|g\| \quad (f, g \in H).$$

Let m_0 be a linear form on a linear subspace H_0 of H such that $|m_0(f)| \leqq \|f\|$ $(f \in H_0)$. Then there is a linear form m on H that extends m_0 and satisfies $|m(f)| \leqq \|f\|$ $(f \in H)$. Taking in particular any $0 \neq f_0 \in H$, $H_0 = \{\alpha f_0 \,|\, \alpha \in \mathbb{R}\}$ and $m_0(\alpha f_0) = \alpha\|f\|$ $(\alpha \in \mathbb{R})$, we get m with $m(f_0) = \|f_0\|$.

6.3. Exercises. Let

$$l^\infty = \left\{x = (x_0, x_1, \ldots) \,\middle|\, x_0, x_1, \ldots \in \mathbb{R},\ \|x\|_\infty = \sup_{n \geqq 0} |x_n| < \infty\right\}$$

6.3.1. Prove that l^∞ is a real linear space and $\|\cdot\|_\infty$ is a norm on l^∞, i.e., a seminorm satisfying $\|x\|_\infty = 0 \Rightarrow x = 0\ (= (0, 0, \ldots))$.

6.3.2. For every $x = (x_0, x_1, \ldots) \in l^\infty$, define

$$p(x) = \inf\left\{\sup_{n \geqq 0} \sum_{j=0}^{r-1} \alpha_j x_{j+n} \,\middle|\, r \geqq 1, \alpha_0, \ldots, \alpha_{r-1} \geqq 0, \alpha_0 + \cdots + \alpha_{r-1} = 1\right\}.$$

Prove that this p satisfies the general requirements imposed in Theorem 6.1, $(p(x + y) \leqq p(x) + p(y)\ (x, y \in l^\infty),\ p(\alpha x) = \alpha p(x)\ (0 \leqq \alpha \in \mathbb{R}, x \in l^\infty))$.

6.3.3. Prove that the $p(\cdot)$ defined in (b) has the representation

$$p(x) = \lim_{r\to\infty} \left[\sup_{n \geqq 0} \frac{1}{r} \sum_{j=0}^{r-1} x_{j+n}\right]$$

6.3.4. Let $l_{\text{conv}} = \{x \,|\, x = (x_0,\ x_1,\ \ldots) \in l^\infty,\ \lim_{n\to\infty} x_n = \text{Lim}(x) \text{ exists}\}$. Prove that $\text{Lim} \leqq p$ on l_{conv}.

6.3.5. Let $l_{\text{Cesáro}} = \{x \,|\, x = (x_0,\ x_1,\ \ldots) \in l^\infty,\ \lim_n (1/n) \sum_{j=0}^{n-1} x_j = \text{Lim}_1(x)$ exists$\}$. Prove that $\text{Lim}_1 \leqq p$ on $l_{\text{Cesáro}}$.

6.3.6. Prove that for every $x \in l^\infty$, $y \in l_{\text{conv}}$,

$$p(x + y) = p(x) + \text{Lim}(y)$$

6.3.7. Prove that for every $x \in l^\infty$, there is a linear form L on l^∞ such that

$$\begin{aligned} L(y) &= \text{Lim}(y) && (y \in l_{\text{conv}}) \\ L(x) &= p(x) && \\ L(z) &\leqq p(z) && (z \in l^\infty). \end{aligned}$$

6.4. Example (Banach limits). Let l^∞, l_{conv}, p, Lim be as in Exercise 6.3. A linear form L on l satisfying

(a) $L(x) \geqq 0$ $(x = (x_0, x_1, \ldots) \in l^\infty,\ x_0, x_1, \ldots \geqq 0)$,

(b) $L((1, 1, \ldots)) = 1$,
(c) $L((x_1, x_2, \ldots)) = L((x_0, x_1, \ldots))$ $((x_0, x_1, \ldots) \in l^\infty)$

is called a *Banach limit*. Let us prove that a linear form L on l^∞ is a Banach limit iff (b) holds and $L(x) \leqq p(x)$ $(x \in l^\infty)$. In fact, if L is a Banach limit, then

$$L((x_0, x_1, \ldots)) = \cdots = L(x_{r-1}, x_r, \ldots))$$
$$= L\left(\left(\frac{1}{r}\sum_{j=0}^{r-1} x_{0+j}, \frac{1}{r}\sum_{j=0}^{r-1} x_{1+j}, \ldots\right)\right) \leqq \sup_{n \geqq 0} \frac{1}{r}\sum_{j=0}^{r-1} x_{n+j}$$

$((x_0, x_1, \ldots) \in l^\infty)$ follows from (a)–(c) and hence $L(x) \leqq p(x)$ $(x \in l^\infty)$ holds by Exercise 6.3.3. In order to prove the converse, we show that $p((x_1, x_2, \ldots) - (x_0, x_1, \ldots)) \leqq 0$ $((x_0, x_1, \ldots) \in l^\infty)$. In fact we have

$$\begin{aligned} &p((x_1, x_2, \ldots) - (x_0, x_1, \ldots)) \\ &\quad = p((x_1 - x_0, x_2 - x_1, \ldots)) \\ &\quad = \lim_{r\to\infty} \sup_{n \geqq 0} \frac{1}{r}\sum_{j=0}^{r-1} (x_{j+1+n} - x_{j+n}) = \lim_{r\to\infty} \sup_{n \geqq 0} \frac{1}{r}(x_{r+n} - x_n) \\ &\quad \leqq \lim_{r\to\infty} \frac{2\|x\|_\infty}{r} = 0 \quad (x = (x_0, x_1, \ldots) \in l^\infty). \end{aligned}$$

Thus $L(x) \leqq p(x)$ $(x \in l^\infty)$ implies $L((x_1, x_2, \ldots)) - L((x_0, x_1, \ldots)) \leqq 0$ and thus (multiply by -1) $L((x_1, x_2, \ldots)) - L(x_0, x_1, \ldots)) \geqq 0$, hence (c). In order to prove (a) we need only observe

$$L(x) \leqq p(x) \leqq 0 \quad (x = (x_0, x_1, \ldots) \in l^\infty, \quad x_0, x_1, \ldots \leqq 0)$$

and multiply by -1. This was to be proved, and Exercise 6.3.7 shows the existence of Banach limits; Exercises 6.3.4 and 5 imply that Banach limits are extensions of the usual limits and Cesáro limits.

7. ELEMENTARY DOMAINS

For any nonempty set Ω, the system $\mathbb{R}^\Omega$ of all real functions on Ω is a vector lattice if we define the linear and lattice operations pointwise. Let us list the most important among the so-called *finite vector lattice operations*, and some of their relations ($\vee$ stands for "maximum" and $\wedge$ for "minimum" as usual):

$$\begin{aligned} f_+ &= f \vee 0 = \tfrac{1}{2}(f + |f|) \qquad \text{(positive part)}; \\ f_- &= -(f \wedge 0) = (-f) \vee 0 \\ &= (-f)_+ = \tfrac{1}{2}(-f + |f|) \qquad \text{(negative part)}; \end{aligned}$$

$$|f| = f \vee (-f) = f_+ + f_- = f_+ + (-f)_+ \qquad \text{(modulus);}$$
$$f \vee g = f + (g - f)_+ \qquad \text{(supremum);}$$
$$f \wedge g = f - (f - g)_+ \qquad \text{(infimum);}$$
$$f + g = (f \wedge g) + (f \vee g);$$
$$f = f_+ - f_-.$$

Observe that the operations $(\cdot)_+$, $(\cdot)_-$, $|\cdot|$, $\vee$, $\wedge$ make sense also in the system $\overline{\mathbb{R}}^\Omega$ of all extended real-valued functions ($\overline{\mathbb{R}} = \mathbb{R} \cup \{+\infty, -\infty\}$ with the usual conventions about ordering), while linear operations suffer under the problem of avoiding $\infty - \infty$, and are thus not always defined in $\overline{\mathbb{R}}^\Omega$. Some relations like $f = f_+ - f_-$ make sense nevertheless.

It is these pointwise defined operations in $\mathbb{R}^\Omega$ or $\overline{\mathbb{R}}^\Omega$ that are tacitly meant whenever we speak of lattices or vector lattices of ($\mathbb{R}$- or $\overline{\mathbb{R}}$-valued functions.

For any $E \subseteq \overline{\mathbb{R}}^\Omega$ we define $\mathscr{E}_+ = \{f \,|\, 0 \leqq f \in \mathscr{E}\}$. For any $f \in \overline{\mathbb{R}}^\Omega$, the set

$$\operatorname{supp}(f) = \{\omega \,|\, f(\omega) \neq 0\}$$

is called the support or carrier of f.

From the above equalities we may deduce:

7.1. Proposition. *A linear subspace of $\mathbb{R}^\Omega$ is a vector lattice iff it is stable under one of the operations $(\cdot)_+$, $|\cdot|$.*

7.2. Definition. *Let $\Omega \neq \varnothing$. A vector lattice $\mathscr{E} \subseteq \mathbb{R}^\Omega$ is called an* **elementary domain** *on (the basic set) Ω if it satisfies*

7.2.1. **Stone's Axiom:** $f \in \mathscr{E} \Rightarrow f \wedge \alpha \in E$ ($\alpha = const > 0$).

In this context, the members of $\mathscr{E}$ are also called **elementary functions.**

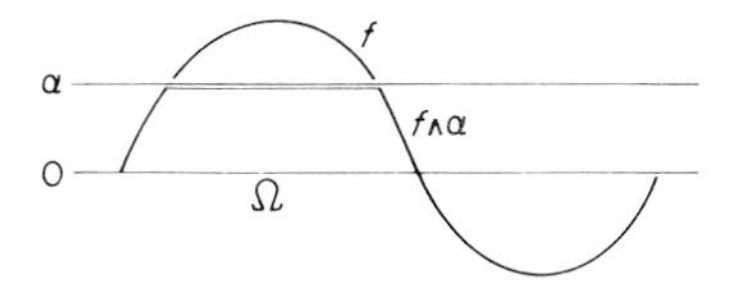

7.3. Remark. Every vector lattice $\mathscr{E} \subseteq \mathbb{R}^\Omega$ contains the constant 0. If it contains any constant $\neq 0$, then it contains all real constants and is an elementary domain. We shall encounter natural examples of elementary domains that contain no constant except 0. Stone's axiom has its name after Marshall Stone (b. 1903).

7.4. Example. Let $\Omega \neq \varnothing$ be arbitrary. Then $\mathscr{R}^\Omega$ is an elementary domain on Ω that contains all constants. $\{f \,|\, f \in \mathscr{R}^\Omega, \operatorname{supp}(f) \text{ is finite}\}$ is an elementary domain that contains no constant $\neq 0$ unless Ω is finite. $\{f \,|\, f \in \mathbb{R}^\Omega, \operatorname{supp}(f) \text{ is at most countable}\}$ is an elementary domain that contains no constant $\neq 0$ unless Ω is at most countable. The system of all bounded real functions on Ω is an elementary domain that contains all constants.

7.5. Example. Let $\Omega = \mathbb{R}$. Then $\mathscr{E} = \mathscr{C}^{00}(\mathbb{R}) = \mathscr{C}^{00}(\mathbb{R}, \mathbb{R}) = \{f \mid f \in \mathbb{R}^{\mathbb{R}},$ f is continuous and vanishes outside some bounded interval$\}$ is an elementary domain not containing any constant $\neq 0$. This easily generalizes to $\mathbb{R}^n$ (see also Example 7.7).

7.6. Exercises. Prove:

7.6.1. For $\Omega = \mathbb{R}_+ = \{x \mid 0 \leqq x \in \mathbb{R}\}$, the system of all functions of the form $f(x) = \alpha x$ with a real α is a vector sublattice of $\mathbb{R}^\Omega$, but does not satisfy Stone's axiom.

7.6.2. For $\Omega = [0, 1] = \{x \mid x \in \mathbb{R}, 0 \leqq x \leqq 1\}$ the system of all piecewise linear real functions on Ω is an elementary domain. Here $f \in \mathbb{R}^\Omega$ is called *piecewise linear* if there are reals $0 = a_0 < a_1 < \cdots < a_n = 1$ such that

$$f(x) = f(a_{k-1}) + \frac{f(a_k) - f(a_{k-1})}{a_k - a_{k-1}}(x - a_{k-1}) \qquad (a_{k-1} \leqq x \leqq a_k, \quad k = 1, \ldots, n).$$

For topologists we include here

7.7. Example. Let $\Omega \neq \varnothing$ be a topological space. Then the following linear subspaces of $\mathbb{R}^\Omega$ are elementary domains:

7.7.1. the system $\mathscr{C}(\Omega, \mathbb{R}) = \mathscr{C}(\Omega)$ of all continuous real functions on Ω;

7.7.2. the system $\mathscr{C}^{\mathrm{b}}(\Omega, \mathbb{R}) = \mathscr{C}^{\mathrm{b}}(\Omega)$ of all bounded real functions on Ω;

7.7.3. the system $\mathscr{C}^0(\Omega, \mathbb{R}) = \mathscr{C}^0(\Omega)$ of all continuous real functions on Ω vanishing at infinity (an $f \in \mathbb{R}^\Omega$ is said to vanish at infinity if for every $\varepsilon > 0$ there is a compact $K \subseteq \Omega$ such that $|f(\omega)| < \varepsilon$ $(\omega \in \Omega \backslash K)$).

7.7.4. the system $\mathscr{C}^{00}(\Omega, \mathbb{R}) = \{f \mid f \in \mathscr{C}(\Omega), \overline{\mathrm{supp}(f)} \text{ is compact}\} = \mathscr{C}^{00}(\Omega)$ (the bar means closure) of all continuous real functions on Ω with a conditionally compact support.

7.7.5. We have

$$\mathscr{C}(\Omega, \mathbb{R}) \supseteq \mathscr{C}^{\mathrm{b}}(\Omega, \mathbb{R}) \supseteq \mathscr{C}^0(\Omega, \mathbb{R}) \supseteq \mathscr{C}^{00}(\Omega, \mathbb{R})$$

with overall equality in case Ω is compact. Generally, $\mathscr{C}^{\mathrm{b}}(\Omega, \mathbb{R})$ contains all real constants.

7.8. Example. Let $\mathscr{R}$ be a set ring in $\Omega \neq \varnothing$. Define

$$\mathscr{E}(\mathscr{R}) = \{\alpha_1 1_{F_1} + \cdots + \alpha_n 1_{F_n} \mid n \geqq 1, \alpha_1, \ldots, \alpha_n \in \mathbb{R}, F_1, \ldots, F_n \in \mathscr{R}\}.$$

The members of $\mathscr{E}(\mathscr{R})$ are called the *step functions* for $\mathscr{R}$. It is obvious that $\mathscr{E}(\mathscr{R})$ is a linear subspace of $\mathbb{R}^{\Omega}$.

$$f = \beta_1 1_{B_1} + \cdots + \beta_r 1_{B_r} \tag{2}$$

is called a *representation* of $f \in \mathscr{E}(\mathscr{R})$ *in disjoint form* if $B_1, \ldots, B_r \in \mathscr{R}$ are pairwise disjoint. Let us show that every $f \in \mathscr{E}(\mathscr{R})$ has a representation in disjoint form. Clearly, every indicator function 1_F of some $F \in \mathscr{R}$ is a step function for $\mathscr{R}$ given in disjoint form. We are through as soon as we can prove that every linear combination of functions given in disjoint form can be represented in disjoint form. Let thus

$$f = \beta_1 1_{B_1} + \cdots + \beta_r 1_{B_r}, \qquad g = \gamma_1 1_{C_1} + \cdots + \gamma_s 1_{C_s}$$

be two functions from $E(R)$ given in disjoint form. Putting

$$B_{r+1} = (C_1 + \cdots + C_s)\backslash(B_1 + \cdots + B_r),$$
$$C_{s+1} = (B_1 + \cdots + B_r)\backslash(C_1 + \cdots + C_s), \qquad \beta_{r+1} = 0 = \gamma_{s+1}$$

we see that we may assume without loss of generality

$$B_1 + \cdots + B_r = C_1 + \cdots + C_s.$$

Put now $D_{jk} = B_j \cap C_k$, $\beta_{jk} = \beta_j$, $\gamma_{jk} = \gamma_k$ $(j = 1, \ldots, r,\ k = 1, \ldots, s)$. Then clearly

$$f = \sum_{j=1}^{r} \sum_{k=1}^{s} \beta_{jk} 1_{D_{jk}}, \qquad g = \sum_{j=1}^{r} \sum_{k=1}^{s} \gamma_{jk} 1_{D_{jk}}$$

are representations of f, g in disjoint form. Now we get, for any reals λ, μ, the following representations of $\lambda f + \mu g$ in disjoint form:

$$\lambda f + \mu g = \sum_{j=1}^{r} \sum_{k=1}^{s} (\lambda \beta_{jk} + \mu \gamma_{jk}) 1_{D_{jk}}.$$

Now that we know that every $f \in \mathscr{E}(\mathscr{R})$ can be represented in disjoint form, it is easy to prove that $\mathscr{E}(\mathscr{R})$ is an elementary domain. Let us use notations like $\vee$, $\wedge$, and $|\cdot|$ for reals as well as for real functions. If $f \in \mathscr{E}(\mathscr{R})$ is given in disjoint form (2), then

$$|f| = \sum_{j=1}^{r} |\beta_j| 1_{B_j}$$

shows $|f| \in \mathscr{E}(\mathscr{R})$ and from Proposition 7.1 $\mathscr{E}(\mathscr{R})$ is a vector lattice.

$$f \wedge \alpha = \sum_{j=1}^{r} (\beta_j \wedge \alpha) 1_{B_j}$$

for every real constant $\alpha > 0$ shows that $\mathscr{E}(\mathscr{R})$ also satisfies Stone's axiom, hence is an elementary domain.

8. MEASURES ON ELEMENTARY DOMAINS

8.1. Definition. *Let $\mathscr{E}$ be an elementary domain on $\Omega \neq \varnothing$. A real function $m: \mathscr{E} \to \mathbb{R}$ on $\mathscr{E}$ is said to be:*

8.1.1. **linear** *or a* **real linear form** *if*

$$m(\alpha_1 f_1 + \cdots + \alpha_n f_n) = \alpha_1 m(f_1) + \cdots + \alpha_n m(f_n)$$
$$(n \geqq 1, \quad \alpha_1, \ldots, \alpha_n \in \mathbb{R}, \quad f_1, \ldots, f_n \in \mathscr{E}).$$

8.1.2. *a positive linear form or a* **(positive) measure** *on $\mathscr{E}$ if it is a linear form and*

$$m(f) \geqq 0 \qquad (0 \leqq f \in \mathscr{E}).$$

8.1.3. **σ-continuous** *(at 0) if*

$$\lim_{k \to \infty} m(f_k) = 0 \quad (\mathscr{E} \ni f_1 \geqq f_2 \geqq \cdots \to 0 \text{ pointwise}).$$

8.1.4. *a* **(positive) σ-measure** *on $\mathscr{E}$ if it is a σ-continuous positive linear form on $\mathscr{E}$.*

8.1.5. **τ-continuous** *(at 0) if for every decreasingly filtered nonempty subset $\mathscr{F}$ of $\mathscr{E}_+$ with*

$$\inf_{f \in \mathscr{F}} f(\omega) = 0 \qquad (\omega \in \Omega)$$

the relation

$$\lim_{\mathscr{F}} m(f) = 0$$

holds, i.e., if for every $\varepsilon > 0$ there is an $f_0 \in \mathscr{F}$ such that $f_0 \geqq f \in \mathscr{F}$ implies $|m(f)| < \varepsilon$.

8.1.6. *a* **(positive) τ-measure** *on $\mathscr{E}$ if it is a τ-continuous positive form on $\mathscr{E}$.*

8.1.7. *If m is a (positive) measure on the elementary domain $\mathscr{E}$ on $\Omega \neq \varnothing$, then the triple $(\Omega, \mathscr{E}, m)$ is also called a* **measure space** *and properties of Ω, $\mathscr{E}$, or m are (sometimes imprecisely) also attributed to the triple. In particular, it is clear what a* **σ-** *or* **τ-measure space** *is.*

8.1.8. *If m is a positive measure on $\mathscr{E}$, we call the extended real number*

$$\|m\| = \sup\{m(f) \mid f \in \mathscr{E}, 0 \leqq f \leqq 1\} \geqq 0$$

the **total mass** *or* **total variation** *of m. If $\|m\| < \infty$, we call m a* **bounded** *(positive) measure; if $\|m\| = 1$, we call m* **normalized**. *If the constant 1 belongs to $\mathscr{E}$ and $m(1) = 1$, then we call m a* **probability measure** *and $(\Omega, \mathscr{E}, m)$ a* **probability (measure) space**. *If $\|m\| = \infty$, we call m an* **unbounded** *(positive) measure.*

8.2. Remarks

8.2.1. The above definition uses some conventions and concepts which should be generally known. In order to ensure complete understanding, let us recall the general conventions about filtered sets and limits. For every set $M \neq \emptyset$ endowed with a partial ordering $\leqq$, a subset F of M is called *increasingly filtered* if for any $n \geqq 1$ and $x_1, \ldots, x_n \in F$, there is an $x \in F$ with $x_1 \leqq x, \ldots, x_n \leqq x$. $F \subseteq M$ is called *decreasingly filtered* if for any $n \geqq 1$ and $x_1, \ldots, x_n \in F$, there is an $x \in F$ with $x \leqq x_1, \ldots, x \leqq x_n$. If $F \subseteq M$ is nonempty and increasingly filtered, if $m: M \to \mathbb{R}$ is any real function and α a real number, we write $\lim_F m(x) = \alpha$ if for every $\varepsilon > 0$, there is a $x_0 \in F$ such that $F \ni x \geqq x_0$ implies $|m(x) - \alpha| < \varepsilon$. An analogous convention is made for decreasingly filtered sets, and a generalization to filters as well as mappings m of M into any topological space obviously makes sense.

8.2.2. A real linear form m on a vector lattice $\mathscr{E} \subseteq \mathbb{R}^\Omega$ (it could be an abstract one as well) is positive iff it is *isotone*, i.e., iff, f, $g \in \mathscr{E}$, $f \leqq g \Rightarrow m(f) \leqq m(g)$ (form $g - f$).

8.2.3. Clearly a probability measure is normalized. We shall, of course, speak of a σ-probability (τ-probability) space if we have a probability measure that is σ-continuous (τ-continuous).

8.2.4. A positive measure m on an elementary domain $\mathscr{E}$ is σ-continuous at 0 iff it is *σ-filter*-continuous at 0, i.e., iff for every countable decreasingly filtered nonempty subset $\mathscr{F}$ of $\mathscr{E}_+$ with $\inf_{f \in \mathscr{F}} f(\omega) = 0$ $(\omega \in \Omega)$, we have $\lim_{\mathscr{F}} m(f) = 0$. In fact it is obvious that σ-filter-continuity implies σ-continuity since every decreasing sequence in $\mathscr{E}$ with pointwise limit 0 gives rise to a countable decreasingly filtered subset of $\mathscr{E}_+$ with pointwise infimum 0. Conversely, for every countable decreasingly filtered subset $\mathscr{F} = \{g_1, g_2, \ldots\}$ of $\mathscr{E}_+$ with pointwise infimum 0, we construct $f_1, f_2, \ldots \in \mathscr{F}$ successively such that $f_1 = g_1$ and $f_n \leqq f_{n-1}$, $f_n \leqq g_n$. We then have $f_1 \geqq f_2 \geqq \cdots \to 0$ pointwise and $\lim_k m(f_k) = 0$ implies

$$\lim_{\mathscr{F}} m(f) = \inf_{f \in \mathscr{F}} m(f) = \lim_{k \to \infty} m(f_k) = 0.$$

8.2.5. A positive measure m on an elementary domain $\mathscr{E}$ is clearly σ-continuous at 0 iff it is σ-continuous at an arbitrary point $f \in \mathscr{E}$, i.e., iff $\mathscr{E} \ni f_1 \geqq f_2 \geqq \cdots \to f$ (pointwise) $\Rightarrow m(f_n) \to m(f)$ (form $f_n - f$). An analogous statement is evident for σ-filter-continuity and for τ-continuity.

8.2.6. Clearly τ-continuity implies σ-continuity.

8.2.7. The real linear forms on an elementary domain $\mathscr{E}$ form a linear subspace of $\mathbb{R}^{\mathscr{E}}$. The positive measures on $\mathscr{E}$ form a set that is stable under finite linear combinations with nonegative real coefficients. The probability measures on an $\mathscr{E}$ containing all constants form a convex set.

8.3. Example. Let $\mathscr{E}$ be an elementary domain on $\Omega \neq \emptyset$. Choose some points $\omega_1, \ldots, \omega_n \in \Omega$ and some reals $\alpha_1, \ldots, \alpha_n$. Then

$$m(f) = \sum_{k=1}^{n} \alpha_k f(\omega_k) \tag{3}$$

defines a τ-continuous linear form on $\mathscr{E}$. If $\alpha_1, \ldots, \alpha_n \geqq 0$, then m is a (positive) τ-measure; in this case it is bounded with total mass $\alpha_1 + \cdots + \alpha_n$. In case $n = 1$, $\alpha_1 = 1$, it is called the *Dirac measure* or *point mass one* at ω_1 and denoted by δ_{ω_1} (some authors prefer ε_{ω_1}). In due course, the m defined by (3) can be written $m = \alpha_1 \delta_{\omega_1} + \cdots + \alpha_n \delta_{\omega_n}$. All this carries easily over to countably many points $\omega_1, \omega_2, \ldots \in \Omega$ if we suppose $|\alpha_1| + |\alpha_2| + \cdots < \infty$ and assume that every $f \in \mathscr{E}$ is bounded.

8.4. Exercises

8.4.1. Let $\mathscr{E}$ be an elementary domain on $\Omega \neq \emptyset$ and assume that $\mathscr{E}$ separates the points of Ω: If $\omega, \eta \in \Omega$, $\omega \neq \eta$, then there is an $f \in \mathscr{E}$ with $f(\omega) \neq f(\eta)$. Assume further that for every $\omega \in \Omega$, there is a $g \in \mathscr{E}$ with $g(\omega) \neq 0$. Prove that then for any $\omega, \eta \in \Omega$ with $\omega \neq \eta$, there is even some $h \in \mathscr{E}$ such that $h(\omega) \neq 0 = h(\eta)$, and

$$\alpha_1 \delta_{\omega_1} + \cdots + \alpha_n \delta_{\omega_n} = \beta_1 \delta_{\eta_1} + \cdots + \beta_s \delta_{\eta_s}$$

in case $0 \notin \{\alpha_1, \ldots, \alpha_n, \beta_1, \ldots, \beta_s\}$, $\omega_j \neq \omega_k$ $(j \neq k)$, $\eta_i \neq \eta_l$ $(i \neq l)$ implies $n = s$ and, after a suitable renumbering of $\eta_1, \ldots, \eta_n$, $\alpha_1 = \beta_1, \ldots, \alpha_n = \beta_n$, $\omega_1 = \eta_1, \ldots, \omega_s = \eta_s$.

8.4.2. Let $\Omega = \mathbb{R}$, $\mathscr{E} = C^{00}(\mathbb{R}, \mathbb{R})$. A positive measure m on $\mathscr{E}$ is called multiplicative if $m(fg) = m(f)m(g)$ $(f, g \in \mathscr{E})$. Prove that for every multiplicative normalized positive measure m on $\mathscr{E} = \mathscr{C}^{00}(\mathbb{R}, \mathbb{R})$, there is a unique point $\omega_0 \in \Omega$ such that $m(f) = f(\omega_0)$ $(f \in \mathscr{E})$. (*Hint:* Call $\omega \in \Omega$ empty if there is an $0 \leqq f \in \mathscr{E}$ with $f(\omega) > 0$ and $m(f) = 0$; show that the set of all empty points in $\mathbb{R}$ is open and $\neq \mathbb{R}$; for any nonempty points $\omega \neq \eta$ form $0 \leqq f, g \in \mathscr{C}^{00}(\mathbb{R}, \mathbb{R})$, such that $f(\omega), g(\eta) > 0$, $fg \equiv 0$ in order to get a contradiction).

8.5. Example. Let $\Omega = \mathbb{R}$, $\mathscr{E} = \mathscr{C}^{00}(\mathbb{R}, \mathbb{R})$, as in Example 7.5. Then every measure m on $\mathscr{E}$ (in particular the Riemann integral defined by $m(f) = \int f(x)\,dx$) is σ- and even τ-continuous. In fact, if $\mathscr{E}_+ \ni f_1 \geqq f_2 \geqq \cdots \to 0$ pointwise, then for every $\varepsilon > 0$, the sets $F_n = \{f_n > \varepsilon\}$ $(n = 1, 2, \ldots)$ are compact and satisfy $F_1 \supseteq F_2 \supseteq \cdots$,

$F_1 \cap F_2 \cap \cdots = \varnothing$. A well-known topological argument shows the existence of some n_0 with $F_{n_0} = \varnothing$, i.e., $f_n \leqq \varepsilon$ $(n \geqq n_0)$. We may now easily find an $f_0 \in \mathscr{E}$ with $1_{\mathrm{supp}(f_1)} \leqq f_0$ and conclude $m(f_n) \leqq m(\varepsilon f_0) = \varepsilon m(f_0)$ $(n \geqq n_0)$. Since $\varepsilon > 0$ was arbitrary, $\lim_{n\to\infty} m(f_n) = 0$ follows. τ-continuity follows in a similar way (exercise), and it is easy to carry over the example to $\mathbb{R}^n$ and $\mathscr{E} = \mathscr{C}^{00}(\mathbb{R}^n, \mathbb{R})$ (exercise). This example can be subsumed under Theorem 10.2. Measures $m: \mathscr{C}^{00}(\mathbb{R}^n, \mathbb{R}) \to \mathbb{R}_+$ are also called *Radon measures* in $\mathbb{R}^n$.

8.6. Example (positive measure that is not σ-continuous). Let $\Omega = \{0, 1, \ldots\}$. Then $\mathbb{R}^\Omega$ is the space of all real sequences $(x_0, x_1, \ldots)$. Consider the set $\mathscr{E} = l^\infty$ of all bounded sequences. It is obviously an elementary domain on Ω. A Banach limit (Example 6.4) m is apparently nothing but a probability measure on l^∞ that satisfies the condition of shift invariance,

$$m((x_1, x_2, \ldots)) = m((x_0, x_1, \ldots)) \quad ((x_0, x_1, \ldots) \in \mathscr{E}).$$

Let now $f_n = (x_0, x_1, \ldots)$ where $x_0 = \cdots = x_n = 0$ and $x_{n+1} = x_{n+2} = \cdots = 1$. Clearly $\mathscr{E} \geqq f_1 \geqq f_2 \geqq \cdots \to 0$ pointwise. On the other hand, normalization plus shift invariance imply

$$m(f_n) = m((1, 1, \ldots)) = 1 \qquad (n = 1, 2, \ldots).$$

Thus m is not σ-continuous.

8.7. Example. Let $(\Omega, \mathscr{R}, m)$ be a content prespace and $\mathscr{E}(\mathscr{R}) = \{\alpha_1 1_{F_1} + \cdots + \alpha_n 1_{F_n} | n \geqq 1, \alpha_1, \ldots, \alpha_n \in \mathbb{R}, F_1, \ldots, F_n \in \mathbb{R}\}$ be the elementary domain of all step functions for $\mathbb{R}$ (Example 7.8). Then there is a unique positive measure (again called) m on $\mathscr{E}(\mathscr{R})$ such that

$$m(1_F) = m(F) \qquad (F \in \mathscr{R}) \tag{4}$$

(the left m denotes the measure and the right m denotes the content). In fact, as shown in Example 7.8, every $f \in \mathscr{E}(\mathscr{R})$ has a representation

$$f = \beta_1 1_{B_1} + \cdots + \beta_r 1_{B_r} \tag{5}$$

in disjoint form.

A representation (5) of an $f \in \mathscr{E}(\mathscr{R})$ is called *standard* if $B_1, \ldots, B_r \neq \varnothing$, $B_j \cap B_k = \varnothing$ $(j \neq k)$, $\beta_1, \ldots, \beta_r \neq 0$, $\beta_j \neq \beta_k$ $(j \neq k)$. It is clear that every $0 \neq f \in \mathscr{E}(\mathscr{R})$ has a unique standard representation: $\{\beta_1, \ldots, \beta_r\} = f(\Omega)\backslash\{0\}$, $B_j = \{f = \beta_j\}$ $(j = 1, \ldots, r)$. It can be obtained from any representation in disjoint form by discarding zero summands and uniting some of the sets involved. Put $m(0) = 0$ and

(6) $$m(f) = \beta_1 m(F_1) + \cdots + \beta_r m(F_r)$$

if $f \neq 0$, and (5) is a standard representation of f. This defines an m: $\mathscr{E}(\mathscr{R}) \to \mathbb{R}$ that clearly satisfies $m(f) \geqq 0$ $(0 \leqq f \in \mathscr{E}(\mathscr{R}))$. Let us now prove that it is a linear form. For this we first observe that (6) holds true also if (5) is only assumed to be in disjoint form (exercise). But now, if we have $f, g \in \mathscr{E}(\mathscr{R})$, we know from Example 7.8 that there are representations (5) of f and

(7) $$g = \gamma_1 1_{B_1} + \cdots + \gamma_r 1_{B_r}$$

in disjoint form with the same sets B_j in (5) and (7) (they were called D_{jk} in Example 7.8). Now for any reals λ, μ, we find that

$$\lambda f + \mu g = (\lambda\beta_1 + \mu\gamma_1)1_{B_1} + \cdots + (\lambda\beta_r + \mu\gamma_r)1_{B_r}$$

is a representation in disjoint form, hence

$$\begin{aligned} m(\lambda f + \mu g) &= (\lambda\beta_1 + \mu\gamma_1)m(B_1) + \cdots + (\lambda\beta_r + \mu\gamma_r)(B_r) \\ &= \lambda m(f) + \mu m(g) \end{aligned}$$

as desired, m is thus a positive measure on $\mathscr{E}(\mathscr{R})$, and it is clear that it satisfies (4) and is the only one satisfying (4). Clearly (6) holds even if (5) is not in disjoint form. We call it the measure on $\mathscr{E}(\mathscr{R})$ derived from the content m on R, and consequently $(\Omega, \mathscr{E}(\mathscr{R}), m)$ the measure space derived from the content prespace $(\Omega, \mathscr{R}, m)$.

8.8. Convention. *Properties, statements, etc. based on a content prespace $(\Omega, \mathscr{R}, m)$ will be attributed, possibly with appropriate modifications of language, to the derived measure space $(\Omega, \mathscr{E}(\mathscr{R}), m)$ and vice versa.*

8.9. Proposition. *Let $(\Omega, \mathscr{E}(\mathscr{R}), m)$ be the measure space derived from the content prespace $(\Omega, \mathscr{R}, m)$. Then the measure m is σ-continuous iff the content m is σ-additive.*

Proof. "if": let $f_n = \sum_{j=1}^{r_n} \beta_{nj} 1_{B_{nj}} \in \mathscr{E}(\mathscr{R})$ be such that $f_1 \geqq f_2 \geqq \cdots \to 0$ pointwise. We may assume $f_1, f_2, \ldots \neq 0$ and all representations standard. With $M = \sup_{1 \leqq j \leqq r_1} \beta_{1j}$ we have $0 \leqq f_n \leqq f_1 \leqq M$ $(n = 1, 2, \ldots)$. Choose any $\varepsilon > 0$ and put $E_n = \{f_n \geqq \varepsilon\} = \bigcup_{\beta_{nj} \geqq \alpha} B_{nj}$. Clearly $E_1, E_2, \ldots \in \mathscr{R}$, $E_1 \supseteq E_2 \supseteq \cdots$, $E_1 \cap E_2 \cap \cdots = \varnothing$. By Proposition 2.2 we can conclude $m(E_n) \to 0$. Now $0 \leqq f_n \leqq \varepsilon 1_{f_1 > 0} + M 1_{E_n}$ shows

$$0 \leqq m(f_n) \leqq \varepsilon m(\{f_1 > 0\}) + M m(E_n).$$

Since $\varepsilon > 0$ was arbitrary, $\lim_{n\to\infty} m(f_n) = 0$ follows.

"only if": Let $\mathscr{R} \ni E_1 \supseteq E_2 \supseteq \cdots$, $E_1 \cap E_2 \cap \cdots = \varnothing$. By Proposition 2.2 it suffices to prove $m(E_n) \to 0$. But $1_{E_n} \to 0$ pointwise, hence $m(E_n) = m(1_{E_n}) \to 0$ follows from (4) and the σ-continuity of m.

8.10. Exercises

8.10.1. Let $A \neq 0$ be a finite set and $\Omega = A \times A \times \cdots$ the Bernoulli space with the alphabet A. A function $f \in \mathbb{R}^\Omega$ is said to be *of order* $r \geqq 0$ if there is a function $f_r \in \mathbb{R}^{A^{r+1}}$ such that

$$f(\omega_0\,\omega_1 \cdots) = f_r(\omega_0 \cdots \omega_r) \qquad (\omega = \omega_0\,\omega_1 \cdots \in \Omega).$$

Let $E_r = \{f \mid f \in \mathbb{R}^\Omega, f \text{ is of order } r\}$ $(r = 0, 1, \ldots)$. Show that every E_r is an elementary domain on Ω, that $E_0 \subseteq E_1 \subseteq \cdots$ and $E = E_0 \cup E_1 \cup \cdots$ is an elementary domain on Ω, too. Prove that every positive measure on $\mathscr{E}$ is τ-continuous. (*Hint:* Use Example 2.5.)

Prove that for every positive measure m on $\mathscr{E}$ the measure space $(\Omega, \mathscr{E}, m)$ is derived from a uniquely determined content prespace $(\Omega, \mathscr{C}, m)$ where $\mathscr{C}$ is defined according to Example 1.6. Prove anew that every content on $\mathscr{C}$ is a σ-content.

8.10.2. Let $\Omega \neq \varnothing$ and $\mathscr{E} \subseteq \mathbb{R}^\Omega$ an elementary domain on Ω. Assume that every $f \in \mathscr{E}$ is bounded. Let $p\colon \mathscr{E} \to \mathbb{R}$ be defined by $p(f) = \sup_{\omega \in \Omega} f(\omega)$. Prove that p is a majorant function and apply Theorem 6.1 (Hahn–Banach) in order to prove the following: Let $\mathscr{E}_0$ be a linear subspace of $\mathscr{E}$ and $m_0\colon \mathscr{E}_0 \to \mathbb{R}$ be linear and such that $f_0 \in \mathscr{E}_0 \Rightarrow m_0(f_0) \leqq p(f_0)$; then there is a positive measure $m\colon \mathscr{E} \to \mathbb{R}$ such that m coincides with m_0 on $\mathscr{E}_0$ (*Hint*: Positivity of m is tantamount to $f \leqq 0 \Rightarrow m(f) \leqq 0$.) We remark that this exercise comprises the following special case: $\mathscr{E}$ contains all constants and $\mathscr{E}_0$ consists of all constants, m_0 carries every constant into itself.

There is still an important class of examples of measure spaces to be described—those with a topological space Ω as a basic set and $\mathscr{C}^{\mathrm{b}}(\Omega, \mathbb{R})$ or $\mathscr{C}^{00}(\Omega, \mathbb{R})$ as an elementary domain. We postpone them until Section 10 and continue now by presenting comparatively simple aspects of the extension problem for measures.

9. *RIEMANN (EUDOXOS)* EXTENSION OF POSITIVE MEASURES

The examples of measure spaces presented so far convey themselves already ample suggestions for extensions of measures from the elementary domain on which they are given to a larger domain.

Let, e.g., $(\mathbb{R}, \mathscr{D}, m)$ be the dyadic content prespace on $\mathbb{R}$ (Examples 1.10, 2.4) and $(\mathbb{R}, \mathscr{E}(\mathscr{D}), m)$ the derived measure space. Apparently $\mathscr{E}(\mathscr{D})$ is nothing but the space of all real functions on $\mathbb{R}$ assuming finitely many real constant values on finitely many dyadic intervals and vanishing elsewhere, and m is the usual elementary (pre-Riemannian, so to speak) integral on $\mathscr{E}(\mathscr{D})$. Every course in calculus provides enough material which makes it desirable to "integrate," i.e., to include into an extended domain for m, functions which are not yet in $\mathscr{E}(\mathscr{D})$, such as functions that vanish outside some interval and coincide with a polynomial, a piecewise linear function, or a step function with nondyadic intervals of constancy, or even a suitable rational function inside, not to mention functions like $1_{\mathbb{Q}} \cdot 1_{[0,1]}$ ($\mathbb{Q}$ = the rationals) which might look pathological at first sight.

In this subsection we present a simple method of extension of measures which is analogous to that given for contents in Section 3 and might well be named after Eudoxos (ca. 408–355 B.C.) again but is usually named after Riemann (1826–1866). We supplement it by considerations about *extension by uniform approximation.*

9.1. Definition. *Let $(\Omega, \mathscr{E}, m)$ be a measure space. A function $g \in \mathbb{R}^\Omega$ is called* **Riemann integrable** *for $(\Omega, \mathscr{E}, m)$ or, briefly, for m, if for every $\varepsilon > 0$ there are $f, h \in \mathscr{E}$ such that*

$$f \leqq g \leqq h, \qquad m(h - f) < \varepsilon,$$

or, equivalently (obey the rule $\inf \varnothing = \infty$, $\sup \varnothing = -\infty$)

$$-\infty < \sup_{\mathscr{E} \ni f \leqq g} m(f) = \inf_{g \leqq h \in \mathscr{E}} m(h) < \infty.$$

Let $\mathscr{E}_{(m)}$ denote the system of all Riemann integrable functions for $(\Omega, \mathscr{E}, m)$.

9.2. Theorem. *Let $(\Omega, \mathscr{E}, m)$ be a measure space and $\mathscr{E}_{(m)}$ the system of the corresponding Riemann measurable functions. Then:*

9.2.1. *$\mathscr{E}_{(m)} \supseteq \mathscr{E}$ and*

$$m(g) = \sup_{\mathscr{E} \ni f \leqq g} m(f) \qquad (g \in \mathscr{E}_{(m)})$$

defines an extension (again denoted by m) of m from $\mathscr{E}$ to $\mathscr{E}_{(m)}$. It is called the **Riemann integral** *for m.*

9.2.2. *$(\Omega, \mathscr{E}_{(m)}, m)$ is a measure space again. It is a σ-measure space iff $(\Omega, \mathscr{E}, m)$ is.*

9.2.3. *Every positive measure on $\mathscr{E}_{(m)}$ that coincides with m on E coincides with the m given in 1. $(\mathscr{E}_{(m)})_{(m)} = \mathscr{E}_{(m)}$.*

Proof. 1 is obvious: For $g \in \mathscr{E}$, take $f = g = h$.

2. Let us first prove that $\mathscr{E}_{(m)}$ is an elementary domain. For this, let us

be content to prove its stability under the passage $g \to g_+ = g \vee 0$ to positive parts; stability under linear operations can be proved in a similar, and then obvious, way; and this will do it after Proposition 7.1. Now, if $g \in \mathscr{E}_{(m)}$ and $\varepsilon > 0$, choose $f, h \in \mathscr{E}$ with $f \leqq g \leqq h$, $m(h - f) < \varepsilon$. Then clearly $\mathscr{E} \ni f_+ \leqq g_+ \leqq h_+ \in \mathscr{E}$, $h_+ - f_+ \leqq h - f$, and thus

$$m(h_+ - f_+) \leqq m(h - f) < \varepsilon.$$

Next we extend m to $\mathscr{E}_{(m)}$ according to 1 and show that $m: \mathscr{E}_{(m)} \to \mathbb{R}$ is linear again. Let us first prove additivity: For any $g, g' \in \mathscr{E}_{(m)}$ we have

$$\begin{aligned} m(g + g') &= \sup_{\mathscr{E} \ni f \leqq g + g'} m(f) \\ &\geqq \sup_{\mathscr{E} \ni f \leqq g,\ \mathscr{E} \ni f' \leqq g'} m(f + f') \\ &= \sup_{\mathscr{E} \ni f \leqq g} m(f) + \sup_{\mathscr{E} \ni f' \leqq g'} m(f') \\ &= m(g) + m(g'). \end{aligned}$$

Now the definition of Riemann integrability, $m(g)$ and $m(g')$ can be computed as infima as well. The corresponding estimates yield

$$m(g + g') \leqq m(g) + m(g')$$

and thus the desired equality

$$m(g + g') = m(g) + m(g') \qquad (g, g' \in \mathscr{E}).$$

The proof of

$$m(\alpha g) = \alpha m(g) \qquad (g \in \mathscr{E}, \quad \alpha \in \mathbb{R})$$

is still simpler (exercise; distinguish the cases $\alpha \geqq 0$ and $\alpha < 0$). Thus $m: \mathscr{E}_{(m)} \to \mathbb{R}$ is linear. It is also positive: For $g \geqq 0$, choose $f \geqq 0$ in $\mathscr{E}$. Let us finally prove that $m: \mathscr{E}_{(m)} \to \mathbb{R}$ is σ-continuous if $m: \mathscr{E} \to \mathbb{R}$ is. Let $\mathscr{E}_{(m)} \ni g_1 \geqq g_2 \geqq \cdots \to 0$ pointwise. For a given $\varepsilon > 0$, choose $f'_1, f'_2, \ldots \in \mathscr{E}$ such that $0 \leqq f'_k \leqq g_k$, $m(f'_k) \geqq m(g_k) - \varepsilon/2^k$ and put $f_k = f'_1 \wedge \cdots \wedge f'_k$ $(k = 1, 2, \ldots)$. Clearly

$$f_1, f_2, \ldots \in \mathscr{E} \qquad \text{and} \qquad 0 \leqq f'_k - f_k \leqq (g_{k-1} - f_{k-1}) + \cdots + (g_1 - f_1)$$

and thus $m(f'_k) \leqq m(f_k) + \varepsilon/2^{k-1} + \cdots + \varepsilon/2 \leqq m(f_k) + \varepsilon$ and hence $m(f_k) \geqq m(g_k) - 2\varepsilon$ $(k = 1, 2, \ldots)$. But we have $f_k \to 0$ pointwise, thus $\lim m(g_k) \leqq 2\varepsilon$, and finally $m(g_k) \to 0$ since $\varepsilon > 0$ was arbitrary. The rest is obvious.

3 is an easy exercise.

9.3. Definition. *For any measure space* $(\Omega, \mathscr{E}, m)$ *the measure space*

$(\Omega, \mathscr{E}_{(m)}, m)$ *defined according to Definition* 9.1. *and Theorem* 9.2 *is called the* **Riemann extension** *of* $(\Omega, \mathscr{E}, m)$.

9.4. Exercises

9.4.1. Let $(\mathbb{R}, \mathscr{E}(\mathscr{D}), m)$ be the σ-measure space derived from the dyadic geometric content prespace $(\mathbb{R}, \mathscr{D}, m)$ (Examples 1.10, 2.4, 8.7). Prove that $(\mathscr{E}(\mathscr{D}))_{(m)}$ contains:

(a) $\mathscr{C}^{00}(\mathbb{R}, \mathbb{R})$
(b) all functions of the form $\alpha_1 1_{I_1} + \cdots + \alpha_n 1_{I_n}$ where $n \geqq 1, \alpha_1, \ldots, \alpha_n \in \mathbb{R}$, $I_1, \ldots, I_n$ arbitrary bounded intervals (including all, none, or some of their endpoints).

and, loosely speaking, all Riemann integrable functions in the usual sense that vanish outside bounded intervals.

9.4.2. Let $(\mathbb{R}, \mathscr{E}(\mathscr{D}), m)$ be as in (a) and $x_1, x_2, \ldots$ a 1-1 enumeration of the rationals in the open interval $]0, 1[$. Choose $g_1', g_2', \ldots \in \mathscr{C}^{00}(\mathbb{R})$ such that

$$0 \leqq g_k' \leqq 1_{]0,1[}, \qquad g_k'(x_k) = 1 \qquad (k = 1, 2, \ldots)$$

but $m(g_1') + m(g_2') + \cdots < 1$. Put $g_k = g_1' \vee \cdots \vee g_k'$ $(k = 1, 2, \ldots)$, $g = \lim g_k$. Prove that $g \notin (\mathscr{E}(\mathscr{D}))_{(m)}$. (*Hint*: If $g \leqq h \in \mathscr{E}(\mathscr{D})$, then $h \geqq 1_{]0,1[}$, thus $g \in (\mathscr{E}(\mathscr{D}))_{(m)}$ would imply $m(g) = 1$; bring this to a contradiction with $\Sigma m(g_k') < 1$.)

9.4.3. Let $(\mathbb{R}, (\mathscr{E}(\mathscr{D}))_{(m)}, m)$ be the σ-measure space defined according to 9.4.1. Prove that this m is not τ-continuous on $(\mathscr{E}(\mathscr{D}))_{(m)}$. (*Hint*: Consider the decreasingly filtered system F of all functions 1_F where $F = [0, 1]$ minus a finite number of points.)

We supplement the above sketch of Riemann integrability by some convenient results about *uniform approximation* (which has, of course, already played a key role in Exercise 9.4.1). We begin with a convenient special case in which we can dispose of constant functions.

9.5. Exercises

9.5.1. Let $(\Omega, \mathscr{E}, m)$ be a measure space such that $\mathscr{E}$ contains all constants. Show that $E_{(m)}$ is closed under uniform approximation. (*Hint*: Use $(\mathscr{E}_{(m)})_{(m)} = \mathscr{E}_{(m)}$ and the fact that uniform approximation makes possible the simple "squeezing-in" procedure which defines Riemann integrability.)

9.5.2. Let $A \neq \varnothing$ be a finite set and $\mathscr{E}$ the elementary domain of all functions of finite order on the compact metric space $\Omega = A \times A \times \cdots$ (Exercise 8.10). Prove that $\mathscr{C}(A \times A \times \cdots)$ is the uniform closure of $\mathscr{E}$, and thus in particular $\mathscr{E}_{(m)} \supseteq \mathscr{C}(A \times A \times \cdots)$ for every positive measure m on $\mathscr{E}$.

A general result on uniform approximation is given by:

9.6. Proposition. *Let* $(\Omega, \mathscr{E}, m)$ *be a measure space and* $\mathscr{E}'$ *an elementary domain of bounded functions on* Ω *such that for every* $f' \in \mathscr{E} \cup \mathscr{E}'$, *there is an* $0 \leqq h \in \mathscr{E}$ *with* $\{f' \neq 0\} \subseteq \{h \geqq 1\}$. *Assume further that* $\mathscr{E}'$ *is contained in the closure of* $\mathscr{E}$ *with respect to uniform approximation, then* $\mathscr{E}' \subseteq \mathscr{E}_{(m)}$.

Proof. For any given $g' \in \mathscr{E}'$ find some $0 \leqq h_0 \in \mathscr{E}$ such that $\{g' \neq 0\} \subseteq \{h_0 \geqq 1\}$ and then some $0 \leqq h_1 \in \mathscr{E}$ with $\{h_0 \neq 0\} \subseteq \{h_1 \geqq 1\}$. Since g' is bounded, we may (after multiplication of h_0 with some constant $\geqq 1$) assume $|g'| \leqq h_0$. Let M be a constant $\geqq h_0$, $\varepsilon > 0$, and $g_0 \in \mathscr{E}$ such that $|g' - g_0| < \varepsilon$ uniformly on Ω. Replacing g_0 by $(g_0 \wedge h_0) \vee (-h_0)$, we may assume $|g_0| \leqq h_0$. Put now $f = g_0 - \varepsilon h_0$, $h = g_0 + \varepsilon h_0$. Clearly, $\mathscr{E} \ni f \leqq g \leqq h \in \mathscr{E}$ and $0 \leqq h - f \leqq 2\varepsilon h_0 \leqq 2\varepsilon M h_1$. We conclude

$$m(h - f) \leqq 2\varepsilon M m(h_1).$$

Since $\varepsilon > 0$ was arbitrary, $g \in \mathscr{E}_{(m)}$ follows.

10. MEASURE SPACES OVER TOPOLOGICAL SPACES

The importance of measure spaces whose basic set Ω is a topological space and whose elementary domain is $\mathscr{C}^{\mathrm{b}}(\Omega) = \mathscr{C}^{\mathrm{b}}(\Omega, \mathbb{R})$, or $\mathscr{C}^{00}(\Omega) = \mathscr{C}^{00}(\Omega, \mathbb{R})$ is a sufficient reason to devote a whole section to them. The reader may consider Example 7.7 and 8.5 and Exercises 8.10, 9.5.2 as a prelude to it. He is now supposed to be informed about the basic concepts and results of general (point set) topology.

10.1. Theorem. *Let* $(\Omega, \mathscr{T})$, $\Omega \neq \varnothing$ *be a compact topological space. Then every positive measure* m *on the elementary domain* $\mathscr{C}(\Omega, \mathbb{R})$ *is* τ*-continuous. Such measures are also called* **Radon measures** *in* Ω.

Proof. We essentially use Dini's theorem and observe that $\mathscr{C}(\Omega, \mathbb{R})$ contains all real constants. Let $\varnothing = \mathscr{F} \subseteq \mathscr{C}(\Omega, \mathbb{R})$ be decreasingly filtered with pointwise infimum 0. For every $\varepsilon > 0$, the family $(\{f \geqq \varepsilon\})_{f \in \mathscr{F}}$ consists of compact sets, is decreasingly filtered, and has an empty intersection. A well-known elementary result in topology (actually not more than the definition of the compactness of Ω applied to the open covering $(\{f < \varepsilon\})_{f \in \mathscr{F}}$ of Ω) says that there is an $f \in \mathscr{F}$ for which $\{f \geqq \varepsilon\}$ is empty, i.e., $0 \leqq f < \varepsilon$, and consequently $0 \leqq m(f) \leqq \varepsilon m(1)$. Since $\varepsilon > 0$ was arbitrary, we find $\inf_{f \in \mathscr{F}} m(f) = 0 = \lim_{\mathscr{F}} m(f)$, and thus the τ-continuity of m.

In more general situations we must try to get some substitute for the constants in $\mathscr{C}(\Omega, \mathbb{R})$.

10.2. Theorem. *Let $(\Omega, \mathscr{T})$, $\Omega \neq \varnothing$ be a locally compact Hausdorff space. Then every positive measure m on the elementary domain $\mathscr{C}^{00}(\Omega, \mathbb{R})$ is τ-continuous. Such measures are also called* **Radon measures** *in Ω.*

Proof. Let $\varnothing \neq \mathscr{F} \subseteq \mathscr{C}^{00}(\Omega, \mathbb{R})$ be decreasingly filtered with pointwise infimum 0. Choose any $f_0 \in \mathscr{F}$ and define $F_0 = \{f \,|\, F \ni f \leqq f_0\}$. Clearly $\inf_{f \in \mathscr{F}_0} f(\omega) = 0$ $(\omega \in \Omega)$ again and $\inf_{f \in \mathscr{F}_0} m(f) = \inf_{f \in \mathscr{F}} m(f)$. Thus we may henceforth assume $f \leqq f_0$ $(f \in \mathscr{F})$ for some $f_0 \in \mathscr{F}$. Consider the compact set $K = \overline{\{f_0 > 0\}}$. For every $\varepsilon > 0$, the Dini-type argument in the proof of Theorem 10.1 yields some $f \in \mathscr{F}$ with $f \leqq \varepsilon$. Basic results in the theory of locally compact Hausdorff spaces (they go with the name of Urysohn) yield the existence of an $f_1 \in \mathscr{C}^{00}(\Omega, \mathbb{R})$ such that $1_K \leqq f_1$. Now we may conclude $f \leqq \varepsilon f_1$ for our above f, and this implies $0 \leqq m(f) \leqq \varepsilon m(f_1)$. Since f_1 is fixed independently of ε and $\varepsilon > 0$ is arbitrary, $\inf_{f \in \mathscr{F}} m(f) = \lim_{\mathscr{F}} m(f) = 0$ follows.

This theorem applies, e.g., to $\Omega = \mathbb{R}^n$ and $m =$ the usual Riemann integral as a positive measure on $\mathscr{C}^{00}(\mathbb{R}^n, \mathbb{R})$.

10.3. Definition. *Let $(\Omega, \mathscr{T})$, $\Omega \neq \varnothing$ be a topological space. A linear form m on an elementary domain $\mathscr{E}$ contained in the space $\mathscr{C}^b(\Omega, \mathbb{R})$ of all bounded continuous real functions on Ω is called* **tight** *if for every $\varepsilon > 0$ there is a compact $K \subseteq \Omega$ and a $\delta > 0$ such that*

$$f \in \mathscr{E}, \ |f| \leqq 1, \ |f(\omega)| < \delta \quad (\omega \in K) \quad \Rightarrow \quad |m(f)| < \varepsilon.$$

If m is a tight (positive) measure, then the measure space $(\Omega, \mathscr{E}, m)$ is also called tight.

Obviously every compact $\Omega \neq \varnothing$ yields every positive measure m on $\mathscr{C}(\Omega, \mathbb{R})$ tight (Theorem 10.1). Examples demonstrating the usefulness of the concept of tightness more convincingly will show up later on in connection with Polish spaces (Chapters V and XI).

10.4. Theorem. *Let $(\Omega, \mathscr{T})$, $\Omega \neq \varnothing$, be a topological space and m a tight linear form on an elementary domain $\mathscr{E} \leqq \mathscr{C}^b(\Omega, \mathbb{R})$. Then m is τ-continuous.*

Proof. Let $\varnothing \neq \mathscr{F} \subseteq \mathscr{E}$ be decreasingly filtered with pointwise infimum 0. For the purpose of proving $\lim_{\mathscr{F}} m(f) = 0$, we may, as in the proof of Theorem 10.2, assume that there exists some $f_0 \in \mathscr{F}$ with $0 \leqq f \leqq f_0$ $(f \in \mathscr{F})$, and even (after multiplication with a strictly positive constant, $0 \leqq f_0 \leqq 1$. For a given $\varepsilon > 0$, we choose a compact $K \subseteq \Omega$ and a $\delta > 0$ according to the definition of tightness. Put $K_f = \{\omega \,|\, \omega \in \Omega, f(\omega) \geqq \delta\}$ $(f \in F)$. Then $(K_f)_{f \in \mathscr{F}}$ is a decreasingly filtered family of compact subsets of K, with an empty intersection. It follows that there is some $f_1 \in \mathscr{F}$ for which $K_f = \emptyset$, i.e., $|f_1(\omega)| < \delta(\omega \in K)$, and consequently $|m(f)| < \varepsilon$ $(\mathscr{F} \ni f \leqq f_1)$ which proves the desired statement.

Next comes a theorem on the preservation of tightness under a certain approximation procedure. We begin with a technical observation:

10.5. Remark. Define generally for any $\Omega \neq \emptyset$,

$$\|f\| = \sup_{\omega \in \Omega} |f(\omega)| \qquad (f \in \mathbb{R}^\Omega)$$

and observe that for any $f, g \in \mathbb{R}^\Omega$ and any constant $M \in \mathbb{R}$ such that $\|f\| \leqq M$ we have

$$|f - [(g \wedge M) \vee (-M)]| \leqq |f - g|.$$

In particular, if $\mathscr{E} \subseteq \mathbb{R}^\Omega$ is an elementary domain and $f \in \mathscr{E}$ satisfies $\|f\| \leqq M$, then for every $g \in \mathscr{E}$ there is a $g' \in \mathscr{E}$ with $\|g'\| \leqq M$ and $|f - g'| \leqq |f - g|$.

10.6. Theorem (HILDENBRAND [1]). *Let $(\Omega, \mathscr{T})$, $\Omega \neq \emptyset$ be a topological space. For every $h \in \mathscr{C}^b(\Omega, \mathbb{R})$, $0 < M < \infty$, $K \subseteq \Omega$ compact, $\delta > 0$ define*

$$\mathscr{U}(h; M, K, \delta) = \{f \mid f \in \mathscr{C}^b(\Omega, \mathbb{R}), \|f\| \leqq M, |f(\omega) - h(\omega)| < \delta\ (\omega \in K)\}.$$

Let $\mathscr{S}$ be the topology in $\mathscr{C}^b(\Omega, \mathbb{R})$ whose basis at h consists of all sets $U(h; M, K, \delta)$ with $M = \|h\| + 1$. Let m be a tight linear form on an elementary domain $\mathscr{E} \subseteq \mathscr{C}^b(\Omega, \mathbb{R})$. Then there is a unique linear form $\overline{m}$ on the $\mathscr{S}$-closure $\overline{\mathscr{E}}$ of $\mathscr{E}$ such that $\overline{m}$ is continuous for the restriction of $\mathscr{S}$ to $\overline{\mathscr{E}}$ and coincides with m on $\mathscr{E}$, $\overline{m}$ is positive if m is, and is tight. If $\mathscr{E}$ separates the points of Ω (i.e., if for any $\omega, \eta \in \Omega$ with $\omega \neq \eta$, there is an $f \in \mathscr{E}$ with $f(\omega) \neq f(\eta)$) and contains all constants, then $\overline{\mathscr{E}} = \mathscr{C}^b(\Omega, \mathbb{R})$.

Proof. Going into the definition of tightness with a multiplicative constant >0, we see that for any $\varepsilon > 0$, $M > 0$, there is a compact $K \subseteq \Omega$ and a $\delta > 0$ such that

$$f \in \mathscr{U}(0; M, K, \delta) \cap \mathscr{E} \quad \Rightarrow \quad |m(f)| < \varepsilon.$$

This implies that for any $\varepsilon > 0$, $M > 0$, there is a compact $K' \subseteq \Omega$ and a $\delta' > 0$ such that for any $h \in \mathscr{C}^b(\Omega, \mathbb{R})$ with $\|h\| \leqq M$

$$\begin{aligned} f, g \in \mathscr{U}(h; M, K', \delta') \cap \mathscr{E} \quad &\Rightarrow \quad f - g \in \mathscr{U}(0; 2M, K', 2\delta') \\ &\Rightarrow \quad |m(f) - m(g)| < \varepsilon. \end{aligned}$$

By mere routine we may now prove: $\overline{\mathscr{E}}$ is an elementary domain again; for every $h \in \overline{\mathscr{E}}$ with $\|h\| \leqq M$ all $\mathscr{U}(h; M, K, \delta) \cap \mathscr{E}$ are nonempty; for every $h \in \overline{\mathscr{E}}$, there is a unique real $\overline{m}(h)$ such that for $\|h\| \leqq M$

$$f \in U(h; M, K', \delta') \cap \mathscr{E} \quad \Rightarrow \quad |m(f) - \overline{m}(h)| < \varepsilon$$

if K', δ' are chosen as above; m is a linear form on $\overline{\mathscr{E}}$, positive if m is.

Let us now prove that $\overline{m}$ is tight. For this, choose an $\varepsilon > 0$ and then a compact $K \subseteq \Omega$ and a $\delta > 0$ such that

$$f \in \mathscr{U}(0; 1, K, \delta) \cap \mathscr{E} \quad \Rightarrow \quad |m(f)| < \varepsilon.$$

Put now $M = 1$ and choose K', δ' as above: For any $h \in \overline{\mathscr{E}}$ with $\|h\| \leqq 1$, we have

$$f \in \mathscr{U}(h; 1, K', \delta') \cap \mathscr{E} \quad \Rightarrow \quad |m(f) - \overline{m}(h)| < \varepsilon.$$

Then for any $h \in \mathscr{U}(0; 1, K \cup K', \frac{1}{2}\min[\delta, \delta']) \cap \mathscr{E}$, we choose some $f \in \mathscr{U}(h; 1, K \cup K', \frac{1}{2}\min[\delta, \delta']) \cap \mathscr{E}$ and get $f \in \mathscr{U}(0; 1, K, \delta) \cap \mathscr{E}$, hence $|m(f)| < \varepsilon$. On the other hand, $f \in \mathscr{U}(h; 1, K', \delta') \cap \mathscr{E}$, hence $|m(f) - m(h)| < \varepsilon$, and thus finally $|\overline{m}(h)| < 2\varepsilon$, proving the tightness of m.

The last statement of the theorem is an easy consequence of the Stone–Weierstrass approximation theorem and of Remark 10.5:

For any $h \in \mathscr{C}^{\mathrm{b}}(\Omega, \mathbb{R})$, $\delta > 0$, $K \subseteq \Omega$ compact, $\mathbb{R} \in M \leqq \|h\| + 1$, choose $f' \in \mathscr{E}$ according to the Stone–Weierstrass theorem such that $|f'(\omega) - h(\omega)| < \delta$ $(\omega \in K)$. By 10.5 the function $f = (f \wedge M) \vee (-M)$ is in $U(h; M, K, \delta)$, as was to be proved.

11. THE EXTENSION PROBLEM FOR MEASURE SPACES

Exercise 9.4.2 more or less indicates the direction that has, by mathematical experience, turned out to be the correct one for a formulation of the

11.1. Extension problem for measure spaces. *Let* $(\Omega, \mathscr{E}, m)$ *be a measure space. Does there exist a measure space* $(\Omega, \mathscr{L}, m)$ *such that* $\mathscr{E} \subseteq \mathscr{L}$, m: $\mathscr{L} \to \mathbb{R}$ *is an extension of* m: $\mathscr{E} \to \mathbb{R}$ *such that the following statement holds:*

Monotone convergence theorem: *If*

$$0 \leqq h_1 \leqq h_2 \leqq \cdots \in \mathscr{L}, \lim_{n\to\infty} m(h_n) < \infty,$$

then the pointwise $\lim_{n\to\infty} h_n = h$ *is in* $\mathscr{L}$ *again and we have*

$$m(h) = \lim_{n\to\infty} m(h_n).$$

11.2. Remarks

11.2.1. From the monotone convergence theorem above there follows immediately a "downward version" of it, which is nothing but the σ-continuity of $m: \mathscr{L} \to \mathbb{R}$ (exercise). This implies, of course, the σ-continuity of $m: \mathscr{E} \to \mathbb{R}$. Thus the extension problem should a priori be considered for σ-measure spaces $(\Omega, \mathscr{E}, m)$ only.

11.2.2. It is easy to provide examples showing that the extension problem for measure spaces in the form 11.1 is not quite correctly posed. Let, e.g., $\Omega = \mathbb{R}$, $\mathscr{E} = \mathscr{C}^{00}(\mathbb{R}, \mathbb{R})$, $m =$ the Riemann integral. Let $f_n(x) = (1 - 2^n|x|) \vee 0$ $(x \in \mathbb{R})$; clearly $m(f_n) = 1/2^n$; put now $h_n = f_1 + \cdots + f_n$; clearly

$$0 \leqq h_1 \leqq h_2 \leqq \cdots \in \mathscr{E}, \qquad \lim_{n \to \infty} m(h_n) = \sum_{k=1}^{\infty} \frac{1}{2^k} = 1 < \infty;$$

but $\lim_{n \to \infty} h_n(0) = \infty$. This indicates that we have to admit functions with values $+\infty$ if we want to obtain the monotone convergence theorem as formulated in 11.1. But with functions attaining values in $\overline{\mathbb{R}} = \mathbb{R} \cup \{\infty, -\infty\}$ we run into difficulties when adding or subtracting them, since $\infty - \infty$ cannot be reasonably defined. This difficulty is one of the *raisons d'être* of the so-called *nullset* business which the reader will encounter in the course of the extension process carried out in Chapter III and which is a characteristic feature of all modern integration (extension) theories; it runs, roughly speaking, as follows: we admit functions with values in $\overline{\mathbb{R}}$ to $\mathscr{L}$ but we shall, with the help of nullsets, define an equivalence relation in $\mathscr{L}$, and it will turn out that every equivalence class contains representatives attaining values in $\mathbb{R}$ only. The linear operations will be defined for the equivalence classes via such finite-valued representatives. The space of equivalence classes will be a vector lattice in an obvious abstract sense.

CHAPTER II

EXTENSION OF σ-CONTENTS AFTER CARATHÉODORY

In this chapter we solve the extension problem for σ-contents (I.5.1) by Carathéodory's (1873–1950) ingenious classical method of additive decomposers (Section 2). This method will, as a matter of fact, yield an extension going, as a rule, far beyond the local σ-ring generated by the set ring from which we start.

A few supplementary extension procedures (extension to the generated σ-ring resp. σ-field, completion) will be presented in Sections 3, 4, and 6. In Section 5 we show how to obtain a σ-content from compatible local data; this is useful, e.g., when dealing with σ-contents on manifolds. The extension theory for σ-measures which is to be presented in Chapter III will lead to another solution of the extension problem for σ-contents (Remark III.6.2.2). A third solution will be given by Theorem XIII.4.1, using Choquet's capacitability theorem and Souslin sets.

1. THE OUTER CONTENT DERIVED FROM A GIVEN CONTENT

1.1. Definition. *Let $(\Omega, \mathscr{R}, m)$ be a content prespace. For every $M \subseteq \Omega$ put*

$$\tilde{m}(M) = \inf\{m(F_1) + m(F_2) + \cdots \mid F_1, F_2, \ldots \in R, M \subseteq F_1 \cup F_2 \cup \cdots\}$$

(with the usual convention $\inf \varnothing = \infty$*). The extended real-valued function $\tilde{m} \geqq 0$ (if we put $\overline{\mathbb{R}} = \mathbb{R} \cup \{\infty, -\infty\}$, it is $\overline{\mathbb{R}}_+$-valued) thus defined on the system $\mathscr{P}(\Omega)$ of all subsets of Ω is called the* **outer content derived** *from the content m on $\mathscr{R}$ (resp. from $(\Omega, \mathscr{R}, m)$).*

1.2. Remark. If $E \in \mathscr{B}^{00}(\mathscr{R})$, then there is some $F \in \mathscr{R}$ with $E \subseteq F$ (Proposition I.4.7.1), hence $\tilde{m}(E)$ is finite. If $E \in \mathscr{R}$, then we may choose $F = E$ and $\tilde{m}(E) = m(E)$ follows. Thus $\tilde{m}$, restricted to $\mathscr{B}^{00}(\mathscr{R})$, provides an $\overline{\mathbb{R}}_+$-valued extension of m from $\mathscr{R}$ to $\mathscr{B}^{00}(\mathscr{R})$. The question is whether it is additive resp. σ-additive there.

1.3. Proposition. *Let $(\Omega, \mathscr{R}, m)$ be a content prespace and $\tilde{m}\colon \mathscr{P}(\Omega) \to \overline{\mathbb{R}}_+$ the outer content derived from $(\Omega, \mathscr{R}, m)$. Then:*

1.3.1. *$\tilde{m}(\varnothing) = 0$ and $\tilde{m}$ is isotone, i.e.,*

$$\tilde{m}(M) \leq \tilde{m}(N) \qquad (M \subseteq N \subseteq \Omega).$$

1.3.2. *$\tilde{m}$ is subadditive, i.e.,*

$$\tilde{m}(M_1 \cup \cdots \cup M_n) \leq \tilde{m}(M_1) + \cdots + \tilde{m}(M_n) \qquad (n > 0, \quad M_1, \ldots, M_n \subseteq \Omega),$$

and even σ-subadditive, i.e.,

$$\tilde{m}(M_1 \cup M_2 \cup \cdots) \leq \tilde{m}(M_1) + \tilde{m}(M_2) + \cdots \qquad (M_1, M_2, \ldots \subseteq \Omega).$$

1.3.3. *If m is σ-additive, then $\tilde{m}$ is lower σ-continuous, i.e.,*

$$\lim_{n\to\infty} \tilde{m}(M_n) = \tilde{m}(M) \qquad (M_1 \subseteq M_2 \subseteq \cdots \subseteq \Omega, \quad M = M_1 \cup M_2 \cup \cdots).$$

Proof. Note that all additions occurring in our proposition make sense although some of the values added may be $= \infty$.

1 is obvious.

2. Let $M_1, \ldots, M_n \subseteq \Omega$. We may assume $\tilde{m}(M_1) + \cdots + \tilde{m}(M_n) < \infty$. Let $\varepsilon > 0$ be arbitrary. For each $k = 1, \ldots, n$, choose $F_{k1}, F_{k2}, \ldots \in \mathscr{R}$ such that $M_k \subseteq F_{k1} \cup F_{k2} \cup \cdots$ and $m(F_{k1}) + m(F_{k2}) + \cdots < \tilde{m}(M_k) + \varepsilon/n$. Clearly

$$M_1 \cup \cdots \cup M_n \subseteq \bigcup_{k=1}^{n} \bigcup_{j=1}^{\infty} F_{kj}$$

and

$$\sum_{k=1}^{n} \sum_{j=1}^{\infty} m(F_{kj}) < \tilde{m}(M_1) + \cdots + \tilde{m}(M_n) + \varepsilon,$$

hence $\tilde{m}(M_1 \cup \cdots \cup M_n) < \tilde{m}(M_1) + \cdots + \tilde{m}(M_n) + \varepsilon$. Since $\varepsilon > 0$ was arbitrary, the claimed subadditivity follows. Working with $\varepsilon/2^n$ in place of ε/n, we can deal with a sequence $M_1, M_2, \ldots \subseteq \Omega$ in the same way and obtain σ-subadditivity (exercise).

3. Let now $m\colon \mathscr{R} \to \mathbb{R}_+$ be σ-additive. We begin our proof of lower σ-continuity of m with some preparations.

(a) We proceed as we did in Definition I.3.1 of Eudoxos measurability and define

$$\underline{m}(M) = \sup_{\mathscr{R} \ni F \subseteq M} m(F) \qquad (M \subseteq \Omega).$$

Clearly $\underline{m}$: $\mathscr{P}(\Omega) \to \overline{\mathbb{R}}$ is $\geqq 0$, extends m: $\mathscr{R} \to R_+$, and is isotone: $\underline{m}(M) \leqq \underline{m}(N)$ $(M \subseteq N \subseteq \Omega)$. Let us introduce the notation

$$\mathscr{R}_\sigma = \{F_1 \cup F_2 \cup \cdots | F_1, F_2, \ldots \in \mathscr{R}\}$$

and show that $\underline{m}$ is lower σ-continuous on $\mathscr{R}_\sigma$ when approximation comes from $\mathscr{R}$, i.e., that $\mathscr{R} \ni F_1 \subseteq F_2 \subseteq \cdots$, $F_1 \cup F_2 \cup \cdots = E$ implies $\lim_{n\to\infty} m(F_n) = \underline{m}(E)$. In fact, if $\mathscr{R} \ni F \subseteq E$, then $\mathscr{R} \ni F \backslash F_1 \supseteq F \backslash F_2 \supseteq \cdots$, $(F\backslash F_1) \cap (F\backslash F_2) \cap \cdots = \varnothing$, which implies $\lim_{n\to\infty} m(F\backslash F_n) = 0$ by Proposition I.2.2. Thus we obtain

$$\lim_{n\to\infty} m(F_n) \geqq \lim_{n\to\infty} m(F \cap F_n) = \lim_{n\to\infty} [m(F) - m(F\backslash F_n)] = m(F).$$

Since $\mathscr{R} \ni F \subseteq E$ was arbitrary, $\lim_{n\to\infty} m(F_n) = \underline{m}(E)$ follows, as desired. Next we show that $\underline{m}$: $\mathscr{R}_\sigma \to \overline{\mathbb{R}}_+$ is lower σ-continuous. Let

$$\mathscr{R}_\sigma \ni E_1 \subseteq E_2 \subseteq \cdots, E_1 \cup E_2 \cup \cdots = E.$$

We prove $E \in \mathscr{R}_\sigma$ and $\lim_{n\to\infty} \underline{m}(E_n) = \underline{m}(E)$. For this let $F_{nj} \in \mathscr{R}$ $(n, j = 1, 2, \ldots)$ be chosen such that $F_{n1} \subseteq F_{n2} \subseteq \cdots$, $F_{n1} \cup F_{n2} \cup \cdots = E_n$ $(n = 1, 2, \ldots)$. From $E = \bigcup_{n,j=1}^{\infty} F_{nj}$ we conclude $E \in \mathscr{R}_\sigma$. Put now $F_k = \bigcup_{n,j \leqq k} F_{nj}$. Clearly $\mathscr{R} \ni F_1 \subseteq F_2 \subseteq \cdots$, $F_1 \cup F_2 \cup \cdots = E$. From what we proved before, we get, for any real $\alpha < \underline{m}(E)$, some k with $m(F_k) > \alpha$. But $E_k \supseteq F_k$, hence $\underline{m}(E_k) \geqq \underline{m}(F_k) = m(F_k) > \alpha$. Varying α, we get $\lim_{n\to\infty} \underline{m}(E_n) \geqq \underline{m}(E)$. Since $\leqq$ is obvious, we are through here.

(b) Next we prove that $E, G \in \mathscr{R}_\sigma \Rightarrow E \cap G, E \cup G \in \mathscr{R}_\sigma$ and

$$\underline{m}(E \cup G) + \underline{m}(E \cap G) = \underline{m}(E) + \underline{m}(G).$$

By additivity, this is true for $E, G \in \mathscr{R}$ (see I.1.9.1). The lower σ-continuity of m: $\mathscr{R}_\sigma \to \overline{\mathbb{R}}_+$ proved in (a) will yield its validity also for $E, G \in \mathscr{R}_\sigma$. The details go as follows. Choose $E_1, G_1, E_2, G_2, \ldots \in \mathscr{R}$ such that $E_1 \cup E_2 \cup \cdots = E$, $G_1 \cup G_2 \cup \cdots = G$. We may assume $E_1 \subseteq E_2 \subseteq \cdots$, $G_1 \subseteq G_2 \subseteq \cdots$. Now clearly $\mathscr{R} \ni E_1 \cup G_1 \subseteq E_2 \cup G_2 \subseteq \cdots$ with union $E \cup G$, and $\mathscr{R} \ni E_1 \cap G_1 \subseteq E_2 \cap G_2 \subseteq \cdots$ with union $E \cap G$, and thus

$$\begin{aligned}\underline{m}(E \cup G) + \underline{m}(E \cap G) &= \lim_{n\to\infty} m(E_n \cup G_n) + \lim_{n\to\infty} m(E_n \cap G_n)\\ &= \lim_{n\to\infty} m(E_n) + \lim_{n\to\infty} m(G_n) = \underline{m}(E) + \underline{m}(G).\end{aligned}$$

(c) Next we prove

$$\tilde{m}(M) = \inf_{M \subseteq E \in \mathscr{R}_\sigma} \underline{m}(E).$$

In fact, let $F_1', F_2', \ldots \in R$ with $F_1' \cup F_2' \cup \cdots \supseteq M$. The first entrance decomposition yields $F_1, F_2, \ldots \in \mathscr{R}$ pairwise disjoint with $F_1 \subseteq F_1'$,

$F_2 \subseteq F'_2, \ldots$ and $F_1 + F_2 + \cdots = F'_1 \cup F'_2 \cup \cdots \supseteq M$. Thus we get

$$\begin{aligned} m(F'_1) + m(F'_2) + \cdots &\geqq m(F_1) + m(F_2) + \cdots \\ &= \lim_{n\to\infty} (m(F_1 + \cdots + F_n)) \\ &= \underline{m}(F_1 + F_2 + \cdots) = \underline{m}(F'_1 \cup F'_2 \cup \cdots). \end{aligned}$$

Thus every term competing in the inf that defines $\tilde{m}(M)$ is minorized by some term which also competes and has the form $\underline{m}(E)$ with $E \in \mathscr{R}_\sigma$. That does it.

(d) Next we prove a general estimate: Let $M \subseteq N \subseteq \Omega$, $\tilde{m}(N) < \infty$, $M \subseteq E \in \mathscr{R}_\sigma$, $\underline{m}(E) < \tilde{m}(M) + \delta$, $N \subseteq G \in \mathscr{R}_\sigma$. Then

$$M \subseteq E \cap G, \underline{m}(E \cap G) < \tilde{m}(M) + \delta, N \subseteq E \cup G, \underline{m}(E \cup G) < \underline{m}(G) + \delta.$$

Clearly the only nontrivial statement is the last inequality in the case $\underline{m}(G) < \infty$. But in this case we have $\tilde{m}(M) \leqq \underline{m}(E \cap G) \leqq \underline{m}(E) < \tilde{m}(M) + \delta$ and (b) yields $\underline{m}(E \cup G) = [\underline{m}(E) - \underline{m}(E \cap G)] + \underline{m}(G) < \underline{m}(G) + \delta$, as desired.

(e) Now we put our achievements together. Let $M_1 \subseteq M_2 \subseteq \cdots \subseteq \Omega$, $M_1 \cup M_2 \cup \cdots = M$. The only interesting case is $\lim_{n\to\infty} \tilde{m}(M_n) < \infty$, and here we have to prove $\tilde{m}(M) \leqq \lim_{n\to\infty} \tilde{m}(M_n)$. Choose any $\varepsilon > 0$ and $M_k \subseteq E_k \in \mathscr{R}_\sigma$ with $\underline{m}(E_k) < \tilde{m}(M_k) + \varepsilon/2^k$ $(k = 1, 2, \ldots)$. Put $G_n = E_1 \cup \cdots \cup E_n$ $(n = 1, 2, \ldots)$. We prove $m(G_k) \leqq \tilde{m}(M_k) + \varepsilon(1 - 1/2^k)$ $(k = 1, 2, \ldots)$ by induction. For $k = 1$, we have

$$\tilde{m}(G_1) = \underline{m}(E_1) < \tilde{m}(M_1) + \varepsilon/2 = \tilde{m}(M_1) + \varepsilon(1 - \tfrac{1}{2})$$

as desired. Assume we are through until k. Apply (d) with $M = M_k$, $N = M_{k+1}$, $E = G_k$, $G = E_{k+1}$, $\delta = \varepsilon(1 - 1/2^k)$. We obtain

$$\begin{aligned} \underline{m}(G_{k+1}) \leqq \underline{m}(E_{k+1}) + \delta &< \tilde{m}(M_{k+1}) + \varepsilon/2^{k+1} + \varepsilon(1 - 1/2^k) \\ &= \tilde{m}(M_{k+1}) + \varepsilon(1 - 1/2^{k+1}) \end{aligned}$$

as desired. Now we consider $G = G_1 \cup G_2 \cup \cdots$ and get $M \subseteq G \in \mathscr{R}_\sigma$,

$$m(G) = \lim_{n\to\infty} \underline{m}(G_n) \leqq \lim_{n\to\infty} \tilde{m}(M_n) + \lim_{n\to\infty} \varepsilon(1 - 1/2^n) = \lim_{n\to\infty} \tilde{m}(M_n) + \varepsilon.$$

Since $\varepsilon > 0$ was arbitrary, the desired inequality follows and the proposition is proved.

1.4. Exercises

1.4.1. Let $(\mathbb{R}, \mathscr{D}, m)$ be as in Example I.2.4 (dyadic geometric σ-content prespace on $\mathbb{R}$). Prove that $\tilde{m}(M) = 0$ for every countable $M \subseteq \mathbb{R}$.

1.4.2. Let $(\mathbb{R}, \mathscr{D}, m)$ be as in Example I.2.4. Prove that $\tilde{m}$ coincides with interval length for any bounded interval.

1.4.3. Carry over 1, 2 to higher dimensions.

1.4.4. Let C be the so-called *Cantor residual set* in the unit interval $[0, 1]$, namely

$$C = \left\{x \,\middle|\, x \in R,\ x = \sum_{k=1}^{\infty} \frac{k}{3^k},\ a_1, a_2, \ldots \in \{0, 2\}\right\}.$$

Prove that C is uncountable and $\tilde{m}(C) = 0$ if $(\mathbb{R}, \mathscr{D}, m)$ is as in 1. (*Hint*: Use 2.)

1.4.5. Prove that the outer content derived from a content space and from its Eudoxos extension are the same.

1.4.6. Let $(\mathbb{R}^2, \mathscr{D}^2, m)$ be the dyadic geometric σ-content prespace on $\mathbb{R}^2$. Prove translation and rotation invariance of the derived outer content, i.e., $\tilde{m}(M) = \tilde{m}(N)$ if $M, N \subseteq \mathbb{R}^2$ coincide after suitable translation and rotation. (*Hint*: Use Exercise I.3.3.5.)

1.4.7. Carry 6 over to higher dimensions.

2. ADDITIVE DECOMPOSERS

2.1. Definition. *Let $(\Omega, \mathscr{R}, m)$ be a content prespace and $\tilde{m}\colon \mathscr{P}(\Omega) \to \overline{\mathbb{R}}_+$ the outer content derived from it. A set $D \subseteq \Omega$ is called an* **additive decomposer** *for $\tilde{m}$ if*

$$\tilde{m}(M) = \tilde{m}(M \cap D) + \tilde{m}(M \backslash D) \qquad (M \subseteq \Omega,\quad m(M) < \infty)$$

The system of all additive decomposers for m is denoted by $\mathscr{D}_{\tilde{m}}$.

2.2. Remarks

2.2.1. The concept of an additive decomposer could as well be defined for an abstract concept of an outer content in a set $\Omega \neq \varnothing$ which need not be the derivative of some content prespace. We resist the temptation to go to such a degree of generality.

2.2.2. It is obvious from the definition that $\varnothing$ and Ω are always additive decomposers and that the complement of an additive decomposer is an additive decomposer again.

2.3. Theorem. *Let $(\Omega, \mathscr{R}, m)$ be a σ-content prespace and $\tilde{m}\colon \mathscr{P}(\Omega) \to \overline{\mathbb{R}}_+$ the outer content derived from it. Then:*

2.3.1. *the system $\mathscr{D}_{\tilde{m}}$ of all additive decomposers for $\tilde{m}$ is a σ-field in Ω. $\tilde{m}\colon \mathscr{D}_{\tilde{m}} \to \overline{\mathbb{R}}_+$ is σ-additive.*

2.3.2. *The set system $\mathscr{B}_m^{00} = \{D \mid D \in \mathscr{D}_{\tilde{m}},\ \tilde{m}(D) < \infty\}$ is a local σ-ring* *containing* $\mathscr{R}$. *The restriction of* $\tilde{m}$ *to* $\mathscr{B}_m^{00}$ *is a* σ*-content that coincides with* m *on* $\mathscr{R}$. *The* σ*-content space* $(\Omega, \mathscr{B}_m^{00}, \tilde{m})$ *is called the* **Carathéodory extension** *of* $(\Omega, \mathscr{R}, m)$. *It is full.*

2.3.3. *The restriction of* m *to the local* σ*-ring* $\mathscr{B}^{00}$ *generated by* $\mathscr{R}$ *is a* σ*-content extending* m.

Proof. (1) We begin by proving that (a) $\mathscr{D}_{\tilde{m}}$ is a set ring. Let $C, D \in \mathscr{D}_{\tilde{m}}$, $M \subseteq \Omega$, $m(M) < \infty$. Consider the disjoint decomposition ($\complement$ denotes the complement)

$$M = M_1 + M_2 + M_3 + M_4$$

where

$$M_1 = M \cap C \cap D, \qquad M_3 = M \cap \complement C \cap D,$$
$$M_2 = M \cap C \cap \complement D, \qquad M_4 = M \cap \complement C \cap \complement D.$$

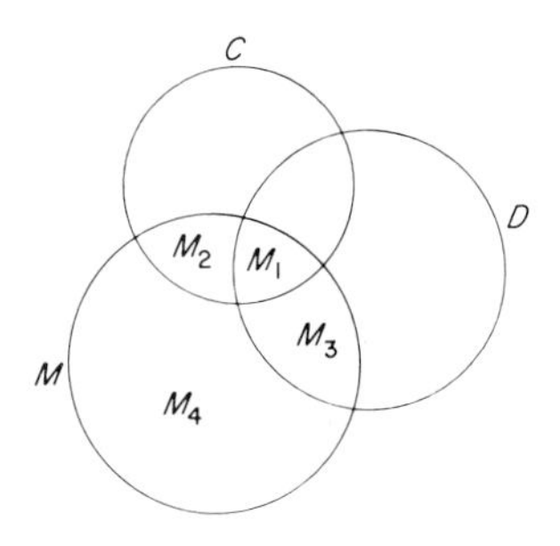

From $C, D \in \mathscr{D}_{\tilde{m}}$ we deduce

$$\tilde{m}(M_1 + M_2) = \tilde{m}(M_1) + \tilde{m}(M_2)$$
$$\tilde{m}(M_1 + M_3) = \tilde{m}(M_1) + \tilde{m}(M_3)$$
$$\vdots$$
$$\tilde{m}(M_3 + M_4) = \tilde{m}(M_3) + \tilde{m}(M_4)$$
$$\tilde{m}(M_2 + M_3 + M_4) = \tilde{m}(M_2) + \tilde{m}(M_3 + M_4) = \tilde{m}(M_2) + \tilde{m}(M_3) + \tilde{m}(M_4)$$
$$\tilde{m}(M_1 + M_3 + M_4) = \tilde{m}(M_1) + \tilde{m}(M_3 + M_4) = \tilde{m}(M_1) + \tilde{m}(M_3) + \tilde{m}(M_4)$$
$$\tilde{m}(M_1 + M_2 + M_4) = \tilde{m}(M_1) + \tilde{m}(M_2 + M_4) = \tilde{m}(M_1) + \tilde{m}(M_2) + \tilde{m}(M_4)$$
$$\tilde{m}(M_1 + M_2 + M_3) = \tilde{m}(M_1 + M_2) + \tilde{m}(M_3) = \tilde{m}(M_1) + \tilde{m}(M_2) + \tilde{m}(M_3)$$
$$\tilde{m}(M) = \tilde{m}(M_1 + M_2) + \tilde{m}(M_3 + M_4)$$
$$= \tilde{m}(M_1) + \tilde{m}(M_2) + \tilde{m}(M_3) + \tilde{m}(M_4).$$

From this we obtain

$$\begin{aligned}\tilde{m}(M) &= \tilde{m}(M_1) + \tilde{m}(M_2 + M_3 + M_4) \\ &= \tilde{m}(M \cap (C \cap D)) + \tilde{m}(M \backslash (C \cap D)) \\ &= \tilde{m}(M_1 + M_2 + M_3) + \tilde{m}(M_4) \\ &= \tilde{m}(M \cap (C \cup D)) + \tilde{m}(M \backslash (C \cup D)) \\ &= \tilde{m}(M_2) + \tilde{m}(M_1 + M_3 + M_4) \\ &= \tilde{m}(M \cap (C \cap D)) + \tilde{m}(M \backslash (C \backslash D)),\end{aligned}$$

proving that $\mathscr{D}_{\tilde{m}}$ is stable under finite unions, intersections, and difference, i.e., is a set ring.

(b) For any $C, D \in \mathscr{D}_{\tilde{m}}$, $C \cap D = \varnothing$, we have

$$\tilde{m}(C + D) = \tilde{m}(C) + \tilde{m}(D).$$

Here $\leqq$ is already known, which turns into $=$ automatically if $\tilde{m}(C + D) = \infty$. If $\tilde{m}(C + D) < \infty$, then $\tilde{m}(C) < \infty > \tilde{m}(D)$ and we need only make the special choice $M = C + D$ in (a) in order to ensure

$$\tilde{m}(C + D) = \tilde{m}(M_2 + M_3) = \tilde{m}(M_2) + \tilde{m}(M_3) = \tilde{m}(C) + \tilde{m}(D).$$

(c) $\mathscr{D}_{\tilde{m}}$ is stable under countable unions. Let $D_1, D_2, \ldots \in \mathscr{D}_{\tilde{m}}$. Since all $D_1 \cup \cdots \cup D_n \in \mathscr{D}_{\tilde{m}}$, we may assume $D_1 \subseteq D_2 \subseteq \cdots$ for our purpose. Choose any $M \subseteq \Omega$ with $\tilde{m}(M) < \infty$. From

$$\tilde{m}(M) = \tilde{m}(M \cap D_n) + \tilde{m}(M \backslash D_n) \geqq \tilde{m}(M \cap D_n) + \tilde{m}\left(M \backslash \bigcup_n D_n\right)$$

and lower continuity of $\tilde{m}$ (Theorem 1.3.3) we get

$$\tilde{m}(M) \geqq \tilde{m}\left(M \cap \bigcup_n D_n\right) + \tilde{m}\left(M \backslash \bigcup_n D_n\right).$$

Since $\leqq$ is already known, we get equality, hence $D_1 \cup D_2 \cup \cdots \in \mathscr{D}_{\tilde{m}}$. Thus we have proved that $\mathscr{D}_{\tilde{m}}$ is a σ-field ($\Omega \in \mathscr{D}_{\tilde{m}}$ is obvious).

The rest of 1 now follows easily (see Exercise I.2.3.3 and Theorem 1.3.3).

2 follows easily from 1.

3 follows from 2.

As a consequence, we obtain the following solution to our extension problem I.5.1 for σ-contents.

2.4. Corollary. *Let $(\Omega, \mathscr{R}, m)$ be a σ-content prespace. Then there is a unique σ-content (again denoted by m) on the local σ-ring $\mathscr{B}^{00}(\mathscr{R})$ generated by $\mathscr{R}$ which coincides with m on $\mathscr{R}$. $(\Omega, \mathscr{B}^{00}(\mathscr{R}), m)$ is called the local σ-content space* **generated** *by the σ-content prespace $(\Omega, \mathscr{R}, m)$. It is a σ-probability space if $(\Omega, \mathscr{R}, m)$ is a σ-probability prespace.*

Remark III.6.2.2 will yield another proof of this corollary. A third proof will be contained in Theorem XIII.4.1.

2.5. Definition

2.5.1. *The σ-content space $(\mathbb{R}^n, \mathscr{B}^{00}, m)$ generated by the dyadic geometric (= Riemann = Jordan = Lebesgue) σ-content prespace (Example I.1.10, Exercise I.1.11.2, Example I.2.4) is called the* **Lebesgue σ-content space** *in $\mathbb{R}^n$. This m is called the* **Lebesgue σ-content** *in $\mathbb{R}^n$.*

2.5.2. *The σ-content space (actually a σ-probability space) generated by the (one-sided) Bernoulli σ-content prespace $(A \times A \times \cdots, \mathscr{C}, m)$ defined by a probability vector p over A (Examples I.1.12, I.2.5) is called the* **Bernoulli σ-probability space** *for p (and A), and so is its "two-sided" analogue (Exercise I.1.13).*

2.5.3. *The σ-content space (actually a σ-probability space) generated by the Markov σ-content prespace $(A \times A \times \cdots, \mathscr{C}, m)$ defined by a stochastic vector p over A and the transition matrix P (Examples I.1.14, I.2.5) is called the* **Markov σ-probability space** *for p, A, and P.*

2.6. Exercises. Let $(\Omega, \mathscr{R}, m)$ be a σ-content prespace and $(\Omega, \mathscr{B}_m^{00}, \tilde{m})$ its Carathéodory extension.

2.6.1. Prove that $\mathscr{R}$ is *dense* in $\mathscr{B}_m^{00}$ in the following sense: For every $E \in \mathscr{B}_m^{00}$ and $\varepsilon > 0$ there is a $F \in \mathscr{R}$ with $m(E \,\Delta\, F) < \varepsilon$, where generally $E \,\Delta\, F = (E \cup F) \backslash (E \cap F)$. (*Hint*: Prove it first for the case where E is a countable union of sets from $\mathscr{R}$.)

2.6.2. Prove that $\mathscr{B}_m^{00}$ is *complete* in the following sense: If $E_1, E_2, \ldots \in \mathscr{B}_m^{00}$, $\lim_{j,k \to \infty} \tilde{m}(E_j \,\Delta\, E_k) = 0$, then there is an $E \in \mathscr{B}_m^{00}$ with $\lim_{n \to \infty} \tilde{m}(E_n \,\Delta\, E) = 0$. (*Hint*: Enforce $m(E_k \,\Delta\, E_{k+1}) < 1/2^k$ by passing to a subsequence and put

$$E = \limsup_{n \to \infty} E_n = \bigcap_{n=1}^{\infty} \bigcup_{k=n}^{\infty} E_k).$$

2.6.3. Prove that $\mathscr{B}_m^{00}$ is the closure of $\mathscr{R}$ in $\mathscr{D}_{\tilde{m}}$ for the metric defined by $\tilde{m}$ and Δ as in (1).

2.6.4. Prove that the σ-content $\tilde{m}\colon \mathscr{B}_m^{00} \to \mathbb{R}_+$ has the same total mass as $m\colon \mathscr{R} \to \mathbb{R}_+$.

2.6.5. Prove the translation and rotation invariance of the Lebesgue σ-content in $\mathbb{R}^n$ (use Exercises 1.4.6, 1.4.7).

2.6.6. Prove: If $E \in \mathscr{B}_m^{00}$ is contained in some $M \in \mathscr{R}$, then there are $F, G \in \mathscr{B}^{00}(\mathscr{R})$ such that $F \subseteq E \subseteq G$, $m(G \backslash F) = 0$.

2.6.7. Let $(\Omega, \mathscr{B}_m^{00}, \tilde{m})$ be the Carathéodory extension of a σ-content prespace. Call $E, F \in \mathscr{B}^{00}$ *equivalent* mod m, and write $E = F \bmod m$ if

$m(E \Delta F) = 0$. Prove that equivalence mod m is an equivalence relation in $\mathscr{B}^{00}$. Write $E \subseteq F \bmod m$ if $E \cap F = E \bmod m$. Prove that this relation "respects" equivalence mod m in that $E = E' \bmod m$, $F = F' \bmod m$, $E \subseteq F \bmod m \Rightarrow E' \subseteq F' \bmod m$. Prove that the system of all equivalence classes mod m in $\mathscr{B}^{00}$ is a *conditionally complete lattice* if endowed with the partial ordering induced by the relation $\subseteq$ mod m. (*Hint*: Show that for any increasingly filtered family $(E_\iota)_{\iota \in I}$ in $\mathscr{B}^{00}$ with $\sup_{\iota \in I} m(E_\iota) < \infty$ there is mod m exactly one $E \in \mathscr{B}^{00}$ such that $E_\iota \subseteq E \bmod m$ $(\iota \in I)$ and $[F \in \mathscr{B}^{00},\ F \supseteq E_\iota \bmod m\ (\iota \in I) \Rightarrow E \subseteq F \bmod m]$; for this, choose $E_{\iota_1} \subseteq E_{\iota_2} \subseteq \cdots$ with $\lim_{n\to\infty} m(E_{\iota_n}) = \sup m(E_\iota)$ and put $E = E_{\iota_1} \cup E_{\iota_2} \cup \cdots$; consult the proof of Theorem III.5.8.5, and 6.1 if necessary.)

3. ROUTINE EXTENSION FROM LOCAL σ-RINGS TO σ-RINGS

In this section we describe an obvious procedure extending a σ-content from a local σ-ring to the generated σ-ring.

3.1. Theorem. *Let $(\Omega, \mathscr{B}^{00}, m)$ be a σ-content space and $\mathscr{B}^0$ the σ-ring generated by $\mathscr{B}^{00}$. Then there is a unique σ-additive mapping $m^0 \colon \mathscr{B}^0 \to \overline{\mathbb{R}}_+ = \mathbb{R}_+ \cup \{\infty\}$ such that m^0 coincides with m on $\mathscr{B}^{00}$.*

$\tilde{\mathscr{B}}^{00} = \{E \mid E \in \mathscr{B}^0,\ m^0(E) < \infty\}$ is a local σ-ring and $(\Omega, \tilde{\mathscr{B}}^{00}, m^0)$ is a full σ-content space. It is called the **full extension** *of $(\Omega, \mathscr{B}^{00}, m)$.*

Proof. By Proposition I.4.7.2.1 we get

$$\mathscr{B}^0 = \{F_1 + F_2 + \cdots \mid F_1, F_2, \cdots \in \mathscr{B}^{00},\ F_j \cap F_k = \varnothing\ (j \neq k)\}.$$

The sole way of defining m^0 is thus

$$m^0(F_1 + F_2 + \cdots) = m(F_1) + m(F_2) + \cdots$$
$$(F_1, F_2, \ldots \in \mathscr{B}^{00},\quad F_j \cap F_k = \varnothing \quad (j \neq k)).$$

The σ-additivity of m on $\mathscr{B}^{00}$ shows that m^0 is indeed well defined that way: If $F_1, G_1, F_2, G_2, \ldots \in \mathscr{B}^{00}$,

$$F_j \cap F_k = \varnothing = G_j \cap G_k \quad (j \neq k), \qquad F_1 + F_2 + \cdots = G_1 + G_2 + \cdots,$$

then

$$\sum_{j=1}^{\infty} m(F_j) = \sum_{j=1}^{\infty} m\left(F_j \cap \sum_{k=1}^{\infty} G_k\right) = \sum_{j,k=1}^{\infty} m(F_j \cap G_k) = \sum_{k=1}^{\infty} m(G_k).$$

If $E_1, E_2, \ldots \in \mathscr{B}^0$, $E_j \cap E_k = \varnothing$ $(j \neq k)$, we choose, for every $j = 1, 2, \ldots$, sets $F_{j1}, F_{j2}, \ldots \in \mathscr{B}^{00}$ such that $F_{jk} \cap F_{jl} = \varnothing$ $(k \neq l)$ and $E_j =$

$F_{j1} + F_{j2} + \cdots$. We can now compute

$$m^0(E_1 + E_2 + \cdots) = m^0\left(\sum_{j,k=1}^{\infty} F_{jk}\right) = \sum_{j,k=1}^{\infty} m(F_{jk})$$

$$= \sum_{j=1}^{\infty} \sum_{k=1}^{\infty} m(F_{jk}) = \sum_{j=1}^{\infty} m^0(E_j).$$

Observe that $E_1 + E_2 + \cdots \in \mathscr{B}^0$ is no problem since $\mathscr{B}^0$ is a σ-ring. The rest is an easy exercise.

3.2. The full extension of the Lebesgue σ-content space (Definition 2.5) is usually called the **full Lebesgue σ-content space** or simply the **Lebesgue σ-content space** again.

3.3. Exercise. Prove the translation and rotation invariance of the full Lebesgue σ-content (use Exercise 2.6.5).

4. MINIMAL EXTENSION TO A σ-FIELD

In this section we prove a general theorem about the σ-additivity of an increasingly filtered supremum of σ-additive functions and use it in order to extend σ-additive functions from a σ-ring to the generated σ-field.

In the next theorem we use the obvious pointwise ordering of $\overline{\mathbb{R}}$-valued functions on an arbitrary domain, e.g., a set system.

4.1. Theorem. *Let $\mathscr{B}$ be a σ-field in $\Omega \neq \varnothing$ and $I = \{\iota, \kappa, \ldots\}$ a non-empty (index) set. Let $(m_\iota)_{\iota \in I}$ be an increasingly filtered family (indexed by I) of σ-additive $\overline{\mathbb{R}}_+$-valued functions on $\mathscr{B}$: for any $\iota, \kappa \in I$ there is a $\lambda \in I$ with*

$$m_\iota \leqq m_\lambda \geqq m_\kappa .$$

Then

$$m(E) = \sup_{\iota \in I} m_\iota(E) \qquad (E \in \mathscr{B})$$

defines a σ-additive function $m\colon \mathscr{B} \to \overline{\mathbb{R}}_+$.

Proof. It is obvious that every m_ι is isotone on $\mathscr{B}$ (Exercise I.1.9.2), and so is m. Let $F_1, F_2, \ldots \in \mathscr{B}$, $F_j \cap F_k = \varnothing$ $(j \neq k)$ and put $F = F_1 + F_2 + \cdots$. We want to prove $m(F) = m(F_1) + m(F_2) + \cdots$. If $m(F) = 0$, then $m(F_1) = m(F_2) = \cdots = 0$ by isotony and we are through. Assume now $m(F) > 0$ (possibly $= \infty$). Choose any $0 \leqq \alpha < m(F)$ and find $\iota \in I$ with $m_\iota(F) > \alpha$. Then

$$\alpha < m_\iota(F) = m_\iota(F_1) + m_\iota(F_2) + \cdots \leqq m(F_1) + m(F_2) + \cdots .$$

Varying α, we find $m(F) \leqq m(F_1) + m(F_2) + \cdots$, and in particular $m(F_1) + m(F_2) + \cdots > 0$. Choose any $0 \leqq \beta < m(F_1) + m(F_2) + \cdots$, find some $n > 0$ with $\beta < m(F_1) + \cdots + m(F_n)$ and $\iota_1, \ldots, \iota_n \in I$ with $\beta < m_{\iota_1}(F_1) + \cdots + m_{\iota_n}(F_n)$. Now this is

$$\leqq m_\iota(F_1) + \cdots + m_\iota(F_n) = m_\iota(F_1 + \cdots + F_n) \leqq m(F)$$

for a suitable $\iota \in I$. Varying β we see $m(F_1) + m(F_2) + \cdots \leqq m(F)$, and the theorem is proved.

4.2. Corollary. *Let $(\Omega, \mathscr{B}^{00}, m)$ be a σ-content space and $\mathscr{B}$ the σ-field generated by $\mathscr{B}^{00}$. Then $E \cap F \in \mathscr{B}^{00}$ $(E \in \mathscr{B}, F \in \mathscr{B}^{00})$ and*

$$\overline{m}(E) = \sup_{F \in \mathscr{B}^{00}} m(E \cap F) \qquad (E \in \mathscr{B})$$

defines a σ-additive function $\overline{m}: \mathscr{B} \to \overline{\mathbb{R}}_+ = \mathbb{R}_+ \cup \{\infty\}$ such that:

(a) *$\overline{m}$ coincides with m on $\mathscr{B}^{00}$*

(b) *If $\overline{\overline{m}}: \mathscr{B} \to \overline{\mathbb{R}}_+$ is σ-additive and satisfies the analogue of (a), then $\overline{\overline{m}} \geqq \overline{m}$, i.e., m is the minimal σ-additive extension of m from $\mathscr{B}^{00}$ to $\mathscr{B}$.*

Proof. Let $I = \mathscr{B}^{00}$. For every $F \in \mathscr{B}^{00} = I$, let $m_F: \mathscr{B} \to \mathbb{R}_+$ be defined by

$$m_F(E) = m(E \cap F) \qquad (E \in \mathscr{B}).$$

By Proposition I.4.7.2.2 this is meaningful. It is practically obvious that m_F is σ-additive. Moreover, $F, G \in \mathscr{B}^{00}$, $F \subseteq G$ implies $m_F \leqq m_G$. Since $\mathscr{B}^{00}$ is increasingly filtered $(F, G \in \mathscr{B}^{00} \Rightarrow F, G \subseteq F \cup G \in \mathscr{B}^{00})$, the hypotheses of Theorem 4.1 are satisfied and we get (a) as an immediate consequence. Let now $\overline{\overline{m}}: \mathscr{B} \to \overline{\mathbb{R}}_+$ be as in (b). Then

$$\overline{\overline{m}}(E) \geqq \overline{\overline{m}}(E \cap F) = m_F(E) \qquad (E \in \mathscr{B}, F \in \mathscr{B}^{00}).$$

Passing to $\sup_{F \in \mathscr{B}^{00}}$, we obtain $\overline{\overline{m}}(E) \geqq \overline{m}(E)$, as desired.

5. DEFINITION OF A σ-CONTENT FROM LOCAL DATA

5.1. Theorem. *Let $(\Omega, \mathscr{B}^{00})$ be a local measurable space. Let $\varnothing \neq \mathscr{U} \subseteq \mathscr{B}^{00}$ be such that every $F \in \mathscr{B}^{00}$ is contained in a finite union of sets from $\mathscr{U}$. Put $\mathscr{B} \cap V = \{F \mid F \in \mathscr{B}^{00}, F \subseteq V\}$ $(V \in \mathscr{B}^{00})$. Assume that for every $U \in \mathscr{U}$, a σ-additive function $m_U: \mathscr{B} \cap U \to \overline{\mathbb{R}}_+$ is given such that the compatibility condition*

$$m_U(F) = m_V(F) \qquad (U, V \in \mathscr{U}, \quad F \in (\mathscr{B} \cap U) \cap (\mathscr{B} \cap V))$$

holds. Then there is a unique σ-additive $m\colon \mathscr{B}^{00} \to \overline{\mathbb{R}}_+$ *such that*

$$m(F) = m_U(F) \qquad (F \in \mathscr{B} \cap U, \quad U \in \mathscr{U}).$$

Proof. We begin by replacing the family $(m_U)_{U \in \mathscr{U}}$ by a family satisfying the hypothesis of Theorem 4.1. Let $\mathscr{V} = \{U_1 \cup \cdots \cup U_n | n > 0,\ U_1, \ldots, U_n \in \mathscr{U}\}$, $\mathscr{B}_V = \{F | F \in \mathscr{B}^{00},\ F \subseteq V\}$ $(V \in \mathscr{V})$. Clearly $\mathscr{V}$ is increasingly filtered, $\mathscr{B}_V$ is a σ-field in V $(V \in \mathscr{V})$ and $V, W \in \mathscr{V}$, $V \subseteq W$ implies $\mathscr{B}_V \subseteq \mathscr{B}_W$. If $\mathscr{B}$ denotes the σ-field generated by $\mathscr{B}^{00}$, then $E \cap V \in \mathscr{B}_V$ $(E \in B,\ V \in \mathscr{V})$ by Proposition I.4.7.2.2. It is an easy exercise to adapt the first entrance decomposition 1.3.3 to our situation so as to give each $F \in \mathscr{B}_V$ with $V \in \mathscr{V}$ at least one representation $F = F_1 + \cdots + F_n$ with suitable $U_1, \ldots, U_n \in \mathscr{U}$ satisfying $U_1 \cup \cdots \cup U_n = V$ and pairwise disjoint $F_1 \in \mathscr{B}_{U_1}, \ldots, F_n \in \mathscr{B}_{U_n}$. We then put

$$m'_V(F) = m_{U_1}(F_1) + \cdots + m_{U_n}(F_n).$$

This definition is independent of the representation of F; if $F = F'_1 + \cdots + F'_r$ with suitable $U'_1, \ldots, U'_r \in \mathscr{U}$, $F'_1 \in \mathscr{B}_{U'_1}, \ldots, F'_r = \mathscr{B}_{U'_r}$ pairwise disjoint, then

$$\begin{aligned} m_{U_1}(F_1) + \cdots + m_{U_n}(F_n) &= m_{U_1}\left(\sum_{j=1}^{r} F_1 \cap F'_j\right) + \cdots + m_{U_n}\left(\sum_{j=1}^{r} F_n \cap F'_j\right) \\ &= \sum_{k=1}^{n} \sum_{j=1}^{r} m_{U_k}(F_k \cap F'_j) = \sum_{j=1}^{r} \sum_{k=1}^{n} m_{U'_j}(F_k \cap F'_j) \\ &= \sum_{j=1}^{r} m_{U'_j}(F'_j). \end{aligned}$$

Here we have exploited the compatibility conditions and the additivity of the m_U $(U \in \mathscr{U})$. Next we put

$$m_V(E) = m'_V(E \cap V) \qquad (E \in \mathscr{B}, \quad V \in \mathscr{V}).$$

It is easy to verify that the σ-additivity of the m_U $(U \in \mathscr{U})$ implies the σ-additivity of the $m_V\colon \mathscr{B} \to \mathbb{R}_+$ $(V \in \mathscr{V})$, and $m_V \leqq m_W$ $(V, W \in \mathscr{V}, V \subseteq W)$. Thus Theorem 4.1 applies, and we get a σ-additive $m'\colon \mathscr{B} \to \overline{\mathbb{R}}_+$ with $m'(E) = \sup_{V \in \mathscr{V}} m_V(E)$. In fact we immediately see $m'(E) = m_V(E)$ for $E \in \mathscr{B}_V$ and in particular $m'(E) = m_U(E)$ $(E \in \mathscr{B}_U)$. If we denote the restriction of m' to $\mathscr{B}^{00}$ by m, we get the first basic statement of our theorem. Its last statement follows similarly to the uniqueness statement of Theorem 3.1 (exercise).

6. COMPLETION

In this section we establish a rather obvious extension procedure for σ-content spaces, called **completion**. We begin with a definition of nullsets compatible with another definition of nullsets to be given later (Chapter III, Section 4).

6.1. Definition. *Let $(\Omega, \mathscr{B}^{00}, m)$ be a σ-content space. A set $N \in \mathscr{B}^{00}$ is called a* **(m-)null set** *if $m(N) = 0$. $(\Omega, \mathscr{B}^{00}, m)$ is called* **complete** *if $\mathscr{B}^{00}$ contains all subsets of countable unions of nullsets, i.e., if*

$$M \subseteq N_1 \cup N_2 \cup \cdots, \qquad N_1, N_2, \ldots \in \mathscr{B}^{00}, \qquad m(N_1) = m(N_2) = \cdots = 0$$
$$\Rightarrow \quad M \in \mathscr{B}^{00}$$

(and, of course, $m(M) = 0$).

6.2. Theorem

6.2.1. *The Carathéodory extension $(\Omega, \mathscr{B}_m^{00}, m)$ of an arbitrary σ-content prespace $(\Omega, \mathscr{R}, m)$ is complete.*

6.2.2. *Let $(\Omega, \mathscr{B}^{00}, m)$ be a σ-content space. Then $\overline{\mathscr{B}}^{00} = \{E \mid$ there is an $F \in \mathscr{B}^{00}$ and m-nullsets $N_1, N_2, \ldots \in \mathscr{B}^{00}$ such that*

$$E \,\Delta\, F = (E \cup F) \backslash (E \cap F) \subseteq N_1 \cup N_2 \cup \cdots\}$$

is a local σ-ring and

$$\overline{m}(E) = m(F) \qquad (F \in \mathscr{B}^{00}, \;\; E \,\Delta\, F \subseteq N_1 \cup N_2 \cup \cdots,$$
$$N_1, N_2, \ldots \in \mathscr{B}^{00}, \;\; m(N_1) = m(N_2) = \cdots = 0)$$

uniquely defines a σ-content $\overline{m}$ on $\overline{\mathscr{B}}^{00}$. The σ-content space $(\Omega, \overline{\mathscr{B}}^{00}, m)$ is complete; $\overline{\mathscr{B}}^{00} = \mathscr{B}^{00}$ iff $(\Omega, \mathscr{B}^{00}, m)$ is complete. $(\Omega, \overline{\mathscr{B}}^{00}, m)$ is called the **completion** *of $(\Omega, \mathscr{B}^{00}, m)$ and $\overline{\mathscr{B}}^{00}$ is called the* **completion** *of $\mathscr{B}^{00}$ for m.*

6.2.3. *Let $(\Omega, \mathscr{R}, m)$ be a σ-content prespace. Then its Carathéodory extension equals the completion of the full extension of the σ-content space $(\Omega, \mathscr{B}^{00}(R), m)$ generated by $(\Omega, \mathscr{R}, m)$.*

Proof. 1. Let $N \subseteq \Omega$, $N_1, N_2, \ldots \in \mathscr{B}_m^{00}$, $\tilde{m}(N_1) = \tilde{m}(N_2) = \cdots = 0$ $N \subseteq N_1 \cup N_2 \cup \cdots$. By Proposition 1.3.2 we get

$$\tilde{m}(N) \leqq \tilde{m}(N_1 \cup N_2 \cup \cdots) \leqq \tilde{m}(N_1) + \tilde{m}(N_2) + \cdots = 0,$$

hence $N \in \mathscr{B}_m^{00}$, $\tilde{m}(N) = 0$.

2. Let us prove that $\overline{\mathscr{B}}^{00}$ is stable under bounded countable unions. Let $\mathscr{N}_\sigma = \{N_1 \cup N_2 \cup \cdots \mid N_1, N_2, \ldots \in \mathscr{B}^{00}, m(N_1) = m(N_2) = \cdots = 0\}$. Let $\overline{F}, \overline{F}_1, \overline{F}_2, \ldots \in \overline{\mathscr{B}}^{00}$, $\overline{F}_1, \overline{F}_2, \ldots \subseteq \overline{F}$ and $F, F_1, F_2, \ldots \in \mathscr{B}^{00}$,

$N, N_1, N_2, \ldots \in \mathcal{N}_\sigma$ such that $\bar{F} \,\Delta\, F \subseteq N$, $\bar{F}_1 \,\Delta\, F_1 \subseteq N_1, \ldots$. Put $N_0 = N \cup N_1 \cup N_2 \cup \cdots$. Clearly $N_0 \in \mathcal{N}_\sigma$, $\bar{F}_1 \backslash N_0 = F_1 \backslash N_0$, $\bar{F}_2 \backslash N_0 = F_2 \backslash N_0, \ldots$ are all in $\mathscr{B}^{00}$ and contained in $F \in \mathscr{B}^{00}$. Hence their union $(\bar{F}_1 \cup \bar{F}_2 \cup \cdots) \backslash N_0$ is in $\mathscr{B}^{00}$ and

$$(\bar{F}_1 \cup \bar{F}_2 \cup \cdots) \,\Delta\, ((F_1 \cup F_2 \cup \cdots) \backslash N_0) \subseteq N_0 \in \mathcal{N}_\sigma,$$

hence $\bar{F}_1 \cup \bar{F}_2 \cup \cdots \in \overline{\mathscr{B}}^{00}$. All other stabilities required for $\overline{\mathscr{B}}^{00}$ to be a local σ-ring are still easier to verify (exercise for the reader). Next let us show that our definition of $\bar{m}$ is unique. In fact, if F, $F' \in \mathscr{B}^{00}$, $N, N' \in \mathcal{N}_\sigma$ are such that $E \,\Delta\, F \subseteq N, E \,\Delta\, F' \subseteq N'$, then clearly $E \backslash (N \cup N') = F \backslash (N \cup N') = F' \backslash (N \cup N') \in \mathscr{B}^{00}$, $m(F) = m(F \backslash (N \cup N')) = m(F')$. The method of subtracting members of $\mathcal{N}_\sigma$ employed twice so far also leads easily to the σ-additivity of m. Every $\bar{m}$-nullset in $\overline{\mathscr{B}}^{00}$ differs from an m-nullset from $\mathscr{B}^{00}$ by a countable union of m-nullsets from $\mathscr{B}^{00}$, hence is contained in such a countable union. It is obvious that this property is shared as well by any countable union of $\bar{m}$-nullsets from $\overline{\mathscr{B}}^{00}$. Thus any set contained in a countable union of $\bar{m}$-nullsets from $\overline{\mathscr{B}}^{00}$ is also contained in a countable union of m-nullsets from $\mathscr{B}^{00}$, hence differs from $\varnothing$ by a subset of such a union and therefore belongs to $\overline{\mathscr{B}}^{00}$ again and is a $\bar{m}$-nullset. Thus $(\Omega, \overline{\mathscr{B}}^{00}, \bar{m})$ is complete. If $(\Omega, \mathscr{B}^{00}, m)$ is complete, then every $E \in \overline{\mathscr{B}}^{00}$ differs from some set in $\mathscr{B}^{00}$ by a set in $\mathscr{B}^{00}$, hence belongs to $\mathscr{B}^{00}$, whence $\overline{\mathscr{B}}^{00} = \mathscr{B}^{00}$. Clearly $\overline{\mathscr{B}}^{00} = \mathscr{B}^{00}$ in turn implies the completeness of $(\Omega, \mathscr{B}^{00}, m)$.

3. It is now obvious that the completion of the full extension $(\Omega, \tilde{\mathscr{B}}^{00}, m^0)$ of $(\Omega, \mathscr{B}^{00}(\mathscr{R}), m)$ is contained in $\mathscr{D}_{\tilde{m}}$. For every $E \in \mathscr{B}^{00}_m$ there is a $G \in \tilde{\mathscr{B}}^{00}$ such that $E \,\Delta\, G$ is contained in a countable union of m^0-nullsets, i.e., an m^0-nullset. Now the very definition of $\tilde{m}(E) < \infty$ implies the existence of countable union $F = F_1 \cup F_2 \cup \cdots$ with F_1, $F_2, \ldots \in \mathscr{R}$, $E \subseteq F_1 \cup F_2 \cup \cdots$, $m(F_1) + m(F_2) + \cdots < \infty$. From this we deduce $F \in \mathscr{B}^0$, $m^0(F) < \infty$. Since E is in $\mathscr{D}_{\tilde{m}}$, we have $\tilde{m}(F) = m^0(F) = \tilde{m}(E) + \tilde{m}(F \backslash E)$. The very definition of $\tilde{m}(E)$ leads to $G_1, G_2, \ldots \in \mathscr{B}^0$,

$$E \subseteq G_1, G_2, \ldots, m^0(G_n) < \tilde{m}(E) + 1/n \; (n = 1, 2, \ldots).$$

$G = G_1 \cap G_2 \cap \cdots$ is in $\mathscr{B}^0$ again and satisfies $E \subseteq G$, $\tilde{m}(E) = m^0(G) < \infty$, thus in particular $G \in \tilde{\mathscr{B}}^{00}$. Similarly, we obtain $F \backslash E \subseteq H \in \mathscr{B}^0$ with $m(F \backslash E) = m^0(H)$. We see now $m^0(F) = m^0(G) + m^0(H)$, hence

$$N = G \backslash (F \backslash H) \in \mathscr{B}^0, m^0(N) = m^0(G) - m^0(F) + m^0(H) = 0.$$

Finally, $E \,\Delta\, G = G \backslash E \subseteq N$, as desired.

6.3. Exercises

6.3.1. Prove that for every σ-content prespace $(\Omega, \mathscr{R}, m)$ with generated

σ-content space $(\Omega, \mathscr{B}^{00}, m)$ and Carathéodory extension $(\Omega, \mathscr{B}_m^{00}, \tilde{m})$, we have $M \in \mathscr{D}_{\tilde{m}}$ iff $M \cap F \in \mathscr{B}_m^{00}$ $(F \in \mathscr{B}^{00})$.

6.3.2. Let $(\Omega, \mathscr{B}^{00}, m)$ be a σ-content space and $(\Omega, \mathscr{B}_m^{00}, \tilde{m})$ its Carathéodory extension. Prove $E \in \mathscr{B}^{00}$, $F \in \mathscr{B}_m^{00} \Rightarrow (E \cap F) \Delta E'$ is an m-nullset for a suitable $E' \in \mathscr{B}^{00}$.

6.4. Example. Let $(\mathbb{R}^n, \mathscr{D}^n, m)$ be a dyadic geometric σ-content prespace (Exercise I.1.11.2) and $(\mathbb{R}^n, \mathscr{B}^{00}, m)$ the σ-content space derived from it, $(\mathbb{R}^n, \mathscr{B}_m^{00}, \tilde{m})$ its Carathéodory extension. Then $\mathscr{B}^0(\mathscr{B}_m^{00}) = \mathscr{D}_{\tilde{m}}$ since $\mathbb{R}$ is a countable union of dyadic intervals. All members of $\mathscr{D}_{\tilde{m}} = \mathscr{B}^0$ $(\mathscr{B}_m^{00})$ are called **Lebesgue measurable sets**. The restriction of $\tilde{m}$ to $\mathscr{B}_m^{00}$ as well as m (on $\mathscr{B}^{00}$) is called the **complete full geometric** or **Lebesgue σ-content in** $\mathbb{R}^n$. With the help of Exercise 3.3 and Theorem 6.2.3 it is easy to show that the Lebesgue σ-content in any dimension is translation and rotation invariant.

6.5. Example. We have already defined the one-sided and two-sided Bernoulli σ-probability space for a probability vector p over a finite set $A \neq \varnothing$ (Definition 2.5.2). The one-sided case is easily generalized to a situation where p may "vary from component to component," i.e., where a sequence $p^0, p^1, \ldots$ of probability vectors over A is given and m is defined on special cylinders by $m\ ([k_0, \ldots, k_r]) = p^0(k_0) \cdots p^r(k_r)$. The resulting content on C is again σ-additive (see Example I.2.5) and its generated σ-probability space is usually again called by the name of **Bernoulli**. The **two-sided Bernoulli σ-probability space** based on a two-sided sequence $\ldots, p^{-1}, p^0, p^1, \ldots$ of probability vectors over A is defined by an obvious analogy. If $(A \times A \times \cdots, \mathscr{C}, m)$ is the Markov probability prespace based on the initial distribution p and the transition matrix P, then the generated σ-probability space $(A \times A \times \cdots, \mathscr{B}^{00}(\mathscr{C}), m)$ is called the **Markov σ-probability space** for p and P, and m: $\mathscr{B}^{00}(\mathscr{C}) = \mathscr{B}(\mathscr{C}) \to [0, 1]$ is called the Markov σ-probability for p and P. The same terminology (and the name Markov in particular) is also employed in the cases treated in Exercise I.1.15 where P was allowed to vary from component to component.

6.6. Exercises

6.6.1. Let $(\Omega, \mathscr{B}^{00}, m)$ be a complete σ-content space and m^0 the unique σ-additive extension of m to $\mathscr{B}^0 = \mathscr{B}^0(\mathscr{B}^{00})$ (Theorem 3.1). Prove that $\mathscr{B}^0$ is complete in the following sense: If $N_1, N_2, \ldots \in \mathscr{B}^0$, $m^0(N_1) = m^0(N_2) = \cdots = 0$, then $N_1 \cup N_2 \cup \cdots \in \mathscr{B}^0$, $m^0(N_1 \cup N_2 \cup \cdots) = 0$; if $M \subseteq \Omega$, E, $N \in \mathscr{B}^0$, $m^0(N) = 0$, $M \Delta E \subseteq N$, then $M \in \mathscr{B}^0$.

6.6.2. Call two reals equivalent if their difference is rational. Let $M \subseteq [0, 1[$ be a set whose intersection with every equivalence class consists of exactly one element. Prove that all rational translates of M are pairwise disjoint and cover $\mathbb{R}$. Use the translation invariance of the Lebesgue content (Example 6.4) in order to prove that M is not Lebesgue measurable. (*Hint*: Assume that M is Lebesgue measurable and use

$$\bigcup_{t \text{ rational}} (M + t) = \mathbb{R}$$

for a proof that $m(M) > 0$; use $\bigcup_{t \text{ rational},\ 0 \leqq t < 1} (M + t) \subseteq [0, 2[$ in order to achieve a contradiction.)

6.6.3. Carry 6.6.2 over to arbitrary dimensions.

CHAPTER III

EXTENSION OF POSITIVE σ- AND τ-MEASURES, AFTER DANIELL

In this chapter we offer a solution to the extension problem I.11.1 for measure spaces for the case of a σ- or τ-measure, applying Daniell's classical method of monotone extension. It is in fact the rudiments of what is usually called integration theory that we present here, including the basic convergence theorems. We implement the theory in sections on the integration of complex-valued functions, on the Hilbert space of square integrable functions, and on stochastic convergence and uniform integrability.

As a by-product we obtain another solution to the extension problem I.5.1 for σ-contents, which turns out to be equivalent to Carathéodory's extension presented in Chapter II.

1. EXTENSION STEP I: σ-UPPER AND σ-LOWER FUNCTIONS, MONOTONE EXTENSION

In this section we follow the suggestion of the monotone convergence statement in the formulation I.11.1 of the extension problem for σ-measure spaces. We extend a given measure along monotone sequences of elementary functions to a larger domain of functions. Our discussion starts with nonnegative functions and "upward" extension.

1.1. Definition. *Let $\Omega \neq \emptyset$ and $\mathscr{E} \subseteq \mathbb{R}^{\Omega}$ an elementary domain (definition I.7.2). A function $u \in \overline{\mathbb{R}}_+^{\Omega}$ (recall that $\overline{\mathbb{R}}_+ = \mathbb{R} \cup \{\infty\}$) is called a* **positive σ-upper function** *for $\mathscr{E}$ if there is a sequence $0 \leqq f_1 \leqq f_2 \leqq \cdots \in \mathscr{E}$ such that $\lim_{n\to\infty} f_n(\omega) = u(\omega)$ $(\omega \in \Omega)$. The set of all positive σ-upper functions for $\mathscr{E}$ is denoted by $\mathscr{E}_+^{\sigma}$.*

The most important properties of positive σ-upper functions are collected in

1.2. Proposition. *Let $\Omega \neq \varnothing$ and $\mathscr{E} \subseteq \mathbb{R}^{\Omega}$ be an elementary domain. Then:*

1.2.1. *$\mathscr{E}_+^{\sigma}$ contains the positive cone $\mathscr{E}_+$ of $\mathscr{E}$.*

1.2.2. *$\mathscr{E}_+^{\sigma}$ is stable under:*

1.2.2.1. *the finite lattice operations:*

$$u, v \in \mathscr{E}_+^{\sigma} \quad \Rightarrow \quad u \vee v, u \wedge v \in \mathscr{E}_+^{\sigma};$$

1.2.2.2. *finite positive linear combinations:*

$$u_1, \ldots, u_n \in \mathscr{E}_+^{\sigma}, \alpha_1, \ldots, \alpha_n \geqq 0 \quad \text{real}$$
$$\Rightarrow \quad \alpha_1 u_1 + \cdots + \alpha_n u_n \in \mathscr{E}_+^{\sigma};$$

1.2.2.3. *countable suprema:*

$$u_1, u_2, \ldots \in \mathscr{E}_+^{\sigma} \quad \Rightarrow \quad u_1 \vee u_2 \vee \cdots \in \mathscr{E}_+^{\sigma};$$

1.2.2.4. *countable positive linear combinations:*

$$u_1, u_2, \ldots \in \mathscr{E}_+^{\sigma}, \alpha_1, \alpha_2, \ldots \geqq 0 \text{ real} \quad \Rightarrow \quad \alpha_1 u_1 + \alpha_2 u_2 + \cdots \in \mathscr{E}_+^{\sigma}.$$

1.2.3.
$$\mathscr{E}_+^{\sigma} = \{f_1 \vee f_2 \vee \cdots | 0 \leqq f_1, f_2, \ldots \in \mathscr{E}\}$$
$$= \left\{\sup_{f \in \mathscr{F}} f \,|\, \varnothing \neq \mathscr{F} \subseteq \mathscr{E}_+, \mathscr{F} \text{ (at most) countable}\right\}.$$

Proof. 3 is an easy exercise.

1 follows from 3.

2.3 follows from 3: If $\mathscr{F}_1, \mathscr{F}_2, \ldots \subseteq \mathscr{E}_+$ are nonempty and at most countable with $\sup_{f \in \mathscr{F}_k} f = u_k$ $(k = 1, 2, \ldots)$, then

$$\varnothing \neq \mathscr{F} = \mathscr{F}_1 \cup \mathscr{F}_2 \cup \cdots \subseteq \mathscr{E}_+$$

is at most countable and $\sup_{f \in \mathscr{F}} f = u_1 \vee u_2 \vee \cdots$.

2.2 follows because $\mathscr{E}_+$ is stable under finite positive linear combinations: If $\mathscr{E}_+ \ni f_{1k} \nearrow u_1, \ldots, \mathscr{E}_+ \ni f_{nk} \nearrow u_n$ pointwise as $k \to \infty$, then

$$\mathscr{E}_+ \ni \alpha_1 f_{1k} + \cdots + \alpha_n f_{nk} \nearrow \alpha_1 u_1 + \cdots + \alpha_n u_n$$

pointwise as $k \to \infty$ and $\alpha_1 u_1 + \cdots + \alpha_n u_n \in \mathscr{E}_+^{\sigma}$ follows.

2.1 is a special case of 2.3 for $\vee$. $\wedge$ stability of $\mathscr{E}_+^{\sigma}$ is deduced from $\wedge$ stability of $\mathscr{E}_+$ in a way similar to the proof of 2.2 (exercise).

2.4 follows from 2.2 and 2.3.

The *monotone convergence theorem* contained in the formulation I.11.1 of the extension problem for measure spaces requires that any extension of m from $\mathscr{E}_+$ to $\mathscr{E}_+^{\sigma}$ be defined according to 2 of the following

1.3. Lemma. *Let* $(\Omega, \mathscr{E}, m)$ *be a* σ*-measure space,* $u \in \mathscr{E}_+^\sigma$ *and*

$$\underline{m}(u) = \sup_{\mathscr{E} \ni f \leqq u} m(f).$$

then the following statements hold:

1.3.1. $\underline{m}(u) = \sup_{\mathscr{E}_+ \ni f \leqq u} m(f)$

1.3.2. *If* $\mathscr{E}_+ \ni f_1 \leqq f_2 \leqq \cdots \nearrow u$ *pointwise, then*

$$\lim_{n \to \infty} m(f_n) = \underline{m}(u)$$

1.3.3. *If* $\varnothing \neq \mathscr{F} \subseteq \mathscr{E}_+$ *is countable and increasingly filtered with pointwise supremum* u*, then*

$$\sup_{f \in \mathscr{F}} m(f) = \underline{m}(u).$$

Proof. 1. $\{f \mid \mathscr{E} \ni f \leqq u\} \supseteq \{f \mid \mathscr{E}_+ \ni f \leqq u\}$ shows $\geqq$. On the other hand, every $\mathscr{E} \ni f \leqq u$ has the majorant $u \geqq f_+ = f \vee 0 \in \mathscr{E}_+$, whence $\leqq$.

2. Let $\mathscr{E}_+ \ni f \leqq u$; then $\mathscr{E}_+ \ni (f - f_n)_+ \searrow (f - u)_+ = 0$ pointwise and the σ-continuity of m (Definition I.8.1.3) implies $m((f - f_n)_+) \searrow 0$. By isotony of m all the following limits exist and we have

$$\begin{aligned} m(f) &= m(f) - \lim m((f - f_n)_+) \leqq m(f) - \lim m(f - f_n) \\ &= \lim m(f_n), \end{aligned}$$

proving $\geqq$ in 2. Since $\leqq$ is obvious, we are through.

3. Let $\mathscr{F} = \{g_1, g_2, \ldots\}$. Since $\mathscr{F}$ is increasingly filtered, we may inductively construct $h_1 = g_1$, $\mathscr{F} \ni h_2 \geqq h_1, g_2$, $\mathscr{F} \ni h_3 \geqq h_2, g_3, \ldots$ and see $g_n \leqq h_n \leqq u$ $(n = 1, 2, \ldots)$, hence $\lim h_n = u$ pointwise. With the help of 2 we get

$$\begin{aligned} \underline{m}(u) \geqq \sup_{f \in \mathscr{F}} m(f) &= \sup_n m(g_n) \geqq \sup_n m(h_n) \\ &= \lim m(h_n) = \underline{m}(u). \end{aligned}$$

1.4. Proposition. *Let* $(\Omega, \mathscr{E}, m)$ *be a* σ*-measure space. The function* $\underline{m}: \mathscr{E}_+^\sigma \to \overline{\mathbb{R}}_+$ *defined in Lemma 1.3. coincides with* m *on* $\mathscr{E}_+$ *and is:*

1.4.1. *isotone, i.e.,*

$$u, v \in \mathscr{E}_+^\sigma, u \leqq v \quad \Rightarrow \quad \underline{m}(u) \leqq \underline{m}(v);$$

1.4.2. *positive homogeneous, i.e.,*

$$\underline{m}(\alpha u) = \alpha \underline{m}(u) \qquad (\alpha \in \mathbb{R}_+, \quad u \in \mathscr{E}_+^\sigma);$$

1.4.3. *σ-additive, i.e.,*

$$\underline{m}(u_1 + u_2 + \cdots) = \underline{m}(u_1) + \underline{m}(u_2) + \cdots \qquad (u_1, u_2, \ldots \in \mathscr{E}_+^\sigma);$$

1.4.4. *lower σ-continuous, i.e.,*

$$u, u_1, u_2, \ldots \in \mathscr{E}_+^\sigma, u_1 \leqq u_2 \leqq \cdots \nearrow u \text{ pointwise}$$
$$\Rightarrow \lim \underline{m}(u_n) = \underline{m}(u).$$

Proof. If $u \in \mathscr{E}_+^\sigma$, then $\{m(f) | \mathscr{E}_+ \ni f \leqq u\}$ contains the maximal element $m(f)$, hence $\underline{m}$ coincides with m on $\mathscr{E}_+$.

1.

$$u \leqq v \quad \Rightarrow \quad \{f | \mathscr{E} \ni f \leqq u\} \subseteq \{f | \mathscr{E} \ni f \leqq v\}$$
$$\Rightarrow \quad \underline{m}(u) \leqq \underline{m}(v).$$

2.

$$u \in \mathscr{E}_+^\sigma, \alpha \in \mathbb{R}_+ \quad \Rightarrow \quad \{f | \mathscr{E}_+ \ni f \leqq \alpha u\} = \{\alpha g | \mathscr{E}_+ \ni g \leqq u\}$$
$$\Rightarrow \quad \underline{m}(\alpha u) = \alpha \underline{m}(u).$$

3. Let $\mathscr{E}_+ \ni f_{n1} \leqq f_{n2} \leqq \cdots \nearrow u_n$ $(n = 1, 2, \ldots)$. Put

$$f_n = f_{1n} + f_{2n} + \cdots + f_{nn}.$$

Clearly, $\mathscr{E}_+ \ni f_1 \leqq f_2 \leqq \cdots \nearrow u_1 + u_2 + \cdots$, hence

$$\underline{m}(u_1 + u_2 + \cdots) = \lim m(f_n) = \lim[m(f_{1n}) + \cdots + m(f_{nn})]$$
$$= \underline{m}(u_1) + \underline{m}(u_2) + \cdots$$

by Lemma 1.3.

4. Let $\mathscr{E}_+ \ni f_{n1} \leqq f_{n2} \leqq \cdots \nearrow u_n$ $(n = 1, 2, \ldots)$. Put $f_n = f_{1n} \vee \cdots \vee f_{nn}$ $(n = 1, 2, \ldots)$. Clearly, $\mathscr{E}_+ \ni f_1 \leqq f_2 \leqq \cdots \nearrow u$ pointwise. On the other hand, $f_n \leqq u_1 \vee \cdots \vee u_n = u_n$ $(n = 1, 2, \ldots)$. Thus we get, by Lemma 1.3, $\underline{m}(u) = \lim m(f_n) \leqq \lim \underline{m}(u_n) \leqq \underline{m}(u)$, the latter $\leqq$ being obvious from 1.

1.5. Exercises

1.5.1. Let $(\mathbb{R}, \mathscr{E}, m)$ be the σ-measure space derived from the dyadic geometric σ-content prespace $(\mathbb{R}, \mathscr{D}, m)$ (Examples I.1.10, I.2.4, I.8.7, Proposition I.8.9) and $x_1, x_2, \ldots$ an enumeration of the set $\mathbb{Q}$ of all rationals. Prove that for every n, the function $f_n = 1_{\mathbb{R}\setminus\{x_1, \ldots, x_n\}}$ is in $\mathscr{E}_+^\sigma$ with $\underline{m}(f) = \infty$ while $\underline{m}(f_n \cdot 1_{[a, b]}) = b - a$ $(-\infty < a < b < \infty)$.

1.5.2. Prove that the function g of Exercise I.9.4.2 is in the $\mathscr{E}_+^\sigma$ of Exercise 1, with $\underline{m}(g) < 1$.

1.5.3. Let Ω be a compact metric space, $E = \mathscr{C}(\Omega, \mathbb{R})$ (Example I.7.7.1). Prove that $\mathscr{E}_+^\sigma$ is the set of all lower semicontinuous functions in $\overline{\mathbb{R}}_+$.

1.5.4. Let Ω be a locally compact Hausdorff space, $\mathscr{E} = \mathscr{C}^{00}(\Omega, \mathbb{R})$ (Example I.7.7.4) and assume that Ω is metrizable and a countable union of compact sets. Prove that $\mathscr{E}_+^\sigma$ is the set of all lower semicontinuous functions in $\overline{\mathbb{R}}_+$.

1.6. Definition. *Let $\Omega \neq \varnothing$ and $\mathscr{E} \subseteq \mathbb{R}^\Omega$ an elementary domain (Definition I.7.2). A function $v \in \overline{\mathbb{R}}^\Omega$ is called:*

1.6.1. *a* **σ-upper function** *for $\mathscr{E}$ if there is a sequence $f_1 \leqq f_2 \leqq \cdots \in \mathscr{E}$ such that $f_n \nearrow v$ pointwise on Ω; the set of all σ-upper functions for $\mathscr{E}$ is denoted by $\mathscr{E}^\sigma$;*

1.6.2. *a* **σ-lower function** *for $\mathscr{E}$ if there is a sequence $g_1 \geqq g_2 \geqq \cdots \in \mathscr{E}$ such that $g_n \searrow v$ pointwise in Ω; the set of all σ-lower functions for $\mathscr{E}$ is denoted by $\mathscr{E}_\sigma$.*

1.7. Proposition. *Let $\Omega \neq \varnothing$ and $\mathscr{E} \subseteq \mathbb{R}^\Omega$ an elementary domain. Then:*

1.7.1. *$\mathscr{E}^\sigma = \mathscr{E} + \mathscr{E}_+^\sigma = \{f + u \mid f \in \mathscr{E}, u \in \mathscr{E}_+^\sigma\}$. Actually,*

$$v \in \mathscr{E}^\sigma, \mathscr{E} \ni f \leqq v \quad \Rightarrow \quad v - f \in \mathscr{E}_+^\sigma.$$

No function in $\mathscr{E}^\sigma$ attains the value $-\infty$ since every function in $\mathscr{E}^\sigma$ has a minorant in $\mathscr{E}$. $\mathscr{E}^\sigma$ is stable under:

1.7.1.1. *positive finite linear combinations, i.e.,*

$$\alpha_1, \ldots, \alpha_n \in \mathbb{R}_+, \quad v_1, \ldots, v_n \in \mathscr{E}^\sigma \quad \Rightarrow \quad \alpha_1 v_1 + \cdots + \alpha_n v_n \in \mathscr{E}^\sigma;$$

1.7.1.2. *countable suprema and finite infima, i.e.,*

$$v_1, v_2, \ldots \in \mathscr{E}^\sigma \quad \Rightarrow \quad v_1 \vee v_2 \vee \cdots \in \mathscr{E}^\sigma$$

$$v_1, \ldots, v_n \in \mathscr{E}^\sigma \quad \Rightarrow \quad v_1 \wedge \cdots \wedge v_n \in \mathscr{E}^\sigma;$$

1.7.1.3. *infima with positive constants, i.e.,*

$$v \in \mathscr{E}^\sigma, \ \alpha \in \mathbb{R}_+ \quad \Rightarrow \quad \alpha \wedge v \in \mathscr{E}^\sigma.$$

1.7.2. *$\mathscr{E}_\sigma = \mathscr{E} - \mathscr{E}_+^\sigma = \{f - u \mid f \in \mathscr{E}, u \in \mathscr{E}_+^\sigma\}$. Actually,*

$$v \in \mathscr{E}_\sigma, \mathscr{E} \ni f \geqq v \quad \Rightarrow \quad f - v \in \mathscr{E}_+^\sigma.$$

No function in $\mathscr{E}_\sigma$ attains the value ∞ since every function in $\mathscr{E}_\sigma$ has a majorant in $\mathscr{E}$. $\mathscr{E}_\sigma$ is stable under:

1.7.2.1. *positive finite linear combinations;*

1.7.2.2. *countable infima and finite suprema;*

1.7.2.3. *infima with positive constants.*

Proof. 1. Let $v \in \mathscr{E}^\sigma, f_1 \leqq f_2 \leqq \cdots \in \mathscr{E}, f_n \nearrow v$ pointwise $(n = 1, 2, \ldots)$. Let $\mathscr{E} \ni f \leqq v$. Put $g_n = (f_n \vee f) - f$. Clearly, $\mathscr{E}_+ \ni g_1 \leqq g_2 \leqq \cdots \nearrow v - f$, proving $v - f \in \mathscr{E}^\sigma_+$. In particular, $\mathscr{E}^\sigma \subseteq \mathscr{E} + \mathscr{E}_+$. The proof of $\supseteq$ is similar and left to the reader.

1.1 is now an easy consequence of the corresponding stability of $\mathscr{E}$ and $\mathscr{E}^\sigma_+$ (Proposition 1.2).

1.2. Choose any $\mathscr{E} \ni f_1 \leqq v_1$, $\mathscr{E} \ni f_2 \leqq v_2$. Put $f = f_1 \wedge f_2$. Then $\mathscr{E} \ni f \leqq v_1, v_2$, $v_1 \vee v_2 = f + (v_1 - f) \vee (v_2 - f)$, $(v_1 - f) \vee (v_2 - f) \in \mathscr{E}^\sigma_+$ (Proposition 1.2.2.1), hence $v_1 \vee v_2 \in \mathscr{E}^\sigma$. By induction we get $v_1 \vee \cdots \vee v_n \in \mathscr{E}^\sigma$. Thus we may henceforth assume $v_1 \leqq v_2 \leqq \cdots$. Choose any $\mathscr{E} \ni f \leqq v_1$. By Proposition 1.2.2.3 we have $(v_1 - f) \vee (v_2 - f) \vee \cdots \in \mathscr{E}^\sigma_+$ and thus $v_1 \vee v_2 \vee \cdots = f + [(v_1 - f) \vee (v_2 - f) \vee \cdots] \in \mathscr{E}^\sigma$. Stability under finite infima follows from Proposition 1.2.2.1 in a similar fashion.

1.3. Choose any $\mathscr{E} \ni f_1 \leqq f_2 \leqq \cdots \nearrow v$ pointwise. Then clearly $\mathscr{E} \ni \alpha \wedge f_1 \leqq \alpha \wedge f_2 \leqq \cdots \nearrow \alpha \wedge v$ pointwise, hence $\alpha \wedge v \in \mathscr{E}^\sigma$.

The proof of 2 is now easy and left to the reader.

1.8. Proposition. *Let $(\Omega, \mathscr{E}, m)$ be a σ-measure space and $\underline{m}$: $\mathscr{E}^\sigma_+ \to \overline{\mathbb{R}}_+$ be defined according to Lemma 1.3. Then:*

1.8.1. *There is a unique function (again denoted by) $\underline{m}$: $\mathscr{E}^\sigma \to \overline{\mathbb{R}}$ such that*

$$\underline{m}(f + u) = m(f) + \underline{m}(u) \qquad (f \in \mathscr{E}, \quad u \in \mathscr{E}^\sigma_+).$$

This function does not attain the value $-\infty$. In fact, every $v \in \mathscr{E}^\sigma$ has a minorant in $\mathscr{E}$ and we have

$$\underline{m}(v) = \sup_{\mathscr{E} \ni f \leqq u} m(f) \qquad (v \in \mathscr{E}^\sigma).$$

$\underline{m}$: $\mathscr{E}^\sigma \to \mathbb{R}$ is isotone and has the following properties:

1.8.1.1. $\underline{m}(\alpha_1 v_1 + \cdots + \alpha_n v_n) = \alpha_1 \underline{m}(v_1) + \cdots + \alpha_n \underline{m}(v_n)$

$$(\alpha_1, \ldots, \alpha_n \in \mathbb{R}_+, \quad v_1, \ldots, v_n \in \mathscr{E}^\sigma).$$

1.8.1.2. *Lower σ-continuity:*

$$\mathscr{E}^\sigma \ni v_1 \leqq v_2 \leqq \cdots \nearrow v \text{ (pointwise)} \quad \Rightarrow \quad \underline{m}(v_n) \nearrow \underline{m}(v).$$

1.8.2. *There is a unique function $\overline{m}$: $\mathscr{E}^\sigma \to \overline{\mathbb{R}}$ such that*

$$\overline{m}(f - u) = m(f) - \underline{m}(u) \qquad (f \in E, \quad u \in \mathscr{E}^\sigma_+).$$

This function does not attain the value ∞. In fact, every $v \in \mathscr{E}_\sigma$ has a majorant in $\mathscr{E}$ and

$$\overline{m}(v) = \inf_{v \leqq f \in \mathscr{E}} m(f) \qquad (v \in \mathscr{E}_\sigma).$$

$\overline{m}$: $\mathscr{E} \to \overline{\mathbb{R}}$ is isotone and has the following properties:

1.8.2.1. $\overline{m}(\alpha_1 v_1 + \cdots + \alpha_n v_n) = \alpha_1 \overline{m}(v_1) + \cdots + \alpha_n \overline{m}(v_n)$

$$(\alpha_1, \ldots, \alpha_n \in \mathbb{R}_+ \quad v_1, \ldots, v_n \in \mathscr{E}_\sigma);$$

1.8.2.2. *Upper σ-continuity:*

$$\mathscr{E}_\sigma \ni v_1 \geqq v_2 \geqq \cdots \searrow v \text{ (pointwise)} \quad \Rightarrow \quad \overline{m}(v_n) \searrow \overline{m}(v).$$

1.8.3. *The following relations between $\underline{m}$ and $\overline{m}$ hold:*

1.8.3.1. $$\overline{m}(-v) = -\underline{m}(v) \qquad (v \in \mathscr{E}^\sigma).$$

1.8.3.2. $$\underline{m}(v) = \overline{m}(v) \qquad (v \in \mathscr{E}^\sigma \cap \mathscr{E}_\sigma).$$

1.8.3.3. $$\overline{m}(v) \leqq \underline{m}(v) \qquad (\mathscr{E}_\sigma \ni v \leqq w \in \mathscr{E}^\sigma).$$

$$\underline{m}(v) \leqq \overline{m}(w) \qquad (\mathscr{E}^\sigma \ni v \leqq w \in \mathscr{E}_\sigma).$$

Proof. 1. If $f, f' \in \mathscr{E}$, $u, u' \in \mathscr{E}_+^\sigma$ are such that $f + u = f' + u'$, we take $g = f \wedge f' \in \mathscr{E}$ and put $f_0 = f - g$, $f'_0 = f' - g$. We get

$$f_0, f'_0 \in \mathscr{E}_+ \subseteq \mathscr{E}_+^\sigma, \; f_0 + u = (f + u) - g = (f' + u') - g = f'_0 + u' \in \mathscr{E}_+^\sigma$$

and thus

$$\begin{aligned} m(f) + \underline{m}(u) &= m(g + f_0) + \underline{m}(u) = m(g) + m(f_0) + \underline{m}(u) \\ &= m(g) + \underline{m}(f_0) + \underline{m}(u) = m(g) + \underline{m}(f_0 + u) \\ &= m(g) + \underline{m}(f'_0 + u') = m(f') + \underline{m}(u'), \end{aligned}$$

using Proposition 1.4.3. By Proposition 1.7.1 every $v \in \mathscr{E}^\sigma$ has a minorant f in $\mathscr{E}$ such that $v = f + u$ for some $u \in \mathscr{E}_+^\sigma$; and by Lemma 1.3 we get

$$\begin{aligned} \underline{m}(v) = m(f) + \underline{m}(u) &= m(f) + \sup_{\mathscr{E} \ni g \leqq u} m(g) \\ &= \sup_{\mathscr{E} \ni g \leqq u} m(f + g) = \sup_{\mathscr{E} \ni h \leqq v} m(h). \end{aligned}$$

From this, isotony follows trivially.

1.1 is now an easy consequence of the corresponding properties of $m: \mathscr{E} \to \mathbb{R}$ and $\underline{m}: \mathscr{E}_+^\sigma \to \overline{\mathbb{R}}_+$.

1.2. Choose any $\mathscr{E} \ni f \leqq v_1$ and put $u_1 = v_1 - f$, $u_2 = v_2 - f$, ... Then $u_1 \leqq u_2 \leqq \cdots \in \mathscr{E}_+^\sigma$ and $v = f + (u_1 \vee u_2 \vee \cdots)$. From Proposition 1.4.4 we conclude now

$$\begin{aligned} \underline{m}(v) = m(f) + \underline{m}(u_1 \vee u_2 \vee \cdots) &= m(f) + \lim_{n\to\infty} \underline{m}(u_n) \\ &= \lim_{n\to\infty} [m(f) + \underline{m}(u_n)] \\ &= \lim_{n\to\infty} \underline{m}(f + u_n) \\ &= \lim_{n\to\infty} \underline{m}(v_n). \end{aligned}$$

2 is proved in an obvious, similar way to 1.

3.1 is obvious.

3.2 is a consequence of 3.3.

3.3. If $\mathscr{E}_\sigma \ni v \leqq w \in \mathscr{E}^\sigma$, find $f, g \in \mathscr{E}$, $t, u \in \mathscr{E}^\sigma_+$ with $v = f - u$, $w = g + t$. We have $f - u \leqq g + t$, hence $f - g \leqq u + t$, and $m(f) - m(g) = m(f - g) = \underline{m}(f - g) \leqq \underline{m}(u + t) = \underline{m}(u) + \underline{m}(t)$ follows from Proposition 1.4. Now $\overline{m}(v) \leqq \underline{m}(w)$ is obvious if $\underline{m}(u) + \underline{m}(t) = \infty$. If $\underline{m}(u) + \underline{m}(t) < \infty$, we get again $\overline{m}(v) = m(f) - \underline{m}(u) \leqq m(g) + \underline{m}(t) = \underline{m}(w)$. The second inequality in 3.3 is proved in a similar way.

1.9. Example. *Let $\Omega = \mathbb{R}$, $\mathscr{E} = \mathscr{C}^{00}(\mathbb{R}, \mathbb{R})$, and $m: \mathscr{E} \to \mathbb{R}$ be the usual Riemann integral. Many functions that are usually called Riemann integrable in the extended sense are in $\mathscr{E}^\sigma$ or $\mathscr{E}_\sigma$, with finite values of $\underline{m}$ resp. $\overline{m}$. If, e.g., $\delta > 0$, then*

$$u(x) = \begin{cases} x^{-1-\delta} & (x > 1) \\ 0 & (x \leqq 1) \end{cases}$$

belongs to $\mathscr{E}^\sigma$ with $\underline{m}(u) = \delta$, and

$$v(x) = \begin{cases} x^{-1+\delta} & (0 < x < 1) \\ 0 & \text{elsewhere on } \mathbb{R} \end{cases}$$

belongs to $\mathscr{E}^\sigma$ with $\underline{m}(v) = 1/\delta$.

2. EXTENSION STEP II: SQUEEZING-IN AND THE DEFINITION OF INTEGRABLE FUNCTIONS

In this section we proceed quickly to a definition of *integrable functions*. A first discussion of the definition will show that it is advisable to invest some labor in a subsidiary concept, the *upper integral*, before proceeding to the proof of basic properties of integrable functions.

2.1. Definition. *Let $(\Omega, \mathscr{E}, m)$ be a σ-measure space and $\underline{m}: \mathscr{E}^\sigma \to \overline{\mathbb{R}}$, $\overline{m}: \mathscr{E}_\sigma \to \overline{\mathbb{R}}$ be defined according to Proposition 1.8. A function $h \in \overline{\mathbb{R}}^\Omega$ is called* **m-(σ-)-integrable** *if for every $\varepsilon > 0$ there is a $u \in \mathscr{E}_\sigma$ and a $v \in \mathscr{E}^\sigma$ such that h is "squeezed between u and v up to ε," more precisely:*

$$u \leqq h \leqq v, \qquad -\infty < \overline{m}(u), \qquad \underline{m}(v) < \infty$$

$$\underline{m}(v) - \overline{m}(u) < \varepsilon$$

The real number

$$\sup_{\mathscr{E}_\sigma \ni u \leqq h} \overline{m}(u) = \inf_{h \leqq v \in \mathscr{E}^\sigma} \underline{m}(v)$$

is then called the **m-integral of h** *and denoted by* $m(h)$, $\int h\,dm$, $\int_\Omega h\,dm$, $\int h(\omega)\,m(d\omega)$, $\int_\Omega h(\omega)\,m(d\omega)$, $\int dm\,h$, $\int m(d\omega)\,h(\omega)$, *etc. The set of all m-σ-integrable functions is denoted by* $\mathscr{L}^1_\sigma(m)$, $\mathscr{L}^1_m$, $\mathscr{L}^1_\sigma(\Omega, \mathscr{E}, m)$ etc., *depending on what specifications are needed. The mapping* $m\colon \mathscr{L}^1_m \to \mathbb{R}$ *is also denoted as the m-integral and symbolized by* $\int dm$ *as well. Also the couple* $\left(\mathscr{L}^1_m, \int dm\right) = (\mathscr{L}^1_m, m) =$ *etc. is called the m-σ-integral (derived from* $(\Omega, \mathscr{E}, m)$ *or, briefly, from m by σ-extension). If* $(\Omega, \mathscr{E}, m)$ *is the σ-measure space derived from a σ-content prespace* $(\Omega, \mathscr{R}, m)$ *(Example I.8.7, Proposition I.8.9), then σ-integrability etc. with respect to the σ-measure m is spoken of synonymously with σ-integrability etc. with respect to the underlying σ-content, and the notation is adapted accordingly, so that it is, e.g., clear what* $\mathscr{L}^1_\sigma(\Omega, \mathscr{R}, m)$ *means.*

2.2. Remarks. The above definition needs some simple explanations.

2.2.1. $\mathscr{E}_\sigma \ni u \leqq v \in \mathscr{E}^\sigma$ implies $\overline{m}(u) \leqq \underline{m}(v)$ by Proposition 1.8.3.3. Hence $\overline{m}(u) \leqq \underline{m}(v)$ in the above context.

2.2.2. The definition of m-integrability of $h \in \mathbb{R}^\Omega$ implies that the equality

$$\sup_{\mathscr{E}_\sigma \ni u \leqq h} \overline{m}(u) = \inf_{h \leqq v \in \mathscr{E}^\sigma} \underline{m}(v)$$

holds and the two members of this equality define the same finite real number, namely the m-integral of h.

2.2.3. If $h \in \mathscr{E}$, then h is a member of $\{u \mid \mathscr{E}_\sigma \ni u \leqq h\}$ majorizing all other members of that set. Similarly, h is a member of $\{v \mid h \leqq v \in \mathscr{E}^\sigma\}$ minorizing all other members of this set. Thus h is m-σ-integrable with $\int h\,dm = m(h)$, the right-hand member denoting the value of $m\colon \mathscr{E} \to \mathbb{R}$ at h. Thus we are justified in denoting the integral by m again: It extends the measure $m\colon \mathscr{E} \to \mathbb{R}$ to a larger set of functions. We shall investigate the question of whether $m\colon \mathscr{L}^1_m \to \mathbb{R}$ in fact solves the extension problem I.11.1 for measures.

2.2.4. If $h \in \mathscr{E}^\sigma$, $\underline{m}(h) < \infty$, then we may choose, for a given $\varepsilon > 0$, some $f \in \mathscr{E}$ with $f \leqq h$ and $m(f) > \underline{m}(h) - \varepsilon$. With $u = f$ and $v = h$ the definition of m-integrability of h is satisfied, and we conclude $h \in \mathscr{L}^1_m$, $\int h\,dm = \underline{m}(h)$. Thus we see that $\{u \mid u \in \mathscr{E}^\sigma,\ \underline{m}(u) < \infty\} \subseteq \mathscr{L}^1_m$ and $\underline{m}$ represents the m-integral on this set. Similarly $\{v \mid v \in \mathscr{E},\ \overline{m}(v) > -\infty\} \subseteq \mathscr{L}^1_m$ and $\overline{m}$ represents the m-integral on this set.

2.2.5. Notice that every $h \in \mathscr{L}^1_\sigma(m)$ lies between a countable supremum and a countable infimum of elementary functions and hence vanishes outside a countable union of supports of elementary functions.

2.2.6. As an immediate consequence of the above definition we observe: If $f, g, h \in \overline{\mathbb{R}}^\Omega, f \leqq g \leqq h, f, h \in \mathscr{L}_\sigma^1(m), \int f\, dm = \int h\, dm$, then $g \in \mathscr{L}_\sigma^1(m)$ and $\int h\, dm = \int f\, dm = \int g\, dm$. We even have the obvious statement: If $g \in \overline{\mathbb{R}}^\Omega$ and if for every $\varepsilon > 0$ there are $f, h \in \mathscr{L}_\sigma^1(m)$ with $f \leqq g \leqq h$, $\int h\, dm - \int f\, dm < \varepsilon$, then $g \in \mathscr{L}_\sigma^1(m)$.

2.2.7. We shall indulge in some liberality concerning the affixes of $\mathscr{L}^1$, tailoring our notation to what we want to emphasize.

2.2.8. If $0 < \alpha \in \mathbb{R}$, then obviously $\alpha m: f \to \alpha m(f)$ $(f \in \mathscr{E})$ is a σ-measure iff m is, $\|\alpha m\| = \alpha\|m\|$, and all the functions $\underline{m}$ etc. take on a factor α when we pass from m to αm. Clearly $\mathscr{L}_\sigma^1(\alpha m) = \mathscr{L}_\sigma^1(m)$ and $\int h\, d(\alpha m) = \alpha \int h\, dm$ $(h \in \mathscr{L}_\sigma^1(m))$. This observation allows us to pass freely between m and αm and hence to assume, e.g., $\|m\| = 1$ if $0 < \|m\| < \infty$.

2.2.9. The letter $\mathscr{L}$ is employed here *in memoriam* Henri Lebesgue (1875–1941) who was one of the founding fathers of modern measure and integration theory (but made important contributions to other branches of mathematics as well. See LEBESGUE [1–4]).

2.2.10. In order to streamline the subsequent investigations it seems feasible to introduce two functions $\tilde{m}: \overline{\mathbb{R}}^\Omega \to \overline{\mathbb{R}}$, $\underset{\sim}{m}: \overline{\mathbb{R}}^\Omega \to \overline{\mathbb{R}}$ by

$$\left.\begin{aligned} \tilde{m}(h) &= \inf_{h \leqq v \in \mathscr{E}_\sigma} \underline{m}(v) \\ \underset{\sim}{m}(h) &= \sup_{\mathscr{E}_\sigma \ni u \leqq h} \overline{m}(u) \end{aligned}\right\} \qquad (h \in \overline{\mathbb{R}}^\Omega)$$

(with the usual convention $\inf \varnothing = \infty$, $\sup \varnothing = -\infty$), to rewrite the definition of m-integrability of $h \in \overline{\mathbb{R}}^\Omega$ as

$$-\infty < \underset{\sim}{m}(h) = \tilde{m}(h) < \infty$$

and to deduce properties of the integral $m: \mathscr{L}_m^1 \to \mathbb{R}$ from properties of $\tilde{m}$ and $\underset{\sim}{m}$, beginning with the observation $\underset{\sim}{m} \leqq \tilde{m}$. It is in fact this device that we shall follow in the sequel.

2.2.11. The whole process leading from the σ-measure space $(\Omega, \mathscr{E}, m)$ to $\mathscr{L}_\sigma^1(m)$ and $m: \mathscr{L}_\sigma^1(m) \to \mathbb{R}$ is called a *(monotone) σ-extension.* As we shall learn in Section 7, τ-measures allow of a "τ-extension" that differs from σ-extension by the absence of countability requirements in certain passages. Since every τ-measure is also a σ-measure, a σ-extension is possible at the same time. As Section 7 will show, the result of both processes is to a certain extent the same.

2.2.12 Stone's axiom I.7.2.1 allows us to squeeze every integrable function that is everywhere below a certain positive constant, between σ-lower and σ-upper functions that are below that constant everywhere as well. The same argument applies to constant minorants. An analogous argument applies to majorants and minorants from the elementary domain $\mathscr{E}$.

2.3. Example. Let $\Omega = \mathbb{R}$, $\mathscr{E} = \mathscr{C}^{00}(\mathbb{R}, \mathbb{R})$, and $m\colon \mathscr{E} \to \mathbb{R}$ be the usual Riemann integral. The functions in $\mathscr{L}^1_\sigma(m)$ are then called **Lebesgue integrable** and $m\colon \mathscr{L}^1_\sigma(m) \to \mathbb{R}$ is called the **Lebesgue integral**. In this case it is customary to write $\int h(x)\,dx$ instead of $\int h(\omega)\,m(d\omega)$.

2.4. Exercises

2.4.1. Let $(\Omega, \mathscr{E}, m)$ be a σ-measure space. Prove that $g \in \overline{\mathbb{R}}^\Omega$ is m-σ-integrable iff there is a countable increasingly filtered set $\varnothing \neq U \subseteq \mathscr{E}_\sigma$ and a countable decreasingly filtered set $\varnothing \neq \mathscr{V} \subseteq \mathscr{E}^\sigma$ such that $f = \sup_{u \in \mathscr{U}} u$ (pointwise) and $h = \inf_{v \in \mathscr{V}} v$ (pointwise) satisfy $f \leqq g \leqq h$ pointwise and

$$-\infty < \lim_{u \in \mathscr{U}} \overline{m}(u) = \lim_{v \in \mathscr{V}} \underline{m}(v) < \infty$$

2.4.2. Let $\Omega = [0, 1]$, $\mathscr{E} = \mathscr{C}([0, 1], \mathbb{R})$, and $m\colon \mathscr{E} \to \mathbb{R}$ be the Riemann integral. Prove that every lower semicontinuous function $u \in \overline{\mathbb{R}}_+^\Omega$ with $\sup_{\mathscr{E} \ni f \leqq u} m(f) < \infty$ is m-σ-integrable and that this sup gives its m-integral.

2.4.3. Let $(\Omega, \mathscr{E}, m)$ be a σ-measure space and $(\Omega, \mathscr{E}_{(m)}, m)$ its Riemann extension. Prove that $\mathscr{E}_{(m)} \subseteq \mathscr{L}^1_\sigma(m)$ and the m-σ-integral coincides with the Riemann integral for m on $\mathscr{E}_{(m)}$. Prove that every $h \in \mathscr{E}_{(m)}$ vanishes outside the support of some $f \in \mathscr{E}$.

2.4.4. Let $(\mathbb{R}, \mathscr{D}, m)$ be the dyadic geometric σ-content prespace. Prove that the σ-integral derived from it is exactly the Lebesgue integral.

3. THE UPPER AND THE LOWER INTEGRAL

In this section we pick up the suggestion made in Remark 2.2.10 and lay the foundations for a quick derivation of the basic results on integrable functions.

3.1. Definition. *Let $(\Omega, \mathscr{E}, m)$ be a σ-measure space and $\underline{m}\colon \mathscr{E}^\sigma \to \mathbb{R} \cup \{\infty\}$, $\overline{m}\colon \mathscr{E}_\sigma \to \mathbb{R} \cup \{-\infty\}$ as in Lemma 1.3 and Proposition 1.8.*

3.1.1. *For any* $h \in \overline{\mathbb{R}}^{\Omega}$, *let*

$$\tilde{m}(h) = \inf_{h \leqq u \in \mathscr{E}_\sigma} \underline{m}(u)$$

(*with the usual convention* $\inf \varnothing = \infty$). *The function* $\tilde{m}: \overline{\mathbb{R}}^{\Omega} \to \overline{\mathbb{R}}$ *thus defined is called the* **upper** $\boldsymbol{m}$**-(**$\boldsymbol{\sigma}$**-)integral** *or the* **upper (**$\boldsymbol{\sigma}$**-)integral derived from** $(\boldsymbol{\Omega}, \mathscr{E}, \boldsymbol{m})$ (*or simply from* m).

3.1.2. *For any* $h \in \overline{\mathbb{R}}^{\Omega}$, *let*

$$\underset{\sim}{m}(h) = \sup_{\mathscr{E}_\sigma \ni v \leqq h} \overline{m}(v)$$

(*with the convention* $\sup \varnothing = -\infty$). *The function* $\underset{\sim}{m}: \overline{\mathbb{R}}^{\Omega} \to \overline{\mathbb{R}}$ *thus defined is called the* **lower** $\boldsymbol{m}$**-(**$\boldsymbol{\sigma}$**-)integral** *or the* **lower (**$\boldsymbol{\sigma}$**-)integral derived from** $(\boldsymbol{\Omega}, \mathscr{E}, \boldsymbol{m})$ (*or simply from* m).

3.2. Remark. As remarked in 2.2.10, $h \in \mathscr{L}^1_\sigma(m)$ is equivalent to $-\infty < \underset{\sim}{m}(h) = \tilde{m}(h) < \infty$, and $\int h\, dm = \underset{\sim}{m}(h) = \tilde{m}(h)$ in this case. The basic properties of the upper and the lower m-σ-integral are collected in

3.3. Proposition. *Let* $(\Omega, \mathscr{E}, m)$ *be a* σ*-measure space and* $\underset{\sim}{m}$, $\tilde{m}$ *the lower and upper* m-σ*-integral. Then:*

3.3.1. m *and* m *are isotone*, i.e.,

$$\underset{\sim}{m}(g) \leqq \underset{\sim}{m}(h), \qquad \tilde{m}(g) \leqq \tilde{m}(h), \qquad (g, h \in \overline{\mathbb{R}}^{\Omega}, \quad g \leqq h).$$

3.3.2. $\underset{\sim}{m}(h) \leqq \tilde{m}(h) \qquad (h \in \overline{\mathbb{R}}^{\Omega}).$

3.3.3. $\tilde{m}(-h) = -\underset{\sim}{m}(h) \qquad (h \in \overline{\mathbb{R}}^{\Omega}).$

3.3.4. $\underset{\sim}{m}(f) = \tilde{m}(f) \qquad (f \in \mathscr{E}).$

3.3.5. $\underset{\sim}{m}$ *and* $\tilde{m}$ *are positive homogeneous*, i.e.,

$$\underset{\sim}{m}(\alpha h) = \alpha \underset{\sim}{m}(h), \qquad \tilde{m}(\alpha h) = \alpha \tilde{m}(h) \qquad (\alpha \in \mathbb{R}_+, \quad h \in \overline{\mathbb{R}}^{\Omega})$$

(*with the usual convention* $0 \cdot \infty = 0$).

3.3.6. $\underset{\sim}{m}$ *is superadditive and* $\tilde{m}$ *is subadditive, i.e., if* $h_1, \ldots, h_n \in \overline{\mathbb{R}}^{\Omega}$ *and if* $h_1 + \cdots + h_n$, $\underset{\sim}{m}(h_1) + \cdots + \underset{\sim}{m}(h_n)$ *and* $\tilde{m}(h_n) + \cdots + \tilde{m}(h_n)$ *make sense (i.e., no case of* $\infty + (-\infty)$ *happens), then*

$$\underset{\sim}{m}(h_1 + \cdots + h_n) \geqq \underset{\sim}{m}(h_1) + \cdots + \underset{\sim}{m}(h_n)$$
$$\tilde{m}(h_1 + \cdots + h_n) \leqq \tilde{m}(h_1) + \cdots + \tilde{m}(h_n).$$

3.3.7. $\tilde{m}$ *is lower* σ*-continuous on nonnegative functions:* $h, h_1, \ldots \in \overline{\mathbb{R}}^{\Omega}_+$, $h_n \nearrow h$ *pointwise* $\Rightarrow \tilde{m}(h_n) \nearrow \tilde{m}(h)$.

3.3.8. *$\underset{\sim}{m}$ is σ-superadditive and $\tilde{m}$ is σ-subadditive on nonnegative functions, i.e.,*

$$h_1, h_2, \ldots \in \overline{\mathbb{R}}_+^{\Omega} \quad \Rightarrow \quad \underset{\sim}{m}(h_1 + h_2 + \cdots) \geqq \underset{\sim}{m}(h_1) + \underset{\sim}{m}(h_2) + \cdots \quad \textit{and}$$
$$\tilde{m}(h_1 + h_2 + \cdots) \leqq \tilde{m}(h_1) + \tilde{m}(h_2) + \cdots.$$

Proof. 1 is obvious: If $g \leqq h$, then

$$\{v \mid v \in \mathscr{E}_\sigma, v \leqq g\} \subseteq \{v \mid v \in \mathscr{E}_\sigma, v \leqq h\}$$

and

$$\underset{\sim}{m}(g) = \sup_{\mathscr{E}_\sigma \ni v \leqq g} \overline{m}(v) \leqq \sup_{\mathscr{E}_\sigma \ni v \leqq h} \overline{m}(v)$$

follows, etc.

2 follows from Proposition 1.8.3.3.

3 follows from Propositions 1.7.2 and 1.8.3.1, and the definition of $\underset{\sim}{m}$, $\tilde{m}$ in an obvious way.

4 follows from Remark 2.2.3.

5 follows from Propositions 1.8.1.1, 1.8.2.1 in an obvious way.

6. Let us be content to prove the first inequality. If $\underset{\sim}{m}(h_1) + \cdots + \underset{\sim}{m}(h_n) = -\infty$, we have nothing to prove. If $m(h_1) + \cdots + m(h_n) > -\infty$, we have $m(h_1), \ldots, m(h_n) > -\infty$. For any choice of finite reals $\alpha_1 < m(h_1), \ldots, \alpha_n < m(h_n)$, we may find $u_1 \leqq h_1, \ldots, u_n \leqq h_n$ in $\mathscr{E}_\sigma$ with $\alpha_1 < \overline{m}(u_1), \ldots, \alpha_n < \overline{m}(u_n)$. Since no $u_1, \ldots, u_n$ attains the value ∞, $u_1 + \cdots + u_n$ makes sense and we get $\overline{m}(u_1 + \cdots + u_n) = \overline{m}(u_1) + \cdots + \overline{m}(u_n)$ by Proposition 1.8.2.1. This sum in turn is $> \alpha_1 + \cdots + \alpha_n$. On the other hand, $h_1 + \cdots + h_n \geqq u_1 + \cdots + u_n \in \mathscr{E}_\sigma$ and we get

$$\underset{\sim}{m}(h_1 + \cdots + h_n) \geqq \overline{m}(u_1 + \cdots + u_n) > \alpha_1 + \cdots + \alpha_n.$$

Varying $\alpha_1, \ldots, \alpha_n$ we get $\underset{\sim}{m}(h_1 + \cdots + h_n) \geqq \underset{\sim}{m}(h_1) + \cdots + \underset{\sim}{m}(h_n)$.

7. Consider $h, h_1, h_2, \ldots \in \overline{\mathbb{R}}_+^{\Omega}$ with $h_1 \leqq h_2 \leqq \cdots \nearrow h$ pointwise. Then $0 \leqq \tilde{m}(h_1) \leqq \tilde{m}(h_2) \leqq \cdots \leqq \tilde{m}(h)$ by 1 and $\lim_{n\to\infty} \tilde{m}(h_n) \leqq \tilde{m}(h)$ follows. If $\lim_{n\to\infty} \tilde{m}(h_n) = \infty$, we are through. Assume now $\lim_{n\to\infty} \tilde{m}(h_n) < \infty$ and choose any $\varepsilon > 0$. For every $n = 1, 2, \ldots$, choose some $u_n' \in \mathscr{E}^\sigma$ with $h_n \leqq u_n'$ and $\underline{m}(u_n') < \tilde{m}(h_n) + \varepsilon/2^n$. Put $u_n = u_1' \vee \cdots \vee u_n'$, $u = \lim_{n\to\infty} u_n$ (pointwise). By Proposition 1.2.2.3 we have $u, u_1, u_2, \ldots \in \mathscr{E}_+^\sigma$, and clearly $u_1 \leqq u_2 \leqq \cdots \nearrow u$ pointwise. We prove $\underline{m}(u_n) \leqq \tilde{m}(h_n) + \varepsilon(1 - 1/2^n)$ by induction. For $n = 1$ this is obviously true. Assume it is true for n. Since $u_{n+1} = u_n \vee u_{n+1}'$, $u_n + u_{n+1}' = (u_n \wedge u_{n+1}') + (u_n \vee u_{n+1}')$, $u_n \wedge u_{n+1}' \geqq h_n$, we find

$$\begin{aligned} \underline{m}(u_n) + \underline{m}(u_{n+1}') &= \underline{m}(u_n + u_{n+1}') \\ &= \underline{m}(u_n \wedge u_{n+1}') + \underline{m}(u_n \vee u_{n+1}') \\ &\geqq \tilde{m}(h_n) + \underline{m}(u_{n+1}) \end{aligned}$$

(we have used Proposition 1.4.3 and the definition of $\tilde{m}$). We now get

$$\begin{aligned}\underline{m}(u_{n+1}) &\leqq (\underline{m}(u_n) - \tilde{m}(h_n)) + \underline{m}(u'_{n+1}) \\ &\leqq \varepsilon\left(1 - \frac{1}{2^n}\right) + \tilde{m}(h_{n+1}) + \frac{\varepsilon}{2^{n+1}} \\ &= \tilde{m}(h_{n+1}) + \varepsilon\left(1 - \frac{1}{2^{n+1}}\right)\end{aligned}$$

as desired. By the lower σ-continuity of $\underline{m}$ (Proposition 1.4.4) we find $\underline{m}(u) = \lim_{n\to\infty} \underline{m}(u_n) \leqq \lim_{n\to\infty} \tilde{m}(h_n) + \varepsilon$. Since

$$h = \lim_{n\to\infty} h_n \leqq \lim_{n\to\infty} u_n = u \in \mathscr{E}^\sigma,$$

this implies $\tilde{m}(h) \leqq \underline{m}(u) \leqq \lim_{n\to\infty} \tilde{m}(h_n) + \varepsilon$. Since $\varepsilon > 0$ was arbitrary, the desired inequality $\tilde{m}(h) \leqq \lim_{n\to\infty} \tilde{m}(h_n)$ follows.

The first part of 8 is an easy consequence of 6 and 7. The second part is obvious in the following way:

$$\begin{aligned}\underset{\sim}{m}(h_1 + h_2 + \cdots) &\geqq \lim_{n\to\infty} \underset{\sim}{m}(h_1 + \cdots + h_n) \geqq \lim_{n\to\infty} [\underset{\sim}{m}(h_1) + \cdots + \underset{\sim}{m}(h_n)] \\ &= \underset{\sim}{m}(h_1) + \underset{\sim}{m}(h_2) + \cdots,\end{aligned}$$

the first $\geqq$ being a consequence of 1 and the second a consequence of 6.

3.4. Exercises

3.4.1. Let $\Omega = \mathbb{R}$, $\mathscr{E} = \{f \mid f \in \mathscr{C}(\mathbb{R}, \mathbb{R}), f(x) = 0\ (x \notin [0, 1])\}$, and $m: \mathscr{E} \to \mathbb{R}$ be the Riemann integral. Let $h_n = 1_{[-1/n,\, 0[}$ $(n = 1, 2, \ldots)$. Show that $\tilde{m}(h_n) = \infty$ $(n = 1, 2, \ldots)$ but $\tilde{m}(\lim h_n) = \tilde{m}(0) = 0$. Thus $\tilde{m}$ is in general not upper σ-continuous on nonnegative functions. Let

$$g_n(x) = \begin{cases} -1/nx & (0 < x < 1) \\ 0 & (x \notin\,]0, 1[). \end{cases}$$

Show that $m(g_n) = -\infty$ $(n = 1, 2, \ldots)$ but $\tilde{m}(\lim g_n) = \tilde{m}(0) = 0$. Thus $\tilde{m}$ is in general not lower σ-continuous on functions that are not nonnegative. Show that $\tilde{m}(-g_n) = \infty$ $(n = 1, 2, \ldots)$.

3.4.2. Let $(\mathbb{R}, \mathscr{D}, m)$ be the dyadic geometric σ-content prespace on $\mathbb{R}$ (Examples I.1.10, I.2.4) and $(\mathbb{R}, \mathscr{E}, m)$ the σ-measure space derived from it (Example I.8.7, Proposition I.8.9). Prove that $\mathscr{E}^\sigma = \mathscr{C}^{00}(\mathbb{R}, \mathbb{R})^\sigma$ and that the $\underline{m}$: $\mathscr{E}^\sigma \to \mathbb{R}$ defined on it on the basis of $(\mathbb{R}, \mathscr{E}, m)$ is the same as the $\underline{m}$ defined on $\mathscr{C}^{00}(\mathbb{R}, \mathbb{R})^\sigma$ on the basis of the ordinary Riemann integral on $\mathscr{C}^{00}(\mathbb{R}, \mathbb{R})$. (*Hint*: Majorize functions in $\mathscr{E}$ by functions in $\mathscr{C}^{00}(\mathbb{R}, \mathbb{R})$ suitably, and vice versa.)

4. NULLFUNCTIONS AND NULLSETS

In this subsection we introduce the "**almost everywhere" terminology** which is so fundamental in measure theory.

4.1. Definition. *Let* $(\Omega, \mathscr{E}, m)$ *be a* σ*-measure space and* m *the upper* σ*-integral derived from* m.

4.1.1. *A function* $h \in \overline{\mathbb{R}}^{\Omega}$ *is called an* **m-(σ-)nullfunction** *if* $\tilde{m}(|h|) = 0$.

4.1.2. *A set* $N \subseteq \Omega$ *is called an* **m-(σ-)nullset** *if its indicator function* 1_N *is an m-σ-nullfunction, i.e., if* $\tilde{m}(1_N) = 0$. *The system of all m-σ-nullsets is denoted by* $\mathscr{N}_{m,\sigma}$ *or simply by* $\mathscr{N}_m$ *if "σ" is clear from the context.*

4.2. Proposition. *Let* $(\Omega, \mathscr{E}, m)$ *be a* σ*-measure space and* $\tilde{m}$ *the upper* σ*-integral derived from it.*

4.2.1. *If* h *is an m-σ-nullfunction and if* $g \in \overline{\mathbb{R}}^{\Omega}$, $|g| \leqq |h|$, *then* g *is an m-σ-nullfunction.*

4.2.2. *Every subset of an m-σ-nullset is an m-σ-nullset.*

4.2.3. *If* $h_1, h_2, \ldots$ *are m-σ-nullfunctions and* $h_1 + h_2 + \cdots$ *makes sense, then* $h_1 + h_2 + \cdots$ *is also an m-σ-nullfunction. In any case* $h_1 \vee h_2 \vee \cdots$ *is an m-σ-nullfunction.*

4.2.4. *Every countable union of m-σ-nullsets is an m-σ-nullset.*

4.2.5. *If* h *is an m-σ-nullfunction, then for every* $\alpha \in \overline{\mathbb{R}}$ αh *is an m-σ-nullfunction.*

4.2.6. $h \in \overline{\mathbb{R}}^{\Omega}$ *is an m-σ-nullfunction iff* $\operatorname{supp}(h) = \{\omega | h(\omega) \neq 0\}$ *is an m-σ-nullset.*

4.2.7. *If* $h \in \overline{\mathbb{R}}^{\Omega}$ *satisfies* $\tilde{m}(h) < \infty$, *then* $\{\omega | h(\omega) = \infty\}$ *is an m-σ-nullset, and if* $\underset{\sim}{m}(h) > -\infty$, *then* $\{\omega | h(\omega) = -\infty\}$ *is an m-σ-nullset.*

Proof. 1. Proposition 3.3.1 implies $0 \leqq \tilde{m}(|g|) \leqq \tilde{m}(|h|) = 0$, which is sufficient.

2 follows from 1 since $M \subseteq N \subseteq \Omega$ is tantamount to $1_M \leqq 1_N$.

3. Propositions 3.3.1, 3.3.7, and 3.3.8 imply

$$0 \leqq \tilde{m}(|h_1 + h_2 + \cdots|) \leqq \tilde{m}(|h_1| + |h_2| + \cdots) \leqq \tilde{m}(|h_1|) + \tilde{m}(|h_2|) + \cdots = 0$$

and

$$0 \leqq m(|h_1 \vee h_2 \vee \cdots|) \leqq m(|h_1| + |h_2| + \cdots) = 0.$$

4 follows from 3 since $1_{N_1 \cup N_2 \cup \cdots} \leqq 1_{N_1} + 1_{N_2} + \cdots$ $(N_1, N_2, \ldots \subseteq \Omega)$.

5. For $\alpha \in \mathbb{R}$, we have $m(|\alpha h|) = m(|\alpha| \cdot |h|) = |\alpha| \cdot m(|h|) = 0$ by Proposition 3.3.5. For $\alpha = \pm\infty$, we have $0 \leqq \tilde{m}(|\alpha h|) = \tilde{m}(\infty |h|) = \tilde{m}(|h| + |h| + \cdots) \leqq \tilde{m}(|h|) + \tilde{m}(|h|) + \cdots = 0$.

6. If h is an m-σ-nullfunction, then $0 \leqq \tilde{m}(1_{\text{supp}(h)}) \leqq \tilde{m}(\infty|h|) = 0$ by 5. If $\text{supp}(h)$ is an m-σ-nullset, then $0 \leqq \tilde{m}(|h|) \leqq \tilde{m}(\infty 1_{\text{supp}(h)}) = 0$ by 5.

7. If $\tilde{m}(h) < \infty$, then there is an $h \leqq u \in \mathscr{E}^\sigma$ with $\tilde{m}(u) = \underline{m}(u) < \infty$. We may put $u = f + v$ with $f \in \mathscr{E}$, $v \in \mathscr{E}_+^\sigma$. Clearly $\{h = \infty\} \subseteq \{v = \infty\}$ and $\underline{m}(v) = \underline{m}(u) - m(f) < \infty$. For every $\alpha \in \mathbb{R}_+$, we have $\alpha 1_{\{v=\infty\}} \leqq v$, hence $0 \leqq \alpha\tilde{m}(1_{\{v=\infty\}}) = \tilde{m}(\alpha 1_{\{v=\infty\}}) \leqq \tilde{m}(v) < \infty$. Varying α, we find that $\{v = \infty\}$ is an m-σ-nullset, hence so is $\{h = \infty\}$. The rest of 7 now follows if we replace h by $-h$.

4.3. Exercises

4.3.1. Let $\Omega \neq \varnothing$ be arbitrary, $\mathscr{E} = \{f \mid f \in \mathbb{R}^\Omega, |\text{supp}(f)| < \infty\}$ and $m(f) = \sum_{f(\omega)\neq 0} f(\omega)$. Prove that $(\Omega, \mathscr{E}, m)$ is a τ-, hence a σ-measure space and that $\varnothing$ is the only m-σ-nullset.

4.3.2. Let $(\Omega, \mathscr{C}^{00}(\mathbb{R}, \mathbb{R}), m)$ be as in Example I.8.5 with $m =$ Riemann integral. Prove that every countable subset of $\mathbb{R}$ is an m-σ-nullset.

4.3.3. Let $(\Omega, \mathscr{R}, m)$ be a σ-content prespace and $(\Omega, \mathscr{E}(\mathscr{R}), m)$ the σ-measure space derived from it. Prove that $N \subseteq \Omega$ is a subset of a countable union of m-nullsets according to Definition II.6.1 (for $(\Omega, \mathscr{R}, m)$) iff it is an m-nullset according to Definition 4.1.2 (for $\Omega, \mathscr{E}(\mathscr{R}), m)$).

4.4. Definition. *Let $(\Omega, \mathscr{E}, m)$ be a σ-measure space.*

4.4.1. *A property P of points $\omega \in \Omega$ is said* **to hold m-(σ-)-almost everywhere** *(m-(σ-)-a.e., mod m, or the like) if the so-called* **exceptional set** *$\{\omega \mid \omega$ does not have the property $P\}$ for P is an m-(σ-)nullset. In particular:*

4.4.2. *Two mappings g, h of Ω into some set are said to coincide m-(σ-) a.e. or mod m if the exceptional set $\{g \neq h\} = \{\omega \mid g(\omega) \neq h(\omega)\}$ is an m-(σ-)nullset. In particular it is clear what equality m-(σ-) a.e. or mod m of functions from $\overline{\mathbb{R}}^\Omega$ means.*

4.4.3. *Two subsets E, F of Ω are said to be m-almost equal or equal mod m if their indicator functions 1_E, 1_F are equal mod m, i.e., if the symmetric difference $E \,\Delta\, F = (E \cup F)\backslash(E \cap F)$ is an m-nullset.*

4.4.4. *If $(\Omega, \mathscr{R}, m)$ is a σ-content prespace and $(\Omega, \mathscr{E}, m)$ is the σ-measure space derived from it, then we adapt the above language in terms of the σ-content m as well.*

4.5. Exercises

4.5.1. Let $(\Omega, \mathscr{E}, m)$ be as in Exercise 4.3.1. Show that equality mod m is strict equality here.

4.5.2. Let $(\Omega, \mathscr{E}, m)$ be as in Exercise 4.3.2. Show that the indicator function of all rationals vanishes m-a.e.

4.6. Proposition. *Let $(\Omega, \mathscr{E}, m)$ be a σ-measure space. Then equality* mod m *of mappings of* Ω *resp. subsets of* Ω *is an equivalence relation. We shall speak of* **m-(σ-)equivalence classes** *or* **equivalence classes of mod** ***m*****.**

Proof. Let us be content to settle the case of mappings and to prove transitivity. But if $f = g \bmod m$, $g = h$ m-a.e., then

$$\{f \neq h\} \subseteq \{f \neq g\} \cup \{g \neq h\}$$

which is an m-σ-nullset again, by Proposition 4.2.4.

Our next definition contains some intentionally not fully precise passages in order to avoid clumsiness of language.

4.7. Definition. *Let* $(\Omega, \mathscr{E}, m)$ *be a* σ*-measure space.*

4.7.1. *Equivalence classes* mod m *of mappings of* Ω *(in particular, functions defined on* Ω*) are also called* **m-(σ-) a.e. defined mappings** *of (functions on)* Ω.

4.7.2. *If* $\mathscr{C}$ *is a system of mappings resp. subsets of* Ω, *then every intersection of an equivalence class* mod m *with* $\mathscr{C}$ *is called an equivalence class* (mod m) *(with)in* $\mathscr{C}$,

4.7.3. *A property* P *of members of a system* $\mathscr{C}$ *of mappings or subsets of* Ω *is called a* **class property** (mod m) *if the system* $\mathscr{C}(P)$ *of all members of* $\mathscr{C}$ *having the property* P *consists of (possibly none) full equivalence classes in* $\mathscr{C}$, *i.e., if* P *holds for a member of* $\mathscr{C}$ *iff it holds for all members of* $\mathscr{C}$ *that are equivalent* (mod m) *to it.*

4.7.4. *A relation between members of a system* $\mathscr{C}$ *of mappings or subsets of* Ω *is called a* **class relation** (mod m) *if passage to equivalent members of* $\mathscr{C}$ *does not lead to a change in its satisfaction.*

4.7.5. *An operation defined for members of a system* $\mathscr{C}$ *of mappings or subsets of* $\mathscr{C}$ *and having results in a set that is also endowed with an equivalence relation is called a* **class operation** (mod m) *if passage to equivalent* (mod m) *members of* $\mathscr{C}$ *yields a passage to an equivalent result.*

4.7.6. *A function defined on a system* $\mathscr{C}$ *of mappings or subset of* Ω *is called a* **class function** *if it is constant on every equivalence class in* $\mathscr{C}$.

4.8. Remarks

4.8.1. Let $(\Omega, \mathscr{E}, m)$ be a σ-measure space and $\mathscr{C}$ a system of mappings of subsets of Ω, then every class property (relation operation, function) in $\mathscr{C}$ immediately gives rise to a property (relation, operation, function) of resp. on full equivalence classes in $\mathscr{C}$, simply by going back to class representatives. In this context we make the

4.8.2. **Convention:** *When dealing with properties (relations, operations, functions) of resp. on equivalence classes which stem from properties etc. of members of* $\mathscr{C}$ *according to* 4.7.1, *we maintain the notations established for the* *latter, possibly with an explanatory* "m-(σ-) a.e." *or* "mod m," *etc. added. We often omit the specification σ or even m if it is clear from the context. Thus it should, e.g., be clear what expressions like*

$$g \leqq h \qquad \textit{for} \quad m\text{-a.e. } \textit{defined functions } f, g \textit{ on } \Omega$$

$$E \cup F \bmod m \qquad (E, F \subseteq \Omega, \quad \bmod m)$$

mean.

Our next proposition again contains some formulations that are at times somewhat imprecise in order to avoid clumsiness of language. The details will become clear in the proof.

4.9. Proposition. *Let* $(\Omega, \mathscr{E}, m)$ *be a* σ*-measure space.*

4.9.1. *The following relations in* $\overline{\mathbb{R}}^\Omega$ *are class relations, operations etc.* mod m:

4.9.1.1. *finite or countable order relations and operations* mod m:

$$\begin{array}{ll} g \leqq h & m\text{-}(\sigma\text{-}) \text{ a.e.} \\ h_1 \leqq h_2 \leqq \cdots & m\text{-}(\sigma\text{-}) \text{ a.e.} \\ h_1 \vee h_2 \vee \cdots & \\ h_1 \wedge h_2 \wedge \cdots; & \end{array}$$

4.9.1.2. *convergence* m-(σ-) a.e. *for sequences, i.e.,*

$$\limsup_{n\to\infty} h_n$$

$$\liminf_{n\to\infty} h_n$$

$$\textit{existence of } \lim_{n\to\infty} h_n \ m\text{-}(\sigma\text{-}) \text{ a.e.}$$

$$\lim_{n\to\infty} h_n;$$

4.9.1.3. *addition, product, multiplication with constants, as far they are meaningful:*

$$g + h, \qquad g \cdot h, \qquad \alpha h;$$

4.9.1.4. *the upper and the lower* m-σ*-integral* $\tilde{m}$, $\underset{\sim}{m}$.

4.9.2. *The following relations, properties, etc.,* mod m *among subsets of* Ω *are class relations etc.* mod m:

4.9.2.1. *finite and countable set order relations:*

$$E \subseteq F \qquad \text{mod } m$$
$$E_1 \subseteq E_2 \subseteq \cdots \qquad \text{mod } m$$
$$E_1 \supseteq E_2 \supseteq \cdots \qquad \text{mod } m.$$

4.9.2.2. *finite and countable set operations:*

$$E \cup F, E \cap F$$
$$E_1 \cup E_2 \cup \cdots$$
$$E_1 \cap E_2 \cap \cdots$$
$$E \backslash F.$$

Proof. 1.1. We have, e.g., to prove the following: If $h_1, h_2, \ldots$ are in $\overline{\mathbb{R}}^\Omega$, with pointwise supremum $h = h_1 \vee h_2 \vee \cdots$, and if $h_1', h_2', \ldots \in \overline{\mathbb{R}}^\Omega$ with pointwise supremum $h' = h_1' \vee h_2' \vee \cdots$ are such that $\{h_1' \neq h_1\}, \{h_2' \neq h_2\}, \ldots$ are m-σ-nullsets, then $\{h' \neq h\}$ is an m-σ-nullset. But clearly

$$\{h' \neq h\} \subseteq \{h_1' \neq h_1\} \cup \{h_2' \neq h_2\} \cup \cdots$$

and thus is an m-σ-nullset by Propositions 4.2.2, 4.2.4. The exercise of making the other statements of 1.1 precise and proving them is now so easy that it can be left to the reader.

1.2 follows from 1.1 since everything here can be expressed in terms of countable suprema and infima, and equality mod m of them, e.g., $\lim_n \sup h_n = g_1 \wedge g_2 \wedge \cdots$ where $g_1 = h_1 \vee h_2 \vee \cdots$, $g_2 = h_2 \vee h_3 \vee \cdots$, and $\lim_{n \to \infty} h_n$ exists m-σ-a.e. iff

$$\limsup_{n \to \infty} h_n = \liminf_{n \to \infty} h_n \qquad \text{holds} \qquad m\text{-}\sigma\text{-a.e.}$$

1.3. If $g, g', h, h' \in \overline{\mathbb{R}}^\Omega$ are such that $g + h$ and $g' + h'$ make sense and $g = g'$ mod m, $h = h'$ mod m, then

$$\{g' + h' \neq g + h\} \subseteq \{g' \neq g\} \cup \{h' \neq h\} \supseteq \{g'h' \neq gh\},$$

and all these sets are m-σ-nullsets by Propositions 4.2.2, 4.2.4. Multiplication with constants is a special case of function multiplication.

1.4. Let $h, h' \in \overline{\mathbb{R}}^\Omega$ and $N = \{h' \neq h\}$ an m-σ-nullset. We have to prove $\underset{\sim}{m}(h') = \underset{\sim}{m}(h)$, $\tilde{m}(h') = \tilde{m}(h)$; and clearly (pass to $-h'$, $-h$) it is sufficient to prove the latter equality. Now, if h has no majorant in $\mathscr{E}^\sigma$, then h' also has none because if $h' \leqq u' \in \mathscr{E}^\sigma$, then we choose some $\infty 1_N \leqq v \in \mathscr{E}_+^\sigma$ (according to Proposition 4.2.5) and get $h \leqq u + v \in \mathscr{E}^\sigma$, a contradiction. Thus we get $\tilde{m}(h') = \tilde{m}(h)$ in this case. If h has a majorant u in $\mathscr{E}$, we choose any majorant v of $\infty 1_N$ in $\mathscr{E}_+$ and get $h' \leqq u + v$, hence

$$\tilde{m}(h') \leqq \tilde{m}(u) + \tilde{m}(v) = \underline{m}(u) + \underline{m}(v).$$

Varying u, v we get $\inf \underline{m}(u) = \tilde{m}(h)$, $\inf \underline{m}(v) = 0$ and thus $m(h') \leqq m(h)$. By symmetry, equality follows: $\tilde{m}$ is a class function.

2 is an obvious consequence of 1 via indicator functions (e.g., $1_{E_1 \cup E_2 \cup \cdots} = 1_{E_1} \vee 1_{E_2} \vee \cdots$). The details are left to the reader.

4.10. Exercises

4.10.1. Let $(\Omega, \mathscr{E}, m)$ be a σ-measure space and $I = \{\iota, \kappa, \ldots\}$ be a countable set endowed with an order relation $\leqq$ such that I is increasingly filtered: For $\iota, \kappa \in I$, there is a $\lambda \in I$ with $\iota, \kappa \leqq \lambda$. Let $(f_\iota)_{\iota \in I}$ be a family of functions from $\overline{\mathbb{R}}^\Omega$ and define

$$\limsup_{\iota \in I} f_\iota = \inf_{\kappa \in I} \left[\sup_{\iota \geqq \kappa} f_\iota\right], \qquad \liminf_{\iota \in I} = \sup_{\kappa \in I} \left[\inf_{\iota \geqq \kappa} f_\iota\right].$$

Prove that $(f_\iota)_{\iota \in I} \to \limsup_{\iota \in I} f_\iota$, $\liminf_{\iota \in I} f_\iota$ are class operations and the existence m-σ-a.e. of $\lim_{\iota \in I} f$ (defined by $\liminf_{\iota \in I} f = \limsup_{\iota \in I} f$ m-σ-a.e.) is a class property.

4.10.2. Let $(\Omega, \mathscr{E}, m)$ be a σ-measure space and let m-σ-a.e. $0 \leqq h_1 \leqq h_2 \leqq \cdots \in \overline{\mathbb{R}}^\Omega$. Prove that $\lim_{n \to \infty} h_n < \infty$ m-σ-a.e. is a class property.

5. BASIC THEOREMS FOR THE INTEGRAL

Let us, for any σ-measure space $(\Omega, \mathscr{E}, m)$, recall definition 2.1 of m-(σ-)integrable functions and its streamlined form 3.2 in terms of the upper and lower m-σ-integrals $\tilde{m}$, $\underset{\sim}{m}$: $h \in \overline{\mathbb{R}}^\Omega$ belongs to the system $\mathscr{L}^1_\sigma(m)$ of all m-σ-integrable functions iff $-\infty < \underset{\sim}{m}(h) = \tilde{m}(h) < \infty$, and we have $\int h\, dm = \underset{\sim}{m}(h) = \tilde{m}(h) = m(h)$ in this case.

In this section we harvest the results of our lengthy preparations and prove the *classical convergence theorems* for the integral along with some other results that round off the picture in the sense of Banach lattice theory, whose terminology we introduce in passing.

5.1. Theorem. *Let $(\Omega, \mathscr{E}, m)$ be a σ-measure space. Then:*

5.1.1. *$\mathscr{L}^1_\sigma(m)$ consists of full equivalence classes* mod m *in $\overline{\mathbb{R}}^\Omega$. The system of all m-σ-equivalence classes in $\mathscr{L}^1_\sigma(m)$ is also denoted by $\mathbf{L}^1_\sigma(m)$, $\mathbf{L}^1_m$, $\mathbf{L}^1_\sigma(\Omega, \mathscr{E}, m)$, etc., depending on what specifications are needed. Every equivalence class $\in \mathbf{L}^1_\sigma(m)$ intersects $\mathbb{R}^\Omega$ in exactly one nonempty equivalence class in $\mathbb{R}^\Omega$; i.e., every m-σ-integrable function is m-σ-a.e. finite.*

5.1.2. *$\mathscr{L}^1_\sigma(m) \cap \mathbb{R}^\Omega$ is an elementary domain, i.e., a vector lattice of real-valued functions (no value $\pm\infty$) satisfying Stone's axiom. Working with class representatives in $\mathscr{L}^1_\sigma(m) \cap \mathbb{R}^\Omega$, we may, and shall, consider $\mathrm{L}^1_\sigma(m)$ as a vector lattice satisfying Stone's axiom in an obvious way. As a matter of fact*

$$h \in \mathscr{L}^1_\sigma(m),\ 0 \leqq \alpha \in \mathbb{R} \quad \Rightarrow \quad h \wedge \alpha \in \mathscr{L}^1_\sigma(m)$$

and every $0 \leqq h \in \mathrm{L}^1_\sigma(m)$ contains a $0 \leqq h \in \mathscr{L}^1_\sigma(m)$.

5.1.3. *$m\colon \mathscr{L}^1_\sigma(m) \to \mathbb{R}$ is a class function and may hence also be considered as a function $m\colon \mathrm{L}^1_\sigma(m) \to \mathbb{R}$. The restriction of $m\colon \mathscr{L}^1_\sigma(m) \to \mathbb{R}$ to $\mathscr{L}^1_\sigma(m) \cap \mathbb{R}^\Omega$ is a positive linear form. Hence $m\colon \mathrm{L}^1_\sigma(m) \to \mathbb{R}$ is a positive linear form.*

5.1.4. *Every $h \in \mathscr{L}^1_\sigma(m)$ vanishes outside some countable union of supports of functions from $\mathscr{E}$.*

5.1.5. *If g, $h \in \mathscr{L}^1_\sigma(m)$, $g \leqq h$ everywhere, and $m(g) = m(h)$, then $g = h$* m-σ-a.e.

5.1.6. *$h \in \overline{\mathbb{R}}^\Omega$ belongs to $\mathscr{L}^1_\sigma(m)$ iff h_+ and h_- belong to $\mathscr{L}^1_\sigma(m)$.*

Proof. 1. By Proposition 4.9.1.4, $\underset{\sim}{m}$ and $\tilde{m}$ are class functions. Since $\mathscr{L}^1_\sigma(m)$ is defined by the behavior of $\underset{\sim}{m}$ and $\tilde{m}$, it consists of full equivalence classes in $\overline{\mathbb{R}}^\Omega$. If $h \in \mathscr{L}^1_\sigma(m)$, then $\tilde{m}(h) < \infty$, hence $\{h = \infty\}$ is an m-σ-nullset, by Proposition 4.2.7. By the same token, $\{h = -\infty\}$ is an m-σ-nullset, and thus h is m-σ-a.e. finite. Put $N = \{|h| = \infty\}$ and $h' = (1 - 1_N)h$. Then $h' = h$ m-σ-a.e. and $h' \in \mathscr{L}^1_\sigma(m) \cap \mathbb{R}^\Omega$.

2 and 3. Let $g, h \in \mathscr{L}^1_\sigma(m) \cap \mathbb{R}^\Omega$. Then

$$\begin{aligned} -\infty < \underset{\sim}{m}(g) + \underset{\sim}{m}(h) \leqq \underset{\sim}{m}(g+h) \leqq \tilde{m}(g+h) &\leqq \tilde{m}(g) + \tilde{m}(h) \\ &= \underset{\sim}{m}(g) + \underset{\sim}{m}(h) < \infty \end{aligned}$$

by Propositions 3.3.6, 3.3.2.

We conclude $-\infty < \underset{\sim}{m}(g+h) = \tilde{m}(g+h) = \tilde{m}(g) + \tilde{m}(h) < \infty$, and thus $g + h \in \mathscr{L}^1_\sigma(m)$, $m(g+h) = m(g) + m(h)$. The proof for $\alpha h \in \mathscr{L}^1_\sigma(m) \cap \mathbb{R}^\Omega$, $m(\alpha h) = \alpha m(h)$ $(\alpha \in \mathbb{R},\ h \in \mathscr{L}^1_\sigma(m) \cap \mathbb{R}^\Omega)$ is still more obvious. Thus $m\colon \mathscr{L}^1_\sigma(m) \cap \mathbb{R}^\Omega \to \mathbb{R}$ is a linear form on a linear subspace of $\mathbb{R}^\Omega$. But this subspace is even a sublattice. Clearly it suffices to prove $h \in \mathscr{L}^1_\sigma(m) \cap \mathbb{R}^\Omega \Rightarrow h_+ = h \vee 0 \in \mathscr{L}^1_\sigma(m) \cap \mathbb{R}^\Omega$. For this we choose some $\varepsilon > 0$ and find $u \in \mathscr{E}_\sigma$, $v \in \mathscr{E}^\sigma$ such that $u \leqq h \leqq v$, $-\infty < \underset{\sim}{m}(u) \leqq \tilde{m}(v) < \infty$, $\tilde{m}(v) - \underset{\sim}{m}(u) < \varepsilon$. By Propositions 1.7 and 1.8 we may conclude $v - u = v + (-u) \in \mathscr{E}^\sigma_+$ and $\underline{m}(v - u) = \underline{m}(v + (-u)) = \underline{m}(v) + \underline{m}(-u) = \underline{m}(v) - \overline{m}(u) < \varepsilon$. Now

$$\mathscr{E}_\sigma \ni u_+ \leqq h_+ \leqq v_+ \in \mathscr{E}^\sigma, \qquad 0 \leqq v_+ - u_+ \leqq v - u$$

and thus

$$\underline{m}(v_+) - \overline{m}(v_-) = \underline{m}(v_+) + \underline{m}(-v_-) = \underline{m}(v_+ - v_-) \leqq \underline{m}(v - u) < \varepsilon.$$

Thus $h_+ \in \mathscr{L}^1_\sigma(m) \cap \mathbb{R}^\Omega$. The proof of Stone's axiom is even more obvious. It is obvious that $m(h) \geqq 0$ if $0 \leqq h \in \mathscr{L}^1_\sigma(m) \cap \mathbb{R}^\Omega$.

4 follows from the fact that every $h \in \mathscr{L}^1_\sigma(m)$ has a minorant that is a countable infimum of functions from $\mathscr{E}$, and a majorant that is a countable supremum of functions from $\mathscr{E}$.

5. We first go back to finite-valued functions: Let

$$N = \{|g| = \infty\} \cup \{h = \infty\}.$$

By 1 this is an m-σ-nullset. Put $g' = (1 - 1_N)g$, $h' = (1 - 1_N)h$. Then $g', h' \in \mathscr{L}^1_\sigma(m) \cap \mathbb{R}^\Omega$, $g' = g \bmod m$, $h' = h \bmod m$, $g' \leqq h'$ everywhere and $m(g') = m(g) = m(h) = m(h')$. We conclude $m(|h' - g'|) = m(h' - g') = m(h') - m(g') = 0$, hence $h' - g'$ is an m-σ-nullfunction, i.e., $h' = g'$ m-σ-a.e. and thus finally $g = h$ m-σ-a.e.

6 is an easy consequence of 2.

5.2. Remark. A careful reader will observe that we have, with the above theorem, started working with a *vector lattice* L^1_m, which is not simply a vector sublattice of $\mathbb{R}^\Omega$ and could, therefore, under the requirement of full mathematical rigor, be treated only in the framework of an abstract theory of vector lattices such as will be presented in Chapter IX, Section 1. We think however that it is not advisable to exaggerate formalism at the present stage and that the relation of L^1_m to $\mathscr{L}^1_m \cap \mathbb{R}^\Omega$ is tight enough to avoid any mistakes. On the contrary, our present medial status of abstraction should provide a good background for the abstract theory given in Section 1 of Chapter IX.

5.3. Theorem (monotone convergence). *Let $(\Omega, \mathscr{E}, m)$ be a σ-measure space. Then the following hold:*

5.3.1. **Monotone convergence theorem, sequential form:**

$$(L^1_\sigma(m) =)\ L^1_m \ni h_1 \leqq h_2 \leqq \cdots, \qquad \lim_{n\to\infty} m(h_n) < \infty$$

$\Rightarrow$ *the m-(σ-)* a.e. *defined function* $h = \lim h_n$ *belongs to L^1_m again and*

$$\lim_{n\to\infty} m(h_n) = m(h).$$

5.3.2. **Monotone convergence theorem, serial form:** $0 \leqq g_2, g_3, \ldots \in L^1_m \ni g_1$, $\sum_{k=1}^\infty m(g_k) < \infty \Rightarrow$ *the m-a.e. defined function $g_1 + g_2 + g_3 + \cdots$ belongs to L^1_m again and*

$$m(g_1 + g_2 + \cdots) = m(g_1) + m(g_2) + \cdots.$$

Proof. 1 and 2 are equivalent via $g_1 = h_1$, $g_2 = h_2 - h_1$, $\ldots$, resp.

$h_1 = g_1, h_2 = g_1 + g_2, \ldots$. It is therefore sufficient to prove 2. But here we see, if $g_2', g_3', \ldots \in \mathbb{R}_+^\Omega$ are representatives of the m-σ-equivalence classes $g_2, g_3, \ldots \geqq 0$, that

$$0 \leqq \underset{\sim}{m}(g_2') + \underset{\sim}{m}(g_3') + \cdots \leqq \underset{\sim}{m}(g_2' + g_3' + \cdots)$$
$$\leqq \tilde{m}(g_2' + g_3' + \cdots) \leqq \tilde{m}(g_2') + \tilde{m}(g_3') + \cdots$$
$$= \underset{\sim}{m}(g_2') + \underset{\sim}{m}(g_3') + \cdots$$

and hence $g_2' + g_3' + \cdots \in \mathscr{L}_\sigma^1(m)$ with

$$m(g_2' + g_3' + \cdots) = m(g_2') + m(g_3') + \cdots,$$

and hence $g_2 + g_3 + \cdots \in L_\sigma^1(m)$ with

$$m(g_2 + g_3 + \cdots) = m(g_2) + m(g_3) + \cdots.$$

Since $m: L_\sigma^1(m) \to \mathbb{R}$ is linear, we get

$$m(g_1 + g_2 + \cdots) = m(g_1) + m(g_2) + \cdots$$

as desired.

5.4. Remarks

5.4.1. It is clear that the above theorem offers a solution to the extension problem I.11.1 for measures in the case of σ-measures.

5.4.2. We have formulated the above theorem in terms of L_m^1 $(= L_\sigma^1(m))$, i.e., of equivalence classes mod m, but had to go back to representatives in $\mathscr{L}_m^1$ for the proof. We could, of course, have formulated the theorem in terms of $\mathscr{L}_m^1$ right away. Throughout this book, it will be used in both forms.

5.4.3. Multiplying by -1, we get a "*downward*" *analogue* of the above theorem, of which we shall make free use throughout the rest of this book.

5.4.4. It is obvious how to formulate and prove a countable "*filter*" *version* of the above theorem: If $\mathscr{F} \subseteq L_m^1$ is countable and increasingly filtered with $\sup_{g \in \mathscr{F}} m(g) < \infty$, then

$$h = \sup_{g \in \mathscr{F}} g \in L_m^1 \qquad \text{and} \qquad m(h) = \sup_{g \in \mathscr{F}} m(g).$$

we shall employ this result as well as its "downward" analogue throughout the rest of this book.

5.5. Theorem (Fatou's lemma). *Let $(\Omega, \mathscr{E}, m)$ be a σ-measure space, $g, h_1, h_2, \ldots \in L_m^1$ $(= L_\sigma^1(m))$, $h_1, h_2, \ldots \leqq g$ and $\limsup_{n\to\infty} m(h_n) > -\infty$. Then the m-a.e. defined function $h = \limsup_{n\to\infty} h_n$ belongs to L_m^1 again and*

$$m(h) = m\left(\limsup_{n\to\infty} h_n\right) \geqq \limsup_{n\to\infty} m(h_n).$$

Proof. Let $g_n = h_n \vee h_{n+1} \vee \cdots = \lim_{k\to\infty} [h_n \vee \cdots \vee h_{n+k}]$ $(n = 1, 2, \ldots)$. Since for every $n = 1, 2, \ldots$ the sequence $h_n \vee \cdots \vee h_{n+k}$ $(k = 1, 2, \ldots)$ is in L_m^1 and majorized by g, we find $h_n \leqq g_n \leqq g$, $g_n \in L_m^1$ $(n = 1, 2, \ldots)$. Clearly $g_1 \geqq g_2 \geqq \cdots$ with $\lim_{n\to\infty} g_n = h$, hence

$$\lim_{n\to\infty} m(g_n) \geqq \limsup_{n\to\infty} m(h_n) > -\infty,$$

and by the "downward" version of Theorem 5.3,

$$m(h) = \lim_{n\to\infty} m(g_n) \geqq \limsup_{n\to\infty} m(h_n).$$

5.6. Remark. It is clear how to formulate and prove a "*downward*" *version* of Fatou's lemma. It is often used with the (not too) special assumption $g = 0$. We shall make free use of such variants throughout the rest of this book.

5.7. Theorem (of Lebesgue, on dominated convergence). *Let* $(\Omega, \mathscr{E}, m)$ *be a* σ*-measure space. Then the following hold:*

5.7.1. **Lebesgue's theorem, sequential form:** *Let* $g, h_1, h_2, \ldots \in L_m^1$ $(= L_\sigma^1(m))$ *and let*

$$|h_1|, |h_2|, \cdots \leqq |g|.$$

Assume that $\lim_{n\to\infty} h_n$ *exists* m-a.e. *and call it* h. *Then* $h \in L_m^1$ *and*

$$m(h) = \lim_n m(h_n).$$

5.7.2. **Lebesgue's theorem, serial form:** *Let* $g, f_1, f_2, \ldots \in L_m^1$ *and let* $|f_1| + |f_2| + \cdots \leqq |g|$. *Then the series* $f_1 + f_2 + \cdots$ *converges to a finite limit* f m-a.e., $f \in L_m^1$ *and*

$$m(f) = m(f_1) + m(f_2) + \cdots.$$

Proof. It is clearly sufficient to prove the sequential form. Applying Fatou's lemma 5.5 "up and downward," we find

$$\liminf_{n\to\infty} h_n = \lim_{n\to\infty} h_n = \limsup_{n\to\infty} h_n \in L_m^1$$

and

$$\begin{aligned} -\infty < m(-g) &\leqq \limsup_{n\to\infty} m(h_n) \leqq m\left(\limsup_{n\to\infty} h_n\right) \\ &= m(h) = m\left(\liminf_{n\to\infty} h_n\right) \leqq \liminf_{n\to\infty} m(h_n) \\ &\leqq m(g) < \infty. \end{aligned}$$

Since $\liminf_n m(h_n) \leqq \limsup_n m(h_n)$, we get all the information needed for the rest of the proof.

5.8. Theorem. *Let $(\Omega, \mathscr{E}, m)$ be a σ-measure space.*

5.8.1. *By*

$$\|h\| = \|h\|_1 = m(|h|) \qquad (h \in \mathscr{L}_m^1 = \mathscr{L}_\sigma^1(m))$$

a class function $\|\cdot\| = \|\cdot\|_1$ with values in $\mathbb{R}_+$ is defined on $\mathscr{L}_m^1$. It can therefore also be considered as a function on L_m^1. As such, it is a norm, i.e., it is:

5.8.1.1. *positive definite:*

$$\|h\| \geqq 0 \qquad (h \in L_m^1)$$

with

$$\|h\| = 0 \quad \text{iff} \quad h = 0 \in L_m^1;$$

5.8.1.2. *absolute homogeneous:*

$$\|\alpha h\| = |\alpha| \cdot \|h\| \qquad (\alpha \in \mathbb{R}, \quad h \in L_m^1);$$

and it satisfies:

5.8.1.3. *the triangle inequality:*

$$\|g + h\| \leqq \|g\| + \|h\| \qquad (g, h \in L_m^1).$$

This norm is also called the 1-norm or L^1-norm for m.

5.8.2. *With the norm $\|\cdot\|$, L_m^1 is a Banach lattice, i.e.:*

5.8.2.1. *the linear operations are norm continuous:*

$$\|(g + h) - (g_0 + h_0)\| \leqq \|g - g_0\| + \|h - h_0\| \qquad (g, h, g_0, h_0 \in L_m^1)$$

$$\|\alpha h - \alpha_0 h_0\| \leqq |\alpha - \alpha_0| \cdot \|h_0\| + |\alpha| \cdot \|h - h_0\| \qquad (\alpha \in \mathbb{R}, \quad h, h_0 \in L_m^1).$$

5.8.2.2. *the finite lattice operations are norm continuous:*

$$\left.\begin{array}{l} \|(g \vee h) - (g_0 \vee h_0)\| \\ \|(g \wedge h) - (g_0 \wedge h_0)\| \end{array}\right\} \leqq \|g - g_0\| + \|h - h_0\| \qquad (g, g_0, h, h_0 \in L_m^1).$$

In particular, a subset of L_m^1 and its norm closure have, if any, the same supremum (= least majorant = l.u.b.) and infimum (= greatest minorant = g.l.b.).

5.8.2.3. *L_m^1 is* **norm complete:** *If $h_1, h_2, \ldots \in L_m^1$ are such that*

$$\lim_{j, k \to \infty} \|h_j - h_k\| = 0,$$

then there is an $h \in L_m^1$ *(and exactly one) such that*

$$\lim_{n \to \infty} \|h_n - h\| = 0.$$

5.8.3. L_m^1 *is even an L-space, i.e., it is a Banach lattice whose norm satisfies,*

$$\|g + h\|_1 = \|g\|_1 + \|h\|_1 \qquad (0 \leqq g, h \in L_m^1).$$

5.8.4. *For every sequence* $h_1, h_2, \ldots \in \mathscr{L}_m^1$ *satisfying* $\lim_{j,k} \|h_j - h_k\| = 0$, *there is a subsequence* $h_{n_1}, h_{n_2}, \ldots$ *(with* $1 \leqq n_1 < n_2 < \cdots$*) that converges* m-a.e.. *By*

$$h = \lim_k h_{n_k} \qquad (m\text{-a.e.})$$

an $h \in \mathscr{L}_m^1$ *is defined uniquely up to equivalence* mod m *such that* $\lim_{n \to \infty} \|h_n - h\| = 0$.

5.8.5. L_m^1 *is a* **conditionally complete lattice:** *Every nonempty subset of* L_m^1 *that has an upper bound in* L_m^1 *has a least upper bound in* L_m^1, *i.e., a majorant that minorizes every majorant of the set, and the analogous statement holds for lower bounds.*

5.8.6. *If* $\varnothing \neq \mathscr{F} \subseteq L_m^1$ *is increasingly filtered with*

$$\sup_{g \in \mathscr{F}} m(g) < \infty.$$

then $\mathscr{F}$ *has a least upper bound* h *in* L_m^1 *and*

$$\lim_{g \in \mathscr{F}} g = h$$

in the norm sense; more precisely, for every $\varepsilon > 0$, *there is a* $g_0 \in \mathscr{F}$ *such that*

$$g_0 \leqq g \in \mathscr{F} \quad \Rightarrow \quad \|g - h\| < \varepsilon.$$

5.8.7. m *is a norm continuous function on* L_m^1*:*

$$|m(g) - m(h)| \leqq \|g - h\| \qquad (g, h \in L_m^1).$$

Proof. 1 is obvious from the properties of the integral stated in Theorem 5.1.

2.1. The first inequality follows from the triangle inequality 1.3 and the second from 1.3 and 1.2:

$$\begin{aligned} \|\alpha h - \alpha_0 h_0\| &\leqq \|\alpha(h - h_0)\| + \|(\alpha - \alpha_0)h_0\| \\ &\leqq |\alpha| \cdot \|h - h_0\| + |\alpha - \alpha_0| \cdot \|h_0\|. \end{aligned}$$

2.2 is an obvious consequence of the well-known inequalities

$$\left.\begin{array}{l}|(x \vee y)-(x_0 \vee y_0)| \\ |(x \wedge y)-(x_0 \wedge y_0)|\end{array}\right\} \leqq |x-x_0|+|y-y_0| \qquad (x, y, x_0, y_0 \in \mathbb{R}).$$

2.3 will follow from 4.
3 is obvious:

$$0 \leqq g, h \in L_m^1 \quad \Rightarrow \quad \|g+h\|_1 = m(g+h) = m(g)+m(h) = \|g\|_1 + \|h\|_1.$$

4. It is easy to find a subsequence $h_{n_1}, h_{n_2}, \ldots$ such that

$$\|h_{n_j} - h_{n_k}\| \leqq 1/2^j \qquad (j \leqq k).$$

Thus the series $|h_{n_1}| + |h_{n_2} - h_{n_1}| + \cdots$ satisfies the hypothesis of the monotone convergence theorem 5.3, serial form:

$$m(|h_{n_1}|) + m(|h_{n_2} - h_{n_1}|) + \cdots \leqq m(|h_{n_1}|) + \frac{1}{2} + \frac{1}{2^2} + \cdots$$

$$= m(|h_{n_1}|) + 1 < \infty.$$

Thus $g = |h_{n_1}| + |h_{n_2} - h_{n_1}| + \cdots \in L_m^1$, in particular $g < \infty$ m-a.e., and thus the series $h_{n_1} + (h_{n_2} - h_{n_1}) + \cdots$ is m-a.e. absolutely convergent, implying the convergence m-a.e. of the sequence $h_{n_1}, h_{n_2}, \ldots$. By $h(\omega) = \lim_{k\to\infty} h_{n_k}(\omega)$ where this limit exists and $h(\omega) = 0$ elsewhere, we get a function h that satisfies $\lim_{k\to\infty} h_{n_k} = h$ m-a.e. Since

$$|h_{n_k}| \leqq |h_{n_1}| + |h_{n_2} - h_{n_1}| + \cdots + |h_{n_k} - h_{n_{k-1}}| \leqq g \in L_m^1 \ (k = 2, 3, \ldots),$$

we may apply Theorem 5.7 (on dominated convergence) by Lebesgue and get $\lim_{k\to\infty} \|h_{n_k} - h\| = 0$. For a given $\varepsilon > 0$, find n_0 such that $\|h_j - h_k\| < \varepsilon/2$ $(j, k \geqq n_0)$ and choose an i such that $n_i \geqq n_0$, $\|h_{n_i} - h\| < \varepsilon/2$. Then $k \geqq n_0$ implies $\|h_k - h\| \leqq \|h_k - h_{n_i}\| + \|h_{n_i} - h\| < \frac{1}{2}\varepsilon + \frac{1}{2}\varepsilon = \varepsilon$.

5 will follow from 6 in the following way: If $\varnothing \neq \mathscr{M} \subseteq L_m^1$ has an upper bound $h_0 \in L_m^1$, then $\mathscr{F} = \{h_1 \vee \cdots \vee h_n \mid n \geqq 1, h_1, \ldots, h_n \in M\}$ is increasingly filtered, $\subseteq L_m^1$, and has the same upper bounds as M, e.g., h_0, which implies $\sup_{h\in\mathscr{F}} m(h) \leqq m(h_0)$.

6. Put $\sup_{g\in\mathscr{F}} m(g) = \alpha$ and choose $f_1, f_2, \ldots \in \mathscr{F}$ with $\lim_{n\to\infty} m(f_n) = \alpha$. Put $g_1 = f_1$, find $\mathscr{F} \ni g_2 \geqq f_2, g_1$, $\mathscr{F} \ni g_3 \geqq f_3, g_2, \ldots$. Clearly now $\mathscr{F} \ni g_1 \leqq g_2 \leqq \cdots$ with $\lim_{n\to\infty} m(g_n) = \alpha$. By Theorem 5.3 on monotone convergence (sequential form) we get an $h \in L_m^1$ such that $\lim_n \|g_n - h\| = 0$. By 7 we see $m(h) = \lim_{n\to\infty} m(g_n) = \alpha$. Let us now prove that $h = \sup_{g\in\mathscr{F}} g$. For this we first prove $h \vee g = h$ $(g \in \mathscr{F})$. Assume there is a $g \in \mathscr{F}$ with $h \vee g \geqq h, \neq h$, then $m(h \vee g) > m(h) = \alpha$, by Theorem 5.1.5. Find $\mathscr{F} \ni h_1 \geqq g_1, g$, $\mathscr{F} \ni h_2 \geqq h_1, g, \ldots$. Clearly $\mathscr{F} \ni h_1 \leqq h_2 \leqq \cdots$. By our previous arguments,

$$\lim_{n\to\infty} h_n = h' \in L_m^1 \qquad \text{with} \qquad \alpha = \lim_{n\to\infty} m(g_n) \leqq \lim_{n\to\infty} m(h_n) = m(h') \leqq \alpha,$$

i.e., $m(h') = \alpha$. On the other hand, $h_1 \geqq g_1, h_2 \geqq g_2, \ldots$ implies $h' \geqq h$; and $h_1, h_2, \ldots \geqq g$ implies $h' \geqq g$; hence $h' \geqq h \vee g$ and thus $m(h') \geqq m(h \vee g) > \alpha$, a contradiction. Thus $h \vee g = h$ $(g \in \mathscr{F})$, which proves that h is an upper bound for $\mathscr{F}$. Every upper bound of $\mathscr{F}$ is an upper bound for the norm closure of $\mathscr{F}$, hence a majorant of h. A reader who has already solved Exercise II.2.6.7 will find himself quite familiar with these arguments.

7 is obvious.

5.9. Remarks

5.9.1. The completeness statement 5.8.2.3 is also called the **Fischer–Riesz theorem** (after E. Fischer (1875–1956) and F. Riesz (1880–1956)).

5.9.2. We have introduced, in passing, notions of abstract *Banach lattice theory* in the above theorem. A reader who feels the necessity of introducing abstract rigor immediately now should consult Chapter IX, Section 1. He will notice that in fact 5.8.6 could be deduced essentially from 5.8.3 alone.

5.9.3. Although $\|\cdot\|: \mathscr{L}_m^1 \to \mathbb{R}_+$ is not a norm in the rigorous sense since $\|h_1\| = 0$ implies only that h is an m-σ-nullfunction, we shall denote it as the 1-norm or L^1-norm on $\mathscr{L}_m^1$ as well.

5.10. Theorem. *Let $(\Omega, \mathscr{E}, m)$ be a σ-measure space. Then $\mathscr{E}$ is 1-norm dense in $\mathscr{L}_m^1$ $(= \mathscr{L}_\sigma^1(m))$.*

Proof. We proceed in two steps.

(a) Every $u \in \mathscr{E}^\sigma$ with $m(u) < \infty$ is 1-norm approximable from $\mathscr{E}$. In fact $\mathscr{E} \ni f_1 \leqq f_2 \leqq \cdots \nearrow u$ implies $\|u - f_n\|_1 = m(u - f_n) = \underline{m}(u) - m(f_n) \to 0$.

(b) Every $h \in \mathscr{L}_m^1$ is 1-norm approximable from $\{u \mid u \in \mathscr{E}^\sigma, \underline{m}(u) < \infty\}$. In fact we may, for any given $\varepsilon > 0$, choose $h \leqq u \in \mathscr{E}^\uparrow$ with $\underline{m}(u) < m(h) + \varepsilon$, hence $\|u - h\| = m(u - h) = \underline{m}(u) - m(h) < \varepsilon$.

Combining (a) and (b) we obtain our theorem.

5.11. Examples

5.11.1. Let $\Omega = \mathbb{R}$, $\mathscr{E} = \mathscr{C}^{00}(\mathbb{R}, \mathbb{R})$, and $m: \mathscr{E} \to \mathbb{R}$ be the Riemann integral. By Theorem 5.10, $\mathscr{C}^{00}(\mathbb{R}, \mathbb{R})$ is 1-norm dense in $\mathscr{L}_m^1$, the space of all Lebesgue integrable functions on $\mathbb{R}$.

5.11.2. Let $A \neq \varnothing$ be finite, $\Omega = A \times A \times \cdots$ the Bernoulli space, $\mathscr{C}$ as in Example I.1.6, and m any content on C. Then m is a σ-content (Example I.2.5.). Let $(\Omega, \mathscr{E}, m)$ be the derived σ-measure space. Here $\mathscr{E}$ is the elementary domain of all functions $\in \mathbb{R}^\Omega$ of finite order (Exercise I.8.10). This domain is 1-norm dense in $\mathscr{L}_\sigma^1(m)$.

5.11.3. If Ω is a locally compact Hausdorff space and $(\Omega, \mathscr{C}^{00}(\Omega, \mathbb{R}), m)$ any measure (hence σ-measure (Theorem I.10.2) space over it, then $\mathscr{C}^{00}(\Omega, \mathbb{R})$ is 1-norm dense in $\mathscr{L}_\sigma^1(m)$.

5.12. Exercises

5.12.1. Let $(\Omega, \mathscr{E}, m)$ be a σ-measure space. Show that for every $h \in \mathscr{L}_\sigma^1(m)$ the functions $h_n = (h \wedge n) \vee (-n)$ $(n = 1, 2, \ldots)$ are in $\mathscr{L}_\sigma^1(m)$ again and $\lim_{n\to\infty} \int h_n \, dm = h \, dm$.

5.12.2. Let $h \in \overline{\mathbb{R}}^{\mathbb{R}}$ be Lebesgue integrable and $g \in \mathscr{C}^{\mathrm{b}}(\mathbb{R}, \mathbb{R})$. Prove that gh is Lebesgue integrable again. (*Hint*: It is sufficient to settle the case $g, h \geqq 0$; prove that $g\mathscr{C}^{00}(\mathbb{R}, \mathbb{R})^\sigma \subseteq \mathscr{C}^{00}(\mathbb{R}, \mathbb{R})^\sigma$, $g\mathscr{C}^{00}(\mathbb{R}, \mathbb{R})_\sigma \subseteq \mathscr{C}^{00}(\mathbb{R}, \mathbb{R})_\sigma$.)

5.12.3. Let $h \in \overline{\mathbb{R}}^{\mathbb{R}}$ be Lebesgue integrable. Prove

$$\lim_{n\to\infty} \int h(x) \cos nx \, dx = 0.$$

(*Hint*: First prove it for $h \in \mathscr{C}^{00}(\mathbb{R}, \mathbb{R})$, then approximate.)

5.13. Theorem (Jensen's inequality). *Let $(\Omega, \mathscr{E}, m)$ be a σ-measure space such that $1 \in \mathscr{L}_\sigma^1(m)$, $m(1) = 1$. Let $I \subseteq \mathbb{R}$ be a closed (possibly infinite) interval and $g\colon I \to \mathbb{R}$ a convex function:*

$$g(\alpha x + (1 - \alpha)y) \leqq \alpha g(x) + (1 - \alpha)g(y) \qquad (0 \leqq \alpha \leqq 1, \quad x, y \in I)$$

Then $h \in \mathscr{L}_\sigma^1(m)$, $h(\omega) \in I$ m-σ-a.e., $g(h) \in \mathscr{L}_\sigma^1(m)$ imply $m(h) \in I$ and **Jensen's inequality**

$$m(g(h)) \geqq g(m(h)).$$

Proof. $1 \in \mathscr{L}_\sigma^1(m)$ implies that $\mathscr{L}_\sigma^1(m)$ contains all finite real constants α, and $m(1) = 1$ implies $m(\alpha) = \alpha$ ($\alpha = \text{const} \in \mathbb{R}$). Let a, b denote the (possibly infinite) left and right endpoints of I. If they are both finite, $a = m(a) \leqq m(h) \leqq m(b) = b$ follows from $a \leqq h \leqq b$ (m-a.e.). In any case $a \leqq h \leqq b \bmod m$ implies $a \leqq m(h) \leqq b$, with $a = m(h)$ iff $h = a \bmod m$ and $m(h) = b \bmod m$ (Theorem 5.1.5). If h is a constant mod m, then we are through. If h is not a constant mod m, then $a < m(h) < b$. In this case we may (this follows from the general elementary theory of convex functions) find a $\gamma \in \mathbb{R}$ with

$$g(m(h)) + \gamma(x - m(h)) \leqq g(x) \qquad (x \in I).$$

We conclude

$$g(m(h)) + \gamma(h - m(h)) \leqq g(h) \qquad m\text{-}\sigma\text{-a.e.}$$

and thus

$$g(m(h)) + \gamma(m(h) - m(h)) \leqq m(g(h)),$$

which yields Jensen's inequality.

5.14. Exercise

5.14.1. Let $(\Omega, \mathscr{E}, m)$ be a σ-measure space such that $1 \in \mathscr{L}_\sigma^1(m)$, $m(1) = 1$. Prove that for any $h \in \mathscr{L}_\sigma^1(m)$

$$m(|h|) \geqq |m(h)|$$

and in case $p \geqq 1$, $|h|^p \in \mathscr{L}_\sigma^1(m)$

$$m(|h|^p) \geqq |m(h)|^p.$$

In case $h \geqq 0$, $\log h \in \mathscr{L}_\sigma^1(m)$ (we put $\log 0 = -\infty$), then $h > 0$ m-a.e. and

$$m(\log h) \leqq \log(m(h)).$$

6. THE σ-CONTENT DERIVED FROM A σ-MEASURE

In this section we use the integral in order to derive a σ-content from it via the integral values of indicator functions. In case the integral stems from a σ-content prespace, the *extension problem* for the latter turns out to be solved thereby once more. Obvious questions, e.g., after the result of one more application of the σ-extension procedure, find well-rounded answers.

6.1. Theorem. *Let $(\Omega, \mathscr{E}, m)$ be a σ-measure space. Then*

$$\mathscr{B}_m^{00} = \mathscr{B}_\sigma^{00}(m) = \mathscr{B}_\sigma^{00}(\Omega, \mathscr{E}, m) = \{F \mid 1_F \in \mathscr{L}_\sigma^1(m)\}$$

is a local σ-ring and

$$m(F) = \int 1_F \, dm \qquad (F \in \mathscr{B}_m^{00})$$

defines a σ-content on $\mathscr{B}_m^{00}$, which we denote by m again. It is called the **σ-content (σ-) derived from the σ-measure m** *(by σ-extension), and $(\Omega, \mathscr{B}_m^{00}, m)$ is called the σ-content space derived from the σ-measure space $(\Omega, \mathscr{E}, m)$. It is full and complete.*

Proof. By Theorem 5.8 we know that $\mathscr{L}_\sigma^1(m) \cap \mathbb{R}^\Omega$ is stable against finite vector lattice operations. This implies stabilities of $\mathscr{B}_m^{00}$ to the extent that it is a ring:

$$\begin{aligned} E, F \in \mathscr{B}_m^{00} \quad &\Rightarrow \quad 1_E, 1_F \in \mathscr{L}_\sigma^1(m) \\ &\Rightarrow \quad 1_{E \cup F} = 1_E \vee 1_F \in \mathscr{L}_\sigma^1(m) \quad \Rightarrow \quad E \cup F \in \mathscr{B}_m^{00}. \\ E, F \in \mathscr{B}_m^{00} \quad &\Rightarrow \quad 1_{E \setminus F} = 1_E - (1_E \wedge 1_F) \in \mathscr{L}_\sigma^1(m) \\ &\Rightarrow \quad E \setminus F \in \mathscr{B}_m^{00}. \end{aligned}$$

The linearity of the integral $m\colon L_m^1 \to \mathbb{R}$ implies that $m\colon \mathscr{B}_m^{00} \to \mathbb{R}_+$ is additive:

$$E, F \in \mathscr{B}_m^{00},\ E \cap F \neq \varnothing \ \Rightarrow\ m(E+F) = m(1_{E+F}) = m(1_E + 1_F) \\ = m(1_E) + m(1_F) = m(E) + m(F).$$

Thus $(\Omega, \mathscr{B}_m^{00}, m)$ is a content prespace. The monotone convergence theorem 5.3 now implies that it is actually a full σ-content space: If $F_1, F_2, \ldots \in \mathscr{B}_m^{00}$, $m(F_1) + m(F_2) + \cdots < \infty$, then we may, by the first entrance decomposition I.1.3.3, construct disjoint $F_1' \subseteq F_1$, $F_2' \subseteq F_2, \ldots$ in $\mathscr{B}_m^{00}$ such that $F_1 \cup \cdots \cup F_n = F_1' + \cdots + F_n'$ $(n = 1, 2, \ldots)$. Clearly,

$$m(1_{F_1'}) + m(1_{F_2'}) + \cdots = m(F_1') + m(F_2') + \cdots \leqq m(F_1) + m(F_2) + \cdots < \infty.$$

By Theorem 5.3 we get $1_{F_1 \cup F_2 \cup \cdots} = 1_{F_1' + F_2' + \cdots} = 1_{F_1'} + 1_{F_2'} + \cdots \in \mathscr{L}_m^1$, hence $F_1 \cup F_2 \cup \cdots \in \mathscr{B}_m^{00}$. If $F_j \cap F_k = \varnothing$ $(j \neq k)$, we get, again by Theorem 5.3,

$$m(F_1 + F_2 + \cdots) = m(1_{F_1 + F_2 + \cdots}) = m(1_{F_1} + 1_{F_2} + \cdots) \\ = m(1_{F_1}) + m(1_{F_2}) + \cdots = m(F_1) + m(F_2) + \cdots.$$

The completeness of $(\Omega, \mathscr{B}_m^{00}, m)$ follows from Proposition 4.2.2.

6.2. Remarks

6.2.1. We shall usually write $\mathscr{B}_m^{00}$ and use the notation $\mathscr{B}_\sigma^{00}(m)$ only if it is necessary to emphasize that we deal with a result of σ- (and not τ-) extension.

6.2.2. *Remark.* Theorem 6.1 yields another solution of the extension problem I.5.1 for σ-contents in the following way: If $(\Omega, \mathscr{R}, m)$ is a σ-content prespace, let $(\Omega, \mathscr{E}(\mathscr{R}), m)$ be the σ-measure space derived from it (Example I.8.7, Proposition I.8.9). Then $F \in \mathscr{R} \Rightarrow 1_F \in \mathscr{E}(\mathscr{R}) \subseteq \mathscr{L}_\sigma^1(m)$, hence $\mathscr{R} \subseteq \mathscr{B}_\sigma^{00}(m)$, in particular the local σ-ring $\mathscr{B}^{00}(\mathscr{R})$ generated by $\mathscr{R}$ is contained in $\mathscr{B}_\sigma^{00}(m)$ and $m(F) = \int 1_F\, dm$ $(F \in \mathscr{R})$, hence we need only restrict the σ-content derived from $(\Omega, \mathscr{E}(\mathscr{R}), m)$ to $\mathscr{B}^{00}(\mathscr{R})$ in order to get a σ-content m on $\mathscr{B}^{00}(\mathscr{R})$ which coincides with the original m on $\mathscr{R}$. We shall in fact see that $(\Omega, \mathscr{B}_m^{00}, m)$ is (as is indicated by our notation) nothing but the Carathéodory extension of $(\Omega, \mathscr{R}, m)$ (Theorem II.2.3, Exercise 6.5.7).

6.3. Theorem. *Let (Ω, E, m) be a σ-measure space, $h \in \mathscr{L}_m^1$ and $0 < \alpha < \beta \in \mathbb{R}$. Then the sets $\{h > \alpha\}$, $\{h \geqq \alpha\}$, $\{\alpha \leqq h < \beta\}$, $\{\alpha \leqq h \leqq \beta\}$, $\{\alpha < h < \beta\}$ are all in $\mathscr{B}_m^{00}$.*

Proof. We employ a device that might be called **blowing up the hat**:

Let

$$1_{\{h>\alpha\}} = \lim_{n\to\infty} [n(h-(\alpha\wedge h))]\wedge 1;$$

the sequence of functions appearing under the lim is an increasing one, and is in $\mathscr{L}^1_\sigma(m)$. Since $[n(h-(\alpha\wedge h))]\wedge 1 \leqq \alpha^{-1}h \in \mathscr{L}^1_\sigma(m)$, we may apply, e.g., Theorem 5.7 (dominated convergence) and get $1_{\{h>\alpha\}} \in \mathscr{L}^1_\sigma(m)$, hence $\{h>\alpha\} \in \mathscr{B}^{00}_\sigma(m)$. Now the sets $\{h>\alpha(1-1/n)\}$ are in $\mathscr{B}^{00}_\sigma(m)$ and decrease toward $\{h \geqq \alpha\}$, which therefore also belongs to $\mathscr{B}^{00}_m$. The rest is an easy exercise, e.g.,

$$\{\alpha \leqq h < \beta\} = \{h \geqq \alpha\}\backslash\{h \geqq \beta\}.$$

6.4. Theorem. *Let $(\Omega, \mathscr{E}, m)$ be a σ-measure space and $\mathscr{R} = \mathscr{B}^{00}_m$, and let $(\Omega, \mathscr{R}, m)$ be the σ-content space derived from $(\Omega, \mathscr{E}, m)$ (by σ-extension) and $(\Omega, \mathscr{E}(\mathscr{R}), m)$ the σ-measure space derived from $(\Omega, \mathscr{R}, m)$. Then*

$$\mathscr{L}^1_\sigma(\Omega, \mathscr{E}, m) = \mathscr{L}^1_\sigma(\Omega, \mathscr{E}(\mathscr{R}), m) = \mathscr{L}^1_\sigma(\Omega, \mathscr{R}, m)$$

and the corresponding integrals coincide.

In particular, if $0 \leqq h \in \mathscr{L}^1_\sigma(\Omega, \mathscr{E}, m)$, then for every $n = 1, 2, \ldots$, the function

$$h_n = \sum_{k=1}^{2^{2n}-1} \frac{k}{2^n} 1_{\{k/2^n \leqq h < (k+1)/2^n\}} + 2^n 1_{\{h \geqq 2^n\}}$$

is in $\mathscr{E}(\mathscr{R}) \subseteq \mathscr{L}^1_\sigma(\Omega, \mathscr{E}, m)$ and $h_1 \leqq h_2 \leqq \cdots \nearrow h$ (pointwise) holds as well as $\lim_{n\to\infty} \|h_n - h\|_1 = 0$. Similarly,

$$h'_n = \sum_{k=1}^{2^{2n}-1} \frac{k}{2^n} 1_{\{k/2^n < h \leqq (k+1)/2^n\}} + 2^n 1_{\{h > 2^n\}}$$

is in $\mathscr{E}(\mathscr{R}) \subseteq \mathscr{L}^1_\sigma(\Omega. \mathscr{E}, m)$ and $h'_n \nearrow h$ (pointwise) holds as well as $\lim_{n\to\infty} \|h'_n - h\|_1 = 0$.

Proof. By the very definition of $\mathscr{R} = \mathscr{B}^{00}_m$ we have $\mathscr{E}(\mathscr{R}) \subseteq \mathscr{L}^1_m$ and coincidence, on $\mathscr{E}(\mathscr{R})$, between the m-σ-integral and the σ-measure derived from the σ-content derived from the σ-measure m on our original $\mathscr{E}$. The last statement of our theorem is an immediate consequence of Theorem 6.3, the monotone convergence theorem 5.3, and the construction of the functions $h_1, h_2, \ldots$ "built under h at scales $1/2^n$." It is now clear that the first step in σ-extension of $(\Omega, \mathscr{E}(\mathscr{R}), m)$ yields exactly the $0 \leqq h \in \mathscr{L}^1_m$ as nonnegative σ-upper functions with finite integral, and thus all σ-upper and σ-lower functions with finite integrals are in $\mathscr{L}^1_m$. It follows (use Remark 2.2.6) that σ-extension leads from $(\Omega, \mathscr{E}(\mathscr{R}), m)$ to exactly $\mathscr{L}^1_m$ and the m-σ-integral derived from $(\Omega, \mathscr{E}, m)$.

6.5. Exercises. The above theorem is an example for the frequent phenomenon that σ-measure spaces on the same basic sets, if properly related, lead to the same spaces $\mathscr{L}_m^1$ by σ-extension. Exercises 6.5.1–6.5.5 display the same phenomenon under various aspects.

6.5.1. Let $(\Omega, \mathscr{E}, m)$, $(\Omega, \mathscr{E}', m')$ be σ-measure spaces with the same basic set $\Omega \neq \varnothing$. Assume that $E \subseteq E'$ and m is the restriction of m' to $\mathscr{E}$. Prove the following statements:

6.5.1.1. $\mathscr{E}^\sigma \subseteq (\mathscr{E}')^\sigma$ and the restriction of $\underline{m}'$: $(\mathscr{E}')^\sigma \to \mathbb{R} \cup \{\infty\}$ to $\mathscr{E}^\sigma$ is $\underline{m}$.

6.5.1.2. $\tilde{m} \geqq \tilde{m}'$.

6.5.1.3. Every m-σ-nullset is also an m'-σ-nullset.

6.5.1.4. $\mathscr{L}_\sigma^1(m) \subseteq \mathscr{L}_\sigma^1(m')$ and every m-equivalence class in $\mathscr{L}_\sigma^1(m)$ is contained in exactly one m'-equivalence class in $\mathscr{L}_\sigma^1(m')$. We may thus write $L_\sigma^1(m) \subseteq L_\sigma^1(m')$.

6.5.1.5. The restriction of m': $L_\sigma^1(m') \to \mathbb{R}$ to $L_\sigma^1(m)$ is m: $L_\sigma^1(m) \to \mathbb{R}$.

6.5.1.6. The σ-content space derived from $(\Omega, \mathscr{E}', m')$ is an extension of the σ-content space derived from $(\Omega, \mathscr{E}, m)$.

6.5.2. Let $(\Omega, \mathscr{E}, m)$ be a σ-measure space.

6.5.2.1. Let $\mathscr{B}^{00}(\mathscr{E}) = \mathscr{B}^{00}(\{\{f > 1\} \mid f \in \mathscr{E}\})$. Let m_0 be the restriction to $\mathscr{B}^{00}(\mathscr{E})$ of the σ-content derived from $(\Omega, \mathscr{E}, m)$. Prove that

$$\mathscr{L}_\sigma^1(\Omega, \mathscr{B}^{00}(\mathscr{E}), m_0) = \mathscr{L}_\sigma^1(\Omega, \mathscr{E}, m)$$

and that the m_0-σ-integral coincides with the m-σ-integral.

6.5.2.2. Let $(\Omega, \mathscr{E}, m)$ be a σ-measure space. Prove that each of the following subsets of $\mathbb{R}^\Omega$ is an elementary domain and the restriction of m: $\mathscr{L}_\sigma^1(m) \to \mathbb{R}$ to it is a σ-measure on it:

$$\mathscr{E}^{\text{step}} = \{f \mid f \in \mathscr{L}_\sigma^1(m), \{f(\omega) \mid \omega \in \Omega\} \text{ is finite and } \subseteq \mathbb{R}\}$$

$$\mathscr{E}^{\text{b}} = \{f \mid f \in \mathscr{L}_\sigma^1(m), \sup_{\omega \in \Omega} |f(\omega)| < \infty\}$$

$$\mathscr{L}_\sigma^1(m) \cap \mathbb{R}^\Omega.$$

Prove that σ-extension of each of the resulting σ-measure spaces $(\Omega, \mathscr{E}^{\text{step}}, m)$, $(\Omega, \mathscr{E}^{\text{b}}, m)$, $(\Omega, \mathscr{L}_m^1 \cap \mathbb{R}^\Omega, m)$ leads exactly to m: $\mathscr{L}_\sigma^1(m) \to \mathbb{R}$.

6.5.3. Let $(\Omega, \mathscr{E}, m)$, $(\Omega, \mathscr{E}', m')$ be σ-measure spaces such that

(a) $\mathscr{E}' \subseteq \mathscr{L}_\sigma^1(m)$ and the restriction of m: $\mathscr{L}_\sigma^1(m) \to \mathbb{R}$ to $\mathscr{E}'$ is m'.
(b) $\mathscr{E} \subseteq \mathscr{L}_\sigma^1(m')$ and the restriction of m': $\mathscr{L}_\sigma^1(m') \to \mathbb{R}$ to $\mathscr{E}$ is m.

Prove that $\mathscr{L}_\sigma^1(m) = \mathscr{L}_\sigma^1(m')$ and the m-σ-integral coincides with the m'-σ-integral.

6.5.4. Let $(\mathbb{R}, \mathscr{C}^{00}(\mathbb{R}, \mathbb{R}), m)$ be a σ-measure (in fact a τ-measure, Theorem I.10.2) space on the basic set $\mathbb{R}$. Prove that $\mathscr{B}_\sigma^{00}(m)$ contains every bounded (closed, half-open, or open) interval and thus the local σ-ring $\mathscr{B}^{00}(\mathbb{R})$ generated by the ring $\mathscr{D}$ of all dyadic intervals (Example I.1.4). Prove that by σ-extension of the σ-measure space derived from the σ-content prespace $(\mathbb{R}, \mathscr{D}, m)$ obtained from the σ-content space $(\mathbb{R}, \mathscr{B}_m^{00}, m)$ derived from $(\mathbb{R}, \mathscr{C}^{00}(\mathbb{R}, \mathbb{R}), m)$ by restriction from $\mathscr{B}_m^{00}$ to $\mathscr{D}$ we obtain exactly $m: \mathscr{L}_\sigma^1(m) \to \mathbb{R}$. Carry over this result to the basic set $\mathbb{R}^n$ $(n = 2, 3, \ldots)$. (*Hint*: Approximate functions $\in \mathscr{C}^{00}(\mathbb{R}^n, \mathbb{R})$ uniformly by functions from $\mathscr{E}(\mathscr{D})$ in a suitable fashion.)

6.5.5. (for topologists) Let Ω be a *locally compact Hausdorff space* and $(\Omega, \mathscr{C}^{00}(\Omega, \mathbb{R}), m)$ a σ-measure (in fact a τ-measure, Theorem I.10.2) space. Let $\mathscr{R}$ be the ring generated by all sets of the form $\{f \geqq 1\}$ $(f \in \mathscr{C}^{00}(\Omega, \mathbb{R}))$ and define the σ-content prespace $(\Omega, \mathscr{R}, m)$ by restriction of the σ-content space $(\Omega, \mathscr{B}_\sigma^{00}(m), m)$ derived from $(\Omega, \mathscr{C}^{00}(\Omega, \mathbb{R}), m)$. Prove that σ-extension of the σ-measure space derived from the σ-content prespace $(\Omega, \mathscr{E}, m)$ leads exactly to $m: \mathscr{L}_\sigma^1(m) \to \mathbb{R}$. (*Hint*: Approximate functions $\in \mathscr{C}^{00}(\Omega, \mathbb{R})$ uniformly by functions from $\mathscr{E}(\mathscr{R})$ in a suitable fashion.) This result is sometimes called *Riesz' theorem*; it can be briefly restated as follows: Every Radon measure on a locally compact Hausdorff space stems from a certain σ-content on the smallest reasonable set ring in that space.

6.5.6. (for topologists) Let Ω be a *Polish space* and $|\cdot, \cdot|$ a complete metric describing the topology. Let $\mathscr{B}$ be the σ-field in Ω generated by the system of all open sets of the topology of Ω.

6.5.6.1. Prove that $\mathscr{B}$ is the σ-field generated by all sets of the form $\{f \geqq 1\}$ $(f \in \mathscr{C}^b(\Omega, \mathbb{R}))$. (*Hint*: Observe that $\mathscr{B}$ is generated by the closed sets as well; for any closed $F \subseteq \Omega$ consider $f(\omega) = [1 - \inf_{\eta \in F} |\omega, \eta|]_+$ $(\omega \in \Omega)$ and prove $\{f \geqq 1\} = F$.)

6.5.6.2. Let $\omega_1, \omega_2, \ldots$ be a dense sequence in Ω and $K_{nj} = \{\omega \,|\, |\omega, \omega_n| \leqq 1/j\}$ $(n, j = 1, 2, \ldots)$. Prove $K_{nj} \in \mathscr{B}$ and $\bigcup_{n=1}^\infty K_{nj} = \Omega$ $(j = 1, 2, \ldots)$.

6.5.6.3. Let m be a positive σ-content on $\mathscr{B}$. Prove

$$\lim_{r \to \infty} m\left(\bigcup_{n=1}^r K_{nj}\right) = m(\Omega) \qquad (j = 1, 2, \ldots).$$

6.5.6.4. For any $\varepsilon > 0$ and $j = 1, 2, \ldots$ choose r_j such that $m(\bigcup_{n=1}^{r_j} K_{nj}) > m(\Omega) - \varepsilon/2^j$ and put $K_\varepsilon = \bigcap_{j=1}^\infty \bigcup_{n=1}^{r_j} \overline{K}_{nj}$. Prove that $K_\varepsilon \in \mathscr{B}$, $m(K_\varepsilon) \geqq m(\Omega) - \varepsilon$ and K_ε is compact.

6.5.6.5. Prove that every $f \in \mathscr{C}^b(\Omega, \mathbb{R})$ is a uniform limit of a sequence

$f_1, f_2, \ldots \in \mathscr{E}(\mathscr{B})$ such that $\sup_{\omega \in \Omega} |f_n(\omega)| \leqq \sup_{\omega \in \Omega} |f(\omega)|$ $(n = 1, 2, \ldots)$, and hence is Riemann integrable for $(\Omega, \mathscr{E}(\mathscr{B}), m)$ (Definition I.9.1).

6.5.6.6. Prove that the σ-measure space $(\Omega, \mathscr{C}^b(\Omega, \mathbb{R}), m)$ obtained by restriction of the Riemann extension $(\Omega, \mathscr{E}(\mathscr{B})_{(m)}, m)$ is tight (Definition I.10.3) and hence a τ-measure space (Theorem I.10.4).

6.5.7. Let $(\Omega, \mathscr{R}, m)$ be a σ-content prespace and $(\Omega, \mathscr{E}(\mathscr{R}), m)$ the σ-measure space derived from it (Example I.8.7, Proposition I.8.9). Prove that the σ-content space $(\Omega, \mathscr{B}_m^{00}, m)$ derived from $(\Omega, \mathscr{E}(\mathscr{R}), m)$ coincides with the Carathéodory extension (Theorem II.2.3) of $(\Omega, \mathscr{R}, m)$. This exercise may be subdivided into the following steps:

6.5.7.1. Prove that $1_F \in \mathscr{L}_\sigma^1(m)$ iff for every $\varepsilon > 0$, there are $\mathscr{R} \ni E_1 \supseteq E_2 \supseteq \cdots$ and $\mathscr{R} \ni G_1 \subseteq G_2 \subseteq \cdots$ such that $E = \bigcap_n E_n$ and $G = \bigcup_n G_n$ satisfy $E \subseteq F \subseteq G$ and

$$\lim_{n\to\infty} m(G_n) - \lim_{n\to\infty} m(E_n) < \varepsilon.$$

6.5.7.2. Let $D \in \mathscr{D}_{\tilde{m}}$ (Theorem II.2.3), i.e., D is an additive decomposer for the outer content $\tilde{m}$ derived from m. Prove that $D \subseteq U$ for some countable union U of sets from $\mathscr{R}$ such that $\tilde{m}(U) < \infty$. Show that U and $U \backslash D$ are additive decomposers for $\tilde{m}$ as well.

6.5.7.3. Prove that D is an additive decomposer for $\tilde{m}$, with $\tilde{m}(D) < \infty$, iff $1_D \in \mathscr{L}_\sigma^1(m)$.

6.5.8. Let $(\Omega, \mathscr{R}', m')$ be a σ-content prespace that is an extension of the σ-content prespace $(\Omega, \mathscr{R}, m)$. Prove that $L_\sigma^1(\Omega, \mathscr{R}, m) \subseteq L_\sigma^1(\Omega, \mathscr{R}', m')$ and the m-σ-integral coincides with the m'-σ-integral on $\mathscr{L}_\sigma^1(\Omega, \mathscr{R}, m)$.

6.5.9. Let $(\Omega, \mathscr{B}^{00}, m)$ be a σ-content space. Call $E, F \in \mathscr{B}^{00}$ equivalent mod m if $1_E = 1_F$ m-a.e., i.e., if $m(E \,\Delta\, F) = 0$. Prove:

6.5.9.1. Countable union, countable intersection, and difference are class operations, $\subseteq$ is a class relation in $\mathscr{B}^{00}$.

6.5.9.2. The set of all equivalence classes in $\mathscr{B}^{00}$ is a conditionally complete lattice with the order defined by $\subseteq$.

6.6. Remark. Theorem 6.4 and Exercises 6.5.1–6.5.3 show that we end up with the same m: $\mathscr{L}_m^1 \to \mathbb{R}$ starting the σ-extension procedure with various σ-measure spaces, partly based on σ-content (pre-) spaces. We are thus, for the rest of this book, entitled to a high flexibility of our hypotheses of definitions, propositions, and theorems. In many situations it will not matter whether we start with a certain σ-content prespace or its generated σ-content space or the full extension of the latter, or with a σ-measure space that leads to the same m: $\mathscr{L}_m^1 \to \mathbb{R}$ after σ-extension. The reader is invited to check currently what the obvious variants of a hypothesis are. We shall choose, as a rule, the one that fits best into the frame of the corresponding chapter or section.

7. τ-EXTENSION OF A τ-MEASURE

Let $(\Omega, \mathscr{E}, m)$ be a τ-measure space. Then we can carry out an analogue of the σ-extension theory differing from the latter by the *omission of a countability assumption* at one point. Since every τ-measure is also a σ-measure, σ-extension can be carried out as well, and we may compare the results of these two procedures. It will turn out that the difference lies essentially only in the scope of nullsets.

We content ourselves here with the following sketch of **τ-extension theory**. Pointwise suprema of arbitrary (not necessarily countable) nonempty subsets of the elementary domain $\mathscr{E}$ are called τ-upper functions, and the set of all these is denoted by $\mathscr{E}^\tau$. Replacing suprema by infima, we arrive at the set $\mathscr{E}_\tau$ of all τ-lower functions. Clearly

$$\mathscr{E}^\sigma \subseteq \mathscr{E}^\tau, \qquad \mathscr{E}_\sigma \subseteq \mathscr{E}_\tau .$$

We define

$$\underline{m}(u) = \sup_{\mathscr{E} \ni f \leqq u} m(f) \qquad (u \in \mathscr{E}^\tau)$$

$$\overline{m}(v) = \inf_{v \leqq f \in \mathscr{E}} m(f) \qquad (v \in \mathscr{E}_\tau)$$

(compare Proposition 1.8) and prove, by τ-continuity, that

$$\underline{m}(u) = \sup_{f \in \mathscr{F}} m(f)$$

holds for every increasingly filtered $\varnothing \neq \mathscr{F} \subseteq \mathscr{E}$ with pointwise supremum $u \in \mathscr{E}^\sigma$. There is an obvious counterpart to this for $v \in \mathscr{E}_\tau$, and we get $\underline{m}(u) = \overline{m}(u)$ $(u \in \mathscr{E}^\sigma \cap \mathscr{E}_\tau)$ along with the analogue of other statements proved in Section 2 for σ-upper and σ-lower functions.

7.1. Definition. *Let $(\Omega, \mathscr{E}, m)$ be a τ-measure space and $\underline{m}$: $\mathscr{E} \to \mathbb{R} \cup \{\infty\}$, $\overline{m}$: $\mathscr{E} \to \mathbb{R} \cup \{-\infty\}$ be defined as above. A function $h \in \overline{\mathbb{R}}^\Omega$ is called* **m-(τ-)integrable** *if for every $\varepsilon > 0$ there is a $u \in \mathscr{E}_\tau$ and a $v \in \mathscr{E}^\tau$ such that*

$$u \leqq h \leqq v$$

$$-\infty < \overline{m}(u), \qquad \underline{m}(v) < \infty, \qquad \underline{m}(v) - \overline{m}(u) < \varepsilon.$$

The real number

$$\sup_{\mathscr{E}_\tau \ni u \leqq h} \overline{m}(u) = \inf_{h \leqq v \in \mathscr{E}^\tau} \underline{m}(v)$$

is then called the **m-integral** *of h and denoted by* $m(h)$, $\int h \, dm$, *etc. The set of all m-τ-integrable functions is denoted by* $L_\tau^1(m)$, L_m^1, $L_\tau^1(\Omega, \mathscr{E}, m)$ *etc. depending on what specifications are needed.*

It is easy to copy the rest of σ-extension theory to the extent that all results on m-σ-nullsets (including the "m-σ-a.e." terminology) can be obtained as well for m-τ-nullsets. We observe that every m-σ-nullset is also an m-τ-nullset and hence m-σ-equivalence implies m-τ-equivalence. The monotone convergence theorem, Fatou's lemma, Lebesgue's theorem on dominated convergence, the norm-denseness of $\mathscr{E}$ in L_m^1, and Jensen's inequality are valid in $\mathscr{L}_\tau^1(m)$ resp. $L_\tau^1(m)$ again. By $\mathscr{B}_\tau^{00}(m) = \{F \mid 1_F \in \mathscr{L}_\tau^1(m)$ and $m(F) = \int 1_F \, dm \; (F \in \mathscr{B}_\tau^{00}(m))$ one defines the **σ-content space** $(\Omega, \mathscr{B}_\tau^{00}(m), m)$ **τ-derived from the τ-measure space** $(\Omega, \mathscr{E}, m)$.

Upon comparing the two extension methods, we get

7.2. Theorem. *Let* $(\Omega, \mathscr{E}, m)$ *be a τ-measure space. Then:*

7.2.1. $\mathscr{L}_\sigma^1(m) \subseteq \mathscr{L}_\tau^1(m)$ *and the m-τ-integral coincides with the m-σ-integral on* $\mathscr{L}_\sigma^1(m)$.

7.2.2. *For every* $h \in \mathscr{L}_\tau^1(m)$ *there is an* $h' \in \mathscr{L}_\sigma^1(m)$ *such that* $h = h'$ *m-τ-a.e.*

7.2.3. *If* $g, h \in \mathscr{L}_\sigma^1(m)$ *and* $g = h$ *m-τ-a.e., then* $g = h$ *m-σ-a.e.*

7.2.4. *Every m-σ-equivalence class in* $L_\sigma^1(m)$ *is contained in exactly one m-τ-equivalence class in* $L_\tau^1(m)$ *and every m-τ-equivalence class in* $L_\tau^1(m)$ *contains exactly one m-σ-equivalence class in* $L_\sigma^1(m)$. *Thus* $L_\sigma^1(m)$ *and* $L_\tau^1(m)$ *are essentially the same space* L_m^1 *(up to a trivial isomorphism).*

Proof. 1 is obvious from $\mathscr{E}^\sigma \subseteq \mathscr{E}^\tau$, $\mathscr{E}_\sigma \subseteq \mathscr{E}_\tau$.

2. Every $h \in \mathscr{L}_\tau^1(m)$ is an m-τ-norm limit of a sequence in $\mathscr{E}$. Passing to a subsequence, we may assume $\mathscr{E} \ni f_1, f_2, \ldots \to h$ m-τ-a.e. as well as $f_1, f_2, \ldots \to h$ m-σ-a.e. The sequence $f_1, f_2, \ldots$ represents a fundamental sequence in $L_\sigma^1(m)$, hence has a norm limit $h' \in \mathscr{L}_\sigma^1(m)$, and we have $f_n \to h'$ m-σ-a.e. This implies $h' = h$ m-τ-a.e.

3. We have $g \wedge h$, $g \vee h \in \mathscr{L}_\sigma^1(m)$, and $g \wedge h = g \vee h$ m-τ-a.e., hence $\int (g \wedge h) \, dm = \int (g \vee h) \, dm$, hence $g \wedge h = g \vee h$ m-σ-a.e. which by $g \wedge h \leqq g$, $h \leqq g \vee h$ clearly implies $g = h$ m-σ-a.e. (Theorem 5.1.5).

4 is a consequence of 2 and 3.

7.3. Exercises

7.3.1. Let m be the ordinary Riemann integral on $\mathscr{C}^{00}(\mathbb{R}, \mathbb{R})$. Prove that its τ-extension coincides with its σ-extension. (*Hint*: Prove that every τ-upper function is a σ-upper function.)

7.3.2. Generalize 7.3.1 to the basic set $\mathbb{R}^n$.

7.3.3. Generalize 7.3.1 to arbitrary $(\Omega, \mathscr{C}^{00}(\Omega, \mathbb{R}), m)$ where Ω is a locally compact metric space.

7.3.4. Carry Exercise 6.5.3 over to τ-measures.

7.3.5. Let $(\Omega, \mathscr{B}_\tau^{00}(m), m)$ be the σ-content space τ-derived from the τ-measure space $(\Omega, \mathscr{E}, m)$. Since the latter is a σ-measure space as well, the σ-content space $(\Omega, \mathscr{B}_\sigma^{00}(m), m)$ is also defined. Prove that $\mathscr{B}_\sigma^{00}(m) \subseteq \mathscr{B}_\tau^{00}(m)$ and that for every $E \in \mathscr{B}_\tau^{00}(m)$ there is an $F \in \mathscr{B}_\sigma^{00}(m)$ such that $m(E \,\Delta\, F) = 0$.

7.3.6. Let $(\Omega, \mathscr{E}, m)$ be a τ-measure space and $\mathscr{F} \subseteq \mathscr{E}^\tau$ be increasingly filtered with pointwise supremum u. Prove that $u \in \mathscr{E}^\tau$ and $\underline{m}(u) = \sup_{v \in \mathscr{F}} \underline{m}(v)$.

8. MEASURABILITY OF REAL AND COMPLEX FUNCTIONS

Let $\mathscr{B}(\mathbb{R})$ be the σ-field of all Borel subsets of $\mathbb{R}$ (Exercise I.4.8.6.2) and $\mathscr{B}(\overline{\mathbb{R}}) = \{F \mid F \subseteq \overline{\mathbb{R}}, F \cap \mathbb{R} \in \mathscr{B}(\mathbb{R})\}$ the σ-field of all Borel subsets (as we call them again) of the extended real line $\overline{\mathbb{R}} = \mathbb{R} \cup \{-\infty, \infty\}$ (Exercise I.4.8.7.1). We shall define, with the help of $\mathscr{B}(\mathbb{R})$ resp. $\mathscr{B}(\overline{\mathbb{R}})$, a class of particularly "nice" functions on every set $\Omega \neq \varnothing$ on which a σ-ring $\mathscr{B}^0$ is given: the so-called **$\mathscr{B}^0$-measurable functions**. It turns out that the special case where $\Omega \in \mathscr{B}^0$, i.e., where $\mathscr{B}^0$ is a σ-field, shows the basic ideas more clearly than the general one, and thus we shall start our presentation with it.

8.1. Definition. *Let $\mathscr{B}$ be a σ-field in $\Omega \neq \varnothing$.*

8.1.1. *A real function $f\colon \Omega \to \mathbb{R}$ is called* **$\mathscr{B}$-measurable** *or* **measurable for $\mathscr{B}$** *if for every $E \in \mathscr{B}(\mathbb{R})$, the set $f^{-1}(E)$ $(= \{\omega \mid f(\omega) \in E\})$ is in $\mathscr{B}$. The set of all $\mathscr{B}$-measurable functions in $\mathbb{R}^\Omega$ is denoted by* mble$(\Omega, \mathscr{B}, \mathbb{R})$.

8.1.2. *A function $f \in \overline{\mathbb{R}}^\Omega$ is called $\mathscr{B}$-measurable if for every $E \in \mathscr{B}(\overline{\mathbb{R}})$ the set $f^{-1}(E) = \{\omega \mid f(\omega) \in E\}$ is in $\mathscr{B}$. The set of all $\mathscr{B}$-measurable functions in $\overline{\mathbb{R}}^\Omega$ is denoted by* mble$(\Omega, \mathscr{B}, \overline{\mathbb{R}})$. *For every $\varnothing \neq \mathscr{E} \subseteq \overline{\mathbb{R}}^\Omega$ the smallest σ-field for which all $f \in \mathscr{E}$ are measurable, i.e., the σ-field*

$$\mathscr{B}(\{f^{-1}E \mid f \in \mathscr{E}, E \in \mathscr{B}(\overline{\mathbb{R}})\}),$$

is denoted by $\mathscr{B}(\mathscr{E})$ and called the **σ-field generated by** $\mathscr{E}$.

8.2. Proposition. *Let $\mathscr{B}$ be a σ-field in $\Omega \neq \varnothing$ and S a dense subset of $\mathbb{R}$.*

8.2.1. *For every $f \in \mathbb{R}^\Omega$ the following statements are equivalent:*

8.2.1.1. $f \in \mathrm{mble}(\Omega, \mathscr{B}, \mathbb{R})$.

8.2.1.2. $\{f < \alpha\}(= \{\omega \mid f(\omega) < \alpha\}) \in \mathscr{B} \quad (\alpha \in S)$.

8.2.1.3. $\{f \leqq \alpha\} \in \mathscr{B} \quad (\alpha \in S)$.

8.2.1.4. $\{f > \alpha\} \in \mathscr{B} \quad (\alpha \in S)$.

8.2.1.5. $\{f \geqq \alpha\} \in \mathscr{B} \quad (\alpha \in S)$.

8.2.1.6. $\{\alpha < f \leqq \beta\} \in \mathscr{B} \quad (\alpha, \beta \in S)$.

8.2.1.7. $\{\alpha < f < \beta\} \in \mathscr{B} \quad (\alpha, \beta \in S)$.

8.2.2. *For every $f \in \overline{\mathbb{R}}^\Omega$ and $N_\infty = \{f = \infty\}$, $N_{-\infty} = \{f = -\infty\}$, the following statements are equivalent:*

8.2.2.1. $f \in \mathrm{mble}(\Omega, \mathscr{B}, \overline{\mathbb{R}})$.

8.2.2.2. $N_\infty, N_{-\infty} \in \mathscr{B}$ *and* $f' = f \cdot (1 - 1_{N_\infty}) \cdot (1 - 1_{N_{-\infty}})$ $(\in \mathbb{R}^\Omega!)$ *is in* $\mathrm{mble}(\Omega, \mathscr{B}, \mathbb{R})$.

8.2.2.3. $\{f < \alpha\} \in \mathscr{B} \quad (\alpha \in S)$.

8.2.2.4. $\{f > \alpha\} \in \mathscr{B} \quad (\alpha \in S)$.

Proof. 1. Clearly 1.1 implies 1.2–1.7 since the sets $]-\infty, \alpha[,]-\infty, \alpha], \ldots,]\alpha, \beta[$ are all in $\mathscr{B}(\mathbb{R})$ (see Exercise I.4.8.6.1). For the converse, let us be content to prove $1.2 \Rightarrow 1.1$. Now, the set system $\mathscr{B}[f] = \{E \mid E \subseteq \mathbb{R}, f^{-1}(E) \in \mathscr{B}\}$ is easily seen to be a σ-field because $\mathscr{B}$ is a σ-field and the operation f^{-1} on sets evidently commutes with all set operations. By hypothesis, $\mathscr{B}[f]$ contains all open half-lines $[-\infty, \alpha[$ with $\alpha \in S$. Let $\alpha \in \mathbb{R}$ be arbitrary and choose $S \ni \alpha_1, \alpha_2, \ldots \nearrow \alpha$. Then $]-\infty, \alpha[=]-\infty, \alpha_1[\cup]-\infty, \alpha_2[\cup \cdots \in \mathscr{B}[f]$, i.e., $\mathscr{B}[f]$ is a σ-field that contains all open half-lines $]-\infty, \alpha[$ $(\alpha \in \mathbb{R})$. It therefore contains also the σ-field generated by the latter, which is nothing but $\mathscr{B}(\mathbb{R})$. But $\mathscr{B}(\mathbb{R}) \subseteq \mathscr{B}[f]$ means 1.1.

2. The stated equivalence is easily deduced from the following list of facts: $\{\infty\}, \{-\infty\} \in \mathscr{B}(\overline{\mathbb{R}})$, $N_\infty = f^{-1}(\{\infty\})$, $N_{-\infty} = f^{-1}(\{-\infty\})$,

$$\{f > \alpha\} = \begin{cases} \{f' > \alpha\} + N_\infty & (\alpha \geqq 0) \\ \{f' > \alpha\} \backslash N_{-\infty} & (\alpha < 0); \end{cases}$$

$$\{f < \alpha\} = \begin{cases} \{f' < \alpha\} + N_{-\infty} & (\alpha \leqq 0) \\ \{f' < \alpha\} \backslash N_\infty & (\alpha > 0); \end{cases}$$

$$f^{-1}(\{\infty\}) = \bigcap_{n=1}^{\infty} \{f > n\}, \qquad f^{-1}(\{-\infty\}) = \bigcap_{n=1}^{\infty} \{f < -n\}.$$

8.3. Exercises

8.3.1. Let $(\Omega, \mathscr{T})$ be a topological space and $\mathscr{B} = \mathscr{B}(\mathscr{T})$.

8.3.1.1. Prove that all $f \in \mathscr{C}(\Omega, \mathbb{R})$ are in $\text{mble}(\Omega, \mathscr{B}, \mathbb{R})$.

8.3.1.2. Let $A_1, A_2, \ldots \in \mathscr{B}$, $A_j \cap A_k = \varnothing$ $(j \neq k)$, $A_1 + A_2 + \cdots = \Omega$. For every $n = 1, 2, \ldots$, let $f_n \in \mathbb{R}^\Omega$ be such that the restriction of f_n to A_n is continuous in the relative topology on A_n. Prove that $f = \sum_n f_n 1_{A_n}$ is in $\text{mble}(\Omega, \mathscr{B}, \mathbb{R})$.

8.3.2. Let $\mathscr{B}$ be a σ-field in $\Omega \neq \varnothing$, $A_1, A_2, \ldots \in \mathscr{B}$, $A_j \cap A_k = \varnothing$ $(j \neq k)$, $\Omega = A_1 + A_2 + \cdots$. Let $f_1, f_2, \ldots \in \mathbb{R}^\Omega$ be such that for every $n = 1, 2, \ldots$, the function $f_n 1_{A_n}$ is in $\text{mble}(\Omega, \mathscr{B}, \mathbb{R})$. Prove that

$$f = \sum_n f_n 1_{A_n} \in \text{mble}(\Omega, \mathscr{B}, \mathbb{R}).$$

8.4. Proposition. *Let $\mathscr{B}$ be a σ-field in $\Omega \neq \varnothing$.*

8.4.1. $\text{mble}(\Omega, \mathscr{B}, \mathbb{R})$ *is a vector sublattice of* $\mathbb{R}^\Omega$ *which contains all constants and is stable under multiplication. If* $f \in \text{mble}(\Omega, \mathscr{B}, \mathbb{R})$, $g \in \text{mble}(\mathbb{R}, \mathscr{B}(\mathbb{R}), \mathbb{R})$, *then* $g(f) \in \text{mble}(\Omega, \mathscr{B}, \mathbb{R})$.

8.4.2. *If* $f_1, f_2, \ldots \in \text{mble}(\Omega, \mathscr{B}, \mathbb{R})$, $f \in \mathbb{R}^\Omega$, *and* $\lim_{n\to\infty} f_n(\omega) = f(\omega)$ $(\omega \in \Omega)$, *then* $f \in \text{mble}(\Omega, \mathscr{B}, \mathbb{R})$.

8.4.3. $\text{mble}(\Omega, \mathscr{B}, \overline{\mathbb{R}})$ *contains all constants from* $\overline{\mathbb{R}}$ *and is stable under:*

8.4.3.1. *addition, as far it is meaningful;*

8.4.3.2. *multiplication;*

8.4.3.3. *finite and countable suprema and infima, lim sup, and lim inf of sequences;*

8.4.3.4. *pointwise limits of sequences;*

8.4.3.5. *superposition of measurable functions, i.e., if* $f \in \text{mble}(\Omega, \mathscr{B}, \overline{\mathbb{R}})$, $g \in \text{mble}(\overline{\mathbb{R}}, \mathscr{B}(\overline{\mathbb{R}}), \overline{\mathbb{R}})$, *then* $g(f) \in \text{mble}(\Omega, \mathscr{B}, \overline{\mathbb{R}})$.

Proof. Keeping Proposition 8.2 in mind we find it sufficient to prove 3.

3.1. Let $f, g \in \overline{\mathbb{R}}^\Omega$ be such that $f(\omega) + g(\omega)$ makes sense for every $\omega \in \Omega$. Let $\mathbb{Q}$ be the set of all rational numbers. Then for every real α

$$\{f + g > \alpha\} = \bigcup_{\beta,\, \gamma \in \mathbb{Q},\, \beta + \gamma > \alpha} \{f > \beta\} \cap \{g > \gamma\}$$

which is in $\mathscr{B}$ again if f, g are $\mathscr{B}$-measurable. Thus $f + g \in \text{mble}(\Omega, \mathscr{B}, \overline{\mathbb{R}})$ by Proposition 8.2.2.4.

3.2. In view of the fact that $fg = f_+ g_+ + f_- g_- - f_+ g_- - f_- g_+$ is meaningful and correct for any $f, g \in \overline{\mathbb{R}}^\Omega$, we may content ourselves to settle the case $f, g \geqq 0$. But here

$$\{fg > \alpha\} = \begin{cases} \Omega \in B & \text{if } \alpha < 0 \\ \bigcup_{0 < \beta, \gamma \in \mathbb{Q}, \beta\gamma > \alpha} \{f > \beta\} \cap \{f > \gamma\} \in \mathscr{B} & \text{if } \alpha \geqq 0 \end{cases}$$

yields the desired result (use Proposition 8.2.2.4).

3.3. Let $f_1, f_2, \ldots \in \mathrm{mble}(\Omega, \mathscr{B}, \overline{\mathbb{R}})$ and $\alpha \in \mathbb{R}$. Then

$$\{f_1 \vee f_2 \vee \cdots > \alpha\} = \{f_1 > \alpha\} \cup \{f_2 > \alpha\} \cup \cdots$$
$$\{f_1 \wedge f_2 \wedge \cdots < \alpha\} = \{f_1 < \alpha\} \cup \{f_2 < \alpha\} \cup \cdots,$$

and thus we are through by Propositions 8.2.2.3, 8.2.2.4.

3.4 follows from 3.3, e.g., by

$$\lim_{n \to \infty} f_n = (f_1 \vee f_2 \vee \cdots) \wedge (f_2 \vee f_3 \vee \cdots) \wedge \cdots.$$

3.5. For every $E \in \mathscr{B}(\overline{\mathbb{R}})$, $\{g(f) \in E\} = f^{-1}(g^{-1}(E)) \in \mathscr{B}$ by our assumptions on f, g.

8.5. Remark. In 8.4.3.5. the reader certainly finds a general principle at work. We shall display it in full generality in Chapter IV, Section 1.

8.6. We now pass to the more general situation where $\mathscr{B}^0$ is an arbitrary σ-ring in $\Omega \neq \varnothing$. We begin with the observation that for every $\varnothing \neq \Omega_0 \in \mathscr{B}^0$, the set system $\mathscr{B}^0 \cap \Omega_0 = \{F \mid \mathscr{B}^0 \ni F \subseteq \Omega_0\} = \{E \cap \Omega_0 \mid E \in \mathscr{B}^0\}$ is a σ-field in Ω_0. Hence it is clear what it means to say that the restriction to Ω_0 of a function from $\overline{\mathbb{R}}^\Omega$ is $(\mathscr{B}^0 \cap \Omega_0)$-measurable.

8.7. Definition. *Let $\mathscr{B}^0$ be a σ-ring in $\Omega \neq \varnothing$.*

8.7.1. *A function $f: \Omega \to \mathbb{R}$ is called* **$\mathscr{B}^0$-measurable**, *or* **measurable for $\mathscr{B}^0$**, *if there is a $\varnothing \neq \Omega_0 \in \mathscr{B}^0$ such that:*

8.7.1.1. *f vanishes in $\Omega \backslash \Omega_0$;*

8.7.1.2. *the restriction of f to Ω_0 is $(\mathscr{B}^0 \cap \Omega_0)$-measurable in the sense of Definition 8.1.1. The set of all $\mathscr{B}^0$-measurable functions in $\mathbb{R}^\Omega$ is denoted by* $\mathrm{mble}(\Omega, \mathscr{B}^0, \mathbb{R})$.

8.7.2. *A function $f \in \overline{\mathbb{R}}^\Omega$ is called $\mathscr{B}^0$-measurable if there is a $\varnothing \neq \Omega_0 \in \mathscr{B}^0$ such that:*

8.7.2.1. *f vanishes in $\Omega \backslash \Omega_0$;*

8.7.2.2. *the restriction of f to Ω_0 is $(\mathscr{B}^0 \cap \Omega_0)$-measurable in the sense of Definition 8.1.2. The set of all $\mathscr{B}^0$-measurable functions in $\overline{\mathbb{R}}^\Omega$ is denoted by* $\mathrm{mble}(\Omega, \mathscr{B}^0, \overline{\mathbb{R}})$.

8.7.3. *For every $\varnothing \neq \mathscr{E} \subseteq \overline{\mathbb{R}}^\Omega$ the smallest σ-ring for which all $f \in \mathscr{E}$ are* *measurable, i.e., the σ-ring*

$$B^0(\{\{f \neq 0\} \cap f^{-1}E \mid f \in \mathscr{E}, E \in \mathscr{B}(\overline{\mathbb{R}})\})$$

is denoted by $\mathscr{B}^0(\mathscr{E})$ and called the **σ-ring generated by** $\mathscr{E}$.

8.8. Remarks

8.8.1. The notations mble$(\Omega, \mathscr{B}^0, \mathbb{R})$ and mble$(\Omega, \mathscr{B}^0, \overline{\mathbb{R}})$ are consistent in view of the fact that for $\Omega \in \mathscr{B}^0$, Definition 8.7 coincides with Definition 8.1.

8.8.2. Observe that constants $\neq 0$ are not $\mathscr{B}^0$-measurable if $\Omega \notin \mathscr{B}^0$.

8.9. Exercises. Let $\mathscr{B}^0$ be a σ-ring in $\Omega \neq \varnothing$.

8.9.1. Prove: An $f \in \overline{\mathbb{R}}^\Omega$ that attains only the at most countably many values $\alpha_1, \alpha_2, \ldots \in \overline{\mathbb{R}}$ is $\mathscr{B}^0$-measurable iff:

8.9.1.1. $\Omega \backslash \{f = \alpha_k\} \in \mathscr{B}^0$ for all k with $\alpha_k = 0$ and

8.9.1.2. $\{f = \alpha_k\} \in \mathscr{B}^0$ for all k with $\alpha_k \neq 0$.

8.9.2. Prove that $f \in$ mble$(\Omega, \mathscr{B}^0, m)$ iff supp$(f) = \{f \neq 0\} \in \mathscr{B}^0$ and either $f = 0$ or the restriction of f to supp(f) is $(\mathscr{B}^0 \cap \text{supp}(f))$-measurable. In particular, any supp$(f) \subseteq \Omega_0 \in \mathscr{B}^0$ would do in Definition 8.7.

8.9.3. Prove that $f \in$ mble$(\Omega, \mathscr{B}^0, \mathbb{R})$ iff $\{\omega \mid f(\omega) \in E\} \in \mathscr{B}^0$ $(0 \notin E \in \mathscr{B}(\mathbb{R}))$ and that the analogous statement holds if we replace $\mathbb{R}$ by $\overline{\mathbb{R}}$.

8.9.4. Formulate and prove criteria for $\mathscr{B}^0$-measurability along the lines of Proposition 8.2.

8.9.5. Let $(\Omega, \mathscr{T})$ be a topological space and $\mathscr{B}^0$ be the σ-ring generated by all compact subsets of Ω. Prove $\mathscr{C}^{00}(\Omega, \mathbb{R}) \subseteq$ mble$(\Omega, \mathscr{B}^0, \mathbb{R})$.

8.9.6. Adopt the convention $\infty - \infty = 0 = (-\infty) - (-\infty)$ and prove: If g, $h \in$ mble$(\Omega, \mathscr{B}^0, \overline{\mathbb{R}})$ and $\varepsilon > 0$, then $\{g - h > \varepsilon\}$, $\{h - g > \varepsilon\}$ and $\{|g - h| > \varepsilon\}$ are in $\mathscr{B}^0$. (*Hint*: Distinguish various cases.)

8.10. Proposition. *Let B^0 be a σ-ring in $\Omega \neq \varnothing$.*

8.10.1. mble$(\Omega, \mathscr{B}^0, \mathbb{R})$ *is a vector sublattice of* $\mathbb{R}^\Omega$ *containing all constants and is stable under multiplication. If* $f \in$ mble$(\Omega, \mathscr{B}^0, \mathbb{R})$, $g \in$ mble$(\mathbb{R}, \mathscr{B}(\mathbb{R}), \mathbb{R})$, $g(0) = 0$, *then* $g(f) \in$ mble$(\Omega, \mathscr{B}^0, \mathbb{R})$.

8.10.2. *If* $f_1, f_2, \ldots \in$ mble$(\Omega, B^0, \mathbb{R})$, $f \in \mathbb{R}^\Omega$, *and* $\lim_n f_n(\omega) = f(\omega)$ $(\omega \in \Omega)$, *then* $f \in$ mble$(\Omega, \mathscr{B}^0, \mathbb{R})$.

8.10.3. mble$(\Omega, \mathscr{B}^0, \overline{R})$ *is stable under:*

8.10.3.1. *addition, as far it is meaningful;*

8.10.3.2. *multiplication;*

8.10.3.3. *finite and countable suprema and infima,* lim sup *and* lim inf *of sequences;*

8.10.3.4. *pointwise limits of sequences;*

8.10.3.5. *restricted superposition—if*

$$f \in \mathrm{mble}(\Omega, \mathscr{B}^0, \overline{\mathbb{R}}),\ g \in \mathrm{mble}(\overline{\mathbb{R}}, \mathscr{B}(\overline{\mathbb{R}}), \overline{\mathbb{R}}),\ g(0) = 0,$$

then $g(f) \in \mathrm{mble}(\Omega, \mathscr{B}^0, \overline{\mathbb{R}})$.

Proof. We use Proposition 8.4 in a way that is clearly visible when displayed in a few cases:

3.1. Let $f, g \in \mathrm{mble}(\Omega, \mathscr{B}^0, \overline{\mathbb{R}})$ and $f + g$ be meaningful. Clearly $\Omega_0 = \mathrm{supp}(f) \cup \mathrm{supp}(g) \in \mathscr{B}^0$, the restrictions of f, g to Ω_0 are in $\mathrm{mble}(\Omega_0, \mathscr{B}^0 \cap \Omega_0, \overline{\mathbb{R}})$, their sum is meaningful and hence in $\mathrm{mble}(\Omega_0, \mathscr{B}^0 \cap \Omega_0, \overline{\mathbb{R}})$.

3.3. If $f_1, f_2, \ldots \in \mathrm{mble}(\Omega, \mathscr{B}^0, \mathbb{R})$, then

$$\Omega_0 = \mathrm{supp}(f_1) \cup \mathrm{supp}(f_2) \cup \cdots \in \mathscr{B}^0$$

and the rest goes as above.

3.5. If $0 \notin E \in \mathscr{B}(\overline{\mathbb{R}})$, then

$$0 \notin g^{-1}(E) \in \mathscr{B}^0,\ g(f)^{-1}(E) = f^{-1}(g^{-1}(E)) \in \mathscr{B}^0,$$

hence $g(f) \in \mathrm{mble}(\Omega, \mathscr{B}^0, \overline{\mathbb{R}})$ by Exercise 8.9.3.

8.11. Proposition. *Let $\mathscr{B}^0$ be a σ-ring in $\Omega \neq \varnothing$ and $f, g \in \mathrm{mble}(\Omega, \mathscr{B}^0, \overline{\mathbb{R}})$. Then the following sets are in $\mathscr{B}^0$:*

$$\{f < g\}, \qquad \{f > g\}, \qquad \{f = g \neq 0\}.$$

Proof. Let $\mathbb{Q} \subseteq \mathbb{R}$ be the set of all rationals. Then

$$\{f < g\} = \bigcup_{0 < \beta \in \mathbb{Q}} [\{g > \beta\} \setminus \{f > \beta\}) \cup [\{f < -\beta\} \setminus \{g < -\beta\}]$$

$$\{f = g \neq 0\} = [\mathrm{supp}(f) \cup \mathrm{supp}(g)] \setminus [\{f < g\} \cup \{f > g\}].$$

8.12. Exercises

8.12.1. Let $\mathscr{B}^0$ be a σ-ring in $\Omega \neq \varnothing$. Let $f_1, f_2, \ldots \in \mathrm{mble}(\Omega, \mathscr{B}^0, \overline{\mathbb{R}})$. Prove:

8.12.1.1. $\{\limsup_{n\to\infty} f_n > \liminf_{n\to\infty} f_n\} \in \mathscr{B}^0$.

8.12.1.2. $\{\omega \mid \lim_{n\to\infty} f_n(\omega) \text{ exists}, |f_1(\omega)| \vee |f_2(\omega)| \vee \cdots > 0\} \in \mathscr{B}^0$.

8.12.1.3. $M = \{\omega \mid \lim_{n\to\infty} f_n(\omega) \text{ exists and is finite}, |f_1(\omega)| \vee |f_2(\omega)| \vee \cdots > 0\} \in \mathscr{B}^0$.

8.12.1.4. $$f(\omega) = \begin{cases} 0 & \text{for } \omega \notin M \quad \text{(from 1.3)} \\ \lim_{n\to\infty} f_n(\omega) & \text{for } \omega \in M \end{cases}$$

defines an $f \in \mathrm{mble}(\Omega, \mathscr{B}^0, \overline{\mathbb{R}})$.

8.12.1.5. Prove: If $f \in \mathrm{mble}(\Omega, \mathscr{B}^0, \overline{\mathbb{R}})$, then $|f|^p \in \mathrm{mble}(\Omega, \mathscr{B}^0, \overline{\mathbb{R}})$ $(0 < p < \infty)$.

8.13. Proposition. *Let $\mathscr{B}^0$ be a σ-ring in $\Omega \neq \varnothing$ and*

$$0 \leqq h \in \mathrm{mble}(\Omega, \mathscr{B}^0, \overline{\mathbb{R}}).$$

Then for every $n = 1, 2, \ldots$, the function with finitely many values in $\mathbb{R}_+$ defined by

$$h_n(\omega) = \begin{cases} 0 & \text{for} \quad 0 \leqq h(\omega) < 1/2^n \\ k/2^n & \text{for} \quad k/2^n \leqq h(\omega) < (k+1)/2^n \ (k = 1, 2, \ldots 2^{2n} - 1) \\ 2^n & \text{for} \quad h(\omega) \geqq 2^n \end{cases}$$

is $\mathscr{B}^0$-measurable. We have

$$0 \leqq h_1 \leqq h_2 \leqq \cdots \nearrow h$$

pointwise on Ω, and even uniformly if h is bounded.

Proof. Apply Exercise 8.9.1 in order to show $h_n \in \mathrm{mble}(\Omega, \mathscr{B}^0, \mathbb{R})$. The rest is direct and obvious.

8.14. Definition. *Let $\mathscr{B}^0$ be a σ-ring in $\Omega \neq \varnothing$. A function $f \in \mathbb{C}^\Omega$ is called a* **$\mathscr{B}^0$-measurable complex-valued function** *on Ω if* $\mathrm{Re}(f)$ *and* $\mathrm{Im}(f)$ *are in* $\mathrm{mble}(\Omega, \mathscr{B}^0, \mathbb{R})$. *The set of all $\mathscr{B}^0$-measurable complex-valued functions on Ω is denoted by* $\mathrm{mble}(\Omega, \mathscr{B}^0, \mathbb{C})$.

8.15. Proposition. *Let $\mathscr{B}^0$ be a σ-ring in $\Omega \neq \varnothing$. Then* $\mathrm{mble}(\Omega, \mathscr{B}^0, \mathbb{C})$ *is stable under:*

8.15.1. *complex linear combinations;*

8.15.2. *pointwise limits of sequences;*

8.15.3. *restricted superposition—let $\mathscr{B}(\mathbb{C})$ be the σ-field of all Borel subsets of $\mathbb{C}$ (Exercise I.4.8.7.3) and $g \in \mathrm{mble}(\mathbb{C}, \mathscr{B}(\mathbb{C}), \mathbb{C})$ with $g(0) = 0$. Then*

$$f \in \mathrm{mble}(\Omega, \mathscr{B}^0, \mathbb{C}) \quad \Rightarrow \quad g(f) \in \mathrm{mble}(\Omega, \mathscr{B}^0, \mathbb{C}).$$

Proof. 1 and 2 follow from Proposition 8.4 in an obvious way.

3. Let $0 \notin E \in \mathscr{B}$. Then $[\mathrm{Re}(g(f))]^{-1}(E) = f^{-1}([\mathrm{Re}(g)]^{-1}(E)) \in \mathscr{B}^0$ since $0 \notin \mathrm{Re}(g)^{-1}(E) \in \mathscr{B}(\mathbb{C})$, hence $\mathrm{Re}(g(f)) \in \mathrm{mble}(\Omega, \mathscr{B}^0, \mathbb{R})$.

$$\mathrm{Im}(g(f)) \in \mathrm{mble}(\Omega, \mathscr{B}^0, \mathbb{R})$$

follows in the same fashion.

8.16. Exercises. Let $\mathscr{B}^0$ be a σ-ring in $\Omega \neq \varnothing$.

8.16.1. Assume $\mathscr{B}^0 = \mathscr{B}^0(\mathscr{B}^{00})$ for some local σ-ring $\mathscr{B}^{00}$. Prove that for every $0 \leqq f \in \mathrm{mble}(\Omega, \mathscr{B}^0, \mathbb{R})$ there is a sequence $0 \leqq f_1 \leqq f_2 \leqq \cdots \nearrow f$

(pointwise) such that $f_1, f_2, \ldots$ are in $\mathscr{E}(\mathscr{B}^{00})$, i.e., step functions for $\mathscr{B}^{00}$ (i.e., of the form $\sum \alpha_k 1_{F_k}$, $F_k \in \mathscr{B}^{00}$ (finite sum)).

8.16.2. Prove that for every $h \in \mathbb{C}^\Omega$, the following statements are equivalent:

8.16.2.1. $h \in \text{mble}(\Omega, \mathscr{B}^0, \mathbb{C})$;

8.16.2.2. $h^{-1}(E) \in \mathscr{B}^0$ for every $0 \notin E \in \mathscr{B}(\mathbb{C})$, where $\mathscr{B}(\mathbb{C})$ denotes the σ-field of all Borel subsets of $\mathbb{C}$ (see Exercises I.4.8.7.3, I.4.8.8).

8.16.3. Extend the list of equivalent statements in 8.16.2.

8.16.4. Prove: If $h \in \text{mble}(\Omega, \mathscr{B}^0, \mathbb{C})$, then

$$|h| = \sqrt{\text{Re}(h)^2 + \text{Im}(h)^2} \in \text{mble}(\Omega, \mathscr{B}^0, \mathbb{R}).$$

8.16.5. Prove: If $h \in \text{mble}(\Omega, \mathscr{B}^0, \mathbb{C})$, then

$$|h|^p \in \text{mble}(\Omega, \mathscr{B}^0, \mathbb{R}) \qquad (0 < p < \infty).$$

9. MEASURABILITY AND INTEGRABILITY

In this section we investigate the relations between the concepts of a measurable and an integrable function. It will turn out that every integrable function is equivalent to one that is measurable with respect to the smallest σ-ring with respect to which one can reasonably expect. In many situations this is a sufficient reason to start the presentation of integrability with measurable functions right away. Furthermore, we shall present a sufficient condition for a measurable function to be integrable.

9.1. Proposition. *Let $\mathscr{E}$ be an elementary domain on $\Omega \neq \varnothing$. Then*

$$\mathscr{B}^0(\mathscr{E}) = B^0(\{\{f > \alpha\} \mid f \in \mathscr{E}, \alpha > 0\})$$

satisfies $\mathscr{E} \subseteq \text{mble}(\Omega, \mathscr{B}^0(\mathscr{E}), \mathbb{R})$ and is contained in every σ-ring $\mathscr{B}^0$ with $\mathscr{E} \subseteq \text{mble}(\Omega, \mathscr{B}^0, \mathbb{R})$. We call $\mathscr{B}^0(\mathscr{E})$ the σ-ring generated by $\mathscr{E}$.

Proof. $\mathscr{E} \subseteq \text{mble}(\Omega, \mathscr{B}^0(\mathscr{E}), \mathbb{R})$ is the only nontrivial statement. Since $f \in \mathscr{E}$ implies $f_+, f_- \in \mathscr{E}, f = f_+ - f_-$, it suffices to settle the case $0 \leqq f \in \mathscr{E}$. Construct now $f_1, f_2, \ldots$ in analogy to the proof of Proposition 8.13. Obviously

$$\left\{f_n = \frac{k}{2^n}\right\} = \left\{f \geqq \frac{k}{2^n}\right\} \Big\backslash \left\{f \geqq \frac{k+1}{2^n}\right\} \qquad (k = 1, \ldots, 2^{2n})$$

$\{f_n = 2^n\} = \{f \geqq 2^n\}$. Generally $\{f \geqq \beta\} = \bigcap_{j=k}^{\infty} \{f > \beta - 1/j\}$ $(\beta > 0,$ $\beta - 1/k > 0)$. This shows that f_n takes on its finitely many values $\neq 0$ on sets that can be obtained from the sets that generate $\mathscr{B}^0(\mathscr{E})$ by definition, by operations under which a σ-ring is stable. Hence they are all in $\mathscr{B}^0(\mathscr{E})$. Thus we find $f \in \text{mble}(\Omega, \mathscr{B}^0(\mathscr{E}), \mathbb{R})$ by Exercise 8.9.1 and Propositions 8.10.2, 8.10.3.4.

9.2. Remark. If $\mathscr{R}$ is a set ring in Ω and $\mathscr{E} = \mathscr{E}_{\mathscr{R}}$, then obviously $\mathscr{B}^0(\mathscr{E}) = \mathscr{B}^0(\mathscr{R})$. In particular, every function from mble$(\Omega, \mathscr{B}^0(\mathscr{E}), \mathbb{R})$ vanishes outside some countable union of sets from $\mathscr{R}$.

9.3. Theorem. *Let $(\Omega, \mathscr{E}, m)$ be a σ-measure space. Then every equivalence class in $L^1_\sigma(\Omega, \mathscr{E}, m)$ contains at least one representative from* mble$(\Omega, \mathscr{B}^0(\mathscr{E}), \mathbb{R})$. *The same holds for τ-measure spaces.*

Proof. By Theorem 7.2.4 it is sufficient to settle the σ-case. We combine Theorem 5.10 with a device employed in the proof of Theorem 5.8.4 and find, for a given $h \in L^1_\sigma(\Omega, \mathscr{E}, m)$, a sequence $f_1, f_2, \ldots \in \mathscr{E}$ such that $\lim f_n = h'$ m-σ-a.e. for every representative h' of h. In particular, $N = \{\omega \mid \lim f_n(\omega)$ does not exist or is infinite$\}$ is an m-σ-nullset. By Exercise 8.12.1.4

$$f(\omega) = \begin{cases} 0 & \text{for} \quad \omega \in N \\ \lim f_n(\omega) & \text{for} \quad \omega \notin N \end{cases}$$

defines a function $f \in \text{mble}(\Omega, \mathscr{B}^0(\mathscr{E}), \mathbb{R})$ which, on the other hand, is equivalent to h'.

9.4. Remark. The above theorem justifies the decision of many authors in measure and integration theory to operate entirely within mble$(\Omega, \mathscr{B}^0(\mathscr{E}), \overline{\mathbb{R}})$.

9.5. Proposition. *Let $(\Omega, \mathscr{E}, m)$ be a σ-measure space and*

$$h \in \text{mble}(\Omega, \mathscr{B}^0(\mathscr{E}), \overline{\mathbb{R}}).$$

Then $h \in \mathscr{L}^1_\sigma(\Omega, \mathscr{E}, m)$ iff there is a $g \in \mathscr{L}^1_\sigma(\Omega, \mathscr{E}, m)$ with $|h| \leqq g$ m-σ-a.e.

Proof. Only the "if" part is nontrivial. If h, g are as supposed, then $h_+, h_- \in \text{mble}(\Omega, \mathscr{B}^0(\mathscr{E}), \overline{\mathbb{R}})$, $h_+ \leqq g$, $h_- \leqq g$. Thus it suffices to consider the case $h \geqq 0$. Construct now $0 \leqq h_1 \leqq h_2 \leqq \cdots$ according to Proposition 8.13. Since every h_n is a finite linear combination of functions of the form $1_{\{h \geqq \alpha\}}$ with $\alpha > 0$, which are all in $\mathscr{L}^1_\sigma(\Omega, \mathscr{E}, m)$ by Theorem 6.3, we find $\mathscr{L}^1_\sigma(\Omega, \mathscr{E}, m) \ni h_1, h_2, \ldots \leqq g$. By Theorem 5.3 (monotone convergence) $h \in \mathscr{L}^1_\sigma(\Omega, \mathscr{E}, m)$ follows.

9.6. Exercises. Let $(\Omega, \mathscr{E}, m)$ be a σ-measure space and $(\Omega, \mathscr{B}^{00}_m, m)$ the σ-content space derived from it.

9.6.1. Use Theorem 9.3 and Proposition 9.5 and prove:

9.6.1.1. $h \in \mathscr{L}^1_m$, $F \in \mathscr{B}^{00}_m \Rightarrow 1_F h \in \mathscr{L}^1_m$.

9.6.1.2. $h \in L_m^1, f \in \mathscr{E}(\mathscr{B}_m^{00}) \Rightarrow fh \in \mathscr{L}_m^1$.

9.6.2. Let $\mathscr{B}^0$ be a σ-ring in Ω. Prove that the set $L_m^1 \cap \mathrm{mble}(\Omega, \mathscr{B}^0, \mathbb{R})$ of all $h \in L_m^1$ containing a function from $\mathrm{mble}(\Omega, \mathscr{B}^0, \mathbb{R})$ is a Banach sublattice of L_m^1.

10. INTEGRATION OF COMPLEX-VALUED FUNCTIONS

In this section we introduce the integral for complex-valued functions in the obvious way: splitting into real and imaginary parts. A manageable criterion for integrability will turn out to make use of measurability.

10.1. Definition. *Let $(\Omega, \mathscr{E}, m)$ be a σ-measure space. A function $h \in \mathbb{C}^\Omega$ is called* **m-(σ-)integrable** *if* Re(h) *and* Im(h) *are m-(σ-)integrable real-valued functions. The space of all m-(σ-)integrable functions from $\mathbb{C}^\Omega$ is denoted by $\mathscr{L}_\sigma^1(m, \mathbb{C})$, $\mathscr{L}_m^1(\mathbb{C})$, $\mathscr{L}_\sigma^1(\Omega, \mathscr{E}, m, \mathbb{C})$ etc., depending on what specifications are needed. In contrast to this, the space $\mathscr{L}_m^1$ of $\overline{\mathbb{R}}$-valued m-σ-integrable functions considered so far is also denoted by $\mathscr{L}_m^1(\overline{\mathbb{R}})$, $\mathscr{L}_\sigma^1(m, \overline{\mathbb{R}})$ etc. For every $h \in \mathscr{L}_m^1(\mathbb{C})$, the complex number*

$$m(h) = m(\mathrm{Re}(h)) + im(\mathrm{Im}(h))$$

is called the **m-integral of h** *and also denoted by $\int h\, dm$, etc. The mapping $m\colon \mathscr{L}_m^1(\mathbb{C}) \to \mathbb{C}$ thus defined is also denoted by $\int dm$ and is called the complex m-(σ-)integral. The same words are also used for the couple $\left(\mathscr{L}_m^1(\mathbb{C}), \int dm\right)$. If $(\Omega, \mathscr{E}, m)$ is the σ-measure space derived from a σ-content prespace $(\Omega, \mathscr{R}, m)$, then σ-integrability etc. of functions in $\mathbb{C}$ with respect to the σ-measure m is spoken of synonymously with σ-integrability etc. with respect to the underlying σ-content (prespace), and the notation is adapted accordingly, such that it is, e.g., clear what $\mathscr{L}_\sigma^1(\Omega, \mathscr{R}, m, \mathbb{C})$ means.*

10.2. Theorem. *Let (Ω, E, m) be a σ-measure space. Then:*

10.2.1. *$\mathscr{L}_\sigma^1(m, \mathbb{C})$ is a complex linear subspace of $\mathbb{C}^\Omega$ and $m\colon \mathscr{L}_\sigma^1(m, \mathbb{C}) \to \mathbb{C}$ is (complex) linear.*

$$m(\bar{h}) = \overline{m(h)},$$

where the bar denotes complex conjugation.

10.2.2. *$f \in \mathrm{mble}(\Omega, \mathscr{B}^0(\mathscr{E}), \mathbb{C})$ is in $\mathscr{L}_\sigma^1(m, \mathbb{C})$ iff $|f| \in \mathscr{L}_\sigma^1(m, \mathbb{R})$.*

10.2.3. *$\mathscr{L}_\sigma^1(m, \mathbb{C})$ consists of full m-σ-equivalence classes in $\mathbb{C}^\Omega$ and m is constant on each such equivalence class, the complex linear operations are* *class operations. Thus m may as well be considered as a (complex) linear form on the complex vector space $L_\sigma^1(m, \mathbb{C}) = L_m^1(\mathbb{C})$ (or whatever affixes are to be written in order to clarify the situation) of all equivalence classes* mod m *in* $\mathscr{L}_\sigma^1(m, \mathbb{C})$.

10.2.4. *By*

$$\|h\|_1 = \int |h| \, dm \qquad (h \in \mathscr{L}_\sigma^1(m, \mathbb{C}))$$

a nonnegative real-valued class function $\|\cdot\|_1$ is defined on $L_\sigma^1(m, \mathbb{C})$. It may thus be considered as a function on $L_\sigma^1(m, \mathbb{C})$. As such it is a norm, and with this norm $L_\sigma^1(m, \mathbb{C})$ is a **complex Banach space**.

Proof. 1 is obvious from the definitions.

2. If $f \in \text{mble}(\Omega, \mathscr{B}^0(\mathscr{E}), \mathbb{C}) \cap \mathscr{L}_\sigma^1(m, \mathbb{C})$, then $|f| \in \text{mble}(\Omega, \mathscr{B}^0(\mathscr{E}), \mathbb{R})$ by Exercise 8.16.4. Now $|f| \leqq |\text{Re}(f)| + |\text{Im}(f)| \in \mathscr{L}_\sigma^1(m, \mathbb{R})$. Thus $|f| \in \mathscr{L}_\sigma^1(m, \mathbb{R})$ by Proposition 9.5. Let now conversely $f \in \text{mble}(\Omega, \mathscr{B}^0(\mathscr{E}), \mathbb{C})$ and $|f| \in \mathscr{L}_\sigma^1(m, \mathbb{R})$. Then clearly $\text{Re}(f) \in \text{mble}(\Omega, \mathscr{B}^0(\mathscr{E}), \mathbb{R})$, $|\text{Re}(f| \leqq |f|$ and thus $\text{Re}(f) \in \mathscr{L}_\sigma^1(m, \mathbb{R})$. Similarly, $\text{Im}(f) \in \mathscr{L}_\sigma^1(m, \mathscr{E})$ follows and we are through.

3 is again obvious from the definitions if we observe that coincidence mod m of two complex-valued functions is tantamount to the coincidence mod m of their real and imaginary parts.

4. Choose any $f, g \in \text{mble}(\Omega, \mathscr{B}^0(\mathscr{E}), \mathbb{R})$ such that $f = \text{Re}(h)$, $g = \text{Im}(h)$ mod m (Theorem 9.3). It follows that $h' = f + ig = h$ mod m and $h' \in \text{mble}(\Omega, \mathscr{B}^0(\mathscr{E}), \mathbb{C})$, hence $|h'| \in \mathscr{L}_m^1(\mathbb{C})$ (by 2) and finally $|h| \in \mathscr{L}_m^1(\mathbb{C})$ by 3. Thus $\|\cdot\|_1$ is well defined. Apparently it is a class function on $\mathscr{L}_m^1(\mathbb{C})$. It is now easy to prove that it is a norm and that $L_m^1(\mathbb{C})$ is a Banach space with this norm, simply by going back to the real case and taking advantage of obvious majorization properties. Let us be content to prove that $L_m^1(\mathbb{C})$ is norm complete. In fact, if $h_1, h_2, \ldots \in L_m^1(\mathbb{C})$, $\lim_{j,k \to \infty} \|h_j - h_k\|_1 = 0$, then

$$\lim_{j,k \to \infty} \|\text{Re}(h_j) - \text{Re}(h_k)\|_1 = 0 = \lim_{j,k \to \infty} \|\text{Im}(h_j) - \text{Im}(h_k)\|_1 = 0$$

since

$$|\text{Re}(h_j) - \text{Re}(h_k)| \leqq |h_j - h_k| \geqq |\text{Im}(h_j) - \text{Im}(h_k)|.$$

Now the existence of $f, g \in L_m^1(\mathbb{R})$ with

$$\lim_{k \to \infty} \|\text{Re}(h_k) - f\|_1 = 0 = \lim_{k \to \infty} \|\text{Im}(h_k) - g\|_1$$

follows, and for $h = f + ig$, we get $h \in L^1_m(\mathbb{C})$,

$$\lim_{k\to\infty} \|h_k - h\|_1 \leq \lim_{k\to\infty} \|\text{Re}(h_k) - f\|_1 + \lim_{k\to\infty} \|\text{Im}(h_k) - g\|_1 = 0.$$

10.3. Remark. If $(\Omega, \mathscr{E}, m)$ is a τ-measure space, then the above theory goes through with obvious minimal modifications. $L^1_\tau(m, \mathbb{C})$ differs from $L^1_\sigma(m, \mathbb{C})$ by a mere "fattening" of equivalence classes.

11. THE REAL AND THE COMPLEX HILBERT SPACE L^2

In this section we introduce the notion of square integrability and prove that the space of all square integrable functions, endowed with a natural scalar product, is a Hilbert space, both for the case of real- and of complex-valued functions. We use a technique of reduction to ordinary integrability here. The notion of measurability plays a useful role in majorant criteria for square integrability.

11.1. Definition. *Let $(\Omega, \mathscr{E}, m)$ be a σ-measure space.*

11.1.1. *A function $h \in \overline{\mathbb{R}}^\Omega$ is called* **square m-(σ-)integrable** *if*

$$h \cdot |h| \in \mathscr{L}^1_\sigma(m, \overline{\mathbb{R}}).$$

The set of all square m-σ-integrable functions $\in \overline{\mathbb{R}}^\Omega$ is denoted by $\mathscr{L}^2(m, \overline{\mathbb{R}})$, $\mathscr{L}^2_m(\overline{\mathbb{R}})$, etc.

11.1.2. *A function $h \in \mathbb{C}^\Omega$ is called a* **square m-(σ-)integrable complex-valued function** *if $h \cdot |h| \in \mathscr{L}^1_\sigma(m, \mathbb{C})$. The set of all square m-σ-integrable complex-valued functions on Ω is denoted by $\mathscr{L}^2(m, \mathbb{C})$, $\mathscr{L}^2_m(\mathbb{C})$, etc.*

11.2. Theorem. *Let $(\Omega, \mathscr{E}, m)$ be a σ-measure space.*

11.2.1. *$\mathscr{L}^2_\sigma(m, \overline{\mathbb{R}})$ consists of full equivalence classes* mod m *in $\overline{\mathbb{R}}^\Omega$. The set of all these equivalence classes is denoted by $L^2_\sigma(m, \overline{\mathbb{R}})$, $L^2_m(\overline{\mathbb{R}})$, etc. Every equivalence class in $L^2_m(\overline{\mathbb{R}})$ has a representative in $\mathbb{R}^\Omega$, and $L^2_m(\overline{\mathbb{R}}) \cap \mathbb{R}^\Omega$ is a vector sublattice of $\mathbb{R}^\Omega$, hence $L^2_m(\overline{\mathbb{R}})$ is a* **vector lattice** *in an obvious sense. Every equivalence class in $L^2_m(\overline{\mathbb{R}})$ has a representative in* mble$(\Omega, \mathscr{B}^0(\mathscr{E}), \mathbb{R})$. *A function $h \in$* mble$(\Omega, \mathscr{B}^0(\mathscr{E}), \overline{\mathbb{R}})$ *is in $\mathscr{L}^2_m(\overline{\mathbb{R}})$ iff $|h|^2 \in \mathscr{L}^1_m(\overline{\mathbb{R}})$.*

11.2.2. *For any $g, h \in \mathscr{L}^2_m(\overline{\mathbb{R}})$ the function gh is in $\mathscr{L}^1_m(\overline{\mathbb{R}})$. By*

$$(g, h) = \int gh \, dm \qquad (g, h \in \mathscr{L}^2_m(\overline{\mathbb{R}}))$$

a function $(\cdot, \cdot)$ is defined on $\mathscr{L}^2_m(\overline{\mathbb{R}}) \times \mathscr{L}^2_m(\overline{\mathbb{R}})$ which is a class function in each variable. It may thus as well be considered as a function on

$L_m^2(\mathbb{R}) \times L_m^2(\mathbb{R})$. As such it is a scalar product and $L_m^2(\mathbb{R})$ is a real **Hilbert space** *with this scalar product, i.e.,*

11.2.2.1. *(g, h) is bilinear; i.e.,*

$$(\alpha_1 g_1 + \alpha_2 g_2, h) = \alpha_1(g_1, h) + \alpha_2(g_2, h),$$
$$(g, \beta_1 h_1 + \beta_2 h_2) = \beta_1(g, h_1) + \beta_2(g, h_2)$$
$$(\alpha_1, \alpha_2, \beta_1, \beta_2 \in \mathbb{R}, \quad g_1, g_2, g, h_1, h_2, h \in L_m^2(\mathbb{R}));$$

11.2.2.2. *(g, h) is symmetric, i.e.,*

$$(g, h) = (h, g) \qquad (g, h \in L_m^2(\mathbb{R}))$$

11.2.2.3. *By $\|h\|_2 = \sqrt{(h, h)}$ a norm is defined on the real vector space $L_m^2(\mathbb{R})$.*

11.2.2.4. *With the norm $\|\cdot\|_2$, $L_m^2(\mathbb{R})$ is* **complete**.

11.2.3. *If $\mathscr{E}$ is stable under $f \to \sqrt{|f|}$, then $\mathscr{E}$ is $\|\cdot\|_2$-norm dense in $L_m^2(\overline{\mathbb{R}})$.*

11.2.4. *Every $\|\cdot\|_2$-norm fundamental sequence in $L_m^2(\mathbb{R})$ has an m-a.e. convergent subsequence and is $\|\cdot\|_2$-convergent to the limit of the latter.*

Proof. 1. If $h = h' \bmod m$, then clearly $h|h| = h'|h'| \bmod m$, proving that $\mathscr{L}_\sigma^2(m, \overline{\mathbb{R}})$ consists of full equivalence classes in $\overline{\mathbb{R}}^\Omega$. If $h \in \mathscr{L}_m^2(\overline{\mathbb{R}})$ and $g = h|h| \bmod m$ and $g \in \text{mble}(\Omega, \mathscr{B}^0(\mathscr{E}), \mathbb{R})$, then

$$h'(\omega) = \begin{cases} \sqrt{g(\omega)} & \text{if } \ g(\omega) \geqq 0 \\ -\sqrt{|g(\omega)|} & \text{if } \ g(\omega) < 0 \end{cases}$$

is in $\text{mble}(\Omega, \mathscr{B}^0(\mathscr{E}), \mathbb{R})$, by Proposition 8.10.3.5. Clearly $h' = h \bmod m$. Let now $h_1, h_2 \in \mathscr{L}_m^2(\overline{\mathbb{R}}) \cap \text{mble}(\Omega, \mathscr{B}^0(\mathscr{E}), \mathbb{R})$. Then

$$\begin{aligned} |(h_1 + h_2)|h_1 + h_2|| &= |h_1 + h_2|^2 \leqq (|h_1| + |h_2|)^2 \\ &\leqq |h_1|^2 + |h_2|^2 + 2|h_1||h_2| \\ &\leqq 4|h_1|^2 \vee |h_2|^2 \end{aligned}$$

proves that $(h_1 + h_2)|h_1 + h_2| \in \mathscr{L}_m^1(\overline{\mathbb{R}})$ (use Propositions 8.10.3.5 and 9.5). But this shows $h_1 + h_2 \in \mathscr{L}_m^2(\overline{\mathbb{R}}) \cap \text{mble}(\Omega, \mathscr{B}^0(\mathscr{E}), \mathbb{R})$. The rest of 1 is now obvious.

In 2 only the following deductions are nontrivial:

2.3. $\|h\|_2 = 0 \Rightarrow \|h\|_2^2 = \int |h|^2 \, dm = 0 \Rightarrow h = 0 \bmod m$.

2.4. For arbitrary $a, b \in \mathbb{R}$, we have

$$\tfrac{1}{2}|a - b|^2 < |a \cdot |a| - b \cdot |b|| \leqq |a - b|(|a| + |b|)$$

(this is easily seen—if $0 < b < a$, then

$$\tfrac{1}{2}|a-b|^2 = \tfrac{1}{2}(a-b)^2 \leqq (a-b)^2 = a^2 - 2ab + b^2$$
$$\leqq a^2 - 2b^2 + b^2 = a^2 - b^2 = \big|a\cdot|a| - b\cdot|b|\big|$$

and

$$\tfrac{1}{2}|a-(-b)|^2 = \tfrac{1}{2}(a^2 + 2ab + b^2) \leqq \tfrac{1}{2}(a^2+b^2) + ab$$
$$\leqq \tfrac{1}{2}(a^2+b^2) + \tfrac{1}{2}(a^2+b^2) = a^2 + b^2 = \big|a\cdot|a| - (-b)\cdot|b|\big|,$$

the rest is obvious). The mapping

$$\varphi : h \quad\to\quad h\cdot|h|$$

sends $\mathscr{L}_m^2(\overline{\mathbb{R}}) \cap \text{mble}(\Omega, \mathscr{B}^0(\mathscr{E}), \mathbb{R})$ onto $\mathscr{L}_m^1(\overline{\mathbb{R}}) \cap \text{mble}(\Omega, \mathscr{B}^0(\mathscr{E}), \mathbb{R})$ in a bijective manner and preserves equivalence mod m. Our above inequalities imply, for $g, h \in \mathscr{L}_m^2(\overline{\mathbb{R}}) \cap \text{mble}(\Omega, \mathscr{B}^0(\mathscr{E}), \mathbb{R})$,

$$\tfrac{1}{2}|g-h|^2 \leqq |\varphi(g) - \varphi(h)| \leqq |g-h|(|g| + |h|)$$

and hence

$$\tfrac{1}{2}\|g-h\|_2^2 \leqq \|\varphi(g) - \varphi(h)\|_1$$
$$\leqq \int |g-h|(|g| + |h|)\, dm = (|g-h|, |g| + |h|)$$
$$\leqq \|g-h\|_2(\|g\|_2 + \|h\|_2)$$

where we have used Schwarz's inequality for the scalar product. Now if $h_1, h_2, \ldots \in \mathscr{L}_m^2(\overline{\mathbb{R}}) \cap \text{mble}(\Omega, \mathscr{B}^0(\mathscr{E}), \mathbb{R})$ is a fundamental sequence in the sense that $\lim_{j,k\to\infty} \|h_j - h_k\|_2 = 0$, then $\|h_j\|_2 \leqq K$ for some constant $K < \infty$; and we may deduce

$$\lim_{j,k\to\infty} \|\varphi(h_j) - \varphi(h_k)\|_1 \leqq \lim_{j,k\to\infty} \|h_j - h_k\|_2 \cdot K = 0.$$

Thus there is some $g' \in \mathscr{L}_m^1(\overline{\mathbb{R}}) \cap \text{mble}(\Omega, \mathscr{B}^0(\mathscr{E}), \mathbb{R})$ with

$$\lim_{k\to\infty} \|\varphi(h_k) - g'\|_1 = 0$$

(use the completeness of L_m^1 (Theorem 5.8.2.3 and Theorem 10.2.4)). Choose $h \in \mathscr{L}_m^1(\overline{\mathbb{R}}) \cap \text{mble}(\Omega, \mathscr{B}^0(\mathscr{E}), \mathbb{R})$ with $\varphi(h) = g'$. Then

$$\lim_{k\to\infty} \|\varphi(h_k) - \varphi(h)\|_1 = 0$$

follows, and by our above general estimate $\lim_{k\to\infty} \|h_j - h\|_2 = 0$ obtains.

3. We use the above mapping φ and see: If

$$h \in \mathscr{L}_m^2(\overline{\mathbb{R}}) \cap \text{mble}(\Omega, \mathscr{B}^0(\mathscr{E}), \mathbb{R}),$$

then $\varphi(h) \in \mathscr{L}_m^1(\overline{\mathbb{R}}) \cap \text{mble}(\Omega, \mathscr{B}^0(\mathscr{E}), \mathbb{R})$; and we may thus find $f_1, f_2, \ldots \in \mathscr{E}$ with $\lim_{k\to\infty} \|f_k - \varphi(h)\|_1 = 0$ (Theorem 5.10). According to the assumed stability of $\mathscr{E}$, we may (going back to positive and negative parts) find $g_1, g_2, \ldots \in \mathscr{E}$ with $\varphi(g_1) = f_1$, $\varphi(g_2) = f_2, \ldots$ and hence

$$\lim_{k\to\infty} \|\varphi(g_k) - \varphi(h)\|_1 = 0,$$

which implies $\lim_{k\to\infty} \|g_k - h\|_2 = 0$ by our previous argument.

4 can easily be deduced from Theorem 5.8.4 by using φ^{-1} and φ (exercise).

11.3. Exercises. Let $(\Omega, \mathscr{E}, m)$ be a σ-measure space.

11.3.1. Prove the monotone convergence theorem for $L_m^2(\overline{\mathbb{R}})$: If $L_m^2(\overline{\mathbb{R}}) \ni h_1 \leqq h_2 \leqq \cdots$ and $\lim_{n\to\infty} \|h_n\|_2 < \infty$, then the m-a.e. defined function $h = \lim_{n\to\infty} h_n$ is in $L_m^2(\overline{\mathbb{R}})$ and $\lim_{n\to\infty} \|h_n - h\|_2 = 0$.

11.3.2. Prove a "filter" analogue of 1.

11.3.3. Formulate and prove a Fatou lemma and Lebesgue's dominated convergence theorem for $L_m^2(\overline{\mathbb{R}})$.

11.3.4. Prove $\mathscr{L}_m^2(\overline{\mathbb{R}}) \subseteq \mathscr{L}_m^1(\overline{\mathbb{R}})$ in case $\mathscr{L}_m^1$ contains all constants. Prove that $L_m^2 \subseteq L_m^1$ is L^1-norm dense in L_m^1. Prove $\|\cdot\|_1 \leqq \|\cdot\|_2$ if in addition $\int 1\, dm = 1$ holds. (*Hint*: If $h \in \text{mble}(\Omega, \mathscr{B}^0(\mathscr{E}), \overline{\mathbb{R}}) \cap \mathscr{L}_m^2(\overline{\mathbb{R}})$, then

$$|h| \leqq |h|^2 \vee 1 \in \mathscr{L}_m^1;$$

if $\int 1\, dm = 1$, then Jensen's inequality applies; for $\mathscr{R} = \mathscr{B}(\mathscr{E})$, $\mathscr{E}(\mathscr{R}) \subseteq L_m^2$ is L^1-norm dense in L_m^1.)

11.3.5. Let $1 \leqq p < \infty$. Call $h \in \overline{\mathbb{R}}^\Omega$ p-integrable for m if there is an $h' \in \text{mble}(\Omega, \mathscr{B}^0(\mathscr{E}), \mathbb{R})$ with $h' = h \bmod m$ and $(h')^p \in \mathscr{L}_m^1(\overline{\mathbb{R}})$. Denote by $L_m^p(\overline{\mathbb{R}})$ the set of all p-integrable (for m) functions in $\mathbb{R}^\Omega$. Formulate and prove results for $L_m^p(\overline{\mathbb{R}})$ analogous to those obtained for $L_m^2(\overline{\mathbb{R}})$ so far. (*Hint*: Use the function mapping $h \to |h|^{p-1} \cdot h$.)

11.3.6. Prove that for every σ-ring $\mathscr{B}^0$ in Ω the set $L_m^2 \cap \text{mble}(\Omega, \mathscr{B}^0, \mathbb{R})$ of all $h \in L_m^2$ containing a function from $\text{mble}(\Omega, \mathscr{B}^0, \mathbb{R})$ is a norm closed linear subspace of the Hilbert space L_m^2.

11.4. Definition. *Let $(\Omega, \mathscr{E}, m)$ be a σ-measure space. A function $h \in \mathbb{C}^\Omega$ is called a* **complex-valued square m-(σ-)integrable function** *if* Re(h) *and* Im(h) *are in $L_m^2(\overline{\mathbb{R}})$. The set of all complex-valued square m-σ-integrable functions on Ω is denoted by $\mathscr{L}_\sigma^2(m, \mathbb{C})$, $\mathscr{L}_m^2(\mathbb{C})$, etc.*

11.5. Proposition. *Let $(\Omega, \mathscr{E}, m)$ be a σ-measure space.*

11.5.1. *$\mathcal{L}^2_\sigma(m, \mathbb{C})$ consists of full equivalence classes in $\mathbb{C}^\Omega$. The set of all these equivalence classes is denoted by $L^2_\sigma(m, \mathbb{C})$, $L^2_m(\mathbb{C})$, etc. Every equivalence class in $L^2_m(\mathbb{C})$ has a representative in the complex linear space* mble $(\Omega, \mathcal{B}^0(\mathcal{E}), \mathbb{C})$ *and $L^2_m(\mathbb{C})$ can be considered as a complex vector space upon calculating with representatives. $h \in$* mble$(\Omega, \mathcal{B}^0(\mathcal{E}), \mathbb{C})$ *is in $L^2_m(\mathbb{C})$ iff* $|h|^2 \in L^2_m(\mathbb{R})$.

11.5.2. *For any $g, h \in \mathcal{L}^2_m(\mathbb{C})$ the function gh is in $\mathcal{L}^1_m(\mathbb{C})$*

$$(g, h) = \int g\bar{h}\, dm \qquad (g, h \in \mathcal{L}^2_m(\mathbb{C}))$$

defines a scalar product on $L^2_m(\mathbb{C})$ (the bar denotes complex conjugation): (g, h) is complex linear in g, $(h, g) = \overline{(g, h)}$ and $\|h\|_2 = \sqrt{(h, h)}$ defines a norm $\|\cdot\|_2$ in $L^2_m(\mathbb{C})$. With this norm, $L^2_m(\mathbb{C})$ is complete, i.e., a **complex Hilbert space**.

The proof (by going back to measurable representatives and real and imaginary parts) is left as an exercise to the reader.

11.6. Exercise. Let $(\Omega, \mathcal{E}, m)$ be a σ-measure space.

11.6.1. Formulate and prove an analogue to Lebesgue's dominated convergence theorem for $L^2_m(\mathbb{C})$.

11.6.2. Assume that m is even a τ-measure. Prove that the obvious definitions of $\mathcal{L}^2_\tau(m, \overline{\mathbb{R}})$ and $\mathcal{L}^2_\tau(m, \mathbb{C})$ lead to a mere "fattening" of equivalence classes, as compared with $\mathcal{L}^2_\sigma(m, \overline{\mathbb{R}})$, $\mathcal{L}^2_\tau(m, \mathbb{C})$. In particular, $L^1_\tau(m, \overline{\mathbb{R}})$ is isomorphic to $L^1_\sigma(m, \overline{\mathbb{R}})$ in an obvious way.

11.6.3. Carry over Exercise 11.3.6 to the complex case.

12. STOCHASTIC CONVERGENCE AND UNIFORM INTEGRABILITY

In connection with content and measure theory we have considered essentially three convergence concepts for functions so far: pointwise convergence, convergence a.e. for a given σ-measure or σ-content, and L^1-norm resp. L^2-norm convergence for a given σ-measure or σ-content. In this section we add one more convergence concept: *stochastic convergence* for a given σ-content. We deem it practical to set out from a σ-content space rather than from a σ-measure space here. The reader is reminded of Remark 6.6 and the flexibility of assumptions to which we are entitled.

12.1. Definition. *Let $(\Omega, \mathcal{B}^{00}, m)$ be a σ-content space and $\mathcal{B}^0 = \mathcal{B}^0(\mathcal{B}^{00})$ the σ-ring generated by $\mathcal{B}^{00}$. Let $f, f_1, f_2, \ldots$* mble$(\Omega, \mathcal{B}^0, \mathbb{R})$. *We* *shall say that the sequence* $f_1, f_2, \ldots$ **converges *m*-stochastically** *toward* f, *and write*

$$f_n \to f \qquad (m\text{-}stochastically)$$

if

$$\lim_{n\to\infty} m(\{|f_n - f| > \varepsilon\} \cap F) = 0 \qquad (\varepsilon > 0, \quad F \in \mathscr{B}^{00}).$$

12.2. Remarks

12.2.1. The reader should observe that $\{|f_n - f| > \varepsilon\} \in \mathscr{B}^0$ (Exercise 8.9.3); by Proposition I.4.7.2.2 we have $\{|f_n - f| > \varepsilon\} \cap F \in \mathscr{B}^{00}$ $(F \in \mathscr{B}^{00})$, hence our definition is in fact meaningful.

12.2.2. In the above definition $\mathscr{B}^{00}$ plays a basic role. If $(\Omega, \tilde{\mathscr{B}}^{00}, m^0)$ is the full extension of $(\Omega, \mathscr{B}^{00}, m)$ (Theorem II.3.1), then clearly $\mathscr{B}^0(\tilde{\mathscr{B}}^{00}) = \mathscr{B}^0$ and one might ask whether the meaning of our definition changes if we replace $\mathscr{B}^{00}$ by $\tilde{\mathscr{B}}^{00}$. But Theorem II.3.1 and its proof show that every $E \in \tilde{\mathscr{B}}^{00}$ has, for any given $\delta > 0$, a representation $E = F + F'$ with $F \in \mathscr{B}^{00}$ and $m^0(F') < \delta$. From this we easily deduce

$$m^0(\{|f_n - f| > \varepsilon\} \cap E \quad \to \quad 0 \qquad (\varepsilon > 0,\ E \in \tilde{\mathscr{B}}^{00})$$

from $m(\{|f_n - f| > \varepsilon\} \cap F) \to 0$ $(\varepsilon > 0,\ F \in \mathscr{B}^{00})$.

12.2.3. Clearly $f_n \to f$ (m-stochastically),

$$f' = f, f'_1 = f_1, \ldots \bmod m \quad \Rightarrow \quad f'_n \to f' \qquad (m\text{-stochastically}),$$

i.e., stochastic convergence is a class relation. On the other hand, $f_n \to f$, $f_n \to f'$ (m-stochastically) implies

$$\begin{aligned} &m(\{|f - f'| > \varepsilon\} \cap F) \\ &\qquad \leq m(\{|f_n - f| > \varepsilon/2\} \cap F) + m(\{|f_n - f'| > \varepsilon/2\} \cap F) \quad (n = 1, 2, \ldots) \end{aligned}$$

and since this becomes arbitrarily small for $n \to \infty$, $\{|f - f'| > \varepsilon\} \cap F$ is an m-nullset, for every $F \in \mathscr{B}^{00}$. Since

$$\{f \neq f'\} \cap F = \bigcup_{n=1}^{\infty} \{|f - f'| > 1/n\} \cap F,$$

we find that $\{f \neq f'\} \cap F$ is an m-nullset for every $F \in \mathscr{B}^{00}$. But $f = f' = 0$ outside some set from $\mathscr{B}^0$, i.e., outside a countable union of sets from $\mathscr{B}^{00}$, hence $\{f \neq f'\}$ is an m-nullset itself. It follows that an m-stochastic limit of a sequence is uniquely determined mod m. It follows as well that the notion of stochastic convergence can easily be carried over to functions that are equivalent mod m to functions with values in $\mathbb{R}$. We carry out our theory for real-valued functions only since we can apply all linear operations freely in that case.

12.2.4. Clearly every subsequence of an m-stochastically convergent sequence is m-stochastically convergent to the same limit.

12.3. Proposition. *Let* $(\Omega, \mathscr{B}^{00}, m)$ *be a* σ*-content space and* $\mathscr{B}^0 = \mathscr{B}^0(\mathscr{B}^{00})$. *Let* $f, f_1, f_2, \ldots \in \text{mble}(\Omega, \mathscr{B}^0, \mathbb{R})$. *Then:*

12.3.1. $f_n \to f$ m-a.e. $\Rightarrow f_n \to f$ *(m-stochastically);*

12.3.2. $f, f_1, f_2, \ldots \in \mathscr{L}_m^1$, $\lim_{n\to\infty} \|f_n - f\|_1 = 0 \Rightarrow f_n \to f$ *(m-stochastically). In fact,* **Tchebyshev's inequality**

$$m(|f_n - f| > \varepsilon) \leqq \varepsilon^{-1}\|f_n - f\|_1 \qquad (\varepsilon > 0, \quad n = 1, 2, \ldots)$$

holds.

12.3.3. $f, f_1, f_2, \ldots \in \mathscr{L}_m^2$, $\lim_{n\to\infty} \|f_n - f\|_2 = 0 \Rightarrow f_n \to f$ *(m-stochastically). In fact,* **Tchebyshev's inequality**

$$m(|f_n - f| > \varepsilon) \leqq \varepsilon^{-2}\|f_n - f\|_2^2 \qquad (\varepsilon > 0, \quad n = 1, 2, \ldots)$$

holds.

Proof. 1. $\{|f_n - f| > \varepsilon\} \subseteq \bigcup_{k \geqq n} \{|f_k - f| > \varepsilon\}$. Call the set on the right-hand side $E_n(\varepsilon)$. Clearly $E_1(\varepsilon) \supseteq E_2(\varepsilon) \supseteq \cdots$ and $f_n \to f$ m-a.e. means that $\bigcap_{n=1}^{\infty} E_n(\varepsilon)$ is an m-nullset $N(\varepsilon)$ $(\varepsilon > 0)$. In fact $\bigcup_{k=1}^{\infty} N(1/k)$ is exactly the exceptional set for $\lim_{n\to\infty} f_n(\omega) = f(\omega)$. Thus we see that for every $F \in \mathscr{B}^{00}$, we have $\lim_{n\to\infty} m(E_n(\varepsilon) \cap F) = m(N(\varepsilon) \cap F) = 0$ (it is important to have some $F \in \mathscr{B}^{00}$ here in order to be sure that $E_n(\varepsilon) \cap F \in \mathscr{B}^{00}$, which is essential for the application of Exercise I.2.3.2).

2. Clearly it is sufficient to prove Tchebyshev's inequality as stated in the theorem. This inequality is a consequence of

$$\varepsilon \cdot 1_{\{|f_n - f| > \varepsilon\}} \leqq |f_n - f|,$$

which is obvious.

3 is proved in way similar to 2.

The rest of this subsection is devoted to conclusions that start essentially with m-stochastic convergence and end up with m-a.e. convergence.

12.4. Definition. *Let* $(\Omega, \mathscr{B}^{00}, m)$ *be a* σ*-content space and* $\mathscr{B}^0 = \mathscr{B}^0(\mathscr{B}^{00})$. *A sequence* $f_1, f_2, \ldots \in \text{mble}(\Omega, \mathscr{B}^0, \mathbb{R})$ *is called an* ***m*-stochastic fundamental sequence** *if*

$$\lim_{j,k\to\infty} m(\{|f_j - f_k| > \varepsilon\} \cap F) = 0 \qquad (\varepsilon > 0, \quad F \in \mathscr{B}^{00}).$$

12.5. Remarks.

12.5.1. $\{|f_j - f_k| > \varepsilon\} \cap F \in \mathscr{B}^{00}$ $(F \in \mathscr{B}^{00})$ follows as in Remark 12.2.1.

12.5.2. The device employed in Remark 12.2.2 shows that we might replace $\mathscr{B}^{00}$ by $\tilde{\mathscr{B}}^{00}$ (from the full extension of $(\Omega, \mathscr{B}^{00}, m)$) without changing the substance of the above definition.

12.5.3. To be an m-stochastic fundamental sequence is clearly a class property for m.

12.5.4. Clearly every subsequence of a m-stochastic fundamental sequence is an m-stochastic fundamental sequence again.

12.5.5. Obviously every norm fundamental sequence in

$$L_m^1 \cap \text{mble}(\Omega, \mathscr{B}^0, \mathbb{R})$$

or $L_m^2 \cap \text{mble}(\Omega, \mathscr{B}^0, \mathbb{R})$ is an m-stochastic fundamental sequence (adapt the corresponding Tchebyshev inequalities).

12.6. Proposition. *Let $(\Omega, \mathscr{B}^{00}, m)$ be a σ-content space and $\mathscr{B}^0 = \mathscr{B}^0(\mathscr{B}^{00})$. Then every m-stochastic fundamental sequence has a subsequence that converges m-a.e. to a $\mathscr{B}^0$-measurable function.*

Proof. Let $f_1, f_2, \ldots \in \text{mble}(\Omega, \mathscr{B}^0, \mathbb{R})$ be an m-stochastic fundamental sequence. Clearly $\text{supp}(f_1) \cup \text{supp}(f_2) \cup \cdots \in \mathscr{B}^0$ again, i.e., there are $F_1, F_2, \ldots \in \mathscr{B}^{00}$ such that $f_1, f_2, \ldots$ vanish outside $F_1 \cup F_2 \cup \cdots$. Thus it is sufficient to prove that for every $F \in \mathscr{B}^{00}$, there is a subsequence of $f_1, f_2, \ldots$ that converges m-a.e. on F since an application of this result to $F_1, F_2, \ldots$ and an obvious diagonal procedure then lead to an m-a.e. convergent subsequence of $f_1, f_2, \ldots$. It is, for a given $F \in \mathscr{B}^{00}$, easy to find $n_1 < n_2 < \cdots$ such that the set $E_k = \{|f_{n_{k+1}} - f_{n_k}| > 1/2^k\} \cap F$ satisfies $m(E_k) < 1/2^k$ $(k = 1, 2, \ldots)$. Let $N = \bigcap_{k=1}^{\infty} \bigcup_{j=k}^{\infty} E_j$. Clearly

$$m(N) \leqq \sum_{j=k}^{\infty} m(E_j) \leqq \frac{1}{2^{k-1}} \qquad (k = 2, 3, \ldots)$$

hence N is an m-nullset. Let $\omega \in E \backslash N$. Then there is a k such that $|f_{n_{j+1}}(\omega) - f_{n_j}(\omega)| < 1/2^j$ $(j \geqq k)$. Thus $\lim_{j \to \infty} f_{n_j}(\omega) \in \mathbb{R}$ exists. The rest is obvious.

12.7. Corollary. *Let $(\Omega, \mathscr{B}^0, m)$ be a σ-content space and $\mathscr{B}^0 = \mathscr{B}^0(\mathscr{B}^{00})$. Then every m-stochastically convergent sequence in* $\text{mble}(\Omega, \mathscr{B}^0, \mathbb{R})$ *has a subsequence that converges m-a.e. to the same limit.*

12.8. Remark. By combination of these results with Theorem 9.3 and Propositions 12.3.2, 12.3.3, we get another proof of the m-a.e. convergence statement of parts of Theorems 5.8.4 and 11.2.4.

12.9. Proposition. *Let $(\Omega, \mathscr{B}^{00}, m)$ be a σ-content space and $\mathscr{B}^0 = \mathscr{B}^0(\mathscr{B}^{00})$. Then for any $f, f_1, f_2, \ldots \in$* $\text{mble}(\Omega, \mathscr{B}^0, \mathbb{R})$ *the following statements are equivalent:*

12.9.1. *$f_n \to f$ (m-stochastically).*

12.9.2. *Every subsequence of $f_1, f_2, \ldots$ has a subsequence that converges m-a.e. to f.*

Proof. $1 \Rightarrow 2$ follows from Corollary 12.7.

$2 \Rightarrow 1$. If $f_n \to f$ (m-stochastically) fails, then there is an $F \in \mathscr{B}^{00}$, an $F \in \mathscr{B}^{00}$, an $\varepsilon > 0$, and a $\delta > 0$ such that

$$m(\{|f_{n_k} - f| > \varepsilon\} \cap F) \geqq \delta \qquad (k = 1, 2, \ldots)$$

for suitable $n_1 < n_2 < \cdots$. By 2 there is $\lim_{\nu\to\infty} f_{n_{k_\nu}} = f$ m-a.e. for suitable $k_1 < k_2 < \cdots$, which by Proposition 12.3.1 implies

$$\lim_{\nu\to\infty} m(\{|f_{n_{k_\nu}} - f| \geqq \varepsilon\} \cap F) = 0,$$

a contradiction.

12.10. Exercises. Let $(\Omega, \mathscr{B}^{00}, m)$ be a σ-content space and $\mathscr{B}^0 = \mathscr{B}^0(\mathscr{B}^{00})$.

12.10.1. Let $f, g \in \text{mble}(\Omega, \mathscr{B}^0, \mathbb{R})$ such that $f = g \bmod m$ fails. Let $f_{2n} = f$, $f_{2n-1} = g$ $(n = 1, 2, \ldots)$. Prove that the sequence $f_1, f_2, \ldots$ has no m-stochastic limit, but every subsequence has a pointwise convergent subsequence.

12.10.2. Let $f, f_1, f_2, \ldots \in \text{mble}(\Omega, \mathscr{B}^0, \mathbb{R})$. Prove that the following two statements are equivalent:

12.10.2.1. $f_n \to f$ (m-stochastically);

12.10.2.2. $\arctan f_n \to \arctan f$ (m-stochastically).

12.10.3. Let $F_0 \in \mathscr{B}^{00}$, $m(F_0) > 0$, and $f_n = n 1_{F_0}$ $(n = 1, 2, \ldots)$. Prove that $\arctan f_n$ $(n = 1, 2, \ldots)$ is an m-stochastic fundamental sequence.

12.10.4. Let $E, F, F_1, F_2, \ldots \in \mathscr{B}^{00}$, $F, F_1, F_2, \ldots \subseteq E$. Prove that $1_{F_n} \to 1_F$ (m-stochastically) iff $m(F_n \,\Delta\, F) \to 0$ as $n \to \infty$.

12.10.5. Let $1 \in \mathscr{L}^1_\sigma(m)$. Prove:

12.10.5.1.

$$d(f, g) = \inf\{\varepsilon \mid \varepsilon > 0, m(|f - g| > \varepsilon) < \varepsilon\} \qquad (f, g \in \text{mble}(\Omega, \mathscr{B}^0, \mathbb{R}))$$

defines a pseudo-metric in $\text{mble}(\Omega, \mathscr{B}^0, \mathbb{R})$, more precisely is nonnegative and symmetric and satisfies the triangle inequality, but $d(f, g) = 0$ implies only $f = g \bmod m$) and $f_n \to f$ (m-stochastically) holds for $f, f_1, f_2, \ldots \in \text{mble}(\Omega, \mathscr{B}^0, \mathbb{R})$ iff $d(f_n, f) = 0$ $(n \to \infty)$.

12.10.5.2.

$$d'(f, g) = \int \frac{|f - g|}{1 + |f - g|}\, dm \qquad (f, g \in \text{mble}(\Omega, \mathscr{B}^0, \mathbb{R}))$$

defines a pseudo-metric in mble$(\Omega, \mathscr{B}^0, \mathbb{R})$ such that for

$$f, f_1, f_2, \ldots \in \text{mble}(\Omega, \mathscr{B}^0, \mathbb{R}), f_n \to f \ (m\text{-stochastically})$$

holds iff $d'(f_n, f) \to 0$ $(n \to \infty)$.

12.10.5.3. Formulate and prove m-stochastic (sequential) continuity of all finite vector lattice operations in mble$(\Omega, \mathscr{B}^0, \mathbb{R})$.

12.10.6. Analyze the proof of Proposition 12.6 and prove the so-called **Egorov theorem:** Let $(\Omega, \mathscr{B}^{00}, m)$ be a σ-content space and f, f_1, $f_2, \ldots \in$ mble$(\Omega, \mathscr{B}^0, \mathbb{R})$ be such that $f_n \to f$ m-a.e. Then for every $F \in \mathscr{B}^{00}$ and every $\varepsilon > 0$, there is a $\mathscr{B}^{00} \ni R \subseteq F$ such that $m(R) < \varepsilon$ and $f_n \to f$ uniformly on $F \backslash R$.

Our next aim is to formulate a condition that, together with stochastic convergence implies L^1-norm convergence.

12.11. Definition. *Let $(\Omega, \mathscr{B}^{00}, m)$ be a σ-content space and $\mathscr{B}^0 = \mathscr{B}^0(\mathscr{B}^{00})$. A subset M of* mble$(\Omega, \mathscr{B}^0, \mathbb{R})$ *is said to be* **uniformly m-(σ-)-integrable** *if for every $\varepsilon > 0$ there is an h, $0 \leqq h \in L^1_\sigma(m)$ such that*

$$|g| - (|g| \wedge h) \in L^1_\sigma(m), \qquad m(|g| - (|g| \wedge h)) < \varepsilon \quad (g \in M).$$

12.12. Remarks

12.12.1. Uniform integrability of a set $M \subseteq$ mble$(\Omega, \mathscr{B}^0, \mathbb{R})$ implies $M \subseteq \mathscr{L}^1_\sigma(m)$. In fact, we have for every $g \in M$, with the above notations, $|g| \leqq h + (|g| - (|g| \wedge h)) \in \mathscr{L}^1_\sigma(m)$, which implies $g \in \mathscr{L}^1_\sigma(m)$ by Proposition 9.5.

12.12.2. We indulge ourselves to allow speaking of uniform integrability of indexed families of functions if the corresponding set of functions is uniformly integrable.

12.12.3. An easy discussion shows the following:

12.12.3.1. The union of two uniformly m-integrable sets is uniformly m-integrable again. In fact, if h serves for M, and h' for M' in the above definition, then $h \vee h'$ serves for $M \cup M'$.

12.12.3.2. Every finite $M \subseteq$ mble$(\Omega, \mathscr{B}^0, \mathbb{R}) \cap L^1(m)$ is uniformly m-integrable (choose $h = |g_1| \vee \cdots \vee |g_n|$ if $M = \{g_1, \ldots, g_n\}$).

12.12.3.3. Every L^1-norm compact $M \subseteq$ mble$(\Omega, \mathscr{B}^0, \mathbb{R}) \cap \mathscr{L}^1_\sigma(m)$ is uniformly integrable. In fact, choose any $\varepsilon > 0$ and find $g_1, \ldots, g_n \in M$ such that for every $g \in M$ there is a $k \in \{1, \ldots, n\}$ such that $\|g - g_k\|_1 < \varepsilon$. $h = |g_1| \vee \cdots \vee |g_n|$ now clearly satisfies the above definition.

12.12.3.4. If $1 \in \mathscr{L}_\sigma^1(m)$, then we may restrict h to constants in the above definition. In fact, if $0 \leqq h \in \mathscr{L}_\sigma^1(m)$, then $h_n = h \wedge n \in L^1(m)$ and

$$\lim_{n \to \infty} \|h_n - h\|_1 = 0.$$

The rest is now essentially an easy consequence of what we know about norm continuity of finite lattice operations in L_m^1 (Theorem 5.8.2.2).

12.13. Proposition. *Let* $(\Omega, \mathscr{B}^{00}, m)$ *be a σ-content space,* $\mathscr{B}^0 = \mathscr{B}^0(\mathscr{B}^{00})$, *and* $f, f_1, f_2, \ldots \in \text{mble}(\Omega, \mathscr{B}^0, \mathbb{R})$ *with*

$$f_n \to f \text{ (}m\text{-stochastically)}.$$

Then the following statements are equivalent

12.13.1. $f, f_1, f_2, \ldots \in L_m^1$ *and*

$$\lim_{n \to \infty} \|f_n - f\|_1 = 0.$$

12.13.2. $f_1, f_2, \ldots$ *is a uniformly m-integrable sequence.*

Proof. $1 \Rightarrow 2$. If $\lim_{n \to \infty} \|f_n - f\|_1 = 0$, then $f, f_1, f_2, \ldots$ is $\|\cdot\|_1$-compact, and Remark 12.12.3.3 implies the uniform m-integrability of $\{f, f_1, f_2, \ldots\}$, hence of $\{f_1, f_2, \ldots\}$.

$2 \Rightarrow 1$. Let $\varepsilon > 0$ and $0 \leqq h \in \mathscr{L}_\sigma^1(m)$ be such that

$$m(|f_n| - (|f_n| \wedge h)) < \varepsilon/2 \qquad (n = 1, 2, \ldots).$$

Let $g_n = (f_n \wedge h) \vee (-h)$. Then $|f_n - g_n| \leqq |f_n| - (|f_n| \wedge h)$, hence

$$\|f_n - g_n\|_1 < \varepsilon/2 \qquad (n = 1, 2, \ldots).$$

If we can prove $\lim_{n \to \infty} \|g_n - g\|_1 = 0$ for $g = (f \wedge h) \vee (-h)$, then we are obviously through. From $|g_n - g| \leqq |f_n - f|$ we see that $g_n \to g$ (m-stochastically). Thus we are through if we can settle the case $|f_n| \leqq h$ $(n = 1, 2, \ldots)$. But this is easily done by Proposition 12.9 and Lebesgue's theorem 5.7 on dominated convergence: Every subsequence of $f_1, f_2, \ldots$ has a subsequence $f_{n_1}, f_{n_2}, \ldots$ with $\lim_{k \to \infty} f_{n_k} = f$ m-a.e., hence $\lim_{k \to \infty} \|f_{n_k} - f\|_1 = 0$. If $\lim_{n \to \infty} \|f_n - f\|_1 = 0$ failed to hold, we could find $n_1 < n_2 < \cdots$ such that $\|f_{n_k} - f\|_1 \geqq \delta > 0$ for some $\delta > 0$ $(k = 1, 2, \ldots)$, a property that is inherited by every subsequence, thus leading to a contradiction.

12.14. Exercises. Let $(\Omega, \mathscr{B}^{00}, m)$ be a σ-content space. For every $h \in \overline{\mathbb{R}}^{\Omega}$, let

$$\text{ess sup } h = \inf\{\alpha \mid \{h > \alpha\} \text{ is an } m\text{-nullset}\}$$
$$\text{ess inf } h = \{\beta \mid \{h < \beta\} \text{ is an } m\text{-nullset}\}$$
$$\|h\|_{\infty} = \text{ess sup } |h|$$

(with the usual conventions $\inf \varnothing = \infty$, $\sup \varnothing = -\infty$). Prove:

12.14.1. For every $h \in \overline{\mathbb{R}}^{\Omega}$, there are $-\infty \leqq \alpha \leqq \beta \leqq \infty$ with $\alpha = \text{ess inf } h$, $\beta = \text{ess sup } h$.

12.14.2. ess sup, ess inf, and $\|\cdot\|_{\infty}$ are class functions (for equivalence mod m) in $\overline{\mathbb{R}}^{\Omega}$.

12.14.3. If $\|h\|_{\infty} < \infty$, then there is an $h' \in \mathbb{R}^{\Omega}$ with $h' = h$ m-a.e.

12.14.4. $\mathscr{F}_m^{\infty}(\mathbb{R}) = \{h \mid h \in \mathbb{R}^{\Omega}, \|h\|_{\infty} < \infty\}$ is a vector sublattice of $\mathbb{R}^{\Omega}$, and the vector lattice operations are class operations on it; thus the set $F_m^{\infty}(\mathbb{R})$ of all equivalence classes in $\mathscr{F}_m^{\infty}(\mathbb{R})$ can be considered as a vector lattice again.

12.14.5. $\|\cdot\|_{\infty}$, considered as a function on $F_m^{\infty}(\mathbb{R})$, is a norm, and $F_m^{\infty}(\mathbb{R})$ is a Banach space with this norm, the finite linear and lattice operations being norm-continuous, i.e., a Banach lattice.

12.14.6. If $\mathscr{B}^0 = \mathscr{B}^0(\mathscr{B}^{00})$, then the set $L_m^{\infty} = L^{\infty}(\Omega, \mathscr{B}^{00}, m)$ etc. of all equivalence classes mod m in

$$\text{mble}(\Omega, \mathscr{B}^0, \mathbb{R}) \cap \mathscr{F}_m^{\infty}(\mathbb{R}) = \mathscr{L}_m^{\infty} = \mathscr{L}^{\infty}(\Omega, \mathscr{B}^{00}, m)$$

etc. is a Banach sublattice of $F_m^{\infty}(\mathbb{R})$.

12.14.7. Carry over the above considerations to H^{Ω}, where H is an arbitrary real Banach lattice. Determine that part of the theory which works if we assume H to be a real Banach space only, in particular if $H = \mathbb{C}$.

12.15. Let $(\Omega, \mathscr{B}, m)$ be a σ-probability space and let $\text{mble}_m(\Omega, \mathscr{B}, \mathbb{R})$ be the set of all equivalence classes mod m of functions from $\text{mble}(\Omega, \mathscr{B}, \mathbb{R})$. Prove that $\leqq$ mod m is a class relation in $\text{mble}(\Omega, \mathscr{B}, \overline{\mathbb{R}})$ and that $\text{mble}_m(\Omega, \mathscr{B}, \overline{\mathbb{R}})$ may thus be considered as partially ordered by $\leqq$. Prove that it is a *complete lattice* with this half-ordering. (*Hint*: Replace every function f by $\arctan f$ and apply Theorem 5.8.5.) Prove that the subsystem of $\text{mble}_m(\Omega, \mathscr{B}, \mathbb{R})$ consisting of all equivalence classes that contain indicator functions of sets from $\mathscr{B}$ is again a complete lattice. Generalize this result to the case where $(\Omega, \mathscr{B}^{00}, m)$ is a σ-content space and Ω is σ-finite.

CHAPTER IV

TRANSFORM OF σ-CONTENTS

In this chapter we define in Section 1, the general concept of a *measurable mapping* and prove some elementary facts pertaining to it. The concept of a *measurable function* introduced in Definition III.8.7 does not strictly fit into our present general scheme, but the deviation is minor. The special case of Definition III.8.1 can be perfectly subsumed under the concept of a measurable mapping. In Section 2 we show how a σ-content can be transformed by a measurable mapping and how the related integrals, norms, etc. behave under this transform. In Section 3 we make the reader acquainted with one of the most important applications of σ-content transform: *the ergodic theorems.* In Section 4 we generalize transform by point mappings to transform by σ-kernels. Contrary to earlier sections of this chapter, we take up the measure viewpoint fully parallel to the content viewpoint. Section 4 concludes with a discussion of a still more general transform method and raises a question concerning under what conditions it can be represented by a kernel. This question will be solved later in Chapter XVI, Section 3.

Before we go into details, a few general observations about point mappings and the transform of functions seem necessary. Let Ω, $\Omega' \neq \varnothing$ and let φ be an arbitrary mapping of Ω into Ω'. Let $X \neq \varnothing$ be arbitrary and consider the sets X^{Ω}, $X^{\Omega'}$ of all X-valued functions on Ω, Ω', i.e., of all mappings of Ω resp. Ω' into X. If $f' \in X^{\Omega'}$, then

$$f' \circ \varphi : \omega \to f'(\varphi(\omega)) \qquad (\omega \in \Omega)$$

uniquely defines a function $f' \circ \varphi \in X^{\Omega}$ which we also call the function f' transformed by φ. This transform of functions is again denoted by φ, namely

$$\varphi : X^{\Omega} \leftarrow X^{\Omega'}$$

(notice the inversion of the arrow). If $X = \{0, 1\}$, then every $f \in \{0, 1\}^\Omega$ is the indicator function $f = 1_F$ of a set $F \subseteq \Omega$:

$$1_F(\omega) = \begin{cases} 1 & \text{if} \quad \omega \in F, \\ 0 & \text{if} \quad \omega \in \Omega \backslash F; \end{cases}$$

and the relation between $\{0, 1\}^\Omega$ and $\mathscr{P}(\Omega)$ thus established is a bijection. The same holds if we replace Ω by Ω'. It is now immediate that

$$\varphi\colon \{0, 1\}^\Omega \leftarrow \{0, 1\}^{\Omega'}$$

is tantamount, by the mentioned bijection, to the so-called (set) *pullback*

$$\varphi^{-1}\colon \mathscr{P}(\Omega) \leftarrow \mathscr{P}(\Omega')$$

defined by

$$\varphi^{-1}(F') = \varphi^{-1}F' = \{\omega \mid \varphi(\omega) \in F'\} \qquad (F' \subseteq \Omega').$$

One of the most simple but important facts in our context is the following: The set pullback commutes with every set operation, finite, countable, or uncountable, and sends $\varnothing$ into $\varnothing$, Ω' into Ω. It is in fact an easy exercise to prove

$$\varphi^{-1}\varnothing = \varnothing, \qquad \varphi^{-1}\Omega' = \Omega$$

$$\varphi^{-1}(E' \backslash F') = (\varphi^{-1}E') \backslash (\varphi^{-1}F') \qquad (E', F' \subseteq \Omega')$$

$$\varphi^{-1}\left(\bigcup_{\iota \in I} F'\right) = \bigcup_{\iota \in I} \varphi^{-1}F', \qquad \varphi^{-1}\left(\bigcap_{\iota \in I} F'\right) = \bigcap_{\iota \in I} \varphi^{-1}F'$$

for any family $(F'_\iota)_{\iota \in I}$ of subsets of Ω indexed by an arbitrary set $I \neq \varnothing$. It should be noticed that φ^{-1} is one-to-one iff $\varphi(\Omega) = \Omega'$. The fact that φ^{-1} is not one-to-one if $\varphi(\Omega) \neq \Omega'$ may turn out to be a serious obstacle for some seemingly straightforward reasonings.

A similar observation holds for the cases $X = \mathbb{R}$ and $X = \overline{\mathbb{R}}$: The mapping

$$\varphi\colon \mathbb{R}^\Omega \leftarrow \mathbb{R}^{\Omega'}$$

commutes with all vector lattice operations (finite linear combinations, finite suprema and infima, taken pointwise); and the mapping

$$\varphi\colon \overline{\mathbb{R}}^\Omega \leftarrow \overline{\mathbb{R}}^{\Omega'}$$

(to choose a convenient situation) commutes with arbitrary (finite, countable, or uncountable) lattice operations, and thus in particular with the operations lim sup and lim inf on sequences. The proofs are easy exercises (and nearly "optical" if one realizes that $f' \circ \varphi$ is constant on those sets of the form $\varphi^{-1}\{\omega'\}$ $(\omega' \in \Omega')$ that are nonempty, and "looks like" f' restricted to $\varphi(\Omega)$).

We have started with a mapping $\varphi\colon \Omega \to \Omega'$ and ended up with mappings $\varphi\colon X^{\Omega} \leftarrow X^{\Omega'}$ resp. $\varphi^{-1}\colon \mathscr{P}(\Omega) \leftarrow \mathscr{P}(\Omega')$. The facts observed in this connection will be considered as obvious henceforth. In Section 1 we shall simply go on in the same way, proceed to the next higher stage of complication, add some more structure, and hence need some more precaution.

1. MEASURABLE MAPPINGS

In this section we present the definition and basic properties of measurable mappings.

1.1. Definition. *Let $\Omega, \Omega' \neq \varnothing$ and $\mathscr{S} \subseteq P(\Omega)$, $\mathscr{S}' \subseteq \mathscr{P}(\Omega')$. A mapping $\varphi\colon \Omega \to \Omega'$ is called $\mathscr{S}$-$\mathscr{S}'$-**measurable** if $\varphi^{-1}\mathscr{S}' \subseteq \mathscr{S}$, i.e., if*

$$\varphi^{-1}F' \in \mathscr{S} \qquad (F' \in \mathscr{S}')$$

1.2. Remark. This very general definition comprises, e.g., the following special case. If $\mathscr{T}$ is a *topology* in Ω (i.e., $\mathscr{T} \subseteq \mathscr{P}(\Omega)$, $\varnothing$, $\Omega \in \mathscr{T}$, $\mathscr{T}$ stable under finite intersections and arbitrary unions) and $\mathscr{T}'$ a topology in Ω', then the $\mathscr{T}$-$\mathscr{T}'$-measurability of a mapping $\varphi\colon \Omega \to \Omega'$ means nothing but the *continuity* of φ in the usual terminology of topology. Throughout this book the accent lies on another special case, namely where $\mathscr{S}$, $\mathscr{S}'$ are σ-fields in Ω, Ω', but attention will be frequently paid to topology.

1.3. Proposition. *Let $\Omega, \Omega' \neq \varnothing$ and $\varphi\colon \Omega \to \Omega'$ be a mapping. Then:*

1.3.1. *For every σ-field $\mathscr{B}$ in Ω,*

$$\{F' \mid F' \subseteq \Omega', \varphi^{-1}F' \in \mathscr{B}\}$$

is a σ-field in Ω'.

1.3.2. *For every σ-field $\mathscr{B}'$ in Ω',*

$$\varphi^{-1}\mathscr{B}' = \{\varphi^{-1}F' \mid F' \in \mathscr{B}'\}$$

is a σ-field in Ω.

1.3.3. *If $\varnothing \neq \mathscr{S}' \subseteq \mathscr{P}(\Omega')$, then*

$$\mathscr{B}(\varphi^{-1}\mathscr{S}') = \varphi^{-1}\mathscr{B}(\mathscr{S}').$$

1.3.4. *If $\varnothing \neq \mathscr{S}' \subseteq \mathscr{P}(\Omega')$, $\mathscr{B}' = \mathscr{B}(\mathscr{S}')$, and $\mathscr{B}$ is a σ-field in Ω, then*

$$\varphi^{-1}\mathscr{S}' \subseteq \mathscr{B} \quad \Rightarrow \quad \varphi \text{ is } \mathscr{B}\text{-}\mathscr{B}'\text{-measurable.}$$

Proof. 1 and 2 are simple consequences of the fact that the pullback φ^{-1} commutes with all set operations and sends $\varnothing$ into $\varnothing$, Ω' into Ω. If, e.g., $F'_1, F'_2, \ldots \subseteq \Omega'$ $\varphi^{-1}F'_1, \varphi^{-1}F'_2, \ldots \in \mathscr{B}$, then $\varphi^{-1}(F'_1 \cup F'_2 \cup \cdots) = \varphi^{-1}F'_1 \cup \varphi^{-1}F'_2 \cup \cdots \in \mathscr{B}$, proving stability under countable unions in 1.

3. $\{F' | F' \subseteq \Omega', \varphi^{-1}F' \in \mathscr{B}\ (\varphi^{-1}\mathscr{S}')$ is a σ-field in Ω' by 1 and contains $\mathscr{S}'$ by definition. Thus $\supseteq$ follows. $\subseteq$ is obvious from $\varphi^{-1}\mathscr{S}' \subseteq \varphi^{-1}\mathscr{B}(\mathscr{S}')$.

4 is an obvious consequence of 3.

1.4. Remarks. Let $\Omega, \Omega' \neq \varnothing$, and $\varphi: \Omega \to \Omega'$ be a mapping.

1.4.1. The methods of proof employed in the above proposition apply to any type of set system and not only to σ-fields, with some precautions which we owe to the fact that the pullback for sets is not one-to-one. Let, e.g., $\mathscr{B}^{00}$ be a local σ-ring in Ω. Then it is easy to prove that $\{F' | F' \subseteq \Omega', \varphi^{-1}F' \in \mathscr{B}^{00}\}$ is a local σ-ring in Ω'; in fact, if $\varphi^{-1}F'$, $\varphi^{-1}F'_1, \ldots \in \mathscr{B}^{00}$, $F'_1, \ldots \subseteq F'$, then $\varphi^{-1}F'_1, \ldots \subseteq \varphi^{-1}F'$ and thus $\varphi^{-1}(F'_1 \cup F'_2 \cup \cdots) = \varphi^{-1}F'_1 \cup \varphi^{-1}F'_2 \cdots \in \mathscr{B}^{00}$. But if $\mathscr{B}^{00\prime}$ is a local σ-ring in Ω', then we have to be cautious when proving that $\varphi^{-1}(\mathscr{B}^{00\prime})$ is a local σ-ring again. The trouble arises in the proof that $\varphi^{-1}(\mathscr{B}^{00\prime})$ is stable under countable bounded unions. If $F', F'_1, \ldots \in \mathscr{B}^{00\prime}$, $\varphi^{-1}F'_1, \ldots \subseteq \varphi^{-1}F'$, then we cannot conclude that $F'_1, \ldots \subseteq F'$. But we can circumvent the problem by the following device:

$$\begin{aligned}\varphi^{-1}(F'_n \cap F') = \varphi^{-1}(F'_n) \cap \varphi^{-1}(F') &= \varphi^{-1}(F'_n) \\ \Rightarrow \quad \varphi^{-1}F'_1 \cup \varphi^{-1}F'_2 &\cup \cdots \\ = \varphi^{-1}((F'_1 \cap F') \cup (F'_2 \cap F') &\cup \cdots) \in \varphi^{-1}\mathscr{B}^{00\prime}.\end{aligned}$$

Thus $\varphi^{-1}\mathscr{B}^{00\prime}$ is in fact a local σ-ring again. Another means of circumventing the mentioned difficulty would consist in taking advantage of Remark I.4.2.2.

1.4.2. The "forward" mapping $F \to \varphi(F) = \{\varphi(\omega) | \omega \in F\}$ of $\mathscr{P}(\Omega) \to \mathscr{P}(\Omega')$ derived from φ does not commute with all set operations in general. It commutes with arbitrary unions, as can be easily seen; but two disjoint sets might have the same image. It is in fact not generally true that $\varphi(\mathscr{B})$ is a σ-field in Ω' if $\mathscr{B}$ is a σ-field in Ω since, e.g., Ω' will not appear in $\varphi(\mathscr{B})$ if $\varphi(\Omega) \neq \Omega'$. Likewise, it is not generally true that $\varphi(\mathscr{R})$ is a ring in Ω' if $\mathscr{R}$ is a ring in Ω. To show this let, e.g., $\Omega = \mathbb{Z} \times \{0, 1\} = \{(n, i) | n \in \mathbb{Z}, i = 0, 1\}$. In Ω we consider the partition ξ into all sets of the form $\{(n, 0), (n + 1, 1)\}$ and the ring $\mathscr{R}$ of all finite unions of members of ξ (Exercise I.1.5.4). Let $\Omega' = \mathbb{Z}$ and $\varphi: \Omega \to \Omega'$ be defined by $\varphi(n, i) = n$. Then clearly $\varphi(\mathscr{R})$ does not contain $\{0\}$ but $\{-1, 0\}$ and $\{0, 1\}$ whose intersection is $\{0\}$. Thus $\varphi(\mathscr{R})$ is not intersection stable. The same $\varphi: \Omega \to \Omega'$ shows that a "forward" analogue of 1.3.3 fails. In fact, let $\mathscr{S} \subseteq \mathscr{P}(\Omega)$ consist of all sets $\{(n, 0), (n + 1, 1)\}$ and $\{(n, 0), (n - 1, 1)\}$ $(n \in \mathbb{Z})$. Clearly $\mathscr{B}(\mathscr{S}) = \mathscr{P}(\Omega)$, but $\{F' | F' \subseteq \Omega', \varphi^{-1}F' \in \mathscr{S} = \varnothing\}$, generating the trivial σ-field $\{\emptyset, \Omega'\}$ in Ω', which is not equal to

$$\{F' \mid \varphi^{-1}F' \in \mathscr{B}(\mathscr{S})\} = \mathscr{P}(\Omega').$$

In view of difficulties of this kind we shall not consider, as a rule, the "forward" mapping of set systems. See, however, Proposition XIII.2.13 on analytic sets and Kuratowski's theorem (XIII.2.18) on Borel sets in Polish spaces.

1.4.3. Definition III.8.1 of $\mathscr{B}$-measurable functions in $\mathbb{R}^\Omega$ or $\overline{\mathbb{R}}^\Omega$ is nothing but a special case of Definition 1.1; and Proposition III.8.2 is essentially Proposition 1.3.4 with various set systems that generate the natural σ-field $\mathscr{B}(\mathbb{R})$ resp. $\mathscr{B}(\overline{\mathbb{R}})$ in place of $\mathscr{S}'$.

1.5. Proposition. *Let $\Omega, \Omega', \Omega'' \neq \emptyset$, and $\varphi: \Omega \to \Omega'$, $\varphi': \Omega' \to \Omega''$ be mappings and $\mathscr{B}, \mathscr{B}', \mathscr{B}''$ σ-fields in $\Omega, \Omega', \Omega''$, respectively. Then:*

$$\begin{aligned} &\varphi \text{ is } \mathscr{B}\text{-}\mathscr{B}'\text{-measurable and } \varphi' \text{ is } \mathscr{B}'\text{-}\mathscr{B}''\text{-measurable} \\ &\Rightarrow \quad \varphi' \circ \varphi: \Omega \to \Omega \text{ is } \mathscr{B}\text{-}\mathscr{B}''\text{-measurable.} \end{aligned}$$

Proof. $(\varphi' \circ \varphi)^{-1}\mathscr{B}'' = \varphi^{-1}((\varphi')^{-1}\mathscr{B}'') \subseteq \varphi^{-1}\mathscr{B}' \subseteq \mathscr{B}$.

The objective of the next proposition is the determination of $\varphi(\mathrm{mble}(\Omega', \mathscr{B}', \mathbb{R}))$ and $\varphi(\mathrm{mble}(\Omega', \mathscr{B}', \overline{\mathbb{R}}))$ for a given measurable φ.

1.6. Proposition. *Let $\Omega, \Omega' \neq \emptyset$, $\varphi: \Omega \to \Omega'$ be a mapping, and $\mathscr{B}'$ a σ-field in Ω'. Then:*

1.6.1. *For every $f \in \mathbb{R}^\Omega$, the following statements are equivalent:*

1.6.1.1. $f \in \mathrm{mble}(\Omega, \varphi^{-1}\mathscr{B}', \mathbb{R})$;

1.6.1.2. $f = f' \circ \varphi$ *for some* $f' \in \mathrm{mble}(\Omega', \mathscr{B}', \mathbb{R})$.

1.6.2. *For every $f \in \overline{\mathbb{R}}^\Omega$, the following statements are equivalent:*

1.6.2.1. $f \in \mathrm{mble}(\Omega, \varphi^{-1}\mathscr{B}', \overline{\mathbb{R}})$;

1.6.2.2. $f = f' \circ \varphi$ *for some* $f' \in \mathrm{mble}(\Omega', \mathscr{B}', \overline{\mathbb{R}})$.

1.6.3. *In both cases, the restriction of f' to $\varphi(\Omega)$ is uniquely determined by f.*

Proof. 1. By Proposition 1.5, only 1.1 $\Rightarrow$ 1.2 is nontrivial. We proceed in four steps.

(a) if $f = 1_F$ for some $F \in \varphi^{-1}\mathscr{B}'$, then $F = \varphi^{-1}F'$ for some $F' \in B$ and $f = f' \circ \varphi$ with $f' = 1_{F'}$.

(b) If $f = \sum_{k=1}^n \alpha_k 1_{F_k}$ with $\alpha_1, \ldots, \alpha_n \in \mathbb{R}$ and $F_1, \ldots, F_n \in \mathscr{B}'$, then $F_1 = \varphi^{-1}F'_1, \ldots, F_n = \varphi^{-1}F'_n$ for suitable $F'_1, \ldots, F'_n \in \mathscr{B}'$ and $f = f' \circ \varphi$ with $f' = \sum_{k=1}^n \alpha_k 1_{F'_k}$.

(c) If $0 \leqq f \in \mathrm{mble}(\Omega, \varphi^{-1}\mathscr{B}', \mathbb{R})$, we construct $0 \leqq f_1 \leqq f_2 \leqq \cdots \nearrow f$ (pointwise) such that f_n is a $\varphi^{-1}\mathscr{B}'$-measurable step function, i.e., a finite linear combination of indicator functions ($n = 1, 2, \ldots$), according to Proposition III.8.13. By (b) we can find $\bar{f}'_n \in \text{mble}(\Omega', \mathscr{B}', \mathbb{R})$ such that $f_n = \bar{f}'_n \circ \varphi$ ($n = 1, 2, \ldots$). Put $f'_n = \bar{f}'_1 \vee \cdots \vee \bar{f}'_n$ ($n = 1, 2, \ldots$). Clearly

$$f'_n \circ \varphi = (\bar{f}'_1 \circ \varphi) \vee \cdots \vee (\bar{f}'_n \circ \varphi) = f_1 \vee \cdots \vee f_n = f_n.$$

($n = 1, 2, \ldots$). Let now

$$f'(\omega) = \begin{cases} 0 & \text{if } \lim f'_n(\omega) = \infty, \\ \lim f'_n(\omega) & \text{otherwise.} \end{cases}$$

Then $(f' \circ \varphi)(\omega) = \lim(f'_n \circ \varphi)(\omega) = \lim f_n(\omega) = f(\omega)$ for all $\omega \in \Omega$ since $f(\omega)$ is always a finite real. Thus $f' \circ \varphi = f$.

(d) If $f \in \text{mble}(\Omega, \varphi^{-1}\mathscr{B}', \mathbb{R})$, then $f = f_+ - f_-$ with $0 \leqq f_+$, $f_- \in \text{mble}(\Omega, \varphi^{-1}\mathscr{B}', \mathbb{R})$. By (c) we find $0 \leqq g', h' \in \text{mble}(\Omega', \mathscr{B}', \mathbb{R})$ with $f_+ = g' \circ \varphi$, $f_- = h' \circ \varphi$, and $f = (g' - h') \circ \varphi$, $g' - h' \in \text{mble}(\Omega', \mathscr{B}', \mathbb{R})$ follows.

2. The proof goes as for 1 with the following modifications. Apply Proposition III.8.13 to f_+, f_- in order to find $f_n \in \text{mble}(\Omega, \mathscr{B}, \mathbb{R})$ such that $|f_n| \leqq |f|$ ($n = 1, 2, \ldots$). $f_n \to f$ pointwise, and f_n is a $\varphi^{-1}\mathscr{B}'$-measurable step function, hence representable in the form $f_n = f'_n \circ \varphi$ with $f'_n \in \text{mble}(\Omega', \mathscr{B}', \mathbb{R})$ ($n = 1, 2, \ldots$), by the same token as in (a), (b) above. Now let $f' = \limsup f'_n$. Then $f' \in \text{mble}(\Omega', \mathscr{B}', \overline{\mathbb{R}})$ $f' \circ \varphi = \limsup (f'_n \circ \varphi) = \limsup f_n = \lim f_n = f$ pointwise.

3 is obvious.

1.7. Exercises

1.7.1. Let $\mathscr{B}^0$ be a σ-ring in $\Omega \neq \varnothing$, $\tilde{\mathscr{B}}^0$ be a σ-ring in $\tilde{\Omega} \neq \varnothing$, and $\varphi: \Omega \to \tilde{\Omega}$ be $\mathscr{B}^0$-$\tilde{\mathscr{B}}^0$-measurable. Prove that

$$h \in \text{mble}(\tilde{\Omega}, \tilde{\mathscr{B}}^0, \overline{\mathbb{R}}) \quad \Rightarrow \quad h \circ \varphi \in \text{mble}(\Omega, \mathscr{B}^0, \overline{\mathbb{R}}).$$

1.7.2. Let $\mathscr{B}$ be a σ-field in $\Omega \neq \varnothing$ and $f_1, \ldots, f_n \in \mathbb{R}^\Omega$. Prove:

1.7.2.1. that the mapping $f: \Omega \to \mathbb{R}^n$ defined by $f(\omega) = (f_1(\omega), \ldots, f_n(\omega))$ is $\mathscr{B}$-$\mathscr{B}(\mathbb{R}^n)$-measurable iff $f_1, \ldots, f_n \in \text{mble}(\Omega, \mathscr{B}, \mathbb{R})$;

1.7.2.2. that for every $g \in \text{mble}(\mathbb{R}^n, \mathscr{B}(\mathbb{R}^n), \mathbb{R})$ the function $g(f_1, \ldots, f_n)$ is in $\text{mble}(\Omega, \mathscr{B}, \mathbb{R})$ if $f_1, \ldots, f_n \in \text{mble}(\Omega, \mathscr{B}, \mathbb{R})$.

1.7.2.3. that $f_1 \wedge \cdots \wedge f_n, f_1 \vee \cdots \vee f_n, f_1 \cdots f_n, f_1 + \cdots + f_n \in \text{mble}(\Omega, \mathscr{B}, \mathbb{R})$ if $f_1, \ldots, f_n \in \text{mble}(\Omega, \mathscr{B}, \mathbb{R})$.

1.7.3. Let $(\Omega, \mathscr{T})$, $(\Omega', \mathscr{T})$ be topological spaces and $\mathscr{B} = \mathscr{B}(\mathscr{T})$, $\mathscr{B}' = \mathscr{B}(\mathscr{T}')$, $\mathscr{B}_0$ resp. $\mathscr{B}'_0$ be the σ-field generated by $\mathscr{C}^b(\Omega, \mathbb{R})$ resp. $\mathscr{C}^b(\Omega', \mathbb{R})$. Let $\varphi: \Omega \to \Omega'$ be continuous. Prove that φ is $\mathscr{B}$-$\mathscr{B}'$-measurable and $\mathscr{B}_0$-$\mathscr{B}'_0$-measurable.

2. TRANSFORM OF σ-ADDITIVE FUNCTIONS

In this section we go one step farther in the direction indicated in the introduction to this chapter. We begin with two sets $\Omega, \Omega' \neq \varnothing$, a ring $\mathscr{R}$ in Ω, a ring $\mathscr{R}'$ in Ω', and a $\mathscr{R}$-$\mathscr{R}'$-measurable mapping. Let pullback yields a mapping $\varphi^{-1}: \mathscr{R} \leftarrow \mathscr{R}'$. This in turn, for any set $X \neq \varnothing$, leads to a mapping, again denoted by the letter φ, sending $X^{\mathscr{R}} \to X^{\mathscr{R}'}$ by the simple formula

$$(1) \qquad m(F') = m(\varphi^{-1}F') \qquad (F' \in \mathscr{R}'; \quad m \in X^{\mathscr{R}}).$$

m is also called the φ-*image* of m, etc. We restrict attention to $X = \mathbb{R}_+$ and σ-additive $m: \mathscr{R} \to \mathbb{R}_+$. By

$$\begin{aligned} F', F_1', F_2', \ldots \in \mathscr{R}', \quad F_j' \cap F_k' = \varnothing\ (j \neq k), \quad F' = F_1' + F_2' + \cdots \\ \Rightarrow \quad \varphi m(F') = m(\varphi^{-1}F') = m(\varphi^{-1}(F_1' + F_2' + \cdots)) \\ = m((\varphi^{-1}F_1') + (\varphi^{-1}F_2') + \cdots) \\ = m(\varphi^{-1}F_1') + m(\varphi^{-1}F_2') + \cdots \\ = \varphi m(F_1') + \varphi m(F_2') + \cdots \end{aligned}$$

we see that φm is then σ-additive again. The reader should be aware that the same argument is valid for σ-additive $m: \mathscr{R} \to \overline{\mathbb{R}}_+$ and that in this case

$$(2) \qquad \varphi^{-1}\{F' \mid F' \in \mathscr{R}', \varphi m(F') < \infty\} \subseteq \{F \mid F \in \mathscr{R}, m(F) < \infty\}$$

$$(3) \qquad \varphi^{-1}\{N' \mid N' \in \mathscr{R}', \varphi m(N') = 0\} \subseteq \{N \mid N \in \mathscr{R}, m(N) = 0\}.$$

The reader should, furthermore, be aware that these statements have obvious consequences for a lot of important special cases, such as:

(a) $\mathscr{R}, \mathscr{R}'$ are σ-fields;
(b) $\mathscr{R}, \mathscr{R}'$ are σ-rings;
(c) $\mathscr{R}, \mathscr{R}'$ are local σ-rings.

σ-extension theory allows us to pass from one of these situations to another by various extension and derivation devices. Proposition 1.3 and Remark 1.4.1. show that measurability of mappings is widely invariant under such passages. We have therefore a high degree of freedom in the choice of the setting for subsequent considerations. I decide to choose, in view of the necessity of speaking of integrable functions, the level of rings.

The main result of our preliminary considerations can now be concentrated into

2.1. Proposition. *Let $\mathscr{R}$ be a ring in $\Omega \neq \emptyset$, $\mathscr{R}'$ a ring in Ω', and $\varphi\colon \Omega \to \Omega'$ $\mathscr{R}$-$\mathscr{R}'$-measurable. Then for every σ-content m on $\mathscr{R}$ the set functions φm defined by* (1) *on $\mathscr{R}'$ is a σ-content again.*

2.2. Remark. Notice that the above proposition covers many seemingly different situations as special cases. We mention the following:

2.2.1. $(\Omega, \mathscr{B}^{00}, m)$ is a full and complete σ-content space, e.g., the Carathéodory extension (Theorem II.2.3) of some σ-content prespace, $\mathscr{S}' \subseteq \mathscr{P}(\Omega')$, φ is $\mathscr{B}^{00}$-$\mathscr{S}'$-measurable. We may then apply Proposition 2.1 with $\mathscr{R} = \mathscr{B}^{00}$ and $\mathscr{R}' = \mathscr{B}^{00}(\mathscr{S}')$.

2.2.2. $\mathscr{B}$ is a σ-field in Ω and $m\colon \mathscr{B} \to \overline{\mathbb{R}}_+$ σ-additive, $\mathscr{B}^{00} = \{F \mid F \in \mathscr{B}, m(F) < \infty, \mathscr{S}' \subseteq \mathscr{P}(\Omega')\}$ and $\varphi\colon \Omega \to \Omega'$ $\mathscr{B}^{00}$-$\mathscr{S}'$-measurable. We may then apply Proposition 2.1 with $\mathscr{R} = \mathscr{B}^{00}$ and $\mathscr{R}' = \mathscr{B}^{00}(\mathscr{S}')$.

2.3. Exercises. Let $(\Omega, \mathscr{R}, m)$, $(\Omega', \mathscr{R}', m')$ be σ-content prespaces and $\varphi\colon \Omega \to \Omega'$ be $\mathscr{R}$-$\mathscr{R}'$-measurable. Let $\mathscr{B}^{00} = \mathscr{B}^{00}(\mathscr{R})$, $\mathscr{B}'^{00} = \mathscr{B}^{00}(\mathscr{R}')$, $\mathscr{B}^{0} = \mathscr{B}^{0}(\mathscr{R})$, $\mathscr{B}'^{0} = \mathscr{B}^{0}(\mathscr{R}')$, $\mathscr{B} = \mathscr{B}(\mathscr{R})$, $\mathscr{B}' = \mathscr{B}(\mathscr{R}')$.

2.3.1. Prove that φ is $\mathscr{B}^{00}$-$\mathscr{B}'^{00}$-, $\mathscr{B}^{0}$-$\mathscr{B}'^{0}$-, and $\mathscr{B}$-$\mathscr{B}'$-measurable.

2.3.2. Let $m = m'$ and denote by m, m' the minimal σ-additive extension of m, m' to $\mathscr{B}$, $\mathscr{B}'$, respectively. Prove $\varphi m = m'$.

2.3.3. Let $\varphi m = m'$. Prove that φ is measurable for the Carathéodory extensions (Theorem II.2.3) of $\mathscr{R}$ resp. $\mathscr{R}'$ with respect to m resp. m' and sends these into one another.

2.3.4. Formulate and prove the analogue of 2.3.2 for the extension steps before the minimal σ-additive extensions to the generated σ-fields as well as for their completions (Theorem II.6.2).

2.4. Theorem. *Let $(\Omega, \mathscr{R}, m)$, $(\Omega', \mathscr{R}', m')$ be σ-content prespaces and $\varphi\colon \Omega \to \Omega'$ be $\mathscr{R}$-$\mathscr{R}'$-measurable with $\varphi m = m'$. Then for every $h \in \overline{\mathbb{R}}^{\Omega'}$*

$$h' \in \mathscr{L}^1_\sigma(\Omega', \mathscr{R}', m') \quad \Leftrightarrow \quad h' \circ \varphi \in \mathscr{L}^1_\sigma(\Omega, \mathscr{R}, m)$$

and if this holds, we have the **integral transport formula**

$$\int h'\, dm' = \int (h' \circ \varphi)\, dm. \tag{4}$$

Proof. Let $(\Omega, \mathscr{B}_m, m)$ be the Carathéodory extension of $(\Omega, \mathscr{R}, m)$, i.e., following Exercise III.6.5.7, $\mathscr{B}^{00}_m = \{F \mid F \subseteq \Omega,\ 1_F \in \mathscr{L}^1_\sigma(m)\}$, $m(F) = \int 1_F\, dm$ $(F \in \mathscr{B}^{00}_m)$. Similarly, let $(\Omega', \mathscr{B}^{00}_{m'}, m')$ be the Carathéodory extension of $(\Omega', \mathscr{R}', m')$. By Exercise 2.3.3, φ is $\mathscr{B}^{00}_m$-$\mathscr{B}^{00}_{m'}$-measurable and sends

$m\colon \mathscr{B}_m^{00} \to \mathbb{R}_+$ into $m'\colon \mathscr{B}_{m'}^{00} \to \mathbb{R}_+$. It follows that (4) is true for $h = 1_{F'}$, $F' \in \mathscr{B}_{m'}^{00}$, and hence for all step functions $h' \in \mathscr{L}_\sigma^1(m')$. Let now $0 \leqq h' \in \overline{\mathbb{R}}^\Omega$ and form step functions $0 \leqq h_n' \nearrow h'$ (pointwise) according to Theorem III.6.4. Then clearly $0 \leqq h_n' \circ \varphi \nearrow h' \circ \varphi$ (pointwise), and the $h_n' \circ \varphi$ are step functions again. If $h' \in \mathscr{L}_\sigma^1(m)$, then $h_n' \in \mathscr{L}_\sigma^1(m)$ and

$$\int h'\,dm' = \lim \int h_n'\,dm' = \lim \int (h_n' \circ \varphi)\,dm < \infty$$

by Theorem III.6.4, and the monotone convergence theorem implies $h' \circ \varphi \in \mathscr{L}_\sigma^1(m)$ and (4). A symmetric argument yields

$$h' \circ \varphi \in \mathscr{L}_\sigma^1(m) \quad \Rightarrow \quad h' \in \mathscr{L}_\sigma^1(m')$$

and (4) again. Thus we are through in the case $h' \geqq 0$. The general case follows now by $h' = h_+' - h_-'$ in an obvious fashion.

2.5. Remark. An alternative proof of part of Theorem 2.4 can be based on Exercise III.6.5.1 as follows. The function mapping $\overline{\mathbb{R}}^{\Omega'} \ni h' \to h' \circ \varphi \in \overline{\mathbb{R}}^\Omega$ sends the elementary domain $\mathscr{E}(\mathscr{R}')$ into an elementary domain $\mathscr{E}_0 \subseteq \mathscr{E}(\mathscr{R})$ and yields $m(h' \circ \varphi) = m'(h')$ $(h' \in \mathscr{E}(\mathscr{R}))$ by linear extension of the definition of $\varphi m = m'$. Going through all stages of σ-extension, both with the σ-measure $m'\colon \mathscr{E}(\mathscr{R}') \to \mathbb{R}$ and the σ-measure $m\colon \mathscr{E}_0 \to \mathbb{R}$, we derive

$$h' \circ \varphi \in \mathscr{L}_\sigma^1(\Omega, \mathscr{E}_0, m) \quad \Leftrightarrow \quad h' \in \mathscr{L}_\sigma^1(\Omega, \mathscr{E}(\mathscr{R}'), m') = \mathscr{L}_\sigma^1(\Omega, \mathscr{R}', m'),$$

plus (4) if these conditions hold. Now Exercise III.6.5.1 shows that

$$h' \circ \varphi \in \mathscr{L}_\sigma^1(\Omega, \mathscr{E}_0, m) \quad \Rightarrow \quad h' \circ \varphi \in \mathscr{L}_\sigma^1(\Omega, \mathscr{E}(\mathscr{R}), m) = \mathscr{L}_\sigma^1(\Omega, \mathscr{R}, m).$$

2.6. Proposition. *Let $(\Omega, \mathscr{R}, m)$, $(\Omega', \mathscr{R}', m')$ be σ-content prespaces and $\varphi\colon \Omega \to \Omega'$ be $\mathscr{R}$-$\mathscr{R}'$-measurable with $\varphi m = m'$. Then for any $g', h' \in \mathbb{R}^{\Omega'}$*

$$g', h' \in \mathscr{L}_\sigma^2(m', \overline{\mathbb{R}}) \quad \Rightarrow \quad g' \circ \varphi, h' \circ \varphi \in \mathscr{L}_\sigma^2(m, \overline{\mathbb{R}});$$

and if this holds, we have $(g'h') \circ \varphi = (g \circ \varphi)(h' \circ \varphi)$ and the transport formulas for the scalar product and the norm:

$$(g', h') = \int g'h'\,dm' = \int (g' \circ \varphi)(h' \circ \varphi)\,dm = (g' \circ \varphi, h' \circ \varphi)$$

$$\|h'\|_2 = \|h' \circ \varphi\|_2 .$$

Proof.

$$\begin{aligned} g' \in \mathscr{L}_\sigma^2(m', \overline{\mathbb{R}}) \quad &\Leftrightarrow \quad g' \cdot |g'| \in \mathscr{L}_\sigma^1(m', \overline{\mathbb{R}}) \\ &\Leftrightarrow \quad (g' \cdot |g'|) \circ \varphi = (g' \circ \varphi)|(g' \circ \varphi)| \in \mathscr{L}_\sigma^1(m, \overline{\mathbb{R}}) \\ &\Leftrightarrow \quad g' \circ \varphi \in \mathscr{L}_\sigma^2(m, \overline{\mathbb{R}}). \end{aligned}$$

The rest is obvious.

2.7. Exercises

2.7.1. Let $(\Omega, \mathscr{R}, m)$, $(\Omega', \mathscr{R}', m')$ be σ-content prespaces and $\varphi: \Omega \to \Omega'$ be such that $\varphi^{-1}N'$ is an m-nullset iff $N' \subseteq \Omega'$ is an m'-nullset. With the notations of Exercise III.12.14, prove that for every $h' \in \overline{\mathbb{R}}^{\Omega'}$ the following relations hold:

$$\text{ess sup}(h' \circ \varphi) = \text{ess sup } h'$$
$$\text{ess inf}(h' \circ \varphi) = \text{ess inf } h'$$
$$\|h' \circ \varphi\|_\infty = \|h'\|_\infty$$

where the symbols on the left-hand side refer to m, and the symbols on the right-hand side to m'.

2.7.2. Let $n \geqq 1$ be an integer and $\varnothing \neq \Omega$, $\Omega' \subseteq \mathbb{R}^n$ be open. Let $\varphi: \Omega \to \Omega'$ be a C^1 diffeomorphism and $\Delta: \Omega \to \mathbb{R}$ its functional determinant. That is, $\varphi: \Omega \to \Omega'$ is a bijection, both φ and φ^{-1} have continuous first derivatives on Ω resp. Ω', and $\Delta(x) = \det((\partial\varphi_j(x)/\partial x_k)_{j,k=1,\ldots,n})$, where φ_j is the jth component of φ $(j = 1, \ldots, n)$. Assume that $\Delta(x) > 0$ $(x \in \Omega)$. For every $f \in \mathscr{C}^{00}(\mathbb{R})$, let $m(f)$ be the usual Riemann integral of f, and for every $f' \in \mathscr{C}^{00}(\Omega', \mathbb{R})$ let $m'(f')$ be the usual Riemann integral of the function $(1/\Delta(x))\, f' \circ \varphi$ on Ω' (which exists since $\Delta(x)$ is bounded away from 0 on the compact support of $f' \circ \varphi$). Prove that for every $h' \in \overline{\mathbb{R}}^{\Omega'}$ we have $h' \in \mathscr{L}_\sigma^1(\Omega', \mathscr{C}^{00}(\Omega', \mathbb{R}), m') \Leftrightarrow h' \circ \varphi \in \mathscr{L}_\sigma^1(\Omega, \mathscr{C}^{00}(\Omega, \mathbb{R}), m)$ and that these conditions imply $\int (h' \circ \varphi)\, dm = \int h'\, dm'$.

2.7.3. Let $(\Omega, \mathscr{E}, m)$ be a σ-measure space and $(\Omega, \mathscr{B}_m^{00}, m)$ the σ-content space derived from it.

2.7.3.1. Let $h \in \mathscr{L}_\sigma^1(\Omega, \mathscr{E}, m)$ and assume $\Omega \in \mathscr{B}_m^{00}$. Prove that, for the σ-field $\mathscr{B}$ of all Borel sets in $\overline{\mathbb{R}}$, the mapping $h: \Omega \to \overline{\mathbb{R}}$ is $\mathscr{B}_m^{00}$-$\mathscr{B}$-measurable and sends m into a σ-content hm on $\mathscr{B}$ such that the identity function $x \to x$ on $\overline{\mathbb{R}}$ is in L_{hm}^1 and $\int h\, dm = \int x(hm)\,(dx)$.

2.7.3.2. Let $h \in \mathscr{L}_\sigma^1(\Omega, \mathscr{E}, m)$ and define $\Omega_0 = \{\omega \mid \omega \in \Omega,\ h(\omega) \neq 0\}$, $\mathscr{B}_0^{00} = \mathscr{B}_m^{00} \cap \Omega_0$, $m_0 =$ the restriction of m: $\mathscr{B}_m^{00} \to \mathbb{R}_+$ to $\mathscr{B}_0^{00}$, $\mathscr{B}^{00} = \{F \mid F \subseteq \overline{\mathbb{R}}\backslash\{0\},\ F \text{ Borel},\ m(h^{-1}F) = m(\{\omega \mid h(\omega) \in F\}) < \infty\}$. Prove that $\mathscr{B}^{00}$ is a local σ-ring in $\overline{\mathbb{R}}\backslash\{0\}$ and $h_0 =$ the restriction of h to Ω_0 is $\mathscr{B}_0^{00}$-$\mathscr{B}^{00}$-measurable, sending m_0 into a σ-content $h_0 m_0$ on $\mathscr{B}^{00}$ such that the identity function $x \to x$ on $\overline{\mathbb{R}}\backslash\{0\}$ is in $L_{h_0 m_0}^1$ and $\int h\, dm = \int x(h_0 m_0)\,(dx)$.

2.7.3.3. Let $\Omega \in \mathscr{B}_m^{00}$, i.e., $1 \in L_m^1$, and $h \in \mathscr{L}_\sigma^2(\Omega, \mathscr{E}, m)$, hence $h \in \mathscr{L}_\sigma^1(\Omega, \mathscr{E}, m)$ (Exercise III.11.3.4). Apply 2.7.3.1 and show in addition that the strictly convex function $x \to x^2$ on R is in L_{hm}^1 and

$$\|h\|_2^2 = \int x^2(hm)\,(dx).$$

2.7.3.4. Let $1 \in L_m^1$ and $g: \overline{\mathbb{R}} \to \overline{\mathbb{R}}$ be such that $g \in L_{hm}^1$. Prove that this implies that $g(h): \Omega \to \overline{\mathbb{R}}$ is in L_m^1 and $\int g(h)\,dm = \int g(x)(hm)\,(dx)$.

2.7.3.5. Try to carry over the result of 2.7.3.4 to the more general situation considered in 2.7.3.2.

2.7.4. Try to carry over Exercise 2.7.3 to σ-measures as far as possible.

2.7.5. Let $(\Omega, \mathscr{B}^{00})$, $(\tilde{\Omega}, \tilde{\mathscr{B}}^{00})$ be local measurable spaces and $\mathscr{S} \subseteq \mathscr{P}(\Omega)$, $\tilde{\mathscr{S}} \subseteq \mathscr{P}(\tilde{\Omega})$ be stable against finite intersections, with $\mathscr{B}^{00} = \mathscr{B}^{00}(\mathscr{S})$, $\tilde{\mathscr{B}}^{00} = \mathscr{B}^{00}(\tilde{\mathscr{S}})$. Let $T: \Omega \to \tilde{\Omega}$ satisfy $T^{-1}\tilde{\mathscr{S}} \subseteq \mathscr{S}$. Prove that $T^{-1}\tilde{\mathscr{B}}^{00} \subseteq \mathscr{B}^{00}$. Let $m: \mathscr{B}^{00} \to \mathbb{R}_+$, $\tilde{m}: \tilde{\mathscr{B}}^{00} \to \mathbb{R}_+$ be σ-contents such that

$$\tilde{S} \in \tilde{\mathscr{S}} \quad \Rightarrow \quad m(T^{-1}\tilde{S}) = \tilde{m}(\tilde{S}) \qquad (\tilde{S} \in \tilde{\mathscr{S}}).$$

Prove that $Tm = \tilde{m}$.

3. ERGODIC THEOREMS

In this section we apply the results of the preceding sections, making the reader acquainted with the rudiments of *ergodic theory* and proving the so-called (v. NEUMANN (1903–1957) [2]) *mean ergodic theorem* and the (G. D. BIRKHOFF (1884–1944) [1]) *individual ergodic theorem*. We refrain from tracing the origins of ergodic theory in statistical mechanics.

1. Dynamical Systems

3.1. Definition. *A quadruple* $(\Omega, \mathscr{R}, m, T)$ *is called a* **dynamical system** *if:*

3.1.1. $(\Omega, \mathscr{R}, m)$ *is a σ-content prespace;*

3.1.2. $T: \Omega \to \Omega$ *is $\mathscr{R}$-$\mathscr{R}$-measurable and preserves m, i.e.,*

$$Tm = m.$$

Properties of $(\Omega, \mathscr{R}, m)$ *are also attributed to* $(\Omega, \mathscr{R}, m, T)$.

3.2. Remarks

3.2.1. If $(\Omega, \mathscr{R}, m, T)$ is a dynamical system, then we say that T *preserves* m or is *m-preserving*, and equivalently, that m is invariant under T or T-invariant.

3.2.2. Exercises 2.3 show that, given a dynamical system $(\Omega, \mathscr{R}, m, T)$, some more dynamical systems are, more or less equivalently, given as well, by obvious extensions of the σ-content prespace $(\Omega, \mathscr{R}, m)$. Thus, e.g., $(\Omega, \mathscr{B}_m^{00}, m, T)$ is a dynamical system where $(\Omega, \mathscr{B}_m^{00}, m)$ denotes the Carathéodory extension of $(\Omega, \mathscr{R}, m)$. In particular the pullback T^{-1} sends m-nullsets into m-nullsets and the function mapping $h \to h \circ T$ sends m-a.e. equal functions into m-a.e. equal functions. We concentrate these observations into the statement that we are always free to assume, e.g., that $(\Omega, \mathscr{R}, m)$ is a full σ-content space. On the other hand, our above definition comprises many seemingly more involved situations as special cases. Let, e.g., $\mathscr{B}$ be a σ-field in $\Omega \neq \varnothing$, $T\colon \Omega \to \Omega$ be $\mathscr{B}$-$\mathscr{B}$-measurable and $m\colon \mathscr{B} \to \overline{\mathbb{R}}_+$ be σ-additive such that $Tm = m$. Put $\mathscr{B}^{00} = \{F \mid F \in \mathscr{B}, m(F) < \infty\}$ and let m also denote the restriction of m to $\mathscr{B}^{00}$. Then clearly $(\Omega, \mathscr{B}^{00}, m, T)$ is a dynamical system.

3.2.3. If $(\Omega, \mathscr{R}, m, T)$ is a dynamical system, then for every $n = 0, 1, \ldots,$ $(\Omega, \mathscr{R}, m, T^n)$ is also a dynamical system (T^0 = the identical mapping in Ω).

3.2.4. If $(\Omega, \mathscr{R}, m, T)$ is a dynamical system such that $T\colon \Omega \to \Omega$ is a bijection whose inverse T^{-1} is $\mathscr{R}$-$\mathscr{R}$-measurable (we shall then say briefly that T is (measurably) invertible), then $(\Omega, \mathscr{R}, m, T^n)$ is a dynamical system for every integer n. In fact, it is sufficient to prove that $T^{-1}m = m$. But this follows from

$$F \in \mathscr{R} \quad \Rightarrow \quad TF = (T^{-1})^{-1}F \in \mathscr{R} \quad \text{and} \quad m(TF) = m(T^{-1}(TF)) = m(F).$$

3.2.5. Given a dynamical system $(\Omega, \mathscr{R}, m, T)$, it is convenient to adopt the following ideas and termini. We interpret Ω as the set of all possible *states* of a physical system, and we assume that the system changes its

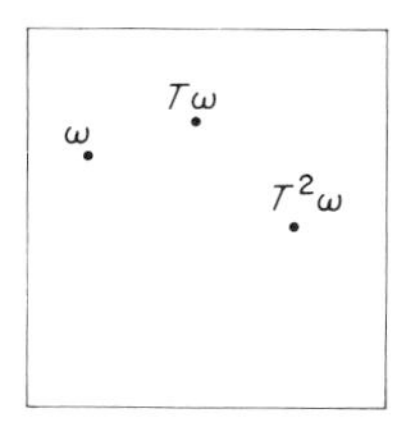

state according to the following rule: If it is in state ω at time t, then it is in state $T\omega$ at time $t + 1$. Thus, if it is in state ω at time 0, then the sequence of its states at times 0, 1, 2, ... is $\omega, T\omega, T^2\omega, \ldots$. The set $\{T^n\omega \mid n = 0, 1, \ldots\}$ is called the (*forward*) *orbit* of ω. If $E \subseteq \Omega$, then $T^{-1}E$ is the set of all ω that *visit* E at time 1, and $T^{-n} = T^{-1}(T^{-(n-1)}E)$ $(n = 1, 2, \ldots)$ is the set of all ω that *visit* E at time n. If T is invertible, then $\{T^n\omega \mid n \in \mathbb{Z}\}$ makes sense as well as $\{T^n\omega \mid 0 \geqq n \in \mathbb{Z}\}$ and are called respectively the (*full*) *orbit* and the *backward orbit* of ω. In this case it makes sense to say that ω visits $E \subseteq \Omega$ at time n also for $0 \geqq n \in \mathbb{Z}$.

3.3. Example. Let $\Omega = \{z | z \in \mathbb{C}, |z| = 1\} = \{e^{ix} | x \in \mathbb{R}\}$. A set of the form $A(a, b) = \{e^{ix} | a \leqq x < b\}$ with reals $a < b < a + 2\pi$ is called a (half-open) arc of length $2\pi(b - a)$. It is obvious how to uniquely define an additive real function $m \geqq 0$ on the field $\mathscr{F}$ of all disjoint unions of half-open arcs, such that $m(A(a, b)) = 2\pi(b - a)$ (compare Exercise I.1.11.4). It is easy to see that $m: \mathscr{F} \to \mathbb{R}_+$ is σ-additive (compare Example I.2.4 and Exercise I.2.6.3). Let now $\alpha \in \mathbb{R}$ and define $T_\alpha: \Omega \to \Omega$ by $T_\alpha z = e^{i\alpha} z$ ($z \in \Omega$), i.e., as a rotation by the angle α. Clearly T_α is $\mathscr{F}$-$\mathscr{F}$-measurable with $T_\alpha m = m$. Let $\mathscr{B}$ be the σ-field generated by $\mathscr{F}$. Then m has a unique σ-additive extension to $\mathscr{B}$ which we denote by m again. Clearly $(\Omega, \mathscr{B}, m, T_\alpha)$ is a dynamical system. It is usually denoted briefly as "*circle rotation.*"

3.4. Example. Let $(\Omega, \mathscr{B}, m) = (A \times A \times \cdots, \mathscr{B}^{00}(\mathscr{C}), m)$ be the Bernoulli σ-content space for the finite alphabet $A \neq \varnothing$ and the probability vector p over A. Define the so-called *shift* $T: \Omega \to \Omega$ by $T(\omega_0 \omega_1 \ldots) = \omega_1 \omega_2 \cdots$ ($\omega = \omega_0 \omega_1 \cdots \in \Omega$). For every special cylinder $[k_0, \ldots, k_r]$ of order r we can write

$$T^{-1}[k_0, \ldots, k_r] = \sum_{k \in A} [k, k_0, \ldots, k_r],$$

i.e., as a disjoint union of special cylinders of order $r + 1$, proving the $\mathscr{B}$-$\mathscr{B}$-measurability of T. Moreover we get

$$\begin{aligned}(Tm)([k_0, \ldots, k_r]) &= m(T^{-1}[k_0, \ldots, k_r]) \\ &= m\left(\sum_{k \in A} [k, k_0, \ldots, k_r]\right) = \sum_{k \in A} m([k, k_0, \ldots, k_r]) \\ &= \sum_{k \in A} p(k)p(k_0) \cdots p(k_r) = p(k_0) \cdots p(k_r) = m([k_0, \ldots, k_r])\end{aligned}$$

since $\sum_{k \in A} p(k) = 1$. This shows that Tm and m agree on all cylinders of finite order. Since the latter form an intersection stable set system generating $\mathscr{B}$, $Tm = m$ follows by Proposition I.5.4. The dynamical system $(\Omega, \mathscr{B}, m, T)$ obtained here is called the *one-sided Bernoulli scheme* or *shift system* for p (and A). If $(\Omega, \mathscr{B}, m)$ is the two-sided Bernoulli σ-probability space for p, then the shift $T: \cdots \omega_{-1} \omega_0 \omega_1 \cdots \to \cdots \omega_0 \omega_1 \omega_2 \cdots$ (we have written out components -1, 0, and 1 of ω and $T\omega$ here) is invertible and defines a dynamical system $(\Omega, \mathscr{B}, m, T)$ which is again called the (*two-sided*) *Bernoulli scheme* or *shift system* for p (and A).

3.5. Exercises

3.5.1. Let $(\Omega, \mathscr{B}, m, T)$ be as in Example 3.3. Prove that the mapping $\varphi: x \to e^{ix}$ of $[0, 2\pi[$ to Ω is one-to-one and $\mathscr{B}([0, 2\pi[)$-$\mathscr{B}$-measurable, where $\mathscr{B}([0, 2\pi[)$ is the σ-field of all Borel subsets of $[0, 2\pi[$. Prove that φ sends the restriction of the Lebesgue σ-content to $\mathscr{B}([0, 2\pi[)$ into m. Prove that $\mathscr{C}(\Omega, \mathbb{R})$ is $\|\cdot\|_1$-norm dense in L_m^1.

3.5.2. Let $\Omega = \{\omega = (z_1, \ldots, z_n) | z_1, \ldots, z_n \in \mathbb{C}, |z_1| = \cdots = |z_n| = 1\} \subseteq \mathbb{C}^n$ and $\mathscr{B}$ the σ-field generated by the natural topology in Ω. Define m on $\mathscr{B}$ as the n-dimensional analogue to Example 3.3 (compare Exercises I.1.11.2, I.2.6.2) and prove that, for any reals $\alpha_1, \ldots, \alpha_n$, the mapping $T: (z_1, \ldots, z_n) \to (e^{i\alpha_1}z_1, \ldots, e^{i\alpha_n}z_n)$ defines a dynamical system $(\Omega, \mathscr{B}, m, T)$ (usually called "torus rotation").

3.5.3. Let $\Omega = \mathbb{R}, \mathscr{B}^{00} = \mathscr{B}^{00}(\mathbb{R}) =$ the system of all bounded Borel subsets of $\mathbb{R}$ and $m: \mathscr{B}^{00} \to \mathbb{R}_+$ the Lebesgue σ-content. Let α be any real and prove that the mapping $T_\alpha: x \to x + \alpha$ defines a dynamical system $(\Omega, \mathscr{B}^{00}, m, T_\alpha)$.

3.5.4. Let $(\Omega, \mathscr{B}, m)$ be the Bernoulli σ-probability space based on a finite alphabet $A \neq \varnothing$ and let $p^0, p^1, \ldots$ be a sequence of probability vectors over A (Example II.6.5), and $T: \Omega \to \Omega$ the shift. Prove that $Tm = m$ iff $p^0 = p^1 = \cdots$.

3.5.5. Prove the analogue of 4 for two-sided Bernoulli spaces.

3.5.6. Let $(\Omega, \mathscr{B}, m)$ be the Markov σ-probability space for the initial distribution p and the transition matrix P (and the finite alphabet $A \neq \varnothing$). Prove that the shift $T: \Omega \to \Omega$ defines a dynamical system iff $pP = p$, i.e., iff $\sum_{k \in A} p(k)P(k, j) = p(j)$ $(j \in A)$. Prove that for every probability vector p^0 over A every limit point p of the conditionally compact sequence $n^{-1} \sum_{k=0}^{n-1} p^0 P^k$ $(n = 1, 2, \ldots;$ P^k denotes the kth power of the square matrix $P)$ satisfies $pP = p$.

3.5.7. Let p, P with $pP = p$ as in 6 and prove that there is a unique dynamical system $(\Omega, \mathscr{B}, m, T)$ where T is the shift in the two-sided sequence space $\cdots \times A \times A \times \cdots$ and

$$m({}_t[k_0, \ldots, k_r]_{t+r}) = p(k_0)P(k_0, k_1) \cdots P(k_{r-1}, k_r)$$

where

$$r \geqq 0,\ t \in \mathbb{Z},\ k_0, \ldots, k_r \in A;$$

$${}_t[k_0, \ldots, k_r]_{t+r} = \{\omega = \cdots \omega_{-1}\omega_0\omega_1 \cdots | \omega_t = k_0, \ldots, \omega_{t+r} = k_r\}.$$

3.6. Remark. The preceding examples and exercises present only the most outstanding ones in a large row of examples of dynamical systems among which one, based on the so-called *Hamiltonian flow* in $\mathbb{R}^{6N}$ (which preserves the $6N$-dimensional Lebesgue σ-content, after a theorem of Liouville), is the historical origin of ergodic theory but by no means the most accessible. We have here intentionally presented only examples that are not too technical but nevertheless play a basic role in present-day ergodic theory, the *two-sided Bernoulli scheme* being the most prominent one.

2. The Mean Ergodic Theorem

Let $(\Omega, \mathscr{R}, m, T)$ be a dynamical system. By Theorem 2.4 and Proposition 2.6 the point mapping $T\colon \Omega \to \Omega$ sends L_m^1 into L_m^1 such that $\int h\, dm = \int (h \circ T)\, dm$, $\|(h \circ T\|_1 = \|h\|_1$ $(h \in L_m^1)$ and also sends L_m^2 into L_m^2 such that $(g, h) = (g \circ T, h \circ T)$, $\|h \circ T\|_2 = \|h\|_2$ $(g, h \in L_m^2)$. In particular, the function mapping $h \to h \circ T$ sends m-a.e. equal functions into m-a.e. equal ones and thus defines a mapping of L_m^1 resp. L_m^2 into itself, and these mappings $T\colon L_m^1 \to L_m^1$ and $T\colon L_m^2 \to L_m^2$ are justly called isometries of the Banach resp. Hilbert spaces L_m^1 resp. L_m^2; they are clearly bijective and invertible if T is. They are, of course, linear. The invariance of the scalar product in L_m^2 is, by the way, logically equivalent to the invariance of the L^2-norm since the latter is defined by $\|h\|_2^2 = (h, h)$ and since the scalar product can be expressed in terms of the norm: $\|g + h\|_2^2 = \|g\|_2^2 + \|h\|_2^2 + 2(g, h)$ $(g, h \in L_m^2)$.

The objective of ergodic theorems is the investigation of the so-called *ergodic means*

$$\frac{1}{t} \sum_{u=0}^{t-1} T^n h$$

either for $h \in L_m^1$ or for $h \in L_m^2$, and their behavior as $t \to \infty$. Actually, the so-called individual ergodic theorem states convergence m-a.e. of these means, and the so-called *mean ergodic theorem* states convergence in the L^1 of L^2 mean, i.e., norm convergence in the Banach space L_m^1 resp. L_m^2. The latter holds for L_m^2 in any case, and for L_m^1 if $1 \in L_m^1$. Of all these statements, norm convergence in L_m^2 is most easy to prove; actually it is nearly a mere geometrical fact, based on the isometric property of $T\colon L_m^2 \to L_m^2$. We therefore begin with

3.7. Theorem (mean ergodic theorem in L^2). *Let $(\Omega, \mathscr{R}, m, T)$ be a dynamical system.*

3.7.1. *There is a linear mapping $\overline{T}\colon L_m^2 \to L_m^2$ such that*

$$\lim_{t\to\infty} \left\| \frac{1}{t} \sum_{u=0}^{t-1} h \circ T^u - \overline{T}h \right\|_2 = 0 \qquad (h \in L_m^2). \tag{1}$$

3.7.2. *$\overline{T}$ has the following properties:*

3.7.2.1. $T \circ \overline{T} = \overline{T} \circ T = \overline{T} = \overline{T} \circ \overline{T}$.

3.7.2.2. $\|\overline{T}h\|_2 \leq \|h\|_2$ $(h \in L_m^2)$.

3.7.2.3. $H_{\text{fix}} = \{h \mid h \in L_m^2,\ h \circ T = h\}$ *equals* $\overline{T}_m^2 L_m^2 = \{\overline{T}h \mid h \in L_m^2\} = \{h \mid h \in L_m^2, \overline{T}h = h\}$ *and is a norm-closed linear subspace of* L_m^2.

3.7.2.4.

$$H_0 = \{h \mid h \in L_m^2, \overline{T}h = 0\} = \left\{ h \mid h \in L_m^2, \lim_{t\to\infty} \left\| \frac{1}{t} \sum_{u=0}^{t-1} h \circ T^u \right\|_2 = 0 \right\}$$

is a norm-closed linear subspace of L_m^2. *If* L *is a linear subspace of* L_m^2 *that is* L^2*-norm dense in* L_m^2, *then* $\{f - f \circ T \mid f \in L\}$ *is a linear subspace of* H_0 *and norm-dense in* H_0.

3.7.2.5. $(g, h) = 0$ $(g \in H_{\text{fix}}, h \in H_0)$.

3.7.2.6. *Every* $h \in L_m^2$ *has a unique representation*

(2) $$h = \bar{h} + h_0, \qquad \bar{h} \in H_{\text{fix}}, \quad h_0 \in H_0.$$

Actually, $\bar{h} = \overline{T}h$.

Proof. The norm we speak of here is always the L^2-norm $\|\cdot\|_2$, of course. It is invariant under T as well as the scalar product in L_m^2, and these facts, together with the linearity of T: $L_m^2 \to L_m^2$, are the only facts used in the proof, which is thus a proof of a much more general theorem.

(a) We begin with a simple observation which is, in a way, the basis of Hilbert space geometry: If $\varnothing \neq K$ is convex (i.e., satisfies $\alpha g + (1 - \alpha)h \in K$ $(g, h \in K, 0 \leqq \alpha \leqq 1)$) and norm-closed, then for every $h_0 \in L_m^2$ there is a unique $h_1 \in K$ such that $\|h_1 - h_0\|_2 \leqq \|h - h_0\|_2$ $(h \in K)$. In fact, let $\beta = \inf_{h \in K} \|h - h_0\|_2$. If $\beta = 0$, then $h_0 \in K$ and we are through, putting $h_1 = h_0$. If $\beta > 0$, we consider the identity

$$\left\| \frac{f+g}{2} \right\|_2^2 + \left\| \frac{f-g}{2} \right\|_2^2 = \left(\frac{f+g}{2}, \frac{f+g}{2} \right) + \left(\frac{f-g}{2}, \frac{f-g}{2} \right)$$
$$= \frac{\|f\|_2^2 + \|g\|_2^2}{2}$$

(the "mixed" scalar products cancel). Choose $g_1, g_2, \ldots \in K$ with $\lim_{n\to\infty} \|g_n - h_0\|_2^2 = \beta$. Then we see

$$\left\| \frac{g_j - g_k}{2} \right\|_2^2 = \left\| \frac{(g_j - h_0) - (g_k - h_0)}{2} \right\|_2^2$$
$$= \frac{\|g_j - h_0\|_2^2 + \|g_j - h_0\|_2^2}{2} - \left\| \frac{(g_j - h_0) + (g_k - h_0)}{2} \right\|_2^2$$
$$= \frac{\|g_j - h_0\|_2^2 + \|g_k - h_0\|_2^2}{2} - \left\| \frac{g_j + g_k}{2} - h_0 \right\|_2^2$$
$$\leqq \frac{\|g_j - h_0\|_2^2 + \|g_k - h_0\|_2^2}{2} - \beta^2 \to \beta^2 - \beta^2 = 0$$

as $j, k \to \infty$. The $\leqq$ here results from the convexity of K, which implies $(g_j + g_k)/2 \in K$ and hence $\|(g_j + g_k)/2 - h_0\|_2 \geqq \beta$. Thus we see that $g_1, g_2 \in K$ is a L^2-norm fundamental sequence in L_m^2 and therefore has a limit h_1 which is in K since K is closed. Since the L^2-norm is an L^2-norm continuous function on L_m^2, we get $\|h_1 - h_0\|_2 = \beta$. If $h_1' \in K$ satisfies $\|h_1' - h_0\|_2 = \beta$, then we use our above identity and $(h_1 + h_1')/2 \in K$ once again in the same way and get $\|h_1 - h_1'\|_2 = 0$, i.e., $h_1 = h_1'$. Let us call h_1 the "point in K next to h_0."

(b) Let H be a closed linear subspace of L_m^2. We show that for every $h_0 \in L_m^2$, the point h_1 in H next to h_0 (it exists by (a) since H is convex) satisfies

$$(h_0 - h_1, g) = 0 \qquad (g \in H).$$

In fact, we may assume $g \neq 0$ and define a quadratic polynomial $x \to p(x)$ on $\mathbb{R}$ by

$$\begin{aligned} p(x) &= \|xg + h_1 - h_0\|_2^2 = (xg + h_1 - h_0, xg + h_1 - h_0) \\ &= x^2\|g\|_2^2 + 2x(g, h_1 - h_0) + \|h_1 - h_0\|_2^2. \end{aligned}$$

Since $xg + h_1 \in H$ ($x \in \mathbb{R}$), $p(x)$ attains its unique minimum for $x = 0$. This clearly implies the vanishing of its first derivative $p'(x)$ for $x = 0$, and this gives exactly the desired result.

(c) Next we prove, for an arbitrary $h \in L_m^2$, the existence of a decomposition (2). To this end we consider the L^2-norm closure $K(h)$ of the set $\{n^{-1}(h \circ T^{r_1} + \cdots h \circ T^{r_n}) | n \geqq 1, r_1, \ldots, r_n \geqq 0\}$. Clearly this latter set equals $\{\alpha_1 h \circ T^{s_1} + \cdots + \alpha_r h \circ T^{s_r} | r \geqq 1, 0 \leqq \alpha_1, \ldots, \alpha_r$ rational with sum 1, $s_1, \ldots, s_r \geqq 0\}$ and $K(h)$ is also the closure of

$$\{\alpha_1 h \circ T^{s_1} + \cdots + \alpha_r h \circ T^{s_r} | 0 \leqq \alpha_1, \ldots, \alpha_r \in \mathbb{R}, \text{ with sum } 1, s_1, \ldots, s_r \geqq 0\}$$

and hence convex (exercise). Since

$$(\alpha_1 h \circ T^{s_1} + \cdots + \alpha_r h \circ T^{s_r}) \circ T = \alpha_1 h \circ T^{s_1 + 1} + \cdots + \alpha_r h \circ T^{s_r + 1}$$

and T carries, being linear and isometric, every L^2-norm convergent sequence into an L^2-norm convergence sequence, it is easily seen that $K(h)$ is T-invariant, i.e., $g \in K(h) \Rightarrow g \circ T \in K(h)$. Let now $\bar{h} \in L_m^2$ be the point in $K(h)$ next to 0. Then $\bar{h} \circ T \in K(h)$ and $\|\bar{h} \circ T\|_2 = \|\bar{h}\|_2$, hence $\bar{h} \circ T$ is next to 0 as well, and $\bar{h} \circ T = \bar{h}$ follows since the point in $K(h)$ next to 0 is unique. Thus we have $\bar{h} \in H_{\text{fix}}$. We put $h_0 = h - \bar{h}$ and will get (2) as soon as we have shown $h_0 \in H_0$. To this end we choose an arbitrary $\varepsilon > 0$ and determine $n \geqq 1, r_1, \ldots, r_n \geqq 0$ such that

$$\left\| \frac{1}{n}(h \circ T^{r_1} + \cdots + h \circ T^{r_n}) - \bar{h} \right\|_2 < \varepsilon.$$

We now get, employing $\bar{h} = \bar{h} \circ T^{r_1} = \cdots = \bar{h} \circ T^{r_n}$,

$$\begin{aligned}
\varepsilon &> \left\| \frac{1}{n}((h - \bar{h}) \circ T^{r_1} + \cdots + (h - \bar{h}) \circ T^{r_n}) \right\|_2 \\
&= \left\| \frac{1}{n}(h_0 \circ T^{r_1} + \cdots + h_0 \circ T^{r_1}) \circ T^u \right\|_2 \qquad (u = 0, \ldots, t) \\
&= \frac{1}{t} \sum_{u=0}^{t-1} \left\| \frac{1}{n}(h_0 \circ T^{r_1 + u} + \cdots + h_0 \circ T^{r_n + u}) \right\|_2 \\
&\geqq \left\| \frac{1}{n}\left(\frac{1}{t}\sum_{u=0}^{t-1} h_0 \circ T^{r_1 + u} + \cdots + \frac{1}{t}\sum_{u=0}^{t-1} h_0 \circ T^{r_n + u}\right) \right\|_2 \\
&= \left\| \frac{1}{n}\left(\frac{1}{t}\sum_{u=r_1}^{r_1+t-1} h_0 \circ T^u + \cdots + \frac{1}{t}\sum_{u=r_1}^{r_n+t-1} h_0 \circ T^u\right) \right\|_2 \\
&= \left\| \frac{1}{n}\left(\frac{n}{t}\sum_{u=0}^{t-1} h_0 \circ T^u\right) - \frac{1}{n}\sum_{k=1}^{n}\left(\frac{1}{t}\left[\sum_{u=0}^{n_k-1} h_0 \circ T^u - \sum_{u=t}^{t+r_k-1} h_0 \circ T^u\right]\right) \right\|_2 \\
&\geqq \left\| \frac{1}{t}\sum_{u=0}^{t-1} h_0 \circ T^u \right\|_2 - \frac{1}{t} \cdot \frac{1}{n}\sum_{k=1}^{n} 2r_k \|h_0\|_2 \\
&> \left\| \frac{1}{t}\sum_{u=0}^{t-1} h_0 \circ T^u \right\|_2 - \varepsilon
\end{aligned}$$

i.e.,

$$\left\| \frac{1}{t}\sum_{u=0}^{t-1} h_0 \circ T^u \right\|_2 < 2\varepsilon$$

if t is sufficiently large, proving $h_0 \in H_0$.

One of the devices used above may justly be called a *cancellation trick*.

(d) Whenever we have (2), then

$$\begin{aligned}
&\lim_{t\to\infty} \left\| \frac{1}{t}\sum_{u=0}^{t-1} h \circ T^u - \bar{h} \right\|_2 \\
&\quad = \lim_{t\to\infty} \left\| \frac{1}{t}\sum_{u=0}^{t-1} h \circ T^u - \bar{h} \circ T^j \right\|_2 \qquad (j = 0, \ldots, t-1) \\
&\quad = \lim_{t\to\infty} \left\| \frac{1}{t}\sum_{u=0}^{t-1} (h - \bar{h}) \circ T^u \right\|_2 = \lim_{t\to\infty} \left\| \frac{1}{t}\sum_{n=0}^{t-1} h_0 \circ T^u \right\|_2 = 0,
\end{aligned}$$

hence $\bar{h}$, h_0 are uniquely determined in (2) and $\overline{T}: h \to \bar{h}$ defines $\overline{T}: L_m^2 \to H_{\text{fix}}$ such that (1) holds.

(e) From

$$\left\| \frac{1}{t} \sum_{u=0}^{t-1} h \circ T^u \right\|_2 \leqq \frac{1}{t} \sum_{u=0}^{t-1} \|h \circ T^u\|_2 = \frac{1}{t} \sum_{u=0}^{t-1} \|h\|_2 = \|h\|_2$$

we get 2.2. We easily get 2.1 from the fact that $\overline{T}$ sends L_m^2 into H_{fix}, clearly leaving every $h \in H_{\text{fix}}$ fixed. The proof that $\overline{T}$ is linear is an easy exercise which we leave to the reader, as well as the proof of 2.3.

(f) Since $\overline{T}$ is norm-contracting and hence norm-continuous, we find that H_0, being the kernel of the linear mapping $\overline{T}$, is a closed subspace of L_m^2.

(g) Next we prove 2.5. Let $g \in H_{\text{fix}}$, $h \in H_0$. Then

$$\begin{aligned} |(g, h)| &= |(g \circ T^u, h \circ T^u)| = |(g, h \circ T^u)| \qquad (u = 0, \ldots, t-1) \\ &= \left| \left(g, \frac{1}{t} \sum_{u=0}^{t-1} h \circ T^u \right) \right| \\ &\leqq \|g\|_2 \cdot \left\| \frac{1}{t} \sum_{u=0}^{t-1} h \circ T^u \right\|_2 (t = 1, 2, \ldots) \\ &\to 0 \qquad (t \to \infty) \end{aligned}$$

and hence $(g, h) = 0$.

(h) Let L be a linear subspace of L_m^2 that is L^2-norm dense in L_m^2. Then $L' = \{f - f \circ T \mid f \in L\}$ is a linear subspace of L_m^2 since T is linear. Our above *cancellation trick*, here used in a simpler form, yields

$$\begin{aligned} \left\| \frac{1}{t} \sum_{u=0}^{t-1} (f - f \circ T) \circ T^u \right\|_2 &= \left\| \frac{1}{t} \sum_{u=0}^{t-1} f \circ T^u - \frac{1}{t} \sum_{u=1}^{t} f \circ T^u \right\|_2 \\ &\leqq \frac{1}{t} (\|f\|_2 + \|f \circ T^t\|_2) \\ &= \frac{2}{t} \|f\|_2 \qquad (f \in L) \end{aligned}$$

which tends to 0 as $t \to \infty$. Thus $L' \subseteq H_0$; and since H_0 is closed, the L^2-norm closure $\overline{L'}$ of L' also satisfies $\overline{L'} \subseteq H_0$. We have to prove equality here. Assume that $\overline{L'} \neq H_0$, choose some $g_0 \in H_0$ not in $\overline{L'}$, and find the point g_1 in $\overline{L'}$ next to g_0. By (b) we find that $h_0 = g_0 - g_1$ satisfies $(h_0, g) = 0$ $(g \in \overline{L'})$ and in particular $(h_0, f - f \circ T) = 0$ $(f \in L)$. By an easy norm approximation we find $(h_0, f - f \circ T) = 0$, i.e., $(h_0, f) = (h_0, f \circ T)$ $(f \in L_m^2)$. Applying this to $f, f \circ T, \ldots$ successively, we see

$$\begin{aligned} (h_0, f) &= (h_0, f \circ T^u) \qquad (u = 0, \ldots, t-1) \\ &= \left(h_0, \frac{1}{t} \sum_{u=0}^{t-1} f \circ T^u \right) \qquad (t = 1, 2, \ldots) \\ &\to (h_0, \overline{T} f), \end{aligned}$$

and this is $=0$ since $h_0 = g_0 - g_1 \in H_0$. Thus we see that $(h_0, f) = 0$ $(f \in L_m^2)$. If we put $f = h_0$, we get $\|h_0\|^2 = (h_0, h_0) = 0$, hence $h_0 = 0$ and 2.4 is proved. The proof of our theorem is now complete.

3.8. Remark. A reader with a bit of background in Hilbert space theory will easily prove the following: Let H be an abstract Hilbert space and G a semigroup of norm-contracting linear mappings of H into itself, i.e., $\|Th\| \leqq \|h\|$ $(T \in G,\ h \in H)$, $T \circ S \in G$ $(T,\ S \in G)$. Put $H_{\text{fix}} = \{h \mid h \in H,\ Th = h\ (T \in G)\}$ and, with $K(h) =$ the norm closure of the convex hull of $\{Th \mid T \in G\}$, $H_0 = \{h \mid h \in H,\ 0 \in K(h)\}$. Then H_{fix} and H_0 are orthogonal complements of each other. A crucial point in the proof is the observation that a $T \in G$ and its so-called adjoint T^* have the same fixed points in H, i.e., $Th = h \Leftrightarrow T^*h = h$ $(h \in H)$. If there is a single $T \in G$ such that $G = \{T^0, T, T^2, \ldots\}$, then (1) holds, where $\overline{T}$ is the so-called orthogonal projection onto H_{fix}.

3.9. Theorem (mean ergodic theorem in L^1). *Let $(\Omega, \mathscr{B}, m, T)$ be a dynamical system such that $1 \in L_m^1$. Then there is a linear mapping $\overline{T}: L_m^1 \to L_m^1$ such that*

$$(3) \qquad \lim_{t \to \infty} \left\| \frac{1}{t} \sum_{u=0}^{t-1} h \circ T^u - \overline{T}h \right\|_1 = 0 \qquad (h \in L_m^1).$$

$\overline{T}$ satisfies $\overline{T} = T \circ \overline{T} = \overline{T} \circ T = \overline{T} \circ \overline{T}$.

Proof. By Exercise III.11.3.4 we have $L_m^2 \subseteq L_m^1$ and L_m^2 is L^1-norm dense in L_m^1. We may and shall assume $\int 1\, dm = 1$ and hence $\|h\|_1 \leqq \|h\|_2$ $(h \in L_m^2)$. Let $h \in L_m^1$ be arbitrary. For an arbitrary $\varepsilon > 0$, we choose some $g \in L_m^2$ such that $\|h - g\|_1 < \varepsilon$. Now we get

$$\left\| \frac{1}{t} \sum_{u=0}^{t-1} h \circ T^u - \frac{1}{t} \sum_{u=0}^{t-1} g \circ T^u \right\|_1 \leqq \frac{1}{t} \sum_{u=0}^{t-1} \|(h - g) \circ T^u\|_1$$
$$= \|h - g\|_1 < \varepsilon \qquad (t = 1, 2, \ldots).$$

On the other hand, we get, for the $\overline{T}$ of Theorem 3.7, $\overline{T}g \in L_m^2 \subseteq L_m^1$ and

$$\lim_{t \to \infty} \left\| \frac{1}{t} \sum_{u=0}^{t-1} g \circ T^u - \overline{T}g \right\|_1 \leqq \lim_{t \to \infty} \left\| \frac{1}{t} \sum_{u=0}^{t-1} g \circ T^u - \overline{T}g \right\|_2 = 0.$$

By combination we find

$$\limsup_t \left\| \frac{1}{t} \sum_{u=0}^{t-1} h \circ T^u - \overline{T}g \right\|_1 \leqq \varepsilon$$

and hence

$$\limsup_{s,t\to\infty}\left\|\frac{1}{s}\sum_{v=0}^{s-1}h\circ T^v-\frac{1}{t}\sum_{u=0}^{t-1}h\circ T^u\right\|_1\leqq\varepsilon.$$

Since $\varepsilon>0$ was arbitrary, $t^{-1}\sum_{u=0}^{t-1}h\circ T^u$ $(t=1,2,\ldots)$ defines an L^1-norm fundamental sequence, hence, by the completeness of L_m^1, an L^1-norm convergent sequence in L_m^1. Define $\overline{T}$ by (3). The proof of the properties of T stated in our theorem is now an easy exercise which we leave to the reader.

3.10. Example. We show that the conclusion of Theorem 3.9 may fail if the hypothesis $1\in L_m^1$ is not fulfilled. Let $(\Omega,\mathscr{B}^{00},m,T)$ be the dynamical system of Exercise 3.5.3, with $\alpha=1$ (translation by 1 in $\mathbb{R}$). Let $h=1_{[0,1[}$. Clearly $h\in L_m^2\subseteq L_m^1$ and $t^{-1}\sum_{u=0}^{t-1}h\circ T^u=t^{-1}1_{[0,t[}$, hence

$$\left\|\frac{1}{t}\sum_{u=0}^{t-1}h\circ T^u\right\|_2^2=\frac{1}{t^2}\int 1_{[0,t[}1_{[0,t[}\,dm$$
$$=\frac{t}{t^2}=\frac{1}{t}$$

which tends to 0 as $t\to\infty$. We have $t^{-1}\sum_{u=0}^{t-1}h\circ T^u\to 0$ uniformly on $\mathbb{R}$ and thus the constant 0 is the only candidate for an L^1-norm limit of this sequence (because an L^1-norm limit is an m-a.e. limit of some subsequence, by Theorem III.5.8.4). But

$$\left\|\frac{1}{t}\sum_{u=0}^{t-1}h\circ T^u\right\|_1=\int\frac{1}{t}1_{[0,t[}\,dm=\frac{t}{t}=1\qquad(t=1,2,\ldots)$$

and thus 0 is not an L^1-norm limit of the sequence in question.

3. The Individual Ergodic Theorem

In the preceding subsection we have operated essentially with isometries in the Banach space L^1 and the Hilbert space L^2 and with the norm convergence of ergodic means. Now we consider *convergence a.e.* of these means, exploiting the fact that the mentioned isometries stem from a point mapping in the basic set. In particular, the lattice properties of L^1 and their preservation under the mapping play an essential role. I should like to mention that there is a theory of a.e. convergence of ergodic means for linear operators in L^1 or L^2 which need not stem from a point mapping (see, e.g., DUNFORD and SCHWARTZ [1], JACOBS [2], AKCOGLU [1]), but we shall not present it here.

We begin with a technical lemma which has proved crucial in the theory of a.e. convergence of ergodic means.

3.11. Lemma (maximal ergodic theorem). *Let* $(\Omega, \mathscr{R}, m, T)$ *be a dynamical system,* $h \in L_m^1$, $n > 1$, *and*

$$F_n = \left\{\omega \,\middle|\, \sup_{1 \leqq t \leqq n} \frac{1}{t} \sum_{u=0}^{t-1} h(T^u\omega) \geqq 0\right\}.$$

Then $h1_{F_n} \in L_m^1$ *and*

$$\int h1_{F_n}\, dm \geqq 0.$$

Proof. If we replace h by a function that is m-a.e. equal to h, then F_n is replaced by a set that differs from F_n by an m-nullset only. In particular, $h1_{F_n}$ is replaced by an m-almost equal function. Thus there is no change in the validity of the statements of our theorem at all, and we may, by Theorem III.9.3, henceforth assume $h \in \text{mble}(\Omega, \mathscr{B}^0(\mathscr{R}), \mathbb{R})$. Now $h1_{F_n} \in L_m^1$ follows by Proposition III.9.5. By Exercise 1.7.1,

$$h \circ T, h \circ T^2, \ldots \in \text{mble}(\Omega, \mathscr{B}^0(\mathscr{R}), \mathbb{R})$$

as well, and hence so are the functions defined by

$$h^0 \equiv 0, \qquad h^n = \sup_{1 \leqq t \leqq n} \sum_{u=0}^{t-1} h \circ T^u \qquad (n = 1, 2, \ldots).$$

Observe now that

$$\begin{aligned}
h &= h + 0 &&\leqq h + h_+^n \circ T \\
h + h \circ T &= h + h \circ T &&\leqq h + h_+^n \circ T \\
h + h \circ T + h \circ T^2 &= h + (h + h \circ T) \circ T &&\leqq h + h_+^n \circ T \\
&\vdots \\
\sum_{u=0}^{n} h \circ T^u &= h + \left(\sum_{u=0}^{n-1} h \circ T^n\right) \circ T &&\leqq h + h_+^n \circ T
\end{aligned}$$

and hence $h^{n+1} \leqq h + h_+^n \circ T$, firstly for $n = 1, 2, \ldots$, but trivially, in view of $h^0 = 0$, also for $n = 0$. We thus have

$$h \geqq h^n - h_+^{n-1} \circ T \qquad (n = 1, 2, \ldots).$$

Now we observe that $F_n = \{h^n \geqq 0\}$ and $h_+^0 \leqq h_+^1 \leqq \cdots$ and obtain

$$\int h 1_{F_n}\, dm \geqq \int h^n 1_{F_n}\, dm - \int (h_+^{n-1} \circ T)\, 1_{F_n}\, dm$$

$$\geqq \int h_+^n\, dm - \int (h_+^{n-1} \circ T)\, dm$$

$$= \int h_+^n\, dm - \int h_+^{n-1}\, dm = \int (h_+^n - h_+^{n-1})\, dm \geqq 0.$$

as desired.

We add one more preparation to the proof of our ergodic theorem.

3.12. Definition. *Let $(\Omega, \mathscr{R}, m, T)$ be a dynamical system.*

3.12.1. *A function $h \in \overline{\mathbb{R}}^\Omega$ is called:*

3.12.1.1. **strictly *T*-invariant** *if*

$$h(T\omega) = h(\omega) \qquad (\omega \in \Omega);$$

3.12.1.2. ***T*-invariant** (mod m) *if*

$$h \circ T = h \qquad (m\text{-a.e.}).$$

3.12.2. *A set $E \subseteq \Omega$ is called:*

3.12.12.1. **strictly *T*-invariant** *if its indicator function 1_E is strictly T-invariant, i.e., if*

$$T\omega \in E \;\Leftrightarrow\; \omega \in E \qquad (\omega \in \Omega);$$

3.12.2.2. ***T*-invariant** (mod m) *if its indicator function 1_E is T-invariant* (mod m), *i.e., if $(T^{-1}E)\,\Delta\, E$ is an m-nullset.*

3.13. Remarks

3.13.1. Since the mapping $h \to h \circ T$ sends m-a.e. equal functions into m-a.e. functions, it may as well be considered as a mapping of the set of all equivalence classes of functions into itself. $h \in \overline{\mathbb{R}}^\Omega$ is T-invariant (mod m) iff it represents a fixed point of the corresponding mapping of equivalence classes. An analogous observation holds for the pullback mapping of subsets of Ω.

3.13.2. In the above definition we put "mod m" in parentheses since in many applications it is omitted, being tacitly understood.

3.14. Examples. Let $(\Omega, \mathscr{R}, m, T)$ be a dynamical system.

3.14.1. Let $h \in \overline{\mathbb{R}}^\Omega$ be arbitrary. Then $\bar{h}(\omega) = \limsup_{n\to\infty} h(T^n\omega)$ clearly defines a strictly invariant function $\bar{h} \in \overline{\mathbb{R}}^\Omega$. In fact,

$$\bar{h}(T\omega) = \limsup_{n\to\infty} h(T^{n+1}\omega) = \limsup_{n\to\infty} h(T^n\omega) = \bar{h}(\omega) \qquad (\omega \in \Omega).$$

3.14.2. Let $E \subseteq \Omega$ be arbitrary. Applying 1 to the indicator function $h = 1_E$ of E, we clearly get a strictly invariant indicator function $\bar{h} = 1_{\bar{E}}$ where $\bar{E} = \bigcap_{n=0}^{\infty} \bigcup_{k=n}^{\infty} T^{-k}E$ is a strictly T-invariant set.

3.15. Proposition. *Let* $(\Omega, \mathscr{R}, m, T)$ *be a dynamical system.*

3.15.1. *A function* $h \in \overline{\mathbb{R}}^{\Omega}$ *is* T-*invariant* (mod m) *iff there is a strictly* T-*invariant function* $\bar{h} \in \overline{\mathbb{R}}^{\Omega}$ *such that* $h = \bar{h}$ m-a.e.

3.15.2. *A set* $E \subseteq \Omega$ *is* T-*invariant* (mod m) *iff there is a strictly* T-*invariant set* $\bar{E} \subseteq \Omega$ *such that* $E \,\Delta\, \bar{E}$ *is an* m-*nullset.*

3.15.3. *The system of all* T-*invariant* (mod m) *functions in* $\overline{\mathbb{R}}$ *is stable under all at most countable lattice operations (in particular,* lim sup *and* lim inf *of sequences), under multiplication and under finite linear combinations as far they are meaningful, and so is the system of all strictly* T-*invariant functions in* $\mathbb{R}$. *Both of them contain all constants from* $\overline{\mathbb{R}}$.

3.15.4. *The system* $\{h \mid h \in \overline{\mathbb{R}}^{\Omega}, h\ T\text{-invariant} \pmod m\} \cap \text{mble}(\Omega, \mathscr{B}(\mathscr{R}), \overline{\mathbb{R}})$ *and* $\{h \mid h \in \overline{\mathbb{R}}^{\Omega}, h \text{ strictly } T\text{-invariant}\} \cap \text{mble}(\Omega, \mathscr{B}(\mathscr{R}), \overline{\mathbb{R}})$ *have the stability properties listed in* 3.

3.15.5. *The set systems*

$$\overline{\mathscr{B}} = \{E \mid E \subseteq \Omega, E\ T\text{-invariant mod } m\},$$
$$\overline{\overline{\mathscr{B}}} = \{E \mid E \subseteq \Omega, E \text{ strictly } T\text{-invariant}\}$$

are σ-*fields as well as their intersections with the* σ-*field* $\mathscr{B}(\mathscr{R})$.

3.15.6. *For every* $h \in \overline{\mathbb{R}}^{\Omega}$ *we have:*

3.15.6.1. h *is* T-*invariant* (mod m) $\Leftrightarrow h$ *is* $\overline{\mathscr{B}}$-*measurable.*

3.15.6.2. h *is strictly* T-*invariant* $\Leftrightarrow h$ *is* $\overline{\overline{\mathscr{B}}}$-*measurable.*

3.15.6.3. h *is* T-*invariant* (mod m) *and* $\mathscr{B}(\mathscr{R})$-*measurable* $\Leftrightarrow h$ *is* $(\overline{\mathscr{B}} \cap \mathscr{B}(\mathscr{R}))$-*measurable.*

3.15.6.4. h *is strictly* T-*invariant and* $\mathscr{B}(\mathscr{R})$-*measurable* $\Leftrightarrow h$ *is* $(\overline{\overline{\mathscr{B}}} \cap \mathscr{B}(\mathscr{R}))$-*measurable.*

3.15.7. *For every* $h \in \overline{\mathbb{R}}^{\Omega}$ *with* $h \circ T = h$ m-a.e., *we have*

$$\lim_{t\to\infty} \frac{1}{t} \sum_{u=0}^{t-1} h \circ T^n = h \qquad (m\text{-a.e.}).$$

Proof. By Remark 3.2.2 the pullback T^{-1} sends m-nullsets into m-nullsets.

1. Let $\bar{h}$ be strictly T-invariant and $h = \bar{h}$ m-a.e. Then

$$\begin{aligned} \{Th \neq h\} &\subseteq \{Th \neq T\bar{h}\} \cup \{T\bar{h} \neq \bar{h}\} \cup \{\bar{h} \neq h\} \\ &= T^{-1}\{h \neq \bar{h}\} \cup \{h \neq \bar{h}\} \end{aligned}$$

is an m-nullset, hence h is T-invariant (mod m). Let conversely h be

T-invariant mod m. Then for $n = 1, 2, \ldots$ the set

$$\begin{aligned}\{T^n h \neq h\} &\subseteq \{T^n h \neq T^{n-1} h\} \cup \cdots \cup \{Th \neq h\} \\ &= T^{n-1}\{Th \neq h\} \cup \cdots \cup T^{-1}\{Th \neq h\} \cup \{Th \neq h\}\end{aligned}$$

is an m-nullset, i.e., the functions h, Th, ... coincide outside some m-nullset, and thus they are all equal mod m to the function $h = \limsup_{n\to\infty} T^n h$ which is strictly T-invariant by Example 3.14.1.

2 is a special case of 1 (use indicator functions).

3. Let us be content to settle the case of a countable supremum of strictly T-invariant functions $h_1, h_2, \ldots$. Clearly

$$\begin{aligned}[T(h_1 \vee h_2 \vee \cdots)](\omega) &= (h_1 \vee h_2 \vee \cdots)(T\omega) \\ &= h_1(T\omega) \vee h_2(T\omega) \vee \cdots \\ &= (Th_1)(\omega) \vee (Th_2)(\omega) \vee \cdots \\ &= (Th_1 \vee Th_2 \vee \cdots)(\omega) \qquad (\omega \in \Omega).\end{aligned}$$

4 follows from 3 and the fact that mble(Ω, $\mathscr{B}$, $\mathbb{R}$) is stable under the operations listed in 3.

5 follows by application of 4 to indicator functions of sets.

6. Let us be content to prove 6.1 (the proofs of 6.2, 6.3, and 6.4 are similar and can be left to the reader). $\Rightarrow$: Since $h \circ T = h$ m-a.e., the set $\{h \circ T > \alpha\} = T^{-1}\{h > \alpha\}$ coincides with the set $\{h > \alpha\}$ up to an m-nullset, hence $\{h > \alpha\}$ is T-invariant (mod m), i.e., in $\overline{\mathscr{B}}$, for every real α; thus h is $\overline{\mathscr{B}}$-measurable. $\Leftarrow$: Assume $h \circ T = h$ m-a.e. does not hold, i.e., $\{h \circ T \neq h\}$ is not an m-nullset. On the other hand,

$$\{h \circ T \neq h\} = \bigcup_{\alpha \text{ rational}} [\{h \circ T > \alpha, h \leqq \alpha\} \cup \{h \circ T \leqq \alpha, h > \alpha\}$$

represents $\{h \circ T \neq h\}$ as a countable union of sets, at least one of which must, therefore, fail to be an m-nullset. But

$$\{h \circ T > \alpha, h \leqq \alpha\} \cup \{h \circ T \leqq \alpha, h > \alpha\} = \{h \leqq \alpha\} \,\Delta\, T^{-1}\{h \leqq \alpha\}$$

shows that the corresponding set $\{h \leqq \alpha\}$ is not T-invariant (mod m).

7 is an easy exercise and is left to the reader. (*Hint:* h, $h \circ T$, ... and $t^{-1} \sum_{u=0}^{t-1} h \circ T^u$ $(t = 1, 2, \ldots)$ are functions that coincide m-a.e.)

3.16. Remark. Let $(\Omega, \mathscr{R}, m, T)$ be a dynamical system and $\mathscr{R} \ni \Omega_0 \neq \varnothing$ a strictly T-invariant set. Put $\mathscr{R}_0 = \mathscr{R} \cap \Omega_0 = \{F \cap \Omega_0 \mid F \in \mathscr{R}\}$, $m_0(F_0) = m(F_0)$ $(F_0 \in \mathscr{R}_0)$, and denote by T_0 the restriction of T to Ω_0. Then clearly $(\Omega_0, \mathscr{R}_0, m_0, T_0)$ is a dynamical system. In fact, $T_0^{-1}(F \cap \Omega_0) = T^{-1}F \cap T^{-1}\Omega_0 = T^{-1}F \cap \Omega_0$ $(F \in \mathscr{R})$ shows $T_0^{-1}\mathscr{R}_0 \subseteq \mathscr{R}_0$, and we get $T_0 m_0 = m_0$ from $Tm = m$ in an obvious way. We call $(\Omega_0, \mathscr{R}_0, m_0, T_0)$ the *restriction* of $(\Omega, \mathscr{R}, m, T)$ to Ω_0. In view of the fact that the construction of a strictly invariant function displayed in the proof of Example 3.14.1 involves countable operations, it is realistic to hope for the existence of strictly invariant sets $\mathscr{R} \ni \Omega_0 \neq \varnothing, \Omega$ only if $\mathscr{R}$ is a σ-ring.

3.17. Theorem (individual ergodic theorem). *Let $(\Omega, \mathscr{R}, m, T)$ be a dynamical system. Then there is a unique linear mapping $\overline{T}: L_m^1 \to L_m^1$ such that the following statements hold:*

3.17.1. *If $h \in \mathscr{L}_m^1$ and $\bar{h}$ is representative of $\overline{T}h'$ where $h' \in L_m^1$ is the equivalence class of h, then*

$$\lim_{t\to\infty} \frac{1}{t} \sum_{u=0}^{t-1} h \circ T^u = \bar{h} \qquad (m\text{-a.e.}). \tag{1}$$

In particular, the limit in question is m-a.e. finite.

3.17.2. $\overline{T} \circ \overline{T} = \overline{T} = \overline{T} \circ T = \overline{T} \circ T$.

3.17.3. $h \geqq 0 \Rightarrow \overline{T}h \geqq 0 \quad (h \in L_m^1)$.

3.17.4. $\|\overline{T}h\|_1 \leqq \|h\|_1 \quad (h \in L_m^1)$.

3.17.5. *If $h \in \mathscr{L}_m^1$, $g \in \overline{\mathscr{R}}^\Omega$, and $g \circ T = g$ m-a.e., then*

$$\lim_{t\to\infty} \frac{1}{t} \sum_{u=0}^{t-1} (hg) \circ T^n = (\overline{T}h)g \qquad (m\text{-a.e.}).$$

In particular, $\overline{T}g = g$ if $g \in \mathscr{L}_m^1$, $g \circ T = g$ m-a.e. and

$$\int h\, 1_F \, dm = \int (\overline{T}h) 1_F \, dm \qquad (h \in \mathscr{L}_m^1,\ F \in \mathscr{B}(\mathscr{R}),\ F\ T\text{-invariant} \pmod m).$$

Proof. In view of Remark 3.2.2 we may, and shall, henceforth assume that $(\Omega, \mathscr{R}, m)$ is a full σ-content space, hence $\mathscr{B}^0(\mathscr{R})$ is the system of all countable unions of sets from $\mathscr{R}$; and if we denote the unique σ-additive extension of m to $\mathscr{B}^0(\mathscr{R})$ by m^0 (Theorem II.3.1), then $\mathscr{R} = \{F \mid F \in \mathscr{B}^0(\mathscr{R}), m^0(F) < \infty\}$. By Theorem III.9.3 we may, and shall, henceforth assume that the functions $h \in \mathscr{L}_m^1$ under consideration are even in mble$(\Omega, \mathscr{B}^0(\mathscr{R}), \mathbb{R})$. Furthermore, for every such function $h \in \mathscr{L}_m^1 \cap$ mble$(\Omega, \mathscr{B}^0(\mathscr{R}), \mathbb{R})$, we have $\{h > \alpha\} \in \mathscr{R} \ni \{h < -\alpha\}$ $(\alpha > 0)$ by Theorem III.6.3, and the same holds if we replace h by $h \circ T^n$ $(n = 1, 2, \ldots)$. It follows that $h, h \circ T, \ldots$ all vanish outside some set $\overline{\Omega}$ that can be written as a countable union $\overline{\Omega} = \Omega_1 \cup \Omega_2 \cup \cdots$ with $\Omega_1, \Omega_2, \ldots \in \mathscr{R}$, $\Omega_1 \subseteq \Omega_2 \subseteq \cdots$. Let us now consider the functions

$$\bar{\bar{h}} = \limsup_{t\to\infty} \frac{1}{t} \sum_{u=0}^{t-1} h \circ T^u, \qquad \underline{\underline{h}} = \liminf_{t\to\infty} \frac{1}{t} \sum_{u=0}^{t-1} h \circ T^u.$$

They are strictly T-invariant. In fact we have, e.g.,

$$\begin{aligned}\overline{h}(T\omega) &= \limsup_{t\to\infty} \frac{1}{t} \sum_{u=0}^{t-1} h(T^{u+1}\omega) \\ &= \limsup_{t\to\infty} \frac{1}{t} \sum_{u=1}^{t} h(T^{u}\omega) \\ &= \limsup_{t\to\infty} \frac{t+1}{t} \left[\frac{1}{t+1} \sum_{u=0}^{t} h(T^{u}\omega) - \frac{1}{t+1} h(\omega)\right] \\ &= \lim_{t\to\infty} \frac{1}{t+1} \sum_{u=0}^{t} h(T^{u}\omega) \\ &= \overline{h}(\omega) \qquad (\omega \in \Omega)\end{aligned}$$

since $h(\omega)$ is always a finite real and $\lim_{t\to\infty} [(t+1)/t] = 1$. Clearly $\overline{h}$ and $\underline{h}$ vanish outside $\overline{\Omega}$ again. We begin by proving that $\overline{h}$ and $\underline{h}$ are m-a.e. finite. For this, we choose any real $\alpha > 0$ and consider the set $\{\overline{h} > \alpha\}$ and the sets

$$H_{nk} = \left\{\omega \,\middle|\, \sup_{1 \leqq t \leqq n} \frac{1}{t} \sum_{u=0}^{t-1} h(T^u\omega) > \alpha\right\} \cap \Omega_k \in \mathscr{R}.$$

Since $s^{-1} \sum_{u=0}^{s-1} \alpha 1_{\Omega_k} \circ T^u \leqq \alpha$ $(s \geqq 1)$, H_{nk} is contained in

$$F_{nk} = \left\{\omega \,\middle|\, \sup_{1 \leqq t \leqq n} \frac{1}{t} \sum_{u=0}^{t-1} (h - \alpha 1_{\Omega_k})(T^u\omega) \geqq 0\right\}.$$

Applying Lemma 3.11 (maximal ergodic theorem) to the function $h - \alpha 1_{\Omega_k} \in \mathscr{L}_m^1$, we find

$$0 \leqq \int (h - \alpha 1_{\Omega_k}) 1_{F_{nk}}\, dm = \int h 1_{F_{nk}}\, dm - \alpha m(F_{nk} \cap \Omega_k),$$

i.e.,

$$m(H_{nk}) \leqq m(F_{nk} \cap \Omega_k) \leqq \frac{1}{\alpha} \int h 1_{F_{nk}}\, dm \leqq \frac{1}{\alpha} \int |h|\, dm < \infty.$$

For $k \to \infty$, we get

$$\left\{\sup_{1 \leqq t \leqq n} \frac{1}{t} \sum_{u=0}^{t-1} h \circ T^u > \alpha\right\} \in \mathscr{R}$$

$$m\left(\left\{\sup_{1 \leqq t \leqq n} \frac{1}{t} \sum_{u=0}^{t-1} h \circ T^u > \alpha\right\}\right) \leqq \frac{1}{\alpha} \int |h|\, dm,$$

and this holds for $n = 1, 2, \ldots$. Passing to $n \to \infty$, we get $\{\bar{h} > a\} \in \mathscr{R}$ and

$$m(\{\bar{h} > \alpha\}) \leqq m\left(\left\{\sup_{t \geqq 1} \frac{1}{t} \sum_{u=0}^{t-1} h \circ T^u > \alpha\right\}\right)$$
$$\leqq \frac{1}{\alpha} \int |h| \, dm.$$

Since the latter term tends to 0 for $\alpha \to \infty$, we get $m(\{\bar{h} = \infty\}) = 0$. Applying this to $-h$, we get $m(\{\underline{h} = -\infty\}) = 0$ and, more precisely, $\{\underline{h} \leqq -\alpha\} \in \mathscr{R}$ and $m(\{\underline{h} < -\alpha\}) \leqq \alpha^{-1} \int |h| \, dm$ $(\alpha > 0)$. We have certainly achieved (1) if we can prove $\bar{h} = \underline{h}$ m-a.e. and by $\bar{h} \geqq \underline{h}$ this is certainly implied by $m(\{\underline{h} < \beta, \bar{h} > \alpha\}) = 0$ (α, β real, $\alpha < \beta$) since

$$\{\underline{h} < \bar{h}\} = \bigcup_{\alpha,\, \beta \text{ rational},\, \beta > \alpha} \{\underline{h} < \beta, \bar{h} > \alpha\}.$$

Our above reasoning shows that either $\beta < 0$ or $\alpha > 0$ (or both), hence $\{\underline{h} < \beta, \bar{h} > \alpha\} \in \mathscr{R}$. Consider now $\Omega_0 = \{\underline{h} < \beta, \bar{h} > \alpha\}$. If $\Omega_0 = \varnothing$, we are through. Assume now $\Omega_0 \neq \varnothing$. By proposition 3.15.6.4, $\Omega_0 \in \mathscr{R}$ is strictly T-invariant. Consider now the restriction $(\Omega_0, \mathscr{R}_0, m_0, T_0)$ of $(\Omega, \mathscr{R}, m, T)$ to Ω_0, and in particular the restriction h_0 of h to Ω_0. Clearly,

$$\limsup_{t \to \infty} \frac{1}{t} \sum_{u=0}^{t-1} h_0 \circ T^u > \alpha$$

and hence

$$\sup_{t \geqq 1} \frac{1}{t} \sum_{u=0}^{t-1} h_0 \circ T^u > \alpha$$

everywhere in Ω_0. Applying Lemma 3.11 (maximal ergodic theorem) to $h_0 - \alpha \in L^1_{m_0}$, we get $\int (h_0 - \alpha) \, dm_0 \geqq 0$, hence $\int h_0 \, dm_0 \geqq \alpha m_0(\Omega_0)$. Applying Lemma 3.11 to $\beta - h_0$, we get $\int (\beta - h_0) \, dm_0 \geqq 0$, hence $\beta m_0(\Omega_0) \geqq \int h_0 \, dm_0 \geqq \alpha m_0(\Omega_0)$. Since $\beta < \alpha$, this is compatible only with $m_0(\Omega_0) = 0$, i.e., $m(\Omega_0) = 0$. Having proved the existence and finiteness m-a.e. in (1), we observe that $L^1_m \ni h \geqq 0$ implies $\bar{h} \geqq 0$; and Fatou's lemma III.5.5 now implies $h \in L^1_m$ and $\int \bar{h} \, dm \leqq \int h \, dm$ in this particular case. If $h \in L^1_m$ is arbitrary, we see that $h = h_+ - h_- \in L^1_m$ and

$$\|\bar{h}\|_1 \leqq \int \bar{h}_+ \, dm + \int \bar{h}_- \, dm \leqq \int h_+ \, dm + \int h_- \, dm = \int |h| \, dm = \|h\|_1.$$

We are now through with the proof of 1, 3, and 4. From the strict invariance of $\underline{\underline{h}}$ and h (see above) we deduce $T(\overline{T}h) = \overline{T}h$ ($h \in L_m^1$), i.e., $T \circ \overline{T} = \overline{T}$; and from the proof leading to the invariance of $\bar{h}$ and $\underline{\underline{h}}$ we see $\overline{T} \circ T = \overline{T}$. From $T \circ \overline{T} = \overline{T}$ we infer

$$\frac{1}{t}\sum_{u=0}^{t-1}(\overline{T}h) \circ T^u = \frac{1}{t}\sum_{u=0}^{t-1}(T^u \circ \overline{T})h = \frac{1}{t}\sum_{u=0}^{t-1}\overline{T}h = \overline{T}h$$

and thus $\overline{T} \circ \overline{T} = \overline{T}$, finishing the proof of 2. For the proof of 5 we observe that from $g \circ T = g$ m-a.e. we get $g \circ T^u = g$ m-a.e. ($u = 1, 2, \ldots$) and thus

$$\frac{1}{t}\sum_{u=0}^{t-1}(hg) \circ T^u = \frac{1}{t}\sum_{u=0}^{t-1}(h \circ T^u)(g \circ T^u) = \left[\frac{1}{t}\sum_{u=0}^{t-1} h \circ T^u\right] g \qquad m\text{-a.e.}$$

which implies the desired statement by (1).

3.18. Exercise. Let $(\Omega, \mathscr{R}, m, T)$ be a dynamical system such that $1 \in L_m^1$. Prove that the $\overline{T}$ of Theorem 3.9 (mean ergodic theorem in L_m^1) coincides with the $\overline{T}$ of Theorem 3.17 (individual ergodic theorem) and thus its restriction to L_m^2 ($\subseteq L_m^1$ here) is the $\overline{T}$ of theorem 3.7 (mean ergodic theorem in L_m^2). (*Hint:* Use Theorems III.5.8.4 and III.11.2.4.) Use Example 3.10 to show that here the hypothesis $1 \in L_m^1$ cannot generally be dispensed with.

4. Ergodicity

Theorem 3.17.2 shows that m-a.e. limit $\bar{h}$ of the ergodic means of a function $h \in L_m^1$ is a T-invariant (mod m) function. In many cases it turns out that every $\mathscr{B}(\mathscr{R})$-measurable and T-invariant (mod m) function is m-a.e. equal to a constant, and hence so are all the $\bar{h}$. We investigate this problem in more detail.

3.19. Definition. *A dynamical system $(\Omega, \mathscr{R}, m, T)$ is said to be* **ergodic** *(and one says then that m is ergodic for T, or T is ergodic for m) if for every T-invariant* (mod m) *set $E \in \mathscr{B}(\mathscr{R})$ either E or $\Omega \backslash E$ is an m-nullset.*

3.20. Proposition. *Let $(\Omega, \mathscr{R}, m, T)$ be a dynamical system. Then $(\Omega, \mathscr{R}, m, T)$ is ergodic $\Leftrightarrow$ every T-invariant* (mod m) *$\mathscr{B}(\mathscr{R})$-measurable function in Ω is m-a.e. equal to a constant.*

Proof. $\Leftarrow$: If $E \in \mathscr{B}(\mathscr{R})$ is T-invariant (mod m), then 1_E is m-a.e. equal to a constant which can at most be 0 or 1. If it is 0, then E is an m-nullset; and if it is 1, then $\Omega \backslash E$ is a m-nullset.

$\Rightarrow$: Let h be a T-invariant (mod m) $\mathscr{B}(\mathscr{R})$-measurable function $\in \overline{\mathbb{R}}^\Omega$.

By Proposition 3.15.6.1 the set $\{h > \alpha\}$ is in $\overline{\mathscr{B}} \cap \mathscr{B}(\mathscr{R})$, and hence either $\{h > \alpha\}$ or $\{h \leqq \alpha\}$ is an m-nullset, for every real α. Let $\alpha_0 = \inf\{\alpha \mid \alpha \in \mathbb{R},\ \{h > \alpha\}$ is an m-nullset$\}$. If $\alpha_0 = -\infty$, then $\{h > -\infty\} = \bigcup_{n=1}^{\infty} \{h > -n\}$ is an m-nullset, hence $h = -\infty$ m-a.e. If $\alpha_0 = \infty$, then $\{h \leqq \alpha\}$ is an m-nullset for every real α, hence $\{h < \infty\} = \bigcup_{n=1}^{\infty} \{h \leqq n\}$ is an m-nullset, hence $h = \infty$ m-a.e. Let now $\alpha_0 \in \mathbb{R}$. By

$$\{h > \alpha_0\} = \bigcup_{n=1}^{\infty} \left\{h > \alpha_0 + \frac{1}{n}\right\},$$

$\{h > \alpha_0\}$ is an m-nullset, and by

$$\{h < \alpha_0\} = \bigcup_{n=1}^{\infty} \left\{h \leqq \alpha_0 - \frac{1}{n}\right\},$$

$\{h < \alpha_0\}$ is an m-nullset, hence $h = \alpha_0$ m-a.e.

3.21. Exercises

3.21.1. Let $(\Omega, \mathscr{R}, m, T)$ be a dynamical system and $\overline{\mathscr{B}}$, $\overline{\overline{\mathscr{B}}}$ as in Proposition 3.15.5. Prove the equivalence of the following statements (and possibly further ones of similar character).

3.21.2. $(\Omega, \mathscr{R}, m, T)$ is ergodic.

3.21.3. For every $E \in \overline{\mathscr{B}} \cap \mathscr{B}^0(\mathscr{R})$, either E or $\Omega \backslash E$ is an m-nullset. (*Hint:* Use Proposition I.4.7.3.1.)

3.21.4. For every $E \in \overline{\overline{\mathscr{B}}} \cap \mathscr{B}(\mathscr{R})$, either E or $\Omega \backslash E$ is an m-nullset.

3.21.5. For every $E \in \overline{\overline{\mathscr{B}}} \cap \mathscr{B}^0(\mathscr{R})$, either E or $\Omega \backslash E$ is an m-nullset.

3.21.6. Every $h \in \text{mble}(\Omega, \overline{\mathscr{B}} \cap \mathscr{B}^0(\mathscr{R}), \overline{\mathbb{R}})$ is m-a.e. constant. (*Hint:* Use Proposition 3.15.6.1.)

3.21.7. Every $h \in \text{mble}(\Omega, \overline{\overline{\mathscr{B}}} \cap \mathscr{B}^0(\mathscr{R}), \overline{\mathbb{R}})$ is m-a.e. constant.

3.21.8. Every $h \in \text{mble}(\Omega, \overline{\overline{\mathscr{B}}} \cap \mathscr{B}^0(\mathscr{R}), \mathbb{R})$ is m-a.e. constant.

3.22. Proposition. *Let $(\Omega, \mathscr{R}, m, T)$ be a dynamical system. Then:*

3.22.1. *If $(\Omega, \mathscr{R}, m, T)$ is ergodic, then for every $h \in L_m^1$ there is a constant $\alpha \in \mathbb{R}$ such that*

$$\lim_{t\to\infty} \frac{1}{t} \sum_{u=0}^{t-1} h \circ T^u = \alpha \qquad (m\text{-a.e.}).$$

3.22.2. *Let $\Omega \in \mathscr{R}$, $m(\Omega) = 1$, and $M \subseteq L_m^1$ be such that the linear span of M in L_m^1 is $\|\cdot\|_1$ norm dense in L_m^1. Then the following statements are equivalent:*

3.22.2.1. *$(\Omega, \mathscr{R}, m, T)$ is ergodic.*

3.22.2.2. $\lim_{t\to\infty} t^{-1} \sum_{u=0}^{t-1} h \circ T^u = \text{const} = \int h\, dm$ m-a.e. $(h \in L_m^1)$.

3.22.2.3. $\lim_{t\to\infty} t^{-1} \sum_{u=0}^{t-1} h \circ T^u = \int h\, dm$ m-a.e. $(h \in M)$.

3.22.2.4. $\lim_{t\to\infty} \left\| t^{-1} \sum_{u=0}^{t-1} h \circ T^u - \int h\, dm \right\|_1 = 0$ $(h \in M)$.

Proof. 1. Define $\bar{h} \in \mathscr{R}^\Omega$ by

$$\bar{h}(\omega) = \begin{cases} \lim\limits_{t\to\infty} \dfrac{1}{t} \sum\limits_{u=0}^{t-1} h(T^u\omega) & \text{if this limit exists and is finite} \\ 0 & \text{otherwise.} \end{cases}$$

Then clearly $\bar{h}$ is $\mathscr{B}(\mathscr{R})$-measurable and T-invariant (mod m), hence m-a.e. equal to a constant α by Proposition 3.20, which can apparently be only a finite real number.

2.1 ⇒ 2.2. By Exercise 3.18 we have (1) both m-a.e. and in $\|\cdot\|_1$-norm (see Theorem 3.9). The latter implies

$$\begin{aligned}\int (\overline{T}h)\, dm &= \lim_{t\to\infty} \int \left[\frac{1}{t} \sum_{u=0}^{t-1} h \circ T^u\right] dm \\ &= \lim_{t\to\infty} \frac{1}{t} \sum_{u=0}^{t-1} \int (h \circ T^u)\, dm \\ &= \lim_{t\to\infty} \frac{1}{t} \sum_{u=0}^{t-1} \int h\, dm = \int h\, dm.\end{aligned}$$

2.2 ⇒ 2.3 is obvious.

2.3 ⇔ 2.4 follows from Exercise 3.18.

2.3 ⇒ 2.2 follows by Theorem 3.9, Exercise 3.18 and norm approximation: Since $\overline{T}$ is linear, we may, and shall, assume that M is $\|\cdot\|_1$ norm dense in L_m^1; a mere repetition of the proof of Theorem 3.9 shows now that for every $h \in L_m^1$, $\overline{T}h$ is a $\|\cdot\|_1$ norm limit of constants and hence a constant itself (exercise).

2.2 ⇒ 2.1. If $F \in \mathscr{B}(M)$ is T-invariant mod m, then $1_F \in L_m^1$ (since $1 \in L_m^1$; use Proposition III.9.5) and $1_F \circ T = 1_F$ m-a.e., hence $\overline{T}1_F = 1_F$ m-a.e. (use Theorem 3.17.5 with $g = 1_F$). By 2.2, 1_F is m-a.e. equal to a constant, hence F or $\Omega \backslash F$ is an m-nullset and the ergodicity of $(\Omega, \mathscr{R}, m, T)$ follows.

3.23. Example. Let $(\Omega, \mathscr{R}, m, T)$ be the one-sided Bernoulli scheme for a finite alphabet A and p be a probability vector on A (example 3.4). Let $M = \{1_F | F \text{ is a special cylinder in } \Omega\}$. Then the linear span of M is the elementary domain $\mathscr{E} = \mathscr{E}_C$ of Example I.8.10.1 and hence $\|\cdot\|_1$ norm dense in L_m^1. We now prove that $(\Omega, \mathscr{R}, m, T)$ is ergodic by verifying 3.22.2.4.

For this we choose arbitrary special cylinders E, F, say, of orders r and s:

$$E = [k_0, \ldots, k_r], \qquad F = [j_0, \ldots, j_s].$$

Now $T^{-u}F = \{\omega \mid \omega_u = j_0, \ldots, \omega_{u+s} = j_s\}$ and

$$E \cap T^{-u}F = \{\omega \mid \omega_0 = k_0, \ldots, \omega_r = k_r, \omega_u = j_0, \ldots, \omega_{u+s} = j_s\}.$$

For $u > r$, we can write

$$E \cap T^{-u}F = \sum_{l_{r+1}, \ldots, l_{u-1} \in A} [k_0, \ldots, k_r, l_{r+1}, \ldots, l_{u-1}, j_0, \ldots, j_s]$$

and hence (by $\sum_{i \in A} p(i) = 1$)

$$\begin{aligned} m(E \cap T^{-u}F) &= \sum_{l_{r+1}, \ldots, l_{u-1} \in A} p(k_0) \cdots p(k_r) p(l_{r+1}) \cdots p(l_{u-1}) p(j_0) \cdots p(j_s) \\ &= p(k_0) \cdots p(k_r) p(j_0) \cdots p(j_s) \\ &= m(E) m(F). \end{aligned}$$

This yields (use Theorem 3.7 (mean ergodic theorem in L_m^2) the scalar product in L_m^2 and its $\|\cdot\|_2$ norm continuity)

$$\begin{aligned} \left(1_E, \lim_{t \to \infty} \frac{1}{t} \sum_{u=0}^{t-1} 1_F \circ T^u\right) &= \lim_{t \to \infty} \frac{1}{t} \sum_{u=0}^{t-1} (1_E, 1_F \circ T^u) \\ &= \lim_{t \to \infty} \frac{1}{t} \sum_{u=0}^{t-1} \int 1_E (1_F \circ T^u) \, dm = \lim_{t \to \infty} \frac{1}{t} \sum_{u=0}^{t-1} m(E \cap T^{-u}F), \end{aligned}$$

and this is clearly $= m(E)m(F) = (1_E, 1)m(F)$ since $m(E \cap T^{-u}F) = m(E)m(F)$ for all but finitely many u. We can rewrite this result as $(g, \overline{T}1_F - m(F) \circ 1) = 0$ for all $g = 1_E$, where E and F are special cylinders. Since the linear span of these $g = 1_E$ is $\|\cdot\|_2$ norm dense in L_m^2 (Theorem III.11.2.3) and since the scalar product is $\|\cdot\|_2$ norm continuous in its first component, we get $(g, \overline{T}1_F - m(F) \cdot 1) = 0$ for all $g \in L_m^2$; and if we put $g = \overline{T}1_F - m(F) \cdot 1$, we have $\|\overline{T}1_F - m(F) \cdot 1\|_2^2 = 0$, i.e., $\overline{T}1_F = m(F) \cdot 1 =$ const for all special cylinders. Taking Theorem III.11.2.3 into account once more (as well as Exercise 3.18), we get 3.22.2.4 as desired. The ergodicity of the two-sided Bernoulli schemes can be proved along the same pattern (exercise).

The result of this example will be generalized in Exercise VI.6.8.2.

3.24. Example. Consider the circle rotation $(\Omega, \mathscr{B}, m, T)$ of Example 3.3 and assume that $\alpha/2\pi$ is irrational. We shall prove that this dynamical system is then ergodic. To this end we shall verify 3.22.2.4 with $M = \mathscr{C}(\Omega, \mathbb{R}) =$ the set of all continuous real functions on the circle line Ω. It is in fact easy to prove that M is $\|\cdot\|_1$ norm dense in L_m^1 (Exercise

3.5.1). Thus let a continuous real function f be given on Ω. We first observe that T is a circle rotation preserving all arc length distances, and that the functions f, $f \circ T$, ..., $t^{-1} \sum_{u=0}^{t-1} f \circ T$ $(t = 1, 2, \ldots)$ form a (countable) uniformly equicontinuous set: there is, for every $\varepsilon > 0$, a $\delta = \delta(\varepsilon) > 0$ such that for any function g from the mentioned set, $|\varphi - \varphi'| < \delta \Rightarrow |g(e^{i\varphi}) - g(e^{i\varphi'})| < \varepsilon$. Next we make the classical observation of KRONECKER that, thanks to the irrationality of $\alpha/2\pi$, the "orbit" $\{T^t\omega | t = 0, 1, \ldots\}$ of any $\omega \in \Omega$ is dense in Ω. In fact, if $\omega = e^{i\varphi}$, then $T^t\omega = e^{i(\varphi + t\alpha)} = e^{i\varphi}e^{it\alpha}$ and $T^t\omega = \omega$ means $e^{it\alpha} = 1$, i.e., $t\alpha/2\pi$ integer, i.e., $\alpha/2\pi$ rational, a contradiction. Subdividing Ω into n disjoint half-open arcs of equal length $\rho > 0$, we find that of the $n + 1$ distinct points ω, $T\omega, \ldots, T^n\omega$ at least two, say $T^j\omega$ and $T^k\omega$ with $0 \leqq j < k \leqq n$ are in the same arc, hence have an arc length distance $< \rho$. Since T^j, T^k are rotations, we see that T^{k-j} is a rotation by an arc length $< \rho$ in the positive or negative sense, hence the sequence ω, $T^{k-j}\omega$, $T^{2(k-j)}\omega$, ... is (even for any $\omega \in \Omega$) a sequence running around the circle in steps of width $< \rho$. Thus every $\eta \in \Omega$ is in the ρ-neighborhood of some $T^{r(k-j)}\omega$. Since $\rho > 0$ was arbitrary, the density of the orbit of ω in Ω follows. Now choose $n_0 > 0$ such that, for a given $\varepsilon > 0$ and the above mentioned $\delta = \delta(\varepsilon)$, there is, for every rotation T^r $(r = 0, 1, \ldots)$, some n with $0 \leqq n \leqq n_0$ such that $T^r\eta$ and $T^n\eta$ are at arc length distance $< \delta$, for every $\eta \in \Omega$. Let $t > n_0$. Then

$$
\begin{aligned}
&\left| \left(\frac{1}{t} \sum_{u=0}^{t-1} h \circ T^u \right)(T^r\eta) - \left(\frac{1}{t} \sum_{u=0}^{t-1} h \circ T^u \right)(\eta) \right| \\
&\quad \leqq \left| \left(\frac{1}{t} \sum_{u=0}^{t-1} h \circ T^u \right)(T^r\eta) - \left(\frac{1}{t} \sum_{u=0}^{t-1} h \circ T^u \right)(T^n\eta) \right| \\
&\qquad + \left| \left(\frac{1}{t} \sum_{u=0}^{t-1} h \circ T^{u+n} \right)(\eta) - \left(\frac{1}{t} \sum_{u=0}^{t-1} h \circ T^u \right)(\eta) \right| \\
&\quad < \varepsilon + \frac{2n_0}{t} \sup_{\eta \in \Omega} |h(\eta)| < 2\varepsilon
\end{aligned}
$$

if t is sufficiently large, uniformly in $\eta \in \Omega$. Since $T^r\eta$ $(r = 0, 1, \ldots)$ runs through a dense subset of Ω, we see that for t sufficiently large, $t^{-1} \sum_{u=0}^{t-1} h \circ T^u$ is constant up to $\varepsilon > 0$. Clearly this proves that $\lim_{t \to \infty} t^{-1} \sum_{u=0}^{t-1} h \circ T^u$ exists even as a uniform limit and is a real constant for every $f \in \mathscr{C}(\Omega, \mathbb{R})$, as was to be proved. Thus a circle rotation with an irrational α is always ergodic.

4. KERNELS

In this section we introduce the notion of a *kernel* both under the aspects of measures and contents. It is shown that a kernel is a generalization of a point mapping. Under suitable assumptions, a kernel can be used to transform a measure resp. a content in a fashion that generalizes transform by a measurable mapping. We shall prove an integral transform equation also in the kernel case. Throughout this section we shall rather freely shift to and fro between the σ-content and the σ-measure aspect.

At the end of this section we pose the problem of kernel representation and present DIEUDONNÉ's [1] famous example which shows that the problem has in general no solution. We display a special case where a solution can be given; the general problem will be tackled again in Chapter XVI.

1. The Concept of a Kernel

4.1. Definition. *Let* $\Omega, \Omega' \neq \emptyset$.

4.1.1. *Let* $\mathscr{E}' \subseteq \mathbb{R}^{\Omega'}$ *be an elementary domain. A function* $P: \Omega \times E' \to \mathbb{R}$ *is called:*

4.1.1.1. *a (positive)* **measure kernel** *from* Ω *to* Ω' *(with* $\mathscr{E}'$*) if for every* $\omega \in \Omega$ *the function* $P(\omega, \cdot)$: $\mathscr{E}' \to \mathbb{R}$ *is a (positive) measure. A measure kernel* P *from* Ω *to* Ω' *is called:*

4.1.1.2. *a* **stochastic kernel** *if* $1 \in \mathscr{E}'$ *and* $P(\omega, 1) = 1$ $(\omega \in \Omega)$;

4.1.1.3. *a* **substochastic kernel** *if* $1 \in \mathscr{E}'$ *and* $P(\omega, 1) \leqq 1$ $(\omega \in \Omega)$;

4.1.1.4. *a* **σ-measure kernel** *if* $P(\omega, \cdot)$ *is a* σ*-measure* $(\omega \in \Omega)$;

4.1.1.5. *a* **τ-measure kernel** *if* $P(\omega, \cdot)$ *is a* τ*-measure* $(\omega \in \Omega)$.

4.1.2. *Let* $\mathscr{R}' \subseteq \mathscr{P}(\Omega')$ *be a set ring. A function* $P: \Omega \times \mathscr{R}' \to \mathbb{R}_+$ *is called:*

4.1.2.1. *a (positive)* **content kernel** *from* Ω *to* Ω' *if for every the function* $P(\omega, \cdot)$: $\mathscr{R}' \to \mathbb{R}_+$ *is a content. A content kernel* P *from* Ω *to* Ω' *is called:*

4.1.2.2. *a* **stochastic content kernel** *if* $\Omega' \in \mathscr{R}'$, *i.e.,* $\mathscr{R}'$ *is a field, and* $P(\omega, \Omega') = 1$ $(\omega \in \Omega)$;

4.1.2.3. *a* **substochastic content kernel** *if* $\Omega' \in \mathbb{R}'$ *and* $P(\omega, \Omega') \leqq 1$ $(\omega \in \Omega)$;

4.1.2.4. *a* **σ-content kernel** *if for every* $\omega \in \Omega$, *the content* $P(\omega, \cdot)$ *is a* σ*-content.*

4.2. Remarks. Let $\Omega, \Omega' \neq \emptyset$.

4.2.1. A measure kernel from Ω to Ω' is nothing but a family of measures on a given elementary domain $\mathscr{E}' \subseteq \mathbb{R}^{\Omega'}$, indexed by Ω; in the same vein, a content kernel from Ω to Ω' is nothing but a family of contents on a given ring $\mathscr{R}' \subseteq \mathscr{P}(\Omega')$, indexed by Ω. The properties of kernels defined above are nothing but the corresponding properties of measures resp. contents holding for all members of the family.

4.2.2. The natural correspondence between σ-measures and σ-contents (Chapter III. Section 6) implies a natural correspondence between σ-measure kernels from Ω to Ω' (and $\mathscr{E}'$) and σ-content kernels from Ω to Ω' (and $\mathscr{R}'$), provided we relate $\mathscr{E}'$ and $\mathscr{R}'$ appropriately. The problem lies in the fact that the σ-extension of σ-measures yields a space of integrable functions strongly depending on the σ-measure under consideration, while we have a universal natural domain for the extension of every σ-content given on a ring $\mathscr{R}'$, namely, the local σ-ring $\mathscr{B}^{00}(\mathscr{R}')$ generated by $\mathscr{R}'$. The following proposals seem to be the most natural ones:

4.2.2.1. Given an elementary domain $\mathscr{E}' \subseteq \mathbb{R}^{\Omega'}$, let $\mathscr{R}'$ be the set ring generated by all sets of the form $\{f' \geqq 1\}$, where $f' \in \mathscr{E}'$. With every σ-measure kernel P from Ω to Ω' (and $\mathscr{E}'$) associate the σ-content kernel P from Ω to Ω' (with $\mathscr{R}'$), where $P(\omega, \cdot)$ is the restriction to $\mathscr{R}'$ of the σ-content derived from the σ-measure $P(\omega, \cdot)$ $(\omega \in \Omega)$. In fact, $F' \in \mathscr{R}'$, $\omega \in \Omega \Rightarrow 1_{F'} \in L^1_{P(\omega, \cdot)}$, making the definition of P meaningful. One could, of course, replace $\mathscr{R}'$ by the local σ-ring $\mathscr{B}^{00}(\mathscr{R}')$. In both cases ($\mathscr{R}'$ and $\mathscr{B}^{00}(\mathscr{R}')$) we call P the σ-content kernel derived from the σ-measure kernel P. Clearly the one P is (sub-)stochastic iff the other P is.

4.2.2.2. Given a ring $\mathscr{R}' \subseteq \mathscr{P}(\Omega')$, let $\mathscr{E}' = \mathscr{E}(\mathscr{R})$ be the elementary domain of all step functions for $\mathscr{R}'$ (Example I.7.8). With every σ-content kernel P from Ω to Ω' (with $\mathscr{R}'$) we associate the σ-measure kernel P from Ω to Ω' (with $\mathscr{E}'$), where $(\Omega', \mathscr{E}', P(\omega, \cdot))$ is the σ-measure space derived from the σ-content prespace $(\Omega', \mathscr{R}', P(\omega, \cdot))$ (Example I.8.7). We call P the σ-measure kernel derived from the σ-content kernel P. In this case it is obvious that $\mathscr{R}' = \{\{f' \geqq 1\} | f' \in \mathscr{E}'\}$ and P is the σ-content kernel derived from the σ-measure kernel P. In particular, the one P is substochastic iff the other P is.

4.2.3. The extension theories for measures and contents presented in Chapters I–III apply in a natural way to kernels: One has to apply them to all measures resp. contents assembled in the kernel simultaneously. Let, e.g., $\mathscr{E}' \subseteq \mathbb{R}^{\Omega'}$ be an elementary domain and P be a σ-measure kernel from Ω to Ω' (with $\mathscr{E}'$). Let

$$\overline{\mathscr{E}}' = \bigcap_{\omega \in \Omega} \mathscr{L}^1_\sigma(\Omega', \mathscr{E}', P(\omega, \cdot)) \cap \mathbb{R}^{\Omega'}.$$

Clearly $\mathscr{E}'$ is an elementary domain and $P(\omega, h') = \int P(\omega, d\omega')h'(\omega')$ $(h' \in E')$ defines a σ-measure kernel from Ω to Ω' (now with $\overline{\mathscr{E}}'$ in the place of $\mathscr{E}'$). In many cases one is content with an elementary domain $\tilde{\mathscr{E}}'$ somewhere between $\mathscr{E}'$ and $\overline{\mathscr{E}}'$. If, e.g., P is stochastic or substochastic and $\mathscr{B}' = \mathscr{B}(\{\{f' \geqq 1\} | f' \in \mathscr{E}'\})$, one may choose for $\tilde{\mathscr{E}}'$ the set of all bounded functions in mble$(\Omega', \mathscr{B}', \mathbb{R})$ (use Proposition III.9.5 in order to prove $\tilde{\mathscr{E}}' \subseteq \mathscr{E}'$). This remark about σ-measure kernels clearly implies an analogous remark on σ-content kernels. It seems, in general, impossible to go beyond $\overline{\mathscr{E}}'$.

4.3. Examples

4.3.1. Let Ω, $\Omega' \neq \emptyset$ and $\varphi\colon \Omega \to \Omega'$ be a mapping. Then $\mathscr{E}' = \mathbb{R}^{\Omega'}$, $P(\omega, f') = f'(\varphi(\omega))$ $(\omega \in \Omega,\ f' \in \mathscr{E}')$ defines a τ-measure kernel from Ω to Ω' which is clearly stochastic. The stochastic σ-content kernel derived from it is clearly given by $\mathscr{R}' = \mathscr{P}(\Omega')$ and $P(\omega, F') = 1_{F'}(\varphi(\omega))$ $(\omega \in \Omega,\ F' \in \mathscr{R}')$. Both kernels are said to be derived from φ.

4.3.2. Let Ω' be finite and $\mathscr{R}' = \mathscr{P}(\Omega')$. Then every positive σ-content kernel P from Ω to Ω' (with $\mathscr{R}'$) is given by a function $p\colon \Omega \times \Omega' \to \mathbb{R}_+$ in the form

$$P(\omega, F') = \sum_{\omega' \in F'} p(\omega, \omega') \qquad (\omega \in \Omega, \quad F' \subseteq \Omega').$$

If also Ω is finite, then p is tantamount to a nonnegative Ω-Ω'-matrix. P is stochastic iff $\sum_{\omega' \in \Omega'} p(\omega, \omega') = 1$ $(\omega \in \Omega)$. This generalizes easily to countable Ω' (and possibly countable Ω).

4.3.3. Let $(\mathbb{R}, \mathscr{B}^{00}, m)$ be the Lebesgue σ-content space on the line (Example II.6.4) where $\mathscr{B}^{00}$ consists of all bounded Borel sets in $\mathbb{R}$. An easy application of the monotone convergence theorem shows that the function $\rho \in \mathscr{C}^{\mathrm{b}}(\mathbb{R}, \mathbb{R})$ defined by

$$\rho(x) = \frac{1}{\sqrt{2\pi}} e^{-x^2/2}$$

is in L_m^1 and that $\int \rho\, dm$ can be evaluated as the classical extended Riemann integral. Classical calculus yields

$$\left[\int_{-\infty}^{\infty} e^{-x^2/2}\, dx\right]^2 = \iint_{R^2} e^{-(x^2+y^2)/2}\, dx\, dy = 2\pi \int_0^{\infty} re^{-r^2/2}\, dr = 2\pi,$$

proving $\int \rho\, dm = 1$. Probabilists call ρ the density of the normal distribution $N(0, 1)$ on $\mathbb{R}$. By the translation invariance of m (see, e.g., Exercise II.2.6.5) we find that for every fixed $y \in \mathbb{R}$ the function $\rho(x + y)$ of $x \in \mathbb{R}$ is in L_m^1 again and has the m-integral 1. Let now $\mathscr{E}' = \mathscr{C}^{00}(\mathbb{R}, \mathbb{R})$ and

$$P(\omega, f') = \int \rho(\omega + \omega') f'(\omega')\, m(d\omega') \qquad (\omega \in \mathbb{R}, \quad f' \in \mathscr{E}').$$

Clearly P is a stochastic τ-measure kernel from $\mathbb{R}$ to $\mathbb{R}$. It is easy to see (exercise) that:

4.3.3.1. $P(\cdot, f') \in \mathscr{C}^b(\mathbb{R}, \mathbb{R})$ $(f' \in \mathscr{C}^{00}(\mathbb{R}, \mathbb{R}))$.

4.3.3.2. $P(\cdot, f') \notin \mathscr{C}^{00}(\mathbb{R}, \mathbb{R})$ if $0 \leqq f' \in \mathscr{C}^{00}(\mathbb{R}, \mathbb{R})$, $f' \neq 0$.

4.3.3.3. $\mathscr{C}^b(\mathbb{R}, \mathbb{R}) \subseteq L^1_{P(\omega,\cdot)}$ $(\omega \in \mathbb{R})$ and $\int P(\omega, d\omega')f'(\omega')$ defines, for every $f' \in \mathscr{C}^b(\mathbb{R}, \mathbb{R})$, a function of ω that is in $\mathscr{C}^b(\mathbb{R}, \mathbb{R})$.

$$P(\omega, f') = \int P(\omega, d\omega')f'(\omega') \qquad (\omega \in \mathbb{R},\ f' \in \mathscr{C}^b(\mathbb{R}, \mathbb{R}))$$

defines a σ-measure kernel from $\mathbb{R}$ to $\mathbb{R}$ (with the new elementary domain $\mathscr{C}^b(\mathbb{R}, \mathbb{R})$).

4.4. Remarks

4.4.1. Example 4.3.1 shows that mappings can be considered as special cases of kernels.

4.4.2. Examples 4.3.2 and 4.3.3 show that there are natural kernels which apparently cannot be given by mappings. (How would you prove that?)

4.4.3. Example 4.3.3 shows that a change in the elementary domain etc. can be the most natural thing. It is in fact a matter of assessment of the nature of a kernel to find out its most natural domain of definition.

4.5. Exercises

4.5.1. Let $(\mathbb{R}, \mathscr{B}^{00}, m)$ be the Lebesgue σ-content space on the line. Let $0 \leqq \rho \in L^1_m \cap \mathscr{C}^b(\mathbb{R}, \mathbb{R})$ be arbitrary. Prove that for every $\omega \in \mathbb{R}$ the function $\rho(\omega + \omega')$ of $\omega' \in \mathbb{R}$ is in $L^1_m \cap \mathscr{C}^b(\mathbb{R}, \mathbb{R})$ and has the same m-integral as ρ. Prove that

$$P(\omega, f') = \int P(\omega + \omega')f'(\omega')\, m(d\omega') \qquad (\omega \in R, \quad f' \in \mathscr{C}^b(\mathbb{R}, \mathbb{R}))$$

defines a σ-measure kernel P from $\mathbb{R}$ to $\mathbb{R}$ (with $\mathscr{C}^b(\mathbb{R}, \mathbb{R})$) such that $P(\cdot, f') \in \mathscr{C}^b(\mathbb{R}, \mathbb{R})$ $(f' \in \mathscr{C}^b(\mathbb{R}, \mathbb{R}))$.

4.5.2. Let $\Omega \neq 0$ and $\Omega' = [0, 1] \subseteq \mathbb{R}$ and P be a measure kernel from Ω to Ω' (with $\mathscr{E}' = \mathscr{C}([0, 1], \mathbb{R})$). Prove that P is a σ-measure kernel. Prove:

4.5.2.1. Let $\varphi\colon \Omega \to \Omega'$ be a mapping. Assume that

$$P(\omega, f') = f'(\varphi(\omega))\ (\omega \in \Omega,\ f' \in \mathscr{C}([0, 1], \mathbb{R})).$$

Then P is multiplicative:

$$P(\omega, f'g') = P(\omega, f')P(\omega, g')\ (\omega \in \Omega,\ f', g' \in \mathscr{C}([0, 1], \mathbb{R}).$$

4.5.2.2. Assume that P is multiplicative. Prove that there is a mapping $\varphi\colon \Omega \to \Omega'$ such that $P(\omega, f') = f'(\varphi(\omega))$ $(\omega \in \Omega,\ f' \in \mathscr{C}([0, 1], \mathbb{R}))$. (*Hint:* Use Exercise I.8.4.2.)

2. Integrability of Kernels. Transform of Measures and Contents by Kernels

4.6. Definition. *Let $\Omega, \Omega' \neq \varnothing$, P be a measure kernel from Ω to Ω' (with an elementary domain $\mathscr{E}' \subseteq \mathbb{R}^{\Omega'}$) and $\hat{P}$ a content kernel from Ω to Ω' (with a ring $\mathscr{R}' \subseteq \mathscr{P}(\Omega')$).*

4.6.1. *Let $(\Omega, \mathscr{E}, m)$ be a σ-measure space. P is said to be m-(σ-)integrable $P(\cdot, f') \in L^1_\sigma(\Omega, \mathscr{E}, m)$ $(f' \in \mathscr{E}')$.*

4.6.2. *Let $(\Omega, \mathscr{E}, m)$ be a τ-measure space. P is said to be m-(τ-)integrable if $P(\cdot, f') \in \mathscr{L}^1_\tau(\Omega, \mathscr{E}, m)$ $(f' \in \mathscr{E}')$.*

4.6.3. *Let $(\Omega, \mathscr{R}, m)$ be a σ-content space. $\hat{P}$ is said to be m-(σ-)integrable if $P(\cdot, F') \in \mathscr{L}^1_\sigma(\Omega, \mathscr{R}, m)$ $(F' \in \mathscr{R}')$.*

4.7. Remarks

4.7.1. The above definition may be viewed in a different way. A kernel from Ω to Ω' is nothing but a family of real functions on Ω indexed with an elementary domain on Ω' or a set ring of subsets of Ω'. Integrability of a kernel means nothing but integrability of all functions of that family.

4.7.2. The above definition could be extended in an obvious way: It is clear how to define, e.g., σ-integrability of a content kernel for a σ-measure, etc. We refrain from listing all possibilities.

4.8. Theorem. *Let $\Omega, \Omega' \neq \varnothing$, $\mathscr{E}' \subseteq \mathbb{R}^{\Omega'}$ be an elementary domain, $\mathscr{R}' \subseteq \mathscr{P}(\Omega')$ a ring, P a measure kernel from Ω to Ω' (with $\mathscr{E}'$) and $\hat{P}$ be a content kernel from Ω to Ω' (with $\mathscr{R}'$).*

4.8.1. *Let $(\Omega, \mathscr{E}, m)$ be a σ-measure space and P an m-integrable kernel. Then*

$$m'(f') = \int m(d\omega)P(\omega, f') \qquad (f' \in \mathscr{E}') \tag{1}$$

defines a σ-measure m' on $\mathscr{E}'$. We write $m' = mP$ for short and say that m' is the **result of transport** *of m by P.*

4.8.2. *Let $(\Omega, \mathscr{E}, m)$ be a τ-measure space and P a τ-measure kernel with $P(\cdot, f') \in \mathscr{E}$ $(f' \in \mathscr{E}')$. Then*

$$m'(f') = m(P(\cdot, f')) \qquad (f' \in \mathscr{E}')$$

defines a τ-measure m' on E'. We write $m' = mP$ and say that m' **results through transport** *of m by P.*

4.8.3. *Let $(\Omega, \mathscr{R}, m)$ be a σ-content space and $\hat{P}$ m-σ-integrable, then*

$$(2) \qquad m'(F') = \int m(d\omega)P(\omega, F') \qquad (F' \in \mathscr{R}')$$

defines a σ-content m' on $\mathscr{R}'$. We write $m' = mP$ for brevity and say that m' **results through transport** *of m by P.*

The proof is an easy exercise (use, e.g., the monotone convergence theorem III. 5.3) and is left to the reader.

4.9. Remarks

4.9.1. The seemingly odd formulation of 4.8.2 is inevitable: Choose, e.g., $\Omega = \Omega' = \mathbb{R}$, $E = \mathscr{C}^{00}(\mathbb{R}, \mathbb{R})$, $\mathscr{B}^{00}$ = all bounded Borel sets in $\mathbb{R}$, $\mathscr{E}' = \mathscr{E}(\mathscr{B}^{00})$ and $P(\omega, f') = f'(\omega)$ $(\omega \in \mathbb{R},\ f' \in \mathscr{E}')$. Let m: $\mathscr{E} \to \mathbb{R}$ be the usual Riemann integral. Clearly P is a m-τ-integrable τ-measure kernel.

It is in fact also an m-σ-integrable σ-measure kernel, hence $m' = mP$ is defined as a σ-measure by 4.8.1. But it is simply the Lebesgue-σ-measure on $\mathscr{E}(\mathscr{B}^{00})$, and this is not a τ-measure (see Exercise I.9.4.3).

4.9.2. Transport of σ-contents by a σ-content kernel generalizes transform of σ-contents by measurable mappings. In fact, if we have the situation of 4.8.3 and P is induced by an $\mathscr{R}$-$\mathscr{R}'$-measurable mapping φ: $\Omega \to \Omega'$, then $m'(F') = m(\varphi^{-1}F')$ $(F' \in \mathscr{R}')$.

4.10. Theorem. *Let $\Omega, \Omega' \neq \emptyset$, $\mathscr{E}' \subseteq \mathbb{R}^{\Omega'}$ be an elementary domain, P a σ-measure kernel from Ω to Ω' (with $\mathscr{E}'$). Let $(\Omega, \mathscr{E}, m)$ be a σ-measure space and P m-σ-integrable, $m' = mP$. Then for every $h' \in \mathscr{L}_\sigma^1(\Omega', \mathscr{E}', m')$, the set*

$$N(h') = \{\omega \mid h' \notin \mathscr{L}_\sigma^1(\Omega', \mathscr{E}', P(\omega, \cdot))\}$$

is an m-σ-nullset and

$$(3) \qquad h(\omega) = \begin{cases} \displaystyle\int P(\omega, d\omega')h'(\omega') & (\omega \in \Omega \backslash N(h')) \\ 0 & (\omega \in N(h')) \end{cases}$$

defines a function $h \in L_\sigma^1(\Omega, \mathscr{E}, m) \cap \mathbb{R}^\Omega$ which we also denote by $h = Ph'$. The mapping P: $h = Ph' \leftarrow h'$ of $\mathscr{L}_\sigma^1(\Omega', \mathscr{E}', m')$ into $\mathscr{L}_\sigma^1(\Omega, \mathscr{E}, m)$ is a class mapping, i.e., sends m'-σ-a.e.-equal functions into m-σ-a.e. equal functions and can thus be considered as a mapping $L_m^1 \leftarrow L_{m'}^1$ which we denote by P again *and call the mapping of integrable functions induced by the kernel P. The latter is linear, positive, and norm-contracting:*

$$Ph' \geqq 0 \qquad (0 \leqq h' \in L^1_{m'})$$
$$\|Ph'\|_1 \leqq \|h'\|_1 \qquad (h' \in \mathscr{L}^1_{m'}).$$

Moreover the **integral transport formula**

$$\int Ph'\,dm = \int h'\,d(mP) \qquad (h' \in L^1_{m'}) \tag{4}$$

holds.

Proof. Let $\mathscr{L}' = \{h' \mid h' \in \mathscr{L}^1_\sigma(\Omega', \mathscr{E}', m'), N(h')$ is an m-σ-nullset, the h defined by (3) is in $\mathscr{L}^1_\sigma(\Omega, \mathscr{E}, m')$ and satisfies (4)$\}$. Clearly (3) well defines $P\colon \mathscr{L}^1_\sigma(\Omega, \mathscr{E}, m) \leftarrow \mathscr{L}'$. By the very definition of the σ-integrability of P and of m' we have $\mathscr{E}' \subseteq \mathscr{L}'$. Let us now show that $\mathscr{L}'$ is stable under the passage to pointwise limits of integral-bounded monotone sequences. It is clearly sufficient to settle the monotone increasing case. Let thus $h'_1, h'_2, \ldots \in \mathscr{L}'$, $h'_1 \leqq h'_2 \leqq \cdots$ everywhere, and define $h' = \lim h'_n$ (pointwise). Let $h_k = Ph'_k$ $(k = 1, 2, \ldots)$ and put $N = N(h'_1) \cup N(h'_2) \cup \cdots$. Clearly N is an m-σ-nullset. For $\omega \in \Omega\backslash N$, we have $h_1(\omega) \leqq h_2(\omega) \leqq \cdots$. Define

$$\bar{h}(\omega) = \begin{cases} \lim h_n(\omega) & (\omega \in \Omega\backslash N) \\ 0 & (\omega \in N). \end{cases}$$

Assume now $\lim \int h'_n\,dm' = \alpha < \infty$. Clearly

$$\lim \int h_n\,dm = \lim \int h'_n\,dm' = \alpha < \infty.$$

By Theorem III.5.3 (monotone convergence) $\bar{h} \in \mathscr{L}^1_\sigma(\Omega, \mathscr{E}, m)$ and $\int \bar{h}\,dm = \alpha = \int h'\,dm'$ follows, and by III.5.1.1 we get an m-σ-nullset $N \subseteq \bar{N} \subseteq \Omega$ such that $\bar{h}(\omega) < \infty$ $(\omega \in \Omega\backslash N)$. Thus for $\omega \in \Omega\backslash\bar{N}$ we get

$$\lim \int P(\omega, d\omega')h'_n(\omega') = \bar{h}(\omega) < \infty$$

and hence $h' \in \mathscr{L}^1_\sigma(\Omega', \mathscr{E}', P(\omega, \cdot))$ and $\int P(\omega, d\omega')h'(\omega') = h(\omega)$ by the monotone integration theorem III.5.3. Consequently, $N(h') \subseteq \dot{\overline{N}}$, hence $N(h')$ is an m-σ-nullset. Moreover (3) defines an h that differs from $\bar{h}$ at most on the m-σ-nullset $\overline{N}$ and hence is in $\mathscr{L}_\sigma^1(\Omega, \mathscr{E}, m)$, satisfying

$$\int h\,dm = \int \bar{h}\,dm = \alpha = \int h'\,dm'.$$

Consequently, we get $h' \in \mathscr{L}'$. Now we show $\mathscr{L}' = \mathscr{L}_\sigma^1(\Omega', \mathscr{E}', m')$ by going through the steps of monotone σ-integration: In the first step we find that σ-upper and σ-lower functions over E' with finite m'-integrals are in $\mathscr{L}'$ again; in the second step we show that functions that can be "squeezed" between σ-upper and σ-lower functions, i.e., all m'-σ-integrable functions, are in $\mathscr{L}'$. The details of the rest of the proof are left to the reader.

4.11. Exercises

4.11.1. Formulate and prove the analogue of Theorem 4.10 for σ-integrable σ-content kernels.

4.11.2. Formulate and prove the analogue of Theorem 4.10 for a τ-measure space $(\Omega, \mathscr{E}, m)$ and a τ-measure kernel P from Ω to Ω' (with $\mathscr{E}' \subseteq \mathbb{R}^{\Omega'}$) such that $P(\cdot, f') \in \mathscr{E}$ $(f' \in \mathscr{E}')$.

4.11.3. Let $\Omega = \mathbb{R}$, $\Omega' = \mathbb{R}^2$, $(\Omega, \mathscr{B}^{00}, m)$ be the Lebesgue σ-content space, $\mathscr{D}'$ the ring of all finite unions of dyadic half-open intervals in $\mathbb{R}^2$. Then

$$P(\omega, F') = m(\{\omega' \mid (\omega, \omega') \in F'\})$$

defines an m-σ-integrable kernel from Ω to Ω', and it is easily seen that $mP = m' =$ the Lebesgue σ-content on $\mathscr{D}'$. Let

$$h'(\omega, \omega') = \begin{cases} 1 & \text{for } |\omega'| < |\omega|, \omega' > 0 \\ -1 & \text{for } |\omega'| < |\omega|, \omega' < 0 \\ 0 & \text{elsewhere.} \end{cases}$$

Prove that $h' \notin \mathscr{L}_\sigma^1(\Omega', \mathscr{D}', m')$, but $h' \in \mathscr{L}_\sigma^1(\Omega', \mathscr{D}', P(\omega, \cdot))$ for every $\omega \in \Omega$ and $\int P(\omega, d\omega')h'(\omega') = 0$ $(\omega \in \Omega)$. This exercise shows that the $\Leftrightarrow$ of Theorem 2.4 does not hold for kernels in general.

4.11.4. Let P be a σ-measure kernel from Ω to Ω' that is σ-integrable for the σ-measure space $(\Omega, \mathscr{E}, m)$. Prove that the σ-content kernel derived from P (Remark 4.2.2.1) is σ-integrable for the σ-content space derived from $(\Omega, \mathscr{E}, m)$. (*Hint*: Show that every $1_{F'}$ with $F' \in \mathscr{R}'$ (of Remark 4.2.2.1) is a pointwise limit of a sequence in $\mathscr{E}'$ that lies between 0 and some $f' \in \mathscr{E}'_+$.)

3. Measurability of Kernels

4.12. Definition. *Let $\Omega, \Omega' \neq \emptyset$, $\mathscr{E}' \subseteq \mathscr{R}^{\Omega'}$ an elementary domain, $\mathscr{R} \subseteq \mathscr{P}(\Omega')$ a ring, P a measure kernel from Ω to Ω' (and $\mathscr{E}'$), $\hat{P}$ a content kernel from Ω to Ω' (and $\mathscr{R}'$) and $\mathscr{B}^0 \subseteq \mathscr{P}(\Omega)$ a σ-ring.*

4.12.1. *P is said to be* **$\mathscr{B}^0$-measurable** *if $P(\cdot, f') \in \text{mble}(\Omega, \mathscr{B}^0, \mathbb{R})$ $(f' \in \mathscr{E}')$.*

4.12.2. *$\hat{P}$ is said to be* **$\mathscr{B}^0$-measurable** *if $P(\cdot, F') \in \text{mble}(\Omega, \mathscr{B}^0, \mathbb{R})$ $(F' \in \mathscr{R}')$.*

4.13. Remarks

4.13.1. Measurability of kernels is nothing but the measurability of the members of the function families that they are according to Remark 4.7.1.

4.13.2. If the content kernel $\hat{P}$ is derived from a point mapping $\varphi: \Omega \to \Omega'$ according to Example 4.3.1, then $\hat{P}$ is $\mathscr{B}^0$-measurable iff φ is $\mathscr{B}^0$-$\mathscr{R}'$-measurable.

4.13.3. If a $\mathscr{B}^0$-measurable σ-measure kernel is given, it is of course interesting to know whether the derived σ-content kernel (Remark 4.2.2.1) is $\mathscr{B}^0$-measurable again. This is in fact true. The proof can be built on the fact that the indicator functions of every set from the ring generated by sets of the form $\{f' \geqq 1\}$, $f' \in \mathscr{E}'$, can be obtained as pointwise limits of sequences in $\mathscr{E}'$, each sequence being bounded by 0 and a function in $\mathscr{E}'_+$. We leave the details as an exercise to the reader. It is also easy to see that the σ-content kernel that we obtain from a given $\mathscr{B}^0$-measurable σ-content kernel from Ω to Ω' (and a ring $\mathscr{R}' \subseteq \mathscr{P}(\Omega')$) by extending every $P(\omega, \cdot)$ uniquely to $\mathscr{B}^{00}(\mathscr{R}')$ is $\mathscr{B}^0$-measurable again. The proof is left to the reader. (*Hint*: $\{E' \mid E' \in \mathscr{B}^{00}(\mathscr{R}'), P(\cdot, E') \in \text{mble}(\Omega, \mathscr{B}^0, \mathbb{R})\}$ is a local σ-ring containing $\mathscr{R}'$, hence equals $\mathscr{B}^{00}(\mathscr{R}')$ (use Proposition I.4.9).)

4.14. Proposition. *Let $\Omega, \Omega' \neq \emptyset$, $\mathscr{E}' \subseteq \mathbb{R}^{\Omega'}$ be an elementary domain, P a measure kernel from Ω to Ω' (and $\mathscr{E}'$). Let $(\Omega, \mathscr{E}, m)$ be a σ-measure space and $\mathscr{B}^0(\mathscr{E})$ the σ-ring generated by $\mathscr{E}$. Assume that P is $\mathscr{B}^0(\mathscr{E})$-measurable. Then P is m-σ-integrable iff for every $f' \in \mathscr{E}'$ there is an $h \in \mathscr{L}^1_\sigma(\Omega, \mathscr{E}, m)$ with $|P(\omega, f')| \leqq h(\omega)$ $(\omega \in \Omega)$.*

Proof. Apply Proposition III.9.5.

4.15. Exercise. Formulate and prove the analogue of the above proposition for σ-contents and content kernels.

4. Products of Kernels

4.16. Theorem. *Let $\Omega, \Omega', \Omega'' \neq \varnothing$, and $\mathscr{E}' \subseteq \mathbb{R}^{\Omega'}$, $\mathscr{E}'' \subseteq \mathbb{R}^{\Omega''}$ be elementary domains, and P a measure kernel from Ω to Ω', P' a measure kernel from Ω' to Ω''. Assume*

$$(5) \qquad P(\cdot, f'') \in \mathscr{E}' \qquad (f'' \in \mathscr{E}'').$$

4.16.1. *Then*

$$\tilde{P}(\omega, f'') = P(\omega, P'(\cdot, f'')) \qquad (\omega \in \Omega, \quad f'' \in \mathscr{E}'')$$

defines a kernel $\tilde{P}$ from Ω to Ω''. It is called the product kernel of P and P' and also denoted by PP'.

4.16.2. *If P, P' are σ-measure kernels, so is PP'.*

4.16.3. *If P, P' are τ-measure kernels, so is PP'.*

The proof is an easy exercise and left to the reader.

4.17. Remarks

4.17.1. We content ourselves with the above theorem on σ-measure kernels here. The analogue for σ-content kernels is an easy corollary to it.

4.17.2. The range of applications for Theorem 4.16 is, given P and $\mathscr{E}'$, clearly the wider the larger $\mathscr{E}'$ is. This provides us with a strong motive to push the extensions given in Remark 4.2.3 to the extreme.

5. The Problem of Kernel Representation

Theorem 4.10 and Exercise 4.11.1 have shown how an integrable σ-measure resp. σ-content kernel P transports a σ-measure resp. σ-content m, for which it is integrable, into another σ-measure resp. σ-content $m' = mP$ and then induces a linear mapping P: $L^1_m \leftarrow L^1_{m'}$ which is positive, integral-preserving, and L^1-norm-contracting. It is obvious to pose the following:

4.18. Problem of kernel representation. *Let $(\Omega, \mathscr{E}, m)$, $(\Omega', \mathscr{E}', m')$ be σ-measure spaces and P: $L^1_m \leftarrow L^1_{m'}$ a positive linear integral-preserving L^1-norm-contracting mapping. Does there exist a σ-measure kernel P from Ω to Ω' (with $\mathscr{E}'$) that is m-σ-integrable and satisfies $mP = m'$, inducing P: $L^1_m \leftarrow L^1_{m'}$ according to Theorem 4.10?*

4.19. Remark. It is clear that there is a "σ-content version" of this problem as well, and that one should be willing to, e.g., vary $\mathscr{E}'$ suitably.

In the σ-measure case, the following ideas offer themselves as helpful toward a solution of 4.18.

The first thing which one should like to do is to pass over from $P\colon L_m^1 \leftarrow L_{m'}^1$ to a positive linear mapping $\tilde{P}\colon \mathscr{L}_m^1 \cap \mathbb{R}^\Omega \leftarrow \mathscr{L}_{m'}^1 \cap \mathbb{R}^{\Omega'}$ which represents P in the obvious sense. It is clear what has to be done in order to achieve this: For every $h' \in L_{m'}^1 \cap \mathbb{R}^\Omega$, the mapping $P\colon L_m^1 \leftarrow L_{m'}^1$ does not give us a single candidate for Ph' but a whole equivalence class mod m of candidates, and the problem is to choose among them in such a fashion that the resulting $\tilde{P}$ is positive and satisfies the uncountably many relations defining its linearity. It is easy to satisfy any countable bundle of linear relations since for any choice of candidates, they are satisfied outside some countable union of exceptional m-nullsets, i.e., outside a single m-nullset, and we can modify all of our candidates to be $=0$ on that nullset and hence to satisfy the countably many linearity relations everywhere. But it remains open whether we can choose our candidates for the Ph' such that the uncountable entirety of all linearity relations is satisfied. We shall see in Chapter XVI, Section 3, how so-called lifting will free us from this problem. Let us nevertheless discuss another obvious idea which should most naturally occur to everyone who tries to solve our problem: Find a countable subset $\{h'_1, h'_2, \ldots\}$ of $\mathscr{L}_{m'}^1 \cap \mathbb{R}^{\Omega'}$ that is "sufficiently representative" for that space and is stable under rational finite linear combinations; choose $\tilde{P}h'_j \in L_m^1 \cap \mathbb{R}^\Omega$ such that all linearity relations

$$\tilde{P}(\alpha_1 h'_1 + \cdots + \alpha_n h'_n) = \alpha_1 \tilde{P}h'_1 + \cdots + \alpha_n \tilde{P}h'_n$$

with rational coefficients $\alpha_1, \ldots, \alpha_n$ $(n = 1, 2, \ldots)$ are satisfied everywhere. The impact of what is "sufficiently representative" should now allow us to choose all other $\tilde{P}h'$ in such a fashion that $\tilde{P}$, now fully defined, satisfies all our wishes.

This program can be carried through, e.g., in the following special case. Let $\Omega' = [0, 1]$ and $\{h'_1, h'_2, \ldots\} \subseteq \mathscr{C}([0, 1], \mathbb{R})$ be dense in the latter space, with respect to uniform approximation. We may and shall assume that $\mathscr{C}([0, 1], \mathbb{R}) \subseteq \mathscr{L}_{m'}^1$ and is L^1-norm dense in $L_{m'}^1$, and that $\{h'_1, h'_2, \ldots\}$ is stable under finite rational linear combinations and finite lattice operations, and contains all rational constants. For every $k = 1, 2, \ldots$, choose any $h_k \in \mathscr{L}_m^1 \cap \mathbb{R}^\Omega$ in the equivalence class mod m $P\hat{h}'_k$ where $\hat{h}'_k$ is the equivalence class mod m' of h'_k. Consider all the countably many inequalities and equalities between finite collections from $\{h_1, h_2, \ldots\}$ that follow from the requirement that $\tilde{P}h'_k = h_k$ defines a positive rational-linear mapping. There is an m-nullset $N \subseteq \Omega$ such that all these relations are true outside N. Define

$$(\tilde{P}h'_k)(\omega) = \begin{cases} h_k(\omega) & (\omega \in \Omega \backslash N), \\ 0 & (\omega \in N). \end{cases}$$

Then $P: \mathscr{L}_m^1 \cap \mathbb{R}^\Omega \leftarrow \{h'_1, h'_2, \ldots\}$ is rational-linear and positive. Since the rational constants are among the h'_k, we have, by positivity and rational linearity,

$$\alpha \leqq h'_k \leqq \beta \quad \Rightarrow \quad \tilde{P}\alpha \leqq \tilde{P}h'_k \leqq \tilde{P}\beta \qquad (-\infty < \alpha < \beta < \infty \text{ rational}).$$

In particular,

$$|h'_k - h'_j| \leqq \delta \quad \Rightarrow \quad |\tilde{P}h'_k - \tilde{P}h'_j| \leqq \delta\, \tilde{P}1 \qquad (\delta > 0 \text{ rational}),$$

i.e., P transforms uniform approximation into pointwise approximation. It is now obvious how to extend the definition of $\tilde{P}$ to $\mathscr{C}([0, 1], \mathbb{R})$ by uniform resp. pointwise approximation, and it is clear that the $\tilde{P}: L_m^1 \cap \mathbb{R}^\Omega \leftarrow \mathscr{C}([0, 1], \mathbb{R})$ thus obtained is positive and real-linear, and sends uniformly convergent sequences into pointwise convergent sequences, then commuting with the passage to limits. Thus we have solved our problem of defining $\tilde{P}$, though not completely, but at least to the extent that we get a reasonable definition on $\mathscr{C}([0, 1], \mathbb{R})$.

Instead of shifting back to the general case, we shall now show how to go on in our special situation. Define a (positive) σ-measure kernel from Ω to $[0, 1]$ by choosing the elementary domains $\mathscr{E} = \tilde{P}(\mathscr{C}([0, 1], \mathbb{R}))$, $\mathscr{E}' = \mathscr{C}([0, 1], \mathbb{R})$ and putting

$$P(\omega, f') = (\tilde{P}f')(\omega) \qquad (\omega \in \Omega, \quad f' \in \mathscr{C}([0, 1], \mathbb{R})).$$

Form now the σ-content kernel derived from this σ-measure kernel, according to Remark 4.2.2, and apply Remark 4.2.3 in order to show that this σ-content kernel actually sends m into m' and induces the given $P: L_m^1 \leftarrow L_{m'}^1$. Every topologist sees that the same conclusions can be drawn if Ω' is an arbitrary compact metric space.

How would one now like to proceed in the general case, once a positive linear $\tilde{P}: \mathscr{L}_m^1 \cap \mathbb{R}^\Omega \leftarrow L_{m'}^1 \cap \mathbb{R}^{\Omega'}$ representing $P: L_m^1 \leftarrow L_{m'}^1$ has been established? One chooses the elementary domains $\mathscr{E} = \mathscr{L}_m^1 \cap \mathbb{R}^\Omega$, $\mathscr{E}' = \mathscr{L}_{m'}^1 \cap \mathbb{R}^{\Omega'}$ and defines a measure kernel P from $\mathscr{E}$ to $\mathscr{E}'$ by

$$P(\omega, f') = (\tilde{P}f')(\omega) \qquad (\omega \in \Omega, f' \in \mathscr{E}').$$

Is this a σ-measure kernel? If $\mathscr{E}' \ni f'_1 \geqq f'_2 \geqq \cdots \searrow 0$ pointwise, we may conclude $\|f'_k\|_1 \searrow 0$, hence $\|\tilde{P}f'_k\|_1 \searrow 0$, but this implies only $P(\omega, f'_k) \searrow 0$ for m-a.e. $\omega \in \Omega$. One could think about modifying the functions $P(\cdot, f')$ $(f' \in \mathscr{E}')$ on m-nullsets such as to ensure $\mathscr{E}' \ni f'_k \searrow 0$ (pointwise) $\Rightarrow P(\omega, f'_k) \searrow 0$ $(\omega \in \Omega)$, and surely this can be done for any given sequence $f'_1, f'_2, \ldots$; but there is, as a rule, an uncountable number of such sequences, and thus one gets uncountably many exceptional m-nullsets. The following celebrated example goes back to Dieudonné [1] and has led to the consequence that to this day no alternative to the use of topological methods similar to those presented above has been found for the solution of the kernel representation problem 4.18 (compare Chapter XVI, Section 3).

4.20. Example. We begin with:

4.20.1. Construction of a σ-content space $(\Omega, \mathscr{B}, m)$ with $\Omega \in \mathscr{B}$, i.e., $\mathscr{B}$ a σ-field, and of a set $M \subseteq \Omega$ such that both M and $\Omega \backslash M$ have outer content $m(\Omega)$ for m. For this we imitate, to a certain extent, the construction of a set that is not Lebesgue measurable given in Example II.6.6.2. For the amusement of the reader, we choose a somewhat different setting now. Let $(\Omega, \mathscr{F}, m)$ be the σ-content space of Example 3.3 (of this chapter), i.e., $\Omega = \{z | z \in \mathbb{C}, |z| = 1\} = \{e^{i\varphi} | \varphi \in \mathbb{R}\} = \{e^{i\varphi} | 0 \leqq \varphi < 2\pi\}$ is the unit circle line, $\mathscr{F}$ the field of all finite unions of half-open arcs in Ω, and $m =$ the σ-content defined by arc length. We know that m is invariant under every circle rotation T: $z \to e^{i\alpha}z$, and the same holds for the Carathéodory extension of m (which we denote by m again) on $\mathscr{B}_m(\mathscr{F})$. Write $\mathscr{B}$ for $\mathscr{B}(\mathscr{F})$. Call $z, z' \in \Omega$ equivalent if $z'z^{-1} = e^{in}$ for some integer n. Clearly this defines an equivalence relation and hence splits Ω into equivalence classes of the form $\{ze^{in} | n \in \mathbb{Z}\}$ which are in fact countably infinite since

$$ze^{in} = ze^{in'} \quad \Rightarrow \quad e^{i(n-n')} = 1 \quad \Rightarrow \quad (n - n')/2\pi \in \mathbb{Z} \quad \Rightarrow \quad n = n'.$$

By the axiom of choice (i.e., Zorn's lemma) there exists a set $M_0 \subseteq \Omega$ that intersects every equivalence class in exactly one point. Clearly $\Omega = \sum_{n \in \mathbb{Z}} M_0 e^{in}$, this being a countable disjoint union. If $M_0 \in \mathscr{B}_m(\mathscr{F})$, then all these sets $M_0 e^{in}$ would have the same content for m; and since $m(\Omega) = 2\pi < \infty$, their content must be 0, which in turn yields $m(\Omega) = 0$, a contradiction. Let now $M = \sum_{n \in \mathbb{Z}} M_0 e^{i2n}$. Clearly $\Omega \backslash M = Me^i$, $(\Omega \backslash M)e^i = M$. Let $\overline{m}$: $\mathscr{P}(\Omega) \to \mathbb{R}_+$ denote the outer content derived from the content m: $\mathscr{F} \to \mathbb{R}_+$ (Definition II.1.1). An obvious analogue to Exercise II.1.4.6 shows that m is rotation invariant. In particular, $\overline{m}(M) = \overline{m}(\Omega \backslash M)$. Assume now that $\overline{m}(M) < m(\Omega)$ $(=2\pi)$. Then the complement F of a suitable countable union of half-open arcs covering M is in $\mathscr{B}(\mathscr{F})$ and has $m(F) > 0$. By extending each of the arcs a little bit over its closed end we may even assume that $\Omega \backslash F$ is open and F is closed. Since $F \subseteq \Omega \backslash M$, we see that $Fe^{ik} \subseteq M$ for every odd integer k. An easy adaptation of Kronecker's argument (Example 3.24) shows that there is sequence $0 < k_1 < k_2 < \cdots$ of odd integers such that $e^{ik_n} \to 1$. This in turn proves

$$\lim_{n \to \infty} 1_{\Omega \backslash F}(e^{ik_n}z) \geqq 1_{\Omega \backslash F}(z) \; (z \in \Omega)$$

and hence

$$\liminf_{n \to \infty} \int 1_{\Omega \backslash F}(e^{ik_n}z)\, 1_{\Omega \backslash F}(z)\, m(dz) \geqq \int 1_{\Omega \backslash F}(z)\, m(dz),$$

i.e.,

$$\liminf_{n\to\infty} m((\Omega\backslash F)e^{ik_n} \cap (\Omega\backslash F)) \geqq m(\Omega\backslash F).$$

Passing to complements, we see $\limsup_{n\to\infty} m(Fe^{ik_n} \cup F) \leqq m(F)$. Since $Fe^{ik_n} \cap F = \varnothing$, this implies $m(F) = \lim_{n\to\infty} m(Fe^{ik_n}) = 0$, a contradiction. Thus we have in fact $\overline{m}(M) = \overline{m}(\Omega\backslash M) = m(\Omega) = 2\pi$.

4.20.2. If M is as above and $\mathscr{B}_m(\mathscr{F}) \ni E \supseteq M$, then $m(E) = \overline{m}(E) \geqq \overline{m}(M) = 2\pi$, i.e., $m(E) = 2\pi$. If $\mathscr{B}_m(\mathscr{F}) \ni F \supseteq M$, then

$$\mathscr{B}_m(\mathscr{F}) \ni E \cap F \supseteq M$$

and $m(E \cap F) = 2\pi$ follows, hence $E \,\Delta\, F \subseteq \Omega\backslash(E \cap F)$ is an m-nullset and $m(E) = m(F)$ $(=m(E \cap F))$ follows. Similarly, we see: If $\mathscr{B}_m(\mathscr{F}) \ni E \subseteq M$, or $\subseteq \Omega\backslash M$, then $m(F) = 0$.

4.20.3. Let M be as above. Then

$$\mathscr{B}' = \{(E \cap M) + F \cap (\Omega\backslash M)) \,|\, E, F \in \mathscr{B}\}$$

is a σ-field in Ω (exercise), namely, the σ-field generated by $\mathscr{F}$ and M. If $E, F, E', F' \in \mathscr{B}$ are such that

$$(E \cap M) + (F \cap (\Omega\backslash M)) = (E' \cap M) + (F' \cap (\Omega\backslash M)),$$

then $E \,\Delta\, E' \subseteq \Omega\backslash M$, hence $m(E \,\Delta\, E') = 0$ and thus $m(E) = m(E')$. It follows that $m'\colon \mathscr{B}' \to \mathbb{R}_+$ is well defined by

$$m'((E \cap M) + (F \cap (\Omega\backslash M))) = m(E) \qquad (E, F \in \mathscr{B}).$$

If $E_1, F_1, E_2, F_2, \ldots \in \mathscr{B}$ are such that the sets $(E_k \cap M) + (F_k \cap (\Omega\backslash M))$ $(k = 1, 2, \ldots)$ are pairwise disjoint, then the $E_k \cap M$ $(k = 1, 2, \ldots)$ are pairwise disjoint and hence $E_k \cap M = E_k' \cap M$ where $E_k' = E_k \backslash \bigcup_{j\neq k} E_j$ $(k = 1, 2, \ldots)$. Now the E_k' $(k = 1, 2, \ldots)$ are still in $\mathscr{B}$ and pairwise disjoint. It follows that we may assume the E_k pairwise disjoint. Now

$$\begin{aligned} m'\left(\sum_{k=1}^{\infty} [(E_k \cap M) + (F_k \cap (\Omega\backslash M))]\right) &= m'\left(\left(\sum_{k=1}^{\infty} E_k\right) \cap M\right) \\ &= m\left(\sum_{k=1}^{\infty} E_k\right) = \sum_{k=1}^{\infty} m(E_k) \\ &= \sum_{k=1}^{\infty} m'((E_k \cap M) + (F_k \cap (\Omega\backslash M))) \end{aligned}$$

and thus $m'\colon \mathscr{B}' \to \mathbb{R}_+$ is a σ-content.

4.20.4. Let now $\Omega' = \Omega$ and let $P\colon L_m^1 \leftarrow L_{m'}^1$ denote the conditional expectation for the σ-field $\mathscr{B} \subseteq \mathscr{B}'$ (see Proposition VIII.6.1). Assume that there is a $\mathscr{B}$-$\mathscr{B}'$-measurable σ-content kernel $\hat{P}$ from Ω to $\Omega' = \Omega$ such that $P: L_m^1 \leftarrow L_{m'}^1$ is the linear mapping derived from $\hat{P}$. Put $N = \{\omega \,|\, P(\omega, M) \neq 1\}$. Since $m'(M) = 2\pi$, i.e., the indicator function of M is $= 1$ m'-a.e., and P sends 1 into 1; we see that N, which is in $\mathscr{B}$, is an m-nullset. For every $F \in \mathscr{B}$, let

$$N(F) = \{\omega \,|\, P(\omega, F) \neq 1_F(\omega)\}.$$

By the very definition of our conditional expectation every $N(F)$ is in $\mathscr{B}$ and an m-nullset. Let now $\Omega \in \mathscr{S} \subseteq \mathscr{B}$ be countable, stable against finite intersections, and such that $\mathscr{B}(\mathscr{S}) = \mathscr{B}$ (take, e.g., all arcs with endpoints in some countable dense subset of Ω, plus Ω itself. Then

$$N_0 = N \cup \bigcup_{F \in \mathscr{S}} N(F)$$

is an m-nullset again. Now

$$\mathscr{B}_0 = \{F \,|\, F \in \mathscr{B},\, N(F) \subseteq N_0\}$$

contains $\mathscr{S}$. But $\mathscr{B}_0$ is a σ-field. In fact, if $E_1, E_2, \ldots \in \mathscr{B}_0$, then for $\omega \notin N_0$ we have $P(\omega, E_k) = 1_{E_k}(\omega)$. If $\omega \in E_1 \cup E_2 \cup \cdots$, then $P(\omega, E_k) = P(\omega, \Omega) = 1$ for some k and $P(\omega, E_1 \cup E_2 \cup \cdots) = 1_{E_1 \cup E_2 \cup \cdots}(\omega)$ follows. If $\omega \notin E_1 \cup E_2 \cup \cdots$, then $P(\omega, E_k) = 0$ $(k = 1, 2, \ldots)$, and

$$P(\omega, E_1 \cup E_2 \cup \cdots) = 0 = 1_{E_1 \cup E_2 \cdots}(\omega)$$

follows. Thus $\mathscr{B}_0$ is stable against countable unions. But it is also stable against passage to complements: If $E \in \mathscr{B}_0$ and $\omega \notin N_0$, then we have $P(\omega, \Omega) = 1$ and $P(\omega, E) = 1_E(\omega)$, hence $P(\omega, \Omega \backslash E) = P(\omega, \Omega) - P(\omega, E) = 1 - P(\omega, E) = 1 - 1_E(\omega) = 1_{\Omega \backslash E}(\omega)$. Consequently, $\mathscr{B}_0 = \mathscr{B}$, i.e., $P(\omega, E) = 1_E(\omega)$ $(E \in \mathscr{B},\ \omega \notin N_0)$. Now we see for any fixed $\omega \in \Omega \backslash N_0$ that $P(\omega, \{\omega\}) = 1_{\{\omega\}}(\omega) = 1$ and $P(\omega, M) = 1$, hence $P(\omega, \{\omega\} \cap M) = 1$ and $\omega \in M$ follows. This implies that $\Omega \backslash N_0 \subseteq M$. But we have shown that this implies $m(\Omega \backslash N_0) = 0$, and this contradicts $m(\Omega \backslash N_0) = m(\Omega) = 2\pi$. Consequently, our P: $L_m^1 \leftarrow L_{m'}^1$ cannot be induced by a $\mathscr{B}$-$\mathscr{B}'$-measurable σ-content kernel.

CHAPTER V

CONTENTS AND MEASURES IN TOPOLOGICAL SPACES. PART I: REGULARITY

The purpose of this chapter is the investigation of one aspect of *σ-contents in topological spaces: regularity.* Loosely speaking, regularity means approximability of measurable sets by other sets, mostly very special measurable sets, which will, in our context, be characterizable by topological properties. The σ-content viewpoint dominates in our presentation. We begin with Section 1, presenting a very abstract definition of regularity for contents in arbitrary basic sets. The abstract level of presentation is justified immediately by an application showing how σ-additivity can be obtained from a suitable regularity property (Proposition 1.6). We prove a prototype regularity theorem (1.7) which shows how to get obvious regularities through the process of σ-extension of a σ-measure and the derivation of a σ-content therefrom. In Section 2 we specialize to the topological situation which is the proper theme of this chapter and prove a general regularity theorem (2.1). In Sections 3–5 we specialize our general result to various cases, exhibiting the discrepancy (and harmony) of Baire and Borel σ-contents in compact and locally compact spaces, and the inner compact-regularity of σ-contents in Polish spaces.

The other major theme of measure and content theory in topological spaces, *weak convergence*, is postponed until Chapter XI.

1. THE GENERAL CONCEPT OF REGULARITY

In this section we introduce a very general concept of regularity, still without alluding to topology, except in an application to so-called abstract σ-compact systems. We prove a very general regularity theorem which shows how regularity arises in a natural fashion if we apply, e.g., the σ-extension process to a σ-measure and derive a σ-content from it.

1.1. Definition. *Let* $\Omega \neq \emptyset$, $\mathscr{F}, \mathscr{G} \subseteq \mathscr{P}(\Omega)$. *Let* $(\Omega, \mathscr{R}, m)$ *be a content prespace and* $\tilde{m}: \mathscr{P}(\Omega) \to \mathbb{R}_+$ *the outer content derived from it (definition* II.1.1). m (*or* $(\Omega, \mathscr{R}, m)$) *is said to be:*

1.1.1. **outer $\mathscr{G}$-regular** *at* $E \in \mathscr{R}$ *if for every* $\varepsilon > 0$ *there is an* $E \subseteq G \in \mathscr{G}$ *such that* $\tilde{m}(G \backslash E) < \varepsilon$;

1.1.2. **inner $\mathscr{F}$-regular** *at* $E \in \mathscr{R}$ *if for every* $\varepsilon > 0$ *there is an* $\mathscr{F} \ni F \subseteq E$ *such that* $\tilde{m}(E \backslash F) < \varepsilon$;

1.1.3. **outer $\mathscr{G}$-regular** *if it is outer* $\mathscr{G}$-*regular at every* $E \in \mathscr{R}$.

1.1.4. **inner $\mathscr{F}$-regular** *if it is inner* $\mathscr{F}$-*regular at every* $E \in \mathscr{R}$.

1.1.5. **$\mathscr{F}$-$\mathscr{G}$-regular** *if it is inner* $\mathscr{F}$- *and outer* $\mathscr{G}$-*regular.*

1.2. Remarks

1.2.1. Observe that we do not require $\mathscr{G}$ or $\mathscr{F}$ to be contained in $\mathscr{R}$ in this definition. A good motive for this will be visible in Example 1.3.2. Nevertheless there are many cases where $\mathscr{F} \subseteq \mathscr{R}$ or $\mathscr{G} \subseteq \mathscr{R}$ or both. It is obvious that

1.2.1.1. $\mathscr{G} \subseteq \mathscr{R}$ and outer G-regularity of $m: \mathscr{R} \to \mathbb{R}_+$ implies

$$m(E) = \inf_{G \in \mathscr{G}} m(G) \qquad (E \in \mathscr{R});$$

in particular, $m: \mathscr{R} \to \mathbb{R}_+$ is uniquely determined by its restriction $m: \mathscr{G} \to \mathbb{R}_+$.

1.2.1.2. $\mathscr{F} \subseteq \mathscr{R}$ and inner F-regularity of $m: \mathscr{R} \to \mathbb{R}_+$ implies

$$m(E) = \sup_{F \in \mathscr{F}} m(F) \qquad (E \in \mathscr{R});$$

in particular, $m: \mathscr{R} \to \mathbb{R}_+$ is uniquely determined by its restriction $m: \mathscr{F} \to \mathbb{R}_+$.

1.2.2. If $\mathscr{G}$ or $\mathscr{F}$ have an easy verbal characterization, we shall often substitute it for $\mathscr{G}$ or $\mathscr{F}$ when talking of regularity. Thus it should, e.g., be clear what *inner compact-regularity* and outer *open-regularity* means.

1.3. Examples

1.3.1. Every content space $(\Omega, \mathscr{R}, m)$ is inner $\mathscr{P}(\Omega)$-regular.

1.3.2. Example I.2.4 displays the dyadic geometric prespace on $\mathbb{R}$ as inner $\mathscr{K}$-regular, where $\mathscr{K}$ is the system of all compact subsets of $\mathscr{R}$ (actually, only finite unions of compact intervals with dyadic rational endpoints were used).

1.3.3. Finite or countable points mass distributions (Example I.1.8) are inner $\mathscr{P}_{\text{fin}}(\Omega)$-regular, where $\mathscr{P}_{\text{fin}}(\Omega)$ denotes the system of all finite subsets of Ω.

1.3.4. Let $\Omega = \{0, 1, \ldots\}$, $L: l^\infty \to \mathbb{R}$ be a Banach limit (Example I.6.4), $\mathscr{R} = \mathscr{P}(\Omega)$, and $m: \mathscr{R} \to \mathbb{R}_+$ be defined by

$$m(F) = L(1_F) \qquad (F \subseteq \Omega).$$

Then m is not inner $\mathscr{P}_{\text{fin}}(\Omega)$-regular, where $\mathscr{P}_{\text{fin}}(\Omega) = \{F \mid F \subseteq \Omega, |F| < \infty\}$.

The next proposition is to show that there is a general idea behind Example 1.3.2 (resp. I.2.4).

1.4. Definition. *Let $\Omega \neq \varnothing$. A set system $\mathscr{K} \subseteq \mathscr{P}(\Omega)$ is called* **σ-compact** *if*

$$K_1, K_2, \ldots \in \mathscr{K}, \quad K_1 \cap K_2 \cap \cdots = \varnothing$$
$$\Rightarrow \quad K_1 \cap \cdots \cap K_n = \varnothing \qquad \textit{for some } n < \infty.$$

1.5. Remark. It is obvious that the compact subsets of a topological space form a σ-compact system. They form even a system which should, by obvious analogy, be called τ-compact. For the purposes of σ-content theory, σ-compactness is sufficient since we are dealing only with countable set operations here. A more systematic discussion of σ-compactness seems unnecessary here and will be postponed until Chapter XIII.

Let us remark that every subsystem of a σ-compact system is σ-compact again.

1.6. Proposition. *Let $(\Omega, \mathscr{R}, m)$ be a content prespace. Let $\mathscr{R}_0 \subseteq \mathscr{R}$ be a ring and $\mathscr{K} \subseteq \mathscr{P}(\Omega)$ a σ-compact set system. Assume that m is inner $\mathscr{K}$-regular at every $E \in \mathscr{R}_0$. Then the restriction of m to $\mathscr{R}_0$ is σ-additive.*

Proof. We use the criterion for σ-additivity given in Proposition I.2.2. Let $\mathscr{R}_0 \ni F_1 \supseteq F_2 \supseteq \cdots \searrow \varnothing$ and assume $\lim m(F_n) = 2\varepsilon > 0$. For every $n = 1, 2, \ldots$, choose $K_n \in \mathscr{K}$ such that $K_n \subseteq F_n$, $\tilde{m}(F_n \backslash K_n) < \varepsilon/2^n$. Clearly, $K_1 \cap \cdots \cap K_n \subseteq F_n$ and

$$\begin{aligned}\tilde{m}(F_n \backslash (K_1 \cap \cdots \cap K_n)) &\leqq \tilde{m}((F_1 \backslash K_1) \cup \cdots \cup (F_n \backslash K_n)) \\ &\leqq \tilde{m}(F_1 \backslash K_1) + \cdots + \tilde{m}(F_n \backslash K_n) < \varepsilon.\end{aligned}$$

Now $\bigcap F_K = \varnothing \Rightarrow \bigcap K_K = \varnothing$, hence there is some n with $K_1 \cap \cdots \cap K_n = \varnothing$. This implies $\varepsilon \geqq \tilde{m}(F_n \backslash (K_1 \cap \cdots \cap K_n)) = \tilde{m}(F_n) = m(F_n) \geqq 2\varepsilon$, a contradiction.

1.7. Theorem. *Let $\Omega \neq \varnothing$, $\mathscr{E} \subseteq \mathbb{R}^\Omega$ an elementary domain,*

$\mathscr{E}^\sigma$ the set of all σ-upper functions

$\mathscr{E}_\sigma$ the set of all σ-lower functions

$\mathscr{E}^\tau$ the set of all τ-upper functions

$\mathscr{E}_\tau$ the set of all τ-lower functions

for $\mathscr{E}$ *and* $m: \mathscr{E} \to \mathbb{R}$ *a* σ*-measure. Let*

$$\mathscr{G}^\sigma = \mathscr{G}^\sigma(\mathscr{E}) = \{\{u > 1\} \,|\, u \in \mathscr{E}^\sigma,\ \underline{m}(u) < \infty\}$$
$$\mathscr{F}_\sigma = \mathscr{F}_\sigma(\mathscr{E}) = \{\{v \geqq 1\} \,|\, v \in \mathscr{E}_\sigma\}$$
$$\mathscr{G}^\tau = \mathscr{G}^\tau(\mathscr{E}) = \{\{u > 1\} \,|\, u \in \mathscr{E}^\tau,\ \overline{m}(u) < \infty\}$$
$$\mathscr{F}_\tau = \mathscr{F}_\tau(\mathscr{E}) = \{\{v \geqq 1\} \,|\, v \in \mathscr{E}_\tau\}.$$

Then:

1.7.1. $\mathscr{G}^\sigma$ *is stable under finite intersections and countable unions*

$\mathscr{F}_\sigma$ *is stable under finite unions and countable intersections*

$\mathscr{G}^\tau$ *is stable under finite intersections and arbitrary unions, i.e., is a topology in* Ω.

$\mathscr{F}_\tau$ *is stable under finite unions and arbitrary intersections.*

1.7.2. *Let* $(\Omega, \mathscr{E}, m)$ *be a* σ*-measure space. Then the* σ*-content space* $(\Omega, \mathscr{B}_\sigma^{00}(m), m)$ σ*-derived from it (Theorem* III.6.1) *is* $\mathscr{F}_\sigma$*-*$\mathscr{G}^\sigma$*-regular and satisfies* $\mathscr{F}_\sigma, \mathscr{G}^\sigma \subseteq \mathscr{B}_\sigma^{00}(m)$.

1.7.3. *Let* $(\Omega, \mathscr{E}, m)$ *be a* τ*-measure space. Then the* σ*-content space* $(\Omega, \mathscr{B}_\tau^{00}(m), m)$ τ*-derived from it (Chapter* III, *Section* 7) *is* $\mathscr{F}_\tau$*-*$\mathscr{G}^\tau$*-regular and satisfies* $\mathscr{F}_\tau, \mathscr{G}^\tau \subseteq \mathscr{B}_\tau^{00}(m)$.

Proof. 1 is, as far as the σ-case is concerned, an easy consequence of Proposition III.1.2, resp. Proposition III.1.7, which states analogous stability properties for $\mathscr{E}_+^\sigma$ resp. $\mathscr{E}^\sigma$. In fact we have, e.g.,

$$\begin{aligned} v_1, v_2, \ldots \in \mathscr{E}_\sigma \quad &\Rightarrow \quad v_n(\omega) = \inf_{f \in \mathscr{E}_n} f(\omega) (\omega \in \Omega) \\ &\qquad \text{for some countable } \mathscr{E}_n \subseteq \mathscr{E} \quad (n = 1, 2, \ldots) \\ &\Rightarrow \quad v_1 \wedge v_2 \wedge \cdots = \inf_{f \in \mathscr{E}_1 \cup \mathscr{E}_n \cup \cdots} f \quad \text{(pointwise)} \\ &\qquad \text{and } \mathscr{E}_1 \cup \mathscr{E}_2 \cup \cdots \subseteq \mathscr{E} \text{ is countable} \\ &\Rightarrow \quad v_1 \wedge v_2 \wedge \cdots \in \mathscr{E}_\sigma \\ &\Rightarrow \quad \bigcap_{n=1}^{\infty} \{v_n \geqq 1\} = \{v_1 \wedge v_2 \wedge \cdots \geqq 1\} \in \mathscr{F}_\sigma. \end{aligned}$$

Dropping countability assumptions, we transform proofs like this into the analogous proofs for the τ-case.

2. $\mathscr{G}^\sigma \in \mathscr{B}_\sigma^{00}(m)$ follows from Theorem III.6.3 since every $u \in \mathscr{E}^\sigma$ with $\underline{m}(u) < \infty$ is in $\mathscr{L}_\sigma^1(m)$. $\mathscr{F}_\sigma \subseteq \mathscr{B}_\sigma^{00}(m)$ follows since $v \in \mathscr{E}_\sigma \Rightarrow v_+ \in \mathscr{E}_\sigma$ (Proposition III.1.7.1.3) and $v_+ \in \mathscr{L}_\sigma^1(m)$ (in fact v_+ lies between some $f \in \mathscr{E}$ and 0), applying Theorem III.6.3 again. Let now $E \in \mathscr{B}_\sigma^{00}(m)$ and $\varepsilon > 0$. Find $u \in \mathscr{E}^\sigma$, $v \in \mathscr{E}_\sigma$ such that $v \leqq 1_E \leqq u, m(u - v) < \varepsilon/4$. Put $F = \{v \geqq 1/4\}, G = \{u > 1/2\}$.

Clearly, $E \supseteq F = \{4v \geqq 1\} \in \mathscr{F}_\sigma$, $E \subseteq G = \{2u > 1\} \in \mathscr{G}^\sigma$ and

$$1_G - 1_F \leqq 4(u - v),$$

hence $m(G - F) \leqq m(4(u - v)) \leqq 4m(u - v) < \varepsilon$. This implies $m(E \backslash F) < \varepsilon$, $m(G \backslash E) < \varepsilon$, and the desired conclusion follows.

3 is proved in the same way as 2.

2. REGULARITY OF σ-CONTENTS IN TOPOLOGICAL SPACES

We now turn to our topological theme and begin by asking the reader to review his basic knowledge in general topology, upon which we shall make increasing demand in the subsequent sections. In particular, the reader should know what G_σ's, F_σ's, and K_σ's are and be familiar with the basic facts about semicontinuous functions, such as: every indicator function of an open set and every supremum of continuous functions is lower semicontinuous. This is a good opportunity to learn something about completely regular spaces, among which there are all locally compact Hausdorff spaces as well as all metrizable spaces.

As a corollary to our general regularity theorem 1.7 we now obtain, with the notations introduced there,

2.1. Theorem. *Let $(\Omega, \mathscr{T})$ be a topological space and $E \subseteq \mathscr{C}(\Omega, \mathbb{R})$ an elementary domain.*

2.1.1. *Let $(\Omega, \mathscr{E}, m)$ be a σ-measure space. Then:*

2.1.1.1. *$\mathscr{F}_\sigma = \mathscr{F}_\sigma(\mathscr{E})$ consists of closed G_δ's.*

2.1.1.2. *$\mathscr{G}^\sigma = \mathscr{G}^\sigma(\mathscr{E})$ consists of open F_σ's.*

2.1.1.3. *The σ-content space $(\Omega, \mathscr{B}_\sigma^{00}(m), m)$ σ-derived from $(\Omega, \mathscr{E}, m)$ is $\mathscr{F}_\sigma$-$\mathscr{G}^\sigma$-regular.*

2.1.2. *Let $(\Omega, \mathscr{E}, m)$ be a τ-measure space. Then:*

2.1.2.1. *$\mathscr{F}_\tau$ consists of closed sets.*

2.1.2.2. *$\mathscr{G}^\tau$ consists of open sets.*

2.1.2.3. *The σ-content space $(\Omega, \mathscr{B}^{00}(m), m)$ τ-derived from $(\Omega, \mathscr{E}, m)$ is $\mathscr{F}_\tau$-$\mathscr{G}^\tau$-regular.*

Proof. 1. Let $v \in \mathscr{E}_\sigma$, say, $v = f_1 \wedge f_2 \wedge \cdots$ with $f_1, f_2, \ldots \in \mathscr{E}$. Then $\{v \geqq 1\} = \bigcap_{n=1} \{f_n > 1 - 1/n\}$ shows that $v \geqq 1$ is a G_σ, and even one of a rather special kind. The proof of 1.2 is achieved in a similar way.

1.3 is a consequence of Theorem 1.7.2.

2 is proved as 1 except that we drop all countability assumptions and

use Theorem 1.7.3 plus the fact that all members of $\mathscr{E}^\tau$ are lower, and all members of $\mathscr{E}_\tau$ are upper semicontinuous, being suprema resp. infima of continuous functions.

3. REGULARITY OF σ-CONTENTS IN COMPACT SPACES

In this section we investigate aspects of regularity in compact Hausdorff spaces. Our main result may, in brief terms, be stated as follows: *On a compact Hausdorff space a positive Radon measure and an inner compact regular Borel σ-content are merely two aspects of the same matter.* The reader is now assumed to be fully familiar with the theory of these spaces. The theorems of Urysohn, Tychonov, and Weierstrass–Stone will be used.

Every compact Hausdorff space is completely regular, by an immediate application of Urysohn's theorem.

The crucial point which makes our theory nontrivial is the quite non-pathological existence of compact Hausdorff spaces in which not every lower semicontinuous function, and even not every indicator function of an open set, is a countable supremum of continuous functions (although it is always an uncountable supremum of such, by complete regularity).

It should be recalled that every measure $m: \mathscr{C}(\Omega, \mathbb{R}) \to \mathbb{R}$ is a τ-measure, hence a σ-measure (Theorem I.10.1).

3.1. Example. Let $\Omega = [0,1]^{[0,1]}$ = the set of all functions on the closed unit interval with values in the closed unit interval. Thus every $\omega \in \Omega$ is of the form $\omega = (\omega(t))_{0 \leqq t \leqq 1}$ with $0 \leqq \omega(t) \leqq 1$. Let $\mathscr{T}$ be the system of all unions of sets of the form

$$\bigcap_{k=1}^{n} \{\omega \mid \omega \in \Omega, a < \omega(t_k) < b\}$$

with $0 \leqq t_1, \ldots, t_n \leqq 1$, $a, b \in \mathbb{R}$. In the terminology of general topology, $\mathscr{T}$ is the product topology in the product Ω of copies of $[0,1]$ over the index set $[0,1]$. By Tychonov's theorem $(\Omega, \mathscr{T})$ is a compact space. For every $0 \leqq t \leqq 1$, let $\varphi_t: \Omega \to [0, 1]$ be given by $\varphi_t(\omega) = \omega(t)$ $(\omega = (\omega(t))_{0 \leqq t \leqq 1} \in \Omega)$, i.e., by evaluation at t. Clearly, the φ_t $(0 \leqq t \leqq 1)$ are in $\mathscr{C}(\Omega, \mathbb{R})$ and separate the points of Ω (this implies that $(\Omega, \mathscr{T})$ is Hausdorff), hence by the Weierstrass–Stone theorem every $f \in \mathscr{C}(\Omega, \mathbb{R})$ can be uniformly approximated by functions obtained from the φ_t $(0 \leqq t \leqq 1)$ by finitely many multiplications and linear combinations (each). It follows that for every $f \in \mathscr{C}(\Omega, \mathbb{R})$ there is a countable set $I \subseteq [0,1]$ such that $f(\omega) = f(\eta)$ $(\omega(t) = \eta(t)$ $(t \in I))$. For every $J \subseteq [0,1]$ and $\omega, \eta \in \Omega$, we write $\omega \sim_J \eta$ if $\omega(t) = \eta(t)$ $(t \in J)$. Clearly

$\sim_I$ is an equivalence relation. A set $M \subseteq \Omega$ is said to be of countable type if there is a countable $I \subseteq [0,1]$ such that M consists of full equivalence classes for $\sim_I$. It is easy to see that the sets of countable type form a σ-field $\mathscr{B}'$ in Ω, that every $f \in \mathscr{C}(\Omega, \mathbb{R})$ is $\mathscr{B}'$-measurable, and that, e.g., the set $\Omega_0 = [0, \frac{1}{2}]^{[0,1]} = \{\omega | 0 \leqq \omega(t) \leqq \frac{1}{2}\ (0 \leqq t \leqq 1)\}$ is compact but not in $\mathscr{B}'$. It follows that $\Omega \backslash \Omega_0$ is open, but its indicator function is not a countable supremum of continuous functions since otherwise $1_{\Omega \backslash \Omega_0}$ would be $\mathscr{B}'$-measurable, i.e., $\Omega \backslash \Omega_0 \in \mathscr{B}'$, hence $\Omega_0 \in \mathscr{B}'$. We conclude with the remark that this example is by no means pathological; the phenomena displayed here clearly appear whenever we form an uncountable product space of compact Hausdorff spaces each of which consists of at least two distinct points.

3.2. Definition. *Let $(\Omega, \mathscr{T})$ be a compact Hausdorff space. We denote by*

3.2.1. *$\mathscr{B}(\mathscr{C}(\Omega, \mathbb{R}))$ the σ-field generated $\mathscr{C}(\Omega, \mathbb{R})$. It is also called the* **Baire σ-field** *for $(\Omega, \mathscr{T})$; and the sets in it are also called the* **Baire subsets** *of Ω. A σ-content on $\mathscr{B}(\mathscr{C}(\Omega, \mathbb{R}))$ is also called a* **Baire σ-content** *in Ω.*

3.2.2. *$\mathscr{B}(\mathscr{T})$ the σ-field generated by $\mathscr{T}$. It is also called the* **Borel σ-field** *for $(\Omega, \mathscr{T})$; and the sets in it are also called the* **Borel subsets** *of Ω. A σ-content on $\mathscr{B}(\mathscr{T})$ is also called a* **Borel σ-content** *in Ω.*

3.3. Remarks

3.3.1. Clearly $\mathscr{B}(\mathscr{C}(\Omega, \mathbb{R})) \subseteq \mathscr{B}(\mathscr{T})$ for every compact Hausdorff space $(\Omega, \mathscr{T})$. The two σ-fields may differ, as is shown by Example 3.1. If, however, $(\Omega, \mathscr{T})$ is compact metric, then clearly $G \in \mathscr{T} \Rightarrow 2 \cdot 1_G = f_1 \vee f_2 \vee \cdots$ for suitable $f_1, f_2, \ldots \in \mathscr{C}(\Omega, \mathbb{R})$

$$\Rightarrow G = \{2 \cdot 1_G > 1\} = \{f_1 > 1\} \cup \{f_2 > 1\} \cup \cdots \in \mathscr{B}(\mathscr{C}(\Omega, \mathbb{R})),$$

hence $\mathscr{B}(\mathscr{T}) = \mathscr{B}(\mathscr{C}(\Omega, \mathbb{R}))$.

3.3.2. Since $\mathscr{C}(\Omega, \mathbb{R})$ contains all constants and Ω is an open set, we shall have to deal exclusively with σ-fields in this section.

3.4. Theorem. *Let $(\Omega, \mathscr{T})$ be a compact Hausdorff space. Then:*

3.4.1. *Let $\mathscr{G}^\sigma$, $\mathscr{F}_\sigma$, $\mathscr{G}^\tau$, $\mathscr{F}_\tau$ be as in Theorems 1.7 and 2.1 (with $\mathscr{E} = \mathscr{C}(\Omega, \mathbb{R})$). Then:*

3.4.1.1. *$\mathscr{G}^\sigma, \mathscr{F}_\sigma \subseteq \mathscr{B}(\mathscr{C}(\Omega, \mathbb{R}))$ and $\mathscr{G}^\sigma = \{G | G \subseteq \Omega$ open, G is a $K_\sigma\}$, $\mathscr{F}_\sigma = \{F | F \subseteq \Omega$ closed (= compact), F is a $G_\sigma\} = \{\Omega \backslash G | G \in G^\sigma\}$.*

3.4.1.2. *$\mathscr{G}^\tau, \mathscr{F}_\tau \subseteq \mathscr{B}(\mathscr{T})$ and $\mathscr{G}^\tau = \mathscr{T}$, $\mathscr{F}_\tau = \{F | F \subseteq \Omega$ closed$\} = \{\Omega \backslash G | G \in \mathscr{T}\}$.*

3.4.2. *Every Baire σ-content in Ω is $\mathscr{F}_\sigma$-$\mathscr{G}^\sigma$-regular, i.e., inner compact-G_σ-regular and outer open-F_σ-regular.*

3.4.3. *Every Baire σ-content in Ω has a unique inner $\mathscr{F}_\tau$-regular (i.e., inner compact-regular) extension to a Borel σ-content in Ω.*

Proof. 1.1.

$$G \in \mathscr{G}^\sigma \quad \Rightarrow \quad G = \bigcup_{n=1} \{f_n > 1\} \qquad \text{for suitable } f_1, f_2, \ldots \in \mathscr{C}(\Omega, \mathbb{R})$$

$$\Rightarrow \quad G = \bigcup_{n,k=1} \{f_n \geqq 1 + 1/k\}, \quad f_1, f_2, \ldots \in \mathscr{C}(\Omega, \mathbb{R})$$

shows that every $G \in \mathscr{G}^\sigma$ is in $\mathscr{B}(\mathscr{C}(\Omega, \mathbb{R}))$ and is an F_σ, hence (since every closed subset of Ω is compact) a K_σ. Let now G be an open K_σ in Ω and $G = K_1 \cup K_2 \cup \cdots$, $K_1, K_2, \ldots$ compact. By Urysohn's theorem there are $f_1, f_2, \ldots \in \mathscr{C}(\Omega, \mathbb{R})$ with $1_{K_n} \leqq f_n \leqq 1_G$, hence $2 \cdot 1_G = \sup 2f_n$, hence $G = \{2 \cdot 1_G > 1\} = \{\sup_n 2 \cdot f_n > 1\} \in \mathscr{G}^\sigma$. Passing to complements we show that every closed G_σ in Ω is of the form $F = \{\inf_n 2f_n \geqq 1\}$, $f_1, f_2, \ldots \in \mathscr{C}(\Omega, \mathbb{R})$, hence in $\mathscr{B}(\mathscr{C}(\Omega, \mathbb{R}))$ and in $\mathscr{F}_\sigma$. The proof that every $F \in \mathscr{F}_\sigma$ is a closed G_σ is now an easy exercise

1.2. $G \in \mathscr{G}^\tau \Leftrightarrow G = \bigcup_{g \in \mathscr{G}} \{g > 1\}$ for some $\varnothing \neq \mathscr{G} \subseteq \mathscr{C}(\Omega, \mathbb{R}) \Leftrightarrow G \in \mathscr{T}$ (the last $\Leftarrow$ follows by complete regularity), $F \in \mathscr{F}_\tau \Leftrightarrow F = \bigcap_{f \in \mathscr{F}} \{f \geqq 1\}$ for some $\varnothing \neq \mathscr{F} \subseteq \mathscr{C}(\Omega, \mathbb{R}) \Leftrightarrow \Omega \backslash F = \bigcup_{f \in \mathscr{F}} \{2f > 1\} \Leftrightarrow \Omega \backslash F \in \mathscr{T}$.

2. Let $m: \mathscr{B}(\mathscr{C}(\Omega, \mathbb{R})) \to \mathbb{R}_+$ be a σ-content. For every $f \in \mathscr{C}(\Omega, \mathbb{R})$ we get

$$f \nwarrow f_n = \sum_{k=1}^{2^{2n}-1} 1_{\{k/2^n < f\} \backslash \{(k+1)/2^n < f\}} + 2^n \cdot 1_{\{2^n < f\}}$$

uniformly as $n \to \infty$. Here $f_1, f_2, \ldots \in \mathscr{E}(\mathscr{B}(\mathscr{C}(\Omega, \mathbb{R})))$ have a common bound (e.g., any bound of f) hence $\mathscr{C}(\Omega, \mathbb{R}) \subseteq \mathscr{L}^1_\sigma(\Omega, \mathscr{B}(\mathscr{C}(\Omega, \mathbb{R})), m)$ follows. The restriction of the m-σ-integral to $\mathscr{C}(\Omega, \mathbb{R})$ is again denoted by m. Consider the σ-measure space $(\Omega, \mathscr{C}(\Omega, \mathbb{R}), m)$, apply the σ-extension process to it, and consider the σ-content space $(\Omega, \mathscr{B}_\sigma(m), m)$ (as we write it here) derived from it. By Theorem III.6.3 every set $\{f > 1\}$, $f \in \mathscr{C}(\Omega, \mathbb{R})$ is in $\mathscr{B}_\sigma(m)$, and in particular $\Omega \in \mathscr{B}_\sigma(m)$ (take $f \equiv 2$), hence $\mathscr{B}_\sigma(m)$ is a σ-field containing $\mathscr{B}(\mathscr{C}(\Omega, \mathbb{R}))$. By combination of Exercises III.6.5.1.6 and III.6.5.8 we see that $(\Omega, \mathscr{B}_\sigma(m), m)$ is an extension of $(\Omega, \mathscr{B}(\mathscr{C}(\Omega, \mathbb{R})), m)$. By combination of 1.1 and Theorem 2.1.1.3 the desired result follows.

3. We proceed as in the proof of 2 but observe that $(\Omega, \mathscr{C}(\Omega, \mathbb{R}), m)$ is a τ-measure space, apply the τ-extension process (Chapter III, Section 7) to it, and consider the σ-content space $(\Omega, \mathscr{B}_\tau(m), m)$ (as we write it here) τ-derived from it. Clearly $\mathscr{B}_\sigma(m) \subseteq \mathscr{B}_\tau(m)$, hence $\mathscr{B}_\tau(m)$ is a σ-field, and $(\Omega, \mathscr{B}_\tau(m), m)$ is an extension of $(\Omega, \mathscr{B}_\sigma(m), m)$. Moreover, by complete regularity, we find $1_G \in \mathscr{C}(\Omega, \mathbb{R})$ $(G \in \mathscr{T})$ and in particular $\mathscr{T} \subseteq \mathscr{B}_\tau(m)$, hence $\mathscr{B}(\mathscr{T}) \subseteq \mathscr{B}_\tau(m)$. This, together with 1.2 and Theorem 2.1.2.3 proves the existence of an $\mathscr{F}_\tau$-$\mathscr{G}^\tau$-regular extension of $m: \mathscr{B}(\mathscr{C}(\Omega, \mathbb{R})) \to \mathbb{R}_+$ to $\mathscr{B}(\mathscr{T})$. Since $\mathscr{F}_\tau \subseteq \mathscr{B}(\mathscr{T})$, we may apply Remark 1.2.1.2 and see that it is sufficient to prove the uniqueness of $m: \mathscr{F}_\tau \to \mathbb{R}_+$. This in turn follows now from the outer $\mathscr{G}^\tau$-regularity

prescribed for our extension. In fact, let $F \in \mathscr{F}_\tau$ be given. Then

$$m(F) = \inf_{F \subseteq G \in \mathscr{T}} m(G),$$

and it will now be sufficient to prove that $\inf_{F \subseteq G \in \mathscr{T}} m(G) = \inf_{F \subseteq G \in \mathscr{G}^\sigma} m(G)$ since the right-hand member here is uniquely determined by

$$m: \mathscr{B}(\mathscr{C}(\Omega, \mathbb{R})) \to \mathbb{R}_+ .$$

For this, it is sufficient to prove that for every $G \in \mathscr{T}$ with $F \subseteq G$, there is some $G' \in \mathscr{G}^\sigma$ with $F \subseteq G' \subseteq G$. But this is obvious: By Urysohn's theorem there is an $f \in \mathscr{C}(\Omega, \mathbb{R})$ such that $1_F \leqq f \leqq 1_G$ and we may take, e.g., $G' = \{2f > 1\}$.

3.5. Remark. There is a compact Hausdorff space $(\Omega, \mathscr{T})$ and a Borel σ-content in Ω that is not outer $\mathscr{G}^\tau$-regular. The known examples are pathological and involve ordinal numbers. We do not present them here (see, e.g., HALMOS [4, p. 231]).

3.6. Exercises

3.6.1. Let $(\Omega, \mathscr{T})$ be a compact Hausdorff space. Prove:

3.6.1.1. For every set $E \in \mathscr{B}(\mathscr{C}(\Omega, \mathbb{R}))$, there is a sequence of compact sets of the form $K_n = \{f_n \geqq 1\}$, $f_n \in \mathscr{C}(\Omega, \mathbb{R})$ $0 \leqq f_n \leqq 1$ $(n = 1, 2, \ldots)$ such that $E \in \mathscr{B}(\{K_1, K_2, \ldots\})$. (*Hint*: Use Proposition I.4.7.5.2.)

3.6.1.2. If $f_1, f_2, \ldots$ are as in 1.1, the mapping $\varphi: \Omega \to [0, 1]^{\mathbb{N}}$ defined by $\varphi(\omega) = (f_1(\omega), f_2(\omega), \ldots)$ is continuous for the product topology in $[0, 1]^{\mathbb{N}}$ which is metrizable.

3.6.1.3. Let $E, f_1, f_2, \ldots$ as in 1.1. Then $E = \varphi^{-1}E'$ for some Baire ($=$ Borel) set $E' \subseteq [0, 1]^{\mathbb{N}}$. If E is compact, then E' can be chosen compact.

3.6.1.3. Every compact Baire set $\subseteq \Omega$ is a G_σ.

3.6.2. Let $(\Omega, \mathscr{T})$, $(\Omega', \mathscr{T}')$ be compact Hausdorff spaces and $\varphi: \Omega \to \Omega'$ continuous (i.e., $\varphi^{-1}\mathscr{T}' \subseteq \mathscr{T}$). Prove $\varphi^{-1}\mathscr{B}(\mathscr{C}(\Omega', \mathbb{R})) \subseteq \mathscr{B}(\mathscr{C}(\Omega, \mathbb{R}))$, $\varphi^{-1}\mathscr{B}(\mathscr{T}') \subseteq \mathscr{B}(\mathscr{T})$. Show that φ transforms every

$$\mathscr{F}_\tau(\mathscr{C}(\Omega, \mathbb{R})) - \mathscr{G}^\tau(\mathscr{C}(\Omega, \mathbb{R}))\text{-regular}$$

Borel σ-content in Ω into an $\mathscr{F}_\tau(\mathscr{C}(\Omega', \mathbb{R}))$–$\mathscr{G}^\tau((\mathscr{C}(\Omega', \mathbb{R}))$-regular Borel σ-content in Ω'.

3.6.3. Let (Ω, T) be a compact Hausdorff space and m an inner $\mathscr{F}_\tau$-regular σ-content on $\mathscr{B}(\mathscr{T})$. Let $\mathscr{K} \subseteq \mathscr{B}(\mathscr{T})$ be a decreasingly filtered system of compact sets. Prove $m(\bigcap_{K \in \mathscr{K}} K) = \inf_{K \in \mathscr{K}} m(K)$ (see Exercise III.7.3.6).

4. REGULARITY OF σ-CONTENTS IN LOCALLY COMPACT SPACES

In this section we investigate aspects of regularity in locally compact Hausdorff spaces. The reader is now assumed to be fully familiar with the theory of these spaces. Again, one of our major results can be briefly stated: *On a locally compact Hausdorff space a positive Radon measure and an inner compact regular Borel σ-content are merely two aspects of the same matter.*

The most characteristic features of locally compact Hausdorff spaces become visible upon comparison with the theory of compact Hausdorff spaces. We mention the following:

4.1. Compact Hausdorff spaces are a special case of locally compact Hausdorff spaces. In fact they are not too special: Alexandroff's one-point compactification—the simplest of all compactification methods—embeds every locally compact Hausdorff space into a compact Hausdorff space, and the latter is metrizable etc. iff the former is.

4.2. In the theory of locally compact Hausdorff spaces, special emphasis is laid on compact subsets. In particular, for any locally compact Hausdorff space $(\Omega, \mathscr{T})$, the space $\mathscr{C}^{00}(\Omega, \mathbb{R})$ of all continuous real functions on Ω for which $\mathrm{supp}(f) = \{\omega \mid f(\omega) \neq 0\}$ has a compact closure is of particular importance. $\mathscr{C}^{00}(\Omega, \mathbb{R})$ is an elementary domain; it contains the constant 1 iff Ω is compact. Thus, if Ω is locally compact but not compact, we have to look for substitutes of constant functions.

4.3. A locally compact Hausdorff space can be viewed as a "compatible" family of compact Hausdorff spaces. The theory of locally compact Hausdorff spaces imitates and uses the compact special case on open subsets with a compact closure, and tries to paste the results together. One of the essential tools in this "pasting" is the fact that every compact set is contained in the interior of another compact set. This makes, e.g., the following construction possible: Let $\Omega_0 \subseteq \Omega_1 \subseteq \Omega_2 \subseteq \Omega$ be open and such that the closures $\overline{\Omega}_i$ are compact ($i = 0, 1, 2,$) and $\overline{\Omega}_i \subseteq \Omega_{i+1}$ ($i = 0, 1$). Apply Urysohn's theorem to the disjoint compact subsets $\overline{\Omega}_0$, $\overline{\Omega}_2 \backslash \Omega_1$ of the compact set $\overline{\Omega}_2$ in order to obtain a function $f \in \mathscr{C}(\Omega_2, \mathbb{R})$ with $1_{\overline{\Omega}_0} \leqq f \leqq 1_{\Omega_2 \backslash \overline{\Omega}_1}$ on $\overline{\Omega}_2$; define $g(\omega) = f(\omega)$ $(\omega \in \overline{\Omega}_2)$, $g(\omega) = 0$ $(\omega \in \Omega \backslash \overline{\Omega}_2)$; then

$$g \in \mathscr{C}^{00}(\Omega, \mathbb{R}), 1_{\overline{\Omega}_0} \leqq g \leqq 1_{\Omega \backslash \overline{\Omega}_1}$$

on Ω; if we deal with functions vanishing outside Ω_0, this g is an appropriate substitute for the constant 1.

4.4 Every locally compact Hausdorff space is completely regular (use 4.3).

4.5. Definition. *Let $(\Omega, \mathscr{T})$ be a locally compact Hausdorff space. We denote by*

4.5.1. $\mathscr{B}^{00}(\mathscr{C}^{00}(\Omega, \mathbb{R}))$ *the local σ-ring generated by all sets of the form $\{f > 1\}$, $f \in \mathscr{C}^0(\Omega, \mathbb{R})$, and by $\mathscr{B}^0(\mathscr{C}^{00}(\Omega, \mathbb{R}))$ resp. $\mathscr{B}(\mathscr{C}^{00}(\Omega, \mathbb{R}))$ the σ-ring resp. σ-field generated by $\mathscr{C}^{00}(\Omega, \mathbb{R})$. The sets in $\mathscr{B}(\mathscr{C}^{00}(\Omega, \mathbb{R}))$ are also called* **Baire subsets of Ω**. *A σ-content on $\mathscr{B}^{00}(\mathscr{C}^{00}(\Omega. \mathbb{R}))$ is also called a* **Baire σ-content** *in Ω.*

4.5.2. $\mathscr{B}_c^{00}(\mathscr{T})$ *the local σ-ring generated by $\{G \mid G \in \mathscr{T}, \overline{G} \text{ compact}\}$ and by $\mathscr{B}_c^0(\mathscr{T})$ resp. $\mathscr{B}_c(\mathscr{T})$ the σ-ring resp. σ-field generated by the same sets. The sets in $\mathscr{B}_c^{00}(\mathscr{T})$ are called the* **bounded Borel sets** *in Ω. The sets in $\mathscr{B}_c^0(\mathscr{T})$ are called the* **σ-bounded Borel sets** *in Ω. A σ-content on $\mathscr{B}_c^{00}(\mathscr{T})$ is called a* **Borel σ-content** *in Ω.*

4.5.3. $\mathscr{B}(\mathscr{T})$ *the σ-field generated by $\mathscr{T}$. The sets in $\mathscr{B}(\mathscr{T})$ (and in particular all those in $\mathscr{B}_c(\mathscr{T})$ ($\subseteq \mathscr{B}(\mathscr{T})$!) are called the* **Borel sets** *in Ω.*

4.6. Remarks

4.6.1. The aim of this definition is systematics and completeness. Only $\mathscr{B}^{00}(\mathscr{C}^{00}(\Omega, \mathbb{R}))$ and $\mathscr{B}_c^{00}(\mathscr{T})$ will play an essential role in the sequel. This selection appears to be most natural if we look at the next two remarks.

4.6.2. Discussion of the relations between $\mathscr{B}^{00}(\mathscr{C}^{00}(\Omega, \mathbb{R})), \ldots, \mathscr{B}(\mathscr{T})$: Clearly $\mathscr{B}^{00}(\mathscr{C}^{00}(\Omega, \mathbb{R})) \subseteq \mathscr{B}^0(\mathscr{C}^{00}(\Omega, \mathbb{R})) \subseteq \mathscr{B}(\mathscr{C}^{00}(\Omega, \mathbb{R}))$,

$$\mathscr{B}^{00}(\mathscr{C}^{00}(\Omega, \mathbb{R})) \subseteq \mathscr{B}_c^{00}(\mathscr{T}) \subseteq \mathscr{B}_c^0(\mathscr{T}) \subseteq \mathscr{B}_c(\mathscr{T}) \subseteq \mathscr{B}(\mathscr{T}).$$

Example 3.1 shows that $\mathscr{B}^{00}(\mathscr{C}^{00}(\Omega, \mathbb{R})) \neq \mathscr{B}_c^{00}(\mathscr{T})$ in general; in fact, Ω compact $\Rightarrow \mathscr{B}^{00}(\mathscr{C}^{00}(\Omega, \mathbb{R})) = \mathscr{B}(\mathscr{C}(\Omega, \mathbb{R}))$, $\mathscr{B}_c^{00}(\mathscr{T}) = \mathscr{B}(\mathscr{T})$, and the example applies. By Proposition I.4.7 every set in $\mathscr{B}^{00}(\mathscr{C}^{00}(\Omega, \mathbb{R}))$ is contained in a compact set of the form $\{f \geqq 1\}$, $f \in \mathscr{C}^{00}(\Omega, \mathbb{R})$, and every set in $\mathscr{B}_c^{00}(\mathscr{T})$ is contained in some compact set; every set in $\mathscr{B}^0(\mathscr{C}^{00}(\Omega, \mathbb{R}))$ is contained in a countable union of compact sets of the form $\{f \geqq 1\}$, $f \in \mathscr{C}^{00}(\Omega, \mathbb{R})$, and every set in $\mathscr{B}_c^0(\mathscr{T})$ is contained in some K_σ; for every set E in $\mathscr{B}_c(\mathscr{T})$ either E or $\Omega \backslash E$ is contained in some K_σ; every set $\subseteq \Omega$ is contained in Ω which is a (in general uncountable) union of compact sets, even such of the form $\{f \geqq 1\}$, $f \in \mathscr{C}^{00}(\Omega, \mathbb{R})$. If $\Omega_0 \in \mathscr{B}^{00}(\mathscr{C}^{00}(\Omega, \mathbb{R}))$, $E \in \mathscr{B}(\mathscr{C}^{00}(\Omega, \mathbb{R}))$, then $\Omega_0 \cap E \in \mathscr{B}^{00}(\mathscr{C}^{00}(\Omega, \mathbb{R}))$. If $\Omega_0 \in \mathscr{B}_c^{00}(\mathscr{T})$, $E \in \mathscr{B}(\mathscr{T})$, then $\Omega_0 \cap E \in \mathscr{B}_c^{00}(\mathscr{T})$.

4.6.3. Every Baire σ-content m in Ω has a unique σ-additive extension to $\mathscr{B}^0(\mathscr{C}^{00}(\Omega, \mathbb{R}))$. If $\Sigma = \{\{f \geqq 1\} \mid f \in \mathscr{C}^{00}(\Omega, \mathbb{R}), \{f \geqq 1\} \neq \varnothing\}$, then m defines a family $(m_{\Omega_0})_{\Omega_0 \in \Sigma}$ of σ-contents on $\mathscr{B}(\mathscr{C}^{00}(\Omega, \mathbb{R}))$ through

$$m_{\Omega_0}(E) = m(\Omega_0 \cap E) \qquad (\Omega_0 \in \Sigma, \quad E \in \mathscr{B}(\mathscr{C}^{00}(\Omega, \mathbb{R})).$$

This family is increasingly filtered since Σ is increasingly filtered

$(\{f \geqq 1\} \cup \{g \geqq 1\} = \{f \vee g \geqq 1\})$ and $\Omega_0, \Omega_1 \in \Sigma$, $\Omega_0 \subseteq \Omega_1 \Rightarrow m_{\Omega_0} \leqq m_{\Omega_1}$. Thus Theorem II.4.1 applies and

$$\overline{m}(E) = \sup_{\Omega_0 \in \Sigma} m_{\Omega_0}(E) \qquad (E \in \mathscr{B}(\mathscr{C}^{00}(\Omega, \mathbb{R}))$$

defines a σ-additive $\overline{m}: \mathscr{B}(\mathscr{C}^{00}(\Omega, \mathbb{R})) \to \mathbb{R}_+$, and it is easy to see (exercise) that this extends the Baire σ-content m with which we started. It is actually a minorant of every other such extension, i.e., it is the unique minimal extension. Its restriction to $\mathscr{B}^0(\mathscr{C}^{00}(\Omega, \mathbb{R}))$ is uniquely determined by m and σ-additivity alone. A similar argument shows that every Borel σ-content m' in Ω has a unique minimal extension $\overline{m}'$ to $\mathscr{B}(\mathscr{T})$ and that $\overline{m}'$ extends $\overline{m}$ if m' extends m (exercise). To sum up, routine extensions allow us to deal with σ-additive functions on the σ-rings and σ-fields introduced in Definition 4.5 as soon as we have studied the Baire and Borel σ-contents on their domains of definition, the local σ-rings introduced in Definition 4.5. We may thus restrict our attention essentially to the latter throughout the rest of this section.

4.7. Theorem. *Let $(\Omega, \mathscr{T})$ be a locally compact Hausdorff space. Then:*

4.7.1. *Let $\mathscr{G}^\sigma$, $\mathscr{F}_\sigma$, $\mathscr{G}^\tau$, $\mathscr{F}_\tau$ be as in Theorems* 1.7 *and* 2.1 *(with $\mathscr{E} = \mathscr{C}^{00}(\Omega, \mathbb{R})$). Put*

$$\mathscr{G}^\sigma_c = \{G \mid G \in \mathscr{G}^\sigma, \overline{G} \text{ compact}\}, \qquad \mathscr{G}^\tau_c = \{G \mid G \in \mathscr{G}^\tau, \overline{G} \text{ compact}\}.$$

Then:

4.7.1.1. $\mathscr{G}^\sigma_c, \mathscr{F}_\sigma \subseteq \mathscr{B}^{00}(\mathscr{C}^{00}(\Omega, \mathbb{R}))$ and

$$\mathscr{G}^\sigma_c = \{G \mid G \subseteq \Omega \text{ open}, \overline{G} \text{ compact}, G \text{ is a } K_\sigma\},$$

$$\mathscr{F}_\sigma = \{F \mid F \subseteq \Omega \text{ compact}, F \text{ is a } G_\sigma\}.$$

4.7.1.2. $\mathscr{G}^\tau_c, \mathscr{F}_\tau \subseteq \mathscr{B}^{00}_c(\mathscr{T})$ *and* $\mathscr{G}^\tau_c = \{G \mid G \subseteq \Omega \text{ open}, \overline{G} \text{ compact}\}$, $\mathscr{F}_\tau = \{F \mid F \subseteq \Omega \text{ compact}\}$.

4.7.2. *Every Baire σ-content in Ω is $\mathscr{F}_\sigma$-$\mathscr{G}^\sigma_c$-regular, i.e., inner compact-G_σ-regular and outer open-F_σ-conditionally-compact-regular.*

4.7.3. *Every Baire σ-content in Ω has a unique $\mathscr{F}_\tau$-$\mathscr{G}^\tau_c$-regular (i.e., inner compact-regular and outer open-conditionally-compact-regular) extension to a Borel σ-content in Ω.*

Proof. 1.1. If $G \in \mathscr{G}^\sigma_c$, then $G = \{f_1 > 1\} \cup \{f_2 > 1\} \cup \cdots$ for suitable $f_1, f_2, \ldots \in \mathscr{C}^{00}(\Omega, \mathbb{R})$, $\overline{G}$ compact. Use 4.3 and construct some $f \in \mathscr{C}^{00}(\Omega, \mathbb{R})$ with $0 \leqq f \leqq 1$, $f(\omega) = 1$ $(\omega \in \overline{G})$. Now clearly

$$G = \{f_1 > 1\} \cup \{f_2 > 1\} \cup \cdots = \{ff_1 > 1\} \cup \{ff_2 > 1\} \cup \cdots \subseteq \{f = 1\}.$$

This shows $G \in \mathscr{B}^{00}(\mathscr{C}^{00}(\Omega, \mathbb{R}))$, and $G = \bigcup_{n,k=1}^{\infty} \{ff_n \geqq 1 + 1/k\}$ shows that G is a K_σ. Let conversely G be a K_σ, say $G = K_1 \cup K_2 \cup \cdots$, $K_1, K_2, \ldots \subseteq \Omega$ compact, and $\overline{G}$ compact. Find $\Omega_0 \supseteq \overline{G}$ open, $\overline{\Omega}_0$ compact. Use 4.3 to construct $f_1, f_2, \ldots \in \mathscr{C}^{00}(\Omega, \mathbb{R})$ such that $1_{K_n} \leqq f_n \leqq 1_{\overline{G}}$. Now clearly $G = \bigcup_n \{2 \cdot 1_{K_n} > 1\} = \bigcup_n \{2 \cdot f_n > 1\}$, hence $G \in \mathscr{G}_c^\sigma$. If $F \in \mathscr{F}_\sigma$, then $F = \{f_1 \geqq 1\} \cap \{f_2 \geqq 1\} \cap \cdots$ for suitable $f_1, f_2, \ldots \in \mathscr{C}^{00}(\Omega, \mathbb{R})$. In particular, $F \subseteq \{f_1 \geqq 1\}$ which is compact, hence F, being closed, is compact. $F = \bigcap_{n,k=1} \{f_n > 1 - 1/k\}$ shows that F is a G_σ. Let now $F \subseteq \Omega$ be any compact G_σ and, say, $F = G_1 \cap G_2 \cap \cdots$ with open $G_1, G_2, \ldots$. Use 4.3 and choose an open $\Omega_0 \supseteq F$ with $\overline{\Omega}_0$ compact. Now clearly $F = (G_1 \cap \Omega_0) \cap (G_2 \cap \Omega_0) \cap \cdots$, $G_1 \cap \Omega_0$, $G_2 \cap \Omega_0$ open, hence we may assume $G_1, G_2, \ldots \subseteq \Omega_0$. Use 4.3 again, choose an open $\Omega_1 \supseteq \overline{\Omega}_0$ such that $\overline{\Omega}_1$ is compact and find $f_1, f_2, \ldots \in \mathscr{C}^{00}(\Omega, \mathbb{R})$ such that $1_F \leqq f_n \leqq 1_{G_n}$ $(n = 1, 2, \ldots)$. Now clearly $F = \{1_F \geqq 1\} = \{f_1 \geqq 1\} \cap \{f_2 \geqq 1\} \cap \cdots$, proving $F \in \mathscr{F}_\sigma$.

1.2. $\mathscr{G}_c^\tau, \mathscr{F}_\tau \subseteq \mathscr{B}_c^{00}(\mathscr{T})$ since the members of $\mathscr{G}_c^\tau$ are open, with compact closures, and the members of $\mathscr{F}_\tau$ are compact, being intersections of compact sets of the form $\{f \geqq 1\}$, $f \in \mathscr{C}^{00}(\Omega, \mathbb{R})$. Let conversely $G \subseteq \Omega$ be open, $\overline{G}$ compact. By complete regularity, 1_G is a (in general uncountable) supremum of functions $f \in \mathscr{C}^{00}(\Omega, \mathbb{R})$, and we may clearly assume $0 \leqq f \leqq 1$ for these. But now $G = \bigcup \{2 \cdot f > 1\}$ (for these f) shows $G \in \mathscr{G}_c^\tau$. Let $F \subseteq \Omega$ be compact. Use 4.3 in order to find, for every $\omega \in \Omega \backslash F$, some $f_\omega \in \mathscr{C}^{00}(\Omega, \mathbb{R})$ such that $1_F \leqq f \leqq 1$, $f_\omega(\omega) = 0$. Now clearly, $F = \bigcap_{\omega \in \Omega \backslash F} \{2 \cdot f_\omega \geqq 1\}$, proving $F \in \mathscr{F}_\tau$.

2. Repeating essentially the arguments for the proof of Theorem 3.4.2 (with $\mathscr{C}^{00}(\Omega, \mathbb{R})$ in place of $\mathscr{C}(\Omega, \mathbb{R})$ and $\mathscr{B}_\sigma^{00}(m)$ in place of $\mathscr{B}_\sigma(m)$), we find that every Baire σ-content m in is $\mathscr{F}_\sigma$-$\mathscr{G}^\sigma$-regular. But every set in $\mathscr{B}^{00}(\mathscr{C}^{00}(\Omega, \mathbb{R}))$ is contained in an open set of the form $\{f > 1\}$, $f \in \mathscr{C}^{00}(\Omega, \mathbb{R})$ which thus has a compact closure. We need therefore, in the definition of outer regularity, only intersect sets from $\mathscr{G}^\sigma$ with sets of the above form in order to ensure outer $\mathscr{G}_c^\sigma$-regularity.

3. Repeating essentially the arguments for the proof of Theorem 3.4.3 (with $\mathscr{C}^{00}(\Omega, \mathbb{R})$ instead of $\mathscr{C}(\Omega, \mathbb{R})$ and $\mathscr{B}_\tau^{00}(m)$ instead of $\mathscr{B}_\tau(m)$), we find that every Baire σ-content m in Ω has an extension m to $\mathscr{B}_c^{00}(\mathscr{T})$ which is inner $\mathscr{F}_\tau$- and outer $\mathscr{G}^\tau$-regular. Intersecting with open sets having compact closures, we find even outer $\mathscr{G}_c^\tau$-regularity. The proof of uniqueness now amounts, in view of $\mathscr{F}_\tau \subseteq \mathscr{B}_c^{00}(\mathscr{T})$ and Remark 1.2.1.2 to a proof that $\overline{m}\colon \mathscr{F}_\tau \to \mathbb{R}_+$ is uniquely determined. The argument used in the proof of 1.2 now shows that every $F \in \mathscr{F}_\tau$ is the intersection of all open sets $G \supseteq F$ in $\mathscr{G}_c^\tau$ and that for every such G there is some G' with $F \subseteq G' \subseteq G$ such that $G' = \{f > 1\}$ for a suitable $f \in \mathscr{C}^{00}(\Omega, \mathbb{R})$, hence, by outer $\mathscr{G}_c^\tau$-regularity,

$$\overline{m}(F) = \inf_{F \subseteq G \in \mathscr{G}_c^\tau} \overline{m}(G) = \inf_{F \subseteq G \subseteq \mathscr{G}_c^\tau} m(G),$$

proving that $\overline{m}$ is uniquely determined by m and its extension and regularity properties.

4.8. Exercises. Let $(\Omega, \mathscr{T})$ be a locally compact Hausdorff space.

4.8.1. Let m be a Baire σ-content in Ω, and let m also denote its unique σ-additive extension to the σ-ring $\mathscr{B}^0(\mathscr{C}^{00}(\Omega, \mathbb{R}))$. Prove that for every $E \in \mathscr{B}^0(\mathscr{C}^{00}(\Omega, \mathbb{R}))$

$$m(E) = \sup_{E \supseteq F \in \mathscr{F}_\sigma} m(F), \qquad m(E) = \inf_{E \subseteq G \in \mathscr{G}^\sigma} m(G).$$

4.8.2. Prove the analogue of 4.8.1 for $\mathscr{F}_\tau$-$\mathscr{G}_c^\tau$-regular Borel σ-contents.

4.8.3. Prove that for every Borel σ-content inner $\mathscr{F}_\tau$-regularity implies outer $\mathscr{G}_c^\tau$-regularity.

5. REGULARITY IN POLISH SPACES

In this section we investigate aspects of regularity in Polish spaces.

We recall that a topological space $(\Omega, \mathscr{T})$ is called *Polish* if there is a countable dense subset of Ω and there exists a metric in Ω that is complete (i.e., satisfies the Cauchy convergence criterion) and describes the topology $\mathscr{T}$ in the usual way. Let us list a few examples of Polish spaces. Every compact metrizable space is Polish. $\mathbb{R}$ with its usual topology is Polish (take the usual metric in $\mathbb{R}$; do not take the metric $|x, y| = |\arctan x - \arctan y|$, since it is not complete, although it describes the topology). For every compact interval $\varnothing \neq [a, b] \subseteq \mathbb{R}$, the space $\mathscr{C}([a, b], \mathbb{R})$ with the usual topology of uniform approximation is Polish (the polynomials with rational coefficients form, when restricted to $[a, b]$, a countable dense set; the metric $|f, g| = \sup_{a \leq x \leq b} |f(x) - g(x)|$ is complete and describes the topology); similarly, if $(X, \mathscr{S})$ is a compact metric space, $\mathscr{C}(X, \mathbb{R})$ is Polish (exercise: find a feasible substitute for the rational polynomials). $\{0, 1\}^{[0, 1]}$ with the usual product topology has no countable dense subset, hence is not Polish, although compact. The same holds for the space $[0, 1]^{[0, 1]}$ of all real functions on the unit interval with values in the unit interval. A countable product space of Polish spaces is Polish again. In particular, $\mathbb{N} = \{1, 2, \ldots\}$ with its discrete topology is Polish, and the space $\mathbb{N}^{\mathbb{N}}$ of all sequences of natural numbers, endowed with the product topology, is Polish.

Moreover, we want to make the reader acquainted with fundamental structural properties of Polish spaces.

5.1. *Let $(\Omega, \mathscr{T})$ be a Polish space. Then there is a topology $\mathscr{T}' \subseteq \mathscr{T}$ such that $(\Omega, \mathscr{T})$ is compact metrizable and $\mathscr{B}(\mathscr{T}') = \mathscr{B}(\mathscr{T})$.*

This means that we may assume every Polish space to be compact metric, as far as the Borel sets are concerned. The proof is intricate and will not be given in this book; see HOFFMAN-JØRGENSEN [1].

5.2. *In a Polish space* $(\Omega, \mathscr{T})$, *every open set can be written in the form* $\{g > 1\}$, $g \in \mathscr{C}^{\mathrm{b}}(\Omega, \mathbb{R}) = \{f \mid f \in \mathscr{C}(\Omega, \mathbb{R}), f \text{ bounded}\}$.

In fact, let $|\cdot, \cdot|$ be any metric defining $\mathscr{T}$, then for every $G \in \mathscr{T}$ $g(\omega) = [1 + \inf_{\eta \notin G} |\omega, \eta|] \wedge 2$ defines a $g \in \mathscr{C}^{\mathrm{b}}(\Omega, \mathbb{R})$ with $G = \{g > 1\}$. Clearly this argument works in every metric space. In particular, $\mathscr{B}(\mathscr{T})$ is the smallest σ-field containing all sets of the form $\{g > 1\}$, $g \in \mathscr{C}^{\mathrm{b}}(\Omega, \mathbb{R})$, hence the dichotomy exhibited by Example 3.1 for compact Hausdorff spaces does not appear in the case of Polish spaces.

5.3. Theorem. *Let* $(\Omega, \mathscr{T})$ *be a Polish space*, $\mathscr{F} = \{F \mid \Omega \backslash F \in \mathscr{T}\}$ *the system of all closed sets, and* $\mathscr{K}$ *the system of all compact sets in* Ω. *Then* $\mathscr{T}$, $\mathscr{F}$, $\mathscr{K} \subseteq \mathscr{B}(\mathscr{T})$ *and then every* σ*-content* m *on* $\mathscr{B}(\mathscr{T})$ *is* $\mathscr{K}$-$\mathscr{T}$*-regular.*

Proof. The conclusions $\mathscr{T}$, $\mathscr{F}$, $\mathscr{K} \subseteq \mathscr{B}(\mathscr{T})$ are trivial. By Exercise III.6.5.6 there is, for every $\varepsilon > 0$, a compact $K_\varepsilon \subseteq \Omega$ such that $m(\Omega \backslash K_\varepsilon) < \varepsilon/2$. If we can prove that m is inner $\mathscr{F}$-regular, inner $\mathscr{K}$-regularity follows: for any $E \in \mathscr{B}(\mathscr{T})$ and any $\varepsilon > 0$, find $E \supseteq F \in \mathscr{F}$ such that $m(E \backslash F) < \varepsilon/2$, then $E \supseteq F \cap K_\varepsilon \in \mathscr{K}$ and $m(E \backslash (F \cap K_\varepsilon)) < \varepsilon$. Moreover, inner $\mathscr{F}$-regularity implies outer $\mathscr{T}$-regularity (pass to complements). Now it is an easy exercise to prove that $\mathscr{C}^{\mathrm{b}}(\Omega, \mathbb{R}) \subseteq \mathscr{L}^1_\sigma(\Omega, \mathscr{B}(\mathscr{T}), m)$ and $m\colon \mathscr{C}^{\mathrm{b}}(\Omega, \mathbb{R}) \to \mathbb{R}$ (obtained by restriction of the m-σ-integral) is a σ-measure (use the monotone convergence theorem III.5.3). Define $\mathscr{F}_\sigma$ as in Theorem 2.1 (for $\mathscr{E} = \mathscr{C}^{\mathrm{b}}(\Omega, \mathbb{R})$). Then $\mathscr{F}_\sigma \subseteq \mathscr{F}$ and m is inner $\mathscr{F}$-regular, by Theorem 2.1.

5.4. Theorem. *Let* $(\Omega, \mathscr{T})$ *be a Polish space. Then every* σ*-measure* m *on* $\mathscr{C}^{\mathrm{b}}(\Omega, \mathbb{R})$ *is* **tight** *and thus a* τ*-measure.*

Proof. We may and shall assume $m(1) = 1$. Use Exercise III.6.5.6 in order to find, for any $\varepsilon > 0$, some compact $K_\varepsilon \subseteq \Omega$ such that $m(\Omega \backslash K_\varepsilon) < \varepsilon/2$. Let $f \in \mathscr{C}^{\mathrm{b}}(\Omega, \mathbb{R})$, $|f(\omega)| < \varepsilon/2$ $(\omega \in K_\varepsilon)$, $|f| \leqq 1$. Then

$$|m(f)| \leqq \int |f| \, dm = \int |f| 1_{K_\varepsilon} \, dm + \int |f| 1_{\Omega \backslash K_\varepsilon} \, dm$$

$$\leqq \tfrac{1}{2}\varepsilon m(1_{K_\varepsilon}) + \tfrac{1}{2}\varepsilon \leqq \tfrac{1}{2}\varepsilon m(1) + \tfrac{1}{2}\varepsilon = \varepsilon.$$

Tightness implies τ-continuity by Theorem I.10.4.

5.5. Exercises

5.5.1. Deduce Theorem 5.3 from the structural property 5.1 of Polish spaces and Theorem 3.4.

5.5.2. Let F be a closed set in a metric space Ω. Construct $f \in \mathscr{C}^b(\Omega, \mathbb{R})$ such that $F = \{f \geqq 1\}$.

5.5.3. Prove that a subset $\mathscr{M}$ of the Polish space $\mathscr{C}([0, 1], \mathbb{R})$ (with the topology of uniform convergence) has a compact closure iff the so-called *Arzela–Ascoli criterion* holds: There is some $x_0 \in [0, 1]$ such that $\{f(x_0) | f \in \mathscr{M}\}$ is a bounded subset of $\mathbb{R}$, and for every $\varepsilon > 0$ there is a $\delta > 0$ such that $f \in M$, $x, y \in [0, 1]$,

$$|x - y| < \delta \quad \Rightarrow \quad |f(x) - f(y)| < \varepsilon.$$

CHAPTER VI

CONTENTS AND MEASURES IN PRODUCT SPACES

In this chapter we study contents and measures in *product spaces* with finitely or infinitely many components. The development of the theory in the latter case was strongly stimulated by probability theory, which is one of the most important fields of application for its results.

We begin by setting up the measurability machinery for product spaces in Section 1. In Section 2 we deal with products of two spaces. We introduce the product of two σ-contents as well as the product of a σ-content and a kernel together with the corresponding Fubini theorems. The concept of a local σ-ring turns out to be very natural in this context; it saves us the traditional restriction to the σ-finite case. The problem of representing any σ-content in a two-factor product space as a product of a σ-content on the first factor and a kernel from the first factor space to the second is not solvable in the positive sense unless some regularity assumptions are made; this is shown by an example (2.6), the positive solution being deferred to the systematic Chapter XVI. In Section 3 we treat the case of finitely many factors; the concept of a local σ-ring is still natural here, but it ceases to be so as soon as we pass to an infinity of factors. This we do in Section 4, and from then on we are under the rule of σ-fields and σ-probability contents on them; Section 4 treats the case of the inde x set $\{1, 2, \ldots\}$, yielding C. Ionescu-Tulcea's general existence theorem. In Section 5 we pass over to arbitrarily, possibly uncountably many factors. The existence and uniqueness of product σ-probabilities is achieved without any additional assumptions, but for *Kolmogorov's theorem on projective* families we have to make some regularity assumptions which we present from the easy stage of compact factors over the common stage of Polish factors to the most abstract case of factors with σ-compact systems approximating from below. Section 6 contains, for future probabilists, the rudiments of the theory of *independence*: and Section 7 presents, so-to-speak as a technical reservoir for more advanced probability theory, the basic results on the *path structure of Markov semigroups* in locally compact spaces.

1. SET SYSTEMS IN PRODUCT SPACES

In this section we consider families of mappings with a common domain Ω of definition. We show how to pull certain types of set systems back from the image spaces to Ω. We shall apply this to the family of all component mappings of a product space and thereby define product topologies, product local σ-rings, etc. Throughout this section, topologies and σ-fields play the main roles, whereas σ-rings and local σ-rings are of interest in somewhat specialized situations only.

1.1. Definition

1.1.1. *Let $I \neq \emptyset$, $\Omega \neq \emptyset$, $\Omega_\iota \neq \emptyset$ $(\iota \in I)$. For every $\iota \in I$, let $\varphi_\iota : \Omega \to \Omega_\iota$ be a mapping.*

1.1.1.1. *Let $\mathscr{T}_\iota$ be a topology in Ω_ι $(\iota \in I)$. The smallest (= coarsest) topology in Ω that makes φ_ι continuous for every $\iota \in I$ is called the* **topology generated by the family** *$(\varphi_\iota)_{\iota \in I}$ (or the set $\{\varphi_\iota | \iota \in I\}$) of mappings (and, of course, the topologies $\mathscr{T}_\iota$ $(\iota \in I)$). It is also denoted by*

$$\mathscr{T}\left(\bigcup_{\iota \in I} \varphi_\iota^{-1}\mathscr{T}_\iota\right) \qquad or \qquad \mathscr{T}((\varphi_\iota)_{\iota \in I}) \qquad or \qquad \mathscr{T}(\{\varphi_\iota | \iota \in I\}).$$

1.1.1.2. *Let $\mathscr{B}_\iota$ be a σ-field in Ω_ι $(\iota \in I)$. The smallest σ-field $\mathscr{B}$ in Ω that makes φ_ι $\mathscr{B}$-$\mathscr{B}_\iota$-measurable for every $\iota \in I$ is called the* **σ-field generated by** *$(\varphi_\iota)_{\iota \in I}$ (or $\{\varphi_\iota | \iota \in I\}$, and, of course, the σ-fields $\mathscr{B}_\iota$ $(\iota \in I)$). It is also denoted by*

$$\mathscr{B}\left(\bigcup_{\iota \in I} \varphi_\iota^{-1}\mathscr{B}_\iota\right) \qquad or \qquad \mathscr{B}((\varphi_\iota)_{\iota \in I}) \qquad or \qquad \mathscr{B}(\{\varphi_\iota | \iota \in I\}).$$

1.2. Remarks

1.2.1. Notice that in the above definition we deal only with topologies and σ-fields. We could easily set up analogous definitions for rings, σ-rings, local σ-rings, etc., but find no point in that, for our purposes.

1.2.2. If it is clear from the context which topologies or σ-fields are to be taken in the spaces Ω_ι (viz. if $\Omega_\iota = \mathbb{R}$ or $|\Omega_\iota| < \infty$), we often refrain from mentioning them at all.

1.2.3. Topologies and σ-fields are finite intersection stable. Hence $\mathscr{T}(\bigcup_{\iota \in I} \varphi_\iota^{-1}\mathscr{T}_\iota)$ contains all sets of the form $\bigcap_{k=1}^{n} \varphi_{\iota_k}^{-1} U_k$, $n \geqq 1$, $\iota_1, \ldots, \iota_n \in I$ pairwise distinct, $U_k \in \mathscr{T}_{\iota_k}$ $(k = 1, \ldots, n)$; it is in fact the system of all unions of such sets (exercise). Similarly, $\mathscr{B}(\bigcup_{\iota \in I} \varphi_\iota^{-1}\mathscr{B}_\iota)$ contains all sets of the same form, with $U_k \in \mathscr{B}_{\iota_k}$ $(k = 1, \ldots, n)$.

1.2.4. Notice that the above definition includes the notion of a topology $\mathscr{T}(\mathscr{E})$ or σ-field $\mathscr{B}(\mathscr{E})$ generated by a set $\mathscr{E}$ (and not a family) of mappings since we may trivially pass from that set to the family obtained by indexing it by (the identity mapping of the set onto) itself.

1.3. Exercises

1.3.1. Let $\Omega = \mathbb{R}^2$ and $\varphi\colon \Omega \to \mathbb{R}$ be given by $\varphi(x, y) = x^2 + y^2$. Prove that the σ-ring generated by φ consists of all Borel sets consisting of circle lines with center (0, 0).

1.3.2. Let $\Omega = \mathbb{R}^n$ and $\varphi_k\colon \omega = (\omega_1, \ldots, \omega_n) \to \omega_k$ be the kth component mapping. Show that the σ-field of all Borel sets in $\mathbb{R}^n$ is the σ-field generated by $(\varphi_k)_{k=1,\ldots,n}$.

1.3.3. Let $\Omega = A \times A \times \cdots$ be the one-sided Bernoulli space over the finite alphabet A. Let $\varphi_k\colon \omega = (\omega_0 \omega_1 \cdots) \to \omega_k$ be the kth component mapping. Let the ring $\mathscr{C} \subseteq \mathscr{P}(\Omega)$ be given as in Example I.1.12. Prove that $\mathscr{B}(\mathscr{C})$ equals the σ-ring generated by $(\varphi_k)_{k=0,1,\ldots}$.

1.3.4. Let $\Omega \neq \varnothing$ and $\varphi_1, \varphi_2, \ldots$ be mappings from Ω to $\{0, 1\}$. Prove that the σ-field generated by $(\varphi_k)_{k=1,2,\ldots}$ equals the σ-field generated by $\varphi = \sum_{k=1}^{\infty} (2/3^k)\varphi_k$.

1.3.5. Generalize 1.3.4 to:

1.3.5.1. $[0, 1]$ instead of $\{0, 1\}$;

1.3.5.2. $\mathbb{R}$ instead of $\{0, 1\}$.

1.3.6. Let $I \neq \varnothing$ and $\mathscr{I} \subseteq \mathscr{P}(I)$ be increasingly σ-filtered (i.e., $J_1, J_2, \ldots \in \mathscr{I} \Rightarrow J_1 \cup J_2 \cup \cdots \subseteq J$ for some $J \in \mathscr{I}$) with union I. Let $\varphi_\iota\colon \Omega \to \Omega_\iota$ and $\mathscr{B}_\iota$ $(\iota \in I)$ be as in Definition 1.1.2. For every $J \in \mathscr{I}$, let $\mathscr{B}_J$ be the σ-field generated by $(\varphi_\iota)_{\iota \in J}$ in Ω. Prove that $\mathscr{B}_I = \bigcup_{J \in I} \mathscr{B}_J$. (*Hint*: Use Proposition I.4.7.5.)

1.3.7. Let $\Omega \neq \varnothing$ and $\varnothing \neq \mathscr{E} \subseteq \mathbb{R}^\Omega$. We investigate $\mathscr{B}(\mathscr{E})$. Prove:

1.3.7.1. $\mathscr{B}(\mathscr{E}) = \mathscr{B}(\{f^{-1}E \mid f \in \mathscr{E}, E \subseteq \mathbb{R} \text{ Borel}\})$.

1.3.7.2. If $\mathscr{E}$ is a vector lattice $\subseteq \mathbb{R}^\Omega$, then $\mathscr{B}(\mathscr{E}) = \mathscr{B}(\{\{f \geqq \alpha\} \mid f \in \mathscr{E}, \alpha \geqq 0\})$.

1.3.7.3. If $\mathscr{E} \subseteq \mathscr{M} \subseteq \text{mble}(\Omega, \mathscr{B}(\mathscr{E}), \mathbb{R})$, $\mathscr{M}$ is an elementary domain, and $\mathscr{M}$ is stable under pointwise convergence of increasing and decreasing sequences of functions (i.e., is what should be called a monotone class of functions), then $\mathscr{M} = \text{mble}(\Omega, \mathscr{B}(\mathscr{E}), \mathbb{R})$. (*Hint*: Prove $f \in \mathscr{M} \Rightarrow 1_{\{f>1\}} \in \mathscr{M}$ by "blowing up the hat," i.e., taking $1_{\{f>1\}} = \lim_n [n(f_+ - (f_+ \wedge 1))] \wedge 1$ into account); make use of monotone classes, i.e., of Exercise I.5.5.1.)

1.3.7.4. If $\mathscr{E} \subseteq \mathscr{M} \subseteq \text{mble}(\Omega, \mathscr{B}(\mathscr{E}), \mathbb{R})$ and $\mathscr{M}$ is an elementary domain such that $\{f \mid f \in \mathscr{M}, 0 \leqq f \leqq 1\}$ is stable under pointwise convergence of increasing and decreasing sequences of functions (i.e., is what should be called a *monotone class* of functions), then $E \in \mathscr{B}(\mathscr{E}) \Leftrightarrow 1_E \in \mathscr{M}$.

1.4. Definition. *Let* $I \neq \varnothing, \Omega_\iota \neq \varnothing \ (\iota \in I)$,

$$\Omega = \prod_{\iota \in I} \Omega_\iota = \{(\omega_\iota)_{\iota \in I} \mid \omega_\iota \in \Omega_\iota \ (\iota \in I)\}$$

and $\varphi_\iota\colon (\omega_k)_{k \in I} \to \omega_\iota$ *be the* ι*th component mapping of* Ω *onto* Ω_ι $(\iota \in I)$.

1.4.1. *Let* $\mathscr{T}_\iota$ *be a topology in* Ω_ι $(\iota \in I)$. *Then the topology* $\mathscr{T} = \mathscr{T}(\bigcup_{\iota \in I} \varphi_\iota^{-1}\mathscr{T}_\iota)$ *is called the* **product topology** *or the topology of componentwise convergence (for the component topologies* $\mathscr{T}_\iota$ $(\iota \in I)$ *in* Ω. *It is also denoted by* $\mathscr{T} = \prod_{\iota \in I} \mathscr{T}_\iota$, *and the topological space* $(\Omega, \mathscr{T})$ *is also called the* **topological product space** *of the component (topological) spaces* $(\Omega_\iota, \mathscr{T}_\iota)$ $(\iota \in I)$ *and is also denoted by* $\prod_{\iota \in I} (\Omega_\iota, \mathscr{T}_\iota)$.

1.4.2. *Let* $\mathscr{B}_\iota$ *be a* σ*-field in* Ω_ι $(\iota \in I)$. *Then the* σ*-field* $\mathscr{B} = \mathscr{B}(\bigcup_{\iota \in I} \varphi_\iota^{-1}\mathscr{B}_\iota)$ *is called the* **product σ-field** *of the* $\mathscr{B}_\iota$ $(\iota \in I)$ *and is also denoted by* $\mathscr{B} = \prod_{\iota \in I} \mathscr{B}_\iota$. *The measurable space* $(\Omega, \mathscr{B})$ *is then also called the* **(measurable) product space** *of the component or factor (measurable) spaces* $(\Omega_\iota, \mathscr{B}_\iota)$ $(\iota \in I)$ *and is also denoted by* $(\Omega, \mathscr{B}) = \prod_{\iota \in I} (\Omega_\iota, \mathscr{B}_\iota)$.

1.4.3. *Let* I *be finite and* $\mathscr{B}^{00}$ *a local* σ*-ring in* Ω_ι $(\iota \in I)$. *Then the local* σ*-ring*

$$\mathscr{B}^{00} = \mathscr{B}^{00}\left(\left\{\prod_{\iota \in I} F_\iota \mid F_\iota \in \mathscr{B}_\iota^{00} \ (\iota \in I)\right\}\right)$$

is called the **local product σ-ring** *of the* $\mathscr{B}_\iota^{00}$ $(\iota \in I)$, *and is also denoted by* $\mathscr{B}^{00} = \prod_{\iota \in I} \mathscr{B}_\iota^{00}$. $(\Omega, \mathscr{B}^{00})$ *is then also called the* **local product measurable space** $\prod_{\iota \in I} (\Omega_\iota, \mathscr{B}_\iota^{00})$.

1.4.4. *Let* I *be finite and* $\mathscr{B}_\iota^0$ *a* σ*-ring in* Ω_ι $(\iota \in I)$. *Then the* σ*-ring* $\mathscr{B}^0 = \mathscr{B}^0(\{\prod_{\iota \in I} F_\iota \mid F_\iota \in \mathscr{B}_\iota^0 \ (\iota \in I)\})$ *is called the* **product σ-ring** *of the component* σ*-rings* $\mathscr{B}_\iota^0$ $(\iota \in I)$. $(\Omega, \mathscr{B}^0)$ *is also called the* **product measurable space** $\prod_{\iota \in I} (\Omega_\iota, \mathscr{B}_\iota^0)$.

1.5. Remarks

1.5.1. Notice that product topologies etc. are not cartesian products of the component topologies etc. (except in very special situations such as $|I| = 1$), but topologies etc. *formed at the occasion* of a cartesian product of the component spaces Ω_ι $(\iota \in I)$. In Chapter XIII we shall use, for $|I| = 2$, say $I = \{1, 2,\}$, $\Omega_1 = \Omega$, $\Omega_2 = X$, $\mathscr{P} \subseteq \mathscr{P}(\Omega)$, $\mathscr{Q} \subseteq \mathscr{P}(X)$, the notation $[\mathscr{P}, \mathscr{Q}]$ for the system $\{E \times F \mid E \in \mathscr{P}, F \in \mathscr{Q}\}$ of all "rectangles" in $\Omega \times X$.

1.5.2. If it is clear from the context what set systems are to be taken in the component spaces Ω_ι $(\iota \in I)$ (viz. if $\Omega_\iota = \mathbb{R}$ or $|\Omega| < \infty$), we shall usually refrain from mentioning them explicitly.

1.5.3. Notice that we consider local product σ-rings and product σ-rings only for for finite I. If all component local σ-rings are σ-rings resp. σ-fields, so is the local product σ-ring.

1.5.4. Since topologies etc. are finite intersection stable, the product topology contains all so-called (finite-dimensional) *cylinders* $\bigcap_{k=1}^{r} \varphi_{\iota_k}^{-1} U_k$, $n \geqq 1$, $\iota_1, \ldots, \iota_n \in I$ pairwise distinct, where the U_k are taken from the topology etc., given in the component space Ω_{ι_k} $(k = 1, \ldots, n)$. In the case of topologies every member of the product topology is a union of such cylinders (exercise).

1.5.5. Observe that Example V.3.1 has shown that there are families of compact metrizable spaces $((\Omega_\iota, \mathscr{T}_\iota))_{\iota \in I}$ such that $\mathscr{B}(\prod_{\iota \in I} \mathscr{B}(\mathscr{T}_\iota))$ is contained in but different from $\mathscr{B}(\mathscr{T}(\prod_{\iota \in I} \mathscr{T}_\iota))$.

1.5.6. If I is finite or countable, we often replace $\prod_{\iota \in I} \mathscr{T}_\iota$ etc., somewhat imprecisely, by $\mathscr{T}_0 \times \mathscr{T}_1, \mathscr{T}_1 \times \cdots \times \mathscr{T}_n, \mathscr{T}_1 \times \mathscr{T}_2 \times \cdots$ etc., according to the choice of I.

1.6. Exercises

1.6.1. Let $I \neq \varnothing$ and $\mathscr{I} \subseteq \mathscr{P}(I)$ be increasingly σ-filtered with union I (see Exercise 1.3.6) Let $\Omega_\iota \neq \varnothing$, $\mathscr{B}_\iota \subseteq \mathscr{P}(\Omega_\iota)$ be a σ-field, $\varphi_\iota\colon \prod_{\kappa \in \Omega} \Omega_\kappa \to \Omega_\iota$ the ιth component mapping $(\iota \in I)$. For every $J \in \mathscr{I}$, let

$$\mathscr{B}_J = \mathscr{B}\left(\bigcup_{\iota \in J} \varphi_\iota^{-1} \mathscr{B}_\iota\right).$$

Prove that $\prod_{\iota \in I} \mathscr{B}_\iota = \bigcup_{J \in I} \mathscr{B}_J$. (*Hint:* Use Proposition I.4.7.5 and Exercise 1.3.6.)

1.6.2. Let $0 < |I| < \infty$, $\Omega_\iota \neq \varnothing$, $\varphi_\iota\colon \prod_{\kappa \in I} \Omega_\kappa \to \Omega_\iota$ be the ιth component mapping, and $\mathscr{B}_\iota^{00} \subseteq \mathscr{P}(\Omega_\iota)$ a local σ-ring $(\iota \in I)$. Prove that every $F \in \prod_{\iota \in I} \mathscr{B}_\iota^{00}$ is contained in some $\prod_{\iota \in I} F_\iota$ with $F_\iota \in \mathscr{B}_\iota^{00}$ $(\iota \in I)$ (see Proposition I.4.7.1). Formulate and prove a similar statement for product σ-rings.

1.6.3. Let $(\Omega_k, \mathscr{T}_k)$ $(k = 1, 2, \ldots)$ be *Polish* spaces. Prove

$$\prod_{k=1}^{\infty} \mathscr{B}(\mathscr{T}_k) = \mathscr{B}\left(\prod_{k=1}^{\infty} \mathscr{T}_k\right)$$

(see the proof of Theorem 5.6).

1.6.4. Let $I, K \neq \varnothing$, $I_\kappa \neq \varnothing$ $(\kappa \in K)$, $I_\lambda \cap I_\kappa = \varnothing$ $(\lambda \neq \kappa)$, $\sum_{\kappa \in K} I_\kappa = I$.

Let $\Omega_\iota \neq \varnothing$, $\mathscr{B}_\iota \subseteq \mathscr{P}(\Omega_\iota)$ be a σ-field, ψ_ι: $\prod_{\kappa \in I} \Omega_\kappa \to \Omega_\iota$ the ιth component mapping ($\iota \in I$). Let $\bar{\Omega}_\kappa = \prod_{\iota \in I_\kappa} \Omega_\iota$, $\bar{\mathscr{B}}_\kappa = \prod_{\iota \in I_\kappa} \mathscr{B}_\iota$. Prove that $\prod_{\iota \in I} \Omega_\iota = \prod_{\kappa \in K} \bar{\Omega}_\kappa$ up to a trivial identification, and $\prod_{\iota \in I} \mathscr{B}_\iota = \prod_{\kappa \in K} \bar{\mathscr{B}}_\kappa$ up to an analogous identification. Formulate and prove analogous statements for topologies, local σ-rings, and σ-rings.

1.6.5. Let $I \neq \varnothing$, $(\Omega_\iota, \mathscr{T}_\iota)$ be a compact Hausdorff space, and $\mathscr{B}_\iota$ the Baire σ-field in Ω_ι ($\iota \in I$). Prove that $\prod_{\iota \in I} \mathscr{B}_\iota$ is the Baire σ-field in the space $\prod_{\iota \in I} \Omega_\iota$ endowed with the product topology $\prod_{\iota \in I} \mathscr{T}_\iota$ (use the fact that the component mappings separate the points in $\prod_{\iota \in I} \Omega_\iota$ and the Weierstrass–Stone theorem; compare the proof of Theorem 5.5).

1.7. Definition. *Let Ω, I, $\Omega_\iota \neq \varnothing$ ($\iota \in I$) and φ_ι: $\Omega \to \Omega_\iota$ be given. Then the mapping φ: $\Omega \to \prod_{\iota \in I} \Omega_\iota$ defined by $\psi_\iota \circ \varphi = \varphi_\iota$ ($\iota \in I$) (where ψ_ι denotes the ιth component mapping $\prod_{\kappa \in I} \Omega_\kappa \to \Omega_\iota$), i.e., by*

$$\varphi(\omega) = (\omega_\iota)_{\iota \in I} \qquad \text{where} \quad \omega_\iota = \varphi_\iota(\omega) \quad (\iota \in I)$$

is called the **direct product** *of $(\varphi_\iota)_{\iota \in I}$ and also denoted by $\varphi = \prod_{\iota \in I} \varphi_\iota$. If I is finite, etc., we often prefer a notation like $\varphi = \varphi_0 \times \varphi_1$, $\varphi = \varphi_1 \times \cdots \times \varphi_n$, $\varphi = \varphi_1 \times \varphi_2 \times \cdots$.*

1.8. Proposition. *Let Ω, I, φ_ι, φ, ψ_ι ($\iota \in I$) be as above.*

1.8.1. *Let $\mathscr{T}$ be a topology in Ω and $\mathscr{T}_\iota$ a topology in Ω_ι ($\iota \in I$). Then $\varphi = \prod_{\iota \in I} \varphi_\iota$: $\Omega \to \prod_{\iota \in I} \Omega_\iota$ is $\mathscr{T}$-$\prod_{\iota \in I} \mathscr{T}_\iota$-continuous iff φ_ι: $\Omega \to \Omega_\iota$ is $\mathscr{T}$-$\mathscr{T}_\iota$-continuous ($\iota \in I$).*

1.8.2. *Let $\mathscr{B}$ be a σ-field in Ω and $\mathscr{B}_\iota$ a σ-field in Ω_ι ($\iota \in I$). Then $\varphi = \prod_{\iota \notin I} \varphi_\iota$: $\Omega \to \prod_{\iota \in I}$ is $\mathscr{B}$-$\prod_{\iota \in I}$ $\mathscr{B}$-measurable iff φ_ι: $\Omega \to \Omega_\iota$ is $\mathscr{B}$-$\mathscr{B}_\iota$-measurable ($\iota \in I$).*

Proof. We begin with the identities

$$\varphi^{-1}(\psi_\iota^{-1}F) = (\psi_\iota \circ \varphi)^{-1}F = \varphi_\iota^{-1}F \qquad (F \subseteq \Omega_\iota, \quad \iota \in I).$$

1. Let F run through all $\mathscr{T}_\iota$, $\iota \in I$. Then we see that

$$\varphi_\iota\colon \Omega \to \Omega_\iota \text{ continuous} \quad (\iota \in I)$$

$$\Leftrightarrow \quad \varphi^{-1}\left(\bigcup_{\iota \in I} \psi_\iota^{-1}\mathscr{T}_\iota\right) \subseteq \mathscr{T}.$$

But $\prod_{\iota \in I} \mathscr{T}_\iota$ is the smallest topology in $\prod_{\iota \in I} \Omega_\iota$ containing $\bigcup_{\iota \in I} \psi_\iota^{-1}\mathscr{T}_\iota$ and $\{M \mid M \subseteq \prod_{\iota \in I} \Omega_\iota,\ \varphi^{-1}M \in \mathscr{T}\}$ is a topology in $\prod_{\iota \in I} \Omega_\iota$ which thus contains $\prod_{\iota \in I} \mathscr{T}_\iota$. This does it.

2 is proved in exactly the same way.

1.9. Proposition. *Let $I, \Omega, \Omega_\iota, \varphi_\iota$ $(\iota \in I)$ be as above. Let $\mathscr{B}_\iota \subseteq \mathscr{P}(\Omega_\iota)$ be a σ-field $(\iota \in I)$ and $\mathscr{B} = \mathscr{B}(\bigcup_{\iota \in I} \varphi_\iota^{-1}\mathscr{B}_\iota)$ the σ-field generated by $(\varphi_\iota)_{\iota \in I}$. Then for every $f \in \overline{\mathbb{R}}^\Omega$ the following statements are equivalent.*

1.9.1. $f \in \text{mble}(\Omega, \mathscr{B}, \overline{\mathbb{R}})$.

1.9.2. *There is an* $f' \in \text{mble}(\prod_{\iota \in I} \Omega_\iota, \prod_{\iota \in I} \mathscr{B}_\iota, \overline{\mathbb{R}})$ *such that* $f = f' \circ \prod_{\iota \in I} \varphi_\iota$.

Proof. Apply Proposition IV.1.6.2 to our present situation.

1.10. Remarks

1.10.1. There is, of course, an analogue to the above proposition with $\mathbb{R}$ in place of $\overline{\mathbb{R}}$ (consult Proposition IV.1.6.1).

1.10.2. It is worthwhile to write out special cases of the above proposition:

Let $I = \{1, \ldots, n\}$, $\varphi_k \colon \Omega \to \Omega_k$ be a mapping $(k = 1, \ldots, n)$, and $\mathscr{B} = \mathscr{B}(\bigcup_{k=1}^n \varphi_k^{-1}\mathscr{B}_k)$ the σ-field generated by $\varphi_1, \ldots, \varphi_n$ in Ω. Then $f \in \overline{\mathbb{R}}^\Omega$ is $\mathscr{B}$-measurable iff $f(\omega) = f'(\varphi_1(\omega), \ldots, \varphi_n(\omega))$ for some $\prod_{k=1}^n \mathscr{B}_k$-measurable $f' \in \overline{\mathbb{R}}^{\prod_{k=1}^n \Omega_k}$. The reader is advised to write out the case $I = \mathbb{N} = \{1, 2, \ldots\}$.

1.11. Definition. *Let $I \neq \varnothing$, $\Omega_\iota \neq \varnothing$ $(\iota \in I)$, and $\Omega = \prod_{\iota \in I} \Omega_\iota$. Let I_1, $I_2 \neq \varnothing$, $I_1 \cap I_2 = \varnothing$, $I = I_1 + I_2$ and $\tilde{\Omega}_i = \prod_{\iota \in I_i} \Omega_\iota$ $(i = 1, 2,)$, $\tilde{\omega}_2 \in \tilde{\Omega}_2$.*

1.11.1. *For every $h \in \overline{\mathbb{R}}^\Omega$ the function $h_{\tilde{\omega}_2} \in \overline{\mathbb{R}}^{\tilde{\Omega}_1}$ defined by*

$$h_{\tilde{\omega}_2}(\tilde{\omega}_1) = h(\tilde{\omega}_1, \tilde{\omega}_2) \qquad (\tilde{\omega}_1 \in \tilde{\Omega}_1)$$

is called the $\tilde{\boldsymbol{\omega}}_2$**-section** *of h.*

1.11.2. *For every $M \subseteq \Omega$, the set $M_{\tilde{\omega}_2} \subseteq \tilde{\Omega}_1$ defined by*

$$M_{\tilde{\omega}_2} = \{\tilde{\omega}_1 \mid \tilde{\omega}_1 \in \tilde{\Omega}_1, (\tilde{\omega}_1, \tilde{\omega}_2) \in M\}$$

is called the $\tilde{\boldsymbol{\omega}}_2$**-section** *of M.*

1.12. Remark. In this definition we have identified Ω with $\tilde{\Omega}_1 \times \tilde{\Omega}_2$ in an obvious way: If $\tilde{\omega}_i = (\omega_\iota)_{\iota \in I_i} \in \tilde{\Omega}_i$ $(i = 1, 2)$, then $(\tilde{\omega}_1, \tilde{\omega}_2) \leftrightarrow (\omega_\iota)_{\iota \in I} \in \Omega$. Clearly $1_{M_{\tilde{\omega}_2}} = (1_M)_{\tilde{\omega}_2}$ $(M \subseteq \Omega)$.

1.13. Proposition. *Let I, $\Omega_\iota (\iota \in I)$, I_1, I_2, $\tilde{\Omega}_1$, $\tilde{\Omega}_2$, Ω be as above, $\tilde{\omega}_2 \in \tilde{\Omega}_2$.*

1.13.1. *Let $\mathscr{T}_\iota$ be a topology in Ω_ι $(\iota \in I)$. Then for every $h \in \mathbb{R}^\Omega$ which is continuous for $\prod_{\iota \in I} \mathscr{T}_\iota$ the section $h_{\tilde{\omega}_2}$ is continuous for $\prod_{\iota \in I_1} \mathscr{T}_\iota$, and for every $U \in \prod_{\iota \in I} \mathscr{T}_\iota$ the section $U_{\tilde{\omega}_2}$ is in $\prod_{\iota \in I_1} \mathscr{T}_\iota$.*

1.13.2. *Let $\mathscr{B}_\iota$ be a σ-field in Ω_ι $(\iota \in I)$. Then-for every*

$$h \in \text{mble}\left(\Omega, \prod_{\iota \in I} \mathscr{B}_\iota, \overline{\mathbb{R}}\right)$$

the section $h_{\tilde{\omega}_2}$ *is in* $\mathrm{mble}(\tilde{\Omega}_1, \prod_{\iota \in I_1} \mathscr{B}_\iota, \overline{\mathbb{R}})$ *and for every* $E \in \prod_{\iota \in I} \mathscr{B}_\iota$ *the section* $E_{\tilde{\omega}_2}$ *is in* $\prod_{\iota \in I_1} \mathscr{B}_\iota$.

1.13.3. *Let* $|I| < \infty$ *and* $\mathscr{B}_\iota^{00}$ *be a local σ-ring in* Ω_ι $(\iota \in I)$. *Then for every* $h \in \mathrm{mble}(\tilde{\Omega}_1, \mathscr{B}^0(\prod_{\iota \in I_1} \mathscr{B}_\iota^{00}), \overline{\mathbb{R}})$ *vanishing outside some* $F \in \prod_{\iota \in I} \mathscr{B}_\iota^{00}$ *the section* $h_{\tilde{\omega}_2}$ *is in* $\mathrm{mble}(\tilde{\Omega}_1, \mathscr{B}^0(\prod_{\iota \in I_1} \mathscr{B}_\iota^{00}), \overline{\mathbb{R}})$, *vanishing outside some* $\tilde{E} \in \prod_{\iota \in I_1} \mathscr{B}_\iota^{00}$; *and for every* $F \in \prod_{\iota \in I} \mathscr{B}_\iota^{00}$, *the section* $F_{\tilde{\omega}_2}$ *is in* $\prod_{\iota \in I} \mathscr{B}_\iota^{00}$.

1.13.4. *Let* $|I| < \infty$ *and* $\mathscr{B}_\iota^0$ *be a σ-ring in* Ω_ι $(\iota \in I)$. *Then for every* $h \in \mathrm{mble}(\Omega, \prod_{\iota \in I} \mathscr{B}_\iota^0, \overline{\mathbb{R}})$ *the section* $h_{\tilde{\omega}_2}$ *is in* $\mathrm{mble}(\tilde{\Omega}_1, \prod_{\iota \in I_1} \mathscr{B}_\iota^0, \overline{\mathbb{R}})$; *and for every* $E \in \prod_{\iota \in I} \mathscr{B}_\iota^0$, *the section* $E_{\tilde{\omega}_2}$ *is in* $\prod_{\iota \in I_1} \mathscr{B}_\iota^0$.

Proof. Let $\tilde{\omega}_2 = (\omega_\iota)_{\iota \in I_2}$ *and define the mapping*

$$\varphi_{\tilde{\omega}_2}\colon \prod_{\iota \in I_1} \Omega_\iota \to \prod_{\iota \in I} \Omega_\iota \qquad \text{by} \qquad \varphi_{\tilde{\omega}_2} = \prod_{\iota \in I} \tau_\iota$$

where $\tau_\iota((\omega_\kappa)_{\kappa \in I_1}) = \omega_\iota$ $(\iota \in I)$, i.e., τ_ι is the ιth component mapping for $\iota \in I_1$, and the constant ω_ι for $\iota \in I_2$. Clearly every τ_ι is continuous resp. measurable with respect to the set systems envisaged in the respective parts of our proposition (use Proposition 1.8).

1. Since $\varphi_{\tilde{\omega}_2}$ is continuous, $\varphi_{\tilde{\omega}_2}^{-1} U = U_{\tilde{\omega}_2}$ is open in $\prod_{\iota \in I_1} \Omega_\iota$ and $h_{\tilde{\omega}_2} = h \circ \varphi_{\tilde{\omega}_2} \in \mathscr{C}(\tilde{\Omega}_1, \mathbb{R})$.

2. Since $\varphi_{\tilde{\omega}_2}$ is $\prod_{\iota \in I_1} \mathscr{B}_\iota$-$\prod_{\iota \in I} \mathscr{B}_\iota$-measurable, $\varphi_{\tilde{\omega}_2}^{-1} E = E_{\tilde{\omega}_2}$ is in $\prod_{\iota \in I_1} \mathscr{B}_\iota$ and $h_{\tilde{\omega}_2} = h \circ \varphi_{\tilde{\omega}_2}$ is in $\mathrm{mble}(\tilde{\Omega}_1, \prod_{\iota \in I_1} \mathscr{B}_\iota, \overline{\mathbb{R}})$.

3 and 4 are proved in almost the same fashion.

2. TWO FACTORS

In this section we consider a cartesian product of two spaces and investigate the possibilities of defining σ-contents in it. A counterexample shows that not every probability σ-content in such a product space can be obtained via a σ-content plus a kernel. A positive solution to this representation problem will be given in Chapter XVI.

2.1. Theorem. *Let* $(\Omega_0, \mathscr{B}_0^{00}, m_0)$, $(\Omega_1, \mathscr{B}_1^{00}, m_1)$ *be σ-content spaces. Then there is a unique σ-content* m *on* $\mathscr{B}^{00} = \prod_{i=0}^{1} \mathscr{B}_i^{00} = \mathscr{B}_0^{00} \times \mathscr{B}_1^{00}$ (*in* $\Omega = \Omega_0 \times \Omega_1$) *such that*

$$m(F_0 \times F_1) = m_0(F_0) m_1(F_1) \qquad (F_i \in \mathscr{B}_i^{00}, \quad i = 0, 1). \tag{1}$$

m is called the **product σ-content** *of* m_0 *and* m_1 *and also denoted by* $m = m_0 \times m_1$. *We shall also write*

$$(\Omega, \mathscr{B}^{00}, m) = (\Omega_0, \mathscr{B}_0^{00}, m_0) \times (\Omega_1, \mathscr{B}_1^{00}, m_1)$$

and call this the **product σ-content space** *of* $(\Omega_0, \mathscr{B}_0^{00}, m_0)$ *and* $(\Omega_1, \mathscr{B}_1^{00}, m_1)$. *For every* $F \in \mathscr{B}^{00}$, *we have* **Fubini's formula**

$$(2) \qquad m(F) = \int m_1(F_{\omega_0})\, m_0(d\omega_0) = \int m_0(F_{\omega_1})\, m_1(d\omega_1).$$

m is a probability σ-content iff m_0, m_1 are probability σ-contents.

Proof. We begin by observing that Fubini's formula (2) is meaningful. $F_{\omega_0} \in \mathscr{B}_1^{00}$ $(\omega_0 \in \Omega_0)$ and $F_{\omega_1} \in \mathscr{B}_0^{00}$ $(\omega_1 \in \Omega_1)$ are special cases of Proposition 1.13.3. Hence $m_1(F_{\omega_0})$ is a well-defined function of $\omega_0 \in \Omega_0$, and $m_0(F_{\omega_1})$ is a well-defined function of $\omega_1 \in \Omega_1$. We shall henceforth concentrate our attention upon the first equality in (2); the second will follow in a symmetric fashion. Let $\mathscr{S} = \{F_0 \times F_1 \mid F_i \in \mathscr{B}_i^{00}\ (i = 0, 1)\}$. Clearly, $\mathscr{S}$ is stable under finite intersections and satisfies $\mathscr{B}^{00}(\mathscr{S}) = \mathscr{B}^{00}$. By Proposition I.5.3, m is uniquely determined by m_0, m_1, and (1). Let us now prove existence. Consider $\tilde{\mathscr{B}}^{00} = \{F \mid F \in \mathscr{B}^{00},\ m_1(F_{\omega_0})$ is an m_0-σ-integrable function of $\omega_0 \in \Omega_0\}$. Let us first notice that by Exercise 1.6.2 every $F \in \mathscr{B}^{00}$ is contained in some $F_0 \times F_1$ with $F_i \in \mathscr{B}_i^{00}$ $(i = 0, 1)$, and hence $F_{\omega_0} \subseteq F_2$ $(\omega_0 \in \Omega_0)$. The real functions $m_1(F_{\omega_0})$ $(\omega_0 \in \Omega_0)$ under consideration are thus bounded. Now clearly $\tilde{\mathscr{B}}^{00} \supseteq \mathscr{S}$. We prove that hypotheses of proposition I.4.9 are satisfied. In fact, $E, F \in \tilde{\mathscr{B}}^{00}$,

$$F \subseteq E \quad \Rightarrow \quad m_1((E \backslash F)_{\omega_0}) = m_1(E_{\omega_0} \backslash F_{\omega_0}) = m_1(E_{\omega_0}) - m_1(F_{\omega_0})$$

$(\omega_0 \in \Omega_0)$ shows that $\tilde{\mathscr{B}}^{00}$ is stable under proper differences. But $\tilde{\mathscr{B}}^{00}$ is also stable under bounded countable disjoint unions. In fact, Let E, E^1, $E^2, \ldots \in \tilde{\mathscr{B}}^{00}$, $E^1, E^2, \ldots \subseteq E$, $E^j \cap E^k = \varnothing$ $(j \neq k)$; find $F_0 \in \mathscr{B}_0^{00}$, $F_1 \in \mathscr{B}_1^{00}$ such that $E \subseteq F_0 \times F_1$. Then for every $\omega_0 \in \Omega_0$, we have $E_{\omega_0}^j \cap E_{\omega_0}^k = \varnothing$ $(j \neq k)$, $E_{\omega_0}^j \subseteq F_1$ $(j = 1, 2, \ldots)$ and thus

$$\begin{aligned} m_1(F_1) 1_{F_0}(\omega_0) &\geqq m_1((E^1 + E^2 + \cdots)_{\omega_0}) = m_1(E_{\omega_0}^1 + E_{\omega_0}^2 + \cdots) \\ &= m_1(E_{\omega_0}^1) + m_1(E_{\omega_0}^2) + \cdots \geqq 0. \end{aligned}$$

Lebesgue's dominated convergence theorem (III.5.7) now implies $E^1 + E^2 + \cdots \in \tilde{\mathscr{B}}^{00}$. By Proposition I.4.9 we get $\tilde{\mathscr{B}} = \mathscr{B}^{00}$ and thus Fubini's formula. The rest is obvious.

2.2. Theorem (Fubini). *Let $(\Omega_i, \mathscr{B}_i^{00}, m_i)$ $(i = 0, 1)$ be σ-content spaces and $(\Omega, \mathscr{B}^{00}, m) = (\Omega_0, \mathscr{B}_0^{00}, m_0) \times (\Omega_1, \mathscr{B}_1^{00}, m_1)$.*

2.2.1. *Let $h \in L_m^1$. Then for m_0-a.e. $\omega_0 \in \Omega_0$ the ω_0-section h_{ω_0} of h belongs* *to* $L^1_{m_1}$. *The* m_0-a.e. *defined function* $h_0(\omega_0) = \int h_{\omega_0}(\omega_1)\, m_1(d\omega_1) = \int h(\omega_0, \omega_1)\, m_1(d\omega_1)$ *is in* $L^1_{m_0}$ *and*

$$(3) \qquad \int h\, dm = \int h_0\, dm_0 = \int m_0\, (d\omega_0)\left[\int h(\omega_0, \omega_1)\, m_1(d\omega_1)\right].$$

2.2.2. *Let* $\mathscr{B}^0 = \mathscr{B}^0(\mathscr{B}^{00})$ *and* $h \in \text{mble}(\Omega, \mathscr{B}^0, \overline{\mathbb{R}})$. *Then* $h \in \mathscr{L}^1_m$ *iff* $h_{\omega_0} \in \mathscr{L}^1_{m_1}$ *for m-a.e.* $\omega_0 \in \Omega_0$ *and* $h_0(\cdot) = \int h(\cdot, \omega_1)\, m_1(d\omega_1) \in \mathscr{L}^1_{m_0}$. *If this is true, then* (3) *holds. Moreover,*

$$\left\{\omega_0 \,|\, h_{\omega_0} \in \mathscr{L}^1_{m_1}, \int h(\omega_0, \omega_1) m_1(d\omega_1) \neq 0\right\} \in \mathscr{B}^0(\mathscr{B}^{00}_0)$$

and the function $h_0 \in \mathbb{R}^{\Omega_0}$ *defined by* $h_0(\omega_0) = 0$ *if* $h_{\omega_0} \notin \mathscr{L}^1_{m_1}$ *and* $h_0(\omega_0) = \int h(\omega_0, \omega_1)\, m_1(d\omega_1)$ *otherwise, is in* $\text{mble}(\Omega_0, \mathscr{B}^0(\mathscr{B}^{00}_0), \mathbb{R})$.

Proof. 1 is true for $h = 1_F$, $F \in \mathscr{B}^{00}$, by Theorem 2.1. If 1 is true for $h_1, \ldots, h_n \in \mathbb{R}^\Omega$, and $\alpha_1, \ldots, \alpha_n \in \mathbb{R}$, then 1 is true for $\alpha_1 h_1 + \cdots + \alpha_n h_n$ (exercise). Assume that $h^1, h^2, \ldots \in \mathscr{L}^1_m$ are such that 1 is true for each of them and $h^1 \leqq h^2 \leqq \cdots$ (pointwise), $\lim \int h^n\, dm < \infty$ holds. Put $h = \lim h^n$ (pointwise). Then $h_{\omega_0} = \lim h^n_{\omega_0}$ pointwise in Ω_1, for every $\omega_0 \in \Omega_0$. For every $n = 1, 2, \ldots$, let $N_n = \{\omega_0 \,|\, h^n_{\omega_0} \notin \mathscr{L}^1_{m_1}\}$. Put $N = \bigcup_n N_n$ and $h^n_0(\omega_0) = \int h^n(\omega_0, \omega_1) m_1(d\omega_1)$ for $\omega_0 \in \Omega_0 \backslash N$, $h^n_0(\omega_0) = 0$ for $\omega_0 \in N$. Clearly $h^1_0 \leqq h^2_0 \leqq \cdots$ pointwise on Ω_0. By assumption we have $h^1_0, h^2_0, \ldots \in \mathscr{L}^1_{m_0}$ and $\lim \int h^n_0\, dm_0 = \lim \int h^n\, dm < \infty$. By Theorems III.5.1.1 and III.5.3 (monotone convergence) we have an m_0-nullset N' such that $\omega_0 \in \Omega_0 \backslash (N' \cup N) \Rightarrow \lim h^n_0(\omega_0) \in \mathbb{R}$ and

$$\begin{aligned}
\lim h^n_0(\omega_0) &= \lim \int h^n(\omega_0, \omega_1) m_1(d\omega_1) \\
&= \int [\lim h^n(\omega_0, \omega_1)] m_1(d\omega_1) \\
&= \int h(\omega_0, \omega_1) m_1(d\omega_1).
\end{aligned}$$

Moreover, the function h_0 defined on Ω_0 by $h_0(\omega_0) = \lim h_0^n(\omega_0)$ if $\omega_0 \in \Omega_0 \setminus (N' \cup N)$ and $h_0(\omega_0) = 0$ otherwise is in $\mathscr{L}^1_{m_0}$. Thus we have proved, by applying Theorem III.5.3 twice, that h again satisfies 1. With this general background, we go through the steps of monotone σ-integration for m and see successively: 1 is true: (a) for all functions of the form $\alpha_1 1_{F_1} + \cdots + \alpha_n 1_{F_n}$ where $n > 0$, $\alpha_1, \ldots, \alpha_n \in \mathbb{R}$, $F_1, \ldots, F_n \in \mathscr{B}^{00}$; (b) for all σ-upper functions $h \in \mathscr{E}^\sigma(\mathscr{B}^{00})$ with a finite m-integral, and likewise (applying our previous monotone convergence argument "downward") for all σ-lower functions with a finite m-integral; (c) for all monotone decreasing pointwise sequence limits of functions from $\mathscr{E}^\sigma(\mathscr{B}^{00})$, with a finite m-integral, and likewise for all monotone increasing pointwise limits of sequences of σ-lower functions, with a finite m-integral. We now turn to the phenomenon of "squeezing-in" and observe first: If $h \in \mathbb{R}_+^\Omega$ is an m-σ-null function, then there is a $g \in \mathscr{L}^1_m$ with $0 \leqq |h| \leqq g$, $\int g\, dm = 0$ and 1 holds for g, by (c) (use also Proposition III.1.7.1.2). We have thus $|h_{\omega_0}| \leqq g_{\omega_0}$ pointwise on Ω_1, for every $\omega_0 \in \Omega_0$, and $\int g_{\omega_0}\, dm_1 = 0$ for m_0-a.e. $\omega_0 \in \Omega_0$, i.e., h_{ω_0} is an m_1-σ-nullfunction for m_0-a.e., $\omega_0 \in \Omega_0$. This is in particular true if h is an indicator function of an m-nullset N, and we conclude that the m_0-a.e. section N_{ω_0} is an m_1-nullset. Let now $h \in \mathscr{L}^1_m$ be arbitrary. Then we may find, according to (c) and the definition of m-σ-integrability, f, $g \in \mathscr{L}^1_m$ satisfying 1 and $f \leqq h \leqq g$ (pointwise), $\int f\, dm = \int g\, dm$ and hence $N' = \{\omega \mid f(\omega) \neq g(\omega)\}$ is an m-nullset, and thus so is $N = \{\omega \mid f(\omega) \neq h(\omega)\}$.

For m_0-a.e. $\omega_0 \in \Omega_0$ we have thus $f_{\omega_0} \in \mathscr{L}^1_{m_1}$ and $f_{\omega_0} = h_{\omega_0}$ m_1-a.e., hence $h_{\omega_0} \in \mathscr{L}^1_{m_1}$, and moreover $f_0 = h_0$ m_0-a.e., proving $h_0 \in \mathscr{L}^1_{m_0}$ and the validity of 1 for h.

2. Taking 1 into account, a proof of the "if" direction of the whole of 2 will suffice. By the decomposition $h = h_+ - h_-$ we may henceforth assume $h \geqq 0$. Combining Propositions I.4.7.2.1 and III.8.13, we may represent h as a monotone increasing pointwise limit of a sequence h^n of functions from $\mathscr{E}(\mathscr{B}^{00})$. For these, 2 and 1 are clearly true, even with empty exceptional sets. Let now $N \subseteq \Omega_0$ be an m_0-nullset such that $h_{\omega_0} \in \mathscr{L}^1_{m_1}$ for $\omega_0 \notin N$, and define $h_0'(\omega_0) = \int h_{\omega_0}\, dm_1$ for $\omega_0 \in \Omega_0 \setminus N$ and $h_0(\omega_0) = 0$ for every $\omega_0 \in N$. By assumption $h_0' \in \mathscr{L}^1_{m_0}$. For every $\omega_0 \in \Omega_0 \setminus N$, we have $h^1_{\omega_0} \leqq h^2_{\omega_0} \leqq \cdots \leqq h_{\omega_0}$ everywhere on Ω_1, thus we obtain $\int h^n_{\omega_0}\, dm_1 \leqq \int h_{\omega_0}\, dm_1$. It follows that

$$\int h^n\, dm = \int m_0(d\omega_0) \int h^n(\omega_0, \omega_1) m_1(d\omega_1) \leqq \int h_0'\, dm_0 < \infty \qquad (n = 1, 2, \ldots).$$

By Theorem III.5.3 (monotone convergence) $h \in \mathscr{L}_m^1$ now follows. Finally, we settle the measurability problem. Clearly $h_0^n \in \text{mble}(\Omega_0, \mathscr{B}^0(\mathscr{B}_0^{00}), \mathbb{R})$. It follows that $N_0 = \{\omega_0 \mid \lim h_0^n(\omega_0) = \infty\} \in \mathscr{B}^0(\mathscr{B}_0^{00})$. This set is contained in N, hence is an m_0-nullset. For $\omega_0 \in \Omega_0 \backslash N_0$, we have $\lim h_0^n(\omega_0) < \infty$, hence $h_{\omega_0} = \lim h_{\omega_0}^n \in \mathscr{L}_{m_1}^1$. For $\omega_0 \in N_0$, we have $h_{\omega_0} \notin \mathscr{L}_{m_1}^1$ because otherwise Theorem III.5.7 (on dominated convergence) would imply $\lim h_0^n(\omega_0) = \int h_{\omega_0}\, dm_1 < \infty$, contradicting $\omega_0 \in N_0$. Thus we have

$$h_0 = (1 - 1_{N_0}) \lim h_0^n = \lim[(1 - 1_{N_0})h_0^n]$$

pointwise in Ω_0. Since clearly $(1 - 1_{N_0})h_0^n \in \text{mble}(\Omega_0, \mathscr{B}^0(\mathscr{B}_0^{00}), \mathbb{R})$, we get $h_0 \in \text{mble}(\Omega_0, \mathscr{B}^0(\mathscr{B}_0^{00}), \mathbb{R})$ by Proposition III.8.10.3.3.

2.3. Theorem. *Let $(\Omega_0, \mathscr{B}^{00}, m_0)$ be a σ-content space, $\mathscr{B}_1^{00}$ a local σ-ring in $\Omega_1 \neq \varnothing$, and $P\colon \Omega_0 \times \mathscr{B}_1^{00} \to \mathbb{R}_+$ an m_0-integrable $\mathscr{B}^0(\mathscr{B}^{00})$-measurable σ-content kernel from Ω_0 to Ω_1 (Definitions* IV.4.1.2.1, IV.4.6.3, IV.4.12.2). *Then there is a unique σ-content m on $\mathscr{B}^{00} = \mathscr{B}_0^{00} \times \mathscr{B}_1^{00}$ such that*

$$(4) \quad m(F_0 \times F_1) = \int m_0(d\omega_0) 1_{F_0}(\omega_0) P(\omega_0, F_1) \qquad (F_i \in \mathscr{B}_i^{00} \quad (i = 0, 1)).$$

m is called the **product σ-content of u_0 and P**, *and is also denoted by $m = m_0 \times P$. For every $F \in \mathscr{B}^{00}$, we have*

$$(5) \qquad m(F) = \int m_0(d\omega_0) P(\omega_0, F_{\omega_0}).$$

If $\mathscr{B}_0^{00}$, $\mathscr{B}_1^{00}$ are σ-fields, m_0 is a probability σ-content and P stochastic, then $m = m_0 \times P$ is a probability σ-content: $m(\Omega) = 1$. If m is a probability σ-content, then so is m_0.

The proof, which is an obvious generalization of the proof of Theorem 2.1, is left as an exercise to the reader.

2.4. Theorem (Fubini). *Let $\Omega_0, \Omega_1, \mathscr{B}_0^{00}, \mathscr{B}_1^{00}, m_0, P$ be as in Theorem 1.3, and $m = m_0 \times P$.*

2.4.1. *Let $h \in \mathscr{L}_m^1$. Then for m_0-a.e. ω_0 the ω_0-section of h is in $\mathscr{L}_{P(\omega_0, \cdot)}^1$. The m_0-a.e. defined function $h_0(\omega_0) = \int P(\omega_0, d\omega_1) h(\omega_0, \omega_1)$ is in $\mathscr{L}_{m_0}^1$ and*

$$(6) \qquad \int h\, dm = \int h_0\, dm_0 = \int m_0(d\omega_0) \left[\int P(\omega_0, d\omega_1) h(\omega_0, \omega_1) \right].$$

2.4.2. *Let* $\mathscr{B}^0 = \mathscr{B}^0(\mathscr{B}^{00})$ *and* $h \in \text{mble}(\Omega, \mathscr{B}^0, \overline{\mathbb{R}})$. *Then* $h \in \mathscr{L}_m^1$ *iff for* m*-a.e.,* $\omega_0 \in \Omega_0$ *the* ω_0*-section of* h *is in* $\mathscr{L}^1_{P(\omega_0, \cdot)}$ *and* $\int P(\cdot, d\omega_1)h(\cdot, \omega_1) \in \mathscr{L}^1_{m_0}$. *In this case* (6) *holds. Moreover,*

$$\left\{\omega_0 \,\middle|\, h_{\omega_0} \in \mathscr{L}^1_{P(\omega_0, \cdot)}, \int h(\omega_0, \omega_1)P(\omega_0, d\omega_1) \neq 0\right\} \in \mathscr{B}^0(\mathscr{B}_0^{00})$$

and the function $h_0 \in \mathbb{R}^{\Omega_0}$ *defined by* $h_0(\omega_0) = 0$ *if* $h_{\omega_0} \notin \mathscr{L}^1_{P(\omega_0, \cdot)}$ *and* $h_0(\omega_0) = \int P(\omega_0, d\omega_1)h(\omega_0, \omega_1)$ *otherwise, is in* $\text{mble}(\Omega_0, \mathscr{B}^0(\mathscr{B}_0^{00}), \mathbb{R})$.

The proof is again left as an exercise to the reader.

We now turn to the obvious question whether every σ-content on a local product σ-ring in a product space can be obtained in the form $m_0 \times P$ as in Theorem 2.3. The next proposition gives necessary conditions in a special case.

2.5. Proposition. *Let* $\Omega_0, \Omega_1 \neq \varnothing$ *and* $\psi_i \colon \Omega_0 \times \Omega_1 \to \Omega_i$ *be the* ith *component mapping. Let* $\mathscr{B}_0^{00} \subseteq \mathscr{P}(\Omega_0)$, $\mathscr{B}_1^{00} \subseteq \mathscr{P}(\Omega_1)$ *be* σ*-fields,* $\mathscr{B}^{00} = \mathscr{B}_0^{00} \times \mathscr{B}_1^{00}$, m_0 *a* σ*-content on* $\mathscr{B}_0^{00}$, P *a* $\mathscr{B}^0(\mathscr{B}_0^{00})$*-measurable stochastic* σ*-content kernel from* Ω_0 *to* Ω_1 *(with* $\mathscr{B}_1^{00}$*). Let* $m = m_0 \times P$. *Then*

$$\psi_0 m = m_0, \qquad \psi_1 m = m_0 P.$$

Proof.

$$(\psi_0 m)(F_0) = m(F_0 \times \Omega_1) = \int m_0(d\omega_0)1_{F_0}(\omega_0)P(\omega_0, \Omega_1)$$
$$= m_0(F_0) \qquad (F_0 \in B_0^{00})$$

and

$$(\psi_1 n)(F_1) = m(\Omega_0 \times F_1) = \int m_0(d\omega_0)P(\omega_0, F_1) = (m_0 P)(F_1)$$
$$(F_1 \in \mathscr{B}_1^{00}).$$

2.6. Example. *(of a* σ*-content in a product space having no representation in the form* $m_0 \times P$*).* Let $(\Omega, \mathscr{B}, m)$, $(\Omega, \mathscr{B}', m')$ be as in Example IV.4.20. Put $(\Omega_0, \mathscr{B}_0, m_0) = (\Omega, \mathscr{B}, m)$, $(\Omega_1, \mathscr{B}_1, m_1) = (\Omega', \mathscr{B}', m')$. Let $(\tilde{\Omega}, \tilde{\mathscr{B}}', \tilde{m}') = (\Omega_0, \mathscr{B}_0, m_0) \times (\Omega_1, \mathscr{B}_1, m_1)$. Define $\delta \colon \tilde{\Omega} \to \tilde{\Omega}$ by $\delta(\omega_0, \omega_1) = (\omega_1, \omega_1)$ $((\omega_0, \omega_1) \in \tilde{\Omega})$. Let φ_i denote the ith component mapping of $\tilde{\Omega}$ $(i = 0, 1)$. Then $\delta = \varphi_1 \times \varphi_1$ shows that δ is $\tilde{\mathscr{B}}$-$\tilde{\mathscr{B}}$-measurable (use Proposition 1.8.2 and notice that φ_1 is $\mathscr{B}'$-$\mathscr{B}$-measurable as well as $\mathscr{B}'$-$\mathscr{B}'$-measurable since it is the latter and $\mathscr{B} \subseteq \mathscr{B}'$). Let now $\tilde{m} = \delta\tilde{m}'$. Clearly, $\varphi_0 \circ \delta = \psi_1$ is $\tilde{\mathscr{B}}'$-$\mathscr{B}$-measurable. Under this equality, φ_0 sends $\tilde{m}$ into $\varphi_1\tilde{m}' = m'$, restricted to $\mathscr{B}$, i.e., into m, φ_1 sends $\tilde{m}$ into m', when considered as a $\tilde{\mathscr{B}}'$-$\mathscr{B}'$-measurable mapping. Assume now that there is a representation $\tilde{m} = m \times P$ with a $\mathscr{B}$-measurable m-integrable kernel P from Ω to Ω (with $\mathscr{B}'$). Then for any $F \in \mathscr{B}$, $F' \in \mathscr{B}'$, we have

$$\begin{aligned}\int 1_F(\omega)P(\omega, F')\, m(d\omega) &= \tilde{m}(F \times F') = (\delta\tilde{m}')(1_{F \times F'}) = \tilde{m}'(1_{F \times F'} \circ \delta) \\ &= \int m(d\omega)\left[\int 1_F(\omega')1_{F'}(\omega')\, m'(d\omega')\right] \\ &= 2\pi \int 1_F(\omega')1_{F'}(\omega')\, m'(d\omega').\end{aligned}$$

A reader who is already familiar with the notion of conditional expectation (Proposition VIII.6.1) will now notice that the kernel $(1/2\pi)P$ represents the conditional expectation with respect to $\mathscr{B} \subseteq \mathscr{B}'$ and apply Example IV.4.20 to achieve a contradiction. A reader who is not familiar with the notion of conditional expectation can carry through the argument of Example IV.4.20.4 with $(1/2\pi)P$ in place of P and get the contradiction as well. We mention, here without proof, that under some regularity conditions every m in a product space is representable in the form $m_0 \times P$, with a kernel P (Chapter XVI, Section 3).

2.7. Exercises. Let $\Omega_0, \Omega_1 \neq \varnothing$ and φ_i be the ith component mapping of $\Omega = \Omega_0 \times \Omega_1$ onto Ω_i ($i = 0, 1$). Let $\mathscr{B}_0^{00} \subseteq \mathscr{P}(\Omega_0)$, $\mathscr{B}_1^{00} \subseteq \mathscr{P}(\Omega_1)$ be local σ-rings and $\mathscr{B}^{00} = \mathscr{B}_0^{00} \times \mathscr{B}_1^{00}$ their local product σ-ring.

2.7.1. Let $\mathscr{B}_i = \mathscr{B}_i^{00}$ be σ-fields ($i = 0, 1$). Let $(\Omega, \mathscr{B}, m)$ be a σ-probability content space and $\psi_i\colon \Omega \to \Omega_i$ $\mathscr{B}$-$\mathscr{B}_i$-measurable ($i = 0, 1$). Put $\mathscr{C}_i = \psi_i^{-1}\mathscr{B}_i$ ($i = 0, 1$). Prove that

$$\begin{aligned} m(F_0 \cap F_1) = m(F_0)m(F_1) \quad (F_0 \in \mathscr{C}_0, F_1 \in \mathscr{C}_1) \\ \Leftrightarrow \quad \psi_0 \times \psi_1 \text{ sends } m \text{ into } (\psi_0 m) \times (\psi_1 m).\end{aligned}$$

2.7.2. Prove that for any representation $\mathbb{R}^n = \mathbb{R}^{n_0} \times \mathbb{R}^{n_1}$ ($n = n_0 + n_1$; $n_0, n_1 > 0$ integral).

2.7.2.1. The Lebesgue σ-content in $\mathbb{R}^n$ is the product σ-content of the Lebesgue σ-contents in $\mathbb{R}^{n_0}$ and $\mathbb{R}^{n_1}$.

2.7.2.2. The *Gaussian* probability σ-content in $\mathbb{R}^n$ is the product σ-content of the Gaussian σ-contents in $\mathbb{R}^{n_0}$ and $\mathbb{R}^{n_1}$. Generally, the *Gaussian* σ-content m_k in $\mathbb{R}^k$, $k > 1$, is defined by

$$m_k(F) = \frac{1}{(2\pi)^{k/2}} \int 1_F(x_1, \ldots, x_k)e^{-(x_1{}^2 + \cdots + x_k{}^2)/2}\, dx_1 \cdots dx_k \qquad (F \subseteq \mathbb{R}^k \text{ Borel}).$$

2.7.3. Let $\Omega_0 = \Omega_1 = \mathbb{R}$ and p_i be a probability σ-content on the σ-field of all Borel sets in $\mathbb{R}$ $(i = 0, 1)$. Consider the mapping φ: $(x, y) \to x + y$ of $\mathbb{R}^2 \to \mathbb{R}$. Prove that

$$(\varphi p)(F) = \int p_0(dx_0)p_1(F - x_0) \qquad (F \subseteq \mathbb{R} \text{ Borel}),$$

where $F - x_0$ stands for $\{x - x_0 \mid x \in F\}$. p is also called the *convolution* of p_0 and p_1 and is written $p = p_0 * p_1$. (*Hint*: Use Fubini's formula (2).)

2.7.4. Let $\mathscr{B}_i = \mathscr{B}_i^{00}$ be a σ-field in $\Omega_i = \mathbb{R}$, $(\Omega, \mathscr{B}, m)$ a σ-probability space, φ_i: $\Omega \to \Omega_i$ $\mathscr{B}$-$\mathscr{B}_i$-measurable, and $p_i = \varphi_i m$ $(i = 0, 1)$. Prove that

$$(\varphi_0 \times \varphi_1)m = p_0 \times p_1 \quad \Rightarrow \quad (\varphi_0 + \varphi_1)m = p_0 * p_1.$$

2.7.5. Implement Example 2.6 by the following observation: If a σ-content m: $\mathscr{B}^{00} \to \mathbb{R}_+$ has a representation $m_0 \times P$ in the sense of Proposition 2.5, then

$$E_0 \in \mathscr{B}_0^{00},\ E_1 \in \mathscr{B}_1^{00} \quad \Rightarrow \quad m(E_0 \times E_1) = \int 1_{E_0}(\omega_0)P(\omega_0, E_1)\, m(d\omega_0).$$

A reader who is familiar with the notion of *conditional expectation* will read this result to the following effect: A representation $m = m_0 \times P$ amounts to the *representation of a certain conditional expectation by a kernel.* This problem will be tackled in Chapter XVI, Section 3.

3. FINITELY MANY FACTORS

In this section we present the theory of σ-contents in a product of finitely many spaces. This is not only for convenience in citing results, but also is good preparation for the case of countably many factors to be treated in the next section.

One of the technical problems of this section is the ensurement of all integrabilities we need. This could be done immediately by assuming that all local σ-rings in question are σ-fields. In view of some applications (viz. Lebesgue σ-content in $\mathbb{R}^n$) we opt for a somewhat more complicated notion.

3.1. Definition. *Let $\Omega \neq \varnothing$ and $\mathscr{B}^{00} \subseteq \mathscr{P}(\Omega)$ be a local σ-ring. A function $f \in \mathbb{R}^\Omega$ is said to be* $\mathscr{B}^{00}$**-bounded** *if its restriction to any $F \in \mathscr{B}^{00}$ is bounded.*

3.2. Theorem. *Let* $0 \leqq t \in \mathbb{Z}$, $\Omega_s \neq \varnothing$ *and* $\mathscr{B}_s^{00} \subseteq \mathscr{P}(\Omega_s)$ *be a local* σ*-ring* $(s = 0, \ldots, t)$. *For any* $0 \leqq u \leqq v \leqq t$, *put*

$$\Omega(u, v) = \prod_{s=u}^{v} \Omega_s, \qquad \mathscr{B}^{00}(u, v) = \prod_{s=u}^{v} \mathscr{B}_s^{00}$$

$$\Omega(0, t) = \Omega, \qquad \mathscr{B}^{00}(0, t) = \mathscr{B}^{00}.$$

Let $m_0 : \mathscr{B}_0^{00} \to \mathbb{R}_+$ *be a* σ*-content and* P_s *a* σ*-content kernel from* $\Omega(0, s-1)$ *to* Ω_s *(with* $\mathscr{B}_s^{00}$*) such that for every* $F \in \mathscr{B}_s^{00}$ *the function* $P_s(\cdot, F)$ *on* $\Omega(0, s-1)$ *is* $\mathscr{B}^{00}(0, s-1)$*-bounded and* $\mathscr{B}^0(\mathscr{B}^{00}(0, s-1))$*-measurable. Then:*

3.2.1. *There is a* σ*-content kernel* Q_s *from* $\Omega(0, s)$ *to* Ω *(with* $\mathscr{B}^{00}$*) such that for every* $F \in \mathscr{B}^{00}$ *the function* $Q_s(\cdot, F)$ *on* $\Omega(0, s)$ *is* $\mathscr{B}^{00}(0, s)$*-bounded and* $\mathscr{B}^0(\mathscr{B}^{00}(0, s))$*-measurable* $(s = 0, \ldots, t)$ *and*

$$\text{(1)} \quad Q_t(\omega_0, \ldots, \omega_t, F) = 1_F(\omega_0, \ldots, \omega_t) \qquad (F \in \mathscr{B}^{00},\ \omega_0 \in \Omega_0, \ldots, \omega_t \in \Omega_t)$$

$$\text{(2)} \quad Q_s(\omega_0, \ldots, \omega_s, F) = \int P_{s+1}(\omega_0, \ldots, \omega_s, d\omega_{s+1})\, Q_{s+1}(\omega_0, \ldots, \omega_s, \omega_{s+1}, F) \qquad (F \in \mathscr{B}^{00},\ \omega_0 \in \Omega_0, \ldots, \omega_s \in \Omega_s).$$

If $P_1, \ldots, P_t$ *are stochastic, so are* $Q_0, \ldots, Q_t$.

3.2.2. *The so-called* **Fubini formula**

$$\text{(3)} \quad m(F) = \int m_0(d\omega_0) Q_0(\omega_0, F) = \int m_0(d\omega_0) \int P_1(\omega_0, d\omega_1) \cdots \int P_t(\omega_0, \ldots, \omega_{t-1}, d\omega_t) 1_F(\omega_0, \ldots, \omega_t) \qquad (F \in \mathscr{B}^{00})$$

defines a σ*-content* m *on* $\mathscr{B}^{00}$ *which we denote by* $m_0 \times P_1 \times \cdots \times P_t$ *or* $m_0 \times \prod_{s=1}^{t} P_s$. *We also write*

$$(\Omega, \mathscr{B}^{00}, m) = (\Omega_0, \mathscr{B}_0^{00}, m_0) \times (\Omega_1, \mathscr{B}_1^{00}, P_1) \times \cdots \times (\Omega_t, \mathscr{B}_t^{00}, P_t).$$

Proof. Fix any $F \in \mathscr{B}^{00}$. By Exercise 1.6.2 there are $F_s \in \mathscr{B}_s^{00} (s = 0, \ldots, t)$ such that $F \subseteq F_0 \times \cdots \times F_t$. We shall see that all functions that we have to integrate in the course of this proof vanish outside a set $F_0 \times \cdots \times F_s$ or F_s with $s \in \{0, \ldots, t\}$. We begin with Q_t, which is clearly an assembly of point masses one and a σ-content kernel with all values in [0, 1]. As a function of ω_t, $Q_t\ (\omega_0, \ldots, \omega_{t-1}, \omega_t, F)$ vanishes outside F_t and is in mble$(\Omega_t, \mathscr{B}^0(\mathscr{B}_t^{00}), \mathbb{R})$, hence, by Proposition III.9.5, in $\mathscr{L}^1_{P_t(\omega_0, \ldots, \omega_{t-1}, \cdot)}$ for every choice of $(\omega_0, \ldots, \omega_{t-1}) \in \Omega(0, t-1)$. Thus (2) makes sense for $s = t - 1$ and Q_{t-1} is well defined. The $\mathscr{B}^0(\mathscr{B}^{00}(0, t-1))$-measurability of Q_{t-1} follows from Theorem 2.4.2. Clearly, $Q_{t-1}(\omega_0, \ldots, \omega_{t-1}, F)$ vanishes outside $F_0 \times \cdots \times F_s$ and is bounded by any bound of $P_t(\omega_0, \ldots, \omega_{t-1}, F_t)$ within that set. This is enough to make (2) meaningful for $s = t - 2$. It is now clear how to proceed down to $s = 0$ and thus to finish the proof of the theorem.

3.3. Remarks

3.3.1. If m_s is a σ-content on $\mathscr{B}^{00}$ $(s = 0, \ldots, t)$, we can apply the above theorem to the σ-content kernels $P_s(\omega_0, \ldots, \omega_s, F) = m_s(F)$ $(F \in \mathscr{B}_s^{00}$, $s = 1, \ldots, t)$, and m_0, of course. The conditions of the theorem are in fact satisfied trivially. The resulting m is also denoted by $m = m_0 \times \cdots \times m_t$ or $\prod_{s=0}^t m_s$ and called the *product σ-content* of the $m_0, \ldots, m_t$ which are then called the *factors* of m. We then write also

$$(\Omega, \mathscr{B}^{00}, m) = (\Omega_0, \mathscr{B}_0^{00}, m_0) \times \cdots \times (\Omega_t, \mathscr{B}_t^{00}, m_t) = \prod_{s=0}^{t} (\Omega_s, \mathscr{B}_s^{00}, m_s).$$

For $F = F_0 \times \cdots \times F_t$, the Fubini formula (3) clearly boils down to

$$m(F) = m_0(F_0) \cdots m_t(F_t) \tag{4}$$

in our special case.

3.3.2. If $\mathscr{B}_s$ is a σ-field in Ω_s $(s = 0, \ldots, t)$ and m_0 a σ-content on $\mathscr{B}_0$, P'_s a $\mathscr{B}_{s-1}$-measurable stochastic kernel from Ω_{s-1} to Ω_s (with $\mathscr{B}_s$), then we may apply the above theorem to $\mathscr{B}_s^{00} = \mathscr{B}_s$ $(s = 0, \ldots, t)$, m_0, and the kernels P_s defined by

$$P_s(\omega_0, \ldots, \omega_{s-1}, F) = P'_s(\omega_{s-1}, F) \quad (\omega_{s-1} \in \Omega_{s-1}, \quad F \in \mathscr{B}_s, \quad s = 1, \ldots, n).$$

The conditions of the theorem are in fact satisfied trivially. Since $\mathscr{B}^{00}$ is a σ-field, we write $\mathscr{B}$ instead. The resulting m is also denoted by $m = m_0 \times P'_1 \times \cdots \times P'_t = m_0 \times \prod_{s=1}^t P'_s$ and called the *Markov σ-content* for the *initial distribution* m_0 and the *transition kernels* $P'_1, \ldots, P'_t$. The *Fubini formula* (3) becomes

(5)

$$m(1_F) = \int m_0(d\omega_0) \int P'_1(\omega_0, d\omega_1) \cdots \int P'_t(\omega_{t-1}, d\omega_t) 1_F(\omega_0, \ldots, \omega_t) \quad (F \in \mathscr{B})$$

in our special case.

3.3.3. It is easy to prove a Fubini theorem like 2.2 or 2.4 for our present situation of a product of finitely many spaces. The reader is advised to carry out the details.

3.3.4. It is an easy exercise to represent the Lebesgue σ-content on, say, the bounded Borel sets in $\mathbb{R}^n$ as a product of n copies of the Lebesgue σ-content in $\mathbb{R}$ $(n = 1, 2, \ldots)$.

4. COUNTABLY MANY FACTORS

In this section we proceed to cartesian products of countably many spaces. It turns out that a setup in terms of local σ-rings or even σ-rings is no longer feasible. We shall from now on deal with σ-fields only. Moreover, we shall assume all σ-contents involved to be normalized, i.e., probabilities.

4.1. Theorem. *Let* $(\Omega_t, \mathscr{B}_t, m_t)$ $(t = 0, 1, \ldots)$ *be σ-probability spaces,* $\Omega = \prod_{t=0}^{\infty} \Omega_t$, $\mathscr{B} = \prod_{t=0}^{\infty} \mathscr{B}_t$. *Then there is a unique σ-content m on $\mathscr{B}$ such that*

(1)
$$m(F_0 \times \cdots \times F_t \times \Omega_{t+1} \times \cdots) = m_0(F_0) \cdots m(F_t)$$
$$(t \geqq 0, \quad F_0 \in \mathscr{B}_0, \ldots, F_t \in \mathscr{B}_t).$$

m is called the **product (σ-probability)** *of (its factors)* $m_0, m_1, \ldots$ *and is also denoted by* $m = m_0 \times m_1 \times \cdots = \prod_{t=0}^{\infty} m_t$. *We call* $(\Omega, \mathscr{B}, m)$ *the* **product (σ-probability) space** *of* $(\Omega_t, \mathscr{B}_t, m_t)$ $(t = 0, 1, \ldots)$ *and write also*

$$(\Omega, \mathscr{B}, m) = (\Omega_0, \mathscr{B}_0, m_0) \times (\Omega_1, \mathscr{B}_1, m_1) \times \cdots = \prod_{t=0}^{\infty} (\Omega_t, \mathscr{B}_t, m_t).$$

A direct proof of this theorem is not less difficult than the proof of the next Theorem 4.3, of which Theorem 4.1 is a special case and to which the reader is referred.

4.2. Remark. Notice that (1) is insensitive against an enlargement of t as long as the set $F_0 \times \cdots \times F_t \times \Omega_{t+1} \times \cdots$ remains unchanged since $m_s(\Omega_s) = 1$ $(s = 0, 1, \ldots)$.

4.3. Theorem (C. IONESCU-TULCEA [1]). *Let $\Omega_t \neq \emptyset$, $\mathscr{B}_t \subseteq \mathscr{P}(\Omega_t)$ be a σ-field $(t = 0, 1, \ldots)$. Let $\Omega = \prod_{t=0}^{\infty} \Omega_t$, $\mathscr{B} = \prod_{t=0}^{\infty} \mathscr{B}_t$. For every $t = 0, 1, \ldots,$ let $\Omega(0, t) = \Omega_0 \times \cdots \times \Omega_t$, $\mathscr{B}(0, t) = \mathscr{B}_0 \times \cdots \times \mathscr{B}_t$. Let m_0 be a σ-content on $\mathscr{B}_0$. For every $t = 1, 2, \ldots,$ let P_t be a stochastic $\mathscr{B}(0, t-1)$-measurable σ-content kernel from $\Omega(0, t-1)$ to Ω_t (with $\mathscr{B}_t$) and $m_t = m_0 \times P_1 \times \cdots \times P_t$. Then there is a unique σ-content m on $\mathscr{B}$ such that*

$$(2) \qquad m(F_0 \times \cdots \times F_t \times \Omega_{t+1} \times \cdots) = m_t(F_0 \times \cdots \times F_t)$$
$$(t \geqq 0,\ F_0 \in \mathscr{B}_0, \ldots, F_t \in \mathscr{B}_t).$$

In particular, $m(\Omega) = m_0(\Omega_0)$ and m is a (σ-)probability iff m_0 is.

Proof. (2) is meaningful since m_t is defined on $\mathscr{B}(0, t)$. The sets $F_0 \times \cdots \times F_t \times \Omega_{t+1} \times \cdots$ considered in (2) form an intersection stable generator for $\mathscr{B}$, hence m is, by Proposition I.5.3, uniquely determined by (2). Let us now prove existence. We begin by noticing that $m_0, m_1, \ldots$ are compatible in the following sense: Choose any $F \in \mathscr{B}(0, t)$; then

$$m_{t+1}(F \times \Omega_{t+1}) = \int m_0(d\omega_0) \int P_1(\omega_0, d\omega_1) \cdots \int P_t(\omega_0, \ldots, \omega_{t-1}, d\omega_t)$$
$$\int P_{t+1}(\omega_0, \ldots, \omega_t, d\omega_{t+1}) 1_F(\omega_0, \ldots, \omega_t) 1_{\Omega_{t+1}}(\omega_{t+1})$$
$$= \int m_0(d\omega_0) \int P_1(\omega_0, d\omega_1)$$
$$\cdots \int P_t(\omega_0, \ldots, \omega_{t-1}, d\omega_t) 1_F(\omega_0, \ldots, \omega_t)$$
$$= m_t(F) \qquad (t = 0, 1, \ldots).$$

These compatibilities imply that a content m is well defined on the field $\mathscr{F} = \bigcup_{t=0}^{\infty} \{F \times \Omega_{t+1} \times \cdots \mid F \in \mathscr{B}(0, t)\}$ by

$$\tilde{m}(F \times \Omega_{t+1} \times \cdots) = m_t(F) \qquad (t \geqq 0, \quad F \in \mathscr{B}(0, t)).$$

$\tilde{m}$ is additive: If $E_1, \ldots, E_n \in \mathscr{F}$, $E_j \cap E_k = \emptyset$ $(j \neq k)$, then there is some t

and $F_1, \ldots, F_n \in \mathscr{B}(0, t)$ such that $E_k = F_k \times \Omega_{t+1} \times \cdots$, hence $F_j \cap F_k = \varnothing$ $(j \neq k)$ and $E_1 + \cdots + E_n = (F_1 + \cdots + F_n) \times \Omega_{t+1} \times \cdots$,

$$\begin{aligned}\tilde{m}(E_1 + \cdots + E_n) &= m_t(F_1 + \cdots + F_n) \\ &= m_t(F_1) + \cdots + m_t(F_n) = \tilde{m}(E_1) + \cdots + \tilde{m}(E_n).\end{aligned}$$

The theorem is proved by Corollary II.2.4 as soon as we have shown that $\tilde{m}: \mathscr{F} \to \mathbb{R}_+$ is σ-additive. For this, we use the criterion of Proposition I.2.2. Let $F_1, F_2, \ldots \in \mathscr{F}$, $F_1 \supseteq F_2 \supseteq \cdots$, $F_1 \cap F_2 \cdots = \varnothing$. We have to prove $\lim \tilde{m}(F_n) = 0$. For this, we first construct, for every $t = 1, 2, \ldots$, $\mathscr{B}(0, s)$-measurable stochastic σ-content kernels Q_s^t from $\Omega(0, s)$ to $\Omega(0, t)$ (with $\mathscr{B}(0, t)$) $(s = 0, \ldots, t)$ according to Theorem 3.2.1. For every $s \geqq 0$, $(\omega_0, \ldots, \omega_s) \in \Omega(0, s)$, the σ-contents $Q_s^t(\omega_0, \ldots, \omega_s, \cdot)$ $(t = s, s+1, \ldots)$ are clearly compatible in the sense that

$$Q_s^t(\omega_0, \ldots, \omega_s, F) = Q_s^{t+1}(\omega_0, \ldots, \omega_s, F \times \Omega_{t+1})$$
$$(F \in \mathscr{B}(0, t), \quad t = s, s+1, \ldots)$$

and thus lead to content kernels Q_s from $\Omega(0, s)$ to $\mathscr{F}$ via

$$Q_s(\omega_0, \ldots, \omega_s, F_t \times \Omega_{t+1} \times \cdots) = Q_s^t(\omega_0, \ldots, \omega_s, F_t)$$
$$(F \in \mathscr{B}(0, t), \quad t = 0, 1, \ldots).$$

We now get

$$\tilde{m}(F_n) = \int m_0(d\omega_0) Q_0(\omega_0, F_n) \qquad (n = 1, 2, \ldots).$$

Assume now that $\lim \tilde{m}(F_n) = \alpha_0 > 0$. Then the monotone convergence theorem III.5.3 yields the existence of some $\omega_0 \in \Omega_0$ and an $\alpha_1 > 0$ with $\lim Q_0(\omega_0, F_n) = \alpha_1$. But $Q_0(\omega_0, F_n) = \int P_1(\omega_0, d\omega_1) Q_1(\omega_0, \omega_1, F_n)$ and thus there is an $\omega_1 \in \Omega_1$ and an $\alpha_2 > 0$ such that $\lim Q_1(\omega_0, \omega_1, F_n) = \alpha_2 > 0$. Continuing in this way, we build up a point $\omega = (\omega_0, \omega_1, \ldots)$ such that $Q_s(\omega_0, \ldots, \omega_s, F_n) \geqq \alpha_{s+1} > 0$ $(s = 0, 1, \ldots; n = 1, 2, \ldots)$. For fixed n and s sufficiently large, we have $F_n = F_{sn} \times \Omega_{s+1} \times \cdots$ with some $F_{sn} \in \mathscr{B}(0, s)$ and $Q_s(\omega_0, \ldots, \omega_s, F_n) = Q_s^s(\omega_0, \ldots, \omega_s, F_{sn}) = 1_{F_{sn}}(\omega_0, \ldots, \omega_s) > 0$, hence $(\omega_0, \ldots, \omega_s) \in F_{sn}$, i.e., $\omega \in F_n$. Thus $\omega \in F_1 \cap F_2 \cap \cdots$, a contradiction.

4.4. Remarks

4.4.1. Extension procedures which can be considered as obvious at this stage of the book prove the following: For every $F \in \mathscr{B}(0, t)$, we have

$$m(F \times \Omega_{t+1} \times \cdots) = \int m_0(d\omega_0) \int P_1(\omega_0, d\omega_1) \cdots \int P_t(\omega_0, \ldots \omega_{t-1}, d\omega_t) 1_F(\omega_0, \ldots, \omega_t).$$

For every bounded $f' \in \text{mble}(\Omega(0, t), \mathscr{B}(0, t), \mathbb{R})$ and f defined on Ω by $f(\omega) = f(\omega_0, \omega_1, \ldots) = f'(\omega_0, \ldots, \omega_t)$, we have

$$\int f \, dm = \int m_0(d\omega_0) \cdots \int P_t(\omega_0, \ldots, \omega_{t-1}, d\omega_t) f'(\omega_0, \ldots, \omega_t).$$

4.4.2. Let us assume in addition that for $t = 1, 2, \ldots$, there is a $\mathscr{B}_{t-1}$-measurable stochastic kernel P' from Ω_{t-1} to Ω_t (with $\mathscr{B}_t$) such that $P_t(\omega_0, \ldots, \omega_{t-1}, F) = P'_t(\omega_{t-1}, F)$ $(\omega_0 \in \Omega_0, \ldots, \omega_{t-1} \in \Omega_{t-1}, F \in \mathscr{B}_t)$. Then there is a unique σ-content m on $\mathscr{B}$ such that

$$m(F_0 \times \cdots \times F_t \times \Omega_{t+1} \times \cdots) = \int m_0(d\omega_0) 1_{F_0}(\omega_0) \cdots \int P'_t(\omega_{t-1}, F_t)$$

$$(F_0 \in \mathscr{B}_0, \ldots, F_t \in \mathscr{B}_t).$$

m is a σ-probability content if m_0 is m is called the Markov σ-content in Ω, for the initial distribution m_0 and the transition kernels $P'_1, P'_2, \ldots$.

4.4.3. Let us assume in addition that $\Omega_0, \Omega_1, \ldots$ are finite and $\mathscr{B}_t = \mathscr{P}(\Omega_t)$ $(t = 0, 1, \ldots)$. In this case the P'_t of exercise 3 boil down to matrices $P'_t = (P'_t(\xi, \eta))_{\xi \in \Omega_{t-1}, \eta \in \Omega_t}$ with $|\Omega_{t-1}|$ rows and $|\Omega_t|$ columns, and the properties

$$P'_t(\xi, \eta) \geqq 0, \qquad \sum_{\tau \in \Omega_t} P'_t(\xi, \tau) = 1 \quad (\xi \in \Omega_{t-1}, \eta \in \Omega_t).$$

In fact we have only to put $P'_t(\omega_{t-1}, \{\omega_t\}) = P'_t(\omega_{t-1}, \omega_t)$ in order to get the desired identification. But now we are back to Examples I.1.14, I.2.5, and Definition II.2.5.3, whose results appear as a special case of the above Theorem 4.3.

The reader is no doubt aware of the important role that the natural ordering of the index set $\{0, 1, \ldots\}$ plays in Theorem 4.3. Actually, the theorem does not even answer the existence question for a "two-sided" analogue of Markov σ-contents. The second half of the next section will be devoted to a theorem of Kolmogorov which allows one to answer even much more general questions. We shall need a considerable but rather obvious notational machinery in order to formulate that theorem appropriately. Moreover, it shows a significant difference from Theorem 4.3 in that it always makes some regularity assumptions on the spaces Ω_ι.

5. ARBITRARILY MANY FACTORS

In this section we set aside all restrictions on the power or other structure of the index set $I = \{\iota, \kappa, \ldots\} \neq \varnothing$ employed in the formation of product spaces. In particular, the case of uncountably many factors is now under consideration. Throughout this section we shall exclusively deal with σ-fields and (σ-)probabilities.

The leading idea of this section is compatibility. We shall set up a considerable but rather obvious machinery for its appropriate formulation. Next to this, the idea of reduction to the case of countably many factors, as prepared by Proposition I.4.7.5 and Exercises 1.3.6, 1.6.1, will be predominant. It will turn out that these two ideas are fully sufficient in order to prove the existence of product σ-contents, i.e., the analogue of Theorem 4.1, for uncountably many factors. As soon as we step away from product σ-contents, regularity assumptions on the component spaces are demanded, and we enter into the realm of what is usually called Kolmogorov's theorem. We shall prove it in a variety of versions.

We now begin to build the notational machinery for the idea of compatibility.

5.1. Definition. *Let $I = \{\iota, \ldots\} \neq \varnothing$. For every $\iota \in I$, let $\mathscr{B}$ be a σ-field in $\Omega_\iota \neq \varnothing$. For every $\varnothing \neq J \subseteq I$, let $\Omega_J = \prod_{\iota \in J} \Omega_\iota$, $\mathscr{B}_J = \prod_{\iota \in J} \mathscr{B}_\iota$; for $\varnothing = J \subseteq K \subseteq I$, let φ_J^K be the* **natural projection** *of Ω_K onto Ω_J: $\omega = (\omega_\kappa)_{\kappa \in K} \in \Omega_K \Rightarrow [\varphi_J^K \omega]_\iota = \omega_\iota \quad (\iota \in J)$. Put $\Omega = \Omega_I$, $\mathscr{B} = \mathscr{B}_I$, $\varphi_J = \varphi_J^I$ $(\varnothing \neq J \subseteq I)$. For $\varphi_{\{\iota\}}^J$ resp. $\varphi_{\{\iota\}}$, we write φ_ι^J, resp. φ_ι for brevity.*

If m is a σ-content on $\mathscr{B}$ and $\varnothing \neq J \subseteq I$, then $\varphi_J m$ is called the **marginal** *(σ-content) of m for J. A marginal for a finite $J \subseteq I$ is also called a* **finite-dimensional marginal.** *Let $\mathscr{J}$ be any family of nonempty subsets of I. A family $(m_J)_{J \in \mathscr{J}}$ of probability σ-contents m_J on $\mathscr{B}_J$ $(J \in \mathscr{J})$ is called* **compatible or projective** *if*

$$\varphi_J^K m_K = m_J \qquad (J, K \in \mathscr{J}, \quad J \subseteq K)$$

holds.

5.2. Remarks

5.2.1. The exact definition of φ_J^K for $\varnothing \neq J \subseteq K \subseteq I$ can be written out with the help of the component mappings $\varphi_\iota^K: (\omega_\kappa)_{\kappa \in K} \to \omega_\iota$ resp. φ_ι^J as follows:

$$\varphi_\iota^J \circ \varphi_J^K = \varphi_\iota^K \qquad (\iota \in J).$$

In view of Definition 1.7 we could write as well

$$\varphi_J^K = \prod_{\iota \in J} \varphi_\iota^K.$$

The effect of φ_J^K may briefly be described as "throwing away all components not in J."

5.2.2. We are now in a position to formulate the main problem treated in this section. Let $I, \Omega_\iota, \mathscr{B}_\iota$ $(\iota \in I)$, etc. be as in the above definition. Let $\mathscr{J}$ be an increasingly filtered system of nonempty subsets of I such that $I = \bigcup_{J \in \mathscr{J}} J$ and $(m_J)_{J \in \mathscr{J}}$ a projective family of σ-probabilities. Does there exist a σ-probability m on Ω such that its family of marginals for $\mathscr{J}$ is exactly $(m_J)_{J \in \mathscr{J}}$? If yes, is m uniquely determined by this condition? One of the most important special cases is the one where $\mathscr{J}$ is the system of all nonempty finite subsets of I. We shall start our presentation with a theorem treating a particularly simple case, namely, the case of product σ-probabilities. This theorem can be formulated without mentioning compatibility, but this concept will appear in its proof. It is therefore a suitable prelude to the more general theory which is to be presented subsequently.

5.3. Theorem. *Let $I = \{\iota, \ldots\} \neq \varnothing$ be arbitrary and $(\Omega_\iota, \mathscr{B}_\iota, m_\iota)$ $(\iota \in I)$ σ-probability spaces. Let $(\Omega, \mathscr{B}) = \prod_{\iota \in I} (\Omega_\iota, \mathscr{B}_\iota)$. Then there is a unique σ-probability m on $\mathscr{B}$ such that for any choice of a natural number $n > 0$, pairwise distinct $\iota_1, \ldots, \iota_n \in I$, $F_{\iota_1} \in \mathscr{B}_{\iota_1}, \ldots, F_{\iota_n} \in \mathscr{B}_{\iota_n}$, $F_\iota = \Omega_\iota$ $(\iota \notin \{\iota_1, \ldots, \iota_n\})$ the equality*

$$(1) \qquad m\left(\prod_{\iota \in I} F_\iota\right) = \prod_{\iota \in I} m_\iota(F_\iota) = \prod_{k=1}^{n} m_{\iota_k}(F_{\iota_k})$$

holds. m is called the **product (σ-probability content)** *of the σ-probabilities m $(\iota \in I)$ and also denoted by $m = \prod_{\iota \in I} m_\iota$. We also write $(\Omega, \mathscr{B}, m) = \prod_{\iota \in I} (\Omega_\iota, \mathscr{B}_\iota, m)$ and call this product the* **σ-probability space** *of the family $((\Omega_\iota, \mathscr{B}_\iota, m_\iota))_{\iota \in I}$ of σ-probability spaces.*

Proof. The case of an at most countable I has already been settled in Remark 3.3.1 and Theorem 4.1. Let now I be uncountable. Let $\mathscr{I}$ denote the set of all countable subsets of I. For any $J \in \mathscr{I}$, let $(\Omega_J, \mathscr{B}_J, m_J) = \prod_{\iota \in J} (\Omega_\iota, \mathscr{B}_\iota, m_\iota)$ and denote by $\mathscr{B}(J)$ the σ-field in Ω generated by $(\varphi_\iota)_{\iota \in J}$. It is an easy exercise to show that $(m_J)_{J \in \mathscr{I}}$ is projective (go back to Theorem 4.1 and see what $(\varphi_J^K)^{-1}$ of a set $\prod_{\iota \in J} F_\iota$ looks like).

Exercise 1.3.6 (see also Proposition I.4.7.5) shows

$$\mathscr{B} = \bigcup_{J \in \mathscr{J}} \mathscr{B}(J). \tag{2}$$

Fix any $J \in \mathscr{J}$. Since $\varphi_J\colon \Omega \to \Omega_J$ is surjective, the associated pullback $\varphi_J^{-1}\colon \mathscr{P}(\Omega) \leftarrow \mathscr{P}(\Omega_J)$ is injective, sends $\mathscr{B}_J$ onto $\mathscr{B}(J)$, and commutes with all set operations. Clearly there is a unique σ-probability m'_J on $\mathscr{B}(J)$ such that $\varphi_J m'_J = m_J$. Let now $J, K \in \mathscr{J}$ be such that $J \subseteq K$ and choose any $F \in \mathscr{B}(J)$. Let $E \in \mathscr{B}_J$ be such that $F = \varphi_J^{-1}E$. Put $G = (\varphi_J^K)^{-1}E$. Then $F = \varphi_K^{-1}G$ and $\varphi_J^K m_K = m_J$ implies

$$m'_K(F) = m_K(G) = m_J(E) = m'_J(F).$$

This shows the following: There is a unique m on $\mathscr{B}$ such that for every $F \in \mathscr{B}$ and for every $J \in \mathscr{J}$ with $F \in \mathscr{B}(J)$, we have $m(F) = m'_J(F)$. In particular, m satisfies (1). We still have to prove that m is σ-additive. But this is a nearly obvious consequence of (2): Choose any pairwise disjoint $F_1, F_2, \ldots \in \mathscr{B}$. For every $k = 1, 2, \ldots$, find $J_k \in \mathscr{J}$ such that $F_k \in \mathscr{B}(J_k)$. Put $J = J_1 \cup J_2 \cup \cdots$. Clearly $J \in \mathscr{J}$ again and $F_1, F_2, \ldots \in \mathscr{B}(J)$. Now $F_1 + F_2 + \cdots \in \mathscr{B}(J)$ follows and

$$\begin{aligned} m(F_1 + F_2 + \cdots) &= m_J(F_1 + F_2 + \cdots) \\ &= m_J(F_1) + m_J(F_2) + \cdots = m(F_1) + m(F_2) + \cdots \end{aligned}$$

The exercise comprised in the proof of this theorem has provided us with a very simple example of a projective family $(m_J)_{J \in \mathscr{J}}$. The rest of this section is devoted to a study of projective families in full generality. We begin with a few general observations, some of which are nothing but explicit statements of observations already made during the proof of the preceding theorem.

5.4. Proposition. *Under the assumptions of Definition* 5.1:

5.4.1. *If m is a probability σ-content on Ω and $\mathscr{J}$ a family of nonempty subsets of I, then the family $(m_J)_{J \in \mathscr{J}}$, where m_J is the marginal σ-content of m for J, is projective.*

5.4.2. *If $\mathscr{J}$ is an increasingly filtered covering of I with nonempty sets and $(m_J)_{J \in \mathscr{J}}$ is a projective family of probability σ-contents, then there is at most one probability σ-content m on Ω such that m_J is the marginal σ-content for J of m $(J \in \mathscr{J})$.*

5.4.3. *If $\mathscr{J}$ is an increasingly filtered family of nonempty subsets of I, then so is $\mathscr{K} = \bigcup_{J \in \mathscr{J}} \mathscr{P}(J)$ and $\mathscr{J}$ covers I iff $\mathscr{K}$ does. If $(m_J)_{J \in \mathscr{J}}$ is a projective family of probability σ-contents, and if for every $K \in \mathscr{K}$, $K \subseteq J \in \mathscr{J}$,*

$$m_K = \varphi_K^J m_J, \tag{3}$$

then every m_K $(K \in \mathscr{K})$ *is well defined (independently of the choice of* J) *and the family* $(m_K)_{K \in \mathscr{K}}$ *is projective.*

5.4.4. *If* $\mathscr{J}$ *is an increasingly* σ*-filtered covering of* I*, then for every projective family* $(m_J)_{J \in \mathscr{J}}$ *of probability* σ*-contents there is exactly one probability* σ*-content* m *on* $\mathscr{B}$ *such that* $\varphi_J m = m_J$ $(J \in \mathscr{J})$.

Proof. The relations

$$\varphi_J^K \circ \varphi_K^L = \varphi_J^L \qquad (\varnothing \neq J \subseteq K \subseteq L \subseteq I) \tag{5}$$

are obvious from the definition of the mappings involved.

1. For $J, K \in \mathscr{J}$, $J \subseteq K$, we have

$$m_J = \varphi_J m = (\varphi_J^K \circ \varphi_K) m = \varphi_J^K m_K ,$$

proving that $(m_J)_{J \in \mathscr{J}}$ is projective.

2. We put $\mathscr{B}(J) = \varphi_J^{-1}\mathscr{B}_J$ $(J \subseteq I)$ as in the proof of Theorem 5.3. $\mathscr{B}(J)$ is, by Proposition 1.8, nothing but the σ-field in Ω generated by $(\varphi_\iota)_{\iota \in J}$. It follows that $\mathscr{B}(J) \subseteq \mathscr{B}(K)$ $(\varnothing \neq J \subseteq K \subseteq I)$. In particular $\bigcup_{J \in \mathscr{J}} J = I$ implies that $\mathscr{R} = \bigcup_{J \in \mathscr{J}} \mathscr{B}(J)$ contains all $\varphi_\iota^{-1}\mathscr{B}_\iota$ and hence generates $\mathscr{B}$. Since $\mathscr{J}$ is increasingly filtered, so is $(\mathscr{B}(J))_{J \in \mathscr{J}}$, and hence $\mathscr{R}$, being an increasingly filtered union of fields in Ω, is a field again (Exercise I.1.5.9). In particular, $\mathscr{R}$ is stable under finite intersections.

By proposition I.5.3 a σ-probability on $\mathscr{B}$ is uniquely determined by its restriction to $\mathscr{R}$. But its restriction to $\mathscr{R}$ is uniquely determined by the marginals m_J $(J \in \mathscr{J})$: If $J \in \mathscr{J}$, $F \in \mathscr{B}(J)$, say $F = \varphi_J^{-1}E$ for a suitable $E \in \mathscr{B}_J$, then $m(F) = m_J(E)$.

3 is an easy exercise which we leave to the reader.

4 has essentially been proved in the proof of Theorem 5.3. The easy task of adapting the ideas displayed there to our present situation is left as an exercise to the reader.

We now settle our Problem 5.2.2 in the case of compact component spaces.

5.5. Theorem (Kolmogorov [1, 2]). *Let* $0 \neq I$ *and* Ω_ι *be a compact Hausdorff space,* $\mathscr{B}_\iota$ *the Baire* σ*-field in* Ω_ι $(\iota \in I)$. *Then*

5.5.1. *For every* $\varnothing \neq J \subseteq I$, $\mathscr{B}_J = \prod_{\iota \in J} \mathscr{B}_\iota$ *is the Baire* σ*-field of the compact Hausdorff space* $\Omega_J = \prod_{\iota \in J} \Omega_\iota$.

5.5.2. *Let* $\mathscr{J}$ *be an increasingly filtered family of finite nonempty subsets of* I *such that* $I = \bigcup_{J \in \mathscr{J}} J$. *With the notations of Definition 5.1, let* $(m_J)_{J \in \mathscr{J}}$ *be a projective family of* σ*-probabilities. Then there is a unique* σ*-probability* m *on the Baire* σ*-field* $\mathscr{B} = \prod_{\iota \in I} \mathscr{B}_\iota$ *in* $\Omega = \prod_{\iota \in I} \Omega_\iota$ *such that* $(m_J)_{J \in \mathscr{J}}$ *is the family of its marginals for* $\mathscr{J}$.

Proof. For 1, see Exercise 1.6.5. We sketch a proof of 1 and 2 at the same time. The uniqueness statement of our theorem is clear from Proposition 5.4.2. Let us now prove existence. By Urysohn's theorem, $\mathscr{C}(\Omega_\iota, \mathbb{R})$ separates the points of Ω_ι $(\iota \in I)$. This implies that $\mathscr{C}(\Omega_J, \mathbb{R})$, which contains all functions of the form $f \circ \varphi_\iota$ $(\iota \in J, f \in \mathscr{C}(\Omega_\iota, \mathbb{R}))$, separates the points of Ω_J $(J \in \mathscr{J})$. Let $\mathscr{C}(J)$ consist of all functions $f \circ \varphi_J$ where $f \in \mathscr{C}(\Omega_J, \mathbb{R})$ $(J \in \mathscr{J})$. Clearly, $(\mathscr{C}(J))_{J \in \mathscr{J}}$ is an increasingly filtered family of subspaces of $\mathscr{C}(\Omega, \mathbb{R})$, and $\mathscr{C} = \bigcup_{J \in \mathscr{J}} \mathscr{C}(J)$ separates the points of Ω and is an algebra of real functions. By the Weierstrass–Stone theorem, $\mathscr{C}$ is dense in $\mathscr{C}(\Omega, \mathbb{R})$ with respect to uniform approximation. At the same time, it is an elementary domain of Ω since it clearly contains all constants and is a vector lattice. Since $\varphi_J\colon \Omega \to \Omega_J$ is surjective, the function mapping $f \circ \varphi_J \leftarrow f$ of $\mathscr{C}(\Omega_J, \mathbb{R})$ on $\mathscr{C}(J)$ is bijective, and hence a (positive) measure m'_J is uniquely defined on $\mathscr{C}(J)$ by $m'_J(f \circ \varphi_J) = m_J(f)$ $(f \in \mathscr{C}(\Omega_J, \mathbb{R}))$.

The compatibility of $(m_J)_{J \in \mathscr{J}}$ now shows: $J, K \in \mathscr{J}$, $J \subseteq K$ implies that m'_J is the restriction of m'_K to $\mathscr{C}(J)$. In fact the integral transform formula (Theorem IV.2.4) shows that for every $f \in \mathscr{C}(\Omega_J, \mathbb{R})$ and $g = f \circ \varphi_J^K$,

$$m'_K(f \circ \varphi_J) = m'_K(g \circ \varphi_K) = \int g \, dm_K = \int f \, dm_J = m'_J(f \circ \varphi_J).$$

It follows that there is a unique $m'\colon \mathscr{C} \to \mathbb{R}$ such that $J \in \mathscr{J}, f \in \mathscr{C}(\Omega_J, \mathbb{R})$ implies $m'(f \circ \varphi_J) = m_J(f)$. Clearly, m is a (positive) measure on $\mathscr{C}$, and Dini's theorem shows that it is σ- (and even τ-) continuous (see the proof of Theorem I.10.2). The local σ-ring $\mathscr{B}_m^{00}$ (Theorem III.6.1) clearly is a σ-field containing $\mathscr{B}$. The restriction to $\mathscr{B}$ of the σ-probability derived from m will again be denoted by m. The corresponding σ-integral coincides, by Theorem III.6.4, with m on $\mathscr{C}$, and this clearly proves all statements of the theorem.

The reader should notice that compactness allowed us a very smooth approach here, avoiding the idea of reduction to the case of a countable index set.

Next we settle our Problem 5.2.2 in the case of Polish component spaces.

5.6. Theorem (KOLMOGOROV [1, 2]). *Let $\emptyset \neq I$ and Ω_ι be a* **Polish** *space, $\mathscr{B}_\iota$ the Baire (= Borel) σ-field in Ω_ι $(\iota \in I)$. Then:*

5.6.1. *For every at most countable $\emptyset \neq J \subseteq I$, $\mathscr{B}_J = \prod_{\iota \in J} \mathscr{B}_\iota$ is the Baire (= Borel) σ-field in the Polish space $\Omega_J = \prod_{\iota \in J} \Omega_\iota$.*

5.6.2. *Let $\mathscr{J}$ be an increasingly filtered family of finite nonempty subsets of I such that $I = \bigcup_{J \in \mathscr{J}} J$. With the notations of Definition 5.1, let $(m_J)_{J \in \mathscr{J}}$ be a projective family of σ-probabilities. Then there is a unique σ-probability*

m on the σ-field $\mathscr{B} = \prod_{\iota \in I} \mathscr{B}_\iota$ in $\Omega = \prod_{\iota \in I} \Omega_\iota$ such that $(m_J)_{J \in \mathscr{J}}$ is the family of its marginals for $\mathscr{J}$.

Proof. 1. Since J is countable here, we may assume $J = \{1, 2, \ldots\}$. If $|\cdot, \cdot|_k$ is a complete metric in Ω_k defining the topology of Ω_k, then

$$|\omega, \eta| = \sum_{k=1}^{\infty} \frac{1}{2^k} [|\omega_k, \eta_k| \wedge 1] \qquad (\omega = (\omega_1, \omega_2, \ldots), \quad \eta = (\eta_1, \eta_2, \ldots) \in \Omega)$$

defines a complete metric in Ω_J which defines the product topology in Ω_J. If $\omega_k^1, \omega_k^2, \ldots$ is a dense sequence in Ω_k $(k = 1, 2, \ldots)$, then the points of the form $(\omega_1^{n_1}, \ldots, \omega_r^{n_r}, \omega_{r+1}^1, \omega_{r+2}^1, \ldots)$ $(r > 0, n_1, \ldots, n_r \geqq 1)$ form a countable dense subset of Ω_J. It is an easy exercise to prove that every open set in Ω_J is a countable union of sets of the form $G_1 \times \cdots \times G_r \times \Omega_{r+1} \times \cdots$ where $r > 0$ and $G_1 \subseteq \Omega_1, \ldots, G_r \subseteq \Omega_r$ are open. This proves that the Baire σ-field in Ω_J is contained in $\prod_{\iota \in J} \mathscr{B}_\iota$. The opposite inclusion is practically obvious.

2. Uniqueness is clear from Proposition 5.4.2. As for existence, we first show that it suffices to consider the case of a countable I. If I is finite, there is nothing to prove. Assume now that the theorem is true in every situation involving a countable index set. Let $\mathscr{K}$ be the family of all unions of countable increasingly filtered subfamilies of $\mathscr{J}$. It is easy to see that $\mathscr{K}$ contains $\mathscr{J}$, is increasingly σ-filtered, and covers I (exercise). By assumption we get, for every $K \in \mathscr{K}$, a unique probability σ-content m_K on $\mathscr{B}_K$ (in Ω_K) such that $\varphi_J^K m_K = m_J$ $(\mathscr{J} \ni J \subseteq K)$. Now $(m_K)_{K \in \mathscr{K}}$ is projective: Let $K, L \in \mathscr{K}$, $K \subseteq L$, then

$$\varphi_J^K m_K = m_J = \varphi_J^L m_L = \varphi_J^K(\varphi_K^L m_L) \qquad (\mathscr{J} \ni J \subseteq K).$$

This means that m_K and $\varphi_K^L m_K$ agree on $\bigcup_{\mathscr{J} \ni J \subseteq K} (\varphi_J^K)^{-1} \mathscr{B}_J$, which is easily seen to be a field generating $\mathscr{B}_K$; Proposition I.5.4 now proves $m_K = \varphi_K^L m_K$. By Proposition 5.4.4 the unique existence of an m on $\mathscr{B}$ with $\varphi_K m = m_K$ $(K \in \mathscr{K})$ follows. In order to complete the proof, we still have to settle the case of a countable I, which we now take under consideration. We may now assume $I = \{1, 2, \ldots\}$. By Proposition 5.4.3 we may also assume that $\mathscr{J}$ consists of all finite nonempty sets $\subseteq I$, and hence contains in particular the one-element subsets of I. As in the proof of Theorem 5.5, we choose an approach dealing with measures and integrals. For every $J \in \mathscr{J}$, let $\mathscr{C}(J) = \{f \circ \varphi_J \mid f \in \mathscr{C}^{\mathrm{b}}(\Omega_J, \mathbb{R})\}$. It is an easy exercise (see the proof of Theorem 5.5) to prove that $\mathscr{C} = \bigcup_{J \in \mathscr{J}} \mathscr{C}(J)$ is an elementary domain on Ω which contains all constants and separates the points of Ω (since $\mathscr{C}^{\mathrm{b}}(\Omega_J, \mathbb{R})$ separates the points of the Polish space Ω_J, for every $J \in \mathscr{J}$) and that there is a unique measure m on $\mathscr{C}$ such that $m(f \circ \varphi_J) = m_J(f)$ $(J \in \mathscr{J}$, $f \in \mathscr{C}^{\mathrm{b}}(\Omega_J, \mathbb{R}))$. Let us now prove that $m\colon \mathscr{C} \to \mathbb{R}$ is tight. For this, we observe that for every $k = 1, 2, \ldots, (\Omega_K, \mathscr{B}_K, m_K)$ is a tight σ-probability space (Exercise III.6.5.6). Choose now any $\varepsilon > 0$ and determine a compact $K_k \subseteq \Omega_k$ such that $m_k(\Omega_k \backslash K_k) \leqq \varepsilon/2^k$ $(k = 1, 2, \ldots)$. The set $K = K_1 \times K_2 \times \cdots \subseteq \Omega$ is now compact again (by Tychonov's theorem, which might be replaced by a simple direct argument here (exercise)). Let now $f \in \mathscr{C}$, $|f| \leqq 1$ everywhere and assume that $|f| \leqq \delta$ on K. Find $n \geqq 1$ such that $f \in \mathscr{C}(\{1, \ldots, n\})$, i.e., $f(\omega) = f'(\omega_1, \ldots, \omega_n)$ $(\omega = (\omega_1, \omega_2, \ldots) \in \Omega)$ for some $f' \in \mathscr{C}^{\mathrm{b}}(\Omega_{\{1,\ldots,n\}}, \mathbb{R})$. If $(\omega_1, \ldots, \omega_n) \in K_1 \times \cdots \times K_n$, we may choose any $\omega_{n+1} \in K_{n+1}, \ldots$ and get $\omega = (\omega_1, \ldots, \omega_n, \omega_{n+1}, \ldots) \in K_1 \times \cdots K_n \times K_{n+1} \times \cdots = K$, hence $|f'(\omega_1, \ldots, \omega_n)| = |f(\omega_1, \ldots, \omega_n, \omega_{n+1}, \ldots)| < \delta$. We conclude

$$
\begin{aligned}
m(f) = \int f' \, dm_{\{1,\ldots,n\}} &\leqq \delta m_{\{1,\ldots,n\}}(K_1 \times \cdots \times K_n) \\
&\quad + 1 \cdot m(\Omega_{\{1,\ldots,n\}} \backslash (K_1 \times \cdots \times K_n)) \\
&\leqq \delta + \sum_{k=1}^{n} m_k(\Omega_k \backslash K_k) < \delta + \sum_{k=1}^{n} \frac{\varepsilon}{2^k} = \delta + \varepsilon.
\end{aligned}
$$

Clearly, this proves the tightness of $m: \mathscr{C} \to \mathbb{R}$. $\mathscr{L}_m^1$ contains all indicator functions of open sets in Ω (use, e.g., the exercise comprised in part 1 of this proof), hence $\mathscr{B}_m^{00}$ contains $\mathscr{B}$. Here we use Theorem III.6.1, and Theorem III.6.4 shows that $m(f) = \int f \, dm$ $(f \in \mathscr{C})$ where m in the right-hand member means the σ-content derived from the σ-measure $m: \mathscr{C} \to \mathbb{R}$. Even its restriction to $\mathscr{B}$ would suffice here. From this it is obvious that $\varphi_J m = m_J$ $(J \in \mathscr{J})$, and the desired existence statement is proved. Uniqueness follows from Proposition 5.4.2. The theorem is proved.

5.7. Exercises

5.7.1. Use Theorem 5.3 on the inner compact regularity of σ-contents in Polish spaces in order to give an alternative proof of theorem 5.6 entirely in terms of σ-contents, avoiding measures (see also the proof of the subsequent theorem 5.10).

The next theorem will solve our Problem 5.2.2 in a rather abstract situation, containing Theorems 5.5 and 5.6 as special cases. Before we formulate it, a few preparatory investigations about abstract σ-compact set systems (Definition V.1.4) are in order. We begin with an exercise:

5.7.2. Let $A \neq \varnothing$ be arbitrary. In the space

$$A^{\mathbb{N}} = \{\omega = (\omega_1, \omega_2, \ldots) | \omega_1, \omega_2, \ldots \in A\}$$

of all sequences in A we consider the so-called *special cylinders*, i.e., all sets of the form

$$[a_1, \ldots, a_n] = \{\omega = (\omega_1, \omega_2, \ldots) | \omega_1 = a_1, \ldots, \omega_n = a_n\}$$
$$(n > 0, a_1, \ldots, a_n \in A).$$

Prove that the system $\mathscr{K}$ of all such special cylinders is σ-compact.

5.8. Proposition. *Let $\mathscr{K}$ be a σ-compact system of subsets of $\Omega \neq \varnothing$.*

5.8.1. *Let $X \neq \varnothing$ and $\varphi: X \to \Omega$ be a surjective mapping. Then the system $\varphi^{-1}\mathscr{K} = \{\varphi^{-1}K | K \in \mathscr{K}\}$ of subsets of X is σ-compact.*

5.8.2. *The set systems*

$$\mathscr{K}_{\cup f} = \{K_1 \cup \cdots \cup K_n | n > 0, K_1, \ldots, K_n \in \mathscr{K}\}$$
$$\mathscr{K}_{\delta} = \{K_1 \cap K_2 \cap \cdots | K_1, K_2, \ldots \in \mathscr{K}\}$$

are σ-compact, and consequently so is the system $(\mathscr{K}_{\delta})_{\cup f}$ of all finite unions of countable intersections of sets from $\mathscr{K}$.

Proof. 1. Let $K_1, K_2, \ldots \in \mathscr{K}$ be such that $\varphi^{-1}K_1 \cap \cdots \cap \varphi^{-1}K_n \neq \varnothing$ $(n = 1, 2, \ldots)$. Since the pullback φ^{-1} commutes with intersections, $\varphi^{-1}(K_1 \cap \cdots \cap K_n) \neq \varnothing$ and hence $K_1 \cap \cdots \cap K_n \neq \varnothing$ $(n = 1, 2, \ldots)$ follows. Since $\mathscr{K}$ is σ-compact, $K_1 \cap K_2 \cap \cdots \neq \varnothing$, and since φ is surjective, $\varphi^{-1}(K_1 \cap K_2 \cap \cdots) \neq \varnothing$, i.e., $\varphi^{-1}K_1 \cap \varphi^{-1}K_2 \cap \cdots \neq \varnothing$ follows.

2. The σ-compactness of $\mathscr{K}_\delta$ is obvious (exercise). Let now $F_1, F_2, \ldots \in K_{\cup f}$ be such that $F_1 \cap \cdots \cap F_n \neq \varnothing$ $(n = 1, 2, \ldots)$. If $F_k = K_{k1} \cup \cdots \cup K_{kr_k}$ with $K_{k1}, \ldots, K_{kr_k} \in \mathscr{K}$, we see that for every n there is a choice $\rho_{n1} \in \{1, \ldots, r_1\}, \rho_{n2} \in \{1, \ldots, r_2\}, \ldots$ such that $K_{1\rho_{n1}} \cap \cdots \cap K_{n\rho_{nn}} \neq \varnothing$. We now apply an obvious diagonal procedure in order to get $\rho_1, \rho_2, \ldots$ such that $\rho_{n_\nu 1} = \rho_1$ for a suitable sequence $n_\nu \to \infty$, $\rho_{n_\nu' 2} = \rho_2$ for a suitable subsequence $n_\nu' \to \infty$ of $n_1, n_2, \ldots$, etc. It is now clear that

$$K_{1\rho_1} \cap \cdots \cap K_{n\rho_n} \neq \varnothing \qquad (n = 1, 2, \ldots).$$

Since all the K_{jk} are in the σ-compact $\mathscr{K}$, $K_{1\rho_1} \cap K_{2\rho_2} \cap \cdots \neq \varnothing$ and consequently $K_1 \cap K_2 \cap \cdots \neq \varnothing$ follows.

5.9. Proposition. *Let $\Omega_1, \Omega_2, \ldots \neq \varnothing$ and $\mathscr{K}_n$ be a σ-compact system of subsets of Ω_n $(n = 1, 2, \ldots)$. Then the system $\mathscr{K}$ of all finite unions of sets of the form $K_1 \times \cdots \times K_r \times \Omega_{r+1} \times \cdots$, where $r > 0, K_1 \in \mathscr{K}_1, \ldots, K_r \in \mathscr{K}_r$, is a σ-compact system of subsets of $\Omega = \Omega_1 \times \Omega_2 \times \cdots$.*

Proof. For every $n = 1, 2, \ldots$ let $\varphi_n: \Omega \to \Omega_n$ be the natural component mapping. Since it is surjective, Proposition 5.8.1 shows that $\{\varphi_n^{-1}K | K \in \mathscr{K}_n\}$ is σ-compact. Let now $K_n = K_{n1} \times \cdots \times K_{nr_n} \times \Omega_{r_n+1} \times \cdots$ $(n = 1, 2, \ldots)$ be a sequence in $\mathscr{K}$ such that $K_1 \cap \cdots \cap K_m \neq \varnothing$ $(m = 1, 2, \ldots)$. If $r = \max\{r_1, \ldots, r_m\}$, we have $K_1 \cap \cdots \cap K_m = K_1' \times \cdots \times K_r' \times \Omega_{r+1} \times \cdots$ where K_ρ' is a finite intersection of some sets in $\mathscr{K}_\rho$ $(\rho = 1, \ldots, r)$. From $K_1 \cap \cdots \cap K_m \neq \varnothing$ we conclude $K_1', \ldots, K_\rho' \neq \varnothing$. It follows that $K_1 \times K_2 \times \cdots = K_1'' \times K_2'' \times \cdots$ where every K_r'' is either $= \Omega_r$ or a finite intersection $\neq \varnothing$ of members from $\mathscr{K}_r$ or a countable intersection of such members such that any finite intersection is nonempty. By the σ-compactness of the $\mathscr{K}_1, \mathscr{K}_2, \ldots$ we conclude $K_1'', K_2'', \ldots \neq \varnothing$ and hence $K_1 \times K_2 \times \cdots \neq \varnothing$. The rest is an obvious consequence of Proposition 5.8.2.

The theorem announced previously is:

5.10. Theorem. *Let $I \neq \varnothing$ and $\mathscr{B}_\iota$ be a σ-field in $\Omega_\iota \neq \varnothing$ $(\iota \in I)$. Let $\mathscr{J}$ be an increasingly filtered covering of I with nonempty finite sets and $(m_J)_{J \in \mathscr{J}}$ a projective family (in the notation of Definition 4.1). Then for every $\iota \in I$, a probability σ-content m is unambiguously defined on $\mathscr{B}_\iota$ by $m_\iota = \varphi_\iota^J m_J$ $(\iota \in J \in \mathscr{J})$. Assume that for every $\iota \in I$, there is a σ-compact set system $\mathscr{K}_\iota \subseteq \mathscr{B}_\iota$ such that m_ι is inner $\mathscr{K}_\iota$-regular, i.e.,*

$$m_\iota(E) = \sup_{\mathscr{K}_\iota \ni K \subseteq E} m_\iota(K) \qquad (E \in \mathscr{B}_\iota).$$

Then there is a unique probability σ-content m on $\mathscr{B}$ such that m_J is the marginal of m for J $(J \in \mathscr{J})$.

Proof. Uniqueness is clear from Proposition 5.4.2. As for existence, we may first proceed as in the beginning of part 2 of Theorem 5.6 in order to reduce our problem to the case $I = \{1, 2, \ldots\}$. It is essentially an application of Proposition 5.4.4. Applying Proposition 5.4.3 twice, we find that we may assume $J = \{\{1, \ldots, n\} \mid n = 1, 2, \ldots\}$. Form the σ-compact system $\mathscr{K}$ of subsets of Ω exactly as in Proposition 5.9. Besides this, we consider the system $\mathscr{F}$ of all finite unions of sets of the form $F_1 \times \cdots \times F_n \times \Omega_{n+1} \times \cdots$, where $n > 0$, $F_1 \in \mathscr{B}_1, \ldots, F_n \in \mathscr{B}_n$. It is easy to see that $\mathscr{F}$ is a field of subsets of Ω (exercise) and the compatibility of $(m_{\{1, \ldots, n\}})_{n=1,2,\ldots}$ shows that there is a unique content m on $\mathscr{F}$ such that

$$(6) \qquad m(F_1 \times \cdots \times F_n \times \Omega_{n+1} \times \cdots) = m_{\{1, \ldots, n\}}(F_1 \times \cdots \times F_n)$$
$$(n > 0, \quad F_1 \in \mathscr{B}_1, \ldots, F_n \in \mathscr{B}_n).$$

Now we pass to σ-compact subsystems $\mathscr{K}^n$ of $\mathscr{K}$ and fields $\mathscr{F}^n \subseteq \mathscr{F}$ in the following obvious way: $\mathscr{K}^n$ consists of all finite unions of sets of the form $K_1 \times \cdots \times K_n \times \Omega_{n+1} \times \cdots$ where $K_1 \in \mathscr{K}_1, \ldots, K_n \in \mathscr{K}_n$, and $\mathscr{F}^n$ consists of all finite unions of sets of the form $F_1 \times \cdots \times F_n \times \Omega_{n+1} \times \cdots$ where $F_1 \in \mathscr{B}_1, \ldots, F_n \in \mathscr{B}_n$.

Next we prove that the restriction of m to $\mathscr{F}^n$ is inner $\mathscr{K}^n$-regular. For this it suffices to consider $F_1 \in \mathscr{B}_1, \ldots, F_n \in \mathscr{B}_n$, to choose some $\varepsilon > 0$, and to find $K_1 \in \mathscr{K}_1, \ldots, K_n \in \mathscr{K}_n$ such that $m_k(F_k \backslash K_k) < \varepsilon/n$ $(k = 1, \ldots, n)$. We then get $K = K_1 \times \cdots \times K_n \times \Omega_{n+1} \times \cdots \in \mathscr{K}^n$, which is a subset of $F = F_1 \times \cdots \times F_n \times \Omega_{n+1} \times \cdots \in \mathscr{F}^n$ and satisfies

$$\begin{aligned} m(F \backslash K) &\leqq m((F_1 \backslash K_1) \times \Omega_2 \times \cdots) + \cdots + m(\Omega_1 \times \cdots \times (F_n \backslash K_n) \times \Omega_{n+1} \times \cdots) \\ &= m_1(F_1 \backslash K_1) + \cdots + m_n(F_n \backslash K_n) \leqq n(\varepsilon/n) = \varepsilon. \end{aligned}$$

From this result it is obvious that $m\colon \mathscr{F} \to \mathbb{R}_+$ is inner $\mathscr{K}$-regular. By Proposition V.1.6, m is a σ-content. By Corollary II.2.4 it has a unique σ-additive extension to $\mathscr{B}^{00}(\mathscr{F})$ which is apparently the σ-field $\mathscr{B} = \mathscr{B}_1 \times \mathscr{B}_2 \times \cdots$. From (6) we deduce that this extension has the prescribed marginals, and the theorem is proved.

Theorems 5.5, 5.6, and 5.10 cover practically all cases of interest.

6. INDEPENDENCE

In this section we treat a phenomenon that is fundamental in probability on the one hand and is closely related to product σ-probabilities on the other: independence. We first investigate independence of set systems and then define the independence of mappings on this basis. Theorem 6.6 relates independence with product σ-probabilities.

6.1. Definition. *Let* $(\Omega, \mathscr{B}, m)$ *be a* σ*-probability space and* $I \neq \varnothing$, $\mathscr{S}_\iota \subseteq \mathscr{B}$ $(\iota \in I)$. *The family* $(\mathscr{S}_\iota)_{\iota \in I}$ *of set systems* $\subseteq \mathscr{B}$ *is called* **independent** *(for* m*) if* $n \geqq 1$, $\iota_1, \ldots, \iota_n \in I$, $\iota_k \neq \iota_j$ $(k \neq j)$, $F_k \in \mathscr{S}_{\iota_k}$ $(k = 1, \ldots, n)$ *implies the* **product** *or* **independence formula**

$$m\left(\bigcap_{k=1}^{n} F_k\right) = \prod_{k=1}^{n} m(F_k).$$

6.2. Remark. Observe that this definition works with arbitrary finite subsets of I. If $I = \{1, 2, \ldots\}$, the subsets $\{1, \ldots, n\}$ would not suffice.

6.3. Example. Let $I \neq \varnothing$ and $((\Omega_\iota, \mathscr{B}_\iota, m_\iota))_{\iota \in I}$ be a family of σ-probability spaces, $(\Omega, \mathscr{B}, m) = \prod_{\iota \in I} (\Omega_\iota, \mathscr{B}_\iota, m_\iota)$. For every $\iota \in I$, let

$$\mathscr{S}_\iota = \left\{\prod_{\kappa} F_\kappa \,\middle|\, F_\iota \in \mathscr{B}_\iota,\ F_\kappa = \Omega_\kappa\ (\kappa \neq \iota)\right\}.$$

Then Theorem 5.3 states the independence of $(\mathscr{S}_\iota)_{\iota \in I}$. We are thus provided with a vast domain of examples for independence. Theorem 6.6 will tell us that this is essentially the most general example.

Our next aim is to establish the independence of enlarged set systems given the independence of given ones.

6.4. Theorem. *Let $(\Omega, \mathscr{B}, m)$ be a σ-probability space and $(\mathscr{S}_\iota)_{\iota \in I}$ an independent family of set systems $\subseteq \mathscr{B}$.*

6.4.1. *For every $\iota \in I$, let*

$$\mathscr{S}'_\iota = \mathscr{S}_\iota \cup \{\Omega\}$$
$$\mathscr{S}''_\iota = \{E \backslash F \mid E, F \in \mathscr{S}_\iota, F \subseteq E\}$$
$$\mathscr{S}'''_\iota = \{F_1 + F_2 + \cdots \mid F_1, F_2, \ldots \mathscr{S}_\iota, F_j \cap F_k = \varnothing\ (j \neq k)\}.$$

Then each of the families $(\mathscr{S}'_\iota)_{\iota \in I}$, $(\mathscr{S}''_\iota)_{\iota \in I}$, $(\mathscr{S}'''_\iota)_{\iota \in I}$ is independent.

6.4.2. *Assume that for every $\iota \in I$, the set system $\mathscr{S}_\iota$ is stable under finite intersections: $\iota \in I$, $E, F \in \mathscr{S}_\iota \Rightarrow E \cap F \in \mathscr{S}_\iota$. Then the family $(\mathscr{B}(\mathscr{S}_\iota))_{\iota \in I}$ of the σ-fields generated by the $\mathscr{S}_\iota$ $(\iota \in I)$ is independent.*

Proof. 1. The independence of $(\mathscr{S}'_\iota)_{\iota \in I}$ is immediate from Definition 6.1. In order to prove the independence of $(\mathscr{S}''_\iota)_{\iota \in I}$, it is clearly sufficient to observe $n \geqq 1, \iota_1, \ldots, \iota_n \in I, \iota_k \neq \iota_j\ (k \neq j), E_1, F_1 \in \mathscr{S}_{\iota_1}, F_1 \subseteq E_1, F_k \in \mathscr{S}_{\iota_k}$ $(k = 2, \ldots, n)$

$$\begin{aligned}
\Rightarrow\quad & m((E_1 \backslash F_1) \cap (F_2 \cap \cdots \cap F_n) \\
&= m((E_1 \cap F_2 \cap \cdots \cap F_n) \backslash (F_1 \cap F_2 \cap \cdots \cap F_n)) \\
&= m(E_1 \cap F_2 \cap \cdots \cap F_n) - m(F_1 \cap F_2 \cap \cdots \cap F_n) \\
&= m(E_1)m(F_2) \cdots m(F_n) - m(F_1)m(F_2) \cdots m(F_n) \\
&= m(E_1 \backslash F_1)m(F_2) \cdots m(F_n).
\end{aligned}$$

The independence of $(\mathscr{S}'''_\iota)_{\iota \in I}$ is proved in an obvious analogous way.

2. It is clearly sufficient to prove $n \geqq 1$, $\iota_1, \ldots, \iota_n \in I$, $\iota_j \neq \iota_k$ $(j \neq k)$, $F_1 \in \mathscr{B}(\mathscr{S}_{\iota_1})$, $F_2 \in \mathscr{S}_{\iota_2}, \ldots, F_n \in \mathscr{S}_{\iota_n}$

$$\Rightarrow\quad m(F_1 \cap F_2 \cap \cdots \cap F_n) = m(F_1)m(F_2) \cdots m(F_n).$$

This is achieved as follows. Choose $\iota_1, \ldots, \iota_n$ as required and fix some $F_2 \in \mathscr{S}_{\iota_2}, \ldots, F_n \in \mathscr{S}_{\iota_n}$. Let $\overline{\mathscr{S}}_{\iota_1} = \{F \mid F \in \mathscr{B}, m(F \cap F_2 \cap \cdots \cap F_n) = m(F)m(F_2) \cdots m(F_n)\}$. By 1 the set system $\overline{\mathscr{S}}_{\iota_1}$ contains Ω and is stable under proper differences and countable disjoint unions. By Proposition I.4.9, $\overline{\mathscr{S}}_{\iota_1} \supseteq \mathscr{B}(\overline{\mathscr{S}}_{\iota_1})$, and we are through.

6.5. Definition. *Let $(\Omega, \mathscr{B}, m)$ be a σ-probability space and $((\Omega_\iota, \mathscr{B}_\iota))_{\iota \in I}$ an arbitrary family of measurable spaces with $\Omega_\iota \in \mathscr{B}_\iota$ $(\iota \in I)$. For every $\iota \in I$, let $f_\iota\colon \Omega \to \Omega_\iota$ be a $\mathscr{B}$-$\mathscr{B}_\iota$-measurable mapping. The family $(f_\iota)_{\iota \in I}$ of mappings* *is said to be* **independent** *(for m) if the family* $(f_\iota^{-1}\mathscr{B}_\iota)_{\iota \in I}$ *of σ-fields* $\subseteq \mathscr{B}$ *is independent.*

6.6. Theorem. *Let* $(\Omega, \mathscr{B}, m)$ *be a σ-probability space and* $((\Omega_\iota, \mathscr{B}_\iota))_{\iota \in I}$ *a family of measurable spaces with* $\Omega_\iota \in \mathscr{B}_\iota$ $(\iota \in I)$. *For every* $\iota \in I$, *let* $f_\iota: \Omega \to \Omega_\iota$ *be a measurable mapping and* $\mathscr{S}_\iota \subseteq \mathscr{B}_\iota$ *a set system that is stable under finite intersections and satisfies* $\mathscr{B}(\mathscr{S}_\iota) = \mathscr{B}_\iota$. *Then the following statements are equivalent:*

6.6.1. *The family* $(f_\iota^{-1}\mathscr{S}_\iota)_{\iota \in I}$ *of set systems is independent (for m).*

6.6.2. *The family* $(f_\iota)_{\iota \in I}$ *of mappings is independent (for m).*

6.6.3. *The product mapping* $f = \prod_{\iota \in I} f_\iota: \Omega \to \prod_{\iota \in I} \Omega_\iota$ *sends m into the product σ-probability* $\prod_{\iota \in I} (f_\iota m)$.

Proof. $1 \Rightarrow 2$. By Proposition IV.1.3.3 we have $\mathscr{B}(f_\iota^{-1}\mathscr{S}_\iota) = f_\iota^{-1}\mathscr{B}_\iota$; and since the pullback f_ι^{-1} of sets commutes with all set operations, we find that $f_\iota^{-1}\mathscr{S}_\iota$ is stable under finite intersections. Theorem 6.4.2 now shows that $(f_\iota^{-1}\mathscr{B}_\iota)_{\iota \in I}$ is an independent family, i.e., $(f_\iota)_{\iota \in I}$ is independent.

$2 \Rightarrow 3$. Put $m_\iota = f_\iota m$ $(\iota \in I)$, $f = \prod_{\iota \in I} f_\iota$ for short. Choose $\iota_1, \ldots, \iota_n \in I$, $\iota_j \neq \iota_k$ $(j \neq k)$ and $F_{\iota_k} \in \mathscr{B}_{\iota_k}$ $(k = 1, \ldots, n)$. Put $F_\iota = \Omega_\iota$ $(\iota \in I \setminus \{\iota_1, \ldots, \iota_n\})$. Then

$$f(m)\left(\prod_{\iota \in I} F_\iota\right) = m\left(f^{-1} \prod_{\iota \in I} F_\iota\right)$$

$$= m\left(\bigcap_{k=1}^{n} f_{\iota_k}^{-1} F_{\iota_k}\right) = \prod_{k=1}^{n} m\left(f_{\iota_k}^{-1} F_{\iota_k}\right) = \prod_{k=1}^{n} m_{\iota_k}(F_{\iota_k}),$$

hence $fm = \prod_{\iota \in I} m_\iota$, by the very definition of a product σ-probability given in Theorem 5.3.

$3 \Rightarrow 1$ follows by rereading the equalities we have just used.

6.7. Remarks

6.7.1. If $\Omega_\iota = \mathbb{R}$ $(\iota \in I)$, then $\mathscr{S}_\iota = \{]-\infty, x] \mid x \in \mathbb{R}\}$ is a set system that is stable under finite intersections and generates the natural σ-field in $\mathbb{R}$. The independence of $(f_\iota^{-1}\mathscr{S}_\iota)_{\iota \in I}$ then reads: For any $n \geqq 1$, $\iota_1, \ldots, \iota_n \in I$, $\iota_j \neq \iota_k$ $(j \neq k)$, $x_1, \ldots, x_n \in \mathbb{R}$ implies

$$m(\{f_{\iota_1} \leqq x_1, \ldots, f_{\iota_n} \leqq x_n\}) = m(\{f_{\iota_1} \leqq x_1\}) \cdots m(\{f_{\iota_n} \leqq x_n\}).$$

In this form, independence occurs frequently in probability theory, in particular in the older literature. A similar statement holds if $\Omega_\iota = \overline{\mathbb{R}}$ $(\iota \in I)$.

6.7.2. If $(\Omega, \mathscr{B}) = \prod_{\iota \in I} (\Omega_\iota, \mathscr{B}_\iota)$ and $f_\iota: \Omega \to \Omega_\iota$ is the ιth component projection, i.e., $f_\iota(\omega_\kappa)_{\kappa \in I} = \omega_\iota$ $((\omega_\kappa)_{\kappa \in I} \in \prod_{\kappa \in I} \Omega_\kappa$, $\iota \in I)$, then $\prod_{\iota \in I} f_\iota$ clearly is the identity mapping of Ω onto itself, and one obvious consequence of the above theorem is: A σ-probability m on Ω is a product σ-probability $\prod_{\iota \in I} m_\iota$ iff $(f_\iota)_{\iota \in I}$ is an independent family; in this case $f_\iota m = m_\iota$ $(\iota \in I)$.

6.8. Exercises

6.8.1. Let $(\Omega, \mathscr{B}, m)$ be a σ-probability space and $(\mathscr{B}_\iota)_{\iota \in I}$ an independent family of σ-fields $\subseteq B$.

6.8.1.1. Let $(\Omega, \overline{\mathscr{B}}, m)$ be the completion of $(\Omega, \mathscr{B}, m)$ (Theorem II.6.2.2); for every $\iota \in I$, let $\overline{\mathscr{B}}_\iota$ be the system of all $F \subseteq \Omega$ for which we can find an $E \in \mathscr{B}_\iota$ and an $N \in \mathscr{B}$ such that $m(N) = 0$, $E \,\Delta\, F \subseteq N$. Prove that $(\overline{\mathscr{B}}_\iota)_{\iota \in I}$ is independent.

6.8.1.2. Let $I = \sum_{\kappa \in K} I_\kappa$ be a disjoint decomposition of I such that $I_\kappa \neq \varnothing$ $(\kappa \in K)$; for every $\kappa \in K$, let $\mathscr{B}_\kappa = \mathscr{B}(\bigcup_{\iota \in I_\kappa} \mathscr{B}_\iota)$. Prove that $(\mathscr{B}_\kappa)_{\kappa \in K}$ is an independent family.

6.8.1.3. Let $((\Omega_\iota, \mathscr{D}_\iota))_{\iota \in I}$ be a family of measurable spaces with $\Omega_\iota \in \mathscr{D}_\iota$ $(\iota \in I)$. Let $(f_\iota)_{\iota \in I}$ be an independent family of mappings $f_\iota: \Omega \to \Omega_\iota$ such that $f_\iota^{-1}\mathscr{D}_\iota \subseteq \mathscr{B}$. Let $I = \sum_{\kappa \in K} I_\kappa$ be a disjoint decomposition of I such that $I_\kappa \neq \varnothing$ $(\kappa \in K)$. Let $(X_\kappa, \mathscr{C}_\kappa)_{\kappa \in K}$ be a family of measurable spaces such that $X_\kappa \in \mathscr{C}_\kappa$ $(\kappa \in K)$. For every $\kappa \in K$, let $g_\kappa: \prod_{\iota \in I_\kappa} \Omega_\iota \to X_\kappa$ be a $(\prod_{\iota \in I_\kappa} \mathscr{D}_\iota)$-$\mathscr{C}_\kappa$-measurable mapping and $h_\kappa: \Omega \to X_\kappa$ be defined by $h_\kappa = g_\kappa \circ (\prod_{\iota \in I_\kappa} f_\iota)$. Prove that $(h_\kappa)_{\kappa \in K}$ is independent.

6.8.1.4. Let in particular $I = \{1, 2, \ldots\}$ and $f_n \in \text{mble}(\Omega, \mathscr{B}, \overline{\mathbb{R}})$ $(n = 1, 2, \ldots)$ such that $(f_n)_{n=1,2,\ldots}$ is independent. Prove that

$$f = \limsup_{n \to \infty} f_{2n} \qquad \text{and} \qquad f = \liminf_{n \to \infty} f_{2n+1}$$

are two independent mappings. Prove that

$$\limsup_{n \to \infty} \frac{1}{n} \sum_{k=1}^{n} f_k,\ f_1,\ f_2, \ldots$$

is an independent family.

6.8.1.5. Let again $I = \{1, 2, \ldots\}$ and $\tilde{\mathscr{B}} = \bigcap_{n=1}^{\infty} \mathscr{B}(\bigcup_{k \geq n} \mathscr{B}_k)$. Prove the so-called 0-1-law: $F \in \tilde{\mathscr{B}} \Rightarrow m(F) = 0$ or $= 1$. (*Hint*: Prove that $\tilde{\mathscr{B}}, \mathscr{B}_1, \mathscr{B}_2, \ldots$ is an independent family and hence so is $(\tilde{\mathscr{B}}, \tilde{\mathscr{B}})$.)

6.8.1.6. Prove that the $\limsup_{n \to \infty} n^{-1} \sum_{k=1}^{n} f_k$ considered in 6.8.1.4 is a constant mod m.

6.8.2. Let $(\Omega, \mathscr{B}, m)$ be a σ-probability space and $(\Omega_t, \mathscr{B}_t, m_t) = (\Omega, B, m)$ $(t = 0, 1, \ldots)$. Let $(\tilde{\Omega}, \tilde{\mathscr{B}}, \tilde{m}) = \prod_{t=0}^{\infty} (\Omega_t, \mathscr{B}_t, m_t)$. In $\tilde{\Omega}$ define the so-called shift as the mapping $\tilde{T}: \tilde{\Omega} \to \tilde{\Omega}$ defined by $\tilde{T}(\omega_0, \omega_1, \ldots) = (\omega_1, \omega_2, \ldots)$ $(\omega_0, \omega_1, \ldots \in \Omega)$. Prove that $(\Omega, \mathscr{B}, m, T)$ is an ergodic dynamical system (Definitions IV.3.1, IV.3.19). (*Hint*: Use the technique of Example IV.3.23.)

6.8.3. Let $(\Omega, \mathscr{B}, m)$ be a σ-probability space and $F_1, F_2, \ldots \in \mathscr{B}$. Prove:

6.8.3.1. **(first Borel–Cantelli lemma).** If $\sum_{k=1}^{\infty} m(F_k) < \infty$, then $\bar{F} = \{\omega \mid \omega \in F_k$ for infinitely many $k\}$ satisfies $m(\bar{F}) = 0$. (*Hint*: $\bar{F} \subseteq \bigcup_{k \geqq n} F_k$ $(n = 1, 2, \ldots)$.)

6.8.3.2. **(second Borel–Cantelli lemma).** If $(\{F_k\})_{k=1,2,\ldots}$ is independent for m and $\sum_{k=1}^{\infty} m(F_k) = \infty$, then the $\bar{F}$ just considered satisfies $m(\bar{F}) = 1$. (*Hint*: $\Omega \backslash \bar{F} \subseteq \bigcup_{n=1}^{\infty} \bigcap_{k \geqq n} (\Omega \backslash F_k)$.)

7. MARKOVIAN SEMIGROUPS AND THEIR PATH STRUCTURE

In this section we treat, mainly for the use of probabilists, the fundamental facts about Markovian semigroups and their paths in compact and locally compact metrizable spaces. Subsection 1 begins with the definition of Markovian and sub-Markovian semigroups in abstract spaces X, gives a general construction for their extension to X plus a "cemetery" ∞ and applies Kolmogorov's theorem 5.10 in order to prove the existence of so-called *canonical Markovian σ-contents* with given initial distribution. These σ-contents live in the continuous cartesian product space $\Omega = \prod_{t \in \mathbb{R}_+} X_t$ $(X_t = X$ $(t \geqq 0))$ whose elements $\omega = (x_t)_{t \geqq 0}$ may justly be called *paths* in X if we interpret $t \in \mathbb{R}_+$ as time. If X bears a topology, one may ask whether, after the possible elimination of a nullset, all paths are continuous, or have right and left limits everywhere (which means, roughly speaking, that they do not oscillate too much). The theory which answers such questions is called *path theory*, here for Markovian semigroups, and we present it in a form originally proposed by NELSON [1] which avoids the explicit presentation of a theory of so-called *separability* of stochastic processes (see DOOB [2], MEYER [2, 3], BLUMENTHAL and GETOOR [1], DYNKIN [1, 2] and the jewel on p. 102 of A. and C. IONESCU–TULCEA [1] (see Chapter XVI, Section 3)) and substitutes for it the regular extension of σ-contents from Baire to Borel σ-fields in *compact Hausdorff spaces* (Theorem V.3.4.3), a construction based mainly on the uncountable case of Dini's theorem. Subsection 2 presents the locally compact case which is, in subsection 3, settled by reduction to the compact case. In brief, the story runs as follows: The sets of interest, like the set of all continuous paths, do not belong to the product (= Baire) σ-field in Ω as a rule; but if X is compact metrizable, they can be shown to belong to the Borel σ-field in the (nonmetrizable) compact Hausdorff space Ω and to have nullsets as their complements, under suitable assumptions about the semigroup. If X is only *locally compact*, we pass over to the *one-point compactification* $\bar{X}$ of X and apply what we know about the compact case, but with the additional precaution of excluding all paths that "reach infinity in finite time"; this can be done under a standard regularity condition for the semigroup.

The content of this chapter provides a small but essential portion of the preliminary theory of continuous time Markov processes which the reader can study after BLUMENTHAL and GETOOR [1], DYNKIN [1, 2], MEYER [2, 3].

1. Markovian Semigroups and Markovian Canonical σ-Probabilities

Let $X \neq \varnothing$ and $\mathscr{B}$ be a σ-field in X. Consider two substochastic $\mathscr{B}$-measurable kernels P, Q from X to X (Definitions IV.4.1, IV.4.12). The appropriate elementary domain to be employed here is the space $\mathscr{E} = \text{mble}^{\text{b}}(X, \mathscr{B}, \mathbb{R})$ of all bounded real-$\mathscr{B}$-measurable functions on X. Clearly, P can be considered as a mapping from $\mathscr{E}$ to $\mathscr{E}$, e.g., by

$$P: f \to Pf$$

where

$$(Pf)(x) = \int P(x, dy) f(y)$$

for any $f \in \mathscr{E}$. This notation puts the σ-content kernel aspects in the foreground, and we shall maintain this aspect in the sequel. The same statements apply to Q and the product kernel PQ in the sense of Theorem IV.4.16 can be written in the form

$$(PQ)(x, F) = \int P(x, dy) Q(y, F) \qquad (x \in X, \quad F \in \mathscr{B}).$$

The reader may check the σ-continuity of PQ under the σ-content kernel aspect by a simple application of the monotone convergence theorem. Clearly PQ is substochastic again and stochastic if P, Q are stochastic. The special kernel I defined by

$$I(x, F) = 1_F(x) \qquad (x \in X, \quad F \in \mathscr{B})$$

serves as a neutral element:

$$IP = PI = P$$

for every substochastic kernel P. If $f \in \mathscr{L}^1_{P(x, \cdot)}$ $(x \in X)$ we shall write $P(x, f)$ in place of $\int P(x, dy) f(y)$ and Pf in place of $P(\cdot, f)$ if we like.

Similarly, for every σ-content m on $\mathscr{B}$, the σ-content mP is defined on $\mathscr{B}$ by

$$(mP)(F) = \int m(dx)P(x, F) \qquad (F \in \mathscr{B})$$

(use the monotone convergence theorem for the proof of its σ-additivity). There is a simple duality between these two actions of P:

$$(mP)(f) = m(Pf)$$

whenever the integrals involved here exist.

7.1. Definition. *Let $\mathscr{B}$ be a σ-field in $X \neq \varnothing$. A one-parameter family $(P_t)_{t \in \mathbb{R}_+}$ of $\mathscr{B}$-measurable (sub-)stochastic kernels from X to X is called a* **(sub-)Markov(ian) semigroup** *in the measurable space $(X, \mathscr{B})$ (or in X) if:*

7.1.1. $$P_0 = I;$$

7.1.2. $$P_s P_t = P_{s+t} \qquad (s, t \geqq 0).$$

The sub-Markovian case is easily linked with the Markovian case. The next theorem will be enough of a pretext for us to restrict attention to the Markovian case in the sequel.

7.2. Proposition. *Let $(P_t)_{t \geqq 0}$ be a sub-Markovian semigroup in $(X, \mathscr{B})$ and $\infty \notin X$. Put*

$$X = X + \{\infty\},$$
$$\overline{\mathscr{B}} = \{E + F \mid E \in \mathscr{B}, F = \varnothing \text{ or } = \{\infty\}\}.$$

Then $\overline{\mathscr{B}}$ is a σ-field in $\overline{X}$. By

$$\overline{P}_t(\overline{x}, E + F) = \begin{cases} P_t(\overline{x}, E) + (1 - P_t(\overline{x}, X))1_F(\infty) & (\overline{x} \in X) \\ 1_F(\overline{x}) & (\overline{x} = \infty) \end{cases}$$

with

$$t \geqq 0, \qquad E \in \mathscr{B}, \qquad F = \varnothing \text{ or } = \{\infty\}$$

a Markovian semigroup $(\overline{P}_t)_{t \geqq 0}$ is defined in $(\overline{X}, \overline{\mathscr{B}})$. It is called the **natural Markovian extension** *of $(P_t)_{t \geqq 0}$.*

Remark. The new point ∞ is often called the "cemetery" for $(P_t)_{t \geqq 0}$, for obvious reasons.

Proof. The proof of the statement concerning $\overline{\mathscr{B}}$ is an easy exercise, and so is the proof of the fact that every $\overline{P}_t$ is a stochastic kernel. For the rest, let us be content to prove $\overline{P}_s \overline{P}_t = \overline{P}_{s+t}$. We distinguish four cases in order to have simpler formulas.

Case I: $\bar{x} \in X$, $E \in \mathscr{B}$. Then

$$\int \bar{P}_s(\bar{x}, d\bar{y})\bar{P}_t(\bar{y}, E) = \int P_s(\bar{x}, dy)P_t(y, E)$$
$$= P_{s+t}(\bar{x}, E) = \bar{P}_{s+t}(x, E).$$

Case II: $\bar{x} = \infty$, $E \in \mathscr{B}$. Then

$$\int \bar{P}_s(\bar{x}, d\bar{y})\bar{P}_t(\bar{y}, E) = \bar{P}_t(\infty, E) = 0$$
$$= \bar{P}_{s+t}(\bar{x}, E).$$

Case III: $\bar{x} \in X$, $F = \{\infty\}$. Then

$$\int \bar{P}_s(\bar{x}, d\bar{y})\bar{P}_t(\bar{y}, F) = \int P_s(\bar{x}, dy)(1 - P_t(y, x)) + (1 - P_s(\bar{x}, X))$$
$$= P_s(\bar{x}, X) - P_{s+t}(\bar{x}, X) + 1 - P_s(\bar{x}, X)$$
$$= 1 - P_{s+t}(\bar{x}, X) = \bar{P}_{s+t}(\bar{x}, F).$$

Case IV: $\bar{x} = \infty$, $F = \{\infty\}$. Then

$$\int \bar{P}_s(\bar{x}, d\bar{y})\bar{P}_t(\bar{y}, F) = 1 = \bar{P}_{s+t}(\bar{x}, F).$$

The general case follows from these four cases by addition.

7.3. Theorem. *Let $\mathscr{B}$ be a σ-field in $X \neq \varnothing$ and $(P_t)_{t \geqq 0}$ a Markovian semigroup in $(X, \mathscr{B})$. Assume that there is a σ-compact set system $\mathscr{K} \subseteq \mathscr{B}$ such that for every $x \in X$, the σ-content $P(x, \cdot)$ is inner $\mathscr{K}$-regular. Then there is a stochastic kernel P from $(X, \mathscr{B})$ to $(\Omega, \tilde{\mathscr{B}}) = \prod_{t \in \mathbb{R}_+} (X_t, \mathscr{B}_t)$ $((X_t, B_t) = (X, \mathscr{B})$ $(t \in \mathbb{R}_+))$, such that (we write $P^x(E)$ instead of $P(x, E)$, and φ_t denotes the* t*th component projection of Ω onto X)*

$$(1) \qquad P^x(\varphi_{t_0}^{-1} E_0 \cap \cdots \cap \varphi_{t_n}^{-1}(E_n))$$
$$= 1_{E_0}(x) \int P_{t_1}(x, dx_1)1_{E_1}(x_1) \int P_{t_2 - t_1}(x_1, dx_2)1_{E_2}(x_2)$$
$$\cdots \int P_{t_n - t_{n-1}}(x_{n-1}, dx_n)1_{E_n}(x_n)$$
$$(0 = t_0 < t_1 < \cdots < t_n, \quad E_0, \ldots, E_n \in \mathscr{B}).$$

Proof. Let $\mathscr{J} = \{J \mid 0 \in J \subseteq \mathbb{R}_+ \text{ finite}\}$. Clearly $\mathscr{J}$ is increasingly filtered and covers $\mathbb{R}_+$. For given $J = \{t_0, \ldots, t_n\}$, $0 = t_0 < t_1 < \cdots < t_n$, and $x \in X$, there is exactly one σ-content m_J on $\mathscr{B} \times \cdots \times \mathscr{B}$ in $X \times \cdots \times X$ ($n+1$ factors) such that for any $E_0, \ldots, E_n \in \mathscr{B}$, $m_J(E_0 \times \cdots \times E_n)$ is given by the right-hand member of (1). This follows from Theorem 3.2 (and Remark 3.3.2) plus the observation that these sets $E_0 \times \cdots \times E_n$ form an intersection stable system generating $\mathscr{B} \times \cdots \times \mathscr{B}$. Let us now show that these m_J (for our fixed choice of $x \in X$) form a projective family. Whenever $J \subseteq K \in \mathscr{J}$, then K arises from J by inserting some indexes between the t_i and adding some larger than t_n (one of these possibilities may not be realized). Thus it seems sufficient to settle the following two cases:

Case 1: $K = \{t_0, t_1, \ldots, t_{n-1}, s, t_n\}$ with $t_{n-1} < s < t_n$. Here (we employ the notational machinery from Section 5)

$$\begin{aligned}\varphi_J^K m_K(E_0 \times \cdots \times E_n) &= m_K(E_0 \times \cdots \times E_{n-1} \times X \times E_n)\\ &= 1_{E_0}(x) \int P_{t_1}(x, dx_1) 1_{E_1}(x_1)\\ &\quad \cdots \int P_{s-t_{n-1}}(x_{n-1}, du) 1_X(u) \int P_{t-s}(u, dx_n) 1_{E_n}(x_n)\end{aligned}$$

but $1_X \equiv 1$ and $\int P_{s-t_{n-1}}(x_{n-1}, du) P_{t_n - s}(u, \cdot) = P_{t_n - t_{n-1}}(x_{n-1}, \cdot)$ by the semigroup property, hence $\varphi_J^K m_K(E_0 \times \cdots \times E_n)$ equals the righthand member of (1), which is $m_J(E_0 \times \cdots \times E_n)$, as desired.

Case 2: $K = \{t_0, \ldots, t_n, t_{n+1}\}$ with $t_{n+1} > t_n$. Here

$$\begin{aligned}\varphi_J^K m_K(E_0 \times \cdots \times E_n) &= m_K(E_0 \times \cdots \times E_n \times X)\\ &= 1_{E_0}(x) \int \cdots \int P_{t_{n+1}-t_n}(x_n, dx_{n+1}) 1_X(x_{n+1}),\end{aligned}$$

and the latter integral equals 1 because $P_{t_{n+1}-t_n}$ is a stochastic kernel.

Theorem 5.10 now proves the rest.

It should be clear how to extend formula (1) to somewhat more complicated sets which will occur, e.g., in the proof of Lemma 7.8.

7.4. Definition. *Under the hypotheses of Theorem 7.3, P^x is called the canonical Markovian σ-content associated with the given Markovian semigroup $(P_t)_{t \geq 0}$ and the starting point $x \in X$. If p is any probability σ-content on $\mathscr{B}$ (in X), the probability σ-content $p \times P$ on $\tilde{\mathscr{B}}$ (in Ω) is called the canonical Markovian σ-content associated with the semigroup and the initial distribution p.*

2. The Compact Case

In this subsection we treat the fundamentals of path theory for Markovian σ-contents in $\Omega = \prod_{t \in \mathbb{R}_+} X_t$ where $X_t = X$ $(t \geqq 0)$ is a compact metric space. This case deserves special attention because the theory is particularly simple and rounded-off here.

Let us recall that by *Tychonov's theorem* $\Omega = \prod_{t \in \mathbb{R}_+} X_t$ is a compact Hausdorff space again. In X the Baire σ-field $\mathscr{B}$ coincides with the Borel σ-field. In Ω, the Baire σ-field $\tilde{\mathscr{B}}$ coincides with the product σ-field of copies of $\mathscr{B}$ (Theorem 5.5.1) but is different from the (larger) Borel σ-field $\hat{\mathscr{B}}$ in Ω whenever X has more than one point (Example V.3.1). Every σ-content m on $\tilde{\mathscr{B}}$ is inner-G_δ-compact- and outer-K_σ-regular, and has a unique inner-compact- and outer-open-regular extension m to the Borel σ-field $\hat{\mathscr{B}}$ (Theorem V.3.4).

Since $\mathscr{B}$ is linked to the compact metric space X in a canonical way, we shall omit it from the notations occasionally and speak, e.g., of Markovian semigroups in X instead of $(X, \mathscr{B})$.

7.5. Definition. *Let X be a compact metric space. A Markovian semigroup $(P_t)_{t \geqq 0}$ in X is called* **Feller** *if:*

7.5.1. $P_t f \in \mathscr{C}(X, \mathbb{R})$ $(f \in \mathscr{C}(X, \mathbb{R}), t \geqq 0)$;

7.5.2. $\lim_{t \to 0+0} \|P_t f - f\| = 0$ $(f \in \mathscr{C}(X, \mathbb{R}))$ *(strong continuity) where $\|\cdot\|$ denotes the uniform (sup) norm in the space $\mathscr{C}(X, \mathbb{R})$ of all continuous real functions on X.*

From the general estimate

$$\|P_t g\| \leqq \|g\| \qquad (t \geqq 0, \quad g \in \mathscr{C}(X, \mathbb{R}))$$

we deduce

$$\|P_{s+t} f - P_s f\| \leqq \|P_t f - f\| \qquad (s, t \geqq 0, \quad f \in \mathscr{C}(X, \mathbb{R}))$$

which shows that 7.5.2 implies the only seemingly more general condition

7.5.3. $$\lim_{t \to s+0} \|P_t f - P_s f\| = 0 \qquad (s \geqq 0, \quad f \in \mathscr{C}(X, \mathbb{R})).$$

7.6. Example. Let X be finite with the discrete topology and

$$A = (A(x, y))_{x, y \in X}$$

be a real matrix with the properties

$$A(x, y) \geqq 0 \qquad (x, y \in X, \quad x \neq y), \tag{2}$$

$$\sum_{y \in X} A(x, y) = 0 \qquad (x \in X). \tag{3}$$

Then

$$P_t = e^{tA} = \sum_{n=0}^{\infty} \frac{t^n A^n}{n!}$$

defines a Markovian semigroup $(P_t)_{t \geqq 0}$ in X which is Feller. The Feller property being obvious, we restrict ourselves to a proof that every P_t is stochastic, i.e., a matrix $(P_t(x, y))_{x, y \in X}$ satisfying

$$P_t(x, y) \geqq 0 \qquad (x, y \in X), \tag{4}$$

$$\sum_{y \in X} P_t(x, y) = 1 \qquad (x \in X). \tag{5}$$

In fact P_t is a differentiable function of t and, for every fixed t, a continuous function of A. Now if $A(x, y) > 0$ $(x \neq y)$, we find that every element of P_t is of the same sign as $I + tA$ if f is sufficiently small, i.e., strictly positive; and this property goes through for arbitrary $t > 0$ since $P_t = (P_{t/n})^n$ $(n = 1, 2, \ldots)$.

An arbitrary A satisfying (1) and (2) can always be written as a limit of matrices A satisfying (2), (3), and $A(x, y) > 0$ $(x \neq y)$, hence (4) follows in the general case, by continuity. (5) holds trivially for $t = 0$. The general case follows by observing

$$\begin{aligned} \frac{d}{dt}\left(\sum_{y \in X} P_t(x, y)\right) &= \sum_{y \in X} (e^{tA} A)(x, y) \\ &= \sum_{z \in X} (e^{tA})(x, z) \sum_{y \in X} A(z, y) = 0. \end{aligned}$$

It is not hard to show, but beyond our scope here, that every Markovian semigroup in X that is Feller can be obtained from a "generator" A in the above fashion.

A criterion for strong continuity is given in

7.7. Proposition. *Let X be a compact metric space with metric $|\cdot, \cdot|$ and*

$(P_t)_{t \geqq 0}$ *a Markovian semigroup in* X *such that* $P_t \mathscr{C}(X, \mathbb{R}) \subseteq \mathscr{C}(X, \mathbb{R})$ $(t \geqq 0)$. *Then* $(P_t)_{t \geqq 0}$ *is strongly continuous iff for every* $\varepsilon > 0$

$$\lim_{t \to 0+0} \sup_{x \in X} P(x, X \backslash U_\varepsilon(x)) = 0 \tag{6}$$

where $U_\varepsilon(x) = \{y \mid |x, y| < \varepsilon\}$.

Proof. 1. Let $(P_t)_{t \geqq 0}$ be strongly continuous. For a given $\varepsilon > 0$, let $f_x \in \mathscr{C}(X, \mathbb{R})$ satisfy

$$1_{X \backslash U_{2\varepsilon}(x)} \leqq f_x \leqq 1_{X \backslash U_\varepsilon(x)}.$$

A construction of f_x with the help of the metric is an easy exercise. Since X is compact, we may find $x_1, \ldots, x_n \in X$ such that $X = U_\varepsilon(x_1) \cup \cdots \cup U_\varepsilon(x_n)$. For a given $\delta > 0$, choose $t_0 > 0$ such that $\|P_t f_{x_k} - f_{x_k}\| < \delta$ $(0 \leqq t \leqq t_0,\ k = 1, \ldots, n)$, according to strong continuity. Let now $x \in X$ be arbitrary, say $x \in U_\varepsilon(x_k)$. Clearly, $1_{X \backslash U_{3\varepsilon}(x)} \leqq f_{x_k}$. From this we deduce

$$\begin{aligned} P_t(x, X \backslash U_{3\varepsilon}(x)) &\leqq P_t(x, f_{x_k}) \\ &< |f_{x_k}(x)| + \delta = \delta \qquad (0 \leqq t \leqq t_0), \end{aligned}$$

as desired.

2. Assume now that (6) holds for every $\varepsilon > 0$ and let $f \in \mathscr{C}(X, \mathbb{R})$ be arbitrary. By uniform continuity of f we can find, for a given $\delta > 0$, some $\varepsilon > 0$ such that $|f(y) - f(x)| < \delta$ $(y \in U_\varepsilon(x))$, and we can estimate

$$\begin{aligned} |P_t(x, f) - f(x)| &\leqq \left| \int P_t(x, dy) f(y) 1_{U_\varepsilon(x)} - \int P_t(x, dy) f(x) 1_{U_\varepsilon(x)} \right| \\ &\quad + \left| \int P_t(x, dy) f(x) 1_{X \backslash U_\varepsilon(x)} - \int P_t(x, dy) f(y) 1_{X \backslash U_\varepsilon(x)} \right| \\ &< \delta + 2\|f\| P_t(x, X \backslash U_\varepsilon(x)), \end{aligned}$$

and by (6) this becomes $< 2\delta$ uniformly in x if $t \geqq 0$ is sufficiently small.

If we interpret t as time, it is natural to call the elements $(x_t)_{t \geqq 0}$ of Ω paths. Theorem 7.3 guarantees the existence of Markovian σ-contents $\tilde{m} = p \times P$ with arbitrary initial distributions p for every Markovian semigroup $(P_t)_{t \geqq 0}$ in X. There are two subsets of the set Ω of all paths of special interest: the set Ω_{cont} of all continuous paths and the set $\Omega_{\frac{1}{2}\text{-cont}}$ of all paths having right and left limits everywhere. Since every set in the

Baire (= product) σ-field $\tilde{\mathscr{B}}$ depends, roughly speaking, of countably many coordinates only, Ω_{cont} and $\Omega_{\frac{1}{2}\text{-cont}}$ are not in $\tilde{\mathscr{B}}$ since the description of, e.g., continuity involves the consideration of x_t for all t from whole intervals and these are uncountable sets. We shall however show that Ω_{cont} and $\Omega_{\frac{1}{2}\text{-cont}}$ are in the Borel σ-field $\hat{\mathscr{B}}$ of the compact Hausdorff space Ω.

Our intention can now be made more precise by saying that we want to establish $\hat{m}(\Omega_{\text{cont}}) = 1$ resp. $\hat{m}(\Omega_{\frac{1}{2}\text{-cont}}) = 1$ for the unique regular extension $\hat{m}$ of m from $\tilde{\mathscr{B}}$ to $\hat{\mathscr{B}}$, under suitable assumptions about the semigroup.

Let us begin with $\Omega_{\frac{1}{2}\text{-cont}}$ and prove that $\hat{m}(\Omega_{\frac{1}{2}\text{-cont}}) = 1$ if the Markovian semigroup $(P_t)_{t \geqq 0}$ is Feller.

7.8. Lemma (KINNEY [1]). *Let X be compact metric and $(P_t)_{t \geqq 0}$ a Markovian semigroup in X. For any $\varepsilon > 0$ and $t \geqq 0$, let*

$$\alpha_\varepsilon(t) = \sup_{\substack{x \in X \\ 0 \leqq u \leqq t}} P_u(x, X \backslash \bar{U}_\varepsilon(x))$$

where $\bar{U}_\varepsilon(x) = \{y \mid |x, y| \leqq \varepsilon\}$ as usual. Let $\varnothing \neq J \subseteq \mathbb{R}_+$ be finite, $0 \leqq a < b$ and $J \subseteq [a, b]$; put

$$A_\varepsilon^n(J) = \{\omega = (x_t)_{t \geqq 0} \mid \text{there are } t_1 < \cdots < t_{2n} \text{ in } J \text{ such that} \\ |x_{t_{2k-1}}, x_{t_{2k}}| > 4\varepsilon \ (k = 1, \ldots, n)\}$$

Then $A_\varepsilon^n(J) \in \tilde{\mathscr{B}}$, and

$$P^x(A_\varepsilon^n(J)) \leqq (2\alpha_\varepsilon(b - a))^n \qquad (x \in X)$$

and consequently

$$m(A_\varepsilon^n(J)) \leqq (2\alpha_\varepsilon(b - a))^n$$

for any initial distribution p in X and $m = p \times P$.

Proof. Let $J = \{s_1, \ldots, s_r\}$ with $a \leqq s_1 < \cdots < s_r \leqq b$. We proceed by induction with respect to n. If $\omega = (x_t)_{t \geqq 0} \in A_\varepsilon^1(J)$, then there is a first j with $|x_{s_j}, x_{s_l}| > 4\varepsilon$ for some $l > j$. Besides this, let i be minimal with $|x_a, x_{s_i}| > 2\varepsilon$. Now the triangle inequality for our metric implies that such an i actually exists, and this in turn implies that either $|x_a, x_b| > \varepsilon$ or $|x_{s_i}, x_b| > \varepsilon$. Thus we obtain (using a *first entrance decomposition*)

$$A_\varepsilon^1(J) \subseteq \{|x_a, x_b| > \varepsilon\} \cup \sum_{i=1}^{r} \{|x_a, x_{s_\nu}| \leqq 2\varepsilon \ (\nu < i), \\ |x_a, x_{s_i}| > 2\varepsilon, |x_{s_i}, x_b| > \varepsilon\}$$

Now by (1) we get (observe $b - s_i \leqq b - a$ $(i = 1, \ldots, r)$)

$$P^x\left(\sum_{i=1}^{r}\{|x_a, x_{s_v}| \leqq 2\varepsilon\ (v < i),\quad |x_a, x_{s_i}| > 2\varepsilon,\quad |x_{s_i}, x_b| > \varepsilon\}\right)$$

$$\leqq \sum_{i=1}^{r} P^x(\{|x_a, x_{s_v}| \leqq 2\varepsilon\ (v < i),\quad |x_a, x_{s_i}| > 2\varepsilon\})\alpha_\varepsilon(b - a)$$

$$\leqq P^x\left(\sum_{i=1}^{r}\{|x_a, x_{s_v}| \leqq 2\varepsilon\ (v < i),\quad |x_a, x_{s_i}| > 2\varepsilon\right)\alpha_\varepsilon(b - a)$$

$$\leqq \alpha_\varepsilon(b - a)$$

and thus

$$P^x(A_\varepsilon^1(J)) \leqq P^x(\{|x_a, x_b| > \varepsilon\}) + \alpha_\varepsilon(b - a)$$
$$\leqq 2\alpha_\varepsilon(b - a),$$

settling the case $n = 1$. Assume now that we are through for the case $n - 1$. If $\omega = (x_t)_{t \geqq 0} \in A_\varepsilon^n(J)$, then there is a first i with $\omega \in A_\varepsilon^{n-1}(\{s_1, \ldots, s_i\})$ and there is some first $j > i$ with $|x_{s_j}, x_{s_l}| > 4\varepsilon$ for some $l > j$. We thus obtain, using a first entrance decomposition,

$$P^x(A_\varepsilon^n(J)) = \sum_{i=1}^{r} P^x(\{\omega \notin A_\varepsilon^{n-1}(\{s_1, \ldots, s_v\})\ (v < i),$$

$$\omega \in A_\varepsilon^{n-1}(\{s_1, \ldots, s_i\}), \omega \in A_\varepsilon^1(\{s_{i+1}, \ldots, s_r\})\}).$$

With the help of (1) we obtain

$$\leqq \sum_{i=1}^{r} P^x(\{\omega \notin A_\varepsilon^{n-1}(\{s_1, \ldots, s_v\})\ (v < i), \omega \in A_\varepsilon^{n-1}(\{s_1, \ldots, s_i\})\})$$

$$\cdot \sup_{y \in X} P^y(A_\varepsilon^1(\{s_{i+1}, \ldots, s_r\}))$$

$$\leqq \alpha_\varepsilon(J) \sum_{i=1}^{r} P^x(\omega \notin A_\varepsilon^{n-1}(\{s_1, \ldots, s_v\})\ (v < i), \omega \in A^{n-1}(\{s_1, \ldots, s_i\})\})$$

$$= \alpha_\varepsilon(J) \cdot P^x(A_\varepsilon^{n-1}(J)) \leqq \alpha_\varepsilon(J) \cdot \alpha_\varepsilon^{n-1}(J)$$

$$= \alpha_\varepsilon^n(J)$$

by the induction hypothesis. The corresponding estimate for m is obtained by integration over x with respect to p.

For any $\varepsilon > 0$, $0 \leqq a < b$, and any integer $n > 0$, we define now

$$\Omega_\varepsilon^n(a, b) = \bigcup_{J \subseteq [a, b] \text{ finite}} A_\varepsilon^n(J).$$

Here we have an uncountable union and hence possibly $\Omega_\varepsilon^n(a, b) \notin \tilde{\mathscr{B}}$. But topology helps: Clearly $A_\varepsilon^n(J)$ is an open set and hence so is $\Omega_\varepsilon^n(a, b)$, which consequently belongs to $\hat{\mathscr{B}}$. Moreover if J increases, so does $A_\varepsilon^n(J)$, hence the $A_\varepsilon^n(J)$ with $J \subseteq [a, b]$ form an increasingly filtered system. Passing to complements (which are compact) and applying Exercise V.3.6.3, we find

$$\hat{m}(\Omega_\varepsilon^n(a, b)) = \sup_{J \subseteq [a, b] \text{ finite}} m(A_\varepsilon^n(J)).$$

We thus obtain the

7.9. Corollary. *With the above notations $\Omega_\varepsilon^n(a, b) \in \hat{\mathscr{B}}$ and*

$$\hat{m}(\Omega_\varepsilon^n(a, b)) \leqq (2\alpha_\varepsilon(b - a))^n \qquad (\varepsilon > 0, \quad 0 \leqq a < b, \quad n = 1, 2, \ldots).$$

As a consequence, we obtain

7.10. Theorem. *Let X be compact metric and $(P_t)_{t \geqq 0}$ a Markovian semigroup in X that is Feller. The set*

$$\Omega_{\frac{1}{2}\text{-cont}} = \left\{ \omega = (x_t)_{t \geqq 0} \;\; \lim_{t \to s+0} x_t \text{ exists for } s \geqq 0 \text{ and } \lim_{t \to s-0} x_t \text{ exists for } s > 0 \right\}$$

belongs to $\hat{\mathscr{B}}$ and satisfies $\hat{m}(\Omega_{\frac{1}{2}\text{-cont}}) = 1$ where $\hat{m}$ denotes the unique inner compact regular outer open regular extension of $m = p \times P$ from $\tilde{\mathscr{B}}$ to $\hat{\mathscr{B}}$, for an arbitrary probability σ-content p in X. In particular,

$$\hat{P}^x(\Omega_{\frac{1}{2}\text{-cont}}) = 1 \qquad (x \in X).$$

Proof. Let us first prove

$$\Omega \backslash \Omega_{\frac{1}{2}\text{-cont}} = \bigcup_{r=1}^{\infty} \bigcap_{N=1}^{\infty} \bigcup_{k=1}^{\infty} \bigcap_{n=1}^{\infty} \Omega_{1/r}^n\left(\frac{k-1}{N}, \frac{k}{N}\right).$$

In fact, if ω belongs to the set on the right-hand side, then there is an $\varepsilon > 0$ and some finite interval $[a, b]$ such that for any $n = 1, 2, \ldots$ there are $a \leqq t_1 < \cdots < t_{2n} \leqq b$ with $|x_{t_{2k-1}}, x_{t_{2k}}| > 4\varepsilon$ $(k = 1, \ldots, n)$. In this case we define $\chi_n(t) = |\{k \mid t_k \leqq t\}|$ $(a \leqq t \leqq b)$ and consider these "oscillation counting" functions for $n = 1, 2, \ldots$. Clearly $\lim \sup_n \chi_n(b) = \infty$ and thus $s = \inf\{t \mid \lim \sup \chi_n(t) = \infty\}$ is well defined. If $\lim \sup \chi_n(s) = \infty$, then $a < s \leqq b$ and $\lim \sup[\chi_n(s) - \chi_n(s - \delta)] = \infty$ $(0 < \delta < s - a)$. This means that oscillations $> 4\varepsilon$ accumulate to s from the left, and thus $\lim_{t \to s-0} x_t$ cannot exist. If $\lim \sup \chi_n(s) < \infty$, then $a \leqq s < b$ and

$$\lim \sup[\chi_n(s + \delta) - \chi_n(s)] = \infty \;(0 < \delta < b - s).$$

This means that oscillations $> 4\varepsilon$ accumulate to s from the right, and thus $\lim_{t \to s+0} x_t$ cannot exist. It is easy to show that, conversely, the nonexistence of, say $\lim_{t \to s+0} x_t$ implies the existence of some $r > 0$ and, for every

$N = 1, 2, \ldots$, of some $k > 0$ such that $\omega \in \Omega^n_{1/r}((k-1)/N, k/N)$ $(n = 1, 2, \ldots)$. Thus we have proved (7) which implies $\Omega_{\frac{1}{2}\text{-cont}} \in \hat{\mathscr{B}}$ since

$$\Omega^n_{1/r}((k - 1/N, k/N) \in \hat{\mathscr{B}}$$

(Corollary 7.9) and only countable set operations are involved. Now the Feller property implies

$$\lim_{N\to\infty} \alpha_\varepsilon\left(\frac{1}{N}\right) \leqq \lim_{N\to\infty} \left[\sup_{\substack{0 \leqq t \leqq 1/N \\ x \in X}} P_t(x, X \backslash U_{\varepsilon/2}(x))\right.$$

$$= 0 \qquad (\varepsilon > 0);$$

and thus we may, for a given $r = 1, 2, \ldots$, find some $N = 1, 2, \ldots$ such that $2\alpha_{1/r}(1/N) < 1$. Corollary 7.9 now implies

$$\hat{m}\left(\bigcap_{n=1}^{\infty} \Omega^n_{1/r}\left(\frac{k-1}{N}, \frac{k}{N}\right)\right) \leqq \lim_{n\to\infty} \left(2\alpha_{1/r}\left(\frac{1}{N}\right)\right)^n = 0 \qquad (k = 1, 2, \ldots)$$

hence $\bigcup_{k=1}^{\infty} \bigcap_{n=1}^{\infty} \Omega^n_{1/r}((k-1)/N, k/N)$ is an m-nullset and so finally is $\Omega \backslash \Omega_{\frac{1}{2}\text{-cont}}$, proving the theorem.

We pass now over to Ω_{cont}. Again we begin with an estimate.

7.11. Lemma. *Let X, $(P_t)_{t \geqq 0}$ and $\alpha_\varepsilon(t)$ $(\varepsilon, t > 0)$ be as in Lemma 7.8. Let $\varnothing \neq J \subseteq \mathbb{R}_+$ be finite, $0 \leqq a < b$ and $a\ J \subseteq [a, b]$; put*

$$B_\varepsilon(J) = \{\omega = (x_t)_{t \geqq 0} \mid \text{there is a } t \in J \text{ such that } |x_a, x_t| > 3\varepsilon\}.$$

Then $B_\varepsilon(J) \in \tilde{\mathscr{B}}$ and

$$P^x(B_\varepsilon(J)) \leqq 2\alpha_\varepsilon(b - a) \qquad (x \in X)$$

and hence

$$m(B_\varepsilon(J)) \leqq 2\alpha_\varepsilon(b - a)$$

for every probability σ-content p in X and $m = p \times P$.

Proof. Clearly $B_\varepsilon(J)$ is an open set depending on a fixed finite number of components, hence $B_\varepsilon(J) \in \tilde{\mathscr{B}}$. If $J = \{t_0, \ldots, t_n\}$ where $a = t_0 < \cdots < t_n \leqq b$, then

$$B_\varepsilon(J) = \sum_{j=1}^{n} \{|x_a, x_{t_i}| \leqq 3\varepsilon\ (i < j),\ |x_a, x_{t_j}| > 3\varepsilon\}$$

$$= \sum_{j=1}^{n} [\{|x_a, x_{t_i}| \leqq 3\varepsilon\ (i < j),\ |x_a, x_{t_j}| > 3\varepsilon,\ |x_a, x_b| > \varepsilon\}$$

$$+ \{|x_a, x_{t_i}| \leqq 3\varepsilon\ (i < j),\ |x_a, x_{t_j}| > 3\varepsilon,\ |x_a, x_b| \leqq \varepsilon\}]$$

Thus

$$m(B_\varepsilon(J)) \leqq \sum_{j=1}^{n} m(|x_a, x_{t_i}| \leqq 3\varepsilon \ (i<j),\ |x_a, x_{t_j}| > 3\varepsilon,\ |x_a, x_b| > \varepsilon)$$

$$+ \sum_{j=1}^{n} m(|x_a, x_{t_i}| \leqq 3\varepsilon \ (i<j),\ |x_a, x_{t_j}| > 3\varepsilon,\ |x_{t_j}, x_b| > 2\varepsilon).$$

$$\leqq m(|x_a, x_b| > \varepsilon)$$

$$+ \sum_{j=1}^{n} m(|x_a, x_{t_i}| \leqq 3\varepsilon \ (i<j),\ |x_a, x_{t_j}| > 3\varepsilon,\ |x_{t_j}, x_b| > 2\varepsilon)$$

Put $E_j = \{|x_a, x_{t_i}| \leqq 3\varepsilon \ (i<j),\ |x_a, x_{t_j}| > 3\varepsilon\}$ and

$$m_j(F) = m(E_j \cap \{x_{t_j} \in F\}).$$

It is easily seen that m_j is a σ-content in X and $m_1 + \cdots + m_n$ a probability σ-content. Moreover an easy Fubini type argument (see the proof of Theorem 1.4) leads from (1) to

$$m(E_j \cap \{|x_{t_j}, x_b| > 2\varepsilon\}) = \int m_j(dx) P_{b-t_j}(x, X \backslash \bar{U}_{2\varepsilon}(x))$$

$$\leqq m_j(X)\alpha_{2\varepsilon}(b-a) \qquad (j = 1, \ldots, n).$$

Summation over j yields

$$\sum_{j=1}^{n} m(|x_a, x_{t_i}| \leqq 3\varepsilon \ (i<j),\ |x_a, x_{t_j}| > 3\varepsilon,\ |x_{t_j}, x_b| > 2\varepsilon)$$

$$\leqq \alpha_{2\varepsilon}(b-a) \leqq \alpha_\varepsilon(b-a).$$

The estimate $m(|x_a, x_b| > \varepsilon) \leqq \alpha_\varepsilon(b-a)$ is still easier to deduce. This proves the lemma.

The argument leading from Lemma 7.8 to Corollary 7.9 also leads to

7.12. Corollary. *Let X, $(P_t)_{t \geqq 0}$ and $\alpha_\varepsilon(t)$ $(\varepsilon, t > 0)$ be as in Lemma 7.8. Then for every $0 \leqq a < b$, the set*

$$\hat{\mathscr{B}}_\varepsilon(a, b) = \{\omega \mid \text{there is a } t \in [a, b] \text{ such that } |x_a, x_t| > 3\varepsilon\}$$

belongs to $\hat{\mathscr{B}}$ and, with the notation of Corollary 7.9,

$$\hat{m}(\hat{\mathscr{B}}_\varepsilon(a, b)) \leqq 2\alpha_{2\varepsilon}(b-a).$$

As a consequence we obtain

7.13. Theorem. *Let X be compact metric and $(P_t)_{t \geqq 0}$ a Markovian semigroup in X that is Feller and even satisfies*

$$\lim_{\delta \to 0+0} \frac{\alpha_\varepsilon(\delta)}{\delta} = 0 \qquad (\varepsilon > 0). \tag{8}$$

The set

$$\Omega_{\text{cont}} = \{\omega = (x_t)_{t \geqq 0} \,|\, t \to x_t \text{ is continuous}\}$$

belongs to $\hat{\mathscr{B}}$ and satisfies $\hat{m}(\Omega_{\text{cont}}) = 1$ (with the notation of Theorem 7.10).

Proof. We first verify

$$\Omega \backslash \Omega_{\text{cont}} = \bigcup_{r=1}^{\infty} \bigcup_{A=1}^{\infty} \bigcap_{N=1}^{\infty} \bigcup_{k=1}^{AN} \hat{\mathscr{B}}_{1/r}\left(\frac{k-1}{N}, \frac{k}{N}\right). \tag{9}$$

In fact if $\omega = (x_t)_{t \geqq 0}$ belongs to the right-hand set, then there is an $\varepsilon > 0$ and an $A > 0$ such that for every $N = 1, 2, \ldots$ there is a $k = 1, \ldots, AN$ with $|x_{(k-1)/N}, x_t| > 3\varepsilon$ for some $(k-1)/N < t \leqq k/N$. Clearly such an $\omega = (x_t)_{t \geqq 0}$ is not continuous even if we restrict t to $[0, A]$. Assume conversely that $\omega = (x_t)_{t \geqq 0}$ is not a continuous function of t. Let $t_0 \geqq 0$ be some point of discontinuity. Then either

$$\limsup_{t \to t_0 - 0} |x_t, x_{t_0}| \qquad \text{or} \qquad \limsup_{t \to t_0 + 0} |x_{t_0}, x_t|$$

is a strictly positive number. Let us deal with the case $t_0 > 0$, $\limsup_{t \to t_0 - 0} |x_t, x_{t_0}| = \gamma > 0$ and leave the other case as an exercise to the reader. Choose an integer $r > 0$ such that $2(3/r) < \gamma$ and an integer $A > 0$ such that $t_0 + 1 < A$. For a given $N = 1, 2, \ldots,$ find k such that $t_0 \in [(k-1)/N, k/N]$. Clearly this is possible with $1 \leqq k \leqq AN$. It follows that there is $t \in](k-1)/N, k/N]$ such that $|x_t, x_{t_0}| > 2(3/r)$. But then either $|x_{(k-1)/N}, x_t| > 3/r$ or $|x_{(k-1)/N}, x_{t_0}| > 3/r$. In any case $\omega = \hat{\mathscr{B}}_{1/r}((k-1)/N, k/N)$. Thus we get (8). Now for fixed r, we get

$$m\left(\bigcup_{k=1}^{AN} \hat{\mathscr{B}}_{1/r}\left(\frac{k-1}{N}, \frac{k}{N}\right)\right) \leqq AN \cdot 2\alpha_{2/r}\left(\frac{1}{N}\right)$$

using Corollary 7.12. Now (8) implies $\lim_{N \to \infty} AN\, 2\alpha_{2/r}(1/N) = 0$, and we see that $\Omega \backslash \Omega_{\text{cont}}$ is a countable union of m-nullsets, which proves the theorem.

The rest of this subsection dealing with a compact metric X is devoted to preparations for the next subsection which will treat the case of a locally compact metric X by reduction to the compact case.

Throughout the rest of this subsection we fix an arbitrary point called ∞ in X and put $\mathscr{C}_0(X, \mathbb{R}) = \{f \mid f \in \mathscr{C}(X, \mathbb{R}), f(\infty) = 0\}$. The open set $X \setminus \{\infty\}$ can clearly be represented as an increasing union of compact sets $\emptyset \neq K_1 \subseteq K_2 \subseteq \cdots$. We put

$$\Omega_0 = \{\omega = (x_t)_{t \geqq 0} \mid x_t \neq \infty \ (0 \leqq t < \infty)\}$$

and observe

$$\Omega_0 = \bigcap_{A=1}^{\infty} \bigcup_{n=1}^{\infty} \Omega_0(A, K_n)$$

where

$$\Omega_0(A, K_n) = \{\omega = (x_t)_{t \geqq 0} \mid x_t \in K_n \ (0 \leqq t \leqq A)\}.$$

This clearly implies $\Omega_0 \in \hat{\mathscr{B}}$.

7.14. Theorem. *Let X be compact metric and $(P_t)_{t \geqq 0}$ be a Markovian semigroup in X that is Feller and satisfies (with the notation just introduced)*

$$P_t \mathscr{C}_0(X, \mathbb{R}) \subseteq \mathscr{C}_0(X, \mathbb{R}) \qquad (t \geqq 0) \tag{10}$$

$$P_t(x, \{\infty\}) = 0 \qquad (t \geqq 0, \quad \infty \neq x \in X). \tag{11}$$

Then for every probability σ-content p in X satisfying $p(\{\infty\}) = 0$, $m = p \times P$ and m the outer open inner compact regular extension of $\hat{m}$ from the Baire σ-field $\tilde{\mathscr{B}}$ to the Borel σ-field $\hat{\mathscr{B}}$ in Ω, we have

$$\hat{m}(\Omega_0) = 1,$$

and in particular

$$\hat{P}^x(\Omega_0) = 1 \qquad (x \in X).$$

Proof. It is clearly sufficient to prove

$$\lim_{n \to \infty} \hat{m}(\Omega_0 (A, K_n)) = 1 \qquad (A > 0).$$

Since the $\Omega_0(A, K_n)$ are compact subsets of Ω, the device leading to Corollary 7.9 shows that it is sufficient to prove

$$\lim_{n \to \infty} m(\Omega_0(J, K_n)) = 1$$

uniformly for all finite sets $\emptyset \neq J \subseteq [0, A]$.

From (11) we derive

$$m(x_A = \infty) = \int p(dx) P_A(x, \{\infty\})$$

$$= \int p(dx) 1_{X \setminus \{\infty\}} P_A(x, \{\infty\}) = 0,$$

which implies $m(x_A \neq \infty) = 1$ and hence $\lim_n m(x_A \in K_n) = 1$. Fix an $\varepsilon > 0$ and choose n_0 such that $m(x_A \in K_{n_0}) > 1 - \frac{1}{2}\varepsilon$. Find an $f \in \mathscr{C}_0(X, \mathbb{R})$ with $0 \leqq 1_{K_{n0}} \leqq f \leqq 1$ and consider $\beta_n(t) = \sup_{x \notin K_n} (P_t f)(x)$. The Feller property implies the continuity of this function of t (sup is a sup norm continuous functional). Moreover, $\beta_1(t) \geqq \beta_2(t) \geqq \cdots \searrow 0$ $(0 \leqq t \leqq A)$ by (10). Dini's theorem implies the existence of some n_1 such that $\beta_{n_1}(t) < \varepsilon/2$ $(0 \leqq t \leqq A)$, hence $(P_t f)(x) < \varepsilon/2$ $(x \in X \setminus K_{n_1}, 0 \leqq t \leqq A)$. Now let $J = \{t_1, \ldots, t_s\}$ with $0 \leqq t_1 < \cdots < t_s \leqq A$. We conclude

$$\left(1 - \frac{\varepsilon}{2}\right) - m(\Omega_0(J, K_{n_1})) < m(x_A \in K_{n_0}) - m(x_{t_1}, \ldots, x_{t_s} \in K_{n_1})$$

$$= m(x_A \in K_{n_0}, x_{t_j} \notin K_{n_1} \text{ for at least one } j = 1, \ldots, s).$$

First entrance decomposition yields

$$= \sum_{j=1}^{s} m(x_{t_i} \in K_{n_1}\ (i < j), x_{t_j} \notin K_{n_1}, x_A \in K_{n_0})$$

$$= \sum_{j=1}^{s} \int p(dx_0) 1_{K_{n_1}}(x_1) P_{t_1}(x_1, dx_2)$$

$$\cdots \int P(x_{j-1}, dx_j) 1_{X \setminus K_{n_1}}(x_j) P(x_j, K_{n_0}).$$

The last factor here is $<\varepsilon/2$ by construction. Thus we may continue

$$< \frac{\varepsilon}{2} \sum_{j=1}^{s} m(x_{t_i} \in K_{n_1}\ (i < j), x_{t_j} \in K_{n_1})$$

$$\leqq \frac{\varepsilon}{2}.$$

This proves $m(\Omega(J, K_{n_1}) > 1 - \varepsilon$. Observe that n_1 was determined independently of J. The theorem is proved.

3. The Locally Compact Case

In this subsection we treat the fundamental facts of path theory for Markovian σ-contents in $\Omega = \prod_{t \in \mathbb{R}_+} X_t$ where $X_t = X$ ($t \geqq 0$) is a locally compact Hausdorff space that has a countable base for the topology, and hence is metrizable. The Baire and Borel σ-fields in X thus coincide and will be denoted by $\mathscr{B}$. Our treatment will consist in a reduction to the compact case (subsection 2): We shall form the (compact metric) one-point compactification $\bar{X}$ of X and then pass over to $\bar{\Omega} = \prod_{t \geqq 0} \bar{X}_t$ ($\bar{X}_t = \bar{X}$ ($t \geqq 0$)).

7.15. Definition. *Let X be a locally compact Hausdorff space with a countably generated topology. A Markovian semigroup $(P_t)_{t \geqq 0}$ in $(X, \mathscr{B})$ (or, for short, in X) is called* **Feller** *if*

(1) $P_t \mathscr{C}_0(X, \mathbb{R}) \subseteq \mathscr{C}_0(X, \mathbb{R})$, where

$$\mathscr{C}_0(X, \mathbb{R}) = \{f \mid f \in \mathscr{C}(X, \mathbb{R}), \{|f| \geqq \varepsilon\} \text{ is compact } (\varepsilon > 0)\}$$

is the space of all continuous real functions on X that "vanish at infinity."

(2) $\lim_{t \to s+0} \|P_t f - P_s f\| = 0$ ($s \geqq 0$, $f \in \mathscr{C}_0(X, \mathbb{R})$) (uniform convergence, i.e., sup norm).

7.16. Theorem. *Let $(P_t)_{t \geqq 0}$ be a Markovian semigroup in the locally compact Hausdorff space X with a countable base for the topology. Let $\bar{X} = X + \{\infty\}$ be the one-point compactification of X (endowed with its topology). Then the construction of Proposition 7.2 leads to a Markovian semigroup $(\bar{P}_t)_{t \geqq 0}$ in $\bar{X}$ that is Feller there iff $(P_t)_{t \geqq 0}$ is Feller in X. In the Feller case we can, identifying $\mathscr{C}_0(X, \mathbb{R})$ with $\mathscr{C}_0(\bar{X}, \mathbb{R}) = \{f \mid f \in \mathscr{C}(\bar{X}, \mathbb{R}), f(\infty) = 0\}$, state*

$$\bar{P}_t \mathscr{C}_0(X, \mathbb{R}) \subseteq \mathscr{C}_0(X, \mathbb{R}), \tag{12}$$

$$\bar{P}_t(x, \{\infty\}) = 0 \qquad (x \in X, t \geqq 0). \tag{13}$$

Proof. Going through the constructions of Proposition 7.2, we immediately verify (13) and $(\bar{P}_t f)(\infty) = f(\infty)$ ($f \in \mathscr{C}(\bar{X}, \mathbb{R})$) which implies (12) plus the fact that the Feller properties for $(P_t)_{t \geqq 0}$ and $(\bar{P}_t)_{t \geqq 0}$ imply each other.

According to this theorem, every Markovian semigroup $(P_t)_{t \geqq 0}$ in X, together with a probability σ-content p in X, gives rise to a probability

σ-content $m = p \times P$ (using obvious notations) on the Baire σ-field $\tilde{\bar{\mathscr{B}}}$ in the compact Hausdorff space $\bar{\Omega}$ and to its unique outer-open inner-compact-regular extension $\hat{\bar{m}}$ on the Borel σ-field $\hat{\bar{\mathscr{B}}}$ in $\bar{\Omega}$. If $p = \delta_x$ (point mass 1 at $x \in X$), we get $\bar{P}^x$ resp. $\hat{\bar{P}}^x$.

7.17. Theorem. *Let $(P_t)_{t \geqq 0}$ be a Markovian semigroup in the locally compact Hausdorff space X with a countable base for the topology, and assume that it is Feller. Then for every increasing sequence $\emptyset \neq K_1 \subseteq K_2 \subseteq \cdots$ of compact subsets of X such that $X = K_1 \cup K_2 \cup \cdots$ and*

$$\hat{\Omega}_{\text{finite}} = \{\omega = (x_t)_{t \geqq 0} \mid x_t \in K_n \textit{ for some } n = 1, 2, \ldots \textit{ whenever } 0 \leqq t \leqq r \ (r = 1, 2, \ldots)\}$$

$$\hat{\Omega}_{\frac{1}{2}\text{-cont}} = \left\{\omega \mid \omega \in \Omega, \ \lim_{t \to s-0} x_t \textit{ exists in } X \textit{ for every } s > 0 \right.$$
$$\left. \textit{and } \lim_{t \to s+0} x_t \textit{ exists in } X \textit{ for every } s \geqq 0 \right\}$$

we have $\hat{\Omega}_{\frac{1}{2}\text{-cont}} \subseteq \hat{\Omega}_{\text{finite}}$, $\hat{\Omega}_{\frac{1}{2}\text{-cont}}$, $\hat{\Omega}_{\text{finite}} \in \hat{\bar{\mathscr{B}}}$ and

$$\hat{\bar{m}}(\hat{\Omega}_{\frac{1}{2}\text{-cont}}) = \hat{\bar{m}}(\hat{\Omega}_{\text{finite}}) = 1$$

Proof. Theorems 7.16 and 7.10 imply that (with obvious notations) $\hat{\bar{\Omega}}_{\frac{1}{2}\text{-cont}} \in \hat{\bar{\mathscr{B}}}$ and $\hat{\bar{m}}(\hat{\bar{\Omega}}_{\frac{1}{2}\text{-cont}}) = 1$. Theorem 7.14 implies $\hat{\bar{\Omega}}_{\text{finite}} \in \hat{\bar{\mathscr{B}}}$ and $\hat{\bar{m}}(\hat{\bar{\Omega}}_{\text{finite}}) = 1$. From this and $\hat{\Omega}_{\frac{1}{2}\text{-cont}} = \hat{\bar{\Omega}}_{\frac{1}{2}\text{-cont}} \cap \hat{\bar{\Omega}}_{\text{finite}}$ we now easily deduce the rest of the theorem.

7.18. Theorem. *Let $(P_t)_{t \geqq 0}$ etc. be as in Theorem 7.17, in particular assume that it is Feller. For any metric $|\cdot, \cdot|$ that describes the topology in X, let*

$$\alpha_\varepsilon(t) = \sup_{x \in X,\ 0 \leqq u \leqq t} P_u(x, X \backslash \bar{U}_\varepsilon(x)) \quad (\varepsilon, t > 0)$$

where $\bar{U}_\varepsilon(x) = \{y \mid y \in X, \ |x, y| \leqq \varepsilon\}$ $(\varepsilon > 0, x \in X)$. Let

$$\hat{\Omega}_{\text{cont}} = \{\omega = (x_t)_{t \geqq 0} \mid \omega \in \Omega, t \to x_t \textit{ is a continuous mapping of } \mathbb{R}_+ \textit{ into } X\}.$$

Then $\hat{\Omega}_{\text{cont}} \in \hat{\mathscr{B}}$. If

$$\lim_{\delta \to 0+0} \frac{\alpha_\varepsilon(\delta)}{\delta} = 0 \qquad (\varepsilon > 0)$$

then

$$\hat{\bar{m}}(\hat{\Omega}_{\text{cont}}) = 1$$

for every probability σ-content p in X. In particular

$$\hat{\hat{P}}^x(\hat{\Omega}_{\text{cont}}) = 1 \qquad (x \in X).$$

Proof. With obvious notations, we have $\hat{\Omega}_{\text{cont}} \in \hat{\hat{\mathscr{B}}}$ and $\hat{\Omega}_{\text{cont}} = \hat{\Omega}_{\text{cont}} \cap \hat{\Omega}_{\text{finite}}$. Our assumptions imply $m(\hat{\Omega}_{\text{finite}}) = 1$ and $m(\hat{\Omega}_{\text{cont}}) = 1$ (use Theorem 7.13). From this our theorem follows.

CHAPTER VII

SET FUNCTIONS IN GENERAL

In this chapter we present an investigation of set functions from a very general viewpoint. Measures are left out completely in this context. The set functions considered here are allowed to take their values in $\overline{\mathbb{R}} = \mathbb{R} \cup \{-\infty, \infty\}$.

Let us review what we have done about set functions so far. We have considered additive and σ-additive functions on rings and local σ-rings (resp. σ-fields), and restricted attention to set functions with finite real values only. We have solved the extension problem I.5.1 for σ-contents in Chapter II (Corollary II.2.4) and once more in Chapter III (Remark III.6.2.2). In Chapter II we have considered set functions that were not necessarily contents, namely:

(1) the outer content derived from a content; it was introduced as a tool for σ-content extensions; it was allowed to attain nonnegative real values and $+\infty$; it was defined for all subsets of the basic set; it was in general not additive, but σ-subadditive (Proposition II.1.3.2);

(2) the σ-additive routine extension of a σ-content from a local σ-ring to its generated σ-ring; it was allowed to attain nonnegative real values and $+\infty$;

(3) the minimal σ-additive extension of a σ-content from a local σ-ring to its generated σ-field; it was σ-additive; values in $\mathbb{R}_+ \cup \{\infty\}$ were allowed.

Thus we have a broad background of set functions that are not σ-contents. We shall now build up systematically a theory of general set functions, clearing up in particular the relations between subadditivity, additivity, and σ-additivity for nonnegative functions on rings, and the advantages of isotony in dealing with convergence of σ-additive functions. In the context of σ-additivity, we prove the *Vitali–Hahn–Saks theorem*. The last section is devoted to a systematic investigation of total variation.

There is one important class of set functions we do not treat here: capacities. Their theory deviates so far from the additivity concept considered in this chapter, and is, on the other hand, so much entangled with so-called *Souslin sets*, that it seems advisable to treat them not here but in a separate section (Chapter XIII, Section 3).

1. BASIC NOTIONS FOR SET FUNCTIONS

The main content of this section is a definition that introduces some basic notions for set functions in general, and thus supersedes earlier definitions, in which such notions have been introduced mostly under more special assumptions. The present definition will however be fully compatible with the earlier definitions and will not contain alterations of notions ex post. We recall the notations $\overline{\mathbb{R}} = \mathbb{R} \cup \{-\infty, \infty\}$, $\overline{\mathbb{R}}_+ = \mathbb{R}_+ \cup \{\infty\}$, $\overline{\mathbb{R}}_- = \mathbb{R}_- \cup \{-\infty\}$, and the obvious total ordering of $\overline{\mathbb{R}}$ along with the usual (compact) topology in $\overline{\mathbb{R}}$. For every $\alpha \in \overline{\mathbb{R}}$, it is clear what $-\alpha$ means; and for every $\overline{\mathbb{R}}$-valued function m on some set, it is clear what the function $-m$ is.

1.1. Definition. *Let $\Omega \neq \varnothing$ and $\varnothing \neq \mathscr{S} \subseteq \mathscr{P}(\Omega)$. A function defined on $\mathscr{S}$ is also called a set function (on $\mathscr{S}$, or, somewhat imprecisely, in Ω). Let a function $m \in \overline{\mathbb{R}}^{\mathscr{S}}$ be given. It is called:*

1.1.1. **monotone increasing** *or* **isotone** *if*

$$E, F \in \mathscr{S},\ E \subseteq F \quad \Rightarrow \quad m(E) \leqq m(F);$$

1.1.2. **monotone decreasing** *or* **antitone** *if*

$$E, F \in \mathscr{S},\ E \subseteq F \quad \Rightarrow \quad m(E) \geqq m(F)$$

1.1.3. **lower σ-continuous** *at $F \in \mathscr{S}$ if*

$$F_1, F_2, \ldots \in \mathscr{S},\ F_1 \subseteq F_2 \subseteq \cdots,\ F_1 \cup F_2 \cup \cdots = F$$
$$\Rightarrow \quad \lim_{n \to \infty} m(F_n) = m(F);$$

1.1.4. **lower σ-continuous** *if it is lower σ-continuous at every $F \in \mathscr{S}$;*

1.1.5. **upper σ-continuous** *at $F \in \mathscr{S}$ if*

$$F_1, F_2, \ldots \in \mathscr{S},\ F_1 \supseteq F_2 \supseteq \cdots,\ F_1 \cap F_2 \cap \cdots = F$$
$$\Rightarrow \quad \lim_{n \to \infty} m(F_n) = m(F);$$

1.1.6. **upper σ-continuous** *if it is upper σ-continuous at every $F \in \mathscr{S}$.*

1.1.7. **monotone σ-continuous** *if it is both upper and lower σ-continuous.*

1.1.8. **subadditive** *if*

$$n \geqq 1,\ F_1, \ldots, F_n \in \mathscr{S},\ F_j \cap F_k = \varnothing\ (j \neq k),$$
$$F_1 + \cdots + F_n \in \mathscr{S},\ m(F_1) + \cdots + m(F_n)\ \text{meaningful}$$
$$\Rightarrow\ m(F_1 + \cdots + F_n) \leqq m(F_1) + \cdots + m(F_n)\ \text{(subadditivity formula);}$$

1.1.9. **σ-subadditive** *if*

$$F_1, F_2, \ldots \in \mathscr{S},\ F_j \cap F_k = \varnothing\ (j \neq k),$$
$$F_1 + F_2 + \cdots \in \mathscr{S},\ m(F_1) + m(F_2) + \cdots\ \text{meaningful}$$
$$\Rightarrow\ m(F_1 + F_2 + \cdots) \leqq m(F_1) + m(F_2) + \cdots$$
$$(\sigma\text{-subadditivity formula});$$

1.1.10. **superadditive** *if the function $-m \in \overline{\mathbb{R}}^{\mathscr{S}}$ is subadditive (it is clear what the superadditivity formula means);*

1.1.11. **σ-superadditive** *if the function $-m$ is σ-subadditive (it is clear what the σ-superadditivity formula means);*

1.1.12. **additive** *if it is both subadditive and superadditive, i.e., if the additivity formula $m(F_1 + \cdots + F_n) = m(F_1) + \cdots + m(F_n)$ holds whenever its right and left member make sense;*

1.1.13. **σ-additive** *if it is both σ-subadditive and σ-superadditive, i.e., if the σ-additivity formula $m(F_1 + F_2 + \cdots) = m(F_1) + m(F_2) + \cdots$ holds whenever its right and left member make sense.*

1.1.14. *A set $F \in \mathscr{S}$ is called **m-finite** if $m(F) \in \mathbb{R}$ (i.e., $|m(F)| < \infty$).*

1.1.15. *A set $F \subseteq \Omega$ is called **m-σ-finite** or **σ-finite (for m)** if it can be represented as a countable union of m-finite sets.*

1.1.16. *m is called* **finite** *if all $F \in \mathscr{S}$ are m-finite, and* **σ-finite** *if all sets in $\mathscr{S}$ are m-σ-finite.*

1.1.17. *A set $M \subseteq \Omega$ is called a* **support** *of or for m, and we say that m is* **carried** *or* **supported** *by M or* **lives on** *M or is* **concentrated on** *M if $F \in \mathscr{S}, M \cap F = \varnothing \Rightarrow m(F) = 0$.*

1.2. Remarks

1.2.1. One basic feature of the above definition is its excessive generality. We have made it nevertheless in order to have a terminology that covers all possible situations. In order to build sound theories with the notions thus introduced, we have primarily to find sound restrictions of generality. One such restriction consists in considering $\mathbb{R}_+$-valued functions only; but even the physical notion of an electric charge distribution is sufficient to give up that restriction for a while. Another sound thing is the assumption that $\mathscr{S}$ is a set ring, and this is the assumption that we shall most frequently make.

1.2.2. It should be noticed that a support of some $m \in \overline{\mathbb{R}}^{\mathscr{S}}$ does not necessarily belong to $\mathscr{S}$.

1.2.3. It is reasonable to assume that $\mathscr{S}$ is at least a set ring if we deal with additivity. Moreover, it is reasonable to assume that m avoids at least one of the values ∞, $-\infty$. Indeed, try it and assume that $\mathscr{S}$ is a ring and that there are $E, F \in \mathscr{S}$ with $m(E) = -\infty, m(F) = \infty$. Then $E = (E \backslash F) + (E \cap F)$, $F = (F \backslash E) + (E \cap F)$. If the additivity formula holds for these decompositions, then the value $+\infty$ is forbidden for $m(E \backslash F)$, $m(E \cap F)$; and the value $-\infty$ is forbidden for $m(E \backslash F)$ and $m(E \cap F)$, hence $m(E \cap F)$ is a finite real and $m(E \backslash F) = -\infty$, $m(F \backslash E) = \infty$. But now the additivity formula for the disjoint union $(E \backslash F) + (F \backslash E)$ would involve $\infty - \infty$, a contradiction. Thus, if we want the additivity formula to hold whenever a disjoint decomposition in $\mathscr{S}$ stands there, m must avoid $-\infty$ or ∞. In the same way additivity "implies" that subsets of m-finite sets are m-finite (if they belong to $\mathscr{S}$). All problems mentioned here are ruled out if $m \geqq 0$ or $m \leqq 0$.

1.2.4. The σ-additivity formula $m(F_1 + F_2 + \cdots) = m(F_1) + m(F_2) + \cdots$ makes sense, in view of the commutative law for set unions, only if the series $m(F_1) + m(F_2) + \cdots$ converges to the same limit in $\mathbb{R}$ no matter how we rearrange it. In the case of a finite limit this means *absolute convergence.* No problems arise if $m \geqq 0$ or $m \leqq 0$.

1.2.5. If $\mathscr{R}$ is a ring and $m \in \overline{\mathbb{R}}^{\mathscr{R}}$ is additive, then so is the restriction m_0 of m to $\mathscr{R}_0 = \{F \mid F \in \mathscr{R}, |m(F)| < \infty\}$, and $\mathscr{R}_0$ is a ring if it is nonempty, provided we make the general requirement: $E, F \in \mathscr{R}, E \cap F = \varnothing \Rightarrow m(E) + m(F)$ is meaningful. In this case it follows that $E \in \mathscr{R}_0$, $F \in \mathscr{R}$, $F \subseteq E \Rightarrow F \in \mathscr{R}_0$, and $m(\varnothing) = m(\varnothing + \varnothing) = m(\varnothing) + m(\varnothing)$ plus the finiteness of $m(\varnothing)$ implies $m(\varnothing) = 0$.

1.2.6. If $m(\varnothing) = 0$, then σ-additivity implies additivity.

1.2.7. For each of the properties of set functions that we have listed in Definition 1.1, the set of all $m \in \overline{\mathbb{R}}^{\mathscr{S}}$ having that property is easily seen, for every $\varnothing \neq \mathscr{S} \subseteq \mathscr{P}(\Omega)$, to be stable under certain operations. Let us here be content to mention the following: The set of all additive $m \in \mathbb{R}^{\mathscr{S}}$ is a linear subspace of the linear space $\mathbb{R}^{\mathscr{S}}$, and the set of all σ-additive $m \in \mathbb{R}^{\mathscr{S}}$ is again a linear subspace of the former; the sum of two nonnegative superadditive functions is superadditive again. Terms like "$\leqq$," "majorant," etc. are to be understood in the natural partial ordering of $\overline{\mathbb{R}}^{\mathscr{S}}$.

1.3. Exercises. Let $\Omega \neq \varnothing$, $\mathscr{R} \subseteq \mathscr{P}(\Omega)$ be a ring, and $m \in \overline{\mathbb{R}}^{\mathscr{R}}$. Prove:

1.3.1. If m is $\geqq 0$ and superadditive, then m is isotone.

1.3.2. If m is $\geqq 0$ and superadditive, then m is σ-superadditive.

1.3.3. If $\Omega \in \mathscr{R}$ (i.e., $\mathscr{R}$ is a field), m is $\geqq 0$ and isotone, $\Omega_1, \Omega_2, \ldots \in \mathscr{R}$, $m(\Omega_1), m(\Omega_2), \ldots < \infty, \Omega = \Omega_1 \cup \Omega_2 \cup \cdots$, then m is σ-finite (the reader feels no doubt tempted to further relax the hypotheses a bit without loosing the result).

1.3.4. If $m \in \mathbb{R}^{\mathscr{R}}$ and m is additive and lives on $E \in \mathscr{R}$ as well as on $F \in \mathscr{R}$, then m lives on $E \cap F$.

1.3.5. If m is $\geqq 0$, isotone, and subadditive, then the relaxed subadditivity formula $m(F_1 \cup \cdots \cup F_n) \leqq m(F_1) + \cdots + m(F_n)$ holds for any $F_1, \ldots, F_n \in \mathscr{R}$. Prove the same about σ-subadditivity.

1.3.6. The sum of finitely many superadditive resp. subadditive resp. isotone functions $\geqq 0$ on $\mathscr{R}$ is superadditive resp. subadditive resp. isotone.

1.4. Example. Let $\Omega \neq \varnothing \neq \mathscr{S} \subseteq \mathscr{P}(\Omega)$ and $0 \leqq m \in \overline{\mathbb{R}}^{\mathscr{S}}$. Define, for every subset M of Ω,

$$\overline{m}(M) = \inf\{m(F_1) + \cdots + m(F_n) | n \geqq 1, F_1, \ldots, F_n \in \mathscr{S}, \\ M \subseteq F_1 \cup \cdots \cup F_n\}.$$

(with the usual convention $\inf \varnothing = \infty$). Then $m \in \overline{\mathbb{R}}_+^{\mathscr{P}(\Omega)}$ is subadditive. In fact, let $E, F \subseteq \Omega$, $E \cap F = \varnothing$. If $\overline{m}(E) + \overline{m}(F) = \infty$, then the desired subadditivity formula $\overline{m}(E + F) \leqq \overline{m}(E) + \overline{m}(F)$ is trivially true. Assume now $\overline{m}(E) + \overline{m}(F) < \infty$, choose any $\varepsilon > 0$, and find $E_1, \ldots, E_r, F_1, \ldots, F_s \in \mathscr{R}$ such that

$$E \subseteq E_1 \cup \cdots \cup E_r, \ F \subseteq F_1 \cup \cdots \cup F_r,$$
$$m(E_1) + \cdots + m(E_r) < \overline{m}(E) + \varepsilon, \ m(F_1) + \cdots + m(F_r) < \overline{m}(F) + \varepsilon.$$

Then $E + F \subseteq E_1 \cup \cdots \cup E_r \cup F_1 \cup \cdots \cup F_s$ and

$$\overline{m}(E + F) \leqq m(E_1) + \cdots + m(F_s) < \overline{m}(E) + \overline{m}(F) + 2\varepsilon.$$

Since $\varepsilon > 0$ was arbitrary, $\overline{m}(E + F) \leqq \overline{m}(E) + \overline{m}(F)$ follows.

1.5. Exercise. Let $\Omega \neq \varnothing \neq \mathscr{S} \subseteq \mathscr{P}(\Omega)$, $m \in \overline{\mathbb{R}}_+^{\mathscr{S}}$, and define

$$\underline{m}(M) = \sup\{m(F_1) + \cdots + m(F_n) | n \geqq 1, \ F_1, \ldots, F_n \in \mathscr{S}, \\ F_j \cap F_k = \varnothing \ (j \neq k), F_1 + \cdots + F_n \subseteq M\}$$

(with the convention $\sup \varnothing = 0$). Prove that the $\underline{m} \in \mathbb{R}_+^{\mathscr{P}(\Omega)}$ thus obtained is superadditive.

2. THE ADDITIVE AND σ-ADDITIVE PARTS OF A SUPERADDITIVE SET FUNCTION

In this section we decompose an arbitrary superadditive nonnegative function on an arbitrary set ring into a sum of a "purely superadditive," a "purely additive," and a σ-additive function in a unique fashion. Although our presentation is in terms of rings, the reader should be aware of the fact that every local σ-ring and every σ-field is a ring, and hence our results apply to the latter types of set systems as well.

2.1. Theorem. *Let $\Omega \neq \varnothing$ and $\mathscr{R}$ be a set ring in Ω. Let $m \in \overline{\mathbb{R}}_+^{\mathscr{R}}$ be superadditive. Then the set function $m^{\mathrm{add}} \in \overline{\mathbb{R}}_+^{\mathscr{R}}$ defined by*

$$m^{\mathrm{add}}(F) = \inf\{m(F_1) + \cdots + m(F_n) \mid n \geqq 1, F_1, \ldots, F_n \in \mathscr{R},$$
$$F_j \cap F_k = \varnothing \ (j \neq k), F = F_1 + \cdots + F_n\} \qquad (F \in \mathscr{R})$$

is additive, $\leqq m$, and majorizes every additive minorant of m in $\overline{\mathbb{R}}_+^{\mathscr{R}}$, i.e., it is the greatest additive minorant of m in $\overline{\mathbb{R}}_+^{\mathscr{R}}$. We call m^{add} the **additive part** *of m.*

Proof. Observe that we have no problems as to the meaningfulness of the members of additivity and superadditivity formulas since $\mathscr{R}$ is a ring and all functions under consideration are nonnegative. In order to prove that m^{add} is additive, choose arbitrary sets $E, F \in \mathscr{R}$ such that $E \cap F = \varnothing$. We first prove $m^{\mathrm{add}}(E + F) \leqq m^{\mathrm{add}}(E) + m^{\mathrm{add}}(F)$. For this, we may assume $m^{\mathrm{add}}(E) + m^{\mathrm{add}}(F) < \infty$. Choose any $\varepsilon > 0$ and $E_1, \ldots, E_r, F_1, \ldots, F_s \in \mathscr{R}$ such that

$$E_j \cap E_k = \varnothing \ (j \neq k),$$
$$E = E_1 + \cdots + E_r, \ m(E_1) + \cdots + m(E_r) < m^{\mathrm{add}}(E) + \varepsilon,$$
$$F_j \cap F_k = \varnothing \ (j \neq k),$$
$$F = F_1 + \cdots + F_s, \ m(F_1) + \cdots + m(F_s) < m^{\mathrm{add}}(F) + \varepsilon.$$

Then $E + F = E_1 + \cdots + F_s$ and $E_1, \ldots, F_s$ are pairwise disjoint. We thus obtain

$$m^{\mathrm{add}}(E + F) \leqq m(E_1) + \cdots + m(E_r) + m(F_1) + \cdots + m(F_2)$$
$$< m^{\mathrm{add}}(E) + m^{\mathrm{add}}(F) + 2\varepsilon.$$

Since $\varepsilon > 0$ was arbitrary, the desired inequality follows. Secondly, we prove $m^{\mathrm{add}}(E + F) \geqq m^{\mathrm{add}}(E) + m^{\mathrm{add}}(F)$. For this, we may assume $m^{\mathrm{add}}(E + F) < \infty$. Choose any $\varepsilon > 0$, and $G_1, \ldots, G_n \in \mathscr{R}$ such that $G_j \cap G_k = \varnothing$ $(j \neq k)$,

$E + F = G_1 + \cdots + G_n$, $m(G_1) + \cdots + m(G_n) < m^{\text{add}}(E + F) + \varepsilon$. Put $E_k = G_k \cap E$, $F_k = G_k \cap F$ $(k = 1, \ldots, n)$. Then $E_1, \ldots, F_n$ are in $\mathscr{R}$, pairwise disjoint, and $E = E_1 + \cdots + E_n$, $F = F_1 + \cdots + F_n$. We thus obtain

$$\begin{aligned} m^{\text{add}}(E + F) + \varepsilon &> m(G_1) + \cdots + m(G_n) \\ &= m(E_1 + F_1) + \cdots + m(E_n + F_n) \\ &\geqq m(E_1) + m(F_1) + \cdots + m(E_n) + m(F_n) \\ &\geqq m^{\text{add}}(E) + m^{\text{add}}(F). \end{aligned}$$

If $m' \in \overline{\mathbb{R}}_+^{\mathscr{R}}$ is an additive minorant of m, then $E, E_1, \ldots, E_n \in \mathscr{R}$, $E_j \cap E_k = \varnothing$ $(j \neq k)$, $E = E_1 + \cdots + E_n$ implies

$$m(E_1) + \cdots + m(E_n) \geqq m'(E_1) + \cdots + m'(E_n) = m'(E),$$

and thus $m^{\text{add}} \geqq m'$ follows. The theorem is proved.

2.2. Theorem. *Let $\Omega \neq \varnothing$ and $\mathscr{R}$ be a set ring in Ω. Let $m \in \overline{\mathbb{R}}_+^{\mathscr{R}}$ be superadditive. Then the set function $m^\sigma \in \overline{\mathbb{R}}_+^{\mathscr{R}}$ defined by*

$$m^\sigma(F) = \inf\{m(F_1) + m(F_2) + \cdots \mid F_1, F_2, \ldots \in \mathscr{R}, F_j \cap F_k = \varnothing \ (j \neq k), \\ F = F_1 + F_2 + \cdots\} \qquad (F \in \mathscr{R})$$

is σ-additive, $\leqq m$, and majorizes every σ-additive minorant of m in $\overline{\mathbb{R}}_+^{\mathscr{R}}$, i.e., it is the greatest σ-additive minorant of m in $\overline{\mathbb{R}}_+^{\mathscr{R}}$. We call m^σ the **σ-additive part** *of m.*

Proof. In order to prove that m^σ is σ-additive, we choose $F, F_1, F_2, \ldots \in \mathscr{R}$ such that $F_j \cap F_k = \varnothing$ $(j \neq k)$, $F = F_1 + F_2 + \cdots$. We first prove $m^\sigma(F) \leqq m^\sigma(F_1) + m^\sigma(F_2) + \cdots$. For this, we may assume $m^\sigma(F_1) + m^\sigma(F_2) + \cdots < \infty$. Choose any $\varepsilon > 0$. For any $j = 1, 2, \ldots$, choose $F_{j1}, F_{j2}, \ldots \in \mathscr{R}$ such that $F_{ji} \cap F_{jk} = \varnothing$ $(i \neq k)$, $F_j = F_{j1} + F_{j2} + \cdots$, $m(F_{j1}) + m(F_{j2}) + \cdots < m^\sigma(F_j) + \varepsilon 2^{-j}$. Then clearly $F = \sum_{j,k} F_{jk}$ and thus

$$\begin{aligned} m^\sigma(F) &\leqq \sum_{j,k} m(F_{jk}) \\ &= (m(F_{11}) + m(F_{12}) + \cdots) + (m(F_{21}) + m(F_{22}) + \cdots) + \cdots \\ &< m^\sigma(F_1) + \varepsilon 2^{-1} + m^\sigma(F_2) + \varepsilon 2^{-2} + \cdots \\ &= m^\sigma(F_1) + m^\sigma(F_2) + \cdots + \varepsilon. \end{aligned}$$

Since $\varepsilon > 0$ was arbitrary, the desired inequality follows. Secondly, we prove $m^\sigma(F) \geqq m^\sigma(F_1) + m^\sigma(F_2) + \cdots$. For this, we may assume $m^\sigma(F) < \infty$. Choose any $\varepsilon > 0$. Find $G_1, G_2, \ldots \in \mathscr{R}$ such that $G_j \cap G_k = \varnothing$ $(j \neq k)$, $F = G_1 + G_2 + \cdots$, and $m(G_1) + m(G_2) + \cdots < m^\sigma(F) + \varepsilon$. Put $F_{jk} = F_j \cap G_k$. Then $F_{jk} \in \mathscr{R}$, all F_{jk} are pairwise disjoint, $F_j = \sum_k F_{jk}$, $G_k = \sum_j F_{jk}$.

We thus get

$$\begin{aligned} m^\sigma(F) + \varepsilon &\geqq m(G_1) + m(G_2) + \cdots = m\left(\sum_j F_{j1}\right) + m\left(\sum_j F_{j2}\right) + \cdots \\ &\geqq \sum_j m(F_{j1}) + \sum_j m(F_{j2}) + \cdots = \sum_{j,k} m(F_{jk}) \\ &= (m(F_{11}) + m(F_{12}) + \cdots) + (m(F_{21}) + m(F_{22}) + \cdots) + \cdots \\ &\geqq m^\sigma(F_1) + m^\sigma(F_2) + \cdots. \end{aligned}$$

Here we have used the fact that m is σ-superadditive (Exercise 1.3.2). Since $\varepsilon > 0$ is arbitrary, the desired inequality follows. If $m' \in \overline{\mathbb{R}}_+^{\mathscr{R}}$ is a σ-additive minorant of m, we get $m \geqq m'$ in nearly the same fashion as the corresponding statement in Theorem 2.1.

2.3. Exercise. Let $\Omega \neq \varnothing$ and $\mathscr{R}$ a set ring in Ω. Let $m \in \overline{\mathbb{R}}_+^{\mathscr{R}}$ be superadditive. Prove that $(m^{\text{add}})^\sigma = m^\sigma$.

2.4. Definition. *Let $\Omega \neq \varnothing$ and $\mathscr{R}$ a set ring in Ω.*

2.4.1. *A superadditive function $m \in \overline{\mathbb{R}}_+^{\mathscr{R}}$ is called* **purely superadditive** *if $m^{\text{add}} \equiv 0$.*

2.4.2. *An additive function $m \in \mathbb{R}_+^{\mathscr{R}}$ is called* **purely additive** *if $m^\sigma \equiv 0$.*

2.5. Exercises. Let $\Omega \neq \varnothing$ and $\mathscr{R}$ a set ring in Ω. Prove:

2.5.1. A superadditive $m \in \overline{\mathbb{R}}_+^{\mathscr{R}}$ is purely superadditive iff for every $F \in \mathscr{R}$ and every $\varepsilon > 0$, there are $F_1, \ldots, F_n \in \mathscr{R}$ such that $F_j \cap F_k = \varnothing$ $(j \neq k)$, $F = F_1 + \cdots + F_n$ and $m(F_1) + \cdots + m(F_n) < \varepsilon$.

2.5.2. An additive $m \in \mathbb{R}_+^{\mathscr{R}}$ is purely additive iff for every $F \in \mathscr{R}$, every σ-additive $m' \in \overline{\mathbb{R}}_+^{\mathscr{R}}$ with $m'(F) < \infty$ and every $\varepsilon > 0$ there are $F_0, F_1 \in \mathscr{R}$ such that $F_0 \cap F_1 = \varnothing$, $F = F_0 + F_1$, $m(F_1) < \varepsilon$, $m'(F_0) < \varepsilon$.

2.5.3. The sum of two purely superadditive functions $\in \overline{\mathbb{R}}_+^{\mathscr{R}}$ is purely superadditive again.

2.5.4. The sum of two purely additive functions $\mathbb{R}_+^{\mathscr{R}}$ is purely additive again.

2.5.5. If $m_1, m_2 \in \overline{\mathbb{R}}_+^{\mathscr{R}}$ are superadditive, then:

2.5.5.1. $(m_1 + m_2)^{\text{add}} = m_1^{\text{add}} + m_2^{\text{add}}$.

2.5.5.2. $(m_1 + m_2)^\sigma = m_1^\sigma + m_2^\sigma$.

2.5.6. If $m_1, m_2 \in \overline{\mathbb{R}}_+^{\mathscr{R}}$ are superadditive, $m_1 \leqq m_2$, then:

2.5.6.1. $m_1^{\text{add}} \leqq m_2^{\text{add}}$.

2.5.6.2. $m_1^\sigma \leqq m_2^\sigma$.

2.5.7. Let $A = \{0, 1\}$, $p = (\frac{1}{2}, \frac{1}{2})$, and $(\Omega, \mathscr{B}, p)$ be the Bernoulli σ-content space for p. Prove that $m'(F) = m(F)^2$ $(F \in \mathscr{B})$ defines a purely superadditive $m' \in \mathbb{R}^{\mathscr{B}}$.

2.5.8. Let $\Omega = \mathbb{N} = \{1, 2, \ldots\}$, $\mathscr{R} = \mathscr{P}(\Omega)$, m: $l^\infty \to \mathbb{R}$ a Banach limit (Example I.6.4). Define $m' \in \mathbb{R}^{\mathscr{R}}$ by $m'(F) = m(1_F)$ $(F \subseteq \Omega)$. Prove that m' is purely additive.

2.6. Theorem. *Let $\Omega \neq \varnothing$ and $\mathscr{R}$ a set ring in Ω. Let $m \in \mathbb{R}_+^{\mathscr{R}}$ be superadditive. Then $m_0 = m - m^{\text{add}}$ is well defined, $m_0 \in \mathbb{R}_+^{\mathscr{R}}$, and m_0 is purely superadditive. Moreover, $m_1^{\text{add}} = m^{\text{add}} - m^\sigma$ is well defined, $m_1^{\text{add}} \in \mathbb{R}_+^{\mathscr{R}}$, and m_1^{add} is purely additive:*

$$m = m_0 + m_1^{\text{add}} + m^\sigma.$$

If $m = m_0' + m_1' + m_2'$ is such that $m_0', m_1', m_2' \geqq 0$, m_0' is purely superadditive, m_1' is purely additive, and m_2' is σ-additive, then $m_0' = m_0$, $m_1' = m_1^{\text{add}}$, $m_2' = m^\sigma$.

Proof. Observe that all function values envisaged in this theorem are *finite real numbers*. The very definition of m_0 proves $0 \leqq m_0 \leqq m$ and the pure superadditivity of m_0 (use Exercise 2.5.1). The very definition of m_1^{add} proves $0 \leqq m_1^{\text{add}} \leqq m^{\text{add}}$ and the pure additivity of m_1^{add} (modify Exercise 2.5.1 appropriately). The very definition of m_0, m_1^{add}, m^σ proves $m = m_0 + m_1^{\text{add}} + m^\sigma$. Let now $m_0', m_1', m_2' \in \mathbb{R}_+^{\mathscr{R}}$ be such that m_0' is purely superadditive, m_1' is purely additive, m_2' is σ-additive, and $m = m_0' + m_1' + m_2'$. By Exercises 1.3.6 and 2.5.5.1 we conclude

$$m^{\text{add}} = (m_0')^{\text{add}} + (m_1') + (m_2')^{\text{add}} = (m_1')^{\text{add}} + (m_2')^{\text{add}} = m_1' + m_2'.$$

By Exercises 2.3 and 2.5.5.2 we conclude $m^\sigma = (m^{\text{add}})^\sigma = (m_1')^\sigma + (m_2')^\sigma = m_2'$. This proves $m_0' = m_0$, $m_1' = m_1^{\text{add}}$, $m_2' = m^\sigma$.

2.7. The reader might amuse himself attempting to carry over the above theorem to $\overline{\mathbb{R}}$-valued functions as far as possible.

3. σ-ADDITIVITY. THE VITALI–HAHN–SAKS THEOREM

Since σ-additivity has been one of the main themes of our book so far, we have only little to do about it at this particular moment. We shall however do two things here: (1) generalize Proposition I.2.2 to some extent in order to obtain a somewhat more general criterion for σ-additivity in terms of σ-continuity, and (2) to prove the so-called Vitali–Hahn–Saks theorem and some related results, observing that the whole bulk is more or less a section of limiting theory.

We emphasize that the results presented here partly in terms of set rings are in particular true when reformulated in terms of local σ-rings or of σ-fields.

3.1. Proposition. *Let $\Omega \neq \varnothing$ and $\mathcal{R}$ be a set ring in Ω. Then for any additive $m \in \mathbb{R}^{\mathcal{R}}$ the following statements are equivalent:*

3.1.1. *m is σ-additive.*

3.1.2. *m is upper σ-continuous at $\varnothing$.*

3.1.3. *m is lower σ-continuous.*

3.1.4. *m is upper σ-continuous.*

Proof. $1 \Rightarrow 2$. Let $F_1, F_2, \ldots \in \mathcal{R}$, $F_1 \supseteq F_2 \supseteq \cdots$, $F_1 \cap F_2 \cap \cdots = \varnothing$. Put $E_n = F_{n-1} \backslash F_n$ $(n = 2, 3, \ldots)$. Clearly $E_2, E_3, \ldots \in \mathcal{R}$, $E_j \cap E_k = \varnothing$ $(j \neq k)$ and $F_1 = E_2 + E_3 + \cdots$. The σ-additivity of m implies

$$m(F_1) = m(E_2) + m(E_3) + \cdots,$$

the series being convergent. Similarly, $m(F_n) = m(E_{n+1}) + m(E_{n+2}) + \cdots$, and this apparently tends to 0 as $n \to \infty$.

$2 \Rightarrow 3$. Let $F, F_1, F_2, \ldots \in \mathcal{R}$, $F_1 \subseteq F_2 \subseteq \cdots$, $F_1 \cup F_2 \cup \cdots = F$. Put $E_n = F \backslash F_n$ $(n = 1, 2, \ldots)$. Clearly $E_1, E_2, \ldots \in \mathcal{R}$, $E_1 \supseteq E_2 \supseteq \cdots$, $E_1 \cap E_2 \cap \cdots = \varnothing$. Since m is upper σ-continuous at $\varnothing$, we obtain $m(E_n) \to 0$ as $n \to \infty$. Since m is additive, we have $F = F_n + E_n \Rightarrow m(F) = m(F_n) + m(E_n) \Rightarrow m(F_n) \to m(F)$ as $n \to \infty$.

$3 \Rightarrow 4$. Let $E, E_1, E_2, \ldots \in \mathcal{R}$, $E_1 \supseteq E_2 \supseteq \cdots$, $E_1 \cap E_2 \cap \cdots = E$. Put $F = E_1 \backslash E$, $F_n = E_1 \backslash E_n$ $(n = 1, 2, \ldots)$. Then $F, F_1, F_2, \ldots \in \mathcal{R}$, $F_1 \subseteq F_2 \subseteq \cdots$, $F_1 \cup F_2 \cup \cdots = F$. Since m is lower σ-continuous, we get $m(F_n) \to m(F)$ as $n \to \infty$. Since m is additive, we have $m(E_1) = m(E) + m(F) = m(E_n) + m(F_n)$ $(n = 1, 2, \ldots)$ and thus $m(E_n) \to m(E)$ as $n \to \infty$.

$4 \Rightarrow 1$. Let $F, F_1, F_2, \ldots \in \mathcal{R}$, $F_j \cap F_k = \varnothing$ $(j \neq k)$, $F = F_1 + F_2 + \cdots$. Put $E_n = F \backslash (F_1 + \cdots + F_n)$ $(n = 1, 2, \ldots)$. Clearly, $E_1, E_2, \ldots \in \mathcal{R}$, $E_1 \supseteq E_2 \supseteq \cdots$, $E_1 \cap E_2 \cap \cdots = \varnothing$. Since m is upper σ-continuous, we get $m(E_n) \to 0$ as $n \to \infty$. But m is additive and hence

$$m(F) = m(E_n) + m(F_1) + \cdots + m(F_n).$$

It follows that $m(F_1) + m(F_2) + \cdots$ is meaningful and equals $m(F)$. Observe that we have even proved the meaningfulness of the series.

3.2. Remarks

3.2.1. The reader may amuse himself attempting to remove or modify the standing hypothesis that m has finite real values from the above proposition still retaining part of the results.

3.2.2. It is obvious that the σ-additive real functions on a set ring $\mathscr{R}$ form a linear subspace of the vector space $\mathbb{R}^{\mathscr{R}}$. The *Vitali–Hahn–Saks theorem* (3.7) will deal with the closedness of the positive cone of this subspace under setwise convergence (i.e., pointwise convergence on the set $\mathscr{R}$).

3.3. Exercise. Let $\Omega \neq \varnothing$ and $\mathscr{R}$ a set ring in Ω. Let $m_1, m_2, \ldots \in \mathbb{R}^{\mathscr{R}}$ be σ-additive and assume that $m(F) = \lim_n m_n(F)$ exists uniformly on $\mathscr{R}$. Then $m \in \mathbb{R}^{\mathscr{R}}$ is σ-additive again.

We now turn to the *Vitali–Hahn–Saks theorem.* We follow RÉNYI [1] and begin by proving an elementary lemma, preceded by a

3.4. Definition. *A matrix* $t = (t_{nk})_{n,\,k=1,\,2,\,\ldots}$ *is said to be* **limiting** *if it has the following three properties:*

3.4.1.
$$\sup_{1 \leq n < \infty} \sum_{k=1}^{\infty} |t_{nk}| < \infty.$$

3.4.2.
$$\lim_{n \to \infty} \sum_{k=1}^{\infty} t_{nk} = 1.$$

3.4.3.
$$\lim_{n \to \infty} t_{nk} = 0 \ (k = 1, 2, \ldots).$$

3.5. Examples

3.5.1. The identity matrix $e = (e_{nk})_{n,\,k=1,\,2,\,\ldots}$ with $e_{nn} = 1$ $(n = 1, 2, \ldots)$ and $e_{nk} = 0$ otherwise, which may be written out as

$$e = \begin{bmatrix} 1 & 0 & 0 & 0 & \cdots \\ 0 & 1 & 0 & 0 & \cdots \\ 0 & 0 & 1 & 0 & \cdots \\ 0 & 0 & 0 & 1 & \cdots \\ \vdots & \vdots & \vdots & \vdots & \cdots \end{bmatrix},$$

is limiting.

3.5.2. The "Cesáro" matrix

$$c = \begin{bmatrix} 1 & 0 & 0 & 0 & 0 & \cdots \\ \frac{1}{2} & \frac{1}{2} & 0 & 0 & 0 & \cdots \\ \frac{1}{3} & \frac{1}{3} & \frac{1}{3} & 0 & 0 & \cdots \\ \frac{1}{4} & \frac{1}{4} & \frac{1}{4} & \frac{1}{4} & 0 & \cdots \\ \vdots & \vdots & \vdots & \vdots & \vdots & \cdots \end{bmatrix}$$

(the reader may amuse himself by writing it down formally) is limiting.

3.5.3. The modified Cesáro matrix

$$\begin{bmatrix} -1 & 1 & 0 & 0 & 0 & \cdots \\ -\frac{1}{2} & \frac{1}{2} & \frac{1}{2} & 0 & 0 & \cdots \\ -\frac{1}{3} & \frac{1}{3} & \frac{1}{3} & \frac{1}{3} & 0 & \cdots \\ -\frac{1}{4} & \frac{1}{4} & \frac{1}{4} & \frac{1}{4} & \frac{1}{4} & \cdots \\ \vdots & \vdots & \vdots & \vdots & \vdots & \cdots \end{bmatrix}$$

is limiting.

3.5.4. The matrix

$$\begin{bmatrix} -1 & 1 & 0 & 0 & 0 & \cdots \\ 0 & -1 & 1 & 0 & 0 & \cdots \\ 0 & 0 & -1 & 1 & 0 & \cdots \\ 0 & 0 & 0 & -1 & 1 & \cdots \\ \vdots & \vdots & \vdots & \vdots & \vdots & \cdots \end{bmatrix}$$

is *not limiting.*

3.6. Lemma. *Let $t = (t_{nk})_{n,\,k=1,\,2,\,\dots}$ be a limiting matrix. Then there is a sequence $s = (s_1, s_2, \dots)$ in $\{0, 1\}$ such that the sequence $u_1, u_2, \dots$ defined by $u_n = s_1 t_{n1} + s_2 t_{n2} + \cdots$ $(n = 1, 2, \dots)$ is bounded but not convergent.*

Proof. Observe that 3.4.1 implies that $u_1, u_2, \dots$ are well defined in any case. In order to construct s successively, we choose a real $0 < \alpha < \frac{1}{2}$ and determine $r_1 > 0$ such that $\sum_{k > r_1} |t_{1k}| < \alpha$ (use 3.4.1). Then we determine $n_1 > 0$ such that

$$\sum_{k \leqq r_1} |t_{nk}| < \frac{\alpha}{2} \qquad \left| \sum_k t_{nk} - 1 \right| < \frac{\alpha}{2}$$

and hence $|\sum_{k > r_1} t_{nk} - 1| < \alpha$ for $n \geqq n_1$ (use 3.4.2, 3.4.3). Next we determine $r_2 > r_1$ such that

$$\left| \sum_{r_1 < k \leqq r_2} t_{n_1 k} - 1 \right| < \alpha.$$

Then we determine $n_2 > n_1$ such that $n \geqq n_2$ $\sum_{k \leqq r_2} |t_{nk}| < \alpha/2$. Then we determine $r_3 > r_2$ such that $\sum_{k > r_3} |t_{n_2 k}| < \alpha/2$. We proceed further in an obvious alternating manner. Details may be written out upon sight of what has to come out in the end: We put $s_1 = \cdots = s_{r_1} = 0$, $s_{r_1+1} = \cdots = s_{r_2} = 1$, $s_{r_2+1} = \cdots = s_{r_3} = 0, \dots$, and it is now an easy exercise to prove $u_{n_1} > 1 - \alpha$, $u_{n_2} < \alpha$, $u_{n_3} > 1 - \alpha$, $u_{n_4} < \alpha$, etc., proving the desired statement.

3.7. Theorem (Vitali–Hahn–Saks). *Let $\Omega \neq \varnothing$ and $\mathscr{B}^{00}$ be a local σ-ring in Ω. Let $m, m_1, m_2, \ldots \in \mathbb{R}_+^{\mathscr{B}^{00}}$ be such that:*

3.7.1. $m_1, m_2, \ldots$ *are σ-additive;*

3.7.2. $\lim_{n\to\infty} m_n(F) = m(F)$ $(F \in \mathscr{R})$.

Then $m(\varnothing) = 0$ and m is σ-additive.

Proof. It is obvious that m is additive and satisfies $m(\varnothing) = 0$. Let now $F, F_1, F_2, \ldots \in \mathscr{B}^{00}$ be such that $F_j \cap F_k = \varnothing$ $(j \neq k)$, $F_1 + F_2 + \cdots = F$. Clearly,

$$
\begin{aligned}
m(F) = \lim_n m_n(F) &\geqq \lim_n m_n(F_1 + \cdots + F_r) \\
&= \lim_n m_n(F_1) + \cdots + \lim_n m_n(F_r) \\
&= m(F_1) + \cdots + m(F_r) \qquad (r = 1, 2, \ldots)
\end{aligned}
$$

and hence $m(F) \geqq m(F_1) + m(F_2) + \cdots$. Assume now that equality fails to hold here. Then $m(F) > 0$ and we may, after multiplying all $m, m_1, m_2, \ldots$ by the same factor $1/m(F) > 0$, assume $1 = m(F) > m(F_1) + m(F_2) + \cdots$. Put now $\Delta = 1 - (m(F_1) + m(F_2) + \cdots)$ and

$$t_{nk} = \frac{m_n(F_k) - m(F_k)}{\Delta}.$$

We verify now the hypotheses 3.4.1–3 of Lemma 3.4 for the matrix $t = (t_{nk})_{n,k=1,2,\ldots}$. Clearly,

$$\sum_k |t_{nk}| \leqq \frac{1}{\Delta}(m_n(F) + 1 - \Delta) \quad \to \quad \frac{2-\Delta}{\Delta} < \infty \qquad \text{as} \quad n \to \infty,$$

whence 3.3.1,

$$
\begin{aligned}
\lim_{n\to\infty} \sum_k t_{nk} &= \lim_{n\to\infty} \frac{m_n(F) - (m(F_1) + m(F_2) + \cdots)}{\Delta} \\
&= \frac{1 - (m(F_1) + m(F_2) + \cdots)}{\Delta} = \frac{\Delta}{\Delta} = 1,
\end{aligned}
$$

whence 3.4.2. Finally,

$$\lim_{n\to\infty} t_{nk} = \frac{m(F_k) - m(F_k)}{\Delta} = 0 \qquad (k = 1, 2, \ldots),$$

whence 3.4.3. By Lemma 3.6 we now get a sequence $s_1, s_2, \ldots$ of symbols 0 and 1 such that $\lim_n \sum_k t_{nk} s_k$ does not exist. Put now $F' = \sum_{s_k=1} F_k$.

It follows that

$$\sum_k t_{nk} s_k = \sum_{s_k=1} t_{nk} s_k = \frac{1}{\Delta} \sum_{s_s=1} (m_n(F_k) - m(F_k))$$
$$= \frac{m_n(F') - \sum_{s_k=1} m(F_k)}{\Delta} = \frac{1}{\Delta} (m_n(F') - \sum_{s_k=1} m(F_k)).$$

The last term is however convergent to $\Delta^{-1}(m(F') - \sum_{s_k=1} m(F_k))$ by the hypothesis of our theorem. This is a contradiction, hence the theorem is proved.

3.8. Exercises

3.8.1. Let Ω, $\mathscr{B}^{00}$, $m, m_1, m_2, \ldots$ be as in Theorem 3.6 (assume in particular that 3.7.1 and 3.7.2 hold). Put $\overline{m}(F) = \sup_n m_n(F)$ $(F \in \mathscr{B}^{00})$. Prove that $\overline{m}$ is in $\mathbb{R}_+^{\mathscr{B}^{00}}$ and is upper σ-continuous at $\varnothing$.

3.8.2. Deduce Theorem 3.7 once more with the help of Exercise 3.8.1.

3.8.3. Prove that the hypotheses of Theorem 3.7 imply that for every $F \in \mathscr{B}^{00}$, $\lim_n m_n(E) = m(E)$ holds even uniformly on $\{E \mid \mathscr{B}^{00} \ni E \subseteq F\}$.

It is much easier to establish the σ-additivity of limits if we have either uniform convergence (Exercise 3.3) or some sort of monotonicity on the way.

3.9. Proposition. *Let $\Omega \neq \varnothing \neq \mathscr{S} \subseteq \mathscr{P}(\Omega)$ and $M \subseteq \overline{\mathbb{R}}_+^{\mathscr{S}}$ be increasingly filtered, Define $\overline{m} \in \overline{\mathbb{R}}_+^{\mathscr{S}}$ by $\overline{m}(F) = \sup\{m(F) \mid m \in M\}$ $(F \in \mathscr{S})$.*

3.9.1. *Assume that every $m \in M$ is additive. Then $\overline{m}$ is additive.*

3.9.2. *Assume that every $m \in M$ is σ-additive. Then $\overline{m}$ is σ-additive.*

Proof. We prove 2 in such a way that the modification into a proof of 1 is obvious. Let $F, F_1, F_2, \ldots \in \mathscr{S}$, $F_j \cap F_k = \varnothing$ $(j \neq k)$, $F = F_1 + F_2 + \cdots$. Choose any $\alpha < \overline{m}(F)$, find $m \in M$ with $\alpha < m(F)$, then

$$\alpha < m(F) = m(F_1) + m(F_2) + \cdots \leqq \overline{m}(F_1) + \overline{m}(F_2) + \cdots,$$

and $\overline{m}(F) \leqq \overline{m}(F_1) + \overline{m}(F_2) + \cdots$ follows. Choose any

$$\beta < \overline{m}(F_1) + \overline{m}(F_2) + \cdots,$$

find n with $\beta < \overline{m}(F_1) + \cdots + \overline{m}(F_n)$ and then $m_1, \ldots, m_n \in M$ with $\beta < m_1(F_1) + \cdots + m_n(F_n)$. Since M is increasingly filtered, we may find some $m \in M$ with $m_1, \ldots, m_n \leqq m$. Thus

$$\beta < m(F_1) + \cdots + m(F_n) \leqq m(F_1) + \cdots + m(F_n) + m(F_{n+1}) + \cdots = m(F)$$
$$\leqq \overline{m}(F),$$

and $\overline{m}(F) \geqq \overline{m}(F_1) + \overline{m}(F_2) + \cdots$ follows.

4. TOTAL VARIATION

In this section we attach a nonnegative set function defined for all subsets of the basic set to any given set function. It will be called the total variation of the latter, and we shall investigate its properties, particularly in their dependence upon properties of the given set function.

4.1. Definition. *Let $\Omega \neq \varnothing \neq \mathscr{S} \subseteq \mathscr{P}(\Omega)$ and $m \in \overline{\mathbb{R}}^{\mathscr{S}}$ be given. The $\overline{\mathbb{R}}_+$-valued function $\|m\|(\cdot)$ defined on $\mathscr{P}(\Omega)$ by*

$$\|m\|(M) = \sup\{|m(F_1)| + \cdots + |m(F_n)| \,|\, n \geqq 1, \mathscr{S} \ni F_1, \ldots, F_n \subseteq M, F_j \cap F_k = \varnothing \ (j \neq k)\} \qquad (M \subseteq \Omega)$$

(with the convention $\sup \varnothing = 0$*) is called the total variation (function) of m. The extended real number $\|m\|(M)$ is called the total variation of m on M. $\|m\|(\Omega)$ is also briefly written as $\|m\|$ and briefly called the total variation or the norm of m. m is called bounded if $\|m\| < \infty$.*

4.2. Remarks

4.2.1. We shall sometimes write $\|m\|$ instead of $\|m\|(\cdot)$ if there is no danger of confusion.

4.2.2. Clearly $\|m\|(M) = 0$ iff $\Omega \backslash M$ is a support of M. In particular, $\|m\| = 0$ iff m vanishes. Clearly $\|m\|(\varnothing) = 0$ if $m(\varnothing) = 0$.

4.3. Exercises. Let $\Omega \neq \varnothing \neq \mathscr{S} \subseteq \mathscr{P}(\Omega)$ and $m \in \mathbb{R}^{\mathscr{S}}$ be given.

4.3.1. Prove that $\|m\| < \infty \Rightarrow m$ is carried by a set that is σ-finite for m.

4.3.2. Assume that $m(\varnothing) = 0$. Prove that

$$\|m\|(M) = \sup\left\{\sum_{k=1}^{\infty} |m(F_k)| \,\middle|\, \mathscr{S} \ni F_1, F_2, \ldots \subseteq M, F_j \cap F_k = \varnothing \ (j \neq k)\right\} \qquad (M \subseteq \Omega).$$

4.3.3. Assume that $\mathscr{S}$ is a ring and $m(\varnothing) = 0$. Prove that

$$\begin{aligned} \|m\|(E) &= \sup\{|m(F_1)| + \cdots + |m(F_n)| \,|\, n \geqq 1, F_1, \ldots, F_n \in \mathscr{S}, F_j \cap F_k = \varnothing \ (j \neq k), F_1 + \cdots + F_n = E\} \\ &= \sup\left\{\sum_{k=1}^{\infty} |m(F_k)| \,\middle|\, \mathscr{S} \ni F_1, F_2, \ldots \subseteq E, F_j \cap F_k = \varnothing \ (j \neq k), F_1 + F_2 + \cdots = E\right\} \qquad (E \in \mathscr{S}). \end{aligned}$$

4.3.4. Assume that $\mathscr{S}$ is a ring and m is superadditive and $\geqq 0$. Prove that $\|m\|(M) = \sup\{m(F) | \mathscr{S} \ni F \subseteq M\}$.

4.3.5. Assume that $(\Omega, \mathscr{S}, m)$ is a content space. Prove that the two definitions I.1.7.3 and 4.1 of $\|m\|$ coincide.

4.3.6. Prove that $\|m\|(\cdot)$: $\mathscr{P}(\Omega) \to \overline{\mathbb{R}}_+$ is isotone and superadditive (and hence, by Exercise 1.3.2, also σ-superadditive).

4.4. Proposition. *Let $\Omega \neq \varnothing$, $\mathscr{R}$ be a set ring in Ω, and $m \in \overline{\mathbb{R}}^{\mathscr{R}}$ additive. Then $\|m\|(\cdot)$, when restricted to $\mathscr{R}$, is additive.*

Proof. By Exercise 4.3.6 we know that $\|m\|(\cdot)$ is superadditive and isotone even on $\mathscr{P}(\Omega)$. Thus we have to prove only that the restriction of $\|m\|(\cdot)$ to $\mathscr{R}$ is subadditive. Let $E, F \in \mathscr{R}$, $E \cap F = \varnothing$. If $\|m\|(E)$ or $\|m\|(F)$ is $= \infty$, then by isotony also $\|m\|(E + F) = \infty$ and $\|m\|(E + F) = \|m\|(E) + \|m\|(F)$ follows. The case $\|m\|(E + F) = 0$ is settled in a similar way. Assume now $\|m\|(E + F) > 0$ and $\|m\|(E)$, $\|m\|(F) < \infty$. Choose any $\alpha < \|m\|(E + F)$ and determine $E + F \supseteq F_1, \ldots, F_n \in \mathscr{R}$ such that $F_j \cap F_k = \varnothing$ $(j \neq k)$ and $\alpha < |m(F_1)| + \cdots + |m(F_n)|$. Clearly, we may continue

$$
\begin{aligned}
&= |m(F_1 \cap E) + m(F_1 \cap F)| + \cdots + |m(F_n \cap E) + m(F_n \cap F)| \\
&\leqq [|m(F_1 \cap E)| + \cdots + |m(F_n \cap E|] + [|m(F_1 \cap F)| + \cdots + |m(F_n \cap F)|] \\
&\leqq \|m\|(E) + \|m\|(F)
\end{aligned}
$$

(under our assumptions all values of m envisaged here are finite reals). Since $\alpha < \|m\|(E + F)$ was arbitrary, the subadditivity formula

$$\|m\|(E + F) \leqq \|m\|(E) + \|m\|(F)$$

follows.

4.5. Exercises

4.5.1. Let $\Omega \neq \varnothing \neq \mathscr{S} \subseteq \mathscr{P}(\Omega)$ and $m \in \overline{\mathbb{R}}^{\mathscr{S}}$. Prove:

4.5.1.1. $\varnothing$ is a support of m iff $m \equiv 0$.

4.5.1.2. $\varnothing \neq M \subseteq \Omega$ is a support of m iff $\|m\|(\Omega \backslash M) = 0$.

4.5.1.3. m and $\|m\|(\cdot)$ have the same supports, more precisely; a $\varnothing \neq M \subseteq \Omega$ is a support of m iff M is a support of $\|m\|(\cdot)$.

4.5.1.4. Assume that $\mathscr{S}$ is a ring. Then the intersection of two supports of m is again a support of m.

4.5.1.5. Assume that $\mathscr{S}$ is a local σ-ring. Then the intersection of countably many supports of m is a support of m again.

4.5.2. Let $\Omega \neq \varnothing$ and $\mathscr{R}$ a set ring in Ω.

4.5.2.1. Let $m \in \mathbb{R}_+^{\mathcal{R}}$ be subadditive. Prove that $\|m\|(\cdot)$ is additive on $\mathcal{R}$.

4.5.2.2. Let $m \in \mathbb{R}^{\mathcal{R}}$ be σ-additive. Prove that $\|m\|(\cdot)$ is σ-additive on $\mathcal{R}$. (*Hint*: Imitate the proof of Proposition 4.4 making use of Exercise 4.3.3.)

4.5.2.3. Prove that $M \subseteq \Omega \Rightarrow \|m\|(M) = \sup\{\|m\|(F) \mid \mathcal{R} \ni F \subseteq M\}$.

4.5.2.4. Let $m, m' \in \overline{\mathbb{R}}_+^{\mathcal{R}}$ be additive. Prove

$$\|m\|(M) = \sup\{|m(F)| \mid \mathcal{R} \ni F \subseteq M\}, \ \|m + m'\|(M) = \|m\|(M) + \|m'\|(M),$$

for all $M \subseteq \Omega$, and $\|m\| = \sup\{m(F) \mid F \in \mathcal{R}\}$, $\|m + m'\| = \|m\| + \|m'\|$.

4.5.2.5. Let $m, m' \in \mathbb{R}^{\mathcal{R}}$ be additive. Prove

$$\begin{aligned} \|m\|(M) &= \sup\{|m(E)| + |m(F)| \mid \mathcal{R} \ni E, F \subseteq M, E \cap F = \emptyset\} \\ &\leqq 2 \sup\{|m(F)| \mid \mathcal{R} \ni F \subseteq M\}, \end{aligned}$$

$$\|m + m'\|(M) \leqq \|m\|(M) + \|m'\|(M) \qquad (M \subseteq \Omega).$$

4.5.3. Let $\Omega \neq \emptyset \neq \mathcal{S} \subseteq \mathcal{P}(\Omega)$ and $m, m' \in \overline{\mathbb{R}}^{\mathcal{S}}$.

4.5.3.1. Prove $0 \leqq m \leqq m' \Rightarrow \|m\|(M) \leqq \|m'\|(M) \qquad (M \subseteq \Omega)$,

$$\|m\| \leqq \|m'\|$$

4.5.3.2. Prove $\|\alpha m\|(M) = |\alpha| \cdot \|m\|(M)$ $(\alpha \in \mathbb{R}, M \subseteq \Omega)$.

CHAPTER VIII
THE VECTOR LATTICE OF SIGNED CONTENTS

This chapter presents, on the basis of the preceding one, a thorough study of additive resp. σ-additive set functions that attain *finite real values* only (but not necessarily only nonnegative ones) and have finite total variation on every set of their domain. These set functions will be called *signed contents* resp. *signed σ-contents.* For the latter, a set ring is adopted for a large part of this chapter, and the reader should be aware once more that local σ-rings are rings, and so are σ-fields. We refrain from repeating what has been achieved in the preceding chapter under more general assumptions, except in the exercises.

We begin by clearing up the vector lattice structure of the set of all signed contents resp. signed σ-contents on a ring (Section 1). Then we proceed to the case of local σ-rings, prove the existence of so-called *Hahn decompositions* and show how the vector lattice business simplifies through this (Section 2). *Absolute continuity* and *Lebesgue decomposition* are the themes of Sections 3 and 4. In Section 5 we prove the *Radon–Nikodym theorem* for σ-contents with finite total variation. In Section 6 we apply it extensively, defining *conditional expectations*; and in Section 7 we present the rudiments of the theory of *martingales* and *semimartingales*. In Section 8 we pose the *Radon–Nikodym problem*; after some preparations on *locally integrable functions* and *disjoint bases* we prove a very general Radon–Nikodym theorem, assuming a disjoint base, and thus covering, e.g., the case of arbitrary locally compact spaces. An example of Halmos shows that the Radon–Nikodym problem has no general solution. See KÖLZOW [4].

Throughout this chapter we speak of *contents* and thus confine ourselves to additive real set functions. The next chapter will show how to translate the results of this chapter into the language of *measures*, i.e., linear forms, and how to prove some of them anew in that context. The reader will, throughout the present chapter, be frequently confronted with the observation that the most basic facts are of a purely vector lattice theoretical character. The need for a systematic vector lattice theoretical investigation will be satisfied, it is hoped in the next chapter. In the present chapter we prefer to sacrifice some systematics in order to stay closer to applications.

1. SIGNED CONTENTS AND SIGNED σ-CONTENTS

In this section we focus attention on *real-valued additive functions on set rings* and define the concepts of a *signed content* and a *signed σ-content*. We prove that both the signed contents and the signed σ-contents on a set ring form a vector lattice and show how to perform the vector lattice operations. The basic formulas are in Theorem 1.3. They show that the vector lattice operations are local in the sense that what happens on the subsets of a given set does not depend but on the values of the set function on these subsets. Having cleared up the vector lattice structure of the space of all signed contents, we shall proceed to a proof that it is even a *conditionally complete lattice*. Moreover, we shall use total variation in order to introduce seminorms; these turn out to be norms, and the resulting normed linear space to be even an L-space in the important special case of fields. Generally, however, we work with set rings in this section. The next section (on Hahn decompositions) will show how matters simplify slightly if we pass over to local σ-rings or to σ-fields.

Recall the definition and basic properties of total variation as presented in Section VII.4.

1.1. Definition. *Let $\Omega \neq \varnothing$, $\mathscr{R}$ a set ring in Ω and $m \in \mathbb{R}^{\mathscr{R}}$ additive.*

1.1.1 *Assume that the restriction of the total variation $\|m\|(\cdot)$ of m to $\mathscr{R}$ takes only finite real values. Then m is called a* **signed content** *or a (electric)* **charge distribution** *on $\mathscr{R}$, and the triple $(\Omega, \mathscr{R}, m)$ is called a* **signed content prespace.**

1.1.2. *If a signed content m on $\mathscr{R}$ is σ-additive, we call it a* **signed σ-content**, *and $(\Omega, \mathscr{R}, m)$ is called a* **signed σ-content prespace**. *It is called a* **signed σ-content space** *if $\mathscr{R}$ is a local σ-ring.*

1.1.3. *The set of all signed contents on $\mathscr{R}$ is denoted by* $\operatorname{cont}(\mathscr{R})$, *and the set of all signed σ-contents on $\mathscr{R}$ is denoted by* $\operatorname{cont}^{\sigma}(\mathscr{R})$. *Also* $\operatorname{cont}(\mathscr{R})_+$ *resp.* $\operatorname{cont}^{\sigma}(\mathscr{R})_+$ *denotes the set of all (positive) contents resp. (positive) σ-contents on $\mathscr{R}$.*

1.1.4. *If $(\Omega, \mathscr{R}, m)$ is a signed content prespace, then the restriction of $\|m\|(\cdot)$ to $\mathscr{R}$ is denoted by $|m|(\cdot)$ and called the* **modulus** *of m.*

1.2. Proposition. *Let $(\Omega, \mathscr{R}, m)$ be a signed content prespace and $|m|(\cdot)$ the modulus of m. Then:*

1.2.1. *$|m|(\cdot)$ is a (positive) content on $\mathscr{R}$.*

1.2.2. *If m is a signed σ-content, then $|m|(\cdot)$ is a (positive) σ-content on $\mathscr{R}$ and vice versa.*

Proof. Use Proposition VII.4.4 and Exercise VII.4.5.2.2 (plus Proposition VII.3.1).

1.3. Theorem. *Let $\Omega \neq \emptyset$ and $\mathscr{R}$ be a set ring in Ω. Then* $\mathrm{cont}(\mathscr{R})$ *is a linear subspace of* $\mathbb{R}^{\mathscr{R}}$ *and* $\mathrm{cont}(\mathscr{R})_+$ *is a convex cone with tip 0 in* $\mathrm{cont}(\mathscr{R})$*. We consider the partial-ordering in* $\mathrm{cont}(\mathscr{R})$ *defined by that cone. Then:*

1.3.1. $\mathrm{cont}(\mathscr{R})$ *is a vector lattice and* $|m| = m \vee (-m)$ $(m \in \mathrm{cont}(\mathscr{R}))$.

1.3.2. $\mathrm{cont}^\sigma(\mathscr{R})$ *is a vector sublattice of* $\mathrm{cont}(\mathscr{R})$*: If* $m \in \mathrm{cont}^\sigma(\mathscr{R})$*, then* $|m| \in \mathrm{cont}^\sigma(\mathscr{R})$.

1.3.3. *The vector lattice operations in* $\mathrm{cont}(\mathscr{R})$ *can be carried out in the following ways:*

1.3.3.1. *If m, m′* $\in \mathrm{cont}(\mathscr{R})$*, then*

$$(m \vee m')(F) = \sup\left\{\sum_{k=1}^{n} \max\{m(F_k), m'(F_k)\} \,\middle|\, n \geqq 1, F_1, \ldots, F_n \in \mathscr{R},\right.$$

$$\left. F_j \cap F_k = \emptyset (j \neq k), F_1 + \cdots + F_n = F\right\}$$

$$= \sup\{m(E) + m'(E') \mid E, E' \in \mathscr{R}, E \cap E' = \emptyset, E + E' = F\}$$

$$(m \wedge m')(F) = \inf\left\{\sum_{k=1}^{n} \min\{m(F_k), m'(F_k)\} \,\middle|\, n \geqq 1, F_1, \ldots, F_n \in \mathscr{R},\right.$$

$$\left. F_j \cap F_k = \emptyset (j \neq k), F_1 + \cdots + F_n = F\right\}$$

$$= \inf\{m(E) + m'(E') \mid E, E' \in \mathscr{R}, E \cap E' = \emptyset, E + E' = F\}$$

for every $F \in \mathscr{R}$.

1.3.3.2. *If* $m \in \mathrm{cont}(\mathscr{R})$*, then*

$$m_+(F) = \sup\left\{\sum_{k=1}^{n} (m(F_k) \vee 0) \,\middle|\, n \geqq 1, F_1, \ldots, F_n \in \mathscr{R}, F_j \cap F_k = \emptyset\right.$$

$$\left. (j \neq k), F_1 + \cdots + F_n \subseteq F\right\}$$

$$= \sup\{m(E) \mid \mathscr{R} \ni E \subseteq F\}$$

$$m_-(F) = \sup\left\{\sum_{k=1}^{n} ((-m(F_k) \vee 0) \,|\, n \geqq 1, F_1, \ldots, F_n \in \mathscr{R}, F_j \cap F_k = \varnothing \right.$$

$$\left. (j \neq k), F_1 + \cdots + F_n \subseteq F \right\}$$

$$= \sup\{-m(E) \,|\, \mathscr{R} \ni E \subseteq F\}.$$

Proof. It is obvious that $\operatorname{cont}(\mathscr{R})$ is a linear subspace of $\mathbb{R}^{\mathscr{R}}$, $\operatorname{cont}(\mathscr{R})_+$ is a convex cone $\subseteq \operatorname{cont}(\mathscr{R})$, with tip 0, and $\sigma\text{-}\operatorname{cont}(\mathscr{R})$ is a linear subspace of $\operatorname{cont}(\mathscr{R})$.

1. Let $m \in \operatorname{cont}(\mathscr{R})$. We write $|m|$ instead of $|m|(\cdot)$ for brevity. Put $m' = \frac{1}{2}(|m| + m)$, $m'' = \frac{1}{2}(|m| - m)$. Clearly $m', m'' \in \operatorname{cont}(\mathscr{R})$ and $m = m' - m''$. From the definition of $|m|$ it is obvious that $|m| \geqq m, -m$, hence $m', m'' \geqq 0$. If $m''' \in \operatorname{cont}(\mathscr{R})$ and satisfies $m''' \leqq m', m''$, then it is clear from the definition of $|m|$ (see also Exercise VII.4.3.3) that $m'''(F) \leqq |m(F)| + m(F)$, $|m(F)| - m(F)$ and hence $m'''(F) \leqq 0$ $(F \in \mathscr{R})$. We conclude that $m' \wedge m''$ exists and is $= 0$. Lemma 0.2.1 now shows that $\operatorname{cont}(\mathscr{R})$ is a vector lattice. Moreover, the proof of Lemma 0.2.1 shows that in terms of this vector lattice structure we have $m' = m_+$, $m'' = m_-$, and thus $m_+ + m_-$ is the restriction of the total variation $|m|(\cdot)$ to $\mathscr{R}$, thus fully justifying the notation $|m|$ for it: $|m| = m \vee (-m)$.

2 follows from Proposition VII.2.5.5.2.

3. It is certainly sufficient to prove 3.1. The first equality sign follows from

$$\begin{aligned} m \vee m' &= m + (m' - m)_+ = m + \tfrac{1}{2}(|m' - m| + m' - m) \\ &= \tfrac{1}{2}(m + m' + |m' - m|), \end{aligned}$$

the definition of

$$|m' - m|(F) = \sup\left\{\sum_{k=1}^{n} |m'(F_k) - m(F_k)| \,\Big|\, n \geqq 1, F_1, \ldots, F_n \in \mathscr{R}, \right.$$

$$\left. F_j \cap F_k = \varnothing, F_1 + \cdots + F_n = F \right\}$$

$(F \in \mathscr{R}$; use Exercise VII.4.3.3), the additivity of m, m', and the obvious equality $\max\{x, y\} = x + y + |x - y|$ $(x, y \in \mathbb{R})$. The second equality sign now follows from the first by breaking each sum $\sum_{k=1}^{n} \max\{m(F_k), m'(F_k)\}$ into

$$\sum_{m(F_k) > m'(F_k)} m(F_k) = m\left(\sum_{m(F_k) > m'(F_k)} F_k\right)$$

and

$$\sum_{m(F_k) \leqq m'(F_k)} m'(F_k) = m'\left(\sum_{m(F_k) \leqq m'(F_k)} F_k\right)$$

(actually this yields $\leqq$ only, but $\geqq$ is obvious).

1.4. Theorem. *Let $\Omega \neq \varnothing$ and $\mathscr{R}$ be a set ring in Ω. Let $\varnothing \neq M \subseteq \mathrm{cont}_+(\mathscr{R})$ be increasingly filtered. Define $\overline{m} \in \mathbb{R}_+^{\mathscr{R}}$ by*

$$\overline{m}(F) = \sup_{m \in M} m(F) \qquad (F \in \mathscr{R})$$

and assume

$$\overline{m}(F) < \infty \qquad (F \in \mathscr{R}).$$

Then:

1.4.1. *M has a l.u.b. in the lattice $\mathrm{cont}(\mathscr{R})$ and $\overline{m}$ is this l.u.b.:*

$$\overline{m} = \bigvee_{m \in M} m.$$

1.4.2. *If $M \subseteq \sigma\text{-}\mathrm{cont}_+(\mathscr{R})$, then $\overline{m} \in \mathrm{cont}_+^{\sigma}(\mathscr{R})$, M has a l.u.b. in the lattice $\mathrm{cont}^{\sigma}(\mathscr{R})$, and $\overline{m}$ is this l.u.b.*

1.4.3. $\mathrm{cont}(\mathscr{R})$ *and* $\mathrm{cont}^{\sigma}(\mathscr{R})$ *are conditionally complete lattices; more precisely, every $\varnothing \neq M \subseteq \mathrm{cont}^{\sigma}(\mathscr{R})$ that has an upper bound in $\mathrm{cont}(\mathscr{R})$, has a l.u.b. in $\mathrm{cont}^{\sigma}(\mathscr{R})$.*

Proof. Apply Proposition VII.3.9 in order to prove 1 and 2. In order to prove 3, let $\varnothing \neq M \subseteq \mathrm{cont}(\mathscr{R})$ have a majorant in $\mathrm{cont}(\mathscr{R})$. Devices already employed in the proof of Theorem 0.2.2 allow us to assume $M \subseteq \mathrm{cont}_+(\mathscr{R})$, M increasingly filtered. Application of 1 now yields the l.u.b. of M in $\mathrm{cont}(\mathscr{R})$. If $M \subseteq \mathrm{cont}^{\sigma}(\mathscr{R})$, then the same reasoning leads to the l.u.b. of M in σ-$\mathrm{cont}(\mathscr{R})$. Multiplication by -1 yields g.l.b.'s for sets with minorants.

1.5. Exercises. Let $\Omega \neq \varnothing$ and $\mathscr{R}$ be a set ring in Ω.

1.5.1. Let $\mathscr{U} \subseteq \mathscr{R}$ be increasingly filtered (for $\subseteq$). For every $U \in \mathscr{U}$, let a content m_U be defined on the ring $\mathscr{R}_U = \{F \cap U \mid F \in \mathscr{R}\} \subseteq \mathscr{R}$ such that the compatibility conditions

$$m_U(F) = m_V(F) \qquad (U, V \in \mathscr{U}, \quad F \in \mathscr{R}_U \cap \mathscr{R}_V)$$

are satisfied.

1.5.1.1. Prove that there is an $m \in \mathrm{cont}(\mathscr{R})_+$ such that for every $U \in \mathscr{U}$ the restriction of m to $\mathscr{R}_U$ coincides with m_U. (*Hint*: Define m'_U on $\mathscr{R}$ by $m'_U(F) = m_U(F \cap U)$ $(F \in \mathscr{R})$ and use Theorem II.5.1.)

1.5.1.2. Prove the analogue of 1.5.1.1 for σ-contents. Prove that in this case m is unique if every $F \in \mathscr{R}$ can be covered with countably many sets from $\bigcup_{U \in \mathfrak{U}} \mathscr{R}_U$.

1.5.2. Let $m \in \text{cont}(\mathscr{R})$.

1.5.2.1. Prove that $|m|$ is purely additive (Definition VII.2.4.2) iff m can be written in the form $m = m' - m''$, where $0 \leqq m'$, $m'' \in \text{cont}(\mathscr{R})$ and m' and m'' are purely additive. Call $m \in \text{cont}(\mathscr{R})$ purely additive if $|m|$ is purely additive.

1.5.2.2. Prove that an $m \in \text{cont}(\mathscr{R})$ is purely additive iff

$$F \in \mathscr{R} \quad \Rightarrow \quad \inf\left\{ \sum_{k=1}^{n} |m(F_k)| \,\middle|\, n \geqq 1, F_1, F_2, \ldots \in \mathscr{R}, F_j \cap F_k = \varnothing (j \neq k), \right.$$

$$\left. F = F_1 + F_2 + \cdots \right\} = 0.$$

1.5.2.3. Prove that $\text{cont}^{\text{p.a.}}(\mathscr{R}) = \{m \mid m \in \text{cont}(\mathscr{R}),\ m \text{ purely additive}\}$ is a vector sublattice of $\text{cont}(\mathscr{R})$.

1.5.2.4. Prove $\text{cont}^{\text{p.a.}}(\mathscr{R}) \cap \text{cont}^{\sigma}(\mathscr{R}) = \{0\}$.

1.5.2.5. Prove that every $m \in \text{cont}(\mathscr{R})$ has a unique decomposition $m = m_1 + m_2$ where $m_1 \in \text{cont}^{\text{p.a.}}(\mathscr{R})$, $m_2 = \text{cont}^{\sigma}(\mathscr{R})$.

1.5.2.6. Prove $\|(|m|)\|(M) = \|m\|(M)$ $(M \subseteq \Omega)$, and in particular $\|(|m|)\| = \|m\|$.

1.5.2.7. Prove $|\alpha m| = |\alpha| \cdot |m|$ $(\alpha \in \mathbb{R})$.

1.5.2.8. Prove

$$\|m\|(M) = \sup\{|m(E)| + |m(F)| \mid \mathscr{R} \ni E, F \subseteq M, \quad E \cap F = \varnothing\} \qquad (M \subseteq \Omega).$$

1.5.2.9. Assume $m \in \text{cont}^{\sigma}(\mathscr{R})$ and prove

$$\|m\|(M) = \sup\left\{ \int f\, dm_+ - \int f\, dm_- \,\middle|\, f \in L^1_{|m|},\ |f| \leqq 1_M \right\}.$$

(*Hint*: Use Theorem III.5.10 and replace step functions by differences of two indicator functions in a suitable way.)

1.5.3. Let $m \in \text{cont}^{\sigma}(\mathscr{R})$.

1.5.3.1. Let $\mathscr{B}^{00} = \mathscr{B}^{00}(\mathscr{R})$ be the local σ-ring generated by $\mathscr{R}$. Prove that m has a unique σ-additive extension m^{00} to $\mathscr{B}^{00}$ and that $m^{00} \geqq 0$ iff $m \geqq 0$. (*Hint*: Decompose $m = m_+ - m_-$ and extend m_+ and m_- according to Corollary II.2.4.)

1.5.3.2. Prove $\|m^{00}\|(\cdot) = \|m\|(\cdot)$ on $\mathscr{R}$ and $m^{00} \in \text{cont}^{\sigma}(\mathscr{B}^{00})$.

1.5.3.3. Let φ: $\text{cont}^\sigma(\mathscr{B}^{00}) \to \text{cont}^\sigma(\mathscr{R})$ be the restriction mapping. Prove that it is a vector lattice isomorphism.

1.5.3.4. Let $\mathscr{B}^0 = \mathscr{B}^0(\mathscr{R})$ be the σ-ring generated by $\mathscr{R}$. Assume that $\|m_-\| < \infty$. Prove that m has a unique σ-additive extension m^0 to $\mathscr{B}^0$.

1.5.3.5. Prove $\|m^0\|(\cdot) = \|m\|(\cdot)$.

We now use the total variation in order to introduce various seminorms in $\text{cont}(\mathscr{R})$ and norms on various subspaces of $\text{cont}(\mathscr{R})$. A variety of L-spaces result.

1.6. Theorem. *Let $\Omega \neq \varnothing$ and $\mathscr{R}$ be a set ring in Ω.*

1.6.1. *Let* $\text{cont}^\text{b}(\mathscr{R}) = \{m \mid m \in \text{cont}(\mathscr{R}), \|m\| < \infty\}$. *Then* $(\text{cont}^\text{b}(\mathscr{R}), \|\cdot\|)$ *is an L-space. Norm convergence in* $\text{cont}^\text{b}(\mathscr{R})$ *amounts to uniform convergence on* $\mathscr{R}$.

1.6.2. *For every $M \in \mathscr{R}$, a seminorm $\|\cdot\|_M$ is defined on* $\text{cont}(\mathscr{R})$ *by* $\|m\|_M = \|m\|(M)$ $(m \in \text{cont}(\mathscr{R}))$.

1.6.3. *For every $M \in \mathscr{R}$, the set*

$$\text{cont}^M(\mathscr{R}) = \{m \mid m \in \text{cont}(\mathscr{R}), \|m\|(\Omega \backslash M) = 0\}$$

is a vector sublattice of $\text{cont}(\mathscr{R})$. *The restriction of $\|\cdot\|_M$ to* $\text{cont}^M(\mathscr{R})$ *is a norm and* $(\text{cont}^M(\mathscr{R}), \|\cdot\|_M)$ *is an L-space.*

1.6.4. *Assume $\Omega \in \mathscr{R}$, i.e., $\mathscr{R}$ is a field. Then* $\text{cont}^\text{b}(\mathscr{R}) = \text{cont}(\mathscr{R})$ *and* $(\text{cont}(\mathscr{R}), \|\cdot\|)$ *is an L-space.*

Proof. Using Exercise VII.4.5.3.2, we prove $m, m' \in \text{cont}^\text{b}(\mathscr{R})$

$$\begin{aligned} \Rightarrow \quad & \|m + m'\| \leqq \|m\| + \|m'\| < \infty, \\ & \|\alpha m\| = |\alpha| \cdot \|m\| < \infty \ (\alpha \in \mathbb{R}), \ \|(|m|)\| = \|m\| < \infty, \\ & \|m_+\| \leqq \|(|m|)\| = \|m\| < \infty, \end{aligned}$$

hence $m + m'$, αm, $|m|$, $m_+ \in \text{cont}^\text{b}(\mathscr{R})$. Clearly, $\|m\| = 0 \Rightarrow m = 0$. It follows that $\text{cont}^\text{b}(\mathscr{R})$ is a vector sublattice of $\text{cont}(\mathscr{R})$, and $(\text{cont}^\text{b}(\mathscr{R}), \|\cdot\|)$ is a normed vector lattice. By Exercise VII.4.5.2.4 it is an L-space, provided we can prove completeness. Before we do this we clear up the relation between uniform approximation and approximation for $\|\cdot\|$. It is convenient to introduce $|||m||| = \sup\{|m(F)| \mid F \in \mathscr{R}\}$ $(m \in \text{cont}(\mathscr{R}))$. It is easy to deduce from Theorem 1.3.3.1 that $|||m||| \leqq \|m\| \leqq 2|||m|||$ $(m \in \text{cont}(\mathscr{R}))$. This shows that $\text{cont}^\text{b}(\mathscr{R}) = \{m \mid m \in \text{cont}(\mathscr{R}), |||m||| < \infty\}$ and $\|\cdot\|$-convergence is tantamount to $|||\cdot|||$-convergence, i.e., to uniform convergence. It is now an easy exercise to prove that $\text{cont}^\text{b}(\mathscr{R})$ is complete for $|||\cdot|||$, and hence for $\|\cdot\|$.

2. For $m, m' \in \text{cont}(\mathscr{R})$ and $M \in \mathscr{R}$, we have

$$\|m + m'\|_M = |m + m'|(M) \leq |m|(M) + |m'|(M) = \|m\|_M + \|m'\|_M$$

and, similarly, $\|\alpha m\|_M = |\alpha| \cdot \|m\|_M$ ($\alpha \in \mathbb{R}$; see Exercise 1.5.2.7).

3. Consider the field $\mathscr{R}^M = \{F | \mathscr{R} \ni F \subseteq M\} \subseteq \mathscr{R}$ in M. For every $m \in \text{cont}^M(\mathscr{R})$, let $\varphi(m) \in \text{cont}(\mathscr{R}^M)$ be defined by $\varphi(m)(F) = m(F)$ ($F \in \mathscr{R}^M$), i.e., by mere restriction. It is an easy exercise to prove that $\varphi\colon \text{cont}^M(\mathscr{R}) \to \text{cont}(\mathscr{R}^M)$ is a vector lattice isomorphism preserving total variation. Now 3 follows from 1.

4 is a special case of 1 as well as of 3.

1.7. Theorem. *Let $\Omega \neq \varnothing$ and $\mathscr{R}$ be a set ring in Ω.*

1.7.1. *Let* $\text{cont}^\sigma(\mathscr{R}) = \text{cont}^\sigma(\mathscr{R}) \cap \text{cont}^b(\mathscr{R})$. *Then* $(\text{cont}^{\sigma, b}(\mathscr{R}), \|\cdot\|)$ *is an L-space. Norm convergence in* $\text{cont}^{\sigma, b}(\mathscr{R})$ *amounts to uniform convergence on* $\mathscr{R}$.

1.7.2. *For every* $M \in \mathscr{R}$, $\|m\|_M = \|m\|(M)$ $(m \in \text{cont}^\sigma(\mathscr{R}))$ *defines a seminorm on* $\text{cont}^\sigma(\mathscr{R})$.

1.7.3. *For every* $M \in \mathscr{R}$, $\text{cont}^{\sigma, M}(\mathscr{R}) = \text{cont}^\sigma(\mathscr{R}) \cap \text{cont}^{\sigma, M}(\mathscr{R})$ *is a vector sublattice of* $\text{cont}^\sigma(\mathscr{R})$. $(\text{cont}^{\sigma, M}(\mathscr{R}), \|\cdot\|)$ *is an L-space.*

1.7.4. *Assume* $\Omega \in \mathscr{R}$, *i.e.,* $\mathscr{R}$ *is a field. Then* $\text{cont}^{\sigma, b}(\mathscr{R}) = \text{cont}^\sigma(\mathscr{R})$ *and* $(\text{cont}^\sigma(\mathscr{R}), \|\cdot\|)$ *is an L-space.*

Proof. Proceed as in the proof of Theorem 1.6, applying also Theorem 1.3.2 and Exercise VII.3.8.1.

1.8. Exercises. Let $\Omega \neq \varnothing$ and $\mathscr{R}$ be a set ring in Ω.

1.8.1. Formulate and prove the analogue of Theorem 1.6.2 for $\text{cont}^M(\mathscr{R}) \cap \sigma\text{-cont}(M)$. (*Hint*: Use Theorem 1.7.)

1.8.2. Combine Theorems 1.6, 1.7, and 0.2.2 into a new proof of Theorem 1.4.3.

1.8.3. Combine Exercises 1.8.1 and 1.8.2.

1.8.4. Let $M \in \mathscr{R}$. Prove that every $\varnothing \neq S \subseteq \text{cont}^M(\mathscr{R})$ that has an upper bound in $\text{cont}(\mathscr{R})$, has a l.u.b. in $\text{cont}^M(\mathscr{R})$. Prove the analogous result for $\sigma\text{-cont}^M(\mathscr{R})$. (*Hint*: Use Theorem 0.2.2.)

1.8.5. Let $m \in \text{cont}^b(\mathscr{R})$. Prove that there are $M_1, M_2, \ldots \in \mathscr{R}$, $M_j \cap M_k = \varnothing$ $(j \neq k)$ such that $\|m\|(\Omega \backslash (M_1 + M_2 + \cdots)) = 0$.

1.8.6. Let in particular $\Omega = \mathbb{R}$, $\mathscr{D}$ the ring generated by all $[k/2^r, (k+1)/2^r[$ $(r = 0, 1, \ldots;\ k$ integer) and m the geometric content on $\mathscr{D}$ (Example I.1.10). Let $[0, 1[= A + B$ where A, B are dense in $[0, 1[$, $A \cap B = \varnothing$. Put $M = A$. Prove that $\{m | m \in \text{cont}(\mathscr{D}), \|m\|(\mathbb{R} \backslash A) = 0\}$ contains (many) signed contents $\neq 0$, but $\|\cdot\|(A)$ vanishes identically on it.

2. HAHN DECOMPOSITIONS

Let $\Omega = \mathbb{R}^3$ and $\mathscr{R}$ be one of the usual set rings in $\mathbb{R}^3$. It is natural to interpret a signed content m on $\mathscr{R}$ as the mathematical model for an *electric charge distribution.* $m_+ = m \vee 0$ then shows how the positive charges are distributed, and $m_- = (-m) \vee 0$ gives the pattern of negative charges. What is still missing in the eyes of a physicist is a proof that the positive and the negative charges assembled in m lie spatially separated, so that it is not by mathematical magic but for solid physical reasons that they do not annihilate one another. This spatial separation can be mathematically proved if $\mathscr{R}$ is a local σ-ring, and the mathematical keyword for it is *Hahn decomposition.*

In this section we define what Hahn decompositions are, prove their existence, and show their relationship with the vector lattice properties of signed σ-contents.

2.1. Definition. *Let* $\Omega \neq \varnothing \neq \mathscr{S} \subseteq \mathscr{P}(\Omega)$ *and* $m \in \overline{\mathbb{R}}^{\mathscr{S}}$

2.1.1. *A set* $P \subseteq \Omega$ *is said to be* **positive** *for* m *or* m*-positive if* $\mathscr{S} \ni F \subseteq P \Rightarrow m(F) \geqq 0$. *A set* $N \subseteq \Omega$ *is said to be* **negative** *for* m *or* m*-negative if* $\mathscr{S} \ni F \subseteq N \Rightarrow m(F) \leqq 0$.

2.1.2. *Let* M, P, $N \subseteq \Omega$, $P \cap N = \varnothing$, $M = P + N$, P m*-positive,* N m*-negative. Then the couple* (P, N) *is said to form a* **Hahn decomposition** *of* M *for* m.

2.2. Remark. Usually we shall take the formula $M = P + N$ as a symbol for a Hahn decomposition, rather than the couple (P, N).

2.3. Theorem. *Let* $(\Omega, \mathscr{B}^{00}, m)$ *be a signed* σ*-content space. Then every* $E \in \mathscr{B}^{00}$ *has at least one Hahn decomposition*

$$E = P + N, \qquad P, N \in \mathscr{B}^{00}, \quad P \cap N = \varnothing, \quad P \text{ } m\text{-positive}, \quad N \text{ } m\text{-negative}$$

for m *in* $\mathscr{B}^{00}$. *For every such decomposition*

$$\mathscr{B}^{00} \ni F \subseteq E \quad \Rightarrow \quad m_+(F) = m(F \cap P), \qquad m_-(F) = -m(F \cap N)$$
$$|m|(F) = m(F \cap P) - m(F \cap N).$$

Proof. Fix $E \in \mathscr{B}^{00}$. Clearly we may, for the rest of the proof of the existence of a Hahn decomposition of E in $\mathscr{B}^{00}$, forget about $\Omega \backslash E$, i.e., assume that $E = \Omega$ and $\mathscr{B}^{00}$ is a σ-field. This assumption implies $\|m\| < \infty$; in particular all values of m lie in a bounded interval. Let $\mathscr{B}_+ = \{F \mid F \in \mathscr{B}^{00}, F \text{ } m\text{-positive}\}$, $\mathscr{B}_- = \{F \mid F \in \mathscr{B}^{00}, F \text{ } m\text{-negative}\}$. It is an easy exercise to prove that $\mathscr{B}_+$ is a σ-ring in Ω. Likewise, $\mathscr{B}_-$ is a σ-ring in Ω. $\alpha = \sup\{m(F) \mid F \in \mathscr{B}_+\} \leqq \sup\{m(F) \mid F \in \mathscr{B}^{00}\}$ is a finite real and if we choose $P_1, P_2, \ldots \in \mathscr{B}_+$ with $m(P_n) \to \alpha$ and put $P = P_1 \cup P_2 \cup \cdots$, we get $P \in \mathscr{B}_+$, $m(P) = \alpha$. Let $N = \Omega \backslash P$. We are through as soon as we have proved $N \in \mathscr{B}_-$. If $N \notin \mathscr{B}_-$, then we may find some $\mathscr{B}^{00} \ni P' \subseteq N$ with $m(P') > 0$. If $P' \in \mathscr{B}_+$, we observe that $P \cap P' = \varnothing$, find $P + P' \in \mathscr{B}_+$ and $m(P + P') = m(P) + m(P') > \alpha$, and get a contradiction to the construction of α. Thus $P' \notin \mathscr{B}_+$, hence $\beta_1 = \inf\{m(F) | \mathscr{B}^{00} \ni F \subseteq P'\}$ is <0 (it is clearly a finite real). Choose $\mathscr{B}^{00} \ni F_1 \subseteq P'$ with $m(F_1) < \frac{1}{2}\beta_1 < 0$ and put $P_1' = P' \backslash F_1$. Then $m(P_1') = m(P') - m(F_1) > m(P')$. We conclude $P_1' \notin \mathscr{B}_+$ as before and define $\beta_2 = \inf\{m(F) | \mathscr{B}^{00} \ni F \subseteq P_1'\}$. Still $\beta_2 < 0$, but obviously $\beta_2 > \frac{1}{2}\beta_1$. Choose $\mathscr{B}^{00} \ni F_1 \subseteq P_1'$ with $m(F_1) < \frac{1}{2}\beta_2 < 0$ and put $P_2' = P_1' \backslash F_1$. Then $m(P_2') = m(P_1') - m(F_1) > m(P_1')$. Continuing in this way we get $P' \supseteq P_1' \supseteq P_2' \supseteq \cdots$ with $m(P') < m(P_1') < m(P_2') < \cdots$ and

$$\lim_{n \to \infty} [\inf\{m(F) | \mathscr{B}^{00} \ni F \subseteq P_n'\}] = 0.$$

Put $P_\infty' = P_1' \cap P_2' \cap \cdots$. Then $\mathscr{B}^{00} \ni P_\infty' \subseteq P'$ and

$$\inf\{m(F) | \mathscr{B}^{00} \ni F \subseteq P_\infty'\} = 0,$$

hence $P_\infty' \in \mathscr{B}_+$. Moreover, $m(P_\infty') = \lim m(P_n') > m(P_1') > 0$, and we get a contradiction as before. Thus we have proved the existence of a Hahn decomposition (for m and in $\mathscr{B}^{00}$) of every $E \in \mathscr{B}^{00}$. Let now $E = P + N$. Choose any $\mathscr{B}^{00} \ni F \subseteq E$. Then clearly $F \cap P$ is m-positive and $F \cap N = F \backslash (F \cap P)$ is m-negative. It is now obvious that $m(F \cap P) = \sup\{m(G) | \mathscr{B}^{00} \ni G \subseteq F\}$ and $-m(F \cap N) = \sup\{-m(G) | \mathscr{B}^{00} \ni G \subseteq F\}$. Comparing this with Theorem 1.3.3.2, we find $m_+(F) = m(F \cap P)$, $m_-(F) = -m(F \cap N)$, which in turn implies the last equality of our theorem.

2.4. Corollary. *Let $(\Omega, \mathscr{B}, m)$ be a signed σ-content space such that $\mathscr{B}$ is a σ-field. Let $\Omega = P + N$ be a Hahn decomposition of Ω for m. Then m_+ lives on P and m_- lives on N.*

2.5. Proposition. *Let $\Omega \neq \varnothing$, $\mathscr{B}^{00}$ be a local σ-ring in Ω, and m', $m'' \in \sigma\text{-cont}(\mathscr{B}^{00})$. Then the following statements are equivalent:*

2.5.1. $|m'| \wedge |m''| = 0$.

2.5.2. *For every $E \in \mathscr{B}^{00}$, there is a decomposition $E = E' + E''$ with E', $E'' \in \mathscr{B}^{00}$, $E' \cap E'' = \varnothing$ such that*

$$\mathscr{B}^{00} \ni F \subseteq E \quad \Rightarrow \quad m'(F) = m'(F \cap E'),\ m''(F) = m''(F \cap E'')$$

(and hence in particular $|m'|(E'') = 0 = |m''|(E')$).

Proof. $1 \Rightarrow 2$. Put $m = |m'| - |m''|$. By Lemma 0.2.1 and its proof we find $m_+ = |m'|$, $m_- = |m''|$. Let $E = E' + E''$, E', $E'' \in \mathscr{B}^{00}$, $E' \cap E'' = \varnothing$ be a Hahn decomposition according to Theorem 2.3. It follows that

$|m'|(E'') = m(E'' \cap E') = m(\varnothing) = 0 = |m''|(E')$. This in turn implies, by Theorem 1.3, $\mathscr{B}^{00} \ni F \subseteq E \Rightarrow m'(F \cap E'') = 0 = m''(F \cap E')$, hence 2.

$2 \Rightarrow 1$. Fix $E \in \mathscr{B}^{00}$ arbitrary and choose a decomposition $E = E' + E''$, $E', E'' \in \mathscr{B}^{00}$, $E' \cap E'' = \varnothing$ according to 2. Then

$$\begin{aligned} 0 \leqq (|m'| \wedge |m''|)(E) &\leqq \min\{|m'|(E'), |m''|(E')\} + \min\{|m'|(E''), |m''|(E'')\} \\ &\leqq |m''|(E') + |m'|(E'') = 0 \end{aligned}$$

by Theorem 1.3, and 1 follows.

2.6. Exercise. Let $(\Omega, \mathscr{B}^{00}, m)$ be a signed σ-content space and $\|m_-\| < \infty$. Let $\mathscr{B}^0 = \mathscr{B}^0(\mathscr{B}^{00})$ be the σ-ring generated by $\mathscr{B}^{00}$ and m the unique σ-additive extension of m to $\mathscr{B}^0$ (Exercise 1.5.3.4). Prove:

2.6.1. Every set $E \in \mathscr{B}^0$ has a Hahn decomposition in $\mathscr{B}^0$ for m.

2.6.2. Ω has a Hahn decomposition in $\mathscr{B}(\mathscr{B}^{00})$ for m.

2.6.3. If $\Omega = P + N = P' + N'$, $P, P', N, N' \in \mathscr{B}(\mathscr{B}^{00})$, and $P \cap N = \varnothing = P' \cap N'$ are two Hahn decompositions of Ω for m in $\mathscr{B}(\mathscr{B}^{00})$, then $m((P \,\Delta\, P') \cap F) = 0 = m((N \,\Delta\, N') \cap F)$ $(F \in \mathscr{B}^{00})$

3. ABSOLUTE CONTINUITY OF SIGNED σ-CONTENTS

In this section we define the concept of *absolute continuity* for σ-contents and investigate its interplay with lattice operations. The results obtained will be proved once again in Chapter IX in terms of "signed measures," and in a form which is more satisfactory from a systematic point of view. Here we remain closer to the applications of content theory.

3.1. Definition. *Let $\Omega \neq \varnothing$ and $\mathscr{R}$ be a set ring in Ω. Let m_0, $m \in \operatorname{cont}(\mathscr{R})$, We say that m is* **absolutely continuous** *for, or with respect to, m_0, in symbols $m \ll m_0$, if*

$$\mathscr{R} \ni F_1 \supseteq F_2 \supseteq \cdots, \qquad \lim_n |m_0|(F_n) = 0 \quad \Rightarrow \quad \lim_n m(F_n) = 0.$$

We define $\operatorname{cont}^{\ll m_0}(\mathscr{R}) = \{m \mid m \in \operatorname{cont}(\mathscr{R}), m \ll m_0\}$.

3.2. Remarks

3.2.1. Tradition has chosen the word *continuous* here in analogy to the classical "sequential" definition of continuity for functions on, say, intervals. A criterion for absolute continuity to be proved later will resemble the usual classical ε-δ-definition of a continuous function.

3.2.2. Notice that it is not m_0 but $|m_0|$ that enters into the essential part of the definition. We may thus always assume $m_0 \geqq 0$ if we like.

3.2.3. Trivially, $|m| \leqq |m_0| \Rightarrow m \ll m_0$.

3.2.4. It will turn out later that absolute continuity can be characterized by vector lattice relations purely intrinsic to $\operatorname{cont}(\mathscr{R})$ (Proposition IX.4.3).

3.3. Exercises

3.3.1. Let $\Omega \neq \varnothing$ and $\mathscr{R}$ be a set ring in Ω. Prove

$$m_0 \in \operatorname{cont}^\sigma(\mathscr{R}) \quad \Rightarrow \quad \operatorname{cont}^{\ll m_0}(\mathscr{R}) \subseteq \operatorname{cont}^\sigma(\mathscr{R}).$$

3.3.2. Let $\Omega = \mathbb{N} = \{1, 2, \ldots\}$, $\mathscr{R} = \mathscr{P}(\Omega)$, and $m' \in \mathbb{R}^{\mathscr{R}}$ be as in Exercise VII.2.5.8. Define m_0 on $\mathscr{R}$ by $m_0(F) = \sum_{n \in F} 1/2^n$ $(F \subseteq \mathbb{N})$. Prove that m_0, $m' \in \operatorname{cont}(\mathscr{R})$ and m' is not absolutely continuous for m_0.

3.4. Proposition. *Let $\Omega \neq \varnothing$ and $\mathscr{R}$ be a set ring in Ω, $m_0 \in \operatorname{cont}(\mathscr{R})$. Then:*

3.4.1. $\operatorname{cont}^{\ll m_0}(\mathscr{R})$ *is a vector sublattice of* $\operatorname{cont}(\mathscr{R})$.

3.4.2. *If* $\varnothing \neq M \subseteq \operatorname{cont}^{\ll m_0}(\mathscr{R})$ *has a majorant in* $\operatorname{cont}(\mathscr{R})$, *then* $\sup_{m \in M} m \in \operatorname{cont}^{\ll m_0}(\mathscr{R})$.

3.4.3. $\operatorname{cont}^{\ll m_0}(\mathscr{R}) \cap \operatorname{cont}^\sigma(\mathscr{R})$ *is a vector sublattice of* $\operatorname{cont}(\mathscr{R})$.

3.4.4. *If* $\varnothing \neq M \subseteq \operatorname{cont}^{\ll m_0}(\mathscr{R}) \cap \operatorname{cont}^\sigma(\mathscr{R})$ *has a majorant in* $\operatorname{cont}(\mathscr{R})$, *then* $\sup_{m \in M} m \in \operatorname{cont}^{\ll m_0}(\mathscr{R}) \cap \operatorname{cont}^\sigma(\mathscr{R})$.

Proof. 1. Clearly $\operatorname{cont}^{\ll m_0}(\mathscr{R})$ is a linear subspace of $\operatorname{cont}(\mathscr{R})$, hence it suffices to choose any $m \in \operatorname{cont}^{\ll m_0}(\mathscr{R})$ and to prove $m_+ \in \operatorname{cont}^{\ll m_0}(\mathscr{R})$. Assume that the latter relation is wrong. Then there is an $\varepsilon > 0$ and a sequence $\mathscr{R} \ni F_1 \supseteq F_2 \supseteq \cdots$ such that $\lim_n |m_0|(F_n) = 0$ but $m_+(F_n) > \varepsilon$ $(n = 1, 2, \ldots)$. By Theorem 1.3.3.2 there is an $\mathscr{R} \ni E_1 \subseteq F_1$ such that $m(E_1) > \varepsilon$. From $\mathscr{R} \ni E_1 \cap F_1 \supseteq E_1 \cap F_2, \ldots$,

$$\lim_n |m_0|(E_1 \cap F_n) \leqq \lim_n |m_0|(F_n) = 0$$

we conclude $\lim_n m(E_1 \cap F_n) = 0$, thus there is an $n_1 > 1$ such that $m(E_1 \backslash F_n) > \varepsilon$ $(n \geqq n_1)$. Choose, again by Theorem 1.3.3.2, some $\mathscr{R} \ni E_{n_1} \subseteq F_{n_1}$ with $m(E_{n_1}) > \varepsilon$. Again we find $n_2 > n_1$ such that $m(E_{n_1} \backslash F_n) > \varepsilon$ $(n \geqq n_2)$. Put now $G_1 = E_1 \backslash F_{n_1}$, $G_2 = E_{n_1} \backslash F_{n_2}, \ldots$. Then

$$G_1, G_2, \ldots \in \mathscr{R}, \quad G_j \cap G_k = \varnothing \; (j \neq k), \quad G_1, G_2, \ldots \subseteq F_1,$$
$$m(G_1), m(G_2), \ldots > \varepsilon.$$

This clearly leads to $\|m\|(F_1) = \infty$, a contradiction to $m \in \operatorname{cont}(\mathscr{R})$.

2. Let $\varnothing \neq M \subseteq \text{cont}^{\ll m_0}(\mathscr{R})$ have a majorant in $\text{cont}(\mathscr{R})$. Simple devices already employed in the proof of Theorem 0.2.2 allow us to assume that $M \subseteq \text{cont}_+^{\ll m_0}(\mathscr{R})$, M increasingly filtered. Consider

$$\overline{m} = \sup_{m \in M} m \in \text{cont}_+(\mathscr{R})$$

according to Theorem 1.4.3. Let now $\mathscr{R} \ni F_1 \supseteq F_2 \supseteq \cdots$ be such that $\lim_n |m_0|(F_n) = 0$. For every $m \in M$, we have

$$\begin{aligned}\overline{m}(F_1) - m(F_1) &\geqq \overline{m}(F_2) - m(F_2) \geqq \cdots \\ &\geqq \lim_n [\overline{m}(F_n) - m(F_n)] = \lim_n \overline{m}(F_n) \geqq 0.\end{aligned}$$

Varying $m \in M$, we can make the leftmost member as small as we like, proving thus $\lim_n \overline{m}(F_n) = 0$. Hence $\overline{m} \in \text{cont}_+^{\ll m_0}(\mathscr{R})$ follows.

3 and 4 are easy consequences of 1, 2, and Theorems 1.3.2, 1.4.2.

The simplest criteria for absolute continuity are obtained in the case where $\mathscr{R}$ is a local σ-ring and m_0, m are σ-additive.

3.5. Proposition. *Let $\Omega \neq \varnothing$ and $\mathscr{B}^{00}$ be a local σ-ring in Ω. Then for any $m_0, m \in \text{cont}^\sigma(\mathscr{B}^{00})$ the following statements are equivalent:*

3.5.1. $m \ll m_0$.

3.5.2. $F \in \mathscr{B}^{00}$, $|m_0|(F) = 0 \Rightarrow m(F) = 0$.

3.5.3. *For every $E \in \mathscr{B}^{00}$ and every $\varepsilon > 0$, there is a $\delta > 0$ such that $\mathscr{B}^{00} \ni F \subseteq E$, $|m_0|(F) < \delta \Rightarrow |m(F)| < \varepsilon$.*

Proof. $1 \Rightarrow 2$ is trivial.

$2 \Rightarrow 3$. Assume 3 is false. Then there is an $E \in \mathscr{B}^{00}$, an $\varepsilon > 0$, and a sequence $\mathscr{B}^{00} \ni E_1, E_2, \ldots \subseteq E$ such that $|m_0|(E_n) \leqq 1/2^n$, $|m(E_n)| \geqq 2\varepsilon$ $(n = 1, 2, \ldots)$.

For every $n = 1, 2, \ldots$, let $E_n = P_n + N_n$, $P_n, N_r \in \mathscr{B}^{00}$, and $P_n \cap N_n = \varnothing$ be a Hahn decomposition of E_n for m (Theorem 2.3). Clearly, $|m(E_n)| = |m(P_n) + m(N_n)| \geqq 2\varepsilon$ implies that $m(P_n) \geqq \varepsilon$ or $-m(N_n) \geqq \varepsilon$. Passing to a subsequence and possibly multiplying m by -1, we may henceforth assume $E_1 = P_1$, $E_2 = P_2, \ldots$. Put $F_n = E_n \cup E_{n+1} \cup \cdots$ $(n = 1, 2, \ldots)$ and $F = F_1 \cap F_2 \cap \cdots$. Clearly F_1 is positive for m, hence $m(F_n) \geqq m(E_n) \geqq \varepsilon$ $(n = 1, 2, \ldots)$, and hence the σ-additivity of m implies $m(F) = \lim_n m(F_n) \geqq \varepsilon$. On the other hand,

$$|m_0|(F) = \lim_n |m_0|(F_n) \leqq \lim_n \left(\frac{1}{2^n} + \frac{1}{2^{n+1}} + \cdots\right) = 0,$$

i.e., $|m_0|(F) = 0$, a contradiction to 2.

$3 \Rightarrow 1$ is trivial.

Next let us see how much we can carry over of these criteria to the more general case of signed σ-contents on a set ring. Our first step is

3.6. Proposition. *Let $\Omega \neq \emptyset$, $\mathcal{R}$ be a set ring in Ω, and $\mathcal{B}^{00} = \mathcal{B}^{00}(\mathcal{R})$ the local σ-ring generated by $\mathcal{R}$. Then the mapping φ: $\mathrm{cont}^\sigma(\mathcal{B}^{00}) \to \mathrm{cont}^\sigma(\mathcal{R})$ defined by restriction of signed contents $\mathcal{B}^{00}$ to $\mathcal{R}$ is a vector lattice isomorphism. Moreover, $m \ll m_0 \Leftrightarrow \varphi(m) \ll \varphi(m_0)$ for any m, $m_0 \in \mathrm{cont}^\sigma(\mathcal{B}^{00})$.*

Proof. The fact that every $m \in \mathrm{cont}^\sigma(\mathcal{R})$ has a unique σ-additive extension m^{00} to $\mathcal{B}^{00}$ and $m \geqq 0 \Leftrightarrow m^{00} \geqq 0$ (Exercise 1.5.3) proves everything except the last statement. In that statement $\Rightarrow$ is trivial. Let us now prove $\Leftarrow$. It is obvious that we may assume m_0, $m \geqq 0$ (use Remark 3.2.2 and Proposition 3.4). If $m \ll m_0$ fails to hold, we have, by Proposition 3.5.2, some $F \in \mathcal{B}^{00}$ and some $\varepsilon > 0$ such that $m_0(F) = 0$ but $m(F) > \varepsilon$. We now apply Exercise II.2.6.1 to $m_0 + m$ and find some $E_1 \in \mathcal{R}$ such that $m_0(F \,\Delta\, E_1) < \varepsilon/4$, $m(F \,\Delta\, E_1) < \varepsilon/4$. From this we easily deduce

$$m_0(E_1) = \varphi(m_0)(E_1) < \varepsilon/4,\ m(E_1) = \varphi(m)(E_1) \geqq m(F \cap E_1) > \varepsilon(1 - \tfrac{1}{4}).$$

Repeating this argument, we find $E_2' \in \mathcal{R}$ such that

$$m_0((F \cap E_1) \,\Delta\, E_2') < \varepsilon/8,\ m((F \cap E_1) \,\Delta\, E_2') < \varepsilon/8.$$

From this we get, for $E_2 = E_1 \cap E_2'$ firstly

$$\varphi(m_0)(E_2) \leqq \varphi(m_0)(E_2') \leqq m_0(F) + \varepsilon/8 = \varepsilon/8,$$

and secondly

$$\varphi(m)(E_2) > m(F \cap E_1) - \varepsilon/8 > \varepsilon(1 - \tfrac{1}{4} - \tfrac{1}{8}).$$

Continuing in this way we find $\mathcal{R} \ni E_1 \supseteq E_2 \supseteq \cdots$ such that $\lim_n \varphi(m_0)(E_n) = 0$, $\lim_n \varphi(m)(E_n) \geqq \varepsilon/2$, a contradiction to $\varphi(m) \ll \varphi(m_0)$.

3.7. Remark. In the proof of the above proposition we have made heavy use of the particular properties of σ-contents. The more systematic investigations of Chapter IX will reveal that absolute continuity is an "intrinsic" property which can be defined in any conditionally complete vector lattice and hence is invariant under vector lattice isomorphism. We shall thus be able to prove Proposition 3.6 once again: Proposition IX.4.4.

3.8. Corollary. *Let $\Omega \neq \emptyset$ and $\mathcal{R}$ be a set ring in Ω. Then for any m_0, $m \in \mathrm{cont}^\sigma(\mathcal{R})$, the following statements are equivalent:*

3.8.1. $m \ll m_0$.

3.8.2. *For every $E \in \mathcal{R}$ and every $\varepsilon > 0$, there is a $\delta > 0$ such that $\mathcal{R} \ni F \subseteq E$, $|m_0|(F) < \delta \Rightarrow |m(F)| < \varepsilon$.*

Proof. $1 \Rightarrow 2$. Let $\psi(m_0)$, $\psi(m)$ be the unique σ-additive extensions of m_0, m to $\mathscr{B}^{00}(\mathscr{R})$. Since $m \ll m_0 \Rightarrow \psi(m) \ll \psi(m_0)$ by Proposition 3.6, we conclude 2 as a special case of Proposition 3.5.3.

$2 \Rightarrow 1$ is trivial.

3.9. Exercises. Let $\Omega \neq \varnothing$ and $\mathscr{R}$ be a set ring in Ω.

3.9.1. Let $m \in \mathrm{cont}(\mathscr{R})$ be such that there exists a countable union Ω_0 of sets from $\mathscr{R}$ such that $\|m\|(\Omega \backslash \Omega_0) = 0$. Prove that there is an $m_0 \in \mathrm{cont}^{\mathrm{b}}(\mathscr{R})$ such that $m \ll m_0 \ll m$. Prove that $m_0 \in \mathrm{cont}^{\sigma}(\mathscr{R})$ iff $m \in \mathrm{cont}^{\sigma}(\mathscr{R})$.

3.9.2. Let $m, m' \in \mathrm{cont}(\mathscr{R})$, $m \ll m' \ll m$. Prove that for every $M \subseteq \Omega$, we have $\|m\|(M) = 0 \Leftrightarrow \|m'\|(M) = 0$.

3.9.3. Assume that $\mathscr{R}$ is a local σ-ring. Let $m_0 \in \mathrm{cont}_+^{\sigma}(\mathscr{R}) \supseteq \mathscr{C}$ be such that $m \in \mathscr{C} \Rightarrow m \ll m_0$ ($\mathscr{C}$ is then said to be dominated by m_0). Prove that there is an $m_1 \in \mathrm{cont}_+^{\sigma}(\mathscr{R})$ such that $m \ll m_1$ $(m \in \mathscr{C})$ and that furthermore $N \in \mathscr{R}$, $m(N) = 0$ $(m \in \mathscr{C}) \Rightarrow m_1(N) = 0$ holds. (*Hint*: Replace $\mathscr{C}$ by $\mathscr{C}' = \{(m_2 \vee \cdots \vee m_n) \wedge m_0 \,|\, n \geqq 2, m_2, \ldots, m_n \in \mathscr{C}\}$ and put $m_1 = \sup \mathscr{C}$; the result of this exercise is known as the HALMOS–SAVAGE [1] theorem.)

3.9.4. In the situation of Exercise 3.9.3, assume $\Omega \in \mathscr{B}^0(\mathscr{R})$ and prove that there is a countable subset $\mathscr{C}_0 \subseteq \mathscr{C}$ such that, for every $F \in \mathscr{B}^{00}$, $m(F) = 0$ $(m \in \mathscr{C}) \Leftrightarrow m(F) = 0$ $(m \in \mathscr{C}_0)$ (also this result goes under the names of Halmos and Savage).

3.9.5. Let m_0, $m \in \mathrm{cont}^{\sigma}(\mathscr{R})$, $m \ll m_0$. Let $(\Omega, \mathscr{B}^{00}_{m_0}, m'_0)$, $(\Omega, \mathscr{B}^{00}_{m_+}, m'_+)$, $(\Omega, \mathscr{B}^{00}_{m_-}, m'_-)$ denote the completions (Chapter I, Section 6) of the unique σ-additive extensions of m_0, m_+, m_- to $\mathscr{B}^{00}(\mathscr{R})$. Prove $\mathscr{B}^{00}_{m_0} \subseteq \mathscr{B}^{00}_{m_+} \cap \mathscr{B}^{00}_{m_-}$ and $m'' \ll m'_0$ for the restriction m'' of $m'_+ - m'_-$ to $\mathscr{B}^{00}_{m_0}$. (*Hint*: Use Proposition 3.5.)

3.9.6. Let $m_0 \in \mathrm{cont}^{\sigma}(\mathscr{R})$. For every $m \in \mathrm{cont}^{\ll m_0}(\mathscr{R})$, let $m'' \in \mathrm{cont}(\mathscr{B}^{00}_{m_0})$ be as in Exercise 3.9.5. Prove that $m \to m''$ is a vector lattice isomorphism of $\mathrm{cont}^{\ll m_0}(\mathscr{R})$ onto $\mathrm{cont}^{\ll m_0''}(\mathscr{B}^{00}_{m_0})$.

4. LEBESGUE DECOMPOSITIONS

In this section we display *orthogonality* and *absolute continuity* of signed contents as "complementary" properties. Given m_0, every content splits into a component $\ll m_0$ and a component orthogonal to m_0. This is called its *Lebesgue decomposition*. The essential tool in the proof will be Theorem 3.4.2. We shall show how our results indicate that an abstract vector lattice theoretical truth stands behind them. In fact, the abstract and systematic theory of Chapter IX will enable us to obtain Lebesgue decompositions once more. Our present approach is justified by being more close to applications of content theory.

The reader should begin with a reminder from Chapter 0, Section 2 on orthogonality in an arbitrary abstract vector lattice H. Two vectors x, $y \in H$ are said to be orthogonal if $|x| \wedge |y| = 0$. We then also write $x \perp y$. For any $M \subseteq H$, the orthogonal complement $M^\perp$ of M is defined by $M^\perp = \{y | y \in H, x \perp y \ (x \in M)\}$. Clearly,

$$\varnothing^\perp = \{0\}^\perp = H, \qquad M \subseteq N \subseteq H \ \Rightarrow \ M^\perp \supseteq N^\perp, \qquad (M^\perp)^\perp \supseteq M.$$

The orthogonal complement $M^\perp$ of a set $M \subseteq H$ is always a vector sublattice of H:

$$\begin{aligned} x \in M,\, y,\, z \in M^\perp \ \Rightarrow \ 0 &\leqq |x| \wedge |y + z| \\ &\leqq (|x| \wedge |y|) + (|x| \wedge |z|) = 0 \end{aligned}$$

should suffice as hints for a full proof. If a set $S \subseteq M^\perp$ has a l.u.b. $\overline{m} \in H$, then $\overline{m} \in M^\perp$. In fact easy devices employed in the proof of Theorem 0.2.2 show that we may assume $M \subseteq H_+$. But now we have $x \in M$ implies

$$\begin{aligned} \overline{m} \wedge |x| &= \overline{m} + |x| - (\overline{m} \vee |x|) \\ &= m + |x| - \sup_{y \in S} (y \vee |x|) \\ &= \overline{m} + |x| - \sup_{y \in S} (y + |x| - (y \wedge |x|)) \\ &= \overline{m} + |x| - \sup_{y \in S} (y + |x|) \\ &= \overline{m} + |x| - (\overline{m} + |x|) = 0. \end{aligned}$$

4.1. Proposition. *Let $(\Omega, \mathscr{R}, m_0)$ be a content prespace. Define*

$$\mathrm{cont}^{\perp m_0}(\mathscr{R}) = \{m | m \in \mathrm{cont}(\mathscr{R}), m \perp m_0\}.$$

Then:

4.1.1. $\mathrm{cont}^{\perp m_0}(\mathscr{R})$ *is a vector sublattice of* $\mathrm{cont}(\mathscr{R})$.

4.1.2. $\mathrm{cont}^{\perp m_0}(\mathscr{R}) \cap \mathrm{cont}^{\ll m_0}(\mathscr{R}) = \{0\}$.

4.1.3. $m^0 \in \mathrm{cont}^{\ll m_0}(\mathscr{R}),\ m^1 \in \mathrm{cont}^{\perp m_0}(\mathscr{R}) \Rightarrow m^0 \perp m^1$.

4.1.4. *Every $0 \neq S \subseteq \mathrm{cont}^{\perp m_0}(\mathscr{R})$ that has a majorant in H has a l.u.b. in* $\mathrm{cont}^{\perp m_0}(\mathscr{R})$.

4.1.5. $\mathrm{cont}^{\perp m_0}(\mathscr{R}) \cap \mathrm{cont}^{\sigma}(\mathscr{R})$ *is a vector sublattice of* $\mathrm{cont}(\mathscr{R})$.

4.1.6. *Every $0 \neq S \subseteq \mathrm{cont}^{\perp m_0}(\mathscr{R}) \cap \mathrm{cont}^{\sigma}(\mathscr{R})$ that has a majorant in H has a l.u.b. in $\mathrm{cont}^{\perp m_0}(\mathscr{R}) \cap \mathrm{cont}^{\sigma}(\mathscr{R})$.*

Proof. 1 and 4 are consequences of our general preliminary observations.

2. Let $m \in \mathrm{cont}^{\perp m_0}(\mathscr{R}) \cap \mathrm{cont}^{\ll m_0}(\mathscr{R})$. From $|m| \wedge m_0 = 0$ and Theorem 1.3.3.1 we deduce, for any $F \in \mathscr{R}$, the existence of two sequences $E_1, E_2, \ldots$ and $E_1', E_2', \ldots$ in $\mathscr{R}$ such that $E_n \cap E_n' = \varnothing$, $E_n + E_n' = F$ $(n = 1, 2, \ldots)$ and $|m|(E_n) + m_0(E_n') \to 0$ as $n \to \infty$. This implies $\lim_n m_0(E_n') = 0$ and thus, by Corollary 3.8, $\lim |m|(E_n') = 0$ since $|m| \ll m_0$. Now

$$|m|(F) = \lim_n |m|(E_n') + \lim_n |m|(E_n') = 0$$

follows. But $|m| = 0 \Rightarrow m = 0$.

3. $|m^0| \wedge |m^1| \in \mathrm{cont}^{\ll m_0}(\mathscr{R}) \cap \mathrm{cont}^{\perp m_0}(\mathscr{R}) \Rightarrow |m^0| \wedge |m^1| = 0$ by 2.

5 and 6 are consequences of 1 and 2 and Theorem 1.4.3.

4.2. Theorem. *Let $(\Omega, \mathscr{R}, m_0)$ be a content prespace. Then every $m \in \mathrm{cont}(\mathscr{R})$ has a unique representation*

$$m = m^0 + m^1, \qquad m^0 \in \mathrm{cont}^{\ll m_0}(\mathscr{R}), \qquad m^1 \in \mathrm{cont}^{\perp m_0}(\mathscr{R}).$$

It is called the **Lebesgue decomposition** *of m for m_0. We have always $m^0 \perp m^1$.*

Proof. Passing to m_+, m_-, we easily see that it suffices for the proof of the existence of a decomposition to settle the case $m \geqq 0$. Let now $M = \{m' \mid \mathrm{cont}^{\ll m_0}(\mathscr{R}) \ni m' \leqq m\}$. Clearly $0 \in M$, hence $M \neq \varnothing$. Moreover, M has the majorant m. Thus $m^0 = \sup_{m' \in M} m'$ exists and is in $\mathrm{cont}^{\ll m_0}(\mathscr{R})$ (Theorem 3.4.2).

Put $m^1 = m - m^0$. Clearly $m = m^0 + m^1$, $m^1 \geqq 0$. Assume $m^1 \wedge m_0 \neq 0$. Then $m^0 + (m^1 \wedge m_0) \leqq m$ and $m^0 + (m^1 \wedge m_0) \ll m_0$ since $m^0 \ll m_0$ and $m^1 \wedge m_0$ is even $\leqq m_0$. Also, $m^0 + (m^1 \wedge m_0) \geqq m^0$, with $\neq$. But this contradicts the definition of m^0. Thus we have proved the existence of Lebesgue decompositions. If $m = m^0 + m^1 = \tilde{m}^0 + \tilde{m}^1$ are two of them, then $m^0 - m^0 = m^1 - m^1 \in \mathrm{cont}^{\ll m_0}(\mathscr{R}) \cap \mathrm{cont}^{\perp m_0}(\mathscr{R})$, hence $m^0 - \tilde{m}^0 = 0$ by Proposition 4.1.2. This proves uniqueness. The last statement is a consequence of Proposition 4.1.3.

4.3. Exercise. Let $(\Omega, \mathscr{R}, m_0)$ be a content prespace and $\mathrm{cont}(\mathscr{R}) \ni m \ll m_0$. Prove $m \ll m_0 \wedge |m|$. (*Hint*: Form the Lebesgue decomposition of m for $m_0 \wedge |m|$ and show that it coincides with the Lebesgue decomposition for m_0.)

4.4. Proposition. *Let $\Omega \neq \varnothing$ and $\mathscr{B}^{00}$ be a local σ-ring. Then for any signed σ-contents m^0, m^1 on $\mathscr{B}^{00}$ the following statements are equivalent:*

4.4.1. $m^0 \perp m^1$.

4.4.2. *For every $E \in \mathscr{B}^{00}$, there is a decomposition $E = E^0 + E^1$, E^0, $E^1 \in \mathscr{B}^{00}$, $E^0 \cap E^1 = \varnothing$ such that $\|m^0\|(E^1) = 0 = \|m^1\|(E^0)$.*

Proof. Clearly we may assume $m^0, m^1 \geqq 0$ throughout the proof.

$1 \Rightarrow 2$. We have $m^0 \wedge m^1 = 0$. Let $E = P + N$, $P, N \in \mathscr{B}^{00}$, $P \cap N = \varnothing$ be a Hahn decomposition for $m = m^0 - m^1$. Put $E^0 = P$, $E^1 = N$. Since $m^0 = m_+$, $m^1 = m_-$, we have $\|m^0\|(E^1) = m_+(N) = 0 = m_-(P) = \|m^1\|(E^0)$.

$2 \Rightarrow 1$. Choose $E \in \mathscr{B}^{00}$. We have to prove $(|m^0| \wedge |m^1|)(E) = 0$. Decompose $E = E^0 + E^1$ according to 2. Then we get

$$0 \leqq (|m^0| \wedge |m^1|)(E^0) \leqq |m^1|(E^0) = \|m^1\|(E^0) = 0.$$

Similarly $(|m^0| \wedge |m^1|)(E^1) = 0$ follows. We conclude

$$(|m^0| \wedge |m^1|)(E) = (|m^0| \wedge |m^1|)(E^0) + (|m^0| \wedge |m^1|)(E^1) = 0.$$

4.5. Corollary. *Let $(\Omega, \mathscr{B}^{00}, m_0)$ be a σ-content space $m \in \sigma\text{-cont}(\mathscr{B}^{00})$ and $m = m^0 + m^1$ its Lebesgue decomposition for m_0. Then for every $E \in \mathscr{B}^{00}$, there are $E^0, E^1 \in \mathscr{B}^{00}$, $E^0 \cap E^1 = \varnothing$ such that $E = E^0 + E^1$, $\|m^0\|(E^1) = 0 = \|m^1\|(E^0)$.*

4.6. Exercises. Let $\Omega \neq \varnothing$ and $\mathscr{B}$ a σ-field in Ω.

4.6.1. For every $t = 0. 1, \ldots$, let $\mathscr{P}_t = \{A_{t1}, \ldots, A_{tn_t}\} \subseteq \mathscr{B}$ be a partition of Ω such that $\mathscr{P}_t \subseteq \mathscr{B}(\mathscr{P}_{t+1})$. Assume that $m, m' \in \text{cont}_+^\sigma(\mathscr{R})$ satisfy

$$\lim_{t \to \infty} [\sup\{m(A_{tk}) \wedge m'(A_{tk}) | k = 1, \ldots, n_t\}] = 0.$$

Prove that $m \perp m'$

4.6.2. Let $(\Omega, \mathscr{B}, m)$, $(\Omega, \mathscr{B}, m')$ be the Bernoulli σ-probability space for two different probability vectors p, p' over one and the same finite alphabet (Definition II.2.5.2). Prove $m \perp m'$. (The problem posed here is of general importance, see Kakutani [1], Hill [1].)

5. THE RADON–NIKODYM THEOREM FOR SIGNED σ-CONTENTS WITH FINITE TOTAL VARIATION

In this section we consider an arbitrary σ-content space $(\Omega, \mathscr{B}^{00}, m_0)$ and prove that every signed σ-content $m \ll m_0$ on $\mathscr{B}^{00}$ having finite total variation is the indefinite m_0-integral of some m_0-integrable function f, its so-called m_0-*density*:

$$m(F) = \int 1_F \, f \, dm_0 \qquad (F \in \mathscr{B}^{00}).$$

This is the most convenient case of the famous *Radon–Nikodym theorem* which thus identifies part of $\text{cont}^{\ll m_0}(\mathscr{R})$ with $L^1_{m_0}$. The subcase that is most important for applications is that with $\Omega \in \mathscr{B}^{00}$, i.e., $\mathscr{B}^{00}$ is a σ-field. The question whether also a $m \in \operatorname{cont}(\mathscr{B}^{00})$ with infinite total variation can be represented by some sort of an integrable function is difficult and will be treated later (Section 8). The next sections will be devoted to applications of the results of this section.

5.1. Proposition. *Let $(\Omega, \mathscr{B}^{00}, m_0)$ be a σ-content space.*

5.1.1. *Let $f \in \mathscr{L}^1_{m_0}$. Then*

$$m(F) = \int 1_F(\omega) f(\omega) m_0(d\omega) \qquad (F \in \mathscr{B}^{00})$$

defines a signed σ-content m on $\mathscr{B}^{00}$. It is denoted by $m = fm_0$. We also write $dm = f\,dm_0$ for this. We have $fm_0 \in \operatorname{cont}^{\sigma,\mathrm{b}}(\mathscr{B}^{00})$.

5.1.2. $$fm_0 \leqq fm_0 \iff f \leqq f' m_0\text{-a.e.} \qquad (f, f' \in \mathscr{L}^1_{m_0}).$$

In particular, $fm_0 = f'm_0 \Leftrightarrow f = f'm_0$-a.e.

5.1.3. $$f, f' \in \mathscr{L}^1_{m_0} \Rightarrow \begin{array}{l} |fm_0| = |f| m_0, \ (fm_0)_+ = f_+ m_0. \\ (fm_0) \vee (f'm_0) = (f \vee f') m_0. \end{array}$$

5.1.4. $$f \in \mathscr{L}^1_{m_0} \Rightarrow \|fm_0\| = \|f\|_1.$$

Proof. 1. It is obvious that $m = fm_0$ is in $\mathbb{R}^{\mathscr{B}^{00}}$ and additive. It is also σ-additive: Let $F, F_1, F_2, \ldots \in \mathscr{B}^{00}$, $F_j \cap F_k = \varnothing$ $(j \neq k)$, $F = F_1 + F_2 + \cdots$. Then the serial form of Lebesgue's dominated convergence theorem III.5.7.2 yields

$$m(F) = \int 1_F\, dm_0 = \int 1_{F_1} f\, dm_0 + \int 1_{F_2} f dm_0 + \cdots = m(F_1) + m(F_2) + \cdots.$$

$m \in \operatorname{cont}^\sigma(\mathscr{B}^{00})$ and even $m \in \operatorname{cont}^{\sigma,\mathrm{b}}(\mathscr{B}^{00})$ will now follow from our proof of 4.

4. Exercise 1.5.2.9 shows that

$$\|fm_0\| = \sup\left\{\left|\int 1_E f\, dm_0\right| + \left|\int 1_F f\, dm_0\right| \,\middle|\, E, F \in \mathscr{B}^{00}, E \cap F = \varnothing\right\}.$$

By Remark III.9.2 and Theorem III.9.3 we may assume that f vanishes outside a set of the form $E_1 \cup E_2 \cup \cdots$ where $E_1, E_2, \ldots \in \mathscr{B}^{00}$, $E_1 \subseteq E_2 \subseteq \cdots$, and $P = \{f > 0\}$, $N = \{f < 0\} \in \mathscr{B}^0(\mathscr{B}^{00})$, hence $E_n \cap P$, $E_n \cap N \in \mathscr{B}^{00}$ (Proposition I.4.7.2.2). It is obvious that the above supremum can be obtained as

$$\lim_n \left[\int 1_{E_n \cap P} f\, dm_0 + \int 1_{E_n \cap N}(-f)\, dm_0\right] = \int |f|\, dm_0 = \|f\|_1$$

(use $\lim 1_{E_n} f = f$ (pointwise) and Lebesgue's dominated convergence theorem, III.5.7).

2. $\Leftarrow$ is obvious in both lines. It suffices to prove

$$fm_0 \leqq f'm_0 \quad \Rightarrow \quad f \leqq f' \ m\text{-a.e.}$$

By Remark III.9.2 and Theorem III.9.3 we may assume f, $f' \in \text{mble}(\Omega, \mathscr{B}^0(\mathscr{B}^{00}), \mathbb{R})$ and there is some set of the form $M_1 \cup M_2 \cup \cdots$, $M_1, M_2, \ldots \in \mathscr{B}^{00}$ such that f, f' vanish outside that set. For any $n = 1, 2, \ldots$, let $N_n = \{\omega | \omega \in M_r, \ f(\omega) > f'(\omega)\}$. Clearly $0 \leqq \int (f - f')1_{N_n} \, dm_0 = fm_0(N_n) - f'm_0(N_n) \leqq 0$ and $(f - f')1_{N_n} \geqq 0$. By Proposition III.4.2.6, $N_n = \{(f - f')1_{N_n} > 0\}$ is an m_0-nullset. Since $\{f' < f\} \subseteq N_1 \cup N_2 \cup \cdots$, we have $f \leqq f'$ m_0-a.e.

3. It is sufficient to prove $(fm_0)_+ = f_+ m_0$, $(fm_0)_- = f_- m_0$. Now clearly 2 implies that $\leqq$ holds here. But it also implies that $0 \leqq (f_+ m_0) \wedge (f_- m_0) \leqq (f_+ \wedge f_-)m_0 = 0$, hence $(fm_0)_+ = f_+ m_0$, $(fm_0)_- = f_- m_0$ follow from Lemma 0.2.1.

5.2. Theorem (Radon–Nikodym). *Let $(\Omega, \mathscr{B}^{00}, m_0)$ be a σ-content space. Let $m \in \text{cont}^{\sigma,\text{b}}(\mathscr{B}^{00}) \cap \text{cont}^{\ll m_0}(\mathscr{B}^{00})$. Then there is an $f \in \mathscr{L}^1_{m_0}$ such that $m = fm_0$.*

Proof. Since m_0 is σ-additive, we have $\text{cont}^{\ll m_0}(\mathscr{B}^{00}) \subseteq \text{cont}^{\sigma}(\mathscr{B}^{00})$ by Exercise 3.3.1. By Proposition 3.4.3 we may assume $m \geqq 0$. Let now $L = \{f | 0 \leqq f \in L^1_{m_0}, \ fm_0 \leqq m\}$. Clearly, $0 \in L$, hence $L \neq \varnothing$. From Proposition 5.1.3 we conclude

$$f, f' \in L \quad \Rightarrow \quad (f \vee f)m_0 = (fm_0) \vee (f'm_0) \leqq m \quad \Rightarrow \quad f \vee f' \in L.$$

Put $\alpha = \sup\left\{\int f \, dm_0 \,|\, f \in L\right\}$. From

$$f \in L \quad \Rightarrow \quad \int f \, dm_0 = \|fm_0\|(\Omega) \leqq \|m\|(\Omega) < \infty$$

we conclude $\alpha \leqq m(\Omega) < \infty$. We may choose $f_1, f_2, \ldots \in L$ such that $\lim_n \int f_n \, dm_0 = \alpha$, and we may clearly assume $f_1 \leqq f_2 \leqq \cdots$ here. Now the monotone convergence theorem III.5.3.1 yields

$$\lim_n f_n = f \in \mathscr{L}^1_{m_0}, \qquad \int f \, dm_0 = \alpha,$$

$$\lim_n \|f_n - f\|_1 = \lim \left[\left|\int f \, dm - \int f_n \, dm\right|\right] = 0.$$

Since in a normed vector lattice the minorants of a vector form a norm closed set, we get $fm_0 \leqq m$ from $\lim_n \|f_n m_0 - fm_0\| = \lim_n \|f_n - f\|_1 = 0$. We conclude $f \in L$. If $f' \in L$, $\int f' \, dm_0 = \alpha$, we get $L \ni f \vee f' \geqq f, f'$,

$$\int [(f \vee f') - f] \, dm_0 \leqq \alpha - \alpha = 0 = \int [(f \vee f') - f'] \, dm_0,$$

hence $f = f \vee f' = f'$ m_0-a.e. (Proposition III.4.2.6). Thus f is (mod m_0) the unique element in L with $\int f \, dm_0 = \alpha$.

Next we prove $m \neq 0 \Rightarrow \alpha > 0$. Now $m \neq 0$ implies the existence of some $\Omega_0 \in \mathscr{B}^{00}$ such that $m(\Omega_0) > 0$. Define $m_{00} \in \sigma\text{-cont}^b(\mathscr{B}^{00})$ by $m_{00}(F) = m_0(F \cap \Omega_0)$ (in fact $\|m_{00}\| = m_{00}(\Omega_0) < \infty$). We can now conclude that $(m - n^{-1}m_{00})_+ \neq 0$ for some integer $n > 0$ since otherwise we would have

$$\begin{aligned} \lim_n \left\| \left(m - \frac{1}{n} m_{00}\right) - m \right\| &= \lim_n \frac{1}{n} \|m_{00}\| = 0 \\ \Rightarrow \quad \lim_n \left\| \left(m - \frac{1}{n} m_{00}\right)_+ - m_+ \right\| &= 0 \\ \Rightarrow \quad m = m_+ &= 0. \end{aligned}$$

Let now $\Omega_0 = P + N$, $P, N \in \mathscr{B}^{00}$, $P \cap N = \varnothing$ be a Hahn decomposition of Ω_0 for $m - n^{-1}m_{00}$ (Theorem 2.3). Corollary 2.4 shows that $0 \neq (m - n^{-1}m_{00})_+(\Omega_0) = m(P) - n^{-1}m_{00}(P)$. Since $m \ll m_0$, $m_{00}(P) = m_0(P)$, we may conclude

$$\begin{aligned} m_0(P) = 0 \quad &\Rightarrow \quad m(P) = 0 \\ &\Rightarrow \quad m(P) - \frac{1}{n} m_0(P) = m(P) - \frac{1}{n} m_{00}(P) = 0, \end{aligned}$$

a contradiction. Thus $m_0(P) > 0$ follows. Let now $g = n^{-1}1_P$. Then $g \in \mathscr{L}^1_{m_0}$ and $\int g \, dm_0 = n^{-1}m_0(P) > 0$. But we have also

$$\begin{aligned} F \in \mathscr{B}^{00} \quad \Rightarrow \quad (m - gm_0)(F) &= m(F) - \frac{1}{n} m_0(F \cap P) \\ &\geqq m(F \cap P) - \frac{1}{n} m_{00}(F \cap P) = \left(m - \frac{1}{n} m_{00}\right)_+ (P) \geqq 0, \end{aligned}$$

hence $gm_0 \leqq m$, and $g \in L$ follows. The rest of the proof of our theorem is now the following short argument. If $fm_0 \neq m$, then $m' = m - fm_0$ is $\geqq 0$, $\neq 0$, and $\ll m_0$. By what we have just proved, we can find some $0 \leqq h \in \mathscr{L}^1_{m_0}$ such that $0 \neq hm_0 \leqq m'$. We conclude

$$(f + h)m_0 \leqq fm_0 + (m - fm_0) = m,$$

i.e., $f + h \in L$; on the other hand, $\int (f + h)\, dm_0 > \int f\, dm_0 = \alpha$, and a contradiction to the definition of α. The theorem is proved. We combine Proposition 5.1 and Theorem 5.2 into

5.3. Theorem (Radon–Nikodym). *Let $(\Omega, \mathscr{B}^{00}, m_0)$ be a σ-content space. Then the mapping ψ: $L^1_{m_0} \to \mathrm{cont}^{\sigma, \mathrm{b}}(\mathscr{B}^{00})$ defined by $\psi(f) = fm_0$ $(f \in L^1_{m_0})$ is a vector lattice isomorphism of $L^1_{m_0}$ onto*

$$\mathrm{cont}^{\sigma, \mathrm{b}}(\mathscr{B}^{00}) \cap \mathrm{cont}^{\ll m_0}(\mathscr{B}^{00}).$$

It preserves the norm, i.e., $\| fm_0 \| = \| f \|_1$ $(f \in L^1_{m_0})$. If $m \in \mathrm{cont}^{\mathrm{b}}(\mathscr{B}^{00}) \cap \mathrm{cont}^{\ll m_0}(\mathscr{B}^{00})$ and $m = f_{m_0}$, then $f \in L^1_{m_0}$ is uniquely determined by m (and m_0). We call f the **m_0-density of m**, *or the* **density of m for m_0**. *It is sometimes denoted by $f = dm/dm_0$.*

The most important special case of the above theorem is

5.4. Theorem (Radon–Nikodym). *Let $(\Omega, \mathscr{B}, m_0)$ be a σ-content space such that $\mathscr{B}$ is a σ-field, i.e., $\Omega \in \mathscr{B}$. Then $\mathrm{cont}^{\sigma, \mathrm{b}}(\mathscr{B}) \cap \mathrm{cont}^{\ll m_0}(\mathscr{B}) = \mathrm{cont}^{\ll m_0}(\mathscr{B})$ and the mapping ψ: $L^1_{m_0} \to \mathrm{cont}^{\ll m_0}(\mathscr{B})$ defined by $f \to fm_0$ $(f \in L^1_{m_0})$ is a vector lattice isomorphism of $L^1_{m_0}$ onto $\mathrm{cont}^{\ll m_0}(\mathscr{B})$ and preserves the norm.*

5.5. Theorem (chain rule). *Let $(\Omega, \mathscr{B}^{00}, m_0)$ be a σ-content space and $m, m' \in \mathrm{cont}^{\sigma, \mathrm{b}}(\mathscr{B}^{00})$ be such that $m \ll m_0$, $m' \ll m$. Then $m' \ll m_0$ and*

$$\frac{dm'}{dm_0} = \frac{dm'}{dm} \cdot \frac{dm}{dm_0} \qquad (m_0\text{-a.e.}).$$

Proof. Clearly, $F \in \mathscr{B}^{00}$, $m_0(F) = 0 \Rightarrow m(F) = 0 \Rightarrow m'(F) = 0$, hence $m' \ll m_0$ by Proposition 3.5.2. In the sequel we may, and shall assume $m, m' \geqq 0$. Let $f = dm/dm_0$, $f' = dm'/dm$. Then $0 \leqq f \in \mathscr{L}^1_{m_0}$, $0 \leqq f' \in \mathscr{L}^1_m$. Passing to representatives, we may asume $f' \in \mathrm{mble}(\Omega, \mathscr{B}^0(\mathscr{B}^{00}), \mathbb{R})$ (use Remark III.9.2) By Exercise III.8.16.1 we can find a sequence $0 \leqq f'_1 \leqq f'_2 \leqq \cdots \nearrow f'$ of step functions for $\mathscr{B}^{00}$. If

$$g = \alpha_1 1_{F_1} + \cdots + \alpha_r 1_{F_r}, 0 \leqq \alpha_1, \ldots, \alpha_r \in \mathbb{R}, F_1, \ldots, F_r \in \mathscr{B}^{00},$$

then

$$gf = \sum_{\rho=1}^{r} \alpha_\rho 1_{F_\rho} f \in \mathscr{L}^1_{m_0}$$

(see also Exercise III.9.6.1.2) and

$$\int gf \, dm_0 = \sum_{\rho=1}^{r} \alpha_\rho \int 1_{F_\rho} f \, dm_0 = \sum_\rho \alpha_\rho \, m(F_\rho) = \int g \, dm.$$

We thus find $0 \leqq f'_1 f \leqq f'_2 f \leqq \cdots \nearrow f'f$ pointwise, $f'_1 f, f'_2 f, \ldots \in \mathscr{L}^1_{m_0}$,

$$\lim_n \int f'_n f \, dm_0 = \lim_n \int f'_n \, dm = \int f' \, dm < \infty$$

by the monotone convergence theorem III.5.3.1. By the same theorem we get $f'f \in \mathscr{L}^1_{m_0}$, and also, using an obvious modification of the above devices, $F \in \mathscr{B}^{00} \Rightarrow \int 1_F \, f'f \, dm_0 = \int 1_F \, f' \, dm = m'(F)$, proving $f'f = dm'/dm_0$, i.e., the theorem.

5.6. Exercises. Let $(\Omega, \mathscr{B}^{00}, m_0)$ be a σ-content space.

5.6.1. Let $m \in \mathrm{cont}^{\ll m_0}(\mathscr{B}^{00})$, $M \in \mathscr{B}^{00}$. Prove that there is an $f \in \mathscr{L}^1_{m_0}$ such that $\mathscr{B}^{00} \ni F \subseteq M \Rightarrow m(F) = \int 1_F \, f \, dm_0$. Show that if another $f' \in \mathscr{L}^1_{m_0}$ has the same properties, then $1_M f = 1_M f'$ m_0-a.e.

5.6.2. Let $m \in \mathrm{cont}^\sigma(\mathscr{B}^{00})$. Prove that $m \ll |m|$ and that, for every $M \in \mathscr{B}^{00}$ and every $|m|$-density f of m, $P = M \cap \{f > 0\}$, $N = M \cap \{f \leqq 0\}$ is a Hahn decomposition of M for m.

5.6.3. Let $0 \leqq m \in \mathrm{cont}^\sigma(\mathscr{B}^{00})$.

5.6.3.1. Prove that $m_0, m \ll m_0 + m$.

5.6.3.2. Let $m_0 = f_0(m_0 + m)$, $m = f(m_0 + m)$, $\Omega_0 = \{f_0 > 0\}$. Show that $m^0(F) = m(F \cap \Omega_0)$, $m^1(F) = m(F \backslash \Omega_0)$ are meaningful for any $F \in \mathscr{B}^{00}$ and define m^0, $m^1 \in \mathrm{cont}^\sigma(\mathscr{B}^{00})$ such that $m = m^0 + m^1$ is the Lebesgue decomposition of m for m_0.

5.6.3.3. Prove $m^0 = \sup\{m \wedge (nm_0) \mid n = 1, 2, \ldots\}$.

6. CONDITIONAL EXPECTATIONS

In this section we apply Theorem 5.4 in order to define a concept of very frequent use in probability theory: *conditional expectation.* The present section is devoted to a thorough study of its basic properties. Among these there is the property of being a positive linear mapping, hence the problem (IV.4.18) of kernel representation arises. It will be settled to some extent in Chapter XVI, Section 3.

Throughout this section the basic space will be a σ-probability space; in particular, the basic local σ-ring is a σ-field. The case of finite σ-contents can be reduced to the σ-probability case by renorming (multiplication by a factor >0).

The next section will be devoted to an important application of conditional expectations: martingale and semimartingale theory.

6.1. Proposition. *Let $(\Omega, \mathscr{B}, m)$ be a σ-probability space and $\mathscr{B}_0 \subseteq \mathscr{B}$ a σ-field.*

6.1.1. *Let $f \in \mathscr{L}^1_m$. Then there is an $f_0 \in \mathscr{L}^1_m \cap \mathrm{mble}(\Omega, \mathscr{B}_0, \overline{\mathbb{R}})$ such that the so-called* **conditional expectation equations**

$$\int 1_{F_0} f \, dm = \int 1_F f_0 \, dm \qquad (F_0 \in \mathscr{B}_0)$$

hold. f_0 is determined by these properties up to equality mod m. *We call f_0 the* **conditional expectation** *of f for (or with respect to) $\mathscr{B}_0$.*

6.1.2. *Denote by $L^1_m \cap \mathrm{mble}(\Omega, \mathscr{B}_0, \overline{\mathbb{R}})$ the set of all equivalence classes* mod m *in L^1_m containing a function from* $\mathrm{mble}(\Omega, \mathscr{B}_0, \overline{\mathbb{R}})$. *There is a unique mapping E_0: $L^1_m \to L^1_m \cap \mathrm{mble}(\Omega, \mathscr{B}_0, \overline{\mathbb{R}})$ such that for every $f \in \mathscr{L}^1_m$ the image under E_0 of its equivalence class* mod m *consists of conditional expectations of f for $\mathscr{B}_0$. We also write $E^{\mathscr{B}_0}$ instead of E_0.*

Proof. 1. Let m_0 denote the restriction of m to $\mathscr{B}_0$. Define m': $\mathscr{B}_0 \to \mathbb{R}$ by $m'(F_0) = \int 1_{F_0} f \, dm$ $(F_0 \in \mathscr{B}_0)$. It is obvious from Proposition 5.1 that $m' \in \mathrm{cont}^\sigma(\mathscr{B}_0) \cap \mathrm{cont}^{\ll m_0}(\mathscr{B}_0)$ $(\subseteq \mathrm{cont}^\mathrm{b}(\mathscr{B}_0))$. By Theorem 5.2 (Radon–Nikodym) there is an $f_0 \in \mathscr{L}^1_{m_0}$, uniquely determined up to equality mod m_0 ($\Leftrightarrow$ equality mod m) such that $m'(F_0) = \int 1_{F_0} f_0 \, dm_0 = \int 1_{F_0} f_0 \, dm_0$ (use Exercise III.6.5.1.4). Theorem III.9.3 (plus Remark III.9.2) shows that we may choose $f_0 \in \mathrm{mble}(\Omega, \mathscr{B}_0, \overline{\mathbb{R}})$.

2 is an easy consequence of 1.

6.2. Remarks

6.2.1. The application of Exercise III.6.5.1.4 has to be made with the understanding that the elementary domains, $\mathscr{E}, \mathscr{E}'$ considered there are $\mathscr{E}(\mathscr{B}_0)$ and $\mathscr{E}(\mathscr{B})$. Clearly $\mathscr{E}(\mathscr{B}_0) \subseteq \mathrm{mble}(\Omega, \mathscr{B}_0, \mathbb{R})$, $\mathscr{E}(\mathscr{B}) \subseteq \mathrm{mble}(\Omega, \mathscr{B}, \mathbb{R})$.

6.2.2. In view of Theorems III.9.3 and III.4.2.7 we may work with $\mathrm{mble}(\Omega, \mathscr{B}_0, \overline{\mathbb{R}})$ or $\mathrm{mble}(\Omega, \mathscr{B}_0, \mathbb{R})$.

6.2.3. Throughout our discussion of conditional expectations and their applications, we shall shift freely from the viewpoint of functions (such as the elements of $\mathscr{L}_m^1$) and m-a.e. defined functions (such as the equivalence classes belonging to L_m^1) and backward, indulging ourselves in some freedom of notation (such as writing $L_m^1 \cap \text{mble}(\Omega, \mathscr{B}_0, \overline{\mathbb{R}})$ for the set of all equivalence classes in $\mathscr{L}_m^1 \cap \text{mble}(\Omega, \mathscr{B}_0, \mathbb{R})$ and the like).

6.3. Exercises. Let $(\Omega, \mathscr{B}, m)$ be a σ-probability space and $\mathscr{B}_0 \subseteq \mathscr{B}$ a σ-field.

6.3.1. Let $\mathscr{B}_0$ be finite. Exercise I.4.8.3 shows that $\mathscr{B}_0$ contains a finite partition $A_1, \ldots, A_n$ of Ω (i.e., $A_1, \ldots, A_n \in \mathscr{B}_0$, $A_j \cap A_k = \varnothing$ $(j \neq k)$, $\Omega = A_1 + \cdots + A_n$) such that every set in $\mathscr{B}_0$ is a union of some A_k. Let $f \in \mathscr{L}_m^1$. Define

$$f_0(\omega) \begin{cases} \dfrac{1}{m(A_k)} \displaystyle\int 1_{A_k} f \, dm & \text{if} \quad \omega \in A_k, m(A_k) > 0 \\ 0 & \text{otherwise.} \end{cases}$$

Prove that $f_0 = E^{\mathscr{B}_0} f$.

6.3.2. Let $\Omega = \mathbb{R}^2$ and $\mathscr{B}$ consist of all Borel sets in $\mathbb{R}^2$. Let $\mathscr{B}_0$ consist of all those Borel sets in $\mathbb{R}^2$ that consist of full circle lines with center $0 \in \mathbb{R}^2$. Let $f \in \mathscr{C}^{00}(\mathbb{R}^2, \mathbb{R})$. Put $f_0(0, 0) = 0$ and

$$f_0(\omega) = \frac{1}{2\pi\sqrt{\omega_1^2 + \omega_2^2}} \int_0^{2\pi} f\left(\sqrt{\omega_1^2 + \omega_2^2} \cos \varphi, \sqrt{\omega_1^2 + \omega_2^2} \sin \varphi\right) d\varphi$$

$$(\omega = (\omega_1, \omega_2) \in \mathbb{R}^2)$$

else. Prove that $f_0 = E^{\mathscr{B}_0} f$.

6.4. Theorem. *Let $(\Omega, \mathscr{B}, m)$ be a σ-probability space and $\mathscr{B}_0 \subseteq \mathscr{B}_1 \subseteq \mathscr{B}$ σ-fields. Put $E_i = E^{\mathscr{B}_i}$ $(i = 0, 1)$.*

6.4.1. $E_0 \colon L_m^1 \to L_m^1 \cap \text{mble}(\Omega, \mathscr{B}_0, \overline{\mathbb{R}})$ is linear, positive, and contracts the norm: $\|E_0 f\|_1 \leqq \|f\|_1$ $(f \in L_m^1)$.

6.4.2. For every $f \in L_m^1$, we have $E_0 f = f \Leftrightarrow f \in L_m^1 \cap \text{mble}(\Omega, \mathscr{B}_0, \overline{\mathbb{R}})$. In particular:

6.4.3. $E_0 \circ E_0 = E_0$.

6.4.4. $f = \text{const} \Rightarrow E_0 f = f$ $(\in L_m^1)$.

6.4.5.

$$L_m^1 \ni f \leqq C = \text{const } (m\text{-a.e.}) \quad \Rightarrow \quad E_0 f \leqq C \ (m\text{-a.e.}).$$
$$L_m^1 \ni f \geqq c = \text{const } (m\text{-a.e.}) \quad \Rightarrow \quad E_0 f \geqq c \ (m\text{-a.e.}).$$

6.4.6. $E_0 \circ E_1 = E_0$.

6.4.7. $f \in L_m^1$, $g_0 \in L_m^1 \cap \text{mble}(\Omega, \mathscr{B}_0, \overline{\mathbb{R}})$, $fg_0 \in L_m^1 \Rightarrow E_0(g_0 f) = g_0 E_0 f$.

Proof. 1. Since the mappings $f \to fm$, restriction to $\mathscr{B}_0$, formation of densities are linear and positive, so is E_0, their composite. Norm contraction now follows by

$$\|E_0 f\|_1 = \int |E_0 f| \, dm \leqq \int 1_\Omega E_0 f_+ \, dm + \int 1_\Omega E_0 f_- \, dm$$

$$= \int 1_\Omega f_+ \, dm + \int 1_\Omega f_- \, dm = \int |f| \, dm = \|f\|_1.$$

Here $\leqq$ follows from

$$f = f_+ - f_-, f_+, f_- \geqq 0 \quad \Rightarrow \quad E_0 f = E_0 f_+ - E_0 f_-, E_0 f_+, E_0 f_- \geqq 0$$
$$\Rightarrow \quad |E_0 f| \leqq E_0 f_+ + E_0 f_-.$$

2 is an immediate consequence of the definition of E_0.

$3 \Leftarrow 2$, $4 \Leftarrow 2$, and $5 \Leftarrow (1 \text{ and } 4)$ are obvious conclusions.

6. Let $f_1 = E_1 f$, $f_0 = E_0 f_1$. Clearly $f_0 \in L_m^1 \cap \text{mble}(\Omega, \mathscr{B}_0, \mathbb{R})$ and $F_0 \in \mathscr{B}_0 \Rightarrow \int 1_{F_0} f_0 \, dm = \int 1_{F_0} f_1 \, dm_0 = \int 1_{F_0} f \, dm$. Here the first equality follows from $f_0 = E_0 f_1$, and the second follows from $f_1 = E_1 f$ in view of $F_0 \in \mathscr{B}_1$. It is now clear that $f_0 = E_0 f$.

7. It is clearly sufficient to settle the case $0 \leqq f \in L_m^1$, $0 \leqq g_0 \in \text{mble}(\Omega, \mathscr{B}_0, \mathbb{R})$. By Exercise III.8.16.1 we can find a sequence $0 \leqq g_1 \leqq g_2 \leqq \cdots \nearrow g_0$ (pointwise) such that every g_n is of the form $\alpha_1 1_{F_1} + \cdots + \alpha_r 1_{F_r}$ with $0 \leqq \alpha_1, \ldots, \alpha_r \in \mathbb{R}$, $F_1, \ldots, F_r \in \mathscr{B}_0$. We conclude

$$F_0 \in \mathscr{B}_0 \quad \Rightarrow \quad F_0 \cap F_1, \ldots, F_0 \cap F_r \in \mathscr{B}_0$$

$$\Rightarrow \quad \int 1_{F_0}(\alpha_1 1_{F_1} + \cdots + \alpha_r 1_{F_r}) E_0 f \, dm$$

$$= \alpha_1 \int 1_{F_0 \cap F_1} E_0 f \, dm + \cdots + \alpha_r \int 1_{F_0 \cap F_r} E_0 f \, dm$$

$$= \alpha_1 \int 1_{F_0 \cap F_1} f \, dm + \cdots + \alpha_r \int 1_{F_0 \cap F_r} f \, dm$$

$$= \int 1_{F_0}(\alpha_1 1_{F_1} + \cdots + \alpha_r 1_{F_r}) f \, dm$$

and thus $E_0(g_n g) = g_n E_0 f$ $(n = 1, 2, \ldots)$. Now the rest is done with the help of Theorem III.5.3 (monotone integration) as follows:

$$0 \leqq g_1 f \leqq g_2 f \leqq \cdots \nearrow g_0 f \in L_m^1$$

(pointwise), hence $\lim_n \|g_n f - g_0 f\|_1 = 0$, hence

$$\lim_n \|E_0(g_n f) - E_0(g_0 f)\|_1 \leqq \lim_n \|g_n f - g_0 f\|_1 = 0$$

by 1 (norm contraction).

$$0 \leqq g_1 E_0 f \leqq g_2 E_0 f \leqq \cdots \nearrow g_0 E_0 f \quad \Rightarrow \quad \lim_n \|g_n E_0 f - g_0 E_0 f\|_1 = 0.$$

Putting our results together, we obtain $E_0(g_0 f) = g_0 E_0 f$.

6.5. Exercises. Let $(\Omega, \mathscr{B}, m)$ be a σ-probability space and $\mathscr{B}_0 \subseteq \mathscr{B}$ a σ-field, $E_0 = E^{\mathscr{B}_0}$.

6.5.1. Analyze 1 of the above proof and show that

$$\|E_0 f\|_1 = \|f\|_1 \quad \Leftrightarrow \quad E_0 f_+ = (E_0 f)_+, \; E_0 f_- = (E_0 f)_- .$$

6.5.2. Prove $E^{\{\varnothing, \Omega\}} f = \text{const} = \int f \, dm$ (m-a.e., $f \in L_m^1$).

6.5.3. Let $F \in \mathscr{B}$. Show that $E_0 1_F > 0$ m-a.e. on F.

6.5.4. Let $F \in B$, $m(F) > 0$, and restrict the whole situation to F, i.e., restrict m to $\mathscr{B} \cap F = \{E \mid \mathscr{B} \ni E \subseteq F\}$, put $\mathscr{B}_0^F = \mathscr{B}_0 \cap F = \{E \cap F \mid E \in \mathscr{B}_0\}$ and denote by E_0^F: $L^1(F, \mathscr{B} \cap F, m) \to L^1(F, \mathscr{B} \cap F, m)$ the conditional expectation for $\mathscr{B}_0^F$. Show that for every $f \in \mathscr{L}^1(F, \mathscr{B} \cap F, m)$, we have $E_0^F f = E_0 f / E_0 1_F$ m-a.e. on F_0 (Exercise 3 shows that the denominator does not vanish; see the proof of Theorem XIV.1.15.)

6.5.5. Let $\varnothing \neq M \subseteq L_m^1$ have an upper bound in L_m^1, and hence, by Theorem III.5.8.5, a supremum $\sup M$ in the lattice L_m^1. Prove that $E_0 M = \{E_0 f \mid f \in M\}$ has a supremum as well and that $E_0(\sup M) = \sup E_0 M$.

6.6. Theorem (Jensen's inequality). *Let $(\Omega, \mathscr{B}, m)$ be a σ-probability space, $\mathscr{B}_0 \subseteq \mathscr{B}$ a σ-field and $E_0 = E^{\mathscr{B}_0}$. Let $\mathbb{R} \ni a < b \in \mathbb{R}$ and $h \in \mathbb{R}^{]a, b[}$ be convex:*

$$h(\alpha x + (1 - \alpha) y) \leqq \alpha h(x) + (1 - \alpha) h(y) \qquad (x, y \in]a, b[, \quad 0 \leqq \alpha \leqq 1).$$

Let $f \in \mathscr{L}_m^1$ and $a < f < b$ m-a.e., $h(f) \in \mathscr{L}_m^1$. Then $a < E_0 f < b$ m-a.e. and **Jensen's inequality**

$$E[h(f)] \geqq h(E_0 f) \qquad m\text{-a.e.}$$

holds.

Proof. It is clear that the "m-a.e." statements of the theorem are meant to include only that $h(f)$ is m-a.e. defined. But this is sufficient to make a statement like $h(f) \in L_m^1$ meaningful. Now $a \leqq E_0 f \leqq b$ follows from

Theorem 6.4. Let $F_0 = \{a = E_0 f\}$. Clearly $F_0 \in \mathscr{B}_0$ and hence $0 = \int 1_{F_0}(E_0 f - a)\, dm = \int 1_{F_0}(f - a)\, dm$. This $1_{F_0}(f - a)$ is an m-nullfunction, hence $a < f$ m-a.e. $\Rightarrow m(F_0) = 0 \Rightarrow a < E_0 f$ m-a.e. In the same way $E_0 f < b$ m-a.e. is proved. For any $a < x < b$, put

$$\gamma(x) = \inf\{e^{-1}(h(x + e) - h(x)) \mid 0 < e < b - x\}.$$

From the convexity of h we conclude that $\gamma \in \mathbb{R}^{]a,\, b[}$ is monotone increasing (and, by the way, right continuous; these are easy exercises). Thus we may choose a countable set $\{x_1, x_2, \ldots\} \subseteq]a, b[$ which is dense and contains all discontinuities of $\gamma(\cdot)$. For every $a < x < b$, we have, by the convexity of h, $h(y) \geqq h(x) + (y - x)\gamma(x)$ $(a < y < b$; exercise). Choose now any everywhere defined representatives f' of f and f'_0 of $E_0 f$ such that $a < f' < b$, $a < f'_0 < b$ holds everywhere. We have

$$h(f'(\omega)) \geqq h(x) + (f(\omega) - x)\gamma(x)$$

for all ω. By Theorem 6.4.1 we have

$$E_0\, h(f')(\omega) \geqq h(x) + ((E_0 f)(\omega) - x)\gamma(x) \tag{1}$$

for all $\omega \in \Omega \backslash N(x)$, where the exceptional set $N(x)$ is in $\mathscr{B}_0$ and satisfies $m(N(x)) = 0$. Put $N = N(x_1) \cup N(x_2) \cup \cdots$. Clearly, $N \in \mathscr{B}_0$, $m(N) = 0$. For $\omega \in \Omega \backslash N$, we have (1) for all $x \in \{x_1, x_2, \ldots\}$. But every $x \in]a, b[\backslash\{x_1, x_2, \ldots\}$ is a continuity point of γ, hence (1) holds for every $a < x < b$, by approximation from $\{x_1, x_2, \ldots\}$ (h is continuous, being a convex function on an open interval (exercise)). Now we choose any $\omega \in \Omega \backslash N$ and put $x = f'_0(\omega)$. We get $E_0\, h(f')(\omega) \geqq h(f'_0(\omega))$. This proves the theorem.

6.7. Remarks

6.7.1. The above proof seems to be lengthy, but I know none which is shorter and correct at the same time.

6.7.2. In view of Exercise 6.5.2, Jensen's inequality for the integral (Theorem III.5.13) is a special case of the above theorem.

6.8. Exercises. Let $(\Omega, \mathscr{B}, m)$ be a σ-content space and $\mathscr{B}_0 \subseteq \mathscr{B}$ a σ-field; put $E_0 = E^{\mathscr{B}_0}$.

6.8.1. Formulate and prove the analogue of Theorem 6.6 for closed and half-open intervals in place of the open interval $]a, b[$, and also for all kinds of intervals of infinite length.

6.8.2. Prove $f \in L^1_m \Rightarrow |E_0 f| \leqq E_0 |f|$ m-a.e. with equality m-a.e. iff $f \geqq 0$ m-a.e. or $f \leqq 0$ m-a.e.

6.8.3. Let $\mathbb{R} \ni a < b \in \mathbb{R}$. Call $h \in \mathbb{R}^{]a, b[}$ strictly convex if

$$a < x < y < b, 0 < \alpha < 1 \quad \Rightarrow \quad h(\alpha x + (1 - \alpha)y) < \alpha h(x) + (1 - \alpha)h(y).$$

Prove $f, h(f) \in L_m^1$, $E_0(h(f)) = h(E_0 f)$ m-a.e. $\Rightarrow f =$ const (m-a.e.)

6.8.4. Let $h \in L_m^1$.

6.8.4.1. Assume $h \in L_m^2$ ($\subseteq L_m^1$ here, see Exercise III.11.3.4) and prove $E_0 h^2 \leqq (E_0 h)^2$ m-a.e. with equality m-a.e. iff $h =$ const (m-a.e.).

6.8.4.2. Assume $h > 0$ m-a.e. and $\log h \in L_m^1$. Prove $\log(E_0 h) \geqq E_0(\log h)$ with equality m-a.e. iff $h =$ const (m-a.e.).

6.9. Theorem. *Let $(\Omega, \mathscr{B}, m)$ be a probability space, $\mathscr{B}_0 \subseteq \mathscr{B}$ a σ-field. Let $(\cdot, \cdot)$ denote the scalar product in the Hilbert space L_m^2. Then:*

6.9.1. $L_m^2 \subseteq L_m^1$, $L_m^2 \cap \text{mble}(\Omega, \mathscr{B}_0, \overline{\mathbb{R}})$ *is a $\|\cdot\|_2$-closed subspace of L_m^2 and a $\|\cdot\|_1$-closed subspace of $L_m^1 \cap \text{mble}(\Omega, \mathscr{B}_0, \overline{\mathbb{R}})$.*

6.9.2. *The restriction of E_0 to L_m^2 is the orthogonal projection of the Hilbert space L_m^2 onto its closed subspace $L_m^2 \cap \text{mble}(\Omega, \mathscr{B}_0, \overline{\mathbb{R}})$:*

$$(f - E_0 f, g) = 0 \qquad (f \in L_m^2, \quad g \in L_m^2 \cap \text{mble}(\Omega, \mathscr{B}_0, \overline{\mathbb{R}}).$$

Proof. 1 follows from Exercise III.11.3.6.

2. Exercises 6.8.4.1 proves that E_0 sends L_m^2 into (and clearly onto) $L_m^2 \cap \text{mble}(\Omega, \mathscr{B}_0, \mathbb{R})$. Let $f \in L_m^2$, $g \in L_m^2 \cap \text{mble}(\Omega, \mathscr{B}_0, \overline{\mathbb{R}})$. Then $gf \in L_m^1$, and we have $E_0(gf) = g\, E_0 f$ by Theorem 6.4.7. Now we get

$$(f - E_0 f, g) = \int gf \, dm - \int gE_0 f \, dm = 0.$$

7. MARTINGALES, SUBMARTINGALES, AND SUPERMARTINGALES

In this section we apply our general results on conditional expectations in order to define the concepts *martingale*, *submartingale*, and *supermartingale* for the case of $\mathbb{N} \cup \{0\} = \{0, 1, \ldots\}$ as a parameter set, and to prove basic *decomposition* and *convergence theorems* for them. Numerous further applications will be given as exercises.

The reader will no doubt observe certain analogies between the convergence theory of submartingales and the ergodic theorems. In fact, efforts have been made to understand those two theories as special case of an umbrella theory (see ROTA [1]).

We refrain from delving into such a general theory in order to remain closer to the basis of our problems.

7.1. Definition. *Let* $(\Omega, \mathscr{B}, m)$ *be a* σ*-probability space and* $\mathscr{B}_0 \subseteq \mathscr{B}_1 \subseteq \cdots \subseteq \mathscr{B}$ *a sequence of* σ*-fields. Put* $E_t = E^{\mathscr{B}_t}$ $(t = 0, 1, \ldots)$. *A sequence* $f_0, f_1, \ldots$ *with* $f_n \in \mathscr{L}_m^1 \cap \text{mble}(\Omega, \mathscr{B}_n . \overline{\mathbb{R}})$ $(n = 0, 1, 2, \ldots)$ *is called:*

7.1.1. *a* **martingale** *for the sequence* $\mathscr{B}_0 \subseteq \mathscr{B}_1 \subseteq \cdots$ *if the so-called* **martingale equations**

$$f_s = E_s f_t \quad m\text{-a.e.} \qquad (0 \leqq s \leqq t)$$

hold.

7.1.2. *a* **supermartingale** *for the sequence* $\mathscr{B}_0 \subseteq \mathscr{B}_1 \subseteq \cdots$ *if the so-called* **supermartingale inequalities**

$$f_s \geqq E_s f_t \quad m\text{-a.e.} \qquad (0 \leqq s \leqq t)$$

hold.

7.1.3. *a* **submartingale** *for the sequence* $\mathscr{B}_0 \subseteq \mathscr{B}_1 \subseteq \cdots$ *if the so-called* **submartingale inequalities**

$$f_s \leqq E_s f_t \quad m\text{-a.e.} \qquad (0 \leqq s \leqq t)$$

hold.

7.2. Remarks

7.2.1. Some authors say "increasing semimartingale' instead of "submartingale" and "decreasing semimartingale" instead of "supermartingale." Clearly, $f_0, f_1, \ldots$ is a submartingale iff $-f_0, -f_1, \ldots$ is a supermartingale.

7.2.2. It is obvious that the above definition is highly generalizable. As soon as one has an ordered set I and an isotone mapping $I \ni \iota \to \mathscr{B}_\iota$ of I into the system of all σ-fields $\subseteq B$, one can reasonably define what a corresponding martingale etc. is to be. There is an extensive literature on martingales with general index sets (Doob [1], Krickeberg [1], Neveu [4]).

7.2.3. It is obvious that the property of being a martingale etc. is a class property mod m, i.e., insensitive to passing over to m-a.e. equal functions.

7.2.4. Let $\Omega \neq \varnothing$ and $f_0, g_0, f_1, g_1, \ldots \in \overline{\mathbb{R}}^\Omega$. If $f_0 \leqq g_0, f_1 \leqq g_1, \ldots$ (possibly a.e. only, with respect to a given σ-content), we shall say that sequence $f_0, f_1, \ldots$ is a minorant of the sequence $g_0, g_1, \ldots$, etc. In particular it should be clear what e.g., a martingale majorant of a martingale is. Further we shall call $f_0 + g_0, f_1 + g_1, \ldots$ the sum and $f_0 - g_0, f_1 - g_1, \ldots$ the difference of the two given sequences. In particular, it should be clear what a sum of two submartingales etc. is. It is, e.g., obvious that the difference of a submartingale and a martingale is a submartingale again.

7.2.5. Going back to the definition of conditional expectation, we obtain:

7.2.5.1. the *extensive form of the martingale equations:*

$$\int 1_F f_s \, dm = \int 1_F f_t \, dm \qquad (0 \leqq s \leqq t, \quad F \in \mathscr{B}_s)$$

with $\int f_s \, dm = \int f_t \, dm$ $(0 \leqq s \leqq t)$ as special cases;

7.2.5.2. the *extensive form of the supermartingale inequalities:*

$$\int 1_F f_s \, dm \geqq \int 1_F f_t \, dm \qquad (0 \leqq s \leqq t, \quad F \in \mathscr{B}_s)$$

with $\int f_s \, dm \geqq \int f_t \, dm$ $(0 \leqq s \leqq t)$ as special cases;

7.2.5.3. the *extensive form of the submartingale inequalities:*

$$\int 1_F f_s \, dm \leqq \int 1_F f_t \, dm \qquad (0 \leqq s \leqq t, \quad F \in \mathscr{B}_s)$$

with $\int f_s \, dm \leqq \int f_t \, dm$ $(0 \leqq s \leqq t)$ as special cases.

7.3. Exercises. Let $(\Omega, \mathscr{B}, m)$ be a σ-probability space, $\mathscr{B}_0 \subseteq \mathscr{B}_1 \subseteq \cdots \subseteq \mathscr{B}$ σ-fields, $\mathscr{B}_\infty = \mathscr{B}(\mathscr{B}_0 \cup \mathscr{B}_1 \cup \cdots)$. Put $E_t = E^{\mathscr{B}_t}$ $(t = \infty, 0, 1, \ldots)$.

7.3.1. Let $f \in L_m^1$. Put $f_t = E_t f$ $(t = 0, 1, \ldots)$.

7.3.1.1. Prove that $f_0, f_1, \ldots$ is a martingale for $\mathscr{B}_0 \subseteq \mathscr{B}_1 \subseteq \cdots$. (It is called the martingale generated by f (and $\mathscr{B}_0 \subseteq \mathscr{B}_1 \subseteq \cdots$)).

7.3.1.2. Prove $\lim_t \|f_t - E_\infty f\|_1 = 0$. (*Hint*: Prove first that

$$\bigcup_{t=0}^{\infty} L_m^1 \cap \text{mble}(\Omega, \mathscr{B}_t, \overline{\mathbb{R}})$$

is norm-dense in $L_m^1 \cap \text{mble}(\Omega, \mathscr{B}_\infty, \overline{\mathbb{R}})$; see Exercise II.2.6.1.)

7.3.2. Let $f_0, f_1, \ldots$ be a martingale for $\mathscr{B}_0 \subseteq \mathscr{B}_1 \subseteq \cdots$

7.3.2.1. Prove that $f_t \in L_m^2$, $s \leqq t \Rightarrow f_s \in L_m^2$.

7.3.2.2. Assume $f_0, f_1, \ldots \in L_m^2$. Prove $(f_{t+1} - f_t, f_s) = 0$ $(s \leqq t)$.

7.3.2.3. Assume $f_0, f_1, \ldots \in L_m^2$, $\sup_t \|f_t\|_2 < \infty$. Prove that there is an $f \in L_m^2$ such that $\lim_t \|f_t - f\|_2 = 0$. (*Hint*: Use 7.3.2.2 and the completeness of the Hilbert space L_m^2.)

7.3.3. Let $\Omega = [0, 1]$, $\mathscr{B}$ be the σ-field of all Borel subsets of $[0, 1]$, and m the restriction of the Lebesgue σ-content to $\mathscr{B}$. For every $t = 0, 1, \ldots,$ let $\mathscr{P}_t = \{a_{t0}, \ldots, a_{tn_t}\} \subseteq [0, 1]$ be such that $0 = a_{t0} < a_{t1} < \cdots < a_{tn_t} = 1$, $\mathscr{P}_{t-1} \subseteq \mathscr{P}_t$ and $\lim_t [\sup\{a_{tk} - a_{t,k-1} \mid 1 \leqq k \leqq n_t\}] = 0$. Put

$$\mathscr{B}_t = \mathscr{B}(\{[a_{t0}, a_{t1}[, \ldots, [a_{t,n_t-2}, a_{t,n_t-1}[, [a_{t,n_t-1}, a_{t,n_t}]\}) \qquad (t = 0, 1, \ldots).$$

7.3.3.1. Prove $\mathscr{B} = \mathscr{B}(\mathscr{B}_0 \cup \mathscr{B}_1 \cup \cdots)$.

7.3.3.2. Prove $f \in \mathscr{C}([0, 1], \mathbb{R}) \Rightarrow E_t f \to f$ uniformly on $[0, 1]$.

7.3.3.3. Assume $n_t = 2^t$, $a_{tk} = k/2^t$ $(t = 0, 1, \ldots; 0 \leqq k \leqq 2^t)$. Put

$$g_t(x) = \sum_{k=1}^{2^t} (-1)^{k-1} 1_{[a_{t,k-1}, a_{tk}[}, \qquad g_t(1) = 0 \qquad (t = 0, 1, \ldots)$$

(these are the so-called *Rademacher* or *Haar* functions). Prove that $f_t = g_0 + \cdots + g_t$ $(t = 0, 1, \ldots)$ defines a martingale for $\mathscr{B}_0 \subseteq \mathscr{B}_1 \subseteq \cdots$, with $\|f_t\|_2^2 = t + 1$ $(t = 0, 1, \ldots)$.

7.3.4. Let $f_0, f_1, \ldots$ be a martingale for $\mathscr{B}_0 \subseteq \mathscr{B}_1 \subseteq \cdots$. Prove that $|f_0|, |f_1|, \ldots$ is a submartingale for $\mathscr{B}_0 \subseteq \mathscr{B}_1 \subseteq \cdots$.

7.3.5. Let $f_0, f_1, \ldots$ be a submartingale with minorant 0 for $\mathscr{B}_0 \subseteq \mathscr{B}_1 \subseteq \cdots$. Prove that $\|f_0\|_1 \leqq \|f_1\|_1 \leqq \cdots$. Prove the same chain of inequalities for a martingale $f_0, f_1, \ldots$ for $\mathscr{B}_0 \subseteq \mathscr{B}_1 \subseteq \cdots$.

7.4. Proposition. *Let $(\Omega, \mathscr{B}, m)$ be a σ-probability space, $\mathscr{B}_0 \subseteq \mathscr{B}_1 \subseteq \cdots \subseteq \mathscr{B}$ σ-fields.*

7.4.1. *Let $f_0, f_1, \ldots$ be a martingale for $\mathscr{B}_0 \subseteq \mathscr{B}_1 \subseteq \cdots$ and $h \in \mathbb{R}^{\mathbb{R}}$ a convex function such that $h(f_0), h(f_1), \ldots \in L_m^1$. Then $h(f_0), h(f_1), \ldots$ is a submartingale for $\mathscr{B}_0 \subseteq \mathscr{B}_1 \subseteq \cdots$.*

7.4.2. *Let $f_0, f_1, \ldots$ be a submartingale for $\mathscr{B}_0 \subseteq \mathscr{B}_1 \subseteq \cdots$ and $h \in \mathbb{R}^{\mathbb{R}}$ an isotone convex function such that $h(f_0), h(f_1), \ldots \in L_m^1$. Then $h(f_0), h(f_1), \ldots$ is a submartingale for $\mathscr{B}_0 \subseteq \mathscr{B}_1 \subseteq \cdots$.*

Proof. The requested measurabilities are obtained by Proposition III.8.4.3.5. Let $E_t = E^{\mathscr{B}_t}$ $(t = 0, 1, \ldots)$.

1. By Theorem 6.6 (Jensen's inequality) resp. Exercise 6.8.1 we have

$$0 \leqq s \leqq t \quad \Rightarrow \quad E_s[h(f_t)] \geqq h(E_s f_t) = h(f_s).$$

2. Using the same tools plus the isotony of h we get

$$0 \leqq s \leqq t \quad \Rightarrow \quad E_s[h(f_t)] \geqq h(E_s f_t) \geqq h(f_s).$$

7.5. Proposition (Krickeberg majorant). *Let $(\Omega, \mathscr{B}, m)$ be a σ-probability space, $\mathscr{B}_0 \subseteq \mathscr{B}_1 \subseteq \cdots \subseteq \mathscr{B}$ σ-fields. Let $f_0, f_1, \ldots$ be a submartingale for $\mathscr{B}_0 \subseteq \mathscr{B}_1 \mathscr{B} \cdots$ such that $\sup_t \|f_t\|_1 < \infty$. Then there is a martingale g_0, $g_1, \ldots$ for $\mathscr{B}_0 \subseteq \mathscr{B}_1 \subseteq \cdots$ that is a majorant to $f_0, f_1, \ldots$ and satisfies $\|g_t - f_t\|_1 \searrow 0$ $(t \to \infty)$. $g_0, g_1, \ldots$ is called the* **Krickeberg (martingale) majorant** *of the submartingale $f_0, f_1, \ldots$.*

Proof. Let $E_t = E^{\mathscr{B}_t}$ $(t = 0, 1, \ldots)$. For any $0 \leqq s \leqq t$, put $g_{st} = E_s f_t$. Then $0 \leqq s \leqq t \leqq u \Rightarrow f_s \leqq g_{st} = E_s f_t \leqq E_s(E_t f_u) = E_s f_u = g_{su}$, i.e., for every $s \geqq 0$, the sequence $g_{ss}, g_{s,s+1}, \ldots$ is monotone increasing. Moreover,

$$\lim_t \int g_{st}\, dm = \lim_t \int E_s f_t\, dm = \lim_t \int f_t\, dm \leqq \sup_t \|f_t\|_1.$$

The monotone convergence theorem III.5.3 now implies the existence of a $g_s \in L_m^1$ with $\lim_t \|g_{st} - g_s\|_1 = 0$. Clearly $g_s \in L_m^1 \cap \text{mble}(\Omega, \mathscr{B}_s, \overline{\mathbb{R}})$ (use Exercise III.9.6.2). Now Theorem 6.4.1 implies $0 \leqq s \leqq t \leqq u \Rightarrow$

$$\begin{aligned} \|E_s g_t - g_s\|_1 &\leqq \|E_s g_t - E_s g_{tu}\|_1 + \|g_{su} - g_s\|_1 \\ &\leqq \|g_t - g_{tu}\|_1 + \|g_{su} - g_s\|_1 \end{aligned}$$

and this tends to 0 as $u \to \infty$. Thus $g_0, g_1, \ldots$ is a martingale for $\mathscr{B}_0 \subseteq \mathscr{B}_1 \subseteq \cdots$, and a majorant of the submartingale $f_0, f_1, \ldots$. Now $s \geqq 0$ implies

$$\begin{aligned} \|g_s - f_s\|_1 &= \int g_s\, dm - \int f_s\, dm \geqq \int g_{s+1}\, dm - \int E_s f_{s+1}\, dm \\ &= \int g_{s+1}\, dm - \int f_{s+1}\, dm = \|g_{s+1} - f_{s+1}\|_1 \end{aligned}$$

and

$$\begin{aligned} \|g_s - f_s\|_1 &= \int g_s\, dm - \int f_s\, dm \\ &= \lim_t \int f_t\, dm - \int f_s\, dm \to 0 \qquad \text{as} \quad s \to \infty. \end{aligned}$$

Here we use $\sup_t \|f_t\|_1 < \infty$ once more.

7.6. Remark. Let the notations be as above and $f_0, f_1, \ldots$ be a submartingale, with $\sup_t \|f_t\|_1 < \infty$. The convexity and isotony of the function $x \to x_+ = x \vee 0$ on $\mathbb{R}$ implies that $(f_0)_+, (f_1)_+, \ldots$ is again a submartingale (Proposition 7.4.2), clearly satisfies $\sup_t \|(f_t)_+\|_1 \leqq \sup_t \|f_t\|_1 < \infty$, and thus a Krickeberg martingale majorant $g_0, g_1, \ldots$. Now $f_t = (f_t - g_t) - (-g_t)$ represents the submartingale $f_0, f_1, \ldots$ as a difference of a submartingale $f_0 - g_0, f_1 - g_1, \ldots \leqq 0$ and a martingale $-g_0, -g_1, \ldots \leqq 0$. This representation is called the *Krickeberg decomposition.* It allows us to assume in many investigations that the given submartingale is $\leqq 0$.

We intend to prove a theorem on a.e. convergence of norm-bounded submartingales. The basic tool for this will be

7.7. Lemma. *Let $(\Omega, \mathscr{B}, m)$ be a probability σ-space, a, b integers such that $0 \leqq a < b$, let $\mathscr{B}_a \subseteq \mathscr{B}_{a+1} \subseteq \cdots \subseteq \mathscr{B}_b$ be σ-fields, $E_t = E^{\mathscr{B}_t}$ $(t = a, a+1, \ldots, b)$ and $f_a, f_{a+1}, \ldots, f_b$ be such that $f_s \in L_m^1 \cap \operatorname{mble}(\Omega, \mathscr{B}_s, \overline{\mathbb{R}})$ $(s = a, \ldots, b)$, $f_s \leqq E_s f_t$ $(a \leqq s \leqq t \leqq b)$. Let $\alpha, \beta \in \mathbb{R}$, $\alpha < \beta$ and*

$$A = \left\{\omega \,\middle|\, \min_{a \leqq s \leqq b} f_s(\omega) < \alpha\right\},$$

$$B = \left\{\omega \,\middle|\, \max_{a \leqq s \leqq b} f_s(\omega) > \beta\right\},$$

$$C = \Omega \backslash A = \left\{\omega \,\middle|\, \min_{a \leqq s \leqq b} f_s(\omega) \geqq \alpha\right\}.$$

Then

7.7.1. $F \in \mathscr{B}_a \Rightarrow \beta m(F \cap B) \leqq \int 1_{F \cap B} f_b \, dm.$

7.7.2. $F \in \mathscr{B}_a \Rightarrow \int 1_F f_a \leqq \alpha m(F \cap A) + \int 1_{F \cap C} f_b \, dm.$

Proof. 1. First entrance decomposition yields $B = B_a + \cdots + B_b$ where $B_s = \{f_s > \beta, f_a, \ldots, f_{s-1} \leqq \beta\} \in \mathscr{B}_s$ $(s = a, \ldots, b)$. The submartingale inequalities (extensive form, Remark 7.2.5.3) yield

$$\int 1_{F \cap B} f_b \, dm = \sum_{s=a}^{b} \int 1_{F \cap B_s} f_b \, dm \geqq \sum_{s=a}^{b} \int 1_{F \cap B_s} f_s \, dm$$

$$\geqq \sum_{s=a}^{b} \int \beta 1_{F \cap B_s} \, dm = \beta \sum_{s=a}^{b} m(F \cap B_s) = \beta m(F \cap B).$$

2. First entrance decomposition yields $A = A_a + \cdots + A_b$ where $A_s = \{f_s < \alpha, f_a, \ldots, f_{s-1} \geqq \alpha\} \in \mathscr{B}_s$ $(s = a, \ldots, b)$. Put $A'_s = \{f_a, \ldots, f_s \geqq \alpha\}$ $(s = a, \ldots, b)$. Then $A'_s \in \mathscr{B}_s, \Omega = A_a + A'_a, A'_{s-1} = A'_s + A_s$ $(s = a+1, \ldots, b)$.

The submartingale inequalities (extensive form, remark 7.2.5.3.) yield now

$$\int 1_F f_a \, dm = \int 1_{F \cap A_a} f_a \, dm + \int 1_{F \cap A'_a} f_a \, dm$$

$$\leqq \alpha m(F \cap A_a) + \int 1_{F \cap A'_a} f_{a+1} \, dm$$

$$= \alpha m(F \cap A_a) + \int 1_{F \cap A_{a+1}} f_{a+1} \, dm + \int 1_{F \cap A'_{a+1}} f_{a+1} \, dm$$

$$= \alpha m(F \cap A_a) + \alpha m(F \cap A_{a+1}) + \int 1_{F \cap A'_{a+1}} f_{a+1} \, dm = \cdots$$

$$= \alpha[m(F \cap A_a) + \cdots + m(F \cap A_b)] + \int 1_{F \cap A'_b} f_b \, dm$$

$$= \alpha m(F \cap A) + \int 1_{F \cap C} f_b \, dm \qquad \text{since} \quad A'_b = C.$$

7.8. Theorem. *Let $(\Omega, \mathscr{B}, m)$ be a σ-probability space, $\mathscr{B}_0 \subseteq \mathscr{B}_1 \cdots \subseteq \mathscr{B}$ σ-fields. Let $f_0, f_1, \ldots$ be a submartingale for $\mathscr{B}_0 \subseteq \mathscr{B}_1 \subseteq \cdots$ such that $\sup_t \|f_t\|_1 < \infty$. Then there is an $f \in \mathscr{L}^1_m$ such that $\lim_{t \to \infty} f_t = f$ m-a.e. We have $\|f\|_1 \leqq \sup_t \|f_t\|_1$.*

Proof. It will be sufficient to prove convergence m-a.e. Fatou's lemma (Theorem III.5.5) will then prove the integrability of the limit and the last inequality. By Remark 7.6 we may assume $f_0, f_1, \ldots \leqq 0$ for the rest of the proof. We begin by proving that $\inf_t f_t > -\infty$ m-a.e. Apply Lemma 7.7 with any $\alpha < 0$, $a = 0$, $b = t > 0$, $F = \Omega$. We get, for $F_t = \{\inf_{0 \leqq s \leqq t} f_t < \alpha\}$, the inequality

$$\alpha m(F_t) \geqq \int f_0 \, dm - \int 1_C f_t \, dm \geqq -2 \sup_t \|f_t\|_1.$$

Put $F(\alpha) = \{\inf_t f_t < \alpha\}$. Clearly, $F_0 \subseteq F_1 \subseteq \cdots$ and $F_0 \cup F_1 \cup \cdots = F(\alpha)$. We get

$$\alpha m(F(\alpha)) = \alpha \lim_t m(F_t) \geqq -2 \sup_t \|f_t\|_1.$$

As $\alpha \searrow -\infty$, $F(\alpha)$ decreases, hence so does $m(F(\alpha))$, and $\alpha m(F(\alpha))$ remains bounded from below. We conclude $\lim_n m(F(-n)) = 0$ i.e.,

$$\left\{\inf_t f_t = -\infty\right\} = \bigcap_n F(-n)$$

is an m-nullset.

Thus $\inf_t f_t > -\infty$ m-a.e. Assume now that $f_0, f_1, \ldots$ does not converge m-a.e. Since

$$\left\{\liminf_t f_t < \limsup_t f_t\right\} = \bigcup_{\alpha,\,\beta \text{ rational}} \left\{\liminf_t f_t < \alpha < \beta < \limsup_t f_t\right\},$$

we get two reals α, $\beta < 0$ such that the set

$$F = \{\liminf f_t < \alpha < \beta < \limsup f_t\}$$

satisfies $m(F) > 0$. Put $\mathscr{K} = \mathscr{B}_0 \cup \mathscr{B}_1 \cup \cdots$ and $\mathscr{B}_\infty = \mathscr{B}(\mathscr{K})$. Then $\mathscr{K}$ is a field that is dense in the σ-field $\mathscr{B}_\infty$ (Exercise II.2.6.1). Clearly, $F \in \mathscr{B}_\infty$. We may thus, for any $\varepsilon > 0$, find some $t_0 > 0$ and an $F_0 \in \mathscr{B}_t$ such that $m(F \,\Delta\, F_0) < \varepsilon$. Since $F \subseteq \{\lim_t \sup f_t > \beta\}$, we may find $t_1 > t_0$ such that

$$m\left(F \cap \left\{\max_{t_0 \leqq t \leqq t_1} f_t > \beta\right\}\right) > m(F) - \varepsilon$$

and hence

$$m\left(F_0 \cap \left\{\max_{t_0 \leqq t \leqq t_1} f_t > \beta\right\}\right) > m(F) - 2\varepsilon.$$

Apply Lemma 7.7.1 with $a = t_0$, $F = F_0$, $b = t_1$. We get

$$\beta m\left(F_0 \cap \left\{\max_{t_0 \leqq t \leqq t_1} f_t > \beta\right\}\right) \leqq \int 1_{F_0 \cap \{\max_{t_0 \leqq t \leqq t_1} f_t > \beta\}} f_t \, dm.$$

Next we find $t_2 > t_1$ such that

$$m\left(F \cap \left\{\min_{t_1 \leqq t \leqq t_2} f_t < \alpha\right\}\right) > m(F) - \varepsilon,$$

hence

$$m\left(F_0 \cap \left\{\max_{t_0 \leqq t \leqq t_1} f_t > \beta\right\} \cap \left\{\min_{t_1 \leqq t \leqq t_2} f_t < \alpha\right\}\right) > m(F) - 3\varepsilon.$$

Apply Lemma 7.7.2 with $a = t_1$, $b = t_2$. We get

$$\int 1_{F_0 \cap \left\{\max_{t_0 \leqq t \leqq t_1} f_t > \beta\right\}} f_{t_2} \, dm \leqq \alpha m\left(F_0 \cap \left\{\max_{t_0 \leqq t \leqq t_1} f_t > \alpha\right\} \cap \left\{\min_{t_0 \leqq t \leqq t_1} f_t < \alpha\right\}\right)$$

(plus an integral which we may omit because it is $\leqq 0$ in view of our (legal) assumption $f_0, f_1, \ldots \leqq 0$).

Now we observe that $F_0 \cap \{\max_{t_0 \leqq t \leqq t_1} f_t > \beta\} \in \mathscr{B}_{t_1}$. Hence the submartingale inequality (extensive form; Remark 7.2.5.3) yields, together with some of the other inequalities obtained meanwhile (take $\alpha < \beta < 0$ into account!)

$$\beta m\left(F_0 \cap \left\{\max_{t_0 \leqq t \leqq t_1} f_t > \beta\right\}\right) \leqq \int 1_{F_0 \cap \left\{\max_{t_0 \leqq t \leqq t_1} f_t > \beta\right\}} f_{t_1}\, dm$$

$$\leqq \int 1_{F_0 \cap \left\{\max_{t_0 \leqq t \leqq t_1} f_t > \beta\right\}} f_{t_2}\, dm$$

$$\leqq \alpha m\left(F_0 \cap \left\{\max_{t_0 \leqq t \leqq t_1} f_t > \beta\right\} \cap \left\{\min_{t_1 \leqq t \leqq t_2} f_t < \alpha\right\}\right)$$

This now leads to $\beta(m(F) + 2\varepsilon) \leqq \alpha(m(F) - 3\varepsilon)$. Since $\varepsilon > 0$ was arbitrary, $\beta m(F) \leqq \alpha m(F)$ follows. In view of $\alpha < \beta < 0$ this contradicts $m(F) > 0$. The theorem is proved.

7.9. Exercises. Let $(\Omega, \mathscr{B}, m)$ be a σ-probability space and $\mathscr{B}_0 \subseteq \mathscr{B}_1 \subseteq \cdots \subseteq \mathscr{B}$ σ-fields, $B_\infty = \mathscr{B}(\mathscr{B}_0 \cup \mathscr{B}_1 \cup \cdots)$, $E_t = E^{\mathscr{B}_t}$ $(0 \leqq t \leqq \infty)$.

7.9.1. Prove that, for every $f \in L_m^1$, $E_t f \to E_\infty f$ m-a.e.

7.9.2. Assume $\mathscr{B}_\infty = \mathscr{B}$. Let $m' \in \sigma\text{-cont}(\mathscr{B})$ and $m' = m_0' + m_1'$, $m_0' \ll m$, $m_1' \perp m$ its Lebesgue decomposition for m. Assume that for every $t = 0, 1, \ldots,$ there is a partition $\mathscr{P}_t = \{A_{t1}, \ldots, A_{tn_t}\} \in \mathscr{B}_t$, $A_{tj} \cap A_{tk} = \varnothing$ $(j \neq k)$, $A_{t1} + \cdots + A_{tn_t} = \Omega$ such that $\mathscr{B}_t = \mathscr{B}(\mathscr{P}_t)$; put

$$f_t(\omega) = \begin{cases} m'(A_{tk})/m(A_{tk}), & \omega \in A_{tk},\quad m(A_{tk}) > 0 \\ 0 & \text{otherwise.} \end{cases}$$

Prove that $f_0, f_1, \ldots$ is a supermartingale, that $\lim_t f_t = f$ exists m-a.e., and $dm_0' = f\, dm$.

8. THE RADON–NIKODYM PROBLEM

Theorems 5.2 and 5.3 (Radon–Nikodym) are in a way satisfactory: They identify, for an arbitrary given σ-content space $(\Omega, \mathscr{B}^{00}, m_0)$, the (equivalence classes mod m_0 of) m_0-integrable functions with those signed σ-contents on $\mathscr{B}^{00}$ having finite total variation and being absolutely continuous with respect to m_0. In another way, the two mentioned theorems are unsatisfactory: We have a full-fledged vector lattice theory about absolute continuity (Proposition 3.4) and orthogonality (Proposition 4.1) with respect to the given m_0, plus Lebesgue decompositions (Theorem 4.2), and nowhere in this theory is the hypothesis of finite total variation needed. Should there not be a theorem that identifies an arbitrary signed σ-content $m \ll m_0$ with some function f on Ω such that $m(F) = \int 1_F f\, dm$ makes sense and is true for all $F \in \mathscr{B}^{00}$, so that we may justly call it an m_0-density for m? We shall loosely call the problem thus sketched the *Radon–Nikodym problem*, and this section is devoted to a discussion of it. Exercise 5.6 has shown that it is easy to find such a function "*locally*," i.e., on a given set $M \in \mathscr{B}^{00}$. We are apparently facing the problem of "pasting together" such "local densities" into a "global density."

In this section we present a natural concept of *locally integrable functions* and show that they solve our problem under an additional assumption: existence of a *disjoint base*. An example due to Halmos shows that not every σ-content space has a disjoint base. The most important result concerning disjoint bases is their existence for every locally compact Hausdorff space.

We thus leave the reader with the Radon–Nikodym problem in a more precise but open form: to exactly determine all those σ-content spaces $(\Omega, \mathscr{B}^{00}, m_0)$ such that for every signed σ-content $m \ll m_0$ on $\mathscr{B}^{00}$ there is a locally integrable function that is an m_0-density for m. See KÖLZOW [3].

8.1. Definition. *Let $(\Omega, \mathscr{B}^{00}, m_0)$ be a σ-content space.*

8.1.1. *A function $h \in \overline{\mathbb{R}}^\Omega$ is said to be* **locally m_0-integrable** *if $F \in \mathscr{B}^{00} \Rightarrow 1_F h \in \mathscr{L}^1_{m_0}$. The set of all locally m_0-integrable functions is denoted by* loc $\mathscr{L}^1_{m_0}$ *(with further affixes depending on what we want to specify).*

8.1.2. *A function $h \in \overline{\mathbb{R}}^\Omega$ is said to be* **locally $\mathscr{B}^{00}$-measurable** *if $F \in \mathscr{B}^{00} \Rightarrow 1_F h \in \text{mble}(\Omega, \mathscr{B}^0(\mathscr{B}^{00}), \overline{\mathbb{R}})$.*

8.1.3. *A set $N \subseteq \Omega$ is said to be a* **local m_0-nullset** *if $F \in \mathscr{B}^{00} \Rightarrow N \cap F$ is an m_0-nullset.*

8.1.4. *A* **"loc mod m_0" terminology** *is introduced in obvious analogy to the* "mod m_0" *terminology introduced in Definitions* III.4.4, III.4.7, *and Remarks* III.4.8. *In particular, it should be clear what* "loc m_0-a.e." *means and what equivalence classes* "loc mod m_0" *are.*

8.1.5. *The set of all equivalence classes* loc mod m_0 *in* loc $\mathscr{L}^1_{m_0}$ *is denoted by* loc $L^1_{m_0}$ *(with further affixes depending on what we want to specify).*

8.2. Remarks

8.2.1. Since we started from a σ-content space $(\Omega, \mathscr{B}^{00}, m)$, σ-integration for such spaces is tacitly assumed in the above definition. This does not prevent us from doing e.g., the following: Start with a locally compact Hausdorff space Ω and a positive measure m_0 on $\mathscr{C}^{00}(\Omega, \mathbb{R})$; it is a τ-measure, hence we may apply τ-extension (Chapter III, Section 7), get $\mathscr{L}^1_\tau(m)$, take the local σ-ring $\mathscr{B}^{00}_\tau = \{F \mid 1_F \in \mathscr{L}^1_\tau(m), F \text{ conditionally compact}\}$, define a σ-content, again denoted by m_0, on $\mathscr{B}^{00}$ by $m_0(F) = \int 1_F \, dm_0$ ($F \in \mathscr{B}^{00}$), and then proceed as in Definition 8.1.

8.2.2. The reader is assumed to verify immediately the analogues of those results about, say, $\mathscr{L}^1_{m_0}$ and $\int dm_0$, which can be verified for, say loc $L^1_{m_0}$, by mere routine, such as: loc $L^1_{m_0}$ is a vector lattice and $L^1_{m_0}$ is a conditionally complete vector sublattice of it.

8.2.3. It is practically obvious how to formulate and prove the analogue of Theorem II.6.2.2 on completion of σ-content spaces, now with local nullsets in the place of nullsets.

8.2.4. Exercise 8.3.4 below shows that there may be local nullsets that are not nullsets.

8.3. Exercises. Let $(\Omega, \mathscr{B}^{00}, m_0)$ be a σ-content space.

8.3.1. Prove that a function $h \in \text{loc } \mathscr{L}^1_{m_0}$ is loc mod m_0 equivalent to a function $h' \in \mathscr{L}^1_{m_0}$ iff $\sup\left\{\int 1_F |f| \, dm_0 \,\middle|\, F \in \mathscr{B}^{00}\right\} < \infty$. Prove that in this case this real number equals $\|h'\|_1$.

8.3.2. Prove that every $h \in \text{loc } \mathscr{L}^1_{m_0}$ vanishing loc mod m_0 outside some set from $\mathscr{B}^0(\mathscr{B}^{00})$ is loc mod m_0 equivalent to some function from mble$(\Omega, \mathscr{B}^0(\mathscr{B}^{00}), \overline{\mathbb{R}})$.

8.3.3. Prove that $\mathscr{B} = \{F \mid F \subseteq \Omega,\ F \cap E \in \mathscr{B}^{00}\ (E \in \mathscr{B}^{00})\}$ is a σ-field in Ω and that a function $f \in \overline{\mathbb{R}}^\Omega$ is locally $\mathscr{B}^{00}$-measurable iff it is $\mathscr{B}$-measurable.

8.3.4. Let I be uncountable and $(\Omega_\iota, \mathscr{B}^{00}_\iota, m_\iota)$ a copy of the Lebesgue σ-content space $(\mathbb{R}, \mathscr{B}^{00}, m)$ $(\iota \in I)$. Assume $\Omega_\iota \cap \Omega_\kappa = \varnothing$ $(\iota \neq \kappa)$ and put $\Omega = \sum_{\iota \in I} \Omega_\iota$, $\mathscr{B}^{00} = \mathscr{B}^{00}(\bigcup_{\iota \in I} \mathscr{B}^{00}_\iota)$. Prove that

$$\mathscr{B}^{00} = \{F \mid F \subseteq \Omega,\ F \cap \Omega_\iota \in \mathscr{B}^{00}_\iota\ (\iota \in I),\ F \cap \Omega_\iota = \varnothing \text{ except for finitely many } \iota \in I\}.$$

Prove that there is a unique σ-content m on $\mathscr{B}^{00}$ that extends every $m_\iota (\iota \in I)$. Prove that $N \subseteq \Omega$ is an m-nullset iff $N \cap \Omega_\iota$ is an m_ι-nullset $(\iota \in I)$ and $N \cap \Omega_\iota = \varnothing$ except for countably many $\iota \in I$. Prove that $N \subseteq \Omega$ is a local m-nullset iff $N \cap \Omega_\iota$ is an m-nullset $(\iota \in I)$. Construct a local m-nullset that is not an m-nullset.

8.4. Definition. *Let $(\Omega, \mathscr{B}^{00}, m)$ be a σ-content space. A set system $\mathscr{D} \subseteq \mathscr{B}^{00}$ is called a* **disjoint base** *for $(\Omega, \mathscr{B}^{00}, m)$ if the following conditions are satisfied:*

8.4.1. $E, F \in \mathscr{D},\ E \neq F \Rightarrow E \cap F = \varnothing$.

8.4.2. *$E \in \mathscr{D} \Rightarrow m(E) > 0$. In particular, all $E \in \mathscr{D}$ are nonempty sets.*

8.4.3. *For every $M \in \mathscr{B}^{00}$, the set system $\mathscr{D}_M = \{E | E \in \mathscr{D}, m(E \cap M) > 0\}$ is at most countable and its union D_M satisfies*

8.4.4. *$M \backslash D_M$ is an m-nullset.*

8.5. Remarks. Let $(\Omega, \mathscr{B}^{00}, m_0)$ be a σ-content space.

8.5.1. A disjoint base $\mathscr{D}$ of $(\Omega, \mathscr{B}^{00}, m_0)$ remains a disjoint base of $(\Omega, \mathscr{B}^{00}, m)$ if we modify each of its members by an m-nullset $\in \mathscr{B}^{00}$, preserving disjointness (8.4.1). This is, e.g., the case if we replace every $E \in \mathscr{D}$ by some $E' \subseteq E$ such that $E' \backslash E$ is an m-nullset $\in \mathscr{B}^{00}$.

8.5.2. Condition 8.4.3 follows automatically from 8.4.1 and $m(M) < \infty$ (exercise).

8.5.3. Condition 8.4.4 is, given 8.4.1, 8.4.2, 8.4.3, equivalent to: For every $M \in \mathscr{B}^{00}$ with $m(M) > 0$, there is a $D \in \mathscr{D}$ such that $m(D \cap M) > 0$. In fact it is obvious that 8.4.4 implies that statement. Given that statement and some $M \in \mathscr{B}^{00}$ with $m(M) > 0$, let $\mathscr{D}_M$ be as in 8.4.3. We know (8.5.2) that $\mathscr{D}_M$ is at most countable, say, $\mathscr{D}_M = \{D_1, D_1, \ldots\}$, $D_j \cap D_k = \varnothing$ $(j \neq k)$. The serial form of the monotone convergence theorem III.5.3.2 implies $1_{M \backslash D_M} = 1_{M \backslash (D_1 + D_2 + \cdots)} \in \mathscr{L}^1_{m_0}$. By Exercise II.2.6.6 there is some $F \in \mathscr{B}^{00}$ such that $F \subseteq M \backslash D_M$ and $(M \backslash D_M) \backslash F$ is an m_0-nullset. Thus

$$m(M \backslash D_M) > 0 \quad \Rightarrow \quad m(F) > 0$$

and the existence of some $D \in \mathscr{D}$ with $m(D \cap F) > 0$, hence $D \in \mathscr{D}_M$ follows from our presupposed statement. But $D \cap D_j = \varnothing$ $(j = 1, 2, \ldots)$, contradicting the definition of $\mathscr{D}_M$.

8.6. Exercises. Let $(\Omega, \mathscr{B}^{00}, m)$ be a σ-content space.

8.6.1. Assume that Ω is a countable union of sets from $\mathscr{B}^{00}$. Prove that $(\Omega, \mathscr{B}^{00}, m)$ has a disjoint base.

8.6.2. Assume that Ω is a countable union of sets $F_1, F_2, \ldots$ such that $1_{F_k} \in \mathscr{L}^1_m$ $(k = 1, 2, \ldots)$. Prove that $(\Omega, \mathscr{B}^{00}, m)$ has a disjoint base.

8.6.3. Assume that there is an $\Omega_0 \in \mathscr{B}^0(\mathscr{B}^{00})$ such that $\|m\|(\Omega \backslash \Omega_0) = 0$. Prove that $(\Omega, \mathscr{B}^{00}, m)$ has a disjoint base.

8.6.4. Assume that $\|m\| < \infty$. Prove that $(\Omega, \mathscr{B}^{00}, m)$ has a disjoint base.

8.7. Theorem. *Let Ω be a locally compact Hausdorff space and $\mathscr{B}^{00}$ the local σ-ring generated by the system of all compact sets in Ω. Then every σ-content space $(\Omega, \mathscr{B}^{00}, m)$ has a disjoint base $\mathscr{D}$ (called* **"concassage"** *by Bourbaki) such that:*

8.7.1. *$\mathscr{D}$ consists of compact sets.*

8.7.2. *For every compact $K \subseteq \Omega$ the set $\{D \mid D \in \mathscr{D}, D \cap K \neq \varnothing\}$ is at most countable.*

Proof. We may choose $\mathscr{D} = \varnothing$ if $m = 0$. Assume now $m \neq 0$. Let $\Omega(m) = \{\omega \mid \omega \in U,\ U \text{ open},\ \bar{U} \text{ compact} \Rightarrow m(\bar{U}) > 0\}$. Assume $\Omega(m) = \varnothing$ and choose any compact $K \subseteq \Omega$; since $\omega \in K \Rightarrow \omega \notin \Omega(m)$, we may choose, for every $\omega \in K$, an open $U_\omega \ni \omega$ such that $\bar{U}$ is compact and $m(\bar{U}_\omega) = 0$; find $\omega_1, \ldots, \omega_n \in \Omega$ such that $K \subseteq U_{\omega_1} \cup \cdots \cup U_{\omega_n}$; then

$$K \subseteq \bar{U}_{\omega_1} \cup \cdots \cup \bar{U}_{\omega_n}$$

and thus

$$0 \leqq m(K) \leqq m(\bar{U}_{\omega_1}) + \cdots + m(\bar{U}_{\omega_n}) = 0;$$

since every set from $\mathscr{B}^{00}$ is contained in some compact set, $m = 0$ follows, a contradiction. Thus $\Omega(m) \neq \varnothing$. It is easy to see that $\Omega(m)$ is closed and $m(E) = m(E \cap \Omega(m))$ $(E \in \mathscr{B}^{00})$ (exercise). Let now Σ be the system of all $\mathscr{K} \subseteq \mathscr{P}(\Omega)$ with the following properties: (1) $J,\ K \in \mathscr{K} \Rightarrow J \cap K = \varnothing$. (2) $K \in \mathscr{K} \Rightarrow K$ is compact. (3) $K \in \mathscr{K} \Rightarrow m_0(K) > 0$, in particular $K \neq \varnothing$. (4) $K \in \mathscr{K}$, G open $\Rightarrow$ either $K \cap G = \varnothing$ or $m_0(K \cap G) > 0$. It is easy to see that Σ is nonempty. In fact, choose some $\omega \in \Omega(m)$ and a compact K' such that ω is an inner point of K'. Then $K = \Omega(m) \cap K'$ satisfies (3), (4) (exercise), and this $\{K\} \in \Sigma$. With the partial-ordering $\subseteq$, Σ satisfies the hypotheses of *Zorn's lemma* (exercise). Let $\mathscr{D}$ be maximal in Σ. Let $M \in \mathscr{B}^{00}$, $m(M) > 0$. Then there is some open $U \supseteq M$ with a compact $\bar{U}$. Put $\mathscr{D}' = \{D \mid D \in \mathscr{D}, D \cap U \neq \varnothing\}$. Then (4) $\Rightarrow \mathscr{D}_M \subseteq \mathscr{D}'$ and $m(D \cap U) > 0$ $(D \in \mathscr{D}')$. By 8.4.3, $\mathscr{D}'$ is at most countable, say $\mathscr{D}' = \{D_1, D_2, \ldots\}$, $D_j \cap D_k = \varnothing$ $(j \neq k)$. It will now be sufficient to prove $m(U \setminus (D_1 + D_2 + \cdots)) = 0$. Assume this is > 0. By inner compact regularity (Theorem V.4.6.2) there is a compact $K \subseteq U \setminus (D_1 + D_2 + \cdots)$ such that $m(K) > 0$. Clearly $\mathscr{D} \cup \{K \cap \Omega(m)\} \in \Sigma$, contradicting the maximality of $\mathscr{D}$.

8.8. Theorem (Radon–Nikodym). *Let $(\Omega, \mathscr{B}^{00}, m_0)$ be a σ-content space that has a disjoint base. Then for every signed content $m \ll m_0$ on $\mathscr{B}^{00}$ there is an $f \in \operatorname{loc} \mathscr{L}^1_{m_0}$ such that $F \in \mathscr{B}^{00} \Rightarrow m(F) = \int 1_F f \, dm_0$. f is uniquely determined* (loc mod m_0) *by m and is called the* **m_0-density of m.** *We write $m = fm_0$ or $dm = f\, dm_0$ in this case.*

Proof. Let $\mathscr{D} \subseteq \mathscr{B}^{00}$ be a disjoint base for $(\Omega, \mathscr{B}^{00}, m_0)$. For every $D \in \mathscr{D}$, we may apply Exercise 5.6.1 in order to obtain some $f_D \in \mathscr{L}^1_{m_0}$ such that f_D vanishes outside D and $D \supseteq F \in \mathscr{B}^{00} \Rightarrow m(F) = \int 1_F f \, dm_0$. Put $f = \sum_{D \in \mathscr{D}} f_D$. This is meaningful since the $D \in \mathscr{D}$ are pairwise disjoint and every f_D vanishes outside D. Let now $M \in \mathscr{B}^{00}$ and $\mathscr{D}_M = \{D_1, D_2, \ldots\}$,

$D_j \cap D_k = \varnothing$ $(j \neq k)$. Then $1_M f = \sum_{k=1}^{\infty} 1_M f_{D_k}$ and the serial version of Lebesgue's dominated convergence theorem III.5.7.2 implies

$$\int 1_M f \, dm_0 = \sum_k \int 1_M f_{D_k} \, dm_0 = \sum_k m(M \cap D_k) = m(M).$$

8.9. Exercises. Let $(\Omega, \mathscr{B}^{00}, m_0)$ be a σ-content space.

8.9.1. Let $\Omega_0 \subseteq \Omega$. Call $\mathscr{D} \subseteq \mathscr{P}(\Omega_0)$ a disjoint base for $(\Omega, \mathscr{B}^{00}, m_0)$ in Ω_0 if (1) $E, F \in \mathscr{D}$, $E \neq F \Rightarrow E \cap F = \varnothing$; (2) $E \in \mathscr{D} \Rightarrow 1_E \in \mathscr{L}^1_{m_0}$ and $\int 1_E \, dm_0 > 0$; (3) for every $\Omega_0 \supseteq M \in \mathscr{B}^{00}$, the set system

$$\mathscr{D}_M = \left\{ E \,\middle|\, \int 1_{E \cap M} \, dm_0 > 0 \right\}$$

is at most countable and its union D_M satisfies (4): $M \backslash D_M$ is an m_0-nullset. Formulate and prove the analogue of Theorem 8.8 for disjoint bases in a set $\Omega_0 \subseteq \Omega$.

8.9.2. Let $\Omega_0 \subseteq \Omega$ be such that $M \in \mathscr{B}^{00} \Rightarrow M \cap \Omega_0 \in \mathscr{B}^{00}$ and $\|m_0\|(\Omega \backslash \Omega_0) = 0$. Prove that every disjoint base for $(\Omega, \mathscr{B}^{00}, m_0)$ in Ω_0 is also a disjoint base for it in Ω.

8.9.3. Let $\mathscr{B}^0 = \mathscr{B}^0(\mathscr{B}^{00})$ and $\text{loc mble}_m(\Omega, \mathscr{B}^0, \overline{\mathbb{R}})$ be the system of all equivalence classes loc mod m in $\text{mble}(\Omega, \mathscr{B}^0, \overline{\mathbb{R}})$. Assume that $(\Omega, \mathscr{B}^{00}, m)$ has a disjoint base and prove that $\text{loc mble}_m(\Omega, \mathscr{B}^0, \overline{\mathbb{R}})$ is a complete lattice. (*Hint*: Apply Exercise III.12.15 "locally.")

8.9.4. Let $(\Omega, \mathscr{T})$ be a Hausdorff space, $\mathscr{B}^{00} = \mathscr{B}^{00}(\mathscr{C}^{00}(\Omega, \mathbb{R}))$, and $m: \mathscr{B}^{00} \to \mathbb{R}_+$ an inner compact regular σ-content. Prove the existence of a concassage, i.e., a disjoint base for $(\Omega, \mathscr{B}^{00}, m)$ that consists of compact sets. (*Hint*: Modify the proof of Theorem 8.7)

We conclude this section with an example due to Halmos, showing that there are σ-content spaces for which the Radon–Nikodym theorem fails; in particular, they do not admit a disjoint base.

8.10. Example.(after HALMOS [4]). Let X, Y be two uncountable sets and assume that the power of X is strictly less than the power of Y. Put $\Omega = X \times Y$ and call a set $C \subseteq \Omega$ a cross if it satisfies the following conditions: (1) the x-section $C_x = \{y \mid (x, y) \in C\}$ of C is empty for all but finitely many $x \in X$; (2) for every $x \in X$, either C_x or $Y \backslash C_x$ is countable; (3) the y-section $C_y = \{x \mid (x, y) \in C\}$ of C is empty for all but finitely many $y \in Y$; (4) for every $y \in Y$, either C_y or $X \backslash C_y$ is countable. It is easy to see that $\mathscr{B}^{00} = \{C \mid C \subseteq \Omega, C \text{ is a cross}\}$ is a local σ-ring in Ω (exercise).

Define m_X, m_Y on $\mathscr{B}^{00}$ by $m_X(C) = |\{x | Y \backslash C_x \text{ is countable}\}|$, $m_Y(C) = |\{y | X \backslash C_y \text{ is countable}\}$. m_X and m_Y are σ-contents on $\mathscr{B}^{00}$ (exercise). Put $m_0 = m_X + m_Y$, $m = m_X$. Clearly $(\Omega, \mathscr{B}^{00}, m_0)$ is a σ-content space and $0 \leqq m \ll m_0$. The m_0-nullsets are exactly those sets $N \subseteq \Omega$ for which both N_x and N_y are at most countable for all $x \in X$, $y \in Y$, and empty for all but at most countably many $x \in X$, $y \in Y$.

Assume that there is an $f \in \overline{\mathbb{R}}_+^\Omega$ such that for every $C \in \mathscr{B}^{00}$ we have $1_C f \in \mathscr{L}^1_{m_0}$, $\int 1_C f \, dm_0 = m(C)$. Choose any $x \in X$ and $C = \{x\} \times Y$. Then either $\{f = 0\} \cap C$ or $C \backslash \{f = 0\}$ is countable (exercise). $m_0(C) = 1 = m(C)$ excludes the second alternative. It follows that $\{f = \} \cap C$ is at most countable; this holds for every $x \in X$, hence the power of $\{f = 0\}$ is at most countable times the power of X, i.e., at most equal to the power of X. Choose any $y \in Y$ and $C = X \times \{y\}$. Then either $\{f = 0\} \cap C$ or $C \backslash \{f = 0\}$ is countable (exercise). $m_0(C) = 1$, $m(C) = 0$ excludes the first alternative. It follows that $C \backslash \{f = 0\}$ is at most countable, hence $\{f = 0\} \cap C$ has the same power as Y, hence $\{f = 0\}$ has at least the power of Y. We have arrived at a contradiction.

CHAPTER IX

THE VECTOR LATTICE OF SIGNED MEASURES

In the preceding chapter we have investigated the vector lattice of all signed *contents* on a given local σ-ring in some fixed basic set. In this section we parallel these investigations by a detailed analysis of what will be called *signed measures*. The parallelism will, however, not be full, for the following reasons. Signed contents are set functions, and the subsets of the basic set involved in their theory, are, so to speak, the feet on which we stand firmly on the earth. The investigation of the vector lattice properties of signed contents has brought us the suspicion that behind quite a portion of that theory there is nothing but a certain branch of general and abstract vector lattice theory. It is one of the objectives of the present chapter to confirm this suspicion. Measure theory is in fact the best domain for such an enterprise since measures are linear forms on a vector lattice, the elementary domain.

We begin therefore with a section presenting abstract vector lattice theory to an extent that will be sufficient for the application to the vector lattice theory of signed measures to be presented in Section 2.

1. ABSTRACT VECTOR LATTICES AND THEIR DUALS

The basic notions of the abstract theory of vector lattices are, throughout this book, assumed to be familiar to the reader. Chapter 0, Section 2. contains a short account of it. In the present section we treat certain somewhat more advanced topics of abstract vector lattice theory needed for a well-rounded theory of signed measures.

1.1. Definition. *Let* $(H, \leqq)$ *be a partially ordered real vector space.*

1.1.1. *A linear form L on H is said to be* **positive** *if* $0 \leqq f \in H \Rightarrow L(f) \geqq 0$.

1.1.2. *A linear form L on H is said to be* **order bounded** *if it is bounded on order bounded subsets of H, i.e., if for every* $\varnothing \neq M \subseteq H$ *that has a minorant and a majorant in H both*

$$\inf_{f \in M} L(f) > -\infty \qquad \text{and} \qquad \sup_{f \in M} L(f) < \infty$$

hold. The set of all order bounded linear forms on H is called the **order dual** *of H (resp.* $(H, \leqq)$*) and is denoted by* H^{ord}.

1.2. Remark. It is obvious that every positive linear form on H is order bounded since it is isotone.

1.3. Lemma. *Let H be a real vector space and* $K \subseteq H$ *a convex cone with tip* 0 *such that* $H = \text{lin } K$. *Then:*

1.3.1. *Every* $h \in H$ *has at least one representation* $h = f - g, f, g \in K$.

1.3.2. *Let* $L\colon K \to \mathbb{R}$ *be such that:*

1.3.2.1. $\alpha \in \mathbb{R}_+,\ f \in K \Rightarrow L(\alpha f) = \alpha L(f)$.

1.3.2.2. $f, g \in K \Rightarrow L(f + g) = L(f) + L(g)$.

Then there is a unique linear form L on H such that

1.3.2.3. $f, g \in K \Rightarrow L(f - g) = L(f) - L(g)$.

The proof is an easy exercise. (*Hint*: Show that 1.3.2.2 implies that 1.3.2.3 well defines L.)

1.4. Lemma. *Let* $(H, \leqq)$ *be a partially ordered real vector space and* H_+ *its positive cone. Let* H_0 *be a linear subspace of H and* L_0 *a linear form on* H_0 *such that* $0 \leqq f \in H_0 \Rightarrow L_0(f) \geqq 0$. *Then there is a positive linear form L on H such that* L_0 *is the restriction of L to* H_0.

Proof. It is easy to see that $p(h) = \inf_{h \leqq f \in H_0} L_0(f)$ defines a majorant function (in the sense of the Hahn–Banach theorem I.6.1) on H such that $L \leqq p$ on H_0. Apply the Hahn–Banach theorem I.6.1 and get a linear form $L \leqq p$ on H. If $H \ni h \leqq 0$, then $h \leqq 0 \in H_0$, hence $p(h) \leqq L_0(0) = 0$ and hence $L(h) \leqq 0$. Multiplying by -1, we see that L is positive.

1.5. Theorem. *Let* $(H, \leqq)$ *be a vector lattice and* H^{ord} *its order dual. Then:*

1.5.1. *The set* H_+^{ord} *of all positive linear forms on H is a sharp convex cone in* H^{ord}, *with tip* 0. *The partial ordering* $\leqq$ *defined in* H^{ord} *by* $m \leqq m' \Leftrightarrow m' - m \in H_+^{\text{ord}}$ $(m, m' \in H^{\text{ord}})$ *is called the* **dual partial ordering** *of the sharp partial ordering* $\leqq$ *given in H. The partially ordered real vector space* $(H^{\text{ord}}, \leqq)$ *is called the* **order dual** *of* $(H, \leqq)$.

1.5.2. *The order dual* $(H^{\text{ord}}, \leqq)$ *of* $(H, \leqq)$ *is a conditional complete vector lattice.*

1.5.3. *The finite vector lattice operations in* H^{ord} *can be carried out as follows: for* $0 \leqq f \in H$ *and* $m, m' \in H^{\text{ord}}$, *we have*

1.5.3.1.

$$(m \vee m')(f)$$
$$= \sup\left\{\sum_{k=1}^{n} \max\{m(f_k), m'(f_k)\} \,|\, n \geqq 1, 0 \leqq f_1, \ldots, f_n \in H, f_1 + \cdots + f_n = f\right\}$$
$$= \sup\{m(g) + m'(h) \,|\, 0 \leqq g,\ h \in H,\ g + h = f\};$$

1.5.3.2.

$$(m \wedge m')(f)$$
$$= \inf\left\{\sum_{k=1}^{n} \min\{m(f_k), m'(f_k)\} \,|\, n \geqq 1, 0 \leqq f_1, \ldots, f_n \in H, f_1 + \cdots + f_n = f\right\}$$
$$= \inf\{m(g) + m'(h) \,|\, 0 \leqq g,\ h \in H,\ g + h = f\};$$

1.5.3.3. $$m_+(f) = \sup\{m(g) \,|\, 0 \leqq g \in H,\ g \leqq f\}.$$

1.5.4. *If* $\varnothing \neq M \subseteq H_+^{\text{ord}}$ *is increasingly filtered and has a majorant in* H^{ord}, *then*

$$(\sup M)(f) = \sup_{m \in M} m(f) \qquad (0 \leqq f \in H). \tag{0}$$

1.5.5. *If* $\varnothing \neq M \subseteq H_+^{\text{ord}}$ *is increasingly filtered and*

$$\sup_{m \in M} m(f) < \infty \qquad (0 \leqq f \in H),$$

then sup M *exists in* H^{ord}, *and* (0) *holds.*

Proof. 1. It is obvious that H_+^{ord} is a convex cone with tip 0 (exercise). Assume that $m, -m \in H_+^{\text{ord}}$. Then $f \in H_+ \Rightarrow m(f) \geqq 0 \leqq -m(f) \Rightarrow m(f) = 0$, i.e., m vanishes on H_+. Since $H = \text{lin}\, H_+$, we have $m = 0$. Hence H_+^{ord} does not contain a linear subspace of dimension > 0, hence it is a sharp cone with tip 0.

2. We begin by proving that H^{ord} is a real vector space. But it is practically obvious that linear combinations of order bounded linear forms are order bounded again (exercise). Next we prove that H^{ord} is a vector lattice. Let $m \in H^{\text{ord}}$ be arbitrary and define m_+ by 1.5.3.3 and Lemma 1.3. In order to prove that this makes sense, let us first observe that for $f \in H_+$, $m_+(f)$ is a finite real $\geqq 0$ since $\{g \,|\, g \in H,\ 0 \leqq g \leqq f\}$ is an order

bounded set containing 0. 1.3.2.1 (with $m = L$) is obvious. Let us now prove 1.3.2.2 (with $m = L$ again). Choose $0 \leqq f, g \in H$. Let us show that every $0 \leqq h \in H$ with $h \leqq f + g$ has a representation $h = f' + g'$ with $0 \leqq f' \leqq f$, $0 \leqq g' \leqq g$. In fact we may put $f' = h \wedge f$, $g' = h - (h \wedge f)$; then we have $0 \leqq f' \leqq f$ and $0 \leqq g' = h + [(-h) \vee (-f)] = 0 \vee (h - f) \leqq 0 \vee g = g$. Now it is easy to conclude that

$$\begin{aligned} m_+(f + g) &= \sup\{m(h) \,|\, 0 \leqq h \leqq f + g\} \\ &= \sup\{m(f') + m(g') \,|\, 0 \leqq f' \leqq f, 0 \leqq g' \leqq g\} \\ &= \sup\{m(f') \,|\, 0 \leqq f' \leqq f\} + \sup\{m(g') \,|\, 0 \leqq g' \leqq g\} \\ &= m_+(f) + m_+(g). \end{aligned}$$

Thus Lemma 1.3 applies and m_+ is well defined as a linear form on H^{ord}, and obviously a positive, hence an order bounded one. If $m' \in H^{\text{ord}}$ satisfies $m' \geqq 0$, $m' \geqq m$, then $0 \leqq g \leqq f \Rightarrow m(g) \leqq m'(g) \leqq m'(f)$ and thus $m_+(f) \leqq m'(f)$, i.e., $m' \geqq m_+$. Thus we see that $m \vee 0$ exists in H^{ord} and is $= m_+$. It is now easy to prove that H^{ord} is a vector lattice (see Chapter 0, Section 2). The conditional completeness of $(H, \leqq)$ will be an easy consequence of 4.

3. We have already proved 1.5.3.3. The formula $m \vee m' = m + (m' - m)_+$ now implies, for any $0 \leqq f \in H^{\text{ord}}$,

$$\begin{aligned} (m \vee m')(f) &= m(f) + \sup\{(m' - m)(g) \,|\, 0 \leqq g \leqq f\} \\ &= \sup\{m(f) + m'(g) - m(g) \,|\, 0 \leqq o \leqq f\} \\ &= \sup\{m(f - g) + m'(g) \,|\, 0 \leqq g \leqq f\} \\ &= \sup\{m(g) + m'(h) \,|\, 0 \leqq g, h, g + h = f\}. \end{aligned}$$

From $m \vee m' \geqq m, m'$ it is obvious that

$$\begin{aligned} (m \vee m')(f) &= \sum_{k=1}^{n} (m \vee m')(f_k) \\ &\geqq \sum_{k=1}^{n} \max\{m(f_k), m'(f_k)\} \qquad (0 \leqq f_1, \ldots, f_n, \quad f_1 + \cdots + f_n = f). \end{aligned}$$

On the other hand, we may split the sum $\sum_{k=1}^{n}$ into a sum $\sum'$ that runs over those k for which $m(f_k) \geqq m'(f_k)$, and the remaining sum $\sum''$, so that $(m \vee m')(f) \geqq \sum' m(f_k) + \sum'' m'(f_k) = m(\sum' f_k) + m'(\sum'' f_k)$ in the mentioned situation. But these expressions run through all $m(g) + m(h)$ with $0 \leqq g, h$, $g + h = f$, as n and the $f_1, \ldots, f_n$ run as mentioned. This finishes the proof of 1.5.3.1. 1.5.3.2 is proved in a similar fashion (exercise).

4. By standard devices we may pass to the assumption $M \subseteq H_+^{\text{ord}}$. Define $m_0 \colon H_+ \to \mathbb{R}_+$ by $m_0(f) = \sup_{m \in M} m(f)$ $(f \in H_+)$. It is easy to see

that m_0 satisfies the hypotheses of Lemma 1.3 (with $m_0 = L$; exercise). Thus m_0 is well defined on H, and clearly a positive linear form, hence in H_+^{ord}. By the very definition of m_0 it is obvious that $m_0 = \sup M$.

5 is proved in the same way as 4 (exercise).

Let $(H, \leqq)$ be a vector lattice. Recall (Chapter 0, Section 2) that $m, m' \in H$ are said to be orthogonal if $|m| \wedge |m'| = 0$. For every set $S \subseteq H$, the orthogonal complement $S^\perp$ was defined by $S^\perp = \{m \mid m' \in S \Rightarrow m \perp m'\}$. Well-known formulas imply that for any $m, m' \in H$ the following equivalences hold:

$$\begin{aligned} m \perp m' \quad &\Leftrightarrow \quad |m| \vee |m'| = |m| + |m'| \\ &\Leftrightarrow \quad \big||m| - |m'|\big| = |m| + |m'| \\ &\Leftrightarrow \quad [m_+ \perp m'_+,\ m_+ \perp m'_-,\ m_- \perp m'_+,\ m_- \perp m'_-] \end{aligned}$$

(exercise). With these preparations in mind, we can prove the

1.6. Lemma. *Let $(H, \leqq)$ be a vector lattice, $S \subseteq H$ and $|S| = \{|m| \mid m \in S\}$. Then:*

1.6.1. $S^\perp = |S|^\perp$.

1.6.2. *$S^\perp$ is a vector sublattice of H.*

1.6.3. $m \in S^\perp$, $|m'| \leqq |m| \Rightarrow m' \in S^\perp$.

1.6.4. *If $\varnothing \neq M \subseteq S^\perp$ and $\sup M$ exists in H, then $\sup M \in S^\perp$.*

1.6.5. *If H is a conditionally complete lattice, then every $\varnothing \neq M \subseteq S^\perp$ that has a majorant in H, has $\sup M \in S^\perp$. In particular, $S^\perp$ is a conditionally complete sublattice of H.*

Proof. 1–3 are easy exercises.

4. Standard devices show that it is sufficient to settle the case $M, S \subseteq H_+$. Now $M \subseteq S^\perp \Rightarrow m \vee m_1 = m + m_1$ $(m \in M,\ m_1 \in S)$, in short:

$$\begin{aligned} M \vee m_1 = M + m_1 \quad (m_1 \in S) \quad \Rightarrow \quad (\sup M) \vee m_1 &= \sup(M \vee m_1) \\ &= \sup(M + m_1) \\ &= (\sup M) + m_1 \quad (m_1 \in S) \\ \Rightarrow \quad \sup M \subseteq S^\perp. & \end{aligned}$$

5 is an easy consequence of 4.

1.7. Definition. *Let $(H, \leqq)$ be a conditionally complete vector lattice. A set $\varnothing \neq H_0 \subseteq H$ is called a* **(Riesz) band** *in H if:*

1.7.1. *H_0 is a vector sublattice of H.*

1.7.2. $m, m' \in H$, $m \in H_0$, $|m'| \leqq |m| \Rightarrow m' \in H_0$.

1.7.3. *If a set $\varnothing \neq M \subseteq H_0$ has a majorant in H, then $\sup M \in H_0$.*

1.8. Examples

1.8.1. Lemma 1.6 shows that for every conditionally complete vector lattice H and every set $S \subseteq H$, the set $S^{\perp}$ is a band in H.

1.8.2. Theorem VIII.1.4.3 implies that for every set ring $\mathscr{R}$ in a set $\Omega \neq \varnothing$, σ-cont$(\mathscr{R})$ is a band in the conditionally complete vector lattice cont$(\mathscr{R})$. If $m_0 \geqq 0$ is a σ-content on $\mathscr{R}$, then $\text{cont}^{\ll m_0}(\mathscr{R})$ is a band in cont$(\mathscr{R})$, after Proposition VIII.3.4.4 (use also Exercise VIII.3.3.1), and likewise $\text{cont}^{\perp m_0}(\mathscr{R})$ and $\text{cont}^{\perp m_0}(\mathscr{R}) \cap \sigma\text{-cont}(\mathscr{R})$ are bands in cont$(\mathscr{R})$, by Proposition VIII.4.1.4, resp. VIII.4.1.6.

1.9. Proposition. *Let $(H, \leqq)$ be a conditionally complete vector lattice and $\varnothing \neq M \subseteq H$. Then the intersection of all bands in H containing M is a band again: the smallest band containing M. It is also called the* **band generated by** M.

The proof is obvious and left to the reader.

1.10. Remark. Let H be a conditionally complete vector lattice and $\varnothing \neq M \subseteq H$. Then $(M^{\perp})^{\perp}$ is a band containing M. Theorem 1.12.2 will show that it is actually the band generated by M.

1.11. Proposition. *Let $(H, \leqq)$ be a conditionally complete vector lattice. Let $\varnothing \neq S \subseteq H_+$ and assume:*

(1) $m, m' \in S \quad \Rightarrow \quad m + m' \in S$.

(2) $m \in H, m' \in S, 0 \leqq m \leqq m' \quad \Rightarrow \quad m \in S$.

Put $\overline{S} = \{\sup M \mid \varnothing \neq M \subseteq S, M$ has a majorant in $H\}$. For every $0 \leqq m \in H$, put $m_0 = \sup\{m' \mid S \ni m' \leqq m\}$, $m_1 = m - m_0$. Then:

1.11.1. $m = m_0 + m_1$.

1.11.2. $0 \leqq m_0 \in \overline{S}$.

1.11.3. $0 \leqq m_1 \in \overline{S}^{\perp}$.

Proof. m_0 is well defined since H is conditionally complete. 1 is obvious, and 2 is a consequence of the definition of S.

3. It suffices to show $\{m_1\}^{\perp} \supseteq \overline{S}$. Since $\{m_1\}^{\perp}$ is a band (Example 1.8.1), it suffices to show $\{m_1\}^{\perp} \supseteq S$, i.e., $m' \in S \Rightarrow m' \wedge m_1 = 0$. Now in fact $m' \in S \Rightarrow$ (by (2))

$$S \ni m' \wedge m_1 \leqq m - m_0 \quad \Rightarrow \quad 0 \leqq m_0 + (m' \wedge m_1) \leqq m.$$

If $S \ni m'' \leqq m$, then $0 \leqq m'' \leqq m_0$, hence

$$0 \leqq m'' + (m' \wedge m_1) \leqq m \quad \Rightarrow \quad 0 \leqq m'' + (m' \wedge m_1) \leqq m_0$$

by the very definition of m_0 and the fact that $m'' + (m' \wedge m_1) \in S$ by (1).

Varying m'' and passing to suprema, $0 \leqq m_0 + (m' \wedge m_1) \leqq m_0$, and hence $m' \wedge m_1 = 0$ follows.

1.12. Theorem (F. Riesz' theorem on band decompositions). *Let H be a conditionally complete vector lattice and $\varnothing \neq S \subseteq H$ arbitrary. Then:*

1.12.1. $H_1 = S^{\perp}$ *is a band.*

1.12.2. $H_0 = H_1^{\perp}$ $(= (S^{\perp})^{\perp})$ *is a band, the band generated by S.*

1.12.3. $H_0 + H_1 = H$.

1.12.4. $H_0 \cap H_1 = 0$.

1.12.5. *Every $m \in H$ has exactly one decomposition $m = m^0 + m^1$, $m^0 \in H_0$, $m^1 \in H_1$. Here $m \geqq 0 \Rightarrow m^0 \geqq 0$, $m^1 \geqq 0$.*

1.12.6. *The decomposition 3 is called the* **(Riesz') band decomposition** *of H, and the decomposition 5 is called the (Riesz') band decomposition of m, for S resp. H_0 resp. H_1.*

Proof. 1 is obvious (Example 1.8.1). Let H_0' denote the band generated by S. Clearly $S \subseteq H_0' \subseteq H_1^{\perp} = H_0$. 4 follows because

$$m \in H_0 \cap H_1 \quad \Rightarrow \quad m \perp m \quad \Rightarrow \quad |m| = |m| \wedge |m| = 0 \quad \Rightarrow \quad m = 0.$$

Let now $m \in H$ be arbitrary. Apply Proposition 1.11, with $(H_0')_+$ in place of S, to m_+ and m_- and observe that $((H_0')_+)^{\perp} \subseteq S^{\perp} = H_1$. We get a decomposition $m = m_0' + m_1$, $m_0' \in H_0'$, $m_1 \in H$, with $m_0', m_1 \geqq 0$ in case $m \geqq 0$. This implies 3 and 5 is now also proved. We still have to accomplish the proof of 2 by showing $H_0 = H_1^{\perp}$. Choose in particular $m \in H_0$. Then $m_1 = m - m_0' \perp H_1 \Rightarrow m_1 = 0 \Rightarrow m = m_0'$. This does it.

The procedure of band generation is described in more detail by

1.13. Theorem. *Let $(H, \leqq)$ be a conditionally complete vector lattice and $\varnothing \neq S \subseteq H$. Let H_0 be the band generated by S. Put*

$$S_{0+} = \bigcup_{n=1}^{\infty} \bigcup_{m_1, \ldots, m_n \in S} \{m \mid 0 \leqq m \leqq |m_1| + \cdots + |m_n|\}$$

and

$$H_{0+} = \{\sup M \mid \varnothing \neq M \subseteq S_{0+}, M \text{ has a majorant in } H\}.$$

Then $H_{0+} = (H_0)_+$, and in particular $H_0 = H_{0+} - H_{0+}$.

Proof. It is obvious that $S^{\perp} = (S_{0+})^{\perp}$ and $H_{0+} \subseteq H_0$. Apply Proposition 1.11 with S_{0+} in place of S. It follows that every $0 \leqq m \in H$ has a decomposition $m = m_0 + m_1$ with $m_0 \in H_{0+}$, $0 \leqq m_1 \in H_1 = S^{\perp}$.

Choose $0 \leqq m \in H_0$ in particular. Then

$$m_1 = m - m_0 \in H_0 \cap H_1 \quad \Rightarrow \quad m_1 = 0,$$

thus $m = m_0 \in H_{0+}$. This proves $(H_0)_+ = H_{0+}$. The rest is obvious.

1.14. Theorem. *Let $(H, \leqq)$ be a conditionally complete vector lattice and $0 \leqq m_0 \in H$. Let H_0 denote the band generated by $\{m_0\}$. Then for any $0 \leqq m \in H$, we have*

$$m \in H_0 \quad \Leftrightarrow \quad m = \sup_n \{(nm_0) \wedge m \,|\, n = 1, 2, \ldots\}.$$

Riesz' band decomposition for $\{m_0\}$ are also called band decompositions for m_0. In every band decomposition $H_+ \ni m = m^0 + m^1$ for m_0, we have

$$m^0 = \sup_n [(nm_0) \wedge m].$$

The proof is easy by specialisation from Theorem 1.13 (exercise).

2. SIGNED MEASURES

In this section we apply the general theory of Section 1 to an elementary domain and its order dual.

2.1. Definition. *Let $\Omega \neq \varnothing$ and $\mathscr{E} \subseteq \mathbb{R}^\Omega$ be an elementary domain (Definition* I.7.2).

2.1.1. *An order bounded linear form m on $\mathscr{E}$ is also called a* **signed measure** *on an* **(electric) charge distribution** *on $\mathscr{E}$, the triple $(\Omega, \mathscr{E}, m)$ is then also called a* **signed measure space**, *and properties of m are (sometimes loosely) attributed to $(\Omega, \mathscr{E}, m)$. The set $\mathscr{E}^{\text{ord}}$ of all signed measures on $\mathscr{E}$ is also denoted by* meas$(\mathscr{E})$.

2.1.2. *A signed measure m on $\mathscr{E}$ is called* **σ-continuous** *or a* **signed σ-measure** *if $\mathscr{E} \ni f_1 \geqq f_2 \geqq \cdots \searrow 0$* (pointwise) $\Rightarrow \lim_n m(f_n) = 0$. *$(\Omega, \mathscr{E}, m)$ is then also called a* **signed σ-measure space**. *The set of all signed σ-measures on $\mathscr{E}$ is denoted by* meas$^\sigma(\mathscr{E})$.

2.1.3. *A signed measure m on $\mathscr{E}$ is called* **τ-continuous** *or a* **signed τ-measure** *if $\varnothing \neq \mathscr{F} \subseteq \mathscr{E}$, $\mathscr{F}$ decreasingly filtered,*

$$\inf_{f \in \mathscr{F}} f(\omega) = 0 \ (\omega \in \Omega) \quad \Rightarrow \quad \lim_{f \in \mathscr{F}} m(f) = 0,$$

i.e., for every $\varepsilon > 0$ *there is an* $f_0 \in \mathscr{F}$ *such that* $|m(f)| < \varepsilon$ $(\mathscr{F} \ni f \leqq f_0)$. $(\Omega, \mathscr{E}, m)$ *is then also called a* **signed τ-measure space**. *The set of all signed τ-measures on* $\mathscr{E}$ *is denoted by* $\text{meas}^\tau(\mathscr{E})$.

2.1.4. *Let* $m \in \text{meas}(\mathscr{E})$ *and* $M \subseteq \Omega$. *Then we define* $\|m\|(M) = \sup\{|m(f)| \mid f \in \mathscr{E}, |f| \leqq 1_M\}$ *and call this the* **total variation** *of m on M.* $\|m\|(\Omega)$ *is called the* **total variation** *of m and is also denoted by* $\|m\|$. *m is called* **bounded** *if* $\|m\| < \infty$.

2.2. Examples. Let $\Omega \neq \varnothing$ and $\mathscr{E} \subseteq \mathbb{R}^\Omega$ be an elementary domain.

2.2.1. Every positive measure, i.e., positive linear form on $\mathscr{E}$, is also a signed measure. The above definitions of σ- and τ-continuity contain the Definitions I.8.1.3 and I.8.1.5 as special cases.

2.2.2. Let $\omega_1, \ldots, \omega_n \in \Omega$, $\alpha_1, \ldots, \alpha_n \in \mathbb{R}$. Then

$$m(f) = \alpha_1 f(\omega_1) + \cdots + \alpha_n f(\omega_n) \quad (f \in \mathscr{E})$$

defines a signed τ-measure on E ("*electric charge* α_k *at point* ω_k").

2.3. Theorem. *Let* $\Omega \neq \varnothing$ *and* $\mathscr{E} \subseteq \mathbb{R}^\Omega$ *be an elementary domain.*

2.3.1. $\text{meas}(\mathscr{E})$ *is a conditionally complete vector lattice (in the dual partial ordering of the natural partial ordering of* $\mathscr{E}$: $m \leqq m' \Leftrightarrow [m(f) \leqq m'(f)$ $(0 \leqq f \in \mathscr{E})]$ *for any* $m, m' \in \text{meas}(\mathscr{E})$) *whose positive cone* $\text{meas}(\mathscr{E})_+$ *is exactly the set of all positive measures (Definition* I.8.1.2).

2.3.2. *The finite vector lattice operations in* $\text{meas}(\mathscr{E})$ *can be carried out as follows: for* $0 \leqq f \in \mathscr{E}$ *and* $m, m' \in \text{meas}(\mathscr{E})$, *we have*

2.3.2.1.

$$\begin{aligned}(m \vee m')(f) &= \sup\left\{\sum_{k=1}^{n} \max\{m(f_k), m'(f_k)\} \,\middle|\, n \geqq 1, 0 \leqq f_1, \ldots, f_n \in \mathscr{E}, f_1 + \cdots + f_n = f\right\} \\ &= \sup\{m(g) + m'(h) \mid 0 \leqq g, h \in \mathscr{E}, g + h = f\}.\end{aligned}$$

2.3.2.2.

$$\begin{aligned}(m \wedge m')(f) &= \inf\left\{\sum_{k=1}^{n} \inf\{m(f_k, m'(f_k)\} \,\middle|\, n \geqq 1, 0 \leqq f_1, \ldots, f_n \in \mathscr{E}, f_1 + \cdots + f_n = f\right\} \\ &= \inf\{m(g) + m'(h) \mid 0 \leqq g, h \in \mathscr{E}, g + h = f\}.\end{aligned}$$

2.3.2.3. $m_+(f) = \sup\{m(g) \mid 0 \leqq g \in \mathscr{E}, g \leqq f\}$

2.3.3. *If* $\emptyset \neq M \subseteq \text{meas}(\mathscr{E})$ *is increasingly filtered and* $\sup_{m \in M} m(f) < \infty$ $(0 \leqq f \in \mathscr{E})$, *then* $\sup M$ *exists in* $\text{meas}(\mathscr{E})$, *and*

$$(\sup M)(f) = \sup_{m \in M} m(f) \qquad (0 \leqq f \in \mathscr{E})$$

holds.

Proof. Apply theorem 1.5.

2.4. Theorem. *Let* $\Omega \neq \emptyset$ *and* $\mathscr{E} \subseteq \mathbb{R}^{\Omega}$ *be an elementary domain. Then:*

2.4.1. $\text{meas}^{\sigma}(\mathscr{E})$ *is a (Riesz') band in* $\text{meas}(\mathscr{E})$, *i.e.,*

2.4.1.1. $\text{meas}^{\sigma}(\mathscr{E})$ *is a vector sublattice of* $\text{meas}(\mathscr{E})$.

2.4.1.2. $m \in \text{meas}\,(\mathscr{E})$, $m' \in \text{meas}(\mathscr{E})$, $|m'| \leqq |m| \Rightarrow m' \in \text{meas}^{\sigma}(\mathscr{E})$.

2.4.1.3. *If a set* $\emptyset \neq M \subseteq \text{meas}^{\sigma}(\mathscr{E})$ *has a majorant in* $\text{meas}(\mathscr{E})$, *then* $\sup M \in \text{meas}^{\sigma}(\mathscr{E})$.

2.4.2. $\text{meas}^{\tau}(\mathscr{E})$ *is a (Riesz') band in* $\text{meas}(\mathscr{E})$, *i.e.,*

2.4.2.1. $\text{meas}^{\tau}(\mathscr{E})$ *is a vector sublattice of* $\text{meas}(\mathscr{E})$.

2.4.2.2. $m \in \text{meas}^{\tau}(\mathscr{E})$, $m' \in \text{meas}(\mathscr{E})$, $|m'| \leqq |m| \Rightarrow m' \in \text{meas}^{\tau}(\mathscr{E})$.

2.4.2.3. *If a set* $\emptyset \neq M \subseteq \text{meas}^{\tau}(\mathscr{E})$ *has a majorant in* $\text{meas}(\mathscr{E})$, *then* $\sup M \in \text{meas}^{\tau}(\mathscr{E})$.

2.4.3. $\text{meas}(\mathscr{E}) \supseteq \text{meas}^{\sigma}(\mathscr{E}) \supseteq \text{meas}^{\tau}(\mathscr{E})$.

Proof. 2.1. Let $m \in \tau\text{-meas}(\mathscr{E})$ and assume that m_+, which is a positive measure at any rate, is not τ-continuous. Then there is a decreasingly filtered $\emptyset \neq \mathscr{F} \subseteq \mathscr{E}_+$ with pointwise infimum 0, and some $\varepsilon > 0$ such that $f \in \mathscr{F} \Rightarrow m_+(f) > \varepsilon$. By the construction of m_+ given in Theorem 2.3.2.3, we may find, for every $f \in \mathscr{F}$, some $h_f \in \mathscr{E}$ such that $0 \leqq h_f \leqq f$, $m(h_f) > \varepsilon$. For $f, g \in \mathscr{F}$, $g \leqq f$, let $h_{fg} = h_f \vee g$. Fix some $f \in \mathscr{F}$. Then

$$\mathscr{F}_f = \{h_{fg} - h_f | 0 \leqq g \leqq f\}$$

is $\neq \emptyset$, $\subseteq \mathscr{E}_+$, decreasingly filtered, and has pointwise infimum 0. Since m is τ-continuous, we get $\lim_{h \in \mathscr{F}_f} m(h) = 0$. Hence for every $f \in \mathscr{F}$ there is a $g \in \mathscr{F}$ with $g \leqq f$ such that $m(h_{fg}) > \varepsilon$. On the other hand, it is not hard to see that $\mathscr{F}' = \{h_{fg} | f \in \mathscr{F},\ 0 \leqq g \in \mathscr{F},\ g \leqq f\}$ is decreasingly filtered: For any $f, g, f', g' \in \mathscr{F}$, $g \leqq f$, $g' \leqq f'$ choose some $\mathscr{F} \ni e \leqq f, g, f', g'$; now $h_{ee} = h_e \vee e \leqq g \leqq h_f \vee g = h_{fg}$, and similarly $h_{ee} \leqq h_{f'g'}$. Clearly $\inf_{f \in \mathscr{F}'} f(\omega) = 0 \quad (\omega \in \Omega)$ and the τ-continuity of m demands $\lim_{f \in \mathscr{F}'} m(f) = 0$, a contradiction. Thus $m_+ \in \text{meas}^{\tau}(\mathscr{E})$.

2.2. Let $\emptyset \neq \mathscr{F} \subseteq \mathscr{E}_+$ be decreasingly filtered with pointwise infimum 0. Since we know already that $\text{meas}^{\tau}(\mathscr{E})$ is a vector lattice, we obtain

$\lim_{f \in \mathscr{F}} |m|(f) = 0$ and consequently $\lim_{f \in \mathscr{F}} |m'|(f) = 0$. But

$$|m'(f)| \leqq |m'|(f)$$

(apply Theorem 2.3.2.1 to m' and $-m'$), hence $\lim_{f \in \mathscr{F}} m'(f) = 0$ follows.

2.3. Well-known devices allow us to assume $M \subseteq \text{meas}^\tau(\mathscr{E})_+$, M increasingly filtered. Put $\bar{m} = \sup M \in \text{meas}(\mathscr{E})$. By Theorem 2.3.3 we have $\bar{m}(f) = \sup_{m \in M} m(f)$ $(0 \leqq f \in \mathscr{E})$. Let now $\varnothing \neq \mathscr{F} \subseteq \mathscr{E}_+$ be decreasingly filtered with pointwise infimum 0. For any $m \in M$ and any $f \in \mathscr{F}$, we get now

$$\begin{aligned}(\bar{m} - m)(f) &\geqq \lim_{g \in \mathscr{F}} (\bar{m} - m)(g) \\ &= \lim_{g \in \mathscr{F}} \bar{m}(g) - \lim_{g \in \mathscr{F}} m(g) = \lim_{g \in \mathscr{F}} \bar{m}(g) \geqq 0.\end{aligned}$$

Varying $m \in M$, we get $\lim_{g \in \mathscr{F}} \bar{m}(g) = 0$, and thus $\bar{m} \in \text{meas}^\tau(\mathscr{E})$ follows.

The proofs of 1.1–3 are still easier and left as an exercise to the reader. 3 is obvious.

2.5. Remark. Theorem 1.12 on Riesz' band decompositions can now be applied twice and leads to a decomposition $\text{meas}(\mathscr{E}) = M + M^\sigma + \text{meas}^\tau(\mathscr{E})$, where $M \cap M^\sigma = M \cap \text{meas}^\tau(\mathscr{E}) = M^\sigma \cap \text{meas}^\tau(\mathscr{E}) = \{0\}$, $M^\sigma + \text{meas}^\tau(\mathscr{E}) = \text{meas}^\sigma(\mathscr{E})$, M, M^σ (and, of course, $\text{meas}^\tau(\mathscr{E})$) are bands, namely, $M = \text{meas}^\sigma(\mathscr{E})^\perp$, and M^σ is the orthogonal complement of $\text{meas}^\tau(\mathscr{E})$ in the conditionally complete vector lattice $\text{meas}^\sigma(\mathscr{E})$. One feels tempted to call the $m \in M$ *purely linear measures*, and the $m \in M^\sigma$ *purely σ-continuous measures*. This is analogous to the situation investigated in Chapter VII, Section 2 for general set functions (see also Exercise VIII.1.5.2.5).

2.6. Theorem. *Let $\Omega \neq \varnothing$ and $\mathscr{E} \subseteq \mathbb{R}^\Omega$ be an elementary domain.*

2.6.1. *Let* $\text{meas}^b(\mathscr{E}) = \{m \mid m \in \text{meas}(\mathscr{E}), \|m\| < \infty\}$. *Then* $(\text{meas}^b(\mathscr{E}), \|\cdot\|)$ *is an L-space. Norm convergence in* $\text{meas}^b(\mathscr{E})$ *amounts to uniform convergence on* $\{f \mid f \in \mathscr{E}, f \leqq 1\}$.

2.6.2. *For every $M \subseteq \Omega$, a seminorm $\|\cdot\|_M$ is defined on* $\text{meas}(\mathscr{E})$ *by* $\|m\|_M = \|m\|(M)$ $(m \in \text{meas}(\mathscr{E}))$.

2.6.3. *Assume $1 \in \mathscr{E}$. Then* $\text{meas}^b(\mathscr{E}) = \text{meas}(\mathscr{E})$ *and* $(\text{meas}(\mathscr{E}), \|\cdot\|)$ *is an L-space.*

Proof. 1. It is an easy exercise to prove

$$m, m' \in \text{meas}^b(\mathscr{E}) \quad \Rightarrow \quad \begin{aligned}&\|m + m'\| \leqq \|m\| + \|m'\| < \infty, \\ &\|\alpha m\| = |\alpha| \cdot \|m\| < \infty \ (\alpha \in \mathbb{R}), \\ &\|(|m|)\| = \|m\| < \infty, \\ &\|m_+\| \leqq \|(|m|)\| = \|m\| < \infty,\end{aligned}$$

hence $m + m'$, αm, $|m|$, $m_+ \in \text{meas}^{\text{b}}(\mathscr{E})$. Clearly $\|m\| = 0 \Rightarrow m = 0$. It follows that $\text{meas}^{\text{b}}(\mathscr{E})$ is a vector sublattice of $\text{meas}(\mathscr{E})$, and $(\text{meas}^{\text{b}}(\mathscr{E}), \|\cdot\|)$ is a normed vector lattice. If $0 \leqq m$, $m' \in \text{meas}(\mathscr{E})$, then $f, f' \in \mathscr{E}$, $0 \leqq f, f' \leqq 1$ $\Rightarrow f \vee f' \in \mathscr{E}$, $0 \leqq f \vee f' \leqq 1$, $m(f) + m'(f') \leqq m(f \vee f') + m'(f \vee f')$. Thus we get

$$\begin{aligned} \|m\| + \|m'\| &= \sup\{m(f) + m'(f') \mid f, f' \in \mathscr{E}, 0 \leqq f, f' \leqq 1\} \\ &\leqq \sup\{m(g) + m'(g) \mid g \in \mathscr{E}, 0 \leqq g \leqq 1\} \\ &= \|m + m'\| \leqq \|m\| + \|m'\|, \end{aligned}$$

hence $(\text{meas}^{\text{b}}(\mathscr{E}), \|\cdot\|)$ turns out to be an L-space, provided we can prove its norm completeness. But it is obvious from the very definition of $\|\cdot\|$ that $\|\cdot\|$ is the norm of uniform convergence on $M_1 = \{f \mid f \in \mathscr{E}, |f| \leqq 1\}$. It is now an easy exercise to prove that every fundamental sequence for $\|\cdot\|$ in $\text{meas}^{\text{b}}(\mathscr{E})$ is, after restriction to M_1, uniformly convergent to some $m_1 \in \mathbb{R}^{M_1}$, that m_1 is the restriction, to M_1, of a unique linear form m on $\mathscr{E}$, that $m \in \text{meas}^{\text{b}}(\mathscr{E})$ and the given fundamental sequence is $\|\cdot\|$ convergent to m. The details are left as an exercise to the reader. Thus $(\text{meas}^{\text{b}}(\mathscr{E}), \|\cdot\|)$ is in fact an L-space.

2 is an easy exercise which we leave to the reader.

3 follows from 1.

2.7. Exercises. Let $\Omega \neq \varnothing$ and $\mathscr{E} \subseteq \mathbb{R}^{\Omega}$ be an elementary domain.

2.7.1. Put

$$\text{meas}^{\sigma,\text{b}}(\mathscr{E}) = \text{meas}^{\sigma}(\mathscr{E}) \cap \text{meas}^{\text{b}}(\mathscr{E}),\ \text{meas}^{\sigma,\text{b}}(\mathscr{E}) = \text{meas}^{\tau}(\mathscr{E}) \cap \text{meas}^{\text{b}}(\mathscr{E}).$$

Prove:

2.7.1.1. $\text{meas}^{\sigma,\text{b}}(\mathscr{E}), \|\cdot\|)$ is an L-space.

2.7.1.2. $\text{meas}^{\tau,\text{b}}(\mathscr{E}), \|\cdot\|)$ is an L-space.

2.7.2. Assume $1 \in \mathscr{E}$. Prove:

2.7.2.1. $\text{meas}^{\sigma}(\mathscr{E})$ is an L-space.

2.7.2.2. $\text{meas}^{\tau}(\mathscr{E})$ is an L-space.

2.7.3. Assume in particular $\Omega = \mathbb{R}$, $\mathscr{E} = \mathscr{C}^{00}(\mathbb{R}, \mathbb{R})$, $m(f) = f(1)$ $(f \in \mathscr{E})$. Put $M = [0, 1]$. Prove $\|m\|(M) = \|m\|(\Omega \backslash M) = 0$.

2.7.4. Let $m \in \text{meas}^{\sigma}(\mathscr{E})$. Prove

$$\|m\|(\Omega) = \sup\left\{\int f\, dm_+ - \int f\, dm_- \,\Big|\, f \in \mathscr{L}_{\|m\|}, |f| \leqq 1\right\}.$$

(*Hint*: Use theorem III.5.10.)

3. ABSOLUTE CONTINUITY OF MEASURES

In this section we define an analogue to Definition VIII.3.1 of *absolute continuity* of contents, now for *measures*. Our definition will be purely in vector lattice terms. We shall obtain a *Lebesgue decomposition* as a special Riesz' band decomposition.

3.1. Definition. *Let* $\Omega \neq \emptyset$ *and* $\mathscr{E} \subseteq \mathbb{R}^\Omega$ *be an elementary domain,* m_0, $m \in \text{meas}(\mathscr{E})$. *We say that* m *is* **absolutely continuous** *for, or with respect to,* m_0, *in symbols* $m \ll m_0$, *if* $\mathscr{E}_+ \ni f_1 \geqq f_2 \geqq \cdots$,

$$\lim_n |m_0|(f_n) = 0 \quad \Rightarrow \quad \lim_n m(f_n) = 0.$$

We define $\text{meas}^{\ll m_0}(\mathscr{E}) = \{m \mid m \in \text{meas}(\mathscr{E}),\ m \ll m_0\}$.

3.2. Remarks

3.2.1. We have defined absolute continuity of measures by means of sequences in $\mathscr{E}_+$. It is, however, nearly trivial to prove the following. $m \ll m_0$ iff for every decreasingly filtered $\emptyset \neq \mathscr{F} \subseteq \mathscr{E}_+$ with $\inf_{f \in \mathscr{F}} |m_0|(f) = 0$, $\lim_{f \in \mathscr{F}} m(f) = 0$ holds (exercise).

3.2.2. Notice that only $|m_0|$ really enters into the above definition. It is thus always allowed to assume $m_0 \geqq 0$ in this context.

3.3. Exercises. Let $(\Omega, \mathscr{E}, m_0)$ be a measure space.

3.3.1. Prove $m \in \text{meas}(\mathscr{E})$, $|m| \leqq |m_0| \Rightarrow m \ll m_0$.

3.3.2. Prove $m_0 \in \text{meas}^\sigma(\mathscr{E})$, $m \ll m_0 \Rightarrow m \in \text{meas}^\sigma(\mathscr{E})$.

3.3.3. Prove $m_0 \in \text{meas}^\tau(\mathscr{E})$, $m \ll m_0 \Rightarrow m \in \text{meas}^\tau(\mathscr{E})$. (*Hint:* Use Remark 3.2.1.)

3.3.4. Let in particular $\Omega = \mathbb{N} = \{1, 2, \ldots\}$, $\mathscr{E} = l^\infty$,

$$m_0(f) = \frac{1}{2} f(1) + \frac{1}{2^2} f(2) + \cdots \qquad (f \in l^\infty),$$

m a Banach limit on l^∞ (Example I.6.4). Prove that m is not absolutely continuous for m_0.

3.4. Theorem. *Let* $(\Omega, \mathscr{E}, m_0)$ *be a measure space. Then:*

3.4.1. $\text{meas}^{\ll m_0}(\mathscr{E})$ *is a Riesz' band in* $\text{meas}(\mathscr{E})$, *i.e.,*

3.4.1.1. $\text{meas}^{\ll m_0}(\mathscr{E})$ *is a vector sublattice of* $\text{meas}(\mathscr{E})$.

3.4.1.2. $m \leqq \text{meas}^{\ll m_0}(\mathscr{E})$, $m' \in \text{meas}(\mathscr{E})$, $|m'| \leqq |m| \Rightarrow m' \in \text{meas}^{\ll m_0}(\mathscr{E})$.

3.4.1.3. *If* $\emptyset \neq M \subseteq \text{meas}^{\ll m_0}(\mathscr{E})$ *and* M *has a majorant in* $\text{meas}(\mathscr{E})$, *then* $\sup M \in \text{meas}^{\ll m_0}(\mathscr{E})$.

3.4.2. $\text{meas}^{\ll m_0}(\mathscr{E}) \cap \text{meas}^{\sigma}(\mathscr{E})$ *is a Riesz' band in* $\text{meas}(\mathscr{E})$.

3.4.3. $\text{meas}^{\ll m_0}(\mathscr{E}) \cap \text{meas}^{\tau}(\mathscr{E})$ *is a Riesz' band in* $\text{meas}(\mathscr{E})$.

Proof. 1.1. It is obvious that $\text{meas}^{\ll m_0}(\mathscr{E})$ is a linear subspace of $\text{meas}(\mathscr{E})$ which is also stable under $m \to |m|$, hence a vector sublattice.

1.2 is Exercise 3.3.1.

1.3. By obvious devices, we may assume $M \subseteq \text{meas}^{\ll m_0}(\mathscr{E})_+$. Let $\bar{m} = \sup M$ and $\mathscr{E}_+ \ni f_1 \geqq f_2 \geqq \cdots$, $\lim_n m_0(f_n) = 0$. For every $m \in M$, we have

$$(\bar{m} - m)(f_1) \geqq (\bar{m} - m)(f_2) \geqq \cdots \geqq \lim_n (\bar{m} - m)(f_n)$$

$$= \lim_n \bar{m}(f_n) - \lim_n m(f_n) = \lim_n \bar{m}(f_n) \geqq 0.$$

Varying $m \in M$, we see that $\lim \bar{m}(f_n) = 0$.

2 follows from 1 and Theorem 2.4.1.

3 follows from 1 and Theorem 2.4.2.

3.5. Theorem. *Let $(\Omega, \mathscr{E}, m_0)$ be a measure space. Then $\text{meas}^{\ll m_0}(\mathscr{E})$ is the Riesz' band generated by m_0.*

Proof. Clearly $\text{meas}^{\ll m_0}(\mathscr{E})$ contains m_0. By Theorem 3.4.1 it is a band. Hence it contains the band generated by m_0. By Theorem 1.12 the latter is nothing but $(\{m_0\}^{\perp})^{\perp}$. In order to prove $\text{meas}^{\ll m_0}(\mathscr{E}) \subseteq (\{m_0\}^{\perp})^{\perp}$ the following lemma is sufficient.

3.6. Lemma. *Let $\Omega \neq \varnothing$, $\mathscr{E} \subseteq \mathbb{R}^{\Omega}$ be an elementary domain and $m_0, m_1, m' \in \text{meas}(\mathscr{E})$. Then $m_0 \perp m_1$, $m' \ll m_0 \Rightarrow m' \perp m_1$.*

Proof. Clearly, we may assume $m_0, m_1, m' \geqq 0$. Let $m' = m'_0 + m'_1$ by the Riesz' band decomposition of m' for the band $(\{m_0\}^{\perp})^{\perp}$. Then $m'_0, m'_1 \geqq 0$ and $m_0 \perp m'_1$, hence $m_0 \wedge m'_1 = 0$. On the other hand, $0 \leqq m'_1 \leqq m' \leqq m_0$ implies $m'_1 \ll m_0$. We have to prove $m'_1 = 0$; in fact this would imply $m' = m'_0 \perp m_1$. For our proof, we exploit the general representation $(m_0 \wedge m'_1)(f) = \inf\{m_0(g) + m'_1(h) | 0 \leqq g, \ h \in \mathscr{E}, \ g + h = f\}$ $(0 \leqq f \in \mathscr{E})$ (Theorem 2.3.2.2) a countable number of times.

Assume $m'_1 \neq 0$. Then there is some $0 \leqq f_1 \in \mathscr{E}$ with $m'_1(f_1) = 1$. By the above representation, applied to $f = f_1$, we find $0 \leqq g, \ h \in \mathscr{E}$ with $g + h = f_1$ and $m_0(g) < 1/2^2$, $m'_1(h) < 1/2^2$. Put $f_2 = g$, then $0 \leqq f_2 \leqq f_1$ and $m_0(f_2) < 1/2^2$, $m'_1(f_2) = m'_1(f_1 - h) = m'_1(f_1) - m'_1(h) > 1 - (1/2^2)$. Now apply the above representation to $f = f_2$. We get $f_3 \in \mathscr{E}$ with $0 \leqq f_3 \leqq f_2$, $m_0(f_3) < 1/2^3$,

$$m'_1(f_3) = m'_1(f_2 - h) > \left(1 - \frac{1}{2^2}\right) - \frac{1}{2^3} = 1 - \frac{1}{2^2} - \frac{1}{2^3}.$$

Continuing in this way, we get $\mathscr{E} \ni f_1 \geqq f_2 \geqq \cdots$ with $\lim_n m_0(f_n) = 0$ and

$$m'_1(f_n) \geqq 1 - \frac{1}{2^2} - \cdots - \frac{1}{2^n} \geqq \frac{1}{2} \qquad (n = 1, 2, \ldots),$$

which contradicts $m'_1 \ll m_0$. The lemma is proved.

3.7. Theorem. *Let $(\Omega, \mathscr{E}, m_0)$ be a measure space, $0 \leqq m \in \text{meas}(\mathscr{E})$. Then $m \ll m_0 \Leftrightarrow m = \sup\{(nm_0) \wedge m \mid n = 1, 2, \ldots\}$.*

Proof. Apply Theorem 1.14.

3.8. Theorem (Lebesgue decomposition of measures). *Let $(\Omega, \mathscr{E}, m_0)$ be a measure space. Then every $m \in \text{meas}(\mathscr{E})$ has a unique decomposition*

$$m = m^0 + m^1, \qquad m^0 \ll m_0, \quad m^1 \perp m_0.$$

It is called the **Lebesgue decomposition** *of m for m_0. Here*

$$m \geqq 0 \quad \Rightarrow \quad m^0, m^1 \geqq 0, m^0 = \sup_n [m \wedge (nm_0)].$$

Proof. Let $m = m^0 + m^1$ be the Riesz' band decomposition of m for the band $\text{meas}^{\ll m_0}(\mathscr{E})$, which is, by Theorem 3.5, nothing but the band generated by $\{m_0\}$. Clearly $m^0 \ll m_0$, $m^1 \perp m_0$. Let $m = m' + m''$ be another decomposition with $m' \ll m_0$, $m'' \perp m_0$. Then $m'' - m^1 = m^0 - m'$ is $\ll m_0$ and $\perp m_0$. Apply Lemma 3.6 with $m_1 = m' = m''_0 - m'_0$. We get $|m''_0 - m'_0| = |m''_0 - m'_0| \wedge |m''_0 - m'_0| = 0$ and thus $m''_0 - m'_0 = 0$, proving the uniqueness of Lebesgue decomposition. The rest is obvious.

3.9. Remarks

3.9.1. The preceding theory is more general than related theories in most textbooks, in that *no σ-continuity is assumed.* One should however be aware of the fact that the most important special case is $m_0 \in \text{meas}^\sigma(\mathscr{E})$, and that this implies all other involved measures to lie in $\text{meas}^\tau(\mathscr{E})$. In the same way, $m_0 \in \text{meas}^\tau(\mathscr{E})$ implies all other involved measures to lie in $\text{meas}^\tau(\mathscr{E})$. This can be seen from Exercises 3.3.2 and 3.3.3 plus a bit of additional reasoning, which is left as an exercise to the reader.

3.9.2. Theorems 3.5 and 3.7 are of fundamental importance. They show that *absolute continuity,* which we had defined with the help of the elementary domain $\mathscr{E}$ whose order dual $\mathscr{E}^{\text{ord}} = \text{meas}(\mathscr{E})$, *is an intrinsic relation* in $\text{meas}(\mathscr{E})$. Thus we could have presented the theory of absolute continuity as an intrinsic theory about $\text{meas}(\mathscr{E})$; i.e., we could have presented the theory of Riesz' bands before speaking of signed contents and signed measures at all, and let then follow the absolute continuity theory of the latter as a more special case of the abstract theory. But then one would have the disadvantage of seeing essential examples only very late; moreover, the usual definitions of absolute continuity would have to be presented anyhow, namely as important criteria. Thus the approach chosen in this book seems to be sufficiently justified.

4. SIGNED CONTENTS AND SIGNED MEASURES

In this section we prove that absolute continuity and orthogonality of signed contents are tantamount to the like-named relations between signed measure derived from them and that a similar result holds in the opposite direction. Once again contents and measure appear as merely different aspects to the same matter.

4.1. Theorem. *Let $\Omega \neq \emptyset$ and $\mathscr{R}$ be a set ring in Ω. Let*

$$\mathscr{E}(\mathscr{R}) = \{\alpha_1 1_{F_1} + \cdots + \alpha_n 1_{F_n} | n \geqq 1, \alpha_1, \ldots, \alpha_n \in \mathbb{R}, F_1, \ldots, F_n \in \mathscr{R}\}$$

be the elementary domain of all step functions for $\mathscr{R}$ (Example I.7.8). *Then:*

4.1.1. *For every additive real function m on $\mathscr{R}$, there is a unique linear form φm on $\mathscr{E}(\mathscr{R})$ such that*

$$\varphi m(1_E) = m(E) \qquad (E \in \mathscr{R}).$$

For every linear form L on $\mathscr{E}(\mathscr{R})$, there is a unique additive $m \in \mathbb{R}^{\mathscr{R}}$ such that $L = \varphi m$.

4.1.2. *An additive $m \in \mathbb{R}^{\mathscr{R}}$ is a signed σ-content iff $\varphi m \in \mathscr{E}(\mathscr{R})^{\text{alg}}$ is a signed measure.*

4.1.3. *The restriction of φ to* $\text{cont}(\mathscr{R})$ *is a vector lattice isomorphism of* $\text{cont}(\mathscr{R})$ *onto* $\text{meas}(\mathscr{E}(\mathscr{R}))$ *which sends* $\text{cont}^\sigma(\mathscr{R})$ *onto* $\text{meas}^\sigma(\mathscr{E}(\mathscr{R}))$.

4.1.4. *Let m_0, $m \in$* $\text{cont}(\mathscr{R})$. *Then $m \ll m_0 \Leftrightarrow \varphi m \ll \varphi m_0$.*

Proof. 1 is obvious (compare Example I.8.7).

2. Let $m' = \varphi m$ be a signed measure and $E \in \mathscr{R}$. Then

$$\begin{aligned}\mathscr{R} \ni F \subseteq E \quad \Rightarrow \quad |m(F)| &= |\varphi m(1_F)| = |m'(1_F)| \\ &\leqq m'_+(1_F) + m'_-(1_F) \leqq m'_+(1_E) + m'_-(1_E) < \infty,\end{aligned}$$

hence $\|m\|(E) < \infty$ (use Exercise VIII.1.5.2.8) and thus $m \in \text{cont}(\mathscr{R})$. Let $m \in \text{cont}(\mathscr{R})$ and $M \subseteq \mathscr{E}(\mathscr{R})$ be an order bounded set, say with minorant f and majorant g. Put $h = |f| \vee |g|$, $E = \text{supp}\, h = \{\omega | h(\omega) \neq 0\}$. Then $E \in \mathscr{R}$. Let $0 < c \in \mathbb{R}$ be such that $h \leqq c$. Now consider any $f \in M$ and let $f = \alpha_1 1_{F_1} + \cdots + \alpha_n 1_{F_n}$ be a representation of f in disjoint form (i.e., with

$F_j \cap F_k = \emptyset$ $(j \neq k)$, see Example I.7.8), with $\alpha_1, \ldots, \alpha_n \neq 0$. Then $F_1, \ldots, F_n \in E$, $|\alpha_1|, \ldots, |\alpha_n| \leqq c$, and we conclude

$$\begin{aligned} |\varphi m(f)| &= |\alpha_1 m(F_1) + \cdots + \alpha_n m(F_n)| \\ &\leqq c(|m(F_1)| + \cdots + |m(F_n)|) \leqq c\|m\|(M) < \infty. \end{aligned}$$

i.e., $\varphi m \in \mathscr{E}(\mathscr{R})^{\text{ord}}$.

3. φ: $\text{cont}(\mathscr{R}) \to \text{meas}(\mathscr{E}(\mathscr{R}))$ is a bijection, after 1 and 2. Moreover, we have $m \leqq m' \Leftrightarrow \varphi m \leqq \varphi m'$ for any $m, m' \in \text{cont}(\mathscr{R})$. Here $\Leftarrow$ is obvious, and $\Rightarrow$ is seen as follows: Let $0 \leqq f \in \mathscr{E}(\mathscr{R})$ and $f = \alpha_1 1_{F_1} + \cdots + \alpha_n 1_{F_n}$ be a representation of f in disjoint form, with $\alpha_1, \ldots, \alpha_n > 0$. Then $\alpha_1, \ldots, \alpha_n \geqq 0$, and hence

$$\varphi m(f) = \alpha_1 m(F_1) + \cdots + \alpha_n m(F_n) \leqq \alpha_1 m'(F_1) + \cdots + \alpha_n m'(F_n) = \varphi m'(f).$$

4. $\Leftarrow$ is trivial if we go back to the definitions. For a proof of $\Rightarrow$, we may assume $m_0 \geqq 0$, hence $\varphi m_0 \geqq 0$. Let now $\mathscr{E}(\mathscr{R})_+ \ni f_1 \geqq f_2 \geqq \cdots$, $\lim_n \varphi m_0(f_n) = 0$. Let $E \in \mathscr{R}$ and $0 < c \in \mathbb{R}$ be such that $f_1 \leqq c 1_E$, choose any $\varepsilon > 0$ and put $F_n = \{f_n > \varepsilon\}$. Then $f_n \leqq \varepsilon 1_E + c 1_{F_n}$. Clearly $\varepsilon m_0(f_n > \varepsilon) \leqq \varphi m_0(f_n)$, hence $m_0(F_n) \to 0$. We conclude $m(F_n) \to 0$. Now we get $|m|(F_n) \to 0$ (use Proposition VIII.3.4.1) and thus

$$\begin{aligned} |\varphi m(f_n)| &\leqq |\varphi m|(f_n) = (\varphi |m|)(f_n) \\ &\leqq (\varphi |m|)(\varepsilon 1_M + c 1_{F_n}) = \varepsilon |m|(E) + c|m|(F_n) \end{aligned}$$

and hence $\lim |\varphi m(f_n)| \leqq \varepsilon |m|(E)$. Since $\varepsilon > 0$ was arbitrary,

$$\lim |\varphi m(f_n)| = 0$$

follows, i.e., we have $\varphi m \ll \varphi m_0$.

4.2. Remark. Theorem 4.1 shows that absolute continuity of signed contents on a ring $\mathscr{R}$ is an intrinsic vector lattice relation within $\text{cont}(\mathscr{R})$ in the same fashion as it is in $\text{meas}(\mathscr{E})$ for an elementary domain $\mathscr{E}$. It follows that vector lattice isomorphisms do not affect absolute continuity relations.

4.3. Proposition. *Let $(\Omega, \mathscr{R}, m_0)$ be a content prespace. Then for any $0 \leqq m \in \text{cont}(\mathscr{R})$, we have $m \ll m_0$ iff $m = \sup\{(nm_0) \wedge m \mid n = 1, 2, \ldots\}$.*

Proof. Combine Theorem 3.7 with Theorems 4.1.3 and 4.1.4.

4.4. Proposition. *Let $\Omega \neq \emptyset$ and $\mathscr{R}$ be a set ring in Ω, $\mathscr{B}^{00} = \mathscr{B}^{00}(\mathscr{R})$. Let m_0, $m \in \text{cont}^\sigma(\mathscr{R})$ and ψm_0, ψm be their unique σ-additive extensions to $\mathscr{B}^{00}$ (Exercise VIII.1.5.3.1). Then $m \ll m_0 \Leftrightarrow \psi m \ll \psi m_0$.*

Proof. By Exercise VIII.1.5.3.3 ψ: $\text{cont}^\sigma(\mathscr{R}) \to \text{cont}^\sigma(\mathscr{B}^{00})$ is a vector lattice isomorphism. The proposition now follows from Proposition 4.3 (see also Remark 4.2).

The preceding results have dealt with contents and the measures derived from them. We now go in the opposite direction, starting from measures.

4.5. Theorem. *Let* $\Omega \neq \varnothing, \mathscr{E} \subseteq \mathbb{R}^{\Omega}$ *be an elementary domain, and* $\mathscr{B}^{00} = \mathscr{B}^{00}(\{\{f > 0\} \mid f \in \mathscr{E}\})$. *Then:*

4.5.1. *for every* $m \in \text{meas}^{\sigma}(\mathscr{E})$, *there is a unique* $\psi m \in \text{cont}^{\sigma}(\mathscr{B}^{00})$ *such that* $F \in \mathscr{B}^{00} \Rightarrow (\psi m)(F) = \int 1_F \, dm_+ - \int 1_F \, dm_-$. *The mapping*

$$\psi : \text{meas}^{\sigma}(\mathscr{E}) \to \text{cont}^{\sigma}(\mathscr{B}^{00})$$

thus defined is a vector lattice isomorphism. Moreover,

4.5.2. $\|\psi m\|(\Omega) = \|m\|(\Omega)$ $(m \in \text{meas}^{\sigma}(\mathscr{E}),\ M \subseteq \Omega)$.

Proof. 1. Let $0 \leqq m \in \text{meas}^{\sigma}(\mathscr{E})$. Exercise III.6.5.2.1 shows that ψ is well defined on $\text{meas}^{\sigma}(\mathscr{E})$ and one-to-one. It is clearly isotone. Moreover, it is clear from $\mathscr{E} \subseteq \mathscr{L}^1_{\sigma}(\Omega, \mathscr{B}^{00}, m')$ $(m' \in \text{cont}^{\sigma}(\mathscr{B}^{00}))$ that ψ sends $\text{meas}^{\sigma}(\mathscr{E})_+$ onto $\text{cont}^{\sigma}(\mathscr{B}^{00})_+$. The full definition of ψ and the proof of 4.5.1 is now easily achieved by decomposition of an arbitrary $m \in \text{meas}^{\sigma}(\mathscr{E})$ into m_+ and m_-.

2 follows from Exercises VIII.1.5.3.1, VIII.1.5.2.9, and 2.7.4.

4.6. Theorem (Radon–Nikodym). *Let* $(\Omega, \mathscr{E}, m_0)$ *be a* σ*-measure space,* $\text{meas}(\mathscr{E}) \ni m \ll m_0$, $\|m\| < \infty$. *Then there is an* $f \in \mathscr{L}^1_{m_0}$ *such that* $g \in \mathscr{E} \Rightarrow gf \in \mathscr{L}^1_{m_0}$, $\int gf \, dm_0 = m(g)$. *We call f an* $\mathbf{m_0}$**-density** *for m.*

Proof. The vector mapping $\psi : \text{meas}^{\sigma}(\mathscr{E}) \to \text{cont}^{\sigma}(\mathscr{B}^{00})$ constructed in Theorem 4.5.1 is a vector lattice isomorphism, hence preserves absolute continuity (Remark 4.2). By Theorem 4.5.2 it also preserves total variation, hence sends $\text{meas}^{\sigma,\,b}(\mathscr{E})$ onto $\text{cont}^{\sigma,\,b}(\mathscr{B}^{00})$. Apply Theorem VIII.5.2 (Radon–Nikodym) to ψm in order to find an $f \in \mathscr{L}^1_{m_0}$ such that $F \in \mathscr{B}^{00} \Rightarrow \int 1_F f \, d(\psi m_0) = (\psi m)(F)$. By linear combination we establish $\int gf \, dm_0 = \int gf \, dm$ for all $g \in \mathscr{E}(\mathscr{B}^{00})$, and by monotone approximation (see the proof of Theorem VIII.5.5 or VIII.6.4.7) for all $g \in \mathscr{E}$.

4.7. Exercises

4.7.1. Prove that the m_0-density f of Theorem 4.6 is uniquely mod m_0 determined by m.

4.7.2. Try to generalize Theorem 4.6 along the lines of Theorem VIII.8.8.

4.7.3. Let $(\Omega, \mathscr{R}, \tilde{m})$ be a content prespace. Let $m \in \operatorname{cont}(\mathscr{R})$ and $\tilde{m}'_0 = \varphi\tilde{m}_0$, $m' = \varphi m \in \operatorname{meas}(\mathscr{E}(\mathscr{R}))$ according to Theorem 4.1.1. Let $m' = m'_0 + m'_1$, $m'_0 \ll \tilde{m}'$, $m'_1 \perp \tilde{m}'$ be the Lebesgue decomposition of the measure m' for the measure $\tilde{m}'$. Let m_0, $m_1 \in \operatorname{cont}(\mathscr{R})$ be such that $\varphi m_0 = m'_0$, $\varphi m_1 = m'_1$. Prove that $m = m_0 + m_1$ holds and is the Lebesgue decomposition of the content m for the content $\tilde{m}$.

CHAPTER X

THE SPACES L^p

In this chapter we investigate, for a given σ-content space and any $1 \leqq p \leqq \infty$, the space L^p of all functions that are "integrable to the power p." The reader knows already the special cases $p = 1$, $p = 2$ from Chapter III; L^1 and L^2 were introduced there on the basis of a given σ- or τ-measure space; we prefer the σ-content space approach here since we shall have to apply the Radon–Nikodym theorem which was presented in Chapter VIII in the σ-content space setting. The reader should be aware that he may always, given a σ- or τ-measure space, deduce a σ-content space from it for which the integrable functions are the same (see Chapter III, Section 6). The theory presented for σ-content spaces in this chapter therefore carries over to σ- or τ-measure spaces most easily.

In Section 1 we introduce the spaces L^p $(1 \leqq p < \infty)$ and deduce their Banach lattice properties. For L^∞, the reader may consult exercise III.12.14. The so-called *Clarkson inequalities* are put into the exercises. They show that the Banach spaces L^p are uniformly convex for $1 < p < \infty$, and hence, by a general theorem, *reflexive*: They can be identified with the dual of their duals (see, e.g., DUNFORD and SCHWARTZ [1]). We shall neither prove nor use this result but, by purely σ-content theoretical means (mainly the Radon–Nikodym theorem), prove a much more precise result: If $1 < p$, $q < \infty$, $p^{-1} + q^{-1} = 1$, then *L^p and L^q are dual spaces of each other* in a natural fashion. For $p = 1$, $q = \infty$, the situation is a bit more complicated; we have to introduce the space loc L^∞ before we can prove that it is the dual of L^1, and we have to assume the existence of a *disjoint base* for this (observe that no such assumption comes in for the L^p case, $1 < p < \infty$). The dual of L^∞ is out of the scope of this book.

1. THE SPACES L_m^p $(1 \leqq p \leqq \infty)$

In this section we define, for a given σ-content space $(\Omega, \mathscr{B}^{00}, m)$ and any real $1 \leqq p < \infty$, the space $\mathscr{L}_m^p$ of all functions that are *m-integrable to the power p* and investigate its basic properties. For $p = 1$, $\mathscr{L}_m^1$ has the usual meaning. We mention, for $p = \infty$, the space $\mathscr{L}_m^\infty$ of all mod m bounded $\mathscr{B}^0(\mathscr{B}^{00})$-measurable functions, as defined in Exercise III.12.14.6. For any $1 \leqq p \leqq \infty$, we obtain a Banach space $(L_m^p, \|\cdot\|_p)$ from $\mathscr{L}_m^p$ by passage to equivalence classes mod m and introduction of an obvious norm $\|\cdot\|_p$. In the next section we shall investigate the duality properties of these Banach spaces.

We begin with a preparatory

1.1. Lemma. *Let* $\Omega \neq \varnothing$ *and* $\tilde{m}: \mathbb{R}_+^\Omega \to \overline{\mathbb{R}}_+$ *satisfy:*

1.1.1. $f \in \mathbb{R}_+^\Omega, 0 \leqq \alpha \in \mathbb{R} \Rightarrow \tilde{m}(\alpha f) = \alpha \tilde{m}(f)$.

1.1.2. $f, g \in \mathbb{R}_+^\Omega \Rightarrow \tilde{m}(f + g) \leqq \tilde{m}(f) + \tilde{m}(g)$.

1.1.3. $f, g \in \mathbb{R}_+^\Omega, f \leqq g \Rightarrow \tilde{m}(f) \leqq \tilde{m}(g)$.

1.1.4. *Assume that* $n > 0$ *and* $\varphi: \mathbb{R}_+^n \to \mathbb{R}$ *is continuous and satisfies:*

1.1.4.1. $\varphi(x_1, \ldots, x_n) > 0$ $(0 < x_1, \ldots, x_n \in \mathbb{R})$;

1.1.4.2. $\varphi(\alpha x_1, \ldots, \alpha x_n) = \alpha\varphi(x_1, \ldots, x_n)$ $(0 \leqq \alpha, x_1, \ldots, x_n \leqq \mathbb{R})$;

1.1.4.3. $K = \{(x_1, \ldots, x_n) | 0 \leqq x_1, \ldots, x_n \in \mathbb{R}, \varphi(x_1, \ldots, x_n) \geqq 1\}$ *is convex.*

Then:

1.1.4.4. *For any* $h_1, \ldots, h_n \in \mathbb{R}_+^\Omega$ *with* $\tilde{m}(h_1), \ldots, \tilde{m}(h_n) < \infty$, *we have*

$$\tilde{m}(\varphi(h_1, \ldots, h_n)) \leqq \varphi(\tilde{m}(h_1), \ldots, \tilde{m}(h_n)).$$

1.1.5. *Let* $0 < \alpha, \beta < 1$, $\alpha + \beta = 1$, $f, g \in \mathbb{R}_+^\Omega$, $\tilde{m}(f), \tilde{m}(g) < \infty$. *Then* **Hölder's inequality**

$$\tilde{m}(f^\alpha g^\beta) \leqq \tilde{m}(f)^\alpha \tilde{m}(g)^\beta$$

holds.

1.1.6. *Let* $1 \leqq p < \infty, f, g \in \mathbb{R}_+^\Omega$. *Then* **Minkowski's inequality**

$$\tilde{m}((f + g)^p)^{1/p} \leqq \tilde{m}(f^p)^{1/p} + \tilde{m}(g^p)^{1/p}$$

holds.

Proof. 4. By well-known elementary statements about convex sets in $\mathbb{R}^n$ we find that K is an intersection of a family of closed half-spaces in $\mathbb{R}^n$, each of which can be written in the form $\{(x_1, \ldots, x_n) | \beta \leqq \alpha_1 x_1 + \cdots + \alpha_n x_n\}$ with reals $\beta, \alpha_1, \ldots, \alpha_n$, not all $\alpha_k = 0$. By 4.1 and 4.2, K contains all $(x_1, \ldots, x_n)$ with sufficiently large positive components. It follows that for each of our half-spaces, none of the relations $\alpha_1 < 0, \ldots, \alpha_n < 0$ can happen, hence $\alpha_1, \ldots, \alpha_n \geqq 0$ and thus $\alpha_1 x_1 + \cdots + \alpha_n x_n \geqq 0$ on $\mathbb{R}^n_+$, in particular on K. Thus we may always assume $\beta \geqq 0$. The special half-spaces $\{x_1 \geqq 0\}, \ldots, \{x_n \geqq 0\}$ all contain K, but since $0 \notin K$, there is at least one half-space containing K, for which $\beta > 0$. If the $(n+1)$-tuple $(\beta, \alpha_1, \ldots, \alpha_n)$ runs through our family, the half-spaces in $\mathbb{R}^{n+1}$ defined by

$$\{(x_0, x_1, \ldots, x_n) \mid \beta x_0 \leqq \alpha_1 x_1 + \cdots + \alpha_n x_n\},$$

together with $\mathbb{R}^{n+1}_+$, form an intersection which is nothing but the closed convex cone

$$\hat{K} = \{(x_0, \ldots, x_n) \mid 0 \leqq x_0, \ldots, x_n \in \mathbb{R}, x_0 \leqq \varphi(x_1, \ldots, x_n)\}.$$

$\hat{K}$ contains $(\varphi(x_1, \ldots, x_n), x_1, \ldots, x_n)$ for any $0 \leqq x_1, \ldots, x_n \in \mathbb{R}$, hence $\beta\varphi(x_1, \ldots, x_n) \leqq \alpha_1 x_1 + \cdots + \alpha_n x_n$ for every $(\beta, \alpha_1, \ldots, \alpha_n)$ from our family. Thus we see $\beta\varphi(h_1, \ldots, h_n) \leqq \alpha_1 h_1 + \cdots + \alpha_n h_n$ $(h_1, \ldots, h_n \in \mathbb{R}^n_+)$, and upon application of $\tilde{m}$ we obtain $\beta\tilde{m}(\varphi(h_1, \ldots, h_n)) \leqq \alpha_1\tilde{m}(h_1) + \cdots + \alpha_n\tilde{m}(h_n)$; letting $(\beta, \alpha_1, \ldots, \alpha_n)$ vary through our family, we see that $(\tilde{m}(\varphi(h_1, \ldots, h_n)), \tilde{m}(h_1), \ldots, \tilde{m}(h_n)) \in \hat{K}$ and thus $\tilde{m}(\varphi(h_1, \ldots, h_n)) \leqq \varphi(\tilde{m}(h_1), \ldots, \tilde{m}(h_n))$ follows.

5. Choose $n = 2$, $\varphi(x_1, x_2) = x_1^\alpha x_2^\beta$. The boundary of

$$K = \{(x_1, x_2) \mid 0 \leqq x_1, x_2 \in \mathbb{R}, x_1^\alpha x_2^\beta \geqq 1\}$$

is $\{(x_1, x_2) \mid x_1 > 0, x_2 = x_1^{-\alpha/\beta}\}$. The function $\sigma(x) = x^{-\alpha/\beta}$ has, for $x > 0$, the derivatives (use $\alpha + \beta = 1$)

$$\sigma'(x) = -\frac{\alpha}{\beta} x^{-(\alpha/\beta)-1} = -\frac{\alpha}{\beta} x^{-1/\beta}, \qquad \sigma''(x) = \frac{\alpha}{\beta^2} x^{-(1/\beta)-1} > 0,$$

hence K is convex and 5 applies, yielding Hölder's inequality.

6. Only the case $\tilde{m}(f), \tilde{m}(g) < \infty$ is of interest. Choose $n = 2$, $\varphi(x_1, x_2) = (x_1^{1/p} + x_2^{1/p})^p$. The boundary of K is $\{(x_1, x_2) \mid x_1, x_2 \geqq 0, x_1^{1/p} + x_2^{1/p} = 1\}$. The function $\tau(x) = (1 - x^{1/p})^p$ has, for $0 < x \leqq 1$, the derivatives

$$\tau'(x) = -(1 - x^{1/p})^{p-1} x^{(1/p)-1},$$

$$\tau''(x) = \left(1 - \frac{1}{p}\right)(1 - x^{1/p})^{p-2} x^{(1/p)-2} \geqq 0,$$

hence K is convex and 4 applies, yielding Minkowski's inequality.

1.2. Definition. *Let $(\Omega, \mathscr{B}^{00}, m)$ be a σ-content space and $1 \leqq p < \infty$. A function $h \in \overline{\mathbb{R}}^\Omega$ is called* **m-integrable to the power p** *if $h \cdot |h|^{p-1} \in \mathscr{L}^1_m$. The set of all functions $h \in \overline{\mathbb{R}}^\Omega$ that are m-integrable to the power p is denoted by $\mathscr{L}^p_m$, $\mathscr{L}^p_m(\mathbb{R})$, $\mathscr{L}^p(\Omega, \mathscr{B}^{00}, m)$, $\mathscr{L}^p(\Omega, \mathscr{B}^{00}, m, \overline{\mathbb{R}})$, etc., depending on the specifications needed.*

1.3. Theorem. *Let $(\Omega, \mathscr{B}^{00}, m)$ be a σ-content space and $1 \leqq p < \infty$. Then:*

1.3.1. *$\mathscr{L}^p_m$ consists of full equivalence classes* mod m.

1.3.2. *Every equivalence class* mod m *in $\mathscr{L}^p_m$ contains a function from $\mathbb{R}^\Omega$, i.e., a finite-valued representative.*

1.3.3. *The set L^p_m $(= L^p_m(\mathbb{R}) = L^p_m(\Omega, \mathscr{B}^{00}) = \cdots)$ of all equivalence classes* mod m *in $\mathscr{L}^p_m$ is a vector lattice if we define the vector lattice operations in L^p_m via finite-valued representatives.*

By

$$\|h\|_p = \left(\int |h|^p \, dm\right)^{1/p} \qquad (h \in \mathscr{L}^p_m)$$

a nonnegative real-valued function $\|\cdot\|_p$ is defined on $\mathscr{L}^p_m$. It is constant on every equivalence class in $\mathscr{L}^p_m$ and may thus be considered as a function on the real vector space L^p_m. As such, it is a norm. This norm (and also $\|\cdot\|_p\colon \mathscr{L}^p_m \to \mathbb{R}$) is called the p-norm or L^p-norm.

1.3.4. *$(L^p_m, \|\cdot\|_p)$ is a conditionally complete Banach lattice.*

1.3.5. *The elementary domain $\mathscr{E}(\mathscr{B}^{00})$ is L^p-norm dense in $\mathscr{L}^p_m$.*

1.3.6. *Every L^p-norm fundamental sequence in $\mathscr{L}^p_m$ has an m-a.e. convergent subsequence and is $\|\cdot\|_p$-convergent to the limit of the latter.*

Proof. Observe that $g|g|^{p-1} = h|h|^{p-1}$ mod m iff $g = h$ mod m. Now 1 and 2 follow from the corresponding statements about $\mathscr{L}^1_m$, i.e., from Theorem III.5.1.1. Observe further

$$h_+|h_+|^{p-1} = (h|h|^{p-1})_+, \qquad h_-|h_-|^{p-1} = (h|h|^{p-1})_-,$$
$$|h|\cdot|h|^{p-1} = |(h|h|^{p-1}).$$

This shows that $h \in \mathscr{L}^p_m$ iff $h_+, h_- \in \mathscr{L}^p_m$ only if $|h| \in \mathscr{L}^p_m$ (use Theorem III.5.1.2). Next we apply Lemma 1.1.6 with $\tilde{m}(h) = \int h \, dm$ for $0 \leqq h \in \mathscr{L}^1_p$ and $\tilde{m}(h) = \infty$ for all other $h \in \mathbb{R}^\Omega_+$, which is apparently possible. We find

$$\tilde{m}((|g| + |h|)^p)^{1/p} \leqq \tilde{m}(|g|^p)^{1/p} + \tilde{m}(|h|^p)^{1/p} \qquad (g, h \in \mathbb{R}^\Omega).$$

Let now $g, h \in \mathscr{L}^p_m \cap \mathbb{R}^\Omega$; we want to prove $g + h \in \mathscr{L}^p_m$. By passage to equivalent mod m functions we may assume $g, h \in \text{mble}(\Omega, \mathscr{B}^0(\mathscr{B}^{00}), \mathbb{R})$ (Theorem III.9.3). Now $|(g+h)|g+h|^{p-1}| \leqq (|g| + |h|)^p \in \mathscr{L}^1_m$ shows $(g+h)|g+h|^{p-1} \in \mathscr{L}^1_m$ (Proposition III.9.5) and thus $g + h \in \mathscr{L}^p_m$. The proof that $\|\cdot\|_p$ defines a norm on L^p_m is now an easy exercise. In order to prove 4 we have first to show that L^p_m is complete for the norm $\|\cdot\|_p$.

An easy discussion of $x \to x|x|^{p-1}$ yields the inequalities

$$\begin{aligned}|x-y|^p &\leqq 2^p\left|x|x|^{p-1} - y|y|^{p-1}\right| \\ &\leqq 2^{p+1}p|x-y|(|x|+|y|)^{p-1} \qquad (x, y \in \mathbb{R}).\end{aligned}$$

This shows that $h \to |h|^{(1/p)-1}h$ sends L_m^1 onto L_m^p so that fundamental (resp. convergent) sequences in the two spaces exactly correspond to each other. The completeness of L_m^p obviously follows from this and the completeness of L_m^1. The $\|\cdot\|_p$-continuity of the vector lattice operations in L_m^p is an easy exercise (use, e.g., Minkowski's inequality). The proof that every order bounded subset of L_m^p has a supremum and an infimum in L_m^p is easily deduced from the corresponding facts about L_m^1 upon observing that $h \to h|h|^{p-1}$ constitutes an isotone bijection between L_m^p and L_m^1. The proof of 5 is achieved in the same way as the proof of Theorem III.11.2.3 (case $p = 2$); we have only to observe that for every $f' \in \mathscr{E}(\mathscr{B}^{00})$ there is an $f \in \mathscr{E}(\mathscr{B}^{00})$ such that $f|f|^{p-1} = f'$. The proof of 6 follows the lines of the proof of Theorem III.11.2.4 resp. III.5.8.4 (exercises).

1.4. Exercises

1.4.1. Let $1 \leqq p < \infty$. An elementary domain $\mathscr{E} \subseteq \mathbb{R}^\Omega$ is called *p*-stable if for every $f' \in \mathscr{E}$ there is an $f \in \mathscr{E}$ such that $f|f|^{p-1} = f'$. Check all elementary domains you know as to which of them are *p*-stable and which are not (practically all of them are *p*-stable).

1.4.2. Let $(\Omega, \mathscr{E}, m)$ be a σ-measure space, $1 \leqq p < \infty$ and $\mathscr{E}$ be *p*-stable. Let $\mathscr{B}^{00} = \mathscr{B}^{00}(\mathscr{E}) = \mathscr{B}^{00}(\{\{f > 1\} \mid f \in \mathscr{E}\})$ and $(\Omega, \mathscr{B}^{00}, m)$ be the σ-content space obtained by restricting the σ-content derived from $(\Omega, \mathscr{E}, m)$ (Theorem III.6.1) to $\mathscr{B}^{00}$. Prove that $\mathscr{E}$ is $\|\cdot\|_p$-dense in $\mathscr{L}_m^p(\Omega, \mathscr{E}, m)$, which we define to be $\mathscr{L}_m^p(\Omega, \mathscr{B}^{00}, m)$.

1.4.3. Let $(\Omega, \mathscr{B}^{00}, m)$ be a σ-content space and $1 \leqq p < \infty$.

1.4.3.1. Let $h_1, h_2, \ldots \in L_m^p$ satisfy $0 \leqq h_1 \leqq h_2 \leqq \cdots$ and $\lim_n \|h_n\|_p < \infty$. Prove the existence of a unique $h \in L_m^p$ such that $\lim_n \|h_n - h\|_p = 0$ (*monotone convergence theorem in L_m^p, upward version*).

1.4.3.2. Formulate and prove a downward analogue of the preceding statement.

1.4.3.3. Let $g, h_1, h_2, \ldots \in L_m^p$, $0 \leqq h_1, h_2, \ldots \leqq g$. Show that

$$h = \limsup_n h_n \in L_m^p \qquad \text{and} \qquad \|h\|_p \geqq \limsup_n \|h_n\|_p$$

(*Fatou's lemma in L_m^p, upward version*).

1.4.3.4. Formulate and prove a downward analogue of the preceding statement.

1.4.3.5. Formulate and prove *filter versions of the monotone convergence theorem in* L_m^p (see Theorem III.5.8.6).

1.4.3.6. Let $g, h_1, h_2, \ldots \in \mathscr{L}_m^p$, $h \in \overline{\mathbb{R}}_+^\Omega$ be such that $|h_1|, |h_2|, \ldots \leqq g$, $\lim_n h_n = h$ m-a.e. Prove that $h \in \mathscr{L}_m^p$, $\lim_n \|h_n - h\|_p = 0$ (*Lebesgue's dominated convergence theorem in* L_m^p). (*Hint*: Use Fatou's lemma in order to prove $\lim_{j,k} \|h_j - h_k\|_p = 0$.)

1.4.3.7. Prove that for every $h \in \text{mble}(\Omega, \mathscr{B}^0(\mathscr{B}^{00}), \overline{\mathbb{R}})$, $h \in \mathscr{L}_m^p \Leftrightarrow |h|^p \in \mathscr{L}_m^1$ and that every equivalence mod m class in $\mathscr{L}_m^p$ contains a function from $\text{mble}(\Omega, \mathscr{B}^0(\mathscr{B}^{00}), \mathbb{R})$.

1.4.3.8. Let $1 \leqq q < \infty$. Prove that for every $h \in \overline{\mathbb{R}}^\Omega$,

$$h \in \mathscr{L}_m^p \quad \Leftrightarrow \quad h|h|^{(p/q)-1} \in L_m^q \qquad \text{and} \qquad \|h\|_p^p = \|h|h|^{(p/q)-1}\|_q^q$$

in this case.

1.4.3.9. Assume in addition that $1 \in \mathscr{L}_m^1$ and $\int 1\, dm = 1$. Prove that

$$1 \leqq p < q < \infty \quad \Rightarrow \quad \mathscr{L}_m^p \supseteq \mathscr{L}_m^q \text{ and } \|h\|_p \leqq \|h\|_q \qquad (h \in \mathscr{L}_m^q).$$

(*Hint*: Use Jensen's inequality for $x^{p/q}$.)

1.4.3.10. Prove that the set $\mathscr{S}_m$ of all step functions contained in $\mathscr{L}_m^p$ coincides with the set of all step functions contained in $\mathscr{L}_m^1$, and that $\mathscr{S}_m$ is $\|\cdot\|_1$-dense in $\mathscr{L}_m^1$ and $\|\cdot\|_p$-dense in $\mathscr{L}_m^p$. Show that even for every $h \in \mathscr{L}_m^p$ there is a sequence $0 \leqq h_1 \leqq h_2 \leqq \cdots$ in $\mathscr{S}_m$ converging to h m-a.e. and in the L_m^p-norm.

1.4.3.11. Replace $\mathscr{S}_m$ by $\mathscr{S}_m \cap \text{mble}(\Omega, \mathscr{B}^0(\mathscr{B}^{00}), \mathbb{R})$ and prove the same density statement.

1.4.3.12. Prove that $\{h \mid h \in \mathscr{L}_m^p,\ \|h\|_p \leqq 1\}$ is the $\|\cdot\|_p$-closure of $\{f \mid f \in \mathscr{S}_m,\ \|f\|_p \leqq 1\}$ as well as of $\{f \mid f \in \mathscr{E}(\mathscr{B}^{00}),\ \|f\|_p \leqq 1\}$.

1.4.4. Let $2 \leqq p < \infty$. Prove

$$\left(\frac{1+x}{2}\right)^p + \left(\frac{1-x}{2}\right)^p \leqq \frac{1}{2}(1 + x^p) \qquad (0 \leqq x \leqq 1).$$

(*Hint*: Put $F(x) =$ left member minus right member and prove that $F(x) \leqq 0$ $(0 \leqq x \leqq 1)$; $F(0) \leqq 0$ is easy; for $0 < x \leqq 1$ consider $G(x) = 2^p x^{-p} F(x)$ and show $G(1) = 0$, $G'(x) \leqq 0$ $(0 < x < 1.)$

1.4.5. Let $1 < p \leqq 2$. Put $p' = p/(p-1)$ and prove

$$(1 + x)^{p'} + (1 - x)^{p'} \leqq 2(1 + x^p)^{1/(p-1)} \qquad (0 \leqq x \leqq 1).$$

(*Hint*: Only the case $p < 2$ is nontrivial; for $x = 0$ and $x = 1$ equality holds; replace x by $(1 - y)/(1 + y)$ for $0 < y < 1$ and prove

$$(1 + y^{p'})^{p-1} \leqq \tfrac{1}{2}[(1 + y)^p + (1 - y)^p]$$

by expanding power series.)

1.4.6. Let $(\Omega, \mathscr{B}^{00}, m)$ be a σ-content space and prove the CLARKSON [1] *inequalities*

1.4.6.1.

$$\left\| \frac{f + g}{2} \right\|_p^p + \left\| \frac{f - g}{2} \right\|_p^p \leqq \frac{1}{2}[\|f\|_p^p + \|g\|_p^p] \qquad (2 \leqq p < \infty, \quad f, g \in L_m^p)$$

1.4.6.2.

$$\left\| \frac{f + g}{2} \right\|_p^{p'} + \left\| \frac{f - g}{2} \right\|_p^{p'} \leqq \left[\frac{1}{2}\|f\|_p^p + \frac{1}{2}\|g\|_p^p\right]^{1/(p-1)}$$

$$\left(1 < p \leqq 2, \quad p' = \frac{p}{p - 1}, \quad f, g \in L_m^p\right).$$

2. DUALITY OF THE SPACES L_m^p $(1 \leqq p \leqq \infty)$

In this section we prove a number of duality statements about the Banach spaces L_m^p $(1 \leqq p \leqq \infty)$ which may roughly be stated as follows: *L_m^∞ is the dual Banach space of L_m^1*; L_m^1 is strictly smaller than the dual space of L_m^∞ unless it is finite-dimensional; if $1 < p, q < \infty$, $p^{-1} + q^{-1} = 1$, then L_m^p is the dual of L_m^q and vice versa. The theorems of this section state carefully what this sketch says, plus some additional information.

We begin with the dual space of L_m^1. For this, we need new spaces of functions. The first part of the following definition defines the concept of a *local nullset* for σ-content spaces.

2.1. Definition. *Let $(\Omega, \mathscr{B}^{00}, m)$ be a σ-content space.*

2.1.1. *A set $N \subseteq \Omega$ is called a* **local m-nullset** *if $F \in \mathscr{B}^{00} \Rightarrow N \cap F$ is an m-nullset.*

2.1.2. *For every $h \in \overline{\mathbb{R}}^\Omega$, we define*

$$\text{loc ess sup } h = \inf\{\alpha \mid \alpha \in \mathbb{R}, \{h > \alpha\} \textit{ is a local m-nullset}\}$$

$$\text{loc ess inf } h = \sup\{\beta \mid \beta \in \mathbb{R}, \{h < \beta\} \textit{ is a local m-nullset}\}$$

$$\|h\|_{\text{loc}\,\infty} = \text{loc ess sup } |h|$$

(*with the usual conventions* $\inf \varnothing = \infty$, $\sup \varnothing = -\infty$).

2.1.3. *We define*

2.1.3.1. $\text{loc}\,\mathscr{L}_m^1 = \{h \mid h \leqq \overline{\mathbb{R}}^\Omega, \quad h1_F \in \mathscr{L}_m^1 \ (F \in \mathscr{B}^{00})\}$. *The members of* $\text{loc}\,\mathscr{L}_m^1$ *are called* **locally *m*-integrable functions.**

2.1.3.2. $\text{loc}\,\mathscr{L}_m^\infty = \{h \mid h \in \text{loc}\,\mathscr{L}_m^1, \ \|h\|_{\text{loc}\,\infty} < \infty\}$.

2.2. Remarks

2.2.1. The above definition should in part be compared to Definition VIII.8.1. It is clear that the local σ-ring $\mathscr{B}^{00}$ is used only as a means to define what "local" means, and that it could, in this function, be replaced by any set system $\subseteq \mathscr{B}^{00}$ such that every set in $\mathscr{B}^{00}$ is contained in some member of that system. For a general pursuit of ideas of this kind, compare SONDERMANN [1].

2.2.2. The above definition of $\text{loc}\,\mathscr{L}_m^\infty$ should be compared with the definition of L_m^∞ given in Exercise III.12.14.6. Clearly we have $\text{loc}\,L_m^\infty = L_m^\infty$ if $\Omega \in \mathscr{B}^0(\mathscr{B}^{00})$, and in particular if $\mathscr{B}^{00}$ is a σ-field.

2.2.3. It should be noticed that in the case of a τ-measure space $(\Omega, \mathscr{E}, m)$ we may first carry out the τ-extension (Chapter III, Section 7) and then choose another local σ-ring, such as $\mathscr{B}_\tau(\mathscr{E})$, in order to obtain a σ-content space $(\Omega, \mathscr{B}^{00}, m)$ which has all achievements of τ-extension already in it (compare Remark VIII.8.2.1).

2.3. Exercises

Let $(\Omega, \mathscr{B}^{00}, m)$ be a σ-content space.

2.3.1. State and prove the analogues of Remark VIII.8.2.2 and Exercise VIII.8.3 in the present context.

2.3.2. Prove that $\text{loc}\,\mathscr{L}_m^1$ as well as $\text{loc}\,\mathscr{L}_m^\infty$ consist of full equivalence classes loc mod m (it is obvious what loc mod m means) and that the set $\text{loc}\,L_m^1$ of all equivalence classes loc mod m in $\text{loc}\,\mathscr{L}_m^1$ forms a vector lattice as well as the set $\text{loc}\,L_m^\infty$ of all equivalence classes loc mod m in $\text{loc}\,\mathscr{L}_m^\infty$. Multiplication is a class operation on $\text{loc}\,\mathscr{S}_m^\infty$, hence may be viewed as an operation in $\text{loc}\,\mathscr{L}_m^\infty$.

2.3.3. Prove that the intersection of a local m-nullset with a countable union of sets from $\mathscr{B}_m^{00}$ (Theorem III.6.1) is an m-nullset.

2.3.4. Prove that for every countable union M of sets from $\mathscr{B}_m^{00}$ and every $h \in \text{loc}\,\mathscr{L}_m^\infty$, there is an $h' \in \text{mble}(\Omega, \mathscr{B}^0(\mathscr{B}^{00}))$ such that $h1_E$ and $h'1_E$ are equivalent mod m (use Theorem III.9.3).

2.3.5. Assume $\Omega \in \mathscr{B}^0(\mathscr{B}^{00})$ and prove

$$\mathscr{L}_m^\infty = \text{loc}\,\mathscr{L}_m^\infty \cap \text{mble}(\Omega, \mathscr{B}^0(\mathscr{B}^{00}), \mathbb{R})$$

(see Exercise III.12.14.6).

2.4. Theorem. *Let $(\Omega, \mathscr{B}^{00}, m)$ be a σ-content space. Then* $\operatorname{loc} \mathscr{L}_m^\infty$ *is the* **dual Banach space** *of* L_m^1. *More precisely:*

2.4.1. *Every $h \in \operatorname{loc} \mathscr{L}_m^\infty$ defines a $\|\cdot\|_1$-continuous linear form L_h on L_m^1 in the following way: for every $f \in \mathscr{L}_m^1$, the function fh is in $\mathscr{L}_m^1$ and $L_h(f) = \int fh\, dm$ depends only on the equivalence* mod m *class of f in $\mathscr{L}_m^1$, hence L_h may be considered as a real function on L_m^1. As such, it is a linear form which we denote by L_h again, and*

$$(1) \qquad \|h\|_{\operatorname{loc} \infty} = \sup\{|L_h(f)| \mid f \in L_m^1, \|f\|_1 \leqq 1\}.$$

L_h depends only on the equivalence class loc mod m *of h.*

2.4.2. *The mapping $h \to L_h$ of* $\operatorname{loc} L_m^\infty$ *into the dual space of L_m^1 is linear and isometric, hence in particular injective.*

2.4.3. *Assume that the σ-content space $(\Omega, \mathscr{B}^{00}, m)$ has a disjoint base (Definition* VIII.8.4*). Then for every $\|\cdot\|_1$-continuous linear form L on L_m^1 there is an $h \in \operatorname{loc} L_m$ such that $L = L_h$.*

Proof. 1. Let $\alpha = \|h\|_{\operatorname{loc} \infty}$. $(h \wedge \alpha) \vee (-\alpha)$ differs from h only by a local m-nullset. A change of h on a local m-nullset and of f on an m-nullset changes fh only on an m-nullset (exercise). Thus we may assume $|h| \leqq \alpha$ everywhere. Since f vanishes outside a countable union of sets from $\mathscr{B}^{00}$ (use Remark III.2.2.5), we may, by Exercise 2.3.4, assume $f, h \in \operatorname{mble}(\Omega, \mathscr{B}^0(\mathscr{B}^{00}), \mathbb{R})$ for the moment (use Theorem III.9.3). Now $|fh| \leqq \alpha |f| \in \mathscr{L}_m^1$, and $fh \in \mathscr{L}_m^1$ follows from Proposition III.9.5. The rest of 1 is obvious except perhaps for (1). But $|fh| \leqq \alpha|f|$ implies $|L_h(f)| = \left|\int fh\, dm\right| \leqq \int |fh|\, dm \leqq \alpha \int |f|\, dm = \|h\|_{\operatorname{loc} \infty} \cdot \|f\|_1$ and thus proves $\geqq$ in (1). In order to prove $\leqq$, choose (we may clearly assume $\alpha > 0$) any $0 < \gamma < \alpha$. Then $\{h > \gamma\}$ or $\{h < -\gamma\}$ is not a local m-nullset. Assume for simplicity that $\{h > \gamma\}$ is not a local m-nullset. But then there is an $F \in \mathscr{B}^{00}$ such that $\{h > \gamma\} \cap F$ is not an m-nullset. Since $h 1_F \in \mathscr{L}_m^1$, we may, for the rest of our present consideration, assume $h \in \operatorname{mble}(\Omega, \mathscr{B}^0(\mathscr{B}^{00}), \mathbb{R})$, hence $\{h > \gamma\} \cap F \in \mathscr{B}^{00}$ (use Proposition I.4.7.2.2) Put $f = \beta 1_{\{h > \gamma\} \cap F}$ where $\beta > 0$ is such that $\int f\, dm = 1$. We get $L_h(f) = \int fh\, dm \geqq \gamma \int f\, dm = \gamma$. Varying γ, we obtain $\leqq$ in (1).

2 is an easy exercise (the (obvious) definition of $\operatorname{loc} L_m^\infty$ was given in Exercise 2.3.2).

3. The restriction of L to $\mathscr{E}(\mathscr{B}^{00})$ is a signed σ-measure since L is bounded on order bounded ($\Rightarrow$ norm bounded) subsets of $\mathscr{E}(\mathscr{B}^{00})$. Thus the $m'\colon \mathscr{B}^{00} \to \mathbb{R}$ defined by $m'(F) = L(1_F)$ $(F \in \mathscr{B}^{00})$ is a signed σ-content

(Theorem IX.4.1). $F \in \mathscr{B}^{00}$, $m(F) = 0 \Rightarrow \|1_F\|_1 = 0 \Rightarrow L(1_F) = 0 \Rightarrow m'(F) = 0$ shows that $m' \ll m$ (Proposition VIII.3.5.2). By Theorem VIII.8.8 there is an $h \in \text{loc } \mathscr{L}_m^1$ such that $F \in \mathscr{B}^{00} \Rightarrow L(1_F) = m'(F) = \int 1_F h \, dm$. By linear extension we find $L(f) = \int fh \, dm$ for all $f \in \mathscr{E}(\mathscr{B}^{00})$.

We want now to prove that $h \in \text{loc } \mathscr{L}_m^\infty$. Assume the contrary. Then none of the sets $\{|h| > 1\}$, $\{|h| > 2\}$, ... is a local m-nullset and we may assume that $\{h > 1\}$, $\{h > 2\}$, ... are not local m-nullsets. Choose any n. We may find some $F \in \mathscr{B}^{00}$ such that $F \cap \{h > n\}$ is not an m-nullset. Combining Exercises II.6.3.2 and III.6.5.7, we may even assume $F \cap \{h > n\} \in \mathscr{B}^{00}$. Choose β such that $\beta m(F \cap \{h > n\}) = 1/\sqrt{n}$. Then $\|\beta 1_{F \cap \{h > n\}}\|_1 = 1/\sqrt{n}$. On the other hand,

$$L(\beta 1_{F \cap \{h>n\}}) = \beta \int 1_{F \cap \{h>n\}} h \, dm \geqq n\beta \int 1_{F \cap \{h>n\}}$$
$$= n \cdot (1/\sqrt{n}) = \sqrt{n}.$$

Doing this for $n = 1, 2, \ldots$, we arrive at a contradiction to the norm continuity of L. Thus $h \in \text{loc } \mathscr{L}_m^\infty$ and the theorem is proved.

It is now natural to ask whether conversely L_m^1 can be identified with the dual Banach space of loc L_m^∞. That this is not the case even in a very simple situation is shown by

2.5. Example. Let $\Omega = \{1, 2, \ldots\}$, $\mathscr{B}^{00} = \{F \mid F \subseteq \Omega, \ |F| < \infty\}$ and $m(F) = |F|$ $(F \in \mathscr{B}^{00})$, the counting σ-content. Then $\text{loc } L_m^\infty = l^\infty = \{f \mid f \in \mathbb{R}^\Omega, f \text{ bounded}\}$ and $\|\cdot\|_{\text{loc } \infty}$ is simply the supremum norm.

$$L_m^1 = l^1 = \left\{h \,\middle|\, \sum_{n=1}^{\infty} |f(n)| < \infty\right\} \quad \text{and} \quad \|h\|_1 = \sum_{n=1}^{\infty} |h(n)| \quad (h \in \mathscr{L}_m^1).$$

Choose L: loc $L_m^\infty \to \mathbb{R}$, i.e., L: $l^\infty \to \mathbb{R}$ as a Banach limit (Example I.6.4) and assume that there is an $h \in L_m^1$, i.e., an $h \in l^1$ such that $L(f) = \int fh \, dm = \sum_{n=1}^\infty f(n)h(n)$ for all $f \in \text{loc } L_m^1$. The shift invariance of L implies

$$|L(f)| = \left| \sum_{n=1}^{\infty} f(n)h(n+k) \right| \leqq \|f\|_{\text{loc } \infty} \sum_{n=1}^{\infty} |h(n+k)|;$$

and this tends to 0 as $k \to \infty$, hence $L = 0$, contradicting the construction of L as a Banach limit.

2.6. Exercises. Let $(\Omega, \mathscr{B}^{00}, m)$ be a σ-content space.

2.6.1. Prove that L_m^1 is of dimension $n < \infty$ iff there are n pairwise disjoint sets $A_1, \ldots, A_n \in \mathscr{B}^{00}$ such that $m(A_1), \ldots, m(A_n) > 0$ and $\Omega \setminus (A_1 + \cdots + A_n)$ is an m-nullset.

2.6.2. Prove that L^1_m can be identified with the dual of loc L_m iff L^1_m is of finite dimension. (*Hint*: Use Example 2.5.)

Undiscouraged by this negative result, some authors have cleared up the structure of the dual of l^∞, L^∞_m, resp. loc L^∞_m to a considerable extent (see, e.g., GÜNZLER [1, 2, 3].)

Let us now turn to the duality of the spaces L^p_m for $1 < p < \infty$. In this case we get a well-rounded result; no recurrence to local integrability is needed here thanks to a strict convexity argument. Disjoint bases are needed, but the same argument shows that they always exist.

2.7. Theorem. *Let $(\Omega, \mathscr{B}^{00}, m)$ be a σ-content space, $1 < p,\ q < \infty$, $p^{-1} + q^{-1} = 1$.*

2.7.1. *Every $h \in \mathscr{L}^q_m$ defines a $\|\cdot\|_p$-continuous linear form L_h on L^p_m in the following way: For every $f \in \mathscr{L}^p_m$, the function fh is in $\mathscr{L}^1_m$ and $L_h(f) = \int fh\, dm$ depends only on the equivalence* mod *m class of h. We have*

$$\|h\|_q = \sup\{|L_h(f)| \mid \|f\|_p \leqq 1\}$$

2.7.2. *The mapping $h \to L_h$ of L^q_m into the dual space of L^p_m is linear and isometric, hence in particular injective.*

2.7.3. *For every $\|\cdot\|_p$-continuous linear form L on L^p_m, there is an $h \in L^q_m$ such that $L = L_h$.*

Proof. 1. Define $\tilde{m}: \overline{\mathbb{R}}_+ \to \mathbb{R}^\Omega_+$ by $\tilde{m}(h) = \int h\, dm$ if $h \in \mathscr{L}^1_m$ and $\tilde{m}(h) = \infty$ otherwise. Then $\tilde{m}$ satisfies the hypotheses of Lemma 1.1.1–3. Let now $f \in \mathscr{L}^p_m$, $h \in \mathscr{L}^q_m$ and apply Lemma 1.1.5 (Hölder's inequality) with $|f|^p$, $|g|^q$ in place of f, g and p^{-1}, q^{-1} in place of α, β. We get

$$\tilde{m}(|f| \cdot |h|) = \tilde{m}((|f|^p)^{1/p}(|h|^q)^{1/q}) \leq \tilde{m}(|f|^p)^{1/p}\tilde{m}(|h|^q)^{1/q} < \infty.$$

In particular, $|fh| = |f| \cdot |h| \in \mathscr{L}^1_m$ follows. Exercise 1.3.3.7 now proves $fh \in L^1_m$. The rest of 1 is now obvious, except for (2). The inequality derived from Hölder's yields $\|fh\|_1 \leqq \|f\|_p \|h\|_q$ $(f \in \mathscr{L}^p_m, h \in \mathscr{L}^q_m)$, proving $\geqq$ in (2). In order to get $\leqq$ it clearly suffices to consider the case $\|h\|_q = 1$. Put now $f = h|h|^{q-2}$. Then $fh = |h|^q$, hence $m(fh) = \|h\|^q_q = 1$; moreover

$$|f|^p = |h|^{p(q-1)} = |h|^q \quad \Rightarrow \quad \|f\|^p_p = \int |f|^p\, dm = \int |h|^q\, dm = \|h\|^q_q = 1.$$

This shows $\leqq$ in (2).

2 is an easy exercise.

3. Assume that $\|L\| = \sup\{|L(f)| \mid f \in L^p_m,\ \|f\| \leqq 1\} = 1$. Since

$\{f \mid f \in \mathscr{E}(\mathscr{B}^{00}), \|f\|_p \leqq 1\}$ is $\|\cdot\|_p$-dense in $\{f \mid f \in L_m^p, \|f\|_p \leqq 1\}$ and L is $\|\cdot\|_p$-continuous, we get $1 = \|L\|_* = \sup\{|L(f)| \mid f \in \mathscr{E}(\mathscr{B}^{00}), \|f\|_p \leqq 1\}$. Let $f_1, f_2, \ldots \in \mathscr{E}(\mathscr{B}^{00})$, $\|f_1\|_p, \|f_2\|_p, \ldots \leqq 1$ be such that $\lim L(f_n) = 1$. Let $F_n = \text{supp}(f_n)$ $(\in \mathscr{B}^{00})$ and put $E = F_1 \cup F_2 \cup \cdots$. Define $m' \colon \mathscr{B}^{00} \to \mathbb{R}$ by $m'(F) = L(1_F)$ $(F \in \mathscr{B}^{00})$. Clearly, m' is additive since L is linear. Let us prove next that m' lives on E (Definition VII.1.1.17). Assume this is false. Then there is some $F \in \mathscr{B}^{00}$ such that $F \cap E = \varnothing$ and $m'(F) \neq \varnothing$. We may thus find a real β such that $f = \beta 1_F$ satisfies $\|f\|_p = 1$, $L(f) = \varepsilon > 0$. For any $n = 1, 2, \ldots$, the following argument holds: $\text{supp}(f_n) \cap \text{supp}(f) = \varnothing$, hence we have an arbitrary $0 \leqq \alpha \leqq 1$,

$$|\alpha f_n + (1 - \alpha) f|^p = \alpha^p |f_n|^p + (1 - \alpha)^p |f|^p$$

and, by integration, $\|\alpha f_n + (1 - \alpha) f\|_p = [\alpha^p + (1 - \alpha)^p]^{1/p}$. If we put

$$f_{n,\alpha} = (\alpha^p + (1 - \alpha)^p)^{-1/p} (\alpha f_n + (1 - \alpha) f),$$

we get $\|f_{n,\alpha}\|_p = 1$ $(n = 1, 2, \ldots; 0 \leqq \alpha \leqq 1)$. On the other hand,

$$L(f_{n,\alpha}) = (\alpha^p + (1 - \alpha)^p)^{-1/p} (\alpha L(f_n) + (1 - \alpha) L(f)).$$

Differentiation with respect to α yields

$$-\frac{1}{p} (\alpha^p + (1 - \alpha)^p)^{-(1/p) - 1} \cdot p(\alpha^{p-1} + (1 - \alpha)^{p-1})(\alpha L(f_n) + (1 - \alpha) L(f))$$

$$+ (\alpha^p + (1 - \alpha)^p)^{-1/p} (L(f_n) - L(f)).$$

For $\alpha \nearrow 1$, this tends to $-L(f_n) + L(f_n) - L(f) = -L(f) = -\varepsilon < 0$, i.e.,

$$\frac{d}{d\alpha} L(f_{n,\alpha}) = -\varepsilon < 0 \qquad \text{as} \quad \alpha = 1.$$

This holds for $n = 1, 2, \ldots$. But $\lim_n L(f_n) = 1$ shows that there is even a common lower bound for the second derivatives of $L(f_{n,\alpha})$ in a left neighborhood of 1. Thus there is a linear function φ of α such that $\varphi(1) = 0$, $\varphi'(1) = -\varepsilon/2$ and $L(f_{n,\alpha}) \geqq L(f_n) + \varphi(\alpha)$ for all α in a sufficiently small left neighborhood of $\alpha = 1$, uniformly for $n = 1, 2, \ldots$. This proves the existence of an n and an $0 < \alpha < 1$ such that $L(f_{n,\alpha}) > 1$. But this contradicts $\|L\|_* = 1$. Thus m' lives on E. Forget about $\Omega \backslash E$, observe that $(E, \mathscr{B}^{00} \cap E, m)$ has a disjoint base in E (Exercise VIII.8.6.2) and apply Theorem VIII.8.8 (Radon–Nikodym) to the effect that there exists a locally m-integrable function h' on E such that $m'(F) = \int 1_F h' \, dm$ $(F \in \mathscr{B}^{00}, F \subseteq E)$. Define $h(\omega) = h'(\omega)$ for $\omega \in E$ and $h(\omega) = 0$ for $\omega \in \Omega \backslash E$. Then clearly $h \in \text{loc}\, \mathscr{L}_m^1$ and $F \in \mathscr{B}^{00} \Rightarrow m'(F) = m'(F \cap E) = \int 1_F h \, dm$. By linear combination, we obtain

$L(f) = \int fh\, dm$ $(f \in \mathscr{E}(\mathscr{B}^{00}))$. Now we prove $h \in L^q_m$. By Exercise VIII.8.3.2 we may assume $h \in \text{mble}(\Omega, \mathscr{B}^0(\mathscr{B}^{00}), \mathbb{R})$. Put $E^+ = \{h > 0\}$. Then for every $f \in \mathscr{E}(\mathscr{B}^{00})$ with $\|f\|_p \leqq 1$, we have $f 1_{E^+} \in \mathscr{E}(\mathscr{B}^{00})$ (use Proposition I.4.7.2.2) and $\|f 1_{E^+}\|_p \leqq \|f\|_p \leqq 1$, hence

$$\int fh_+\, dm = \int f 1_{E^+} h\, dm = L(f 1_{E^+}) \in [-\|L\|_*, \|L\|_*] = [-1, 1].$$

We conclude $\sup\left\{\int fh_+\, dm \,\middle|\, f \in \mathscr{E}(\mathscr{B}^{00}),\ \|f\|_p \leqq 1\right\} \leqq 1 < \infty$, and a similar relation for h_-. It follows that we may assume $h \geqq 0$. The construction in the proof of Proposition III.8.13 plus the fact that h vanishes outside a countable union of sets from $\mathscr{B}^{00}$ leads to a sequence $0 \leqq h_1 \leqq h_2 \leqq \cdots \nearrow h$ of functions $h_n \in \mathscr{E}(\mathscr{B}^{00})$. From (2) we deduce

$$\begin{aligned}
\|h_n\|_q &= \sup\left\{\int h_n f\, dm \,\middle|\, f \in \mathscr{L}^p_m,\ \|f\|_p \leqq 1\right\} \\
&\leqq \sup\left\{\int hf\, dm \,\middle|\, f \in L^p_m,\ \|f\|_p \leqq 1\right\} \\
&= \sup\{L(f) \mid f \in L^p_m,\ \|f\|_p \leqq 1\} < \infty.
\end{aligned}$$

Now the monotone convergence theorem in $\mathscr{L}^q_m$ (Exercise 1.4.3.1) yields $h \in \mathscr{L}^q_m$.

CHAPTER XI

CONTENTS AND MEASURES IN TOPOLOGICAL SPACES. PART II: THE WEAK TOPOLOGY

In this chapter we continue our study of contents and measures in topological spaces, begun in Chapter V, with a study of regularity phenomena. Now we treat the so-called *weak topology*.

The general ideas which act behind the scene can be summed up as follows: Let H be a real vector space and H' a real linear space of linear forms on H. Then H induces in H' the so-called *w*-topology*. It is by definition the coarsest topology on H' such that for every $f \in H$, the linear form $h' \to h'(f)$ on H' is continuous. A typical w*-neighborhood of some $h'_0 \in H'$ is thus of the form $\{h' \mid h' \in H',\ |h'(f_k) - h'_0(f_k)| < \varepsilon\ (k = 1, \ldots, n)\}$, where $f_1, \ldots, f_n \in H$ and $\varepsilon > 0$ are arbitrarily chosen. w*-convergence of a sequence $h'_1, h'_2, \ldots \in H'$ to some $h' \in H'$ means $\lim_n h'_n(f) = h'(f)$ $(f \in H)$. The most well-rounded special case of this general setup is the following: $(H, \|\cdot\|)$ is a Banach space, $(H', \|\cdot\|) = (H^*, \|\cdot\|) =$ the dual Banach space of $(H, \|\cdot\|)$, i.e., H^* is the set of all $\|\cdot\|$-continuous linear forms on H and $\|h'\| = \sup\{|h'(f)| \mid f \in H, \|f\| \leqq 1\}$; the basic theorem on the w*-topology in H is: *For every real* $K > 0$, *the set* $B_k^* = \{h' \mid h' \in H^*,\ \|h'\| \leqq K\}$ *is* w**-compact*. The proof is of enormous simplicity: Take any ultrafilter $\mathscr{F}^*$ in B_K^* and define $h' \in \mathbb{R}^H$ by $h'(f) = \lim g'(f)$ where g' runs "along $\mathscr{F}^*$" $(f \in H)$; clearly $f \in H$, $\|f\| \leqq 1 \to |h'(f)| \leqq K$; and clearly h' is a linear form; thus $h' \in B_K^*$ and $\lim g' = h'$ where g' runs "along $\mathscr{F}^*$."

In the applications of our general idea to particular situations, H' is often replaced by a vector space H^0 each of whose elements defines a linear form on H in some obvious way. We shall speak of the *w*-topology induced by* H^0 in H also in this situation and give it the obvious meaning.

In many cases H is a vector lattice, and one often considers the restriction of the w*-topology to a subset of H like H_+. We then call it simply the *w*-topology in the subset*.

The situation envisaged in this chapter is $H =$ some vector lattice of continuous real functions on a topological space $(\Omega, \mathscr{T})$, and $H^0 = \text{meas}^\sigma(\mathscr{E})$ or $H^0 = \text{meas}^\tau(\mathscr{E})$ resp. $H^0 = \text{cont}^\sigma(\mathscr{R})$ for some elementary domain $\mathscr{E} \subseteq \mathbb{R}^\Omega$ resp. some set ring $\mathscr{R} \subseteq \mathscr{P}(\Omega)$. This means that the H, H^0 under consideration are always vector lattices. In most cases we shall restrict attention to the positive cone H_+^0 of the vector lattice H^0 and we shall always assume that things fit together in such a fashion that $f \in H$, $m \in H_+^0 \Rightarrow f \in L_m^1$, so that every $m = m_+ - m_- \in H^0$ defines a linear form on H by $\int f\,dm_+ - \int f\,dm_-$ $(f \in H)$. In the special case $H = \mathscr{C}^b(\Omega, \mathbb{R})$ = the vector lattice of all bounded real continuous functions on Ω, we shall give the w*-topology the special name "weak topology." The simplest consequence of our above general statement on w*-compactness is the following: If $(\Omega, \mathscr{T})$ is compact, then the set of all probability measures in Ω is weakly compact (Exercise 2.8.6).

The easiest situation of this kind is the following: Let $(\Omega, \mathscr{T})$ be a locally compact Hausdorff space, put $H = \mathscr{E} = \mathscr{C}^{00}(\Omega, \mathbb{R})$ = the vector lattice of all continuous real functions on Ω each vanishing outside some compact set, and put $H' = H^0 = \text{meas}(\mathscr{E})$. Since $\text{meas}(\mathscr{E})$ is always a (even conditionally complete) vector lattice (Theorem IX.2.3.1) and every positive measure on $\mathscr{E} = \mathscr{C}^{00}(\Omega, \mathbb{R})$ is τ-continuous (Theorem I.10.2), we have $\text{meas}(\mathscr{E}) = \text{meas}^\tau(\mathscr{E})$. In particular there is, whenever we deal with positive measures on $\mathscr{E} = \mathscr{C}^{00}(\Omega, \mathbb{R})$, never any question about σ- or τ-continuity: It is always automatically satisfied. This makes it unnecessary to deal with locally compact spaces in particular, after the establishment of so many general results in Section 1. The *Polish case* however deserves intensive separate treatment and receives it in Section 2.

For reasons that will become plausible in the sequel, we prefer the σ-content space approach. The reader should be aware of the fact that from every σ-measure on an elementary domain $\mathscr{E}$ a σ-content space can be derived such that the σ-integral built from the latter coincides with the σ-integral built from the former (Theorems III.6.1, III.6.4). We shall repeat this in some detail in Section 1. Thus measure theoretical data result in content theoretical data which fit into the development of this chapter.

In Section 1 we deal with general topological spaces, and in Section 2 we display the peculiarities of *Polish spaces*, under the aspect of weak topology of measures. See, e.g., VARADARAJAN [1, 2] for further studies.

1. THE WEAK TOPOLOGY FOR σ-CONTENTS IN ARBITRARY TOPOLOGICAL SPACES

In this section we define, for an arbitrary topological space $(\Omega, \mathscr{T})$, the *weak topology* for σ-contents on the Baire σ-field $\mathscr{B}$ generated by the vector lattice $\mathscr{C}^b(\Omega, \mathbb{R})$ of all bounded real continuous functions on Ω. We start by proving that things fit together here: $\mathscr{C}^{\mathrm{b}}(\Omega, \mathbb{R}) \subseteq \mathscr{L}_m^1$ for every σ-content m on $\mathscr{B}$. After this we derive some basic facts about weak convergence, beginning with criteria for weak convergence and ending up with conditions for conditional *weak compactness* in terms of *uniform σ-continuity* and *uniform tightness*. The reader should keep in mind that the most important special cases are compact, locally compact, and Polish spaces.

1.1. Proposition. *Let $(\Omega, \mathscr{T})$ be a topological space and $\mathscr{B}$ be the σ-field generated by $\mathscr{C}^{\mathrm{b}}(\Omega, \mathbb{R})$. Then:*

1.1.1. *For every (positive) σ-content m on $\mathscr{B}$, $\mathscr{C}^{\mathrm{b}}(\Omega, \mathbb{R}) \subseteq \mathscr{L}_m^1$ holds, and the restriction of the m-σ-integral to $\mathscr{C}^{\mathrm{b}}(\Omega, \mathbb{R})$ is a (positive) σ-measure. In particular, every $f \in \mathscr{C}^{\mathrm{b}}(\Omega, \mathbb{R})$ induces a linear form on* $\mathrm{cont}^\sigma(\mathscr{B})$ by $m \to \int f\, dm$.

1.1.2. *For every (positive) σ-measure m on $\mathscr{C}^{\mathrm{b}}(\Omega, \mathbb{R})$, $\mathscr{B} \subseteq \mathscr{B}_\sigma^{00}(m)$ (Theorem* III.6.1) *and the restriction to $\mathscr{B}$ of the σ-content derived from m is a (positive) σ-content.*

Proof. 1. By the definition of $\mathscr{B}$ we have $\mathscr{C}^{\mathrm{b}}(\Omega, \mathbb{R}) \subseteq \mathrm{mble}(\Omega, \mathscr{B}, \mathbb{R})$. Since $\Omega \in \mathscr{B}$, we have every constant in $\mathscr{L}_m^1$. Since every $h \in \mathscr{C}^{\mathrm{b}}(\Omega, \mathbb{R})$ has a constant majorant, we conclude $\mathscr{C}^{\mathrm{b}}(\Omega, \mathbb{R}) \subseteq \mathscr{L}_m^1$ from Theorem III.9.5. The rest of 1 is obvious (use, e.g., the monotone convergence theorem III.5.3.1 and Remark III.5.4.3).

2. Since $\mathscr{C}^{\mathrm{b}}(\Omega, \mathbb{R})$ and hence $\mathscr{L}_m^1$ contain the constant 1, we deduce from Theorem III.6.1 that $\mathscr{B}^{00}(m)$ is a σ-field for which, by Theorem III.6.4 and Proposition III.8.2, $\mathscr{C}^{\mathrm{b}}(\Omega, \mathbb{R}) \subseteq \mathrm{mble}(\Omega, \mathscr{B}_\sigma^{00}(m), \mathbb{R})$, and hence $\mathscr{B} \subseteq \mathscr{B}_\sigma^{00}(m)$. The rest of 2 is now obvious.

1.2. Definition. *Let $(\Omega, \mathscr{T})$ be a topological space and $\mathscr{B}$ the σ-field generated by $\mathscr{C}^{\mathrm{b}}(\Omega, \mathbb{R})$. Consider the linear forms defined by the $f \in \mathscr{C}^{\mathrm{b}}(\Omega, \mathbb{R})$ on* $\mathrm{cont}^\sigma(\mathscr{B})$ *according to Proposition* 1.1.1. *The w*-topology thus induced by $\mathscr{C}^{\mathrm{b}}(\Omega, \mathbb{R})$ in* $\mathrm{cont}^\sigma(\mathscr{B})$ *is called the* **weak topology** *in* $\mathrm{cont}^\sigma(\mathscr{B})$. *In particular, a sequence $m_1, m_2, \ldots \in \mathrm{cont}^\sigma(\mathscr{B})$ is said* **to converge weakly** *to $m \in \mathrm{cont}^\sigma(\mathscr{B})$, and we write*

$$m_n \to m \quad (\textit{weakly}) \qquad \textit{or} \qquad \lim_n m_n = m \quad (\textit{weakly})$$

if $\lim_n \int f\, dm_n = \int f\, dm \ (f \in \mathscr{C}^{\mathrm{b}}(\Omega, \mathbb{R}))$.

Our next aim is to show that it is not always necessary to check through all functions $f \in \mathscr{C}^{\mathrm{b}}(\Omega, \mathbb{R})$ when trying to prove weak approximation.

1.3. Proposition. *Let $(\Omega, \mathscr{T})$ be a topological space, $\mathscr{B}$ the σ-field generated by $\mathscr{C}^{\mathrm{b}}(\Omega, \mathbb{R})$, and $\mathscr{M} \subseteq \mathrm{mble}^{\mathrm{b}}(\Omega, \mathscr{B}, \mathbb{R})$ (= the vector lattice of all* *bounded* $\mathscr{B}$*-measurable real functions on* Ω) *be such that* $\mathscr{C}^b(\Omega, \mathbb{R})$ *is in the uniform closure of* $\mathscr{M}$. *Then* $\mathscr{M} \subseteq \mathscr{L}^1_m$ *for every* $0 \leqq m \in \text{cont}^\sigma(\mathscr{B})$ *and every* $f \in \mathscr{M}$ *defines a linear form* $m \to \int f\, dm$ *on* $\text{cont}^\sigma(\mathscr{B})$. *The* w*-*topology thus induced on* $\text{cont}^\sigma(\mathscr{B})_+$ *by* $\mathscr{M}$ *is finer than (or equal to) the weak topology in* $\text{cont}^\sigma(\mathscr{B})_+$. *In particular,* $0 \leqq m, m_1, m_2, \ldots \in \text{cont}^\sigma(\mathscr{B})$,

$$\lim_n \int f\, dm_n = \int f\, dm \quad (f \in M) \qquad \Rightarrow \qquad \lim_n m_n = m \quad (\textit{weakly}).$$

Proof. The proof of $\mathscr{M} \subseteq \mathscr{L}^1_m$ $(0 \leqq m \in \text{cont}^\sigma(\mathscr{B}))$ is an easy exercise (compare the proof of Proposition 1.1.1) and is left to the reader. Let $0 \leqq m_0 \in \text{cont}^\sigma(\mathscr{B})$, $\varepsilon > 0$, $f_1, \ldots, f_r \in \mathscr{C}^b(\Omega, \mathbb{R})$ and consider the weak neighborhood

$$U = \left\{ m \,|\, 0 \leqq m \in \text{cont}^\sigma(\mathscr{B}),\ \left| \int f_j\, dm - \int f_j\, dm_0 \right| < \varepsilon \quad (j = 1, \ldots, r) \right\}$$

of m_0 in $\text{cont}^\sigma(\mathscr{B})_+$. For a $\delta > 0$ which we shall determine later, choose $g_0, \ldots, g_r \in \mathscr{M}$ such that each of the inequalities $|g_0 - 1| < \delta$, $|g_1 - f_1| < \delta, \ldots, |g_r - f_r| < \delta$ holds uniformly on Ω. Put

$$V = \left\{ m \,|\, 0 \leqq m \in \text{cont}^\sigma(\mathscr{B}),\ \left| \int g_j\, dm - \int g_j\, dm_0 \right| < \delta \ (j = 0, \ldots, r) \right\}.$$

Let $m \in V$. Then

$$\begin{aligned} \int 1\, dm &\leqq \int (g_0 + \delta)\, dm = \int g_0\, dm + \delta \cdot \int 1\, dm < \int g_0\, dm_0 + \delta + \delta \cdot \int 1\, dm \\ &\leqq \int (1 + \delta)\, dm_0 + \delta + \delta \cdot \int 1\, dm = (1 + \delta) \int 1\, dm_0 + \delta + \delta \cdot \int 1\, dm \end{aligned}$$

and hence

$$\int 1\, dm \leqq \frac{1}{1 - \delta} \left[(1 + \delta) \int 1\, dm_0 + \delta \right] = c \qquad \text{where} \quad c \leqq 4 \int 1\, dm_0 + \frac{1}{2}$$

if $0 < \delta \leqq \frac{1}{2}$, which we shall assume henceforth. Now

$$\left| \int f_j\, dm - \int f_j\, dm_0 \right| \leqq \left| \int g_j\, dm - \int g_j\, dm_0 \right| + 2\delta c < \delta(1 + 2c) < \varepsilon$$

$(j = 1, \ldots, r)$ if we put $\delta = \min\{\frac{1}{2}, \varepsilon/(1 + 2c)\}$, and thus $m \in U$ follows; i.e., we have $V \subseteq U$ and our proposition is proved.

Let us apply this proposition, employing also the set systems similar to the set systems $\mathscr{F}^\sigma$, $\mathscr{G}^\sigma$, etc. appearing in Chapter V, Section 1.

1.4. Proposition. *Let $(\Omega, \mathscr{T})$ be a topological space and $\mathscr{B}$ the σ-field generated by $\mathscr{C}^b(\Omega, \mathbb{R})$.*

1.4.1. *Let*

$$\mathscr{G} = \{\{f > 0\} \mid f \in \mathscr{C}^b(\Omega, \mathbb{R})\}, \qquad \mathscr{F} = \{\{f \geqq 0\} \mid f \in \mathscr{C}^b(\Omega, \mathbb{R})\}$$

Then:

1.4.1.1. $\mathscr{G} \cup \mathscr{F} \subseteq \mathscr{B}$.

1.4.1.2. $\mathscr{C}^b(\Omega, \mathbb{R})$ *is in the uniform closure of* $\mathscr{E}(\mathscr{R}(\mathscr{G}))$.

1.4.1.3. $\mathscr{C}^b(\Omega, \mathbb{R})$ *is in the uniform closure of* $\mathscr{E}(\mathscr{R}(\mathscr{F}))$.

1.4.2. *Let $0 \leqq m \in \mathrm{cont}^\sigma(\mathscr{B})$ and (recall that ∂M denotes the boundary of $M \subseteq \Omega$; clearly, $M \in \mathscr{G} \cup \mathscr{F} \Rightarrow \partial M \in \mathscr{B}$)*

$$\mathscr{G}_m = \{G \mid G \in \mathscr{G}, m(\partial G) = 0\}, \qquad \mathscr{F}_m = \{F \mid F \in \mathscr{F}, m(\partial F) = 0\}.$$

Then:

1.4.2.1. $\mathscr{C}^b(\Omega, \mathbb{R})$ *is in the uniform closure of* $\mathscr{E}(\mathscr{R}(\mathscr{G}_m))$.

1.4.2.2. $\mathscr{C}^b(\Omega, \mathbb{R})$ *is in the uniform closure of* $\mathscr{E}(\mathscr{R}(\mathscr{F}_m))$.

Proof. 1.1 is obvious from the definition of $\mathscr{B}$.

1.2. For any $f \in \mathscr{C}^b(\Omega, \mathbb{R})$ and any reals $\alpha_0 < \alpha_1 < \cdots < \alpha_r$, $\mathscr{E}(\mathscr{R}(\mathscr{G}))$ contains the function

$$\begin{aligned}
&\alpha_0 \cdot [1_{\{f-\alpha_0>0\}} - 1_{\{f-\alpha_1>0\}}] \\
&\quad + \alpha_1 \cdot [1_{\{f-\alpha_1>0\}} - 1_{\{f-\alpha_2>0\}}] + \cdots + \alpha_{r-1} \; [1_{\{f-\alpha_{r-1}>0\}} - 1_{\{f-\alpha_r>0\}}] \\
&= \alpha_0 \, 1_{\{\alpha_0 < f \leqq \alpha_1\}} + \cdots + \alpha_{r-1} 1_{\{\alpha_{r-1} < f \leqq \alpha_r\}}
\end{aligned}$$

which clearly approximates f uniformly as closely as we like, upon appropriate choice of $\alpha_0, \ldots, \alpha_r$,

1.3 is proved in an analogous fashion.

2.1. For any $f \in \mathscr{C}^b(\Omega, \mathbb{R})$ and reals $\alpha \neq \beta$, we have

$$\{f = \alpha\} \cap \{f = \beta\} = \varnothing.$$

Hence $m(\{f = \alpha\}) = 0$ except for countably many values of α (otherwise we could get a contradiction to $m(\Omega) < \infty$). Since

$$\partial\{\alpha < f \leqq \beta\} \subseteq \{f = \alpha\} \cup \{f = \beta\},$$

we see that we can carry over the proof of 1.2 with the precaution that $\alpha_0, \ldots, \alpha_r$ always avoid some countable set.

2.2 is proved in an analogous fashion.

An easy application of this proposition is

1.5. Proposition. *Let $(\Omega, \mathscr{T})$ be a topological space and $\mathscr{B}$ the σ-field generated by $\mathscr{C}^b(\Omega, \mathbb{R})$. Then the set*

$$\{\alpha_1\delta_{\omega_1} + \cdots + \alpha_n\delta_{\omega_n} | n \geqq 1, \alpha_1, \ldots, \alpha_n \in \mathbb{R}, \omega_1, \ldots, \omega_n \in \Omega\}$$

of all "discrete" signed σ-contents on $\mathscr{B}$ (recall $\delta_\omega(F) = 1_F(\omega)$ $(\omega \in \Omega, F \in \mathscr{B})$) is weakly dense in $\text{cont}^\sigma(\mathscr{B})$.

Proof. It is clearly sufficient to prove that $\{\alpha_1\delta_{\omega_1} + \cdots + \alpha_n\delta_{\omega_n} | n \geqq 1, 0 \leqq \alpha_1, \ldots, \alpha_n \in \mathbb{R}, \omega_1, \ldots, \omega_n \in \Omega\}$ is weakly dense in $\text{cont}^\sigma(\mathscr{B})_+$, and for this we simply combine Propositions 1.3 and 1.4.1.2. Let $0 \leqq m_0 \in \text{cont}^\sigma(\mathscr{B})$, and choose $f_1, \ldots, f_r \in \mathscr{E}(\mathscr{R}(\mathscr{G}))$. Clearly there are $A_1, \ldots, A_n \in \mathscr{R}(\mathscr{G})$ such that $A_i \cap A_j = \varnothing$ $(i \neq j)$ and f_k is constant on A_i $(i = 1, \ldots, n)$ and vanishes outside $A_1 + \cdots + A_n$. Put $\alpha_j = m(A_j)$ and choose $\omega_j \in A_j$ if $A_j \neq \varnothing$, and ω_j arbitrary if $A_j = \varnothing$ $(j = 1, \ldots, n)$. Put $m = \alpha_1\delta_{\omega_1} + \cdots + \alpha_n\delta_{\omega_n}$. Then clearly $\int f_i\, dm = \int f_i\, dm_0$ $(i = 1, \ldots, r)$. Hence m is in the neighborhood of m_0 described by $f_1, \ldots, f_r$ and any $\varepsilon > 0$. This does it.

Next we prove criteria for weak convergence.

1.6. Proposition. *Let $(\Omega, \mathscr{T})$ be a topological space and $\mathscr{B}$ the σ-field generated by $\mathscr{C}^b(\Omega, \mathbb{R})$. Let $0 \leqq m, m_1, m_2, \ldots \in \text{cont}^\sigma(\mathscr{B})$. Then the following statements are equivalent:*

1.6.1. $m_n \to m$ *(weakly).*

1.6.2. $\lim_n m_n(\Omega) = m(\Omega)$ *and* $\lim_n \inf m_n(G) \geqq m(G)$

$$(G \in \mathscr{G} = \{\{f > 0\} | f \in \mathscr{C}^b(\Omega, \mathbb{R})\}).$$

1.6.3. $\lim_n m_n(\Omega) = m(\Omega)$ *and* $\lim_n \sup m_n(F) \leqq m(F)$

$$(F \in \mathscr{F} = \{\{f \geqq 0\} | f \in \mathscr{C}^b(\Omega, \mathbb{R})\}).$$

1.6.4. $\lim_n m_n(E) = m(E)$ $(E \in \mathscr{B},\ G \subseteq E \subseteq F,\ m(F) = m(G)$ *for suitable* $F \in \mathscr{F}, G \in \mathscr{G})$.

Proof. $1 \Rightarrow 2.$

$$\lim_n m_n(\Omega) = \lim_n \int 1\, dm_n = \int 1\, dm = m(\Omega).$$

From $1_G = 1_{\{f>0\}} = \lim_{k\to\infty} [1 \wedge (kf)_+]$ (pointwise) and the monotone convergence theorem III.5.3 for m, we find

$$\lim_k \int [1 \wedge (kf)_+]\, dm = \int 1_{\{f>0\}}\, dm = m(G).$$

Thus we may, for every $\varepsilon > 0$, find some $0 \leqq g \in \mathscr{C}^{b}(\Omega, \mathbb{R})$ such that $g \leqq 1_G$, $\int g\, dm > m(G) - \varepsilon$. We conclude

$$\liminf_n m_n(G) \geqq \lim_n \int g\, dm_n > m(G) - \varepsilon,$$

and now the rest of 2 follows because $\varepsilon > 0$ is arbitrary.

$2 \Rightarrow 3$ follows by passage to complements.

$3 \Rightarrow 4$ follows by combination of 2 and 3:

$$\begin{aligned}\limsup_n m_n(E) &\leqq \limsup_n m_n(F) \leqq m(F) = m(G)\\ &\leqq \liminf_n m_n(G) \leqq \liminf_n m_n(E).\end{aligned}$$

$4 \Rightarrow 1$ follows by combining Propositions 1.4.2.1 and 1.3: Let us first show that the sets $E \in \mathscr{B}$, which are as in 4, form a set ring. In fact, if $\{f > 0\} \subseteq E \subseteq \{g \geqq 0\}$, $m(\{f > 0\}) = m(\{g \geqq 0\})$, $\{f' > 0\} \subseteq E' \subseteq \{g' \geqq 0\}$, $m(\{f' > 0\}) = m(\{g' \geqq 0\})$, then

$$\{f \vee f' > 0\} \subseteq E \cup E' \subseteq \{g \vee g' \geqq 0\},$$

$$m(\{g \vee g' \geqq 0\} \backslash \{f \vee f' > 0\}) \leqq m([\{g \geqq 0\} \backslash \{f > 0\}] \cup [\{g' \geqq 0\} \backslash \{f' > 0\}]) = 0,$$

$$\{f \vee (-g') > 0\} \subseteq E \backslash E' \subseteq \{g \vee (-f') \geqq 0\},$$

$$\begin{aligned}&m(\{g \vee (-f') \geqq 0\} \backslash \{f \vee (-g') > 0\})\\ &\leqq m([\{g \geqq 0\} \backslash \{f > 0\}] \cup [\{g' \geqq 0\} \backslash \{f' > 0\}]) = 0\end{aligned}$$

shows that $E \cup E'$, $E \backslash E'$ have the same property. The $E \in \mathscr{F}_m$ have the desired property since $E = \{f \geqq 0\} \Rightarrow \partial E \subseteq \{f = 0\}$ and hence $G = \{f > 0\}$, $F = E$ do the job.

We conclude $\lim_n m_n(E) = m(E)$ for all $E \in \mathscr{R}(\mathscr{F}_m)$, hence, by linear combination,

$$\lim_n \int f\, dm_n = \int f\, dm \qquad (f \in \mathscr{E}(\mathscr{R}(\mathscr{F}_m))).$$

The conclusion 1 now follows from the cited propositions.

1.7. Proposition. *Let $(\Omega, \mathscr{T})$, $(\Omega', \mathscr{T}')$ be topological spaces and $\mathscr{B}$ resp. $\mathscr{B}'$ the σ-field generated by $\mathscr{C}^{b}(\Omega, \mathbb{R})$ resp. $\mathscr{C}^{b}(\Omega', \mathbb{R})$.*

1.7.1. *Let $\varphi: \Omega \to \Omega'$ be continuous. Then the induced σ-content transform mapping φ: $\operatorname{cont}^{\sigma}(\mathscr{B}) \to \operatorname{cont}^{\sigma}(\mathscr{B}')$ (defined by $\varphi m(F') = m(\varphi^{-1}F')$ ($m \in \operatorname{cont}^{\sigma}(\mathscr{B})$, $F' \in \mathscr{B}'$)) is weakly continuous.*

1.7.2. *Let P: $\mathscr{C}^{b}(\Omega, \mathbb{R}) \leftarrow \mathscr{C}^{b}(\Omega', \mathbb{R})$ be linear and positive. Then the* *induced mapping* $P\colon \operatorname{cont}^\sigma(\mathscr{B}) \to \operatorname{cont}^\sigma(\mathscr{B}')$ $\left(\textit{defined by } \int f'\, d(Pm) = \int (Pf')\, dm\right.$ $\left.(m \in \operatorname{cont}^\sigma(\mathscr{B}),\ f' \in \mathscr{C}^{\mathrm{b}}(\Omega, \mathbb{R}))\right)$ *is weakly continuous.*

Proof. Clearly 1 is a special case of 2 (define P by $(Pf')(\omega) = f'(\varphi\omega)$ $(f' \in \mathscr{C}^{\mathrm{b}}(\Omega', \mathbb{R}),\ \omega \in \Omega)$). But a proof of 2 is immediately read from the definition of P: $\operatorname{cont}^\sigma(\mathscr{B}) \to \operatorname{cont}^\sigma(\mathscr{B}')$.

The rest of this section is devoted to investigations of weakly compact subsets of $\operatorname{cont}^\sigma(\mathscr{B})_+$.

1.8. Proposition. *Let* $(\Omega, \mathscr{T})$ *be a topological space and* $\mathscr{B}$ *the* σ*-field generated by* $\mathscr{C}^{\mathrm{b}}(\Omega, \mathbb{R})$. *Let* $\varnothing \neq M \subseteq \operatorname{cont}^\sigma(\mathscr{B})_+$. *Then the following statements are equivalent:*

1.8.1. *M is conditionally weakly compact.*

1.8.2. *M is uniformly σ-continuous in the following sense:*

$$\mathscr{C}^{\mathrm{b}}(\Omega, \mathbb{R}) \ni f_1 \geqq f_2 \geqq \cdots \searrow 0 \quad (\textit{pointwise})$$
$$\Rightarrow \quad \lim_n \left[\sup\left\{\int f_n\, dm \,\middle|\, m \in M\right\}\right] = 0.$$

Proof. $1 \Rightarrow 2$. Let $\varepsilon > 0$ and $U_n = \left\{m \,\middle|\, 0 \leqq m \in \operatorname{cont}^\sigma(\mathscr{B}),\ \int f_n\, dm < \varepsilon\right\}$. Clearly U_n is open in the weak topology in $\operatorname{cont}^\sigma(\mathscr{B})_+$, and $U_1 \subseteq U_2 \subseteq \cdots$, $\bigcup_n U_n = \operatorname{cont}^\sigma(\mathscr{B})_+$. Thus there is some n_0 with $M \subseteq U_n$, i.e.,

$$\sup\left\{\int f_n\, dm \,\middle|\, m \in M\right\} \leqq \varepsilon \qquad (n \geqq n_0).$$

Thus 2 follows.

$2 \Rightarrow 1$. Let $\mathscr{F}$ be an ultrafilter in M and define $m^*(f) = \lim_{\mathscr{F}} \int f\, dm$ $(f \in \mathscr{C}^{\mathrm{b}}(\Omega, \mathbb{R}))$. One sees immediately that m^* is a positive linear form on the elementary domain $\mathscr{C}^{\mathrm{b}}(\Omega, \mathbb{R})$, i.e., a (positive) measure. Let $\mathscr{C}^{\mathrm{b}}(\Omega, \mathbb{R}) \ni f_1 \geqq f_2 \geqq \cdots \searrow 0$ (pointwise). For an arbitrary $\varepsilon > 0$, choose n such that $\sup\left\{\int f_n\, dm \,\middle|\, m \in M\right\} \leqq \varepsilon$. Clearly $m^*(f_n) \leqq \varepsilon$ follows. Thus m is a σ-measure. Let m_0 be the corresponding σ-content on $\mathscr{B}$. Then $\lim_{\mathscr{F}} \int f\, dm = \int f\, dm_0$ follows, hence M is conditionally weakly compact.

1.9. Theorem. *Let* $(\Omega, \mathscr{T})$ *be a topological space and* $\mathscr{B}$ *the* σ*-field generated by* $\mathscr{C}^{\mathrm{b}}(\Omega, \mathbb{R})$. *Let M be a subset of* $\operatorname{cont}^\sigma(\mathscr{B})_+$ *that is* **uniformly tight** *in the following sense: For every* $\varepsilon > 0$, *there is a* $\delta > 0$ *and a compact* *set* $K \subseteq \Omega$ *such that* $m \in M$, $f \in \mathscr{C}^b(\Omega, \mathbb{R})$, $|f| \leqq 1$ *on* Ω, $|f| < \delta$ *on* K

$$\Rightarrow \left| \int f \, dm \right| < \varepsilon.$$

Then M *is* **conditionally weakly compact**.

Proof. By Proposition 1.8 is suffices to prove that M is uniformly σ-continuous. Let therefore $\mathscr{C}^b(\Omega, \mathbb{R}) \ni f_1 \geqq f_2 \geqq \cdots \searrow 0$ pointwise. We may assume $f_1 \leqq 1$. For any $\varepsilon > 0$, determine $\delta > 0$ and a compact $K \subseteq \Omega$ such that $|f| \leqq 1$, $|f| < \delta$ on $K \Rightarrow \left| \int f \, dm \right| < \varepsilon$ $(m \in M)$. By Dini's theorem there is an n_0 such that $f_n < \delta$ uniformly on K for all $n \geqq n_0$, and $\sup\left\{ \int f_n \, dm \,|\, m \in M \right\} \leqq \varepsilon$ $(n \geqq n_0)$ follows, as desired.

1.10. Exercises

1.10.1. Let $(\Omega, \mathscr{T})$ be a topological space and $\mathscr{B}$ the σ-field generated by $\mathscr{C}^b(\Omega, \mathbb{R})$. An $m \in \operatorname{cont}^\sigma(\mathscr{B})$ is called tight if for every $\varepsilon > 0$, there is a $\delta > 0$ and a compact $K \subseteq \Omega$ such that $|f| \leqq 1$, $|f| < \delta$ on $K \Rightarrow \left| \int f \, dm_+ - \int f \, dm_- \right| < \varepsilon$. Denote by $\operatorname{cont}^t(\mathscr{B})$ the system of all tight signed σ-contents on $\mathscr{B}$.

1.10.1.1. Prove that $\operatorname{cont}^t(\mathscr{B})$ is a Riesz band in $\operatorname{cont}^\sigma(\mathscr{B})$ (see Definition IX.1.7).

1.10.1.2. Prove that every order bounded subset of $\operatorname{cont}^t(\mathscr{B})$ is conditionally weakly compact.

1.10.1.3. For every $m \in \operatorname{cont}^\sigma(\mathscr{B})$, let $\|m\|$ denote the total variation of m (Definition VII.4.1). Prove that $(\{m \,|\, m \in \operatorname{cont}^t(\mathscr{B}), \ \|m\| < \infty\}, \ \|\cdot\|)$ is an L-space.

1.10.1.4. Carry over the preceding exercises to measure theory (see Definition I.10.3).

1.10.1.5. Prove: If $M \subseteq \operatorname{cont}^t(\mathscr{B})$ is weakly compact, then M is uniformly tight. (*Hint*: Assume that M is not uniformly tight and consider, for a suitable $\varepsilon > 0$, all sets $M_{f,k} = \left\{ m \,|\, m \in M, \int f \, dm \geqq \varepsilon \right\}$ with $K \subseteq \Omega$ compact, $f \in \mathscr{C}^b(\Omega, \mathbb{R})$, $0 \leqq f \leqq 1, f > 0$ on K.)

1.10.2. Carry over Proposition 1.6 to nets resp. filters of positive σ-contents, instead of sequences.

1.10.3. Let $(\Omega, \mathscr{T})$ be a topological space and $\mathscr{B}$ the σ-field generated by $\mathscr{C}^b(\Omega, \mathbb{R})$. Let $\mathscr{S} \subseteq \mathscr{B}$ be such that every $G \in \mathscr{G} = \{\{f > 0\} \,|\, f \in \mathscr{C}^b(\Omega, \mathbb{R})\}$

is a countable union of sets from $\mathscr{S}$. Prove that, for every $0 \leqq m, m_1, m_2, \ldots \in \operatorname{cont}^\sigma(\mathscr{B})$,

$$\lim_n m_n(E) = m(E) \ (E \in \mathscr{S}) \quad \Rightarrow \quad \lim_n m_n = m \ \text{(weakly)}.$$

Carry over this result to nets and filters.

1.10.4. Let $(\Omega, \mathscr{T})$ be a locally compact space and $\mathscr{B}$ the σ-field generated by $\mathscr{C}^b(\Omega, \mathbb{R})$, $\mathscr{B}^{00} = \mathscr{B}^{00}(\{\{f > 1\} \mid f \in \mathscr{C}^{00}(\Omega, \mathbb{R})\})$.

1.10.4.1. Prove that every finite σ-content $m \geqq 0$ on $\mathscr{B}^{00}$ has a unique σ-additive extension $\overline{m} \geqq 0$ to $\mathscr{B}$ such that $\|m\| = \|\overline{m}\|$. (*Hint:* Use Corollary II.4.2.)

1.10.4.2. Let $m, \overline{m}$ be as in 1.10.4.1. Prove that for every $0 \leqq f \in \mathscr{C}^b(\Omega, \mathbb{R})$,

$$\int f \, d\overline{m} = \sup\left\{\int g \, dm \,\middle|\, g \in \mathscr{C}^{00}(\Omega, \mathbb{R}), 0 \leqq g \leqq f\right\}.$$

1.10.4.3. Assume that Ω is a K_σ and prove that there is a sequence $K_1, K_2, \ldots$ of compact sets such that $K_1 \subseteq K_2 \subseteq \cdots$, $K_1 \cup K_2 \cup \cdots = \Omega$, and every compact set $\subseteq \Omega$ is contained in some K_n.

1.10.4.4. Assume that Ω is a K_σ and $K_1, K_2, \ldots$ is as in 1.10.4.2. Choose $\omega_n \in \Omega \backslash K_n$ and put $m_n = \delta_{\omega_n}$ $(n = 1, 2, \ldots)$. Prove that $\int f \, dm_n \to 0$ $(f \in \mathscr{C}^{00}(\Omega, \mathbb{R}))$, but $m_1, m_2, \ldots$ is not weakly convergent.

1.10.4.5. Prove that every finite subset of $\operatorname{cont}^\sigma(\mathscr{B})_+$ is uniformly tight.

1.10.4.6. Let $0 \leqq m, m_1, m_2, \ldots \in \operatorname{cont}^\sigma(\mathscr{B})$ be such that

$$\lim_n \int f \, dm_n = \int f \, dm \qquad (f \in \mathscr{C}^{00}(\Omega, \mathbb{R})).$$

Prove that $\lim_n m_n = m$ (weakly) iff $\{m_1, m_2, \ldots\}$ is uniformly tight.

2. THE WEAK TOPOLOGY FOR σ-CONTENTS IN POLISH SPACES

In this section we prove a group of results on weak convergence of σ-contents for the particular case where the underlying topological space is *Polish*, i.e., first countable and metrizable with a complete metric. We have already dealt with this situation in Chapter V under the aspect of regularity and know, e.g., that every σ-content in a Polish space is *tight* (Theorem V.5.4).

We begin by specializing to the Polish case some criteria for weak convergence known from Section 1. Then we deal with the weak continuity of transforms by m-a.e. continuous maps. We show how *stochastic con-*

vergence of maps into Polish spaces entail the weak convergence of image σ-contents. We conclude with *Skorokhod's theorem* showing that some a.e. convergence of mappings can be found behind any weak convergence of probability σ-contents in a Polish space.

The reader is reminded that locally compact second countable spaces are Polish.

We begin with a purely topological

2.1. Proposition. *Let $(\Omega, \mathscr{T})$ be a Polish space. Then*

$$\mathscr{G} = \{\{f > 0\} \mid f \in \mathscr{C}^{\mathrm{b}}(\Omega, \mathbb{R})\}$$

coincides with the system $\mathscr{T}$ of all open sets and $F = \{\{f \geqq 0\} \mid f \in \mathscr{C}^{\mathrm{b}}(\Omega, \mathbb{R})\}$ is the system of all closed sets in Ω.

Proof. Let $G \in \mathscr{T}$. Choose $\omega_1, \omega_2, \ldots \in G$ to be dense in G. Use a metric $|\cdot, \cdot|$ that defines the topology in order to define

$$r_n = \min\{1, \inf\{|\eta, \omega_n| \mid \eta \notin G\}\} \qquad \text{and} \qquad f_n(\omega) = [r_n - |\omega, \omega_n|] \vee 0$$

$(\omega \in \Omega)$, $n = 1, 2, \ldots$. Put $f = \sum_{n=1}^{\infty} 2^{-n} f_n$. Clearly, $f \in \mathscr{C}^{\mathrm{b}}(\Omega, \mathbb{R})$ and f vanishes outside G. If $\omega \in G$, put $r = \inf\{|\eta, \omega| \mid \eta \notin G\}$ and choose n such that $|\omega, \omega_n| < \frac{1}{2}\min\{1, r\}$. We can then be sure $|\omega, \omega_n| < r_n$, hence of $f_n(\omega) > 0$. Thus we get $f(\omega) > 0$, and $G = \{f > 0\}$ follows. If $F \subseteq \Omega$ is closed, $\Omega \backslash F$ is open, hence we may find an $f \in \mathscr{C}^{\mathrm{b}}(\Omega, \mathbb{R})$ with $\{f > 0\} = \Omega \backslash F$, i.e., $F = \{-f \geqq 0\}$.

2.2. Remark. It is obvious that the proposition holds for arbitrary first countable metrizable spaces.

2.3. Proposition. *Let $(\Omega, \mathscr{T})$ be a Polish space and $\mathscr{B} = \mathscr{B}(\mathscr{T})$ the σ-field of all Borel sets in Ω. Let $0 \leqq m, m_1, m_2, \ldots \in \operatorname{cont}^{\sigma}(\mathscr{B})$. Then the following statements are equivalent:*

2.3.1. $\lim_n m_n = m$ *(weakly).*

2.3.2. $\lim_n m_n(\Omega) = m(\Omega)$, $\liminf_n m_n(G) \geqq m(G)$ *($G \subseteq \Omega$ open).*

2.3.3. $\lim_n m_n(\Omega) = m(\Omega)$, $\limsup_n m_n(F) \leqq m(F)$ *($F \subseteq \Omega$ closed).*

2.3.4. $\lim_n m_n(E) = m(E)$ $(E \in \mathscr{B}, m(\partial E) = 0)$.

Proof. Combine Propositions 1.6 and 2.1.

2.4. Exercise. Carry the above proposition over to nets resp. filters of σ-contents instead of sequences.

2.5. Proposition. *Let* $(\Omega, \mathcal{T})$ *be a Polish space and* $\mathcal{B} = \mathcal{B}(\mathcal{T})$, $0 \leqq m \in \text{cont}^\sigma(\mathcal{B})$. *Then there is a countable* $\mathcal{S}_m \subseteq \mathcal{B}$ *such that:*

2.5.1. *Every open* $G \subseteq \Omega$ *is a union of sets from* $\mathcal{S}_m$.

2.5.2. $E \in \mathcal{S}_m \Rightarrow m(\partial E) = 0$.

Proof. Let $\omega_1, \omega_2, \ldots \in \Omega$ be dense in Ω. For every $n = 1, 2, \ldots$ and $r > 0$, let $U_r(\omega_n) = \{\eta | \eta \in \Omega, \ |\eta, \omega_n| < r\}$. Clearly, $U_r(\omega_n)$ is open with boundary $\partial U_r(\omega_n) \subseteq \{\eta | \eta \in \Omega, \ |\eta, \omega_n| = r\}$. It follows that

$$0 < r < s \quad \Rightarrow \quad \partial U_r(\omega_n) \cap \partial U_s(\omega_n) = 0.$$

$m(\Omega) < \infty$ now implies that $m(\partial U_r(\omega_n)) > 0$ for at most countably many values of r. Thus there is a countable dense subset R of $\{r | 0 < r \in \mathbb{R}\}$ such that $m(\partial U_r(\omega_n)) = 0$ $(n = 1, 2, \ldots; r \in R)$. Put $\mathcal{S}_m = \{U_r(\omega_n) | n = 1, 2, \ldots; r \in R\}$. Let $G \subseteq \Omega$ be open and $\omega \in G$. Then there is an $s > 0$ such that $\{\eta | \eta \in \Omega, \ |\eta, \omega| < s\} \subseteq G$. Choose n such that $|\omega_n, \omega| < s/2$, and some $r \in R \cap]s/2, s[$. Then $\omega \in U_r(\omega_n) \subseteq G$. Thus $\mathcal{S}_m$ has all the properties required.

2.6. Exercise. Let $(\Omega, \mathcal{T})$ be a Polish space, $\mathcal{B} = \mathcal{B}(\mathcal{T})$, $\Omega_0 \subseteq \Omega$ be countable and dense, and $R \subseteq]0, \infty[$ countable and dense. Let $m, m_1, m_2, \ldots \in \text{cont}^\sigma(\mathcal{B})$ and assume $\lim_n m_n(E) = m(E)$ for every E of the form $E = \{\eta | \eta \in \Omega, \ |\eta, \omega| < r\}$, where $\omega \in \Omega_0$, $r \in R$. Prove that $\lim_n m_n = m$ (weakly). (*Hint*: Use Exercise 1.10.3.)

2.7. Theorem. *Let* $(\Omega, \mathcal{T})$ *be a Polish space and* $\mathcal{B} = \mathcal{B}(\mathcal{T})$. *Let* $m, m_1, m_2, \ldots \in \text{cont}^\sigma(\mathcal{B})$ *be such that* $\lim_n m_n = m$ (weakly). *Let* $f \in \mathbb{R}^\Omega$ *be bounded and* m-a.e. *continuous. Then*

$$\lim_n \int f \, dm_n = \int f \, dm.$$

Proof. Apparently it is sufficient to treat the case $0 \leqq f \leqq 1$, $m(\Omega) = 1$. Let now $\Omega_0 \in \mathcal{B}$ be such that $m(\Omega \backslash \Omega_0) = 0$ and f is continuous on Ω_0. By Theorem V.5.3 there is, for a given $\varepsilon > 0$, some compact $K \subseteq \Omega_0$ such that $m(\Omega_0 \backslash K) < \varepsilon/2$. For every $\omega \in K$, we choose some $\delta(\omega)$ such that $|\eta, \omega| < \delta(\omega) \Rightarrow |f(\eta) - f(\omega)| < \varepsilon/2$ and $m(\{\eta | \ |\eta, \omega| = \delta(\omega)\}) = 0$. This is possible by the continuity of f at every $\omega \in K$, and by an argument used in the proof of Proposition 2.5. Since K is compact, we may find $\omega_1, \ldots, \omega_r \in K$ such that

$$U_k = \{\eta | \ |\eta, \omega_k| < \delta(\omega_k)\} \quad \Rightarrow \quad K \subseteq U_1 \cup \cdots \cup U_r.$$

Write $V_1 = U_1$, $V_2 = U_2 \backslash U_1, \ldots,$ $V_r = U_r \backslash (U_1 \cup \cdots \cup U_{r-1})$. Clearly $V_j \cap V_k = \varnothing$ $(j \neq k)$, $K \subseteq V_1 + \cdots + V_r$, and $m(\partial V_1) = \cdots = m(\partial V_r) = 0$.

Moreover, $\eta \in V_k \Rightarrow |f(\eta) - f(\omega_k)| < \varepsilon/2$ ($k = 1, \ldots, r$). Put

$$g = f(\omega_1)1_{V_1} + \cdots + f(\omega_r)1_{V_r}.$$

Then clearly

$$\lim_n \int g\, dm_n = \sum_{k=1}^{r} f(\omega_k) \lim_n m_n(V_k) = \sum_{k=1}^{r} f(\omega_k) m(V_k) = \int g\, dm.$$

On the other hand, $|g - f| < \varepsilon/2$ on $V_1 + \cdots + V_r$ and thus

$$\left| \int g\, dm_n - \int f\, dm_n \right| < \frac{\varepsilon}{2} m_n(V_1 + \cdots + V_r) + m_n(\Omega \backslash (V_1 + \cdots + V_r))$$

$$< \frac{\varepsilon}{2} m_n(\Omega) + \left[m_n(\Omega) - \sum_{k=1}^{r} m_n(V_k) \right].$$

Clearly this tends to $(\varepsilon/2)m(\Omega) + [1 - m(V_1 + \cdots + V_r)]$ as $n \to \infty$, hence is $< \varepsilon$ for n sufficiently large. Similarly, $\left| \int g\, dm - \int f\, dm \right| < \varepsilon$ follows. Putting our achievements together, we arrive at

$$\limsup_n \left| \int f\, dm_n - \int f\, dm \right| \leqq 2\varepsilon.$$

Since $\varepsilon > 0$ was arbitrary, $\lim_n \int f\, dm_n = \int f\, dm$ follows.

2.8. Exercise. Let $(\Omega, \mathscr{T})$ be a Polish space and $\mathscr{B} = \mathscr{B}(\mathscr{T})$.

2.8.1. Carry Theorem 2.7 over to nets resp. filters of σ-contents.

2.8.2. Let $(\Omega', \mathscr{T}')$ be a topological space and $\mathscr{B}'$ generated by $\mathscr{C}^b(\Omega', \mathbb{R})$. Let φ: $\Omega \to \Omega'$ be $\mathscr{B}$-$\mathscr{B}'$-measurable and m-a.e. continuous. Prove: If $0 \leqq m, m_1, m_2, \ldots \in \text{cont}^\sigma(\mathscr{B})$, $\lim_n m_n = m$ (weakly), then $\varphi m, \varphi m_1, \varphi m_2, \ldots \in \text{cont}^\sigma(\mathscr{B}')_+$, $\lim_n \varphi m_n = \varphi m$ (weakly) (see Chapter IV, Section 2, (1)).

2.8.3. Let in particular $\Omega = \mathbb{R}$ with the usual topology. For any $0 \leqq m, m_1, m_2, \ldots \in \text{cont}^\sigma(\mathscr{B})$, let $F(t) = m(]-\infty, t])$, $F_n(t) = m_n(]-\infty, t])$ ($t \in \mathbb{R}$, $n = 1, 2, \ldots$). Prove:

2.8.3.1. F is an isotone right continuous function, hence has at most countably many discontinuities.

2.8.3.2. $\lim_n m_n = m$ (weakly) iff $\lim_n F_n(t) = F(t)$ for every point t of continuity of F.

2.8.3.3. Let $m_n = n^{-1} \sum_{k=1}^{n} \delta_{k/n}$ ($n = 1, 2, \ldots$) and m be such that $\int f\, dm = \int_0^1 f(x)\, dx$ ($f \in \mathscr{C}^b(\mathbb{R}, \mathbb{R})$). Prove that $m_n \to m$ (weakly).

2.8.3.4. Let $m_n = \delta_{1/n}$ $(n = 1, 2, \ldots)$, $m = \delta_0$, $f = 1_{]-\infty, 0]}$. Prove $m_n \to m$ (weakly), $\lim \int f\, dm_n \neq \int f\, dm$.

2.8.4. Carry over Exercise 2.8.3 to $\Omega = \mathbb{R}^n$.

2.8.5. Let $(\Omega', \mathscr{T}')$ be a Polish space, $\mathscr{B}' = \mathscr{B}(\mathscr{T}')$. Let $0 \leqq m, m_1, m_2, \ldots \in \operatorname{cont}^\sigma(\mathscr{B})$, $\lim_n m_n = m$ (weakly) and $0 \leqq m', m'_1, m'_2, \ldots \in \operatorname{cont}^\sigma(\mathscr{B}')$, $\lim_n m'_n = m'$ (weakly). Prove that $\lim_n (m_n \times m'_n) = m \times m'$. (*Hint*: Construct $\mathscr{S}_m \subseteq \mathscr{B}$, $\mathscr{S}_{m'} \subseteq \mathscr{B}'$ according to Proposition 2.5 and apply Exercise 1.10.3 with $\{E \times E' \mid E \in \mathscr{S}_m, E' \in \mathscr{S}_{m'}\}$.)

2.8.6. Prove that the weak topology in $W = \{m \mid 0 \leqq m \in \operatorname{cont}^\sigma(\mathscr{B}), m(\Omega) = 1\}$ is metrizable and W is compact iff Ω is. (*Hint*: Use Proposition 2.5 and Exercise 1.10.3; see Theorem XV.3.2 and its proof.)

Next we investigate the possibilities of weak convergence for sequences of σ-contents that arise from a single σ-content by transforms with a sequence of measurable mappings.

2.9. Theorem. *Let $(\Omega, \mathscr{T})$ be a Polish space, $\mathscr{B} = \mathscr{B}(\mathscr{T})$. Let $(X, \mathscr{S}, m)$ be a σ-content space, $X \in \mathscr{S}$. Let $\varphi_1, \psi_1, \varphi_2, \psi_2, \ldots: X \to \Omega$ be $\mathscr{S}$-$\mathscr{B}$-measurable and assume that the sequences $\varphi_1, \varphi_2, \ldots$ and $\psi_1, \psi_2, \ldots$ are* ***m*-stochastically equivalent** *in the following sense:*

$$\lim_n m(\{x \mid |\varphi_n(x), \psi_n(x)| \geqq \varepsilon\}) = 0 \qquad (\varepsilon > 0).$$

(We shall prove that the sets in question belong to $\mathscr{S}$.) Assume further $\lim_n \varphi_n m = p$ (weakly) for some $0 \leqq p \in \operatorname{cont}^\sigma(\mathscr{B})$. Then $\lim_n \psi_n m = p$ (weakly).

Proof. By Proposition VI.1.8 the mappings $\varphi_n \times \psi_n: X \to \Omega \times \Omega$ defined by $x \to (\varphi_n(x), \psi_n(x))$ $(x \in X)$ are $\mathscr{S}$-$(\mathscr{B} \times \mathscr{B})$-measurable. The mapping $|\cdot, \cdot|: \Omega \times \Omega \to \mathbb{R}$ defined by $(\omega, \eta) \to |\omega, \eta|$ $(\omega, \eta \in \Omega)$ is continuous and hence $\mathscr{B} \times \mathscr{B}$-measurable (Exercises III.8.3.1 and IV.1.7.3). By Proposition IV.1.5 the function $x \to |\varphi_n(x), \psi_n(x)|$ is in $\operatorname{mble}(X, \mathscr{S}, \mathbb{R})$. Thus $m(\{x \mid |\varphi_n(x), \psi_n(x)| \geqq \varepsilon)$ is a meaningful term. By Proposition 2.3 it is sufficient to prove $\limsup_n (\psi_n m)(F) \leqq p(F)$ for every closed $F \subseteq \Omega$.

For any $\varepsilon > 0$, let $F_\varepsilon = \{\eta \mid \inf\{|\omega, \eta| \mid \omega \in F\} \leqq \varepsilon\}$. Clearly F_ε is closed, hence in $\mathscr{B}$. Since $\psi_n(x) \in F \Rightarrow \varphi_n(x) \in F_\varepsilon$ or $|\varphi_n(x), \psi_n(x)| \geqq \varepsilon$, we have

$$\begin{aligned}\limsup_n (\psi_n m)(F) &= \limsup_n m(\psi_n \in F)\\ &\leqq \limsup_n m(\varphi_n \in F_\varepsilon) + \limsup_n m(|\varphi_n, \psi_n| \geqq \varepsilon)\\ &= \limsup_n (\varphi_n m)(F_\varepsilon) \leqq p(F_\varepsilon).\end{aligned}$$

This holds for every $\varepsilon > 0$. But $F_{1/n} \searrow F$ as $n \to \infty$, hence

$$\limsup(\psi_n m)(F) \leqq p(F)$$

follows.

2.10. Theorem. *Let* $(\Omega, \mathscr{T})$ *be a Polish space,* $\mathscr{B} = \mathscr{B}(\mathscr{T})$. *Let* $(X, \mathscr{S}, m)$ *be a σ-content space,* $X \in \mathscr{S}$. *Let* $\varphi_1, \varphi_2, \ldots: X \to \Omega$ *be* $\mathscr{S}$-$\mathscr{B}$*-measurable. Then:*

2.10.1. *If* $\varphi: X \to \Omega$ *is* $\mathscr{S}$-$\mathscr{B}$*-measurable such that* $\lim_n \varphi_n = \varphi$ m-a.e., *then* $\lim_n \varphi_n m = \varphi m$ *(weakly).*

2.10.2. *If* $\varphi_1, \varphi_2, \ldots$ *is a* **stochastic fundamental sequence** *in the sense that*

$$\lim_{j,k \to \infty} m(|\varphi_j, \varphi_k| \geqq \varepsilon) = 0 \qquad (\varepsilon > 0),$$

then there is an $\mathscr{S}$-$\mathscr{B}$*-measurable* $\varphi: X \to \Omega$ *such that:*

2.10.2.1. $\lim_{j \to \infty} \varphi_{n_j} = \varphi$ m-a.e. *for a suitable sequence* $0 < n_1 < n_2 < \cdots$ *of integers.*

2.10.2.2. $\lim_n \varphi_n m = \varphi m$ *(weakly).*

Proof. 1. Put $\psi_1 = \psi_2 = \cdots = \varphi$. Then the sequences $\varphi_1, \varphi_2, \ldots$ and $\psi_1, \psi_2, \ldots$ are m-stochastically equivalent. In fact $\lim |\varphi_n, \psi_n| = \lim_n |\varphi_n, \varphi| = 0$ m-a.e. implies $\lim |\varphi_n, \psi_n| = 0$ m-stochastically. Theorem 2.9 now does the job.

2. We shall be through as soon as we have proved 2.1. (Theorem 2.9 resp. 1 above settles the rest.) Choose $0 < n_1 < n_2 < \cdots$ in such a fashion that

$$m\left(|\varphi_n, \varphi_{n_k}| \geqq \frac{1}{2^k}\right) < \frac{1}{2^k} \qquad (n \geqq n_n).$$

Put $E_k = \{|\varphi_{n_{k+1}}, \varphi_{n_k}| \geqq 1/2^k\}$. Clearly $N = \bigcap_j \bigcup_{k \geqq j} E_k$ is an m-nullset. If $x \in X \backslash N$, then $\omega \notin \bigcup_{k \geqq j} E_k$ for some j, hence

$$|\varphi_{n_{k+1}}(x), \varphi_{n_k}(x)| < \frac{1}{2^k} \qquad (k \geqq j).$$

Since $(\Omega, |\cdot, \cdot|)$ is complete, $\lim_k \varphi_{n_k}(x)$ exists in X. Choose any $\omega_0 \in \Omega_0$ and put $\varphi(x) = \omega_0$ if $x \in N$ and $\varphi(x) = \lim_k \varphi_{n_k}(x)$ if $x \in \Omega \backslash N$. This does it.

The last topic to be treated in this section is a kind of converse to Theorem 2.10.1: In a suitable model, convergence a.e. is behind a given weak convergence in a Polish space.

The next proposition is a prelude to the final theorem (2.12).

2.11. Proposition. *Let m denote the restriction of the Lebesgue σ-content to the σ-field $\mathscr{S}$ of all Borel subsets of* $[0, 1]$. *Let* $(\Omega, \mathscr{T})$ *be a Polish space and* $\mathscr{B} = \mathscr{B}(\mathscr{T})$. *Then for every probability σ-content p on $\mathscr{B}$ there is an $\mathscr{S}$-$\mathscr{B}$-measurable* $\varphi\colon [0, 1] \to \Omega$ *such that* $\varphi m = p$.

Proof. Consider a complete metric $|\cdot, \cdot|$ describing $\mathscr{T}$. Let us construct, for every $r = 1, 2, \ldots$, a countable partition $\zeta_r \subseteq \mathscr{B}$ of Ω such that $A \in \zeta_r \Rightarrow \operatorname{diam} A = \sup\{|\omega, \eta| \mid \omega, \eta \in A\} < 1/r$ and ζ_{r+1} is a refinement of ζ_r, i.e., every $A \in \zeta_r$ is a union of some members of ζ_{r+1}. This is easily achieved with the help of a dense sequence $\omega_1, \omega_2, \ldots \in \Omega$ and the sets of the form $\{\omega \mid |\omega, \omega_n| < 1/3r\}$ $(r, n = 1, 2, \ldots)$. For every $r = 1, 2, \ldots$, we choose some $\omega_{r, A}$ in every $A \in \zeta_r$ (we assume all members of ζ_r to be nonempty) and put $p_r = \sum_{A \in \zeta_r} p(A)\delta_{\omega_{r, A}}$, which is meaningful in the sense of total variation norm convergence since $\sum_{A \in \zeta_r} p(A) = 1$. Now we show $\lim_r p_r = p$ (weakly). For this we choose any closed $F \subseteq \Omega$ and consider, for an arbitrary $\varepsilon > 0$, the closed set $F_\varepsilon = \{\omega \mid \inf\{|\eta, \omega| \mid \eta \in F\} \leqq \varepsilon\}$. Clearly $\omega_A \in F \Rightarrow A \subseteq F_\varepsilon$ if $1/r < \varepsilon$, and thus

$$\limsup_r p_r(F) = \limsup_r \sum_{\omega_{r, A} \in F} p(A) \leqq p(F_\varepsilon).$$

Since $\lim_{k \to \infty} p(F_{1/k}) = p(F)$, we get $\limsup_r p_r(F) \leqq p(F)$ and thus $\lim_r p_r = p$ (weakly) follows by Proposition 2.3.3. For every $r = 1, 2, \ldots$, we may write $[0, 1]$ as a disjoint union of half-open and possibly empty intervals $I_{r, A}$ $(A \in \zeta_r)$ such that $m(I_{r, A}) = p(A)$ $(A \in \zeta_r)$. Moreover, we may assume $A \in \zeta_r$, $B \in \zeta_{r+1}$, $B \subseteq A$, $p(B) > 0 \Leftrightarrow \varnothing \neq I_{r+1, B} \subseteq I_{r, A}$. Define $\varphi_r\colon [0, 1] \to \Omega$ by $\varphi_r(x) = \omega_{r, A}$ $(x \in I_{r, A})$ and $\varphi_r(1) = \omega_0$ for some fixed $\omega_0 \in \Omega$. It is obvious that $\varphi_1, \varphi_2, \ldots$ are $\mathscr{S}$-$\mathscr{B}$-measurable with $\varphi_r m = p_r$ $(r = 1, 2, \ldots)$. Let $0 < r < s$ and $x \in I_{s, B} \subseteq I_{r, A}$ for $A \in \zeta_r$, $B \in \zeta_s$. Then $\varphi_r(x) = \omega_{r, A} \in A$ and $\varphi_s(x) = \omega_{s, B} \in B \subseteq A$. Since A has diam $A < r$, we get $|\varphi_r(x), \varphi_s(x)| < 1/r$. This inequality holds triviallv for $x = 1$. Thus the sequence $\varphi_1(x), \varphi_2(x), \ldots$ is a $|\cdot, \cdot|$-fundamental sequence, and hence convergent to some $\varphi(x) \in \Omega$, for every $x \in [0, 1]$. From Theorem 2.10.1 we now deduce $m = p$.

2.12. Theorem. *Let m denote the restriction of the Lebesgue σ-content to the σ-field $\mathscr{S}$ of all Borel subsets of* $[0, 1]$. *Let* $(\Omega, \mathscr{T})$ *be a Polish space,* $\mathscr{B} = \mathscr{B}(\mathscr{T})$ *and* $0 \leqq p, p_1, p_2, \ldots \in \operatorname{cont}^\sigma(\mathscr{B})$ *such that* $p(\Omega) = p_1(\Omega) = p_2(\Omega) = \cdots = 1$ *and* $\lim_n p_n = p$ *(weakly). Then there are $\mathscr{S}$-$\mathscr{B}$-measurable mappings* $\varphi, \varphi_1, \varphi_2, \ldots\colon [0, 1]$ *such that* $\varphi m = p$, $\varphi_1 m = p_1$, $\varphi_2 m = p_2, \ldots$ *and* $\lim_{n \to \infty} \varphi_n = \varphi$ *m*-a.e.

Proof. Let $|\cdot, \cdot|$ be a complete metric defining $\mathscr{T}$. It is easy to construct countable disjoint decompositions $\zeta_1, \zeta_2, \ldots$ of Ω into nonempty parts such that $A \in \zeta_r \Rightarrow \operatorname{diam} A < r$ and $m(\partial A) = 0$ (for the latter, slightly modify the "radii" occurring in the construction of the preceding proof), and ζ_{r+1} is a refinement of ζ_r $(r = 1, 2, \ldots)$. For every $r = 1, 2, \ldots$, we construct: (a) a decomposition of $[0, 1]$ into half-open (possibly empty) intervals $I_{r,A}$ $(A \in \zeta_r)$ such that $m(I_{r,A}) = p(A)$ $(A \in \zeta_r)$; and (b) for every $n = 1, 2, \ldots$, a decomposition of $[0, 1]$ into half-open intervals $I^n_{r,A}$ $(A \in \zeta_r)$ such that $m(I^n_{r,A}) = p_n(A)$ $(A \in \zeta_r)$. We may assume that for $A \in \zeta_r$, $B \in \zeta_{r+1}$, the equivalences

$$B \subseteq A,\ p(B) > 0 \quad \Leftrightarrow \quad \varnothing \neq I_{r+1,B} \subseteq I_{r,A}$$
$$B \subseteq A,\ p_n(B) > 0 \quad \Leftrightarrow \quad \varnothing \neq I^n_{r+1,B} \subseteq I^n_{r,A} \qquad (n = 1, 2, \ldots)$$

hold, and moreover that for $n = 1, 2, \ldots, 0 < r < s$, $A \in \zeta_r$, $B \in \zeta_s$, the interval $I^n_{r,A}$ lies left of $I^n_{s,B}$ iff $I_{r,A}$ lies left of $I_{s,B}$. From $p(\partial A) = 0$ we deduce

$$\lim_n m(I^n_{r,A}) = \lim_n p_n(A) = p(A) = m(I_{r,A}) \qquad (A \in \zeta_r,\quad r = 1, 2, \ldots).$$

We thus see that the lengths of the $I^n_{r,A}$ tend to the lengths of the corresponding $I_{r,A}$. Fix any r, $A \in \zeta_r$. The $I^n_{r,B}$, $B \in \zeta_r$, lying left of $I^n_{r,A}$ have a sum of lengths tending, for $n \to \infty$, to a value $\geqq$ the left endpoint of $I_{r,A}$. The length of $I^n_{r,A}$ tends to the length of $I_{r,A}$. Adding a similar argument "from the right," we see that the left endpoint of $I^n_{r,A}$ tends to the left endpoint of $I_{r,A}$, and the right endpoint of $I^n_{r,A}$ tends to the right endpoint of $I_{r,A}$. Choose now $\omega_{r,A} \in A$ $(r = 1, 2, \ldots, A \in \zeta_r)$ and put $\psi_{rn}(x) = \omega_{r,A}$ $(x \in I^n_{r,A},\ r = 1, 2, \ldots, A \in \zeta_r)$, $\psi_{rn}(1) = \omega_0$ for some fixed $\omega_0 \in \Omega$, and similarly $\psi_r(x) = \omega_{r,A}$ $(x \in I_{r,A}, r = 1, 2, \ldots, A \in \zeta_r)$, $\psi_r(1) = \omega_0$. Then we get $\lim_n \psi_{rn}(x) = \psi_r(x)$ for $x = 1$ and every x that is not an endpoint of some $I_{r,A}$, even with equality for n large. Clearly $\lim_r \psi_r(x) = \varphi(x)$ exists for every $x \in [0, 1]$, and also $\lim_r \psi_{rn}(x) = \varphi_n(x)$ exists for every $x \in [0, 1]$. It is now easy to deduce $\lim_n \varphi_n(x) = \varphi(x)$ m-a.e., namely for $x = 1$ and every x that is not an endpoint of some $I_{r,A}$.

In fact, fix such an x and an $\varepsilon > 0$ and choose some r with $1/r < \varepsilon$. Find $A \in \zeta_r$ with $x \in I_{r,A}$ and find n_0 such that $n \geqq n_0 \Rightarrow x \in I^n_{r,A}$. Now we have $|\psi_r(x), \varphi(x)| \leqq 1/r < \varepsilon$, $|\psi_{rn}(x), \varphi_n(x)| \leqq 1/r < \varepsilon$, $\psi_{rn}(x) = \omega_{r,A} = \psi_r(x)$ and hence $|\varphi_n(x), \varphi(x)| < 2\varepsilon$ $(n \geqq n_0)$. By construction we have $\varphi m = p$, $\varphi_1 m = p_1$, $\varphi_2 m = p_2, \ldots$.

CHAPTER XII

THE HAAR MEASURE ON LOCALLY COMPACT GROUPS

From Exercises I.3.3.5, I.3.3.6, and II.3.3 we know that the Lebesgue σ-content on the system of, say, all bounded Borel subsets of $\mathbb{R}^n$ is *translation* (and even rotation) invariant. It is even easy to see that it is *uniquely determined* by translation invariance, provided we prescribe the content of the unit cube to equal 1: If we partition that cube into the 2^r dyadic cubes of order r, then the content must have the same value on all of these, hence that value is $1/2^r$. Now $\mathbb{R}^n$ is a simple example of a *locally compact group*. The purpose of the present chapter is to carry over the ideas just displayed to the general case of an arbitrary locally compact topological group $(G, \mathscr{T})$. Since we do not in general assume commutativity, we have to be aware of the difference of left and right translation. We shall in fact obtain both a left invariant and a right invariant measure on G, each of them uniquely determined up to a positive factor (Theorem 3.10). The existence of such measures was first achieved by the Hungarian mathematician ALFRED HAAR [3] in 1933. Uniqueness was first proved by A. WEIL [1] in 1941. Both authors had several predecessors who settled special cases. An account of the history of Haar measure may be found in HEWITT and ROSS [1] whose proof (which in turn is based on a paper by H. CARTAN [1]) we essentially follow in Sections 2 and 3.

The special case of a compact group is of a particular simplicity and beauty. It can actually be viewed as an introductory chapter to the theory of almost periodic functions on arbitrary groups (see MAAK [2]). For this reason we treat it separately, as an easier prelude to the general theory. The so-called *marriage theorem* of PH. HALL [1] and MAAK [1, 2], which is a purely combinatorial result, plays a central role here, and basic ideas of Riemann integration theory lie transparently at the base of the story. In his unpublished Princeton lectures on measure theory (notes taken by P. R. HALMOS, 1940–1941), V. NEUMANN [3] showed that these ideas can also be used in order to settle the general locally compact case if one uses a generalization, due to KAKUTANI, of the marriage theorem. The resulting theory looks however somewhat clumsy, and I have therefore resisted the temptation to incorporate it into this text. There is, as E. HEWITT told me orally, no easy way to Haar measure in the general case.

Section 1, which deals with the compact case, is followed by a Section 2 which contains the general machinery of locally compact groups, as far as it is used in Section 3, where we prove the *existence and uniqueness of Haar measure* in the general locally compact case. The reader is advised to consult HEWITT and ROSS [1] for further information.

Throughout this chapter we work essentially in the frame of measures, leaving the applications to content theory as exercises.

1. THE HAAR MEASURE ON COMPACT GROUPS

In this section we treat the existence and uniqueness of Haar measure on *compact topological groups* by proving first that every continuous real function on such a group is *almost periodic*, and then proving the existence and uniqueness of *invariant averages* for almost periodic functions via the so-called *marriage theorem.* We do this in order to achieve a certain transparence of methods. It should be noted that we obtain no real gain in generality since every abstract group G can be densely imbedded into a compact topological group $\tilde{G}$ (called its Bohr compactification) such that the almost periodic functions on G are nothing but the restrictions to G of the continuous functions on $\tilde{G}$ after passage to the factor group of G modulo the subgroup of those elements of G that cannot be separated from the neutral element of G by almost periodic functions (see LOOMIS [2]).

1.1. Definition. *Let G be a group whose neutral element we denote by* 1. *Let $\mathcal{T} \subseteq \mathcal{P}(G)$ be a Hausdorff topology on G such that:*

1.1.1. *$(G, \mathcal{T})$ is a compact topological space.*

1.1.2. *The mapping $G \times G \to G$ defined by $(x, y) \to xy$ (group multiplication in G) is continuous if the product topology $\mathcal{T} \times \mathcal{T}$ is employed in $G \times G$.*

1.1.3. *The mapping $x \to x^{-1}$ of G onto G is continuous.*

Then the couple $(G, \mathcal{T})$ is called a **compact (topological) group**.

1.2. Remarks. Let G be a group and $\mathcal{T} \subseteq \mathcal{P}(G)$ a compact Hausdorff topology on G.

1.2.1. The continuity condition of Definition 1.1.2 reads in detail as follows: For any $x_0, y_0 \in G$ and every neighborhood W of $x_0 y_0 \in G$, there is a neighborhood U of x_0 and a neighborhood V of y_0 such that $x \in U, y \in V \Rightarrow xy \in W$.

1.2.2. The continuity condition of Definition 1.1.2 implies that for every $x \in G$ the left translation map L_x: $y \to xy$ of $G \to G$ as well as the right translation map R_x: $y \to yx$ is a bijection that is continuous for $\mathcal{T}$, as well as their inverses $L_{x^{-1}}$ resp. $R_{x^{-1}}$. Clearly $L_x R_y = R_y L_x$ $(x, y \in G)$. Writing, for any subset M of G, xM for $\{xy \mid y \in M\} = L_x M$, and Mx for $R_x M$, as usual in group theory, we thus see, upon considering the homeomorphisms L_x, R_x, that xM is a neighborhood of $x \in G$ iff M is a neighborhood of $1 \in G$ iff Mx is a neighborhood of $x \in G$. Using the L_s, R_t for the transform of functions f on G in the obvious way (e.g., $(L_s f)(x) = f(L_s x) = f(sx)$ $(s, x \in G)$), we find

$$L_s \mathscr{C}(G, \mathbb{R}) = \mathscr{C}(G, \mathbb{R}) = R_t \mathscr{C}(G, \mathbb{R}) \qquad (s, t \in G).$$

1.2.3. The continuity condition of Definition 1.1.2 implies that for every neighborhood U of $1 \in G$, there is a neighborhood V of G such that $VV = \{xy \mid x, y \in V\} \subseteq U$.

1.2.4. The continuity condition of Definition 1.1.3 will not be used in this section. We mention in passing that it can be deduced from 1.1.1 and 1.1.2 (Exercise 1.11). Actually, even the continuity of all L_x and all R_x would suffice (Ellis [1]).

1.2.5. The continuity conditions 1.1.2 and 1.1.3 of the above definition, plus the Hausdorff property of $(G, \mathcal{T})$, make up the *usual definition of the concept of a topological group* (compact or not). Clearly remarks 1–3 apply to general topological groups.

We need an elementary fact about uniform continuity.

1.3. Proposition. *Let $(G, \mathcal{T})$ be a compact topological group and $f \in \mathscr{C}(G, \mathbb{R})$ (i.e., a continuous real function on G). Then for every $\varepsilon > 0$, there is a neighborhood U of $1 \in G$ such that*

1.3.1. $\quad x, y \in G, y \in xU \quad \Rightarrow \quad |f(sx) - f(sy)| < \varepsilon \qquad (s \in G)$

1.3.2. $\quad x, y \in G, y \in Ux \quad \Rightarrow \quad |f(xt) - f(yt)| < \varepsilon \qquad (t \in G)$.

Proof. 1. For every $x \in G$, choose a neighborhood $U(x)$ of $1 \in G$ such that $y \in xU_x \Rightarrow |f(x) - f(y)| < \varepsilon/2$, and choose a neighborhood V_x of $1 \in G$ such that $V_x V_x \subseteq U_x$. Clearly $G = \bigcup_{x \in G} xV_x$, and by compactness we can find $x_1, \ldots, x_n \in G$ with $G = x_1 V_{x_1} \cup \cdots \cup x_n V_{x_n}$. Let $W = V_{x_1} \cap \cdots \cap V_{x_n}$. For any $x, y \in G$ with $y \in xW$, we determine k such that $x \in x_k V_{x_k}$. We then have $y \in xW \subseteq xV_{x_k} \subseteq x_k V_{x_k} V_{x_k} \subseteq x_k V_{x_k}$ and obviously $x \in x_k U_{x_k}$. We conclude

$$|f(x) - f(y)| \leqq |f(x) - f(x_k)| + |f(x_k) - f(y)| < \tfrac{1}{2}\varepsilon + \tfrac{1}{2}\varepsilon = \varepsilon.$$

Clearly $y \in xW$, $s \in G \Rightarrow sy \in sxW \Rightarrow |f(sx) - f(sy)| < \varepsilon$.

2. In a symmetric fashion, we obtain a neighborhood W' of $1 \in G$ such that $x \in W'y \Rightarrow |f(xt) - f(yt)| < \varepsilon$ $(t \in G)$. Upon putting $U = W \cap W'$, we arrive at the full result of our proposition.

There is no patent topology in the following

1.4. Definition. *Let G be any group and f a bounded real function on G.*

1.4.1. *A finite covering $G = M_1 \cup \cdots \cup M_n$ of G by nonempty subsets $M_1, \ldots, M_n$ of G is called an* **ε-covering** *for f if for every $k = 1, \ldots, n$,*

$$x, x' \in M_k, \; s, t \in G \quad \Rightarrow \quad |f(sxt) - f(sx't)| < \varepsilon.$$

1.4.2. *f is called* **almost periodic** *if for every $\varepsilon > 0$ there is an ε-covering for f.*

1.5. Proposition. *Let $(G, \mathscr{T})$ be a compact topological group. Then every $f \in \mathscr{C}(G, \mathbb{R})$ is an almost periodic function.*

Proof. Let $f \in \mathscr{C}(G, \mathbb{R})$. Clearly f is bounded. Choose any $\varepsilon > 0$. Find, by Proposition 1.3, some neighborhood U of $1 \in G$ such that $t, t' \in G$, $t \in t'U \Rightarrow |f(sxt) - f(sxt')| < \varepsilon/4$ $(s, x \in G)$. By compactness we may find $t_1, \ldots, t_r \in G$ such that $G = t_1 U \cup \cdots \cup t_r U$. Let $f_j \in \mathscr{C}(G, \mathbb{R})$ be defined by $f_j(x) = f(xt_j)$ $(j = 1, \ldots, r)$. Find, again by Proposition 1.3, some neighborhoods $V_1, \ldots, V_r$ of $1 \in G$ such that $y \in xV_j \Rightarrow |f_j(sx) - f_j(sy)| < \varepsilon/4$ $(j = 1, \ldots, r)$. Put $V = V_1 \cap \cdots \cap V_r$. Clearly

$$y \in xV \Rightarrow |f_j(sx) - f_j(sy)| < \varepsilon/4 \qquad (j = 1, \ldots, r).$$

By compactness we may find $y_1, \ldots, y_n \in G$ with $G = y_1 V \cup \cdots \cup y_n V$. Put $M_k = y_k V$ $(k = 1, \ldots, n)$. We show that $G = M_1 \cup \cdots \cup M_n$ is an ε-covering for f. Let $k \in \{1, \ldots, n\}$ be arbitrary. Choose $x, x' \in M_k$, $s, t \in G$. Find $j \in \{1, \ldots, s\}$ such that $t \in t_j U$. We conclude

$$\begin{aligned} |f(sxt) - f(sx't)| &\leqq |f(sxt) - f(sxt_j)| + |f_j(sx) - f_j(sy_k)| \\ &\quad + |f_j(sy_k) - f_j(sx')| + |f(sx't_j) - f(sx't)| \\ &< \tfrac{1}{4}\varepsilon + \tfrac{1}{4}\varepsilon + \tfrac{1}{4}\varepsilon + \tfrac{1}{4}\varepsilon = \varepsilon. \end{aligned}$$

Our next result is a purely combinatorial one.

1.6. Proposition (marriage theorem). *Let M, W be two finite nonempty sets. For every $k \in M$, let $F(k) \subseteq W$ be such that*

$$(1) \qquad C \subseteq M \quad \Rightarrow \quad \left| \bigcup_{k \in C} F(k) \right| \geqq |C|,$$

where $|F|$ denotes the cardinality of the set F, as usual. Then there is a one-to-one map

$$\text{ⴵ}: M \to W \qquad \text{such that} \qquad \text{ⴵ}(k) \in F(k) \quad (k \in M).$$

Remark. This proposition goes back to PH. HALL [1], and in a slightly specialized form, to MAAK [1]. KÖNIG [1] had proved an equivalent theorem as early as 1916. The name "marriage theorem" was introduced by WEYL [2] and is based on the following interpretation: M is a set of men, W is a set of women; for every man $k \in M$, the set $F(k)$ is the set of all girl friends of k; condition (1) says that there is never a lack of ladies whenever some clan C of men invites all their girl friends for a party; the marriage mapping ⴵ sends every man k into the arms of one of his girl friends in a one-to-one way (monogamy). The following proof is due to HALMOS and VAUGHAN [1].

Proof of Proposition 1.6. We use induction over the number $|M|$ of men. If $|M| = 1$, then (1) states that the only man has at least one girl friend, and ⴵ is established trivially. Let now $|M| > 1$ and assume that in all cases with fewer than $|M|$ men the proposition is valid.

Case I: Condition (1) is oversatisfied in the sense that

$$\varnothing \neq C \subseteq M, \ C \neq M \quad \Rightarrow \quad \left| \bigcup_{k \in C} F(k) \right| \geqq |C| + 1.$$

In this case we choose some $k_0 \in M$ and some $j_0 \in F(k_0)$, define ⴵ$(k_0) = j_0$ and apply the induction hypothesis to the situation where $M_0 = M \backslash \{k_0\}$, $W_0 = W \backslash \{j_0\}$, $F_0(k) = F(k) \backslash \{j_0\}$; we get the rest of ⴵ since the analogue of (1) is now trivially satisfied again.

Case II. There is some $\varnothing \neq C_0 \neq M$ with $|\bigcup_{k \in C_0} F(k)| = |C_0|$. Clearly we may apply the induction hypothesis in order to obtain ⴵ on C_0. Put now

$$M_1 = M \backslash C_0, \qquad W_1 = W \Big\backslash \Big(\bigcup_{k \in C_0} F(k) \Big), \qquad F_1(k) = F(k) \Big\backslash \Big(\bigcup_{k \in C_0} F(k) \Big).$$

Clearly this scheme satisfies (1) again: If there were some $C_1 \subseteq M_1$ with $|\bigcup_{k \in C_1} F_1(k)| < |C_1|$, the original scheme would fail to satisfy (1) for

$C = C_0 + C_1$. Thus we may once more apply the induction hypothesis and define ∞ also on M_1.

We shall need the following

1.7. Corollary. *Let G be a nonempty set and $G = A_1 \cup \cdots \cup A_n = B_1 \cup \cdots \cup B_n$ be two coverings of G such that*

$$(2) \qquad C \subseteq \{1, \ldots, n\} \;\Rightarrow\; \left| \left\{ j \,\middle|\, \left(\bigcup_{k \in C} A_k \right) \cap B_j \neq \varnothing \right\} \right| \geqq |C|.$$

Then there is a one-to-one mapping $\tau\colon \{1, \ldots, n\} \to \{1, \ldots, n\}$ such that $A_k \cap B_{\tau(k)} \neq \varnothing$ $(k = 1, \ldots, n)$.

Proof. Let $M = \{1, \ldots, n\} = W$. For every $k \in M$, let $F(k) = \{j \mid j \in W,\ A_k \cap B_j \neq \varnothing\}$. Clearly (2) implies (1). Let $\tau = \infty$. Then $\infty(k) \in F(k)$ reads $A_k \cap B_{\tau(k)} \neq \varnothing$.

We apply these results now to a theory of mean values for almost periodic functions.

1.8. Proposition. *Let G be a group and f a bounded real function on G.*

1.8.1. *For any $x_1, \ldots, x_n \in G$, we define*

$$\underline{m}(f; x_1, \ldots, x_n) = \inf\left\{ \frac{1}{n} \sum_{k=1}^{n} f(s x_k t) \,\middle|\, s, t \in G \right\}$$

$$\overline{m}(f; x_1, \ldots, x_n) = \sup\left\{ \frac{1}{n} \sum_{k=1}^{n} f(s x_k t) \,\middle|\, s, t \in G \right\}.$$

and consider the closed interval

$$J(f; x_1, \ldots, x_n) = [\underline{m}(f; x_1, \ldots, x_n), \overline{m}(f; x_1, \ldots, x_n)].$$

Then

1.8.2. *for any $x_1, \ldots, x_n, y_1, \ldots, y_r \in G$, we have*

$$J(f; x_1, \ldots, x_n) \cap J(f; y_1, \ldots, y_r) \neq \varnothing.$$

1.8.3. *Assume now that f is almost periodic. Then for every $\varepsilon > 0$, there are $x_1, \ldots, x_n \in G$ such that $J(f; x_1, \ldots, x_n)$ has a length $< \varepsilon$.*

Proof. 2. Clearly $(nr)^{-1} \sum_{k=1}^{n} \sum_{j=1}^{r} f(x_k y_j)$ is in $J(f; x_1, \ldots, x_n)$ because it can be written as an arithmetic mean

$$\frac{1}{r} \sum_{j=1}^{r} \left(\frac{1}{n} \sum_{k=1}^{n} f(x_k y_j) \right)$$

of reals belonging to $J(f;\, x_1, \ldots, x_n)$; for symmetric reasons, it is in $J(f;\, y_1, \ldots, y_r)$.

3. Let $G = M_1 \cup \cdots \cup M_n$ be an $(\varepsilon/6)$-cover of G for f which is minimal in the sense that there is no $(\varepsilon/6)$-cover of G for f that has strictly fewer than n members. We choose $x_1 \in M_1, \ldots, x_n \in M_n$ and show that $J(f;\, x_1, \ldots, x_n)$ is contained in the $(\varepsilon/3)$-neighborhood of $n^{-1} \sum_{k=1}^{n} f(x_k)$. For this, it is sufficient to establish, for any choice of $s, t \in G$, the inequality

$$\left| \frac{1}{n} \sum_{k=1}^{n} f(sx_k t) - \frac{1}{n} \sum_{k=1}^{n} f(x_k) \right| < \frac{\varepsilon}{3}.$$

It is nearly obvious that $G = (sM_1 t) \cup \cdots \cup (sM_n t)$ is an $(\varepsilon/6)$-cover of G for f again, and obviously a minimal one. We now apply Corollary 1.7 with $A_k = M_k$, $B_k = sM_k t$ $(k = 1, \ldots, n)$. Condition (2) is in fact verified: Assume that some union of, say, p of the M_k intersects only $p' < p$ of the $sM_j t$, then these p' of the $sM_j t$ cover that union of p of the M_k. These p' of the $sM_j t$ then form, together with the remaining $n - p$ of the M_k, an $(\varepsilon/6)$-covering of G for f having $n - p + p' < n$ members, contradicting minimality. From Corollary 1.7 we deduce now the existence of a permutation τ of $\{1, \ldots, n\}$ such that $M_k \cap sM_{\tau(k)} t \neq 0$; choose some y_k in this set $(k = 1, \ldots, n)$. Then

$$\begin{aligned} \frac{1}{n} \sum_{k=1}^{n} f(sx_k t) - \frac{1}{n} \sum_{k=1}^{n} f(x_k) &\leqq \frac{1}{n} \sum_{k=1}^{n} [\,|f(sx_k t) - f(sy_k t)| \\ &\qquad + |f(sy_{\tau(k)} t) - f(x_{\tau(k)})|\,] \\ &< \frac{1}{n} \sum_{k=1}^{n} \left[\frac{\varepsilon}{6} + \frac{\varepsilon}{6}\right] = \frac{\varepsilon}{3}. \end{aligned}$$

1.9. Theorem (mean value theorem for almost periodic functions). *Let G be any group. Denote by $\mathscr{A}(G)$ the set of all almost periodic functions on G. Then:*

1.9.1. *$\mathscr{A}(G)$ is a vector lattice of bounded real functions on G and contains all constants.*

1.9.2. *There is a unique positive linear form m: $\mathscr{A}(G) \to \mathbb{R}$, called the* **mean value**, *such that:*

1.9.2.1. $m(L_s R_t f) = m(f)$ $(f \in \mathscr{A}(G),\ s, t \in G)$

1.9.2.2. *m attains the value 1 on the constant function 1.*

Proof. 1. Let $f, g \in \mathscr{A}(G)$, $\varepsilon > 0$, and $G = A_1 \cup \cdots \cup A_r$ be an ε-covering for f and $G = B_1 \cup \cdots \cup B_s$ an ε-covering for g. It is obvious that

$$G = \bigcup_{k=1}^{r} \bigcup_{j=1}^{s} A_k \cap B_j$$

defines an ε-covering for both f and g, and hence an ε-covering for $f \vee g$ and $f \wedge g$, and a 2ε-covering for $f + g$. The rest of 1 is still more trivial.

2. Propositions 1.8.2 and 3 imply that for every $f \in \mathscr{A}(G)$ the set

$$\bigcap_{n \geqq 1,\, x_1, \ldots, x_n \in G} J(f; x_1, \ldots, x_n)$$

consists of exactly one point, which we denote by $m(f)$; and it is clear that for any $\varepsilon > 0$ there are $x_1, \ldots, x_n \in G$ such that

$$\left| \frac{1}{n} \sum_{k=1}^{n} f(sx_k t) - m(f) \right| < \varepsilon \qquad (s, t \in G). \tag{3}$$

If we choose a further $g \in \mathscr{A}(G)$ and $y_1, \ldots, y_r \in G$ with

$$\left| \frac{1}{r} \sum_{j=1}^{r} g(sy_j t) - m(g) \right| < \varepsilon \qquad (s, t \in G), \tag{4}$$

we get the obvious conclusion

$$\left| \frac{1}{nr} \sum_{k=1}^{n} \sum_{j=1}^{r} (f + g)(sx_k y_j t) - (m(f) + m(g)) \right| < 2\varepsilon \qquad (s, t \in G).$$

This shows that $J(f + g; x_1 y_1, \ldots, x_n y_r)$ is of length $\leqq 4\varepsilon$ and contains $m(f) + m(g)$. It is obvious how to deduce $m(f + g) = m(f) + m(g)$ from this argument. With a few more of similar arguments we reach the conclusion that $m\colon \mathscr{A}(G) \to \mathbb{R}$ is a positive linear form. Since $J(L_s R_t f; x_1, \ldots, x_n) = J(f; x_1, \ldots, x_n)$ holds for any bounded function f on G and any $x_1, \ldots, x_n, s, t \in G$, we obtain $m(L_s R_t f) = m(f)$ ($f \in \mathscr{A}(G)$, $s, t \in G$). We have now established the existence of an $m\colon \mathscr{A}(G) \to \mathbb{R}$ with the required properties. In order to prove uniqueness, we take any $m'\colon \mathscr{A}(G) \to \mathbb{R}$ with the same properties. Fix some $f \in \mathscr{A}(G)$, choose some $\varepsilon > 0$, and find $x_1, \ldots, x_n \in G$ such that (3) holds. We may rewrite this as

$$m(f) - \varepsilon < \frac{1}{n} \sum_{k=1}^{n} f(sx_k t) < m(f) + \varepsilon \qquad (s, t \in G).$$

The assumed properties of m' ensure

$$m(f) - \varepsilon < \frac{1}{n} \sum_{k=1}^{n} m'(L_s R_t f) = m'(f) < m(f) + \varepsilon.$$

Since $\varepsilon > 0$ was arbitrary, $m'(f) = m(f)$ follows.

We now turn back to compact groups. Our next aim is a direct application of Theorem 1.9 to a proof of the existence and uniqueness of Haar measure.

1.10 Theorem. *Let $(G, \mathcal{T})$ be a compact topological group. There is a unique positive τ-measure m: $\mathcal{C}(G, \mathbb{R}) \to \mathbb{R}$ such that:*

1.10.1. $m(L_s R_t f) = m(f)$ $(f \in \mathcal{C}(G, \mathbb{R}),\ s, t \in G)$.

1.10.2. *m attains the value 1 on the constant 1.*

1.10.3. *We call m the (normalized)* **Haar measure** *on G or $(G, \mathcal{T})$.*

Property 1 above is worded as left and right (i.e., bilateral) invariance of m.

Proof. Since $\mathcal{C}(G, \mathbb{R}) \subseteq \mathcal{A}(G)$ by Proposition 1.5, the existence of an m with the stated properties follows immediately from Theorem 1.9. Uniqueness is obtained by copying the uniqueness proof for theorem 1.9. By theorem I.10.1 m is a τ-measure.

1.11. Exercises

1.11.1. Let $(G, \mathcal{T})$ be a compact Hausdorff space, and G a group such that the mapping $(x, y) \to xy$ (group multiplication) of $G \times G \to G$ is continuous for the product topology in $G \times G$. Prove that the mapping $x \to x^{-1}$ is also continuous. (*Hint*: Show first that it suffices to prove that $x \to x^{-1}$ is continuous at $1 \in G$; then prove this indirectly.)

1.11.2. Let $(G. \mathcal{T})$ be a compact topological group and m: $\mathcal{C}(G, \mathbb{R}) \to \mathbb{R}$ the normalized Haar measure on G. Let $\mathcal{B} = \mathcal{B}(\mathcal{T})$ be the σ-field of all Borel subsets of G, and $\mathcal{B}_0 = \mathcal{B}(\mathcal{C}(G, \mathbb{R}))$ (Definition V.3.2.1) the σ-field of all Baire subsets of G.

1.11.2.1. Prove $L_s \mathcal{B} = \mathcal{B} = R_t \mathcal{B}$, $L_s \mathcal{B}_0 = \mathcal{B}_0 = R_t \mathcal{B}_0$ $(s, t \in G)$.

1.11.2.2. Let m denote also the restriction to $\mathcal{B}$ of the σ-content τ-derived from the τ-measure m (Chapter III, Section 7) and let m_0 denote the restriction to $\mathcal{B}_0$ of the σ-content σ-derived from the τ-measure m (Theorem III.6.1). Prove $L_s m \fallingdotseq m = R_t m$, $L_s m_0 = m_0 = R_t m_0$ $(s, t \in G)$. (*Remark*: The transforms of σ-contents considered here are meaningful in view of 1.11.2.1.) Show that m_0 is the restriction of m to $\mathcal{B}_0$ and that $m(U) > 0$ for every nonempty open set $U \subseteq G$. The σ-content m on $\mathcal{B}$ is called the (normalized) Haar σ-content on G.

1.11.2.3. Let $(G, \mathcal{B}, m)$ and $(G, \mathcal{B}_0, m)$ be as defined in 1.11.2.2. Prove that m is the only normalized inner compact regular σ-content on $\mathcal{B}$ that satisfies $L_s m = m = R_t m$ $(s, t \in G)$. (*Hint*: Use the techniques and results of Chapter V, Section 3.)

1.11.2.4. Let m': $\mathcal{C}(G, \mathbb{R}) \to \mathbb{R}$ be a normalized positive τ-measure such that $m'(L_s f) = m'(f) = m'(R_t f)$ for all $f \in \mathcal{C}(G, \mathbb{R})$ and all s, t from a dense subset of G. Prove that $m' = m$.

1.11.2.5. Assume in particular that G has only finitely many, say n, elements. Show that the σ-content derived from m puts mass $1/n$ on every $x \in G$, i.e.,

$$m(f) = \frac{1}{n} \sum_{x \in G} f(x) \qquad (f \in \mathscr{C}(G, \mathbb{R}) = \mathbb{R}^G).$$

1.11.2.6. Let $T: G \to G$ be a continuous endomorphism such that TG is a dense subset of G. Prove that $Tm = m$.

2. DEFINITION AND BASIC MACHINERY OF LOCALLY COMPACT GROUPS

2.1. Definition. *Let G be a group whose neutral element we denote by* 1. *Let $\mathscr{T} \subseteq \mathscr{P}(G)$ be a Hausdorff topology on G such that:*

2.1.1. *$(G, \mathscr{T})$ is a locally compact space.*

2.1.2. *The mapping $(x, y) \to xy^{-1}$ (group multiplication of x with the inverse of y) of $G \times G$ (with the product topology $\mathscr{T} \times \mathscr{T}$) into G (with topology $\mathscr{T}$) is continuous.*

Then the couple $(G, \mathscr{T})$ is called a **locally compact (topological) group**.

2.2. Remarks

2.2.1. The simplest examples of locally compact groups are, of course, the additive group $\mathbb{R}^n$ with its usual topology and an arbitrary group with the discrete topology.

2.2.2. Notice that condition 2.1.2 implies (for $x = 1$) the continuity of the map $y \to y^{-1}$ of G onto itself, and hence, by composition of $(x, y) \to (x, y^{-1})$ and $(x, y^{-1}) \to xy = x(y^{-1})^{-1}$, also the continuity of the mapping $(x, y) \to xy$ of $G \times G$ onto G.

Condition 2.1.2 alone is the *usual definition of a topological group*. Remarks 1.2.1–1.2.3 thus apply also in our present case.

2.2.3. Since the mapping $x \to x^{-1}$ of G onto G is continuous and equals its own inverse, it is a homeomorphism. Thus if V is an open neighborhood of $1 \in G$, so are V^{-1} (which is our obvious notation for $\{x^{-1} | x \in V\}$) and $V \cap V^{-1}$.

2.2.4. If $C, D \subseteq G$ are compact sets, then $C \times D$ is a compact subset of $G \times G$ which is sent, by the continuous mapping $(x, y) \to xy$, into the compact set CD (which is our obvious notation for $\{xy | x \in C, y \in D\}$).

2.2.5. Since $(G, \mathscr{T})$ is also a locally compact Hausdorff space, all results on such topological spaces are available to us. We may, e.g., apply, for any compact $C \subseteq G$ and any open $U \supseteq C$, Urysohn's theorem and obtain a function $F \in \mathscr{C}^{00}(G, \mathbb{R})$ (recall that this denotes the vector lattice of all real-valued continuous functions on G, each of which vanishes in the complement of some compact set), which attains only values between 0 and 1, vanishes on $G \backslash U$, and is $= 1$ on C. We may, e.g., for any covering $C \subseteq U_1 \cup \cdots \cup U_n$ of a compact set C with open sets $U_1, \ldots, U_n$, find $0 \leqq h_1, \ldots, h_n \in \mathscr{C}^{00}(G, \mathbb{R})$ such that $h_1(x) + \cdots + h_n(x) = 1$ $(x \in C)$ and $h_k(x) = 0$ $(x \in G \backslash U_k$, $k = 1, \ldots, n)$.

We shall frequently need

2.3. Proposition. *Let $(G, \mathscr{T})$ be a locally compact topological group. Then for every $\varepsilon > 0$ and every $f \in \mathscr{C}^{00}(G, \mathbb{R})$ there is a neighborhood U of $1 \in G$ such that*

$$x^{-1}y \in U \quad \Rightarrow \quad |f(sx) - f(sy)| < \varepsilon \qquad (s \in G).$$

Proof. Let $C_0 \subseteq G$ be compact such that $x \in G \backslash C_0 \Rightarrow f(x) = 0$. Let W be a neighborhood of $1 \in G$ such that the closure $\overline{W}$ of W is compact. Passing to $W \cap W^{-1}$, we may assume $W = W^{-1}$. Put $C = C_0 \overline{W}$. Clearly C is compact. For every $x \in G$, determine a neighborhood $U_x \subseteq W$ of $1 \in G$ such that $y \in xU_x \Rightarrow |f(x) - f(y)| < \varepsilon/2$, and find a neighborhood V_x of $1 \in G$ with $V_x V_x \subseteq U_x$. Clearly we can find $x_1, \ldots, x_n \in G$ with $C \subseteq x_1 V_{x_1} \cup \cdots \cup x_n V_{x_n}$. Put $U = V_{x_1} \cap \cdots \cap V_{x_n}$. Let now $x^{-1}y \in U$. If $x \in G \backslash C$, then $xU \subseteq xW \subseteq G \backslash C_0$ because a $z \in C_0 \cap xW$ would imply the existence of some $w \in W$ with $z = xw$, i.e., $x = zw^{-1} \in C_0 W \subseteq C$, a contradiction. Thus $x^{-1}y \in U \Rightarrow y \in xU \subseteq G \backslash C_0 \Rightarrow f(y) = f(x) = 0$ in this case. Let now $x \in C$ and find k with $x \in x_k V_{x_k}$. We conclude

$$y \in xU \subseteq x_k V_{x_k} U \subseteq x_k V_{x_k} V_{x_k} \subseteq x_k U_{x_k},$$

and obviously $x \in x_k U_{x_k}$, and thus

$$|f(x) - f(y)| \leqq |f(x) - f(x_k)| + |f(x_k) - f(y)| < \tfrac{1}{2}\varepsilon + \tfrac{1}{2}\varepsilon = \varepsilon.$$

Since $(sx)^{-1}sy = x^{-1}y$, the proposition follows.

2.4. Remark. It is obvious that there is a "right" analogue of Proposition 2.3 and that for any choice of finitely many $f_1, \ldots, f_r \in \mathscr{C}^{00}(G, \mathbb{R})$ and any $\varepsilon > 0$, there is a neighborhood U of $1 \in G$ with

$$x^{-1}y \in U \quad \Rightarrow \quad |f_j(sx) - f_j(sy)| < \varepsilon \qquad (s \in G; j = 1, \ldots, r).$$

3. THE HAAR MEASURE ON LOCALLY COMPACT GROUPS

In this section we prove the *existence* and, in a way, *uniqueness of Haar measure,* now for the *general case of a locally compact group.* Clearly this comprises the compact case once again, but the situation is so much more complicated that a more sophisticated technique has to be developed, and the result is less simple: We get only the existence of a *left invariant* (and symmetrically, a right invariant) Haar measure, and it is uniquely determined only up to a constant factor. In many cases the left invariant Haar measure is not a constant multiple of the right invariant one. As to the construction technique, the basic idea is to take one function e as a meter for all others, so to speak, and to eliminate the influence of the special shape of that function e by measuring it against functions θ vanishing outside smaller and smaller neighborhoods of the neutral element 1 of G. For most of the entire proof we shall deal only with nonnegative functions.

3.1. Definition. *Let $(G, \mathcal{T})$ be a locally topological group. For any two functions $0 \leqq \theta,\ f \in \mathscr{C}^{00}(G, \mathbb{R})$ with $\theta \neq 0$ (i.e., $\theta(x) \neq 0$ for at least one $x \in G$), we define*

$$[f : \theta] = \inf\{\beta_1 + \cdots + \beta_n \,|\, n \geqq 1, \beta_1, \ldots, \beta_n \geqq 0, x_1, \ldots, x_n \in G,$$
$$f \leqq \beta_1 L_{x_1}\theta + \cdots + \beta_n L_{x_n}\theta\}.$$

Here $[f : \theta]$ might, in principle, be an infimum over an empty set, and hence $= \infty$, but the fact that f vanishes outside some compact set, and $\theta \neq 0$, will immediately ensure that this is not the case. We have, in fact, the

3.2. Lemma. *Let $(G, \mathcal{T})$ be a locally compact topological group, $0 \leqq f, f_1, f_2, \theta, \eta \in \mathscr{C}^{00}(G, \mathbb{R})$, $\theta, \eta \neq 0$, $0 \leqq \alpha \in \mathbb{R}$. Then:*

3.2.1. $0 \leqq [f : \theta] < \infty$.

3.2.2. $[\theta : \theta] = 1$.

3.2.3. $f_1 \leqq f_2 \Rightarrow [f_1 : \theta] \leqq [f_2 : \theta]$.

3.2.4. $[f : \theta] = 0 \Rightarrow f = 0$.

3.2.5. $[L_x f : \theta] = [f : L_x \theta] = [f : \theta]$ $(x \in G)$.

3.2.6. $[\alpha f : \theta] = \alpha[f : \theta]$.

3.2.7. $[f_1 + f_2 : \theta] \leqq [f_1 : \theta] + [f_2 : \theta]$.

3.2.8. $[f : \eta] \leqq [f : \theta] \cdot [\theta : \eta]$.

Proof. 1. Let $C \subseteq G$ be compact such that f vanishes on $G \backslash C$. Choose a real constant $A > 0$ such that $f \leqq A$ everywhere. Since $\theta > 0$ at least at one point $x_0 \in G$ and θ is continuous, there is an open $x_0 \in U \subseteq G$

and some real $\gamma > 0$ such that $\theta \geqq \gamma$ on U. From the open covering $C \subseteq \bigcup_{y \in C} L_{yx_0^{-1}} U$ we select a finite subcovering

$$C \subseteq \bigcup_{k=1}^{n} L_{y_k x_0^{-1}} U.$$

Put $x_k = x_0 y_k^{-1}$; clearly $z \in L_{y_k x_0^{-1}} U$ implies $x_k z \in U$; Thus $L_{x_k} \theta$ is $\geqq \gamma$ on $L_{y_k x_0^{-1}} U$ $(k = 1, \ldots, n)$ and consequently $L_{x_1}\theta + \cdots + L_{x_n}\theta \geqq \gamma$ on C, hence for $\beta_1 = \cdots = \beta_n = A/\gamma$, we get $\beta_1 L_{x_1}\theta + \cdots + \beta_n L_{x_n}\theta \geqq f$ everywhere on G, and $[f : \theta] \leqq nA/\gamma < \infty$ follows. In 2 the relation $\leqq$ is obvious, and $\geqq$ will follow from 3 and 7 (put $f = \theta$ there).

4. Let b be a real with $\theta \leqq b$. Then

$$\beta_1 L_{x_1}\theta + \cdots + \beta_n L_{x_n}\theta \leqq b(\beta_1 + \cdots + \beta_n)$$

for any reals $\beta_1, \ldots, \beta_n$ and any $x_1, \ldots, x_n \in G$. Thus $[f : \theta] = 0$ implies that f has an arbitrarily small positive constant as a majorant, hence $f = 0$.

5 is obvious from the translation invariant character of the definition of $[f : \theta]$. 6 is obvious, and 7 is an easy exercise.

8. Choose any reals $\beta_1, \ldots, \beta_n \geqq 0$ and elements $x_1, \ldots, x_n$ of G such that $f \leqq \beta_1 L_{x_1}\theta + \cdots + \beta_n L_{x_n}\theta$. Choose reals $\gamma_1, \ldots, \gamma_r \geqq 0$ and elements $y_1, \ldots, y_r$ of G such that $\theta \leqq \gamma_1 L_{y_1}\eta + \cdots + \gamma_r L_{y_r}\eta$. Then clearly

$$f \leqq \sum_{k=1}^{n} \sum_{j=1}^{r} \beta_k \gamma_j L_{x_k} L_{y_j} \eta,$$

hence

$$[f : \eta] \leqq \sum_{k,\, j} \beta_k \gamma_j = \left(\sum_{k=1}^{n} \beta_k\right) \cdot \left(\sum_{j=1}^{r} \gamma_j\right).$$

Varying the $\beta_k, x_k, \gamma_j, y_j$, we get $[f : \eta] \leqq [f : \theta] \cdot [\theta : \eta]$.

3.3. Definition. *Let $(G, \mathcal{T})$ be a locally compact topological group and $0 \leqq f, e, \theta \in \mathscr{C}^{00}(G, \mathbb{R})$, $e, \theta \neq 0$. We put*

$$m_\theta^e(f) = [f : \theta]/[e : \theta].$$

3.4. Remarks

3.4.1. Lemma 3.2.4 guarantees that the above quotient is always meaningful.

3.4.2. $m_\theta^e(f)$ need not, as a function of f, be additive. We shall be able to establish only that it becomes "more and more additive, the finer θ becomes."

As an immediate consequence of Lemma 3.2, we obtain

3.5. Lemma. *Let* $(G, \mathscr{T})$ *be a locally compact topological group and* $0 \leqq f, e, f_1, f_2, \theta \in \mathscr{C}^{00}(G, \mathbb{R})$, $e, \theta \neq 0$, $0 \leqq \alpha \in \mathbb{R}$. *Then:*

3.5.1. $0 \leqq m_\theta^e(f) < \infty$.

3.5.2. $m_\theta^e(e) = 1$.

3.5.3. $f_1 \leqq f_2 \Rightarrow m_\theta^e(f_1) \leqq m_\theta^e(f_2)$.

3.5.4. $m_\theta^e(f) = 0 \Rightarrow f = 0$.

3.5.5. $m_\theta^e(L_x f) = m_{L_x\theta}^e(f) = m_\theta^e(f)$.

3.5.6. $m_\theta^e(\alpha f) = \alpha m_\theta^e(f)$.

3.5.7. $m_\theta^e(f_1 + f_2) \leqq m_\theta^e(f_1) + m_\theta^e(f_2)$.

3.5.8. $f \neq 0 \Rightarrow 1/[e : f] \leqq m_\theta^e(f) \leqq [f : e]$.

The next lemma contains the result announced in Remark 3.4.2.

3.6. Lemma. *Let* $(G, \mathscr{T})$ *be a locally compact topological group and* $0 \leqq e, g_1, \ldots, g_n \in \mathscr{C}^{00}(G, \mathbb{R})$, $e \neq 0$. *Then for any reals* $\varepsilon, \alpha > 0$, *there is a neighborhood* U *of the neutral element* 1 *of* G *such that the following holds:*

$$0 \leqq \theta \in \mathscr{C}^{00}(G, \mathbb{R}),\ \theta \neq 0,\ \theta \text{ vanishes in } G\backslash U,\ \gamma_1, \ldots, \gamma_n \in \mathbb{R},\ 0 \leqq \gamma_1, \ldots, \gamma_n \leqq \alpha$$
$$\Rightarrow \quad |m_\theta^e(\gamma_1 g_1 + \cdots + \gamma_n g_n) - (\gamma_1 m_\theta^e(g_1) + \cdots + \gamma_n m_\theta^e(g_n))| \leqq \varepsilon.$$

Proof. By Lemmas 3.5.6 and 3.5.7 we always have

$$m_\theta^e(\gamma_1 g_1 + \cdots + \gamma_n g_n) \leqq \gamma_1 m_\theta^e(g_1) + \cdots + \gamma_n m_\theta^e(g_n).$$

It will thus be sufficient to end up with the inequality

$$\gamma_1 m_\theta^e(g_1) + \cdots + \gamma_n m_\theta^e(g_n) \leqq m_\theta^e(\gamma_1 g_1 + \cdots + \gamma_n g_n) + \varepsilon. \tag{1}$$

Find a compact $C \subseteq G$ such that $g_1, \ldots, g_n$ vanish in $G\backslash C$. Let V be a neighborhood of $1 \in G$ whose closure $\overline{V}$ is compact. Choose some $0 \leqq F \in \mathscr{C}^{00}(G, \mathbb{R})$ with $x \in C\overline{V} \Rightarrow F(x) = 1$, and $0 \leqq F \leqq 1$ (Remark 2.2.5). Let $\delta > 0$ be a real that we shall specify later. By Proposition 2.3 we can find a neighborhood $U \subseteq V \cap V^{-1}$ of $1 \in G$ such that

$$x^{-1}y \in U \quad \Rightarrow \quad |g_k(x) - g_k(y)| < \delta^3 \ (k = 1, \ldots, n)$$

and $|F(x) - F(y)| < \delta^3$. We shall prove that this U satisfies the conclusion of our lemma, provided we have chosen δ sufficiently small. Thus we choose some $0 \leqq \theta \in \mathscr{C}^{00}(G, \mathbb{R})$ with $\theta \neq 0$ and $\theta(x) = 0$ $(x \in G\backslash U)$. Consider the function $g = \gamma_1 g_1 + \cdots + \gamma_n g_n + \delta F$. Clearly $g \geqq \delta$ on $C\overline{V}$ and, for any $0 < \gamma \in \mathbb{R}$ with $g_1, \ldots, g_n \leqq \gamma$, also $g \leqq (\gamma_1 + \cdots + \gamma_n)\gamma + \delta \leqq n\alpha\gamma + \delta$. If we prescribe $\delta \leqq n\alpha\gamma$, we get $g \leqq 2n\alpha\gamma$. Moreover, $x^{-1}y \in U$ implies $|g(x) - g(y)| \leqq (\gamma_1 + \cdots + \gamma_n + \delta)\delta^3 \leqq n\gamma\delta^3 + \delta^4$. Define

$$h_k(x) = \begin{cases} \dfrac{\gamma_k g_k(x)}{g(x)} & (x \in C) \\ 0 & (x \in G \backslash C). \end{cases}$$

Clearly $0 \in h_k \in \mathscr{C}^{00}(G, \mathbb{R})$, $gh_k = \gamma_k g_k$ $(k = 1, \ldots, n)$ and $h_1 + \cdots + h_n \leqq 1$. For $x^{-1}y \in U$, $x, y \in C\overline{V}$, we get $g(x), g(y) \geqq \delta$ and

$$\begin{aligned} |h_k(x) - h_k(y)| &= \left| \frac{\gamma_k g_k(x)}{g(x)} - \frac{\gamma_k g_k(y)}{g(y)} \right| = \gamma_k \frac{|g_k(x)g(y) - g_k(y)g(x)|}{g(x)g(y)} \\ &\leqq \frac{\alpha}{\delta^2} (|g_k(x)g(y) - g_k(y)g(y)| + |g_k(y)g(y) - g_k(y)g(x)|) \\ &\leqq \frac{2n\alpha^2\gamma}{\delta^2} |g_k(x) - g_k(y)| + \frac{\alpha\gamma}{\delta^2} |g(x) - g(y)| \\ &\leqq \frac{2n\alpha^2\gamma}{\delta^2} \delta^3 + \frac{\alpha\gamma}{\delta^2} (n\gamma\delta^3 + \delta^4), \end{aligned}$$

and this will become $< \delta/n$ $(k = 1, \ldots, n)$ if δ has been chosen sufficiently small. For $x^{-1}y \in U$, we get $y^{-1}x \in U^{-1} \subseteq V$. Thus $y \in C$ would imply $x = yy^{-1}x \in C\overline{V}$. Similarly, $x \in C$ plus $x^{-1}y \in U \subseteq V$ implies $y \in C\overline{V}$. Hence, if one of x, y is not in $C\overline{V}$, the other is not in C, and hence $h_k(x) = 0 = h_k(y)$ in all these cases. Thus we have generally

$$|h_k(x) - h_k(y)| \leqq \delta/n \qquad (k = 1, \ldots, n, \quad x^{-1}y \in U)$$

This now allows us to estimate $[g : \theta]$ in the following fashion: If $0 \leqq \beta_1, \ldots, \beta_r \in \mathbb{R}$, $x_1, \ldots, x_r \in G$ satisfy $g \leqq \beta_1 L_{x_1}\theta + \cdots + \beta_r L_{x_r}\theta$, i.e.,

$$g(x) \leqq \sum_{j=1}^{r} \beta_j \theta(x_j x) \qquad (x \in G),$$

then we remark that $\theta(x_j x) = 0$ whenever $x \in G \backslash x_j^{-1}U$. Denoting by $\sum_j^{(x)}$ the sum over all those j for which $x \in x_j^{-1}U$, we get $g(x) \leqq \sum_j^{(x)} \beta_j \theta(x_j x)$ $(x \in G)$, hence

$$\gamma_k g_k(x) = g(x)h_k(x) \leqq \sum_j^{(x)} \beta_j \theta(x_j x) \left[h_k(x_j^{-1}) + \frac{\delta}{n} \right] \leqq \sum_{j=1}^{s} \beta_j \theta(x_j x) \left[h_k(x_j^{-1}) + \frac{\delta}{n} \right]$$

This implies $[\gamma_k g_k : \theta] \leqq \sum_{j=1}^{s} \beta_j (h_k(x_j^{-1}) + \delta/n)$, hence

$$\sum_{k=1}^{n} [\gamma_k g_k : \theta] \leqq (1 + \delta) \sum_{j=1}^{s} \beta_j .$$

Varying over the β_j, x_j, we find

$$\sum_{k=1}^{n} [\gamma_k g_k : \theta] \leqq (1 + \delta)[g : \theta].$$

Division by $[e : \theta]$ yields

$$\begin{aligned}\sum_{k=1}^{n} \gamma_k m_\theta^e(g_k) &= \sum_{k=1}^{n} m_\theta^e(\gamma_k g_k) \\ &\leqq (1 + \delta)m_\theta^e(g) = (1 + \delta)m_\theta^e\left(\sum_{k=1}^{n} \gamma_k g_k + \delta F\right) \\ &\leqq (1 + \delta)\left[m_\theta^e\left(\sum_{k=1}^{n} \gamma_k g_k\right) + \delta m_\theta^e(F)\right]\end{aligned}$$

It is now obvious that (1) holds if δ was sufficiently small (notice that we have the estimates

$$m_\theta^e\left(\sum_{k=1}^{n} \gamma_k g_k\right) \leqq \alpha \sum_{k=1}^{n} [g_k : e], \qquad m_\theta^e(F) \leqq [F : e]$$

which are independent of δ and θ).

Our next aim is uniform approximation of a "coarse" function by linear combinations of translates of a "fine" function. More precisely, we have the

3.7. Lemma. *Let $(G, \mathscr{T})$ be a locally compact topological group and $0 \leqq f, g \in \mathscr{C}^{00}(G, \mathbb{R})$, $g \neq 0$. Let $\varepsilon_0 > 0$ and let an open neighborhood U of $1 \in G$ be chosen such that*

3.7.1. $y^{-1}x \in U \Rightarrow |f(x) - f(y)| < \varepsilon_0$.

3.7.2. $x \in G \backslash U \Rightarrow g(x) = 0$.

Then for very $\varepsilon > \varepsilon_0$, there exist $0 \leqq \gamma_1, \ldots, \gamma_n \in \mathbb{R}$, $x_1, \ldots, x_n \in G$ such that $\gamma_1 + \cdots + \gamma_n > 0$ and

3.7.3. $$\left| f - \sum_{k=1}^{n} \gamma_k L_{x_k} g \right| \leqq \varepsilon.$$

3.7.4. $\gamma_k \leqq [f : g^*]$ $(k = 1, \ldots, n)$, where $g^*(x) = g(x^{-1})$ $(x \in G)$.

3.7.5. *More precisely, if $C \subseteq G$ is compact such that f vanishes on $G \backslash C$, then we may choose $x_1, \ldots, x_n \in C^{-1}$.*

Proof. Let us observe first that 1 is a realistic assumption in view of Proposition 2.3. Next we observe that $g(y^{-1}x) = 0$ for $y^{-1}x \in G \backslash U$, and $f(x) - \varepsilon_0 \leqq f(y) \leqq f(x) + \varepsilon_0$ for $y^{-1}x \in U$ by assumption. Thus we get

(2) $$(f(x) - \varepsilon_0)g(y^{-1}x) \leqq f(y)g(y^{-1}x) \leqq (f(x) + \varepsilon_0)g(y^{-1}x) \qquad (x, y \in G)$$

Choose now any $\varepsilon > \varepsilon_0$ and consider the function g^* defined by $g^*(x) = g(x^{-1})$. Since $x \to x^{-1}$ is a homeomorphism, we have $g \in \mathscr{C}^{00}(G, \mathbb{R})$ again. Find $\beta > 0$ such that $\beta[f : g^*] < \varepsilon - \varepsilon_0$. Next find an open neighborhood V of $1 \in G$ such that the closure of V is compact and $t^{-1}s \in V \Rightarrow |g^*(s) - g^*(t)| \leqq \beta$ (Proposition 2.3). Or, equivalently $st^{-1} \in V \Rightarrow |g(s) - g(t)| \leqq \beta$. Let $C \subseteq G$ be compact such that f vanishes on $G \backslash C$. Find $s_1, \ldots, s_n \in C$ with $C \subseteq s_1 V \cup \cdots \cup s_n V$. Find

$$0 \leqq h_1, \ldots, h_n \in \mathscr{C}^{00}(G, \mathbb{R})$$

such that h_k vanishes on $G \backslash s_k V$ and $h_1(y) + \cdots + h_n(y) = 1$ for $y \in C$ (Remark 2.2.5). Observe now that $s_k^{-1}y \in G \backslash V$ implies $y \in G \backslash s_k V$ and hence $h_k(y) = 0$. And $s_k^{-1}y \in V$ implies $(s_k^{-1}x)(y^{-1}x)^{-1} \in V$ and hence $|g(y^{-1}x) - g(s_k^{-1}x)| \leqq \beta$, i.e., $g(y^{-1}x) - \beta \leqq g(s_k^{-1}x) \leqq g(y^{-1}x) + \beta$. Thus we get

$$\text{(3)} \quad \begin{aligned} h_k(y)f(y)(g(y^{-1}x) - \beta) &\leqq h_k(y)f(y)g(s_k^{-1}x) \\ &\leqq h_k(y)f(y)(g(y^{-1}x) + \beta) \qquad (x, y \in G). \end{aligned}$$

Summation over k yields

$$f(y)(g(y^{-1}x) - \beta) \leqq \sum_{k=1}^{n} h_k(y)f(y)g(s_k^{-1}x) \leqq f(y)(g(y^{-1}x) + \beta);$$

and upon combining this with (2), we get

$$\begin{aligned} (f(x) - \varepsilon_0)g(y^{-1}x) - \beta f(y) &\leqq \sum_{k=1}^{n} h_k(y)f(y)g(s_k^{-1}x) \\ &\leqq (f(x) + \varepsilon_0)g(y^{-1}x) + \beta f(y), \end{aligned}$$

or in other terms

$$(*) \quad \begin{aligned} (f(x) - \varepsilon_0)L_{x^{-1}}g^* - \beta f &\leqq \sum_{k=1}^{n} g(s_k^{-1}x)h_k f \\ &\leqq (f(x) + \varepsilon_0)L_{x^{-1}}g^* + \beta f \qquad (x \in G). \end{aligned}$$

Choose now some $0 \leqq e, \theta \in \mathscr{C}^{00}(G, \mathbb{R})$, $e, \theta \neq 0$, which will be subject to some further specification later. From the properties of m_θ^e specified in Lemma 3.5 we conclude now that

$$\text{(4)} \quad \begin{aligned} (f(x) - \varepsilon_0)m_\theta^e(g^*) - \beta m_\theta^e(f) &\leqq m_\theta^e\left(\sum_{k=1}^{n} g(s_k^{-1}x)h_k f\right) \\ &\leqq (f(x) + \varepsilon_0)m_\theta^e(g^*) + \beta m_\theta^e(f). \end{aligned}$$

From Lemmas 3.2 and 3.5 we know that

$$\frac{m_\theta^e(f)}{m_\theta^e(g^*)} \leqq [f : g^*] = \frac{\varepsilon' - \varepsilon_0}{\beta}$$

for a suitable $\varepsilon_0 < \varepsilon' < \varepsilon$. Thus we get, after dividing (4) by $m_\theta^e(g^*)$,

$$f(x) - \varepsilon_0 - (\varepsilon' - \varepsilon_0) \leqq m_\theta^e\left(\sum_{k=1}^{n} \frac{g(s_k^{-1}x)}{m_\theta^e(g^*)} h_k f \right) \leqq f(x) + \varepsilon'.$$

Applying Lemma 3.6 with the appropriate specifications, we get

$$f(x) - \varepsilon \leqq \sum_{k=1}^{n} \frac{g(s_k^{-1}x)}{m_\theta^e(g^*)} m_\theta^e(h_k f) \leqq f(x) + \varepsilon$$

if we only make sure that θ vanishes outside a suitable open neighborhood W of $1 \in G$. But now we have only to put

$$\gamma_k = m_\theta^e(h_k f)/m_\theta^e(g^*), \qquad x_k = s_k^{-1} \in C^{-1} \qquad (k = 1, \ldots, n)$$

in order to achieve the conclusion of our lemma (note that $0 \leqq h_k f \leqq f$ and hence

$$\gamma_k < m_\theta^e(f)/m_\theta^e(g^*) \leqq [f : g^*]).$$

As an easy consequence, we obtain

3.8. Lemma. *Let $(G, \mathscr{T})$ be a locally compact topological group and $0 \leqq f \in \mathscr{C}^{00}(G, \mathbb{R})$. Let W be a neighborhood of $1 \in G$ such that its closure $\overline{W}$ is compact. Let $C \subseteq G$ be compact such that f vanishes on $G \backslash C$. Let $0 \leqq F \in \mathscr{C}^{00}(G, \mathbb{R})$ be such that $x \in C\overline{W} \Rightarrow F(x) = 1$. Let $\varepsilon_0 > 0$ and $U \subseteq W$ a neighborhood of $1 \in G$ such that:*

3.8.1. $x^{-1}y \in V \Rightarrow |f(x) - f(y)| \leqq \varepsilon_0$.

Let $0 \leqq \theta \in \mathscr{C}^{00}(G, \mathbb{R})$, $\theta \neq 0$ be such that:

3.8.2. $x \in G \backslash U \Rightarrow \theta(x) = 0$.

Then for every $\varepsilon > \varepsilon_0$ there are $0 \leqq \gamma_1, \ldots, \gamma_n \in \mathbb{R}$, $x_1, \ldots, x_n \in G$ such that $\gamma_1 + \cdots + \gamma_n > 0$ and

3.8.3. $$\left| f - \sum_{k=1}^{n} \gamma_k L_{x_k} \theta \right| \leqq \varepsilon F.$$

3.8.4. $\gamma_k \leqq [f : \theta^*]$ $(k = 1, \ldots, n)$ where $\theta^*(x) = \theta(x^{-1})$.

For the proof, we need only apply Lemma 3.7 with θ in place of g and observe that $x_1, \ldots, x_n \in C^{-1}$ implies that not only f but also every one of the $L_{x_k}\theta$ vanishes on $G \backslash C\overline{W}$ since $x_k^{-1} \in C \Rightarrow L_{x_k}\theta$ vanishes on $G \backslash x_k^{-1}U \supseteq G \backslash C\overline{W}$ $(k = 1, \ldots, n)$.

3.9. Lemma. *Let $(G, \mathscr{T})$ be a locally compact topological group and $0 \leqq e, f \in \mathscr{C}^{00}(G, \mathbb{R})$, $e \neq 0$. Then for every $\varepsilon > 0$, there is an open neighborhood U of $1 \in G$ such that for any $0 \leqq \theta, \eta \in \mathscr{C}^{00}(G, \mathbb{R})$ which both vanish in $G \backslash U$, the inequality*

$$|m_\theta^e(f) - m_\eta^e(f)| \leqq \varepsilon$$

holds.

Proof. Choose a compact $C \subseteq G$ such that both e and f vanish on $G \backslash C$, and let W be any neighborhood of $1 \in G$ whose closure $\overline{W}$ is compact. Let $0 \leqq F \in \mathscr{C}^{00}(G, \mathbb{R})$ be such that $x \in C\overline{W} \Rightarrow F(x) = 1$. Choose some $\delta > 0$ that will be specified later. Find a neighborhood $V \subseteq W$ of $1 \in G$ such that $y^{-1}x \in V \Rightarrow |e(x) - e(y)| < \delta/2$, $|f(x) - f(y)| < \delta/2$, and let

$$0 \leqq g \in \mathscr{C}^{00}(G, \mathbb{R}), g \neq 0$$

vanish in $G \backslash V$. Apply Lemma 3.8 in order to obtain $0 \leqq \gamma_1, \ldots, \gamma_n \leqq [f : g^*]$ and $x_1, \ldots, x_n \in G$ such that

$$\left| f = \sum_{k=1}^{n} \gamma_k L_{x_k} g \right| \leqq \delta F \qquad \text{and} \qquad \gamma_1 + \cdots + \gamma_n > 0.$$

Let $0 \leqq \theta \in \mathscr{C}^{00}(G, \mathbb{R})$, $\theta \neq 0$. Clearly we get

$$\left| m_\theta^e(f) - m_\theta^e\left(\sum_{k=1}^{n} \gamma_k L_{x_k} g \right) \right| \leqq \delta m_\theta^e(F) \leqq \delta [F : e].$$

By Lemma 3.6 we may find a neighborhood $X \subseteq V$ of $1 \in G$ such that $\theta(x) = 0$ $(x \in G \backslash X)$ implies

$$\left| m_\theta^e\left(\sum_{k=1}^{n} \gamma_k L_{x_k} g \right) - \sum_{k=1}^{n} \gamma_k m_\theta^e(g) \right| < \delta.$$

(Here we use the fact that we had an a priori bound $[f : g^*]$ for $\gamma_1, \ldots, \gamma_n$.) Put $\gamma_1 + \cdots + \gamma_n = \gamma$. We have $\gamma > 0$. Now we obtain

$$(5) \qquad |m_\theta^e(f) - \gamma m_\theta^e(g)| < \delta(1 + [F : e]).$$

By the same token, we get, upon replacing X by a still smaller neighborhood U of $1 \in G$, and upon requiring that θ vanish in $G \backslash U$, both (5) and (recall that $m_\theta^e(e) = 1$)

$$(6) \qquad |1 - \gamma_e m_\theta^e(g)| < \delta(1 + [F : e]),$$

with a suitable real $\gamma_e > 0$. Combining (5) and (6), we get

$$\left| m_\theta^e(f) - \frac{\gamma}{\gamma_e} \right| < \delta(1 + [F:e])\left(1 + \frac{\gamma}{\gamma_e}\right). \tag{7}$$

This implies

$$\frac{\gamma}{\gamma_e}(1 - \delta(1 + [F:e])) < m_\theta^e(f) + \delta(1 + [F:e])$$

and thus

$$\frac{\gamma}{\gamma_e} + 1 < \frac{m_\theta^e(f) + 1}{1 - \delta(1 + [F:e])} \leqq 2(1 + [F:e])$$

if we only choose $\delta > 0$ sufficiently small. This again yields, upon combination with (7),

$$\left| m_\theta^e(f) - \frac{\gamma}{\gamma_e} \right| < \frac{\varepsilon}{2}$$

if $\delta > 0$ is sufficiently small. Observe that γ, γ_e, and δ have been chosen independently of U and θ, hence the same estimate holds for any other $0 \leqq \eta \in \mathscr{C}^{00}(G, \mathbb{R})$, $\eta \neq 0$ as well, provided it vanishes on $G \backslash U$. This does it.

We have now enough lemmas in order to prove the fundamental

3.10. Theorem. *Let $(G, \mathscr{T})$ be a locally compact topological group and $0 \leqq e \in \mathscr{C}^{00}(G, \mathbb{R})$, $e \neq 0$. There exists a unique positive linear form $m^e: \mathscr{C}^{00}(G, \mathbb{R}) \to \mathbb{R}$ such that:*

3.10.1. $m^e(L_x f) = m^e(f)$ $(f \in \mathscr{C}^{00}(G, \mathbb{R}), x \in G)$.

3.10.2. $m^e(e) = 1$.

It satisfies

3.10.3. $m^e(f) > 0$ $(0 \leqq f \in \mathscr{C}^{00}(G, \mathbb{R}), f \neq 0)$.

We call m^e, which is actually a τ-measure (Theorem I.10.1), *the* **left (invariant) Haar measure** *that* **normalizes on** ***e***.

Remark. If m^d, m^e are the left Haar measures normalizing on d, e, respectively, then the theorem says that each of them is a constant multiple of the other:

$$m^e(f) = \frac{m^e(e)}{m^d(e)} m^d(f).$$

This is meaningful because Theorem 3.10.3 implies $m^d(e) > 0$.

Proof. For every open neighborhood U of $1 \in G$ and every $0 \leqq f \in \mathscr{C}^{00}(G, \mathbb{R})$, let

$$\overline{m}_U^e(f) = \sup\{m_\theta^e(f) | 0 \leqq \theta \in \mathscr{C}^{00}(G, \mathbb{R}), \theta \neq 0, \theta(x) = 0 \ (x \in G \backslash U)\},$$
$$\underline{m}_U^e(f) = \inf\{m_\theta^e(f) | 0 \leqq \theta \in \mathscr{C}^{00}(G, \mathbb{R}), \theta \neq 0, \theta(x) = 0 \ (x \in G \backslash U)\}.$$

Generally, $0 \leqq f \in \mathscr{C}^{00}(G, \mathbb{R}) \Rightarrow 0 \leqq \underline{m}_U^e(f) \leqq \overline{m}_U^e(f)$. Clearly

$$f = 0 \quad \Rightarrow \quad \overline{m}_U^e(f) = 0 = \underline{m}_U^e(f) = 0.$$

Lemmas 3.2 and 3.5 imply

$$0 < 1/[e : f] \leqq \underline{m}_U^e(f) \leqq \overline{m}_U^e(f) \leqq [f : e] \qquad (0 \leqq f \in \mathscr{C}^{00}(G, \mathbb{R}), \quad f \neq 0) \tag{8}$$

Lemma 3.9 implies that for every $0 \leqq f \in \mathscr{C}^{00}(G, \mathbb{R})$ and every $\varepsilon > 0$ there is a U such that

$$0 \leqq \overline{m}_U^e(f) - \underline{m}_U^e(f) \leqq \varepsilon.$$

Thus for every $0 \leqq f \in \mathscr{C}^{00}(G, \mathbb{R})$, there is a unique real number $m^e(f)$ such that $0 \leqq \underline{m}_U^e(f) \leqq m^e(f) \leqq \overline{m}_U^e(f)$ for every neighborhood U of $1 \in G$, and (8) implies 3 of our theorem.

Lemma 3.9 says, in other terms, that for every $0 \leqq f \in \mathscr{C}^{00}(G, \mathbb{R})$ and every $\varepsilon > 0$, there is a neighborhood U of $1 \in G$ such that

$$0 \leqq \theta \in \mathscr{C}^{00}(G, \mathbb{R}), \ \theta \neq 0, \ \theta(x) = 0 \ (x \in G \backslash U) \quad \Rightarrow \quad |m^e(f) - m_\theta^e(f)| < \varepsilon;$$

and it is clear that we may, by passing to finite intersections of such neighborhoods U of $1 \in G$, obtain such approximations for any finite number of functions $0 \leqq f \in \mathscr{C}^{00}(G, \mathbb{R})$ simultaneously. Choosing any $x \in G$ and applying this reasoning to f and $L_x f$, we get

$$|m^e(f) - m^e(L_x f)| < |m_\theta^e(f) - m_\theta^e(L_x f)| + 2\varepsilon = 2\varepsilon$$

(Lemma 3.5.5). Since $\varepsilon > 0$ was arbitrary, $m^e(f) = m^e(L_x f)$ follows. Choose any $0 \leqq f_1, \ldots, f_n \in \mathscr{C}^{00}(G, \mathbb{R})$, $0 \leqq \alpha_1, \ldots, \alpha_n \in \mathbb{R}$, $\varepsilon > 0$, and determine the neighborhood U of $1 \in G$ according to the above reasoning, for $f_1, \ldots, f_n$, $\alpha_1 f_1 + \cdots + \alpha_n f_n$ simultaneously, plus in accordance with Lemma 3.6. Then $0 \leqq \theta \in \mathscr{C}^{00}(G, \mathbb{R})$, $\theta \neq 0$, $\theta(x) = 0$ $(x \in G \backslash U)$ imply

$$\begin{aligned}
&|m^e(\alpha_1 f_1 + \cdots + \alpha_n f_n) - (\alpha_1 m^e(f_1) + \cdots + \alpha_n m^e(f_n))| \\
&\quad < |m_\theta^e(\alpha_1 f_1 + \cdots + \alpha_n f_n) - (\alpha_1 m_\theta^e(f_1) + \cdots + \alpha_n m_\theta^e(f_n))| \\
&\qquad + \varepsilon + (\alpha_1 + \cdots + \alpha_n)\varepsilon \\
&\quad < \varepsilon + \varepsilon + (\alpha_1 + \cdots + \alpha_n)\varepsilon.
\end{aligned}$$

Since $\varepsilon > 0$ was arbitrary, $m^e(\alpha_1 f_1 + \cdots + \alpha_n f_n) = \alpha_1 m^e(f_1) + \cdots + \alpha_n m^e(f_n)$

follows. Now we are through with linearity as far as nonnegative functions and coefficients are concerned.

Standard reasoning as displayed in Lemma IX.1.3.2 now shows that there is a unique positive linear form $m^e\colon \mathscr{C}^{00}(G, \mathbb{R}) \to \mathbb{R}$ that extends the $m^e\colon \mathscr{C}_+^{00}(G, \mathbb{R}) \to \mathbb{R}_+$ constructed above. By

$$m^e(L_x f) = m^e(L_x f_+) - m^e(L_x f_-) = m^e(f_+) - m^e(f_-) = m^e(f)$$

we get 1 of our theorem.

2 is an obvious consequence of (8) for $f = e$. It remains to prove the uniqueness statement. Let thus $m\colon \mathscr{C}^{00}(G, \mathbb{R}) \to \mathbb{R}$ be any positive linear form with the properties corresponding to 1 and 2 of our theorem. We begin by showing that m also satisfies 3. Choose any $0 \leqq f_1, \theta \in \mathscr{C}^{00}(G, \mathbb{R})$, $f_1, \theta \neq 0$. Since $[g : \theta] < \infty$, we may find $x_1, \ldots, x_n \in G$ and $0 < \alpha_1, \ldots, \alpha_n \in \mathbb{R}$ with $f_1 \leqq \alpha_1 L_{x_1}\theta + \cdots + \alpha_n L_{x_n}\theta$. From this

$$m(f_1) \leqq \alpha_1 m(L_{x_1}\theta) + \cdots + \alpha_n m(L_{x_n}\theta) = (\alpha_1 + \cdots + \alpha_n)m(\theta)$$

follows, whence $m(\theta) > 0$. Actually, we have proved the general estimate

$$m(\theta)/m(f_1) \geqq 1/[f_1 : \theta].$$

Fix now, for any given $0 \leqq f \in \mathscr{C}^{00}(G, \mathbb{R})$, $f \neq 0$, C, W, and F as in Lemma 3.8. Clearly $F \neq 0$, hence $[F : f] > 0$. Choose any $\varepsilon > 0$ with $\varepsilon[F : f] < 1$. We shall deal now with neighborhoods $U \subseteq W$ of $1 \in G$ and functions $0 \leqq \theta \in \mathscr{C}^{00}(G, \mathbb{R})$, $\theta \neq 0$ that vanish on $G \backslash U$. If we choose U in such a fashion that $x^{-1}y \in U \Rightarrow |f(x) - f(y)| \leqq \varepsilon/2$, then we may, by Lemma 3.8, determine $0 \leqq \gamma_1, \ldots, \gamma_n \in \mathbb{R}$ and $x_1, \ldots, x_n \in G$ such that

$$\left| f - \sum_{k=1}^{n} \gamma_k L_{x_k}\theta \right| \leqq \varepsilon F.$$

This implies

$$f \leqq \varepsilon F + \sum_{k=1}^{n} \gamma_k L_{x_k}\theta, \qquad f + \varepsilon F \geqq \sum_{k=1}^{n} \gamma_k L_{x_k}\theta,$$

and thus

$$[f : \theta] \leqq \varepsilon[F : \theta] + \sum_{k=1}^{n} \gamma_k$$

$$\begin{aligned} m(f) + \varepsilon m(F) &\geqq \left(\sum_{k=1}^{n} \gamma_k \right) m(\theta) \\ &\geqq ([f : \theta] - \varepsilon[F : \theta])m(\theta) \\ &= \left(1 - \varepsilon \frac{[F : \theta]}{[f : \theta]} \right) [f : \theta] m(\theta) \\ &\geqq (1 - \varepsilon[F : f])[f : \theta] m(\theta). \end{aligned}$$

Choose now any $0 \leqq f_1 \in \mathscr{C}^{00}(G, \mathbb{R}), f_1 \neq 0$, hence $[f_1 : \theta], m(f_1) > 0$. Divide our above result

$$m(f) + \varepsilon m(F) \geqq (1 - \varepsilon[F : f])[f : \theta]m(\theta)$$

by $m(f_1)$ and use our previous estimate $m(\theta)/m(f_1) \geqq 1/[f_1 : \theta]$. We get

$$\begin{aligned} \frac{m(f)}{m(f_1)} + \varepsilon \frac{m(F)}{m(f_1)} &\geqq (1 - \varepsilon[F : f])[f : \theta] \frac{m(\theta)}{m(f_1)} \\ &\geqq (1 - \varepsilon[F : f]) \frac{[f : \theta]}{[f_1 : \theta]} \\ &= (1 - \varepsilon[F : f]) \frac{m_\theta^e(f)}{m_\theta^e(f_1)} \end{aligned}$$

Now we let U and θ run and have in the limit

$$\frac{m(f)}{m(f_1)} + \varepsilon \frac{m(F)}{m(f_1)} \geqq (1 - \varepsilon[F : f]) \frac{m^e(f)}{m^e(f_1)}$$

Since $\varepsilon > 0$ was arbitrary, we obtain

$$\frac{m(f)}{m(f_1)} \geqq \frac{m^e(f)}{m^e(f_1)}.$$

Upon interchanging f_1 and f here, we obtain equality. Put now $f_1 = e$. Then the dominators equal 1 and we get $m(f) = m^e(f)$.

3.11. Exercises. Let $(G, \mathscr{T})$ be a locally compact topological group.

3.11.1. Carry over Exercise 1.11.2 to the locally compact case. Prove in particular: There is, on the local σ-ring $\mathscr{B}_c^{00}(\mathscr{T})$ generated by all open sets with a compact closure (Definition V.4.4.2), a left invariant inner compact regular σ-content m, uniquely determined by these properties up to a multiplicative constant > 0, and satisfying $m(U) > 0$ $(\varnothing \neq U \in \mathscr{B}_c^{00}(\mathscr{T})$ open); this σ-content is called the left invariant Haar σ-content; symmetrically, there is a right invariant Haar σ-content on $\mathscr{B}_c^{00}(\mathscr{T})$, with analogous properties, and again determined up to a constant factor > 0.

3.11.2. Prove that the left and right Haar measures on G (and the corresponding σ-contents as well) coincide (up to a constant factor, of course) in at least the following cases:

(a) G is abelian;
(b) $\mathscr{T}$ is the discrete topology;
(c) $\mathscr{T}$ is compact.

Write out an explicit expression for $m(f)$ $(f \in \mathscr{C}^{00}(G, \mathbb{R}))$ in the case of a discrete $\mathscr{T}$.

3.11.3. Shorten the proof of the existence part of Theorem 3.10 by a skillful application of Lemmas 3.6, 3.9, and Tychonov's theorem. (*Hint*: Consult HEWITT and ROSS [1], if necessary.)

3.11.4. For any $U, M \subseteq G$ such that U is an open neighborhood of $1 \in G$ and U and M have compact closures, define $[M : U] = \inf\{n \mid n \geqq 0,$ there are $x_1, \ldots, x_n \in G$ with $M \subseteq x_1 U \cup \cdots \cup x_n U\}$. Prove that $[M : U] < \infty$ and $[M_1 + M_2 : U] = [M_1 : U] + [M_2 : U]$ whenever $M_1, M_2 \subseteq G$ have compact closures and $M_1 U \cap M_2 U = \varnothing$. Carry through an existence proof for a left invariant Haar σ-content on $\mathscr{B}_c^{00}(\mathscr{T})$, on the basis of the ideas of proof of Theorem 3.10, Exercise XIII.4.2.2, and the Tychonov theorem. (*Hint*: Consult HALMOS [4].)

3.11.5. For every $f \in \mathscr{C}^{00}(G, \mathbb{R})$, let f^* denote the function defined by $f^*(x) = f(x^{-1})$ $(x \in G)$. Notice that $f^* \in \mathscr{C}^{00}(G, \mathbb{R})$ again. Let m be a left invariant Haar measure on G. Prove that $m^*(f) = m(f^*)$ $(f \in \mathscr{C}^{00}(G, \mathbb{R}))$ defines a right invariant Haar measure on G.

3.12. Theorem. *Let $(G, \mathscr{T})$ be a locally compact topological group. Then there is a unique function $\Delta \in \mathbb{R}^G$ such that for every left invariant Haar measure m on G*

$$0 \leqq f \in \mathscr{C}^{00}(G, \mathbb{R}), f \neq 0 \quad \Rightarrow \quad m(R_{x^{-1}} f)/m(f) = \Delta(x) \qquad (x \in G).$$

Δ is called the **modular function** *of $(G, \mathscr{T})$, and $(G, \mathscr{T})$ is called unimodular if $\Delta = const = 1$. The modular function Δ of $(G, \mathscr{T})$ has the following properties:*

3.12.1. *Δ is continuous.*

3.12.2. *$\Delta(x) > 0$ $(x \in G)$.*

3.12.3. *$\Delta(xy) = \Delta(x)\Delta(y)$ $(x, y \in G)$.*

Proof. Fix any $x \in G$ and some left invariant Haar measure m, and define $m_x \colon \mathscr{C}^{00}(G, \mathbb{R}) \to \mathbb{R}$ by $m_x(f) = m(R_{x^{-1}} f)$. Clearly

$$m_x(L_s f) = m(R_{x^{-1}} L_s f) = m(L_s R_{x^{-1}} f) = m(R_{x^{-1}} f) = m_x(f)$$

since left and right translations commute. Thus m_x is a left invariant Haar measure again, and by Theorem 3.10 there is a real number $\Delta(x) > 0$ such that $m_x(f) = \Delta(x) m(f)$ $(f \in \mathscr{C}^{00}(G, \mathbb{R}))$. This establishes the existence and uniqueness statement of our theorem as well as 2. Moreover, we see that $0 \leqq f \in \mathscr{C}^{00}(G, \mathbb{R}), f \neq 0, x, y \in G$ imply

$$\begin{aligned}\Delta(xy) m(f) = m_{xy}(f) &= m(R_{(xy)^{-1}} f) = m(R_{y^{-1}} R_{x^{-1}} f) \\ &= m_y(R_{x^{-1}} f) = \Delta(y) m(R_{x^{-1}} f) = \Delta(y) m_x(f) = \Delta(y)\Delta(x) m(f),\end{aligned}$$

proving 3. In order to establish the continuity of Δ it suffices to prove the following

3.13. Proposition. *Let* $(G, \mathscr{T})$ *be a locally compact topological group and* $f \in \mathscr{C}^{00}(G, \mathbb{R})$. *Then* $s \to m(R_s f)$, $t \to m(L_t f)$ *are continuous real functions on* G.

Proof. It clearly suffices to prove the continuity of $t \to m(L_t f)$ in the case $|f| \leqq 1$. For this, let $C \subseteq G$ be a compact set such that f vanishes on $G \backslash C$. Let W be any neighborhood of $1 \in G$ whose closure $\overline{W}$ is compact. Let $0 \leqq F \in \mathscr{C}^{00}(G, \mathbb{R})$ be such that $x \in \overline{W}C \Rightarrow F(x) = 1$. By Proposition 2.3 and Remark 2.4 we may find, for any $0 < \varepsilon < \frac{1}{2}$, a neighborhood $U \subseteq W$ of $1 \in G$ such that $U = U^{-1}$ and

$$st^{-1} \in U \quad \Rightarrow \quad |f(sx) - f(tx)| < \varepsilon/m(F) \qquad (x \in G),$$

i.e., $|L_s f - L_t f| < \varepsilon/m(F)$. Moreover $sx \in G \backslash \overline{W}C$ implies $tx \in G \backslash C$ because $tx = z \in C$ would imply $sx = st^{-1}tx \in UC \subseteq \overline{W}C$, a contradiction; similarly $tx \in G \backslash \overline{W}C$ implies $sx \in G \backslash C$ because $sx = u \in C$ would imply $tx = ts^{-1}sx \in U^{-1}C = UC \subseteq \overline{W}C$, again a contradiction. Thus we see that $f(sx) = 0 = f(tx)$ or sx, $tx \in \overline{W}C$, but here we have $F(sx) = 1$. We may thus conclude $|L_s f - L_t f| \leqq \varepsilon L_s F$, and now we get

$$|m(L_s f) - m(L_t f)| \leqq m(|L_s f - L_t f|) \leqq \frac{\varepsilon}{m(F)} m(L_s F) = \varepsilon.$$

3.14. Proposition. *Let* $(G, \mathscr{T})$ *be a locally compact topological group and* Δ *its modular function. For every* $f \in \mathbb{R}^G$ *define* $f^* \in \mathbb{R}^G$ *by* $f^*(x) = f(x^{-1})$ $(x \in G)$. *Then for every left invariant Haar measure* m *on* G, *we have*

$$m(f) = m(f^*/\Delta) \qquad (f \in \mathscr{C}^{00}(G, \mathbb{R})).$$

Proof. Notice that $f \in \mathscr{C}^{00}(G, \mathbb{R}) \Rightarrow f^*/\Delta \in \mathscr{C}^{00}(G, \mathbb{R})$ since Δ is continuous and everywhere > 0. Let $m'(F) = m(f^*/\Delta)$ $(f \in \mathscr{C}^{00}(G, \mathbb{R}))$. Clearly this defines a positive measure m' on $\mathscr{C}^{00}(G, \mathbb{R})$. Moreover, we have

$$\begin{aligned} m'(L_s f) &= m\left(\frac{(L_s f)^*}{\Delta}\right) = m\left(\frac{R_{s^{-1}} f^*}{\Delta}\right) \\ &= m\left(\frac{R_{s^{-1}} f^*}{\Delta(s) R_{s^{-1}} \Delta}\right) = \frac{1}{\Delta(s)} m\left(R_{s^{-1}} \frac{f^*}{\Delta}\right) \\ &= \frac{\Delta(s)}{\Delta(s)} m\left(\frac{f^*}{\Delta}\right) = m\left(\frac{f^*}{\Delta}\right) = m'(f). \end{aligned}$$

Hence m' is a left invariant Haar measure, and there is a real $\alpha > 0$ with $m' = \alpha m$. In order to show that $\alpha = 1$, we notice that $\Delta(1) = 1$, hence $1/\Delta(x)$

is nearly $=1$ if x remains in a suitable neighborhood U of $1 \in G$. Let $0 \leqq g \in \mathscr{C}^{00}(G, \mathbb{R})$, $g(1) \neq 0$ vanish in $G \backslash U$ and put $f = gg^*$, clearly $f(1) \neq 0$, f vanishes in $G \backslash U$, and $f^* = f$. Now $m'(f) = m(f/\Delta)$ is approximately $= m(f)$, hence $\alpha = 1$. A reader who dislikes this qualitative argument is invited to replace it by rigorous epsilontics.

3.15. Proposition. *Let $(G, \mathscr{T})$ be a locally compact topological group and Δ its modular function. Then for every left invariant Haar measure m on G*

$$m'(f) = m(f/\Delta) \qquad (f \in \mathscr{C}^{00}(G, \mathbb{R}))$$

defines a right invariant Haar measure m' on G. We have

$$m'(f) = m(f^*) \qquad (f \in \mathscr{C}^{00}(G, \mathbb{R})).$$

Proof. Notice that $f \in \mathscr{C}^{00}(G, \mathbb{R}) \Rightarrow f/\Delta \in \mathscr{C}^{00}(G, \mathbb{R})$ because Δ is continuous and everywhere > 0. By Exercise 3.11.5, $m''(f) = m(f^*)$ defines a right invariant Haar measure on G. Now clearly Proposition 3.14 implies

$$m''(f) = m(f) = m(f/\Delta) = m'(f) \qquad (f \in \mathscr{C}^{00}(G, \mathbb{R})).$$

3.16. Remarks

3.16.1. Passing from Haar measures to derived σ-contents (Exercises 1.11.2, 3.11.1), we can interpret Proposition 3.15 as follows: If m is a left invariant and m' is a right invariant Haar σ-content on G, then $m \ll m'$ with density Δ (for the technicalities of absolute continuity of σ-contents and measures see Chapter VIII, Sections 3 and 4, and Chapter IX, Section 3).

3.16.2. Exercise 3.11.2 implies that Δ is constant $= 1$ if G is abelian, or the topology $\mathscr{T}$ is compact or discrete. For the compact case, a very easy proof of $\Delta \equiv 1$ runs as follows: $\Delta\colon G \to \mathbb{R}_+ \backslash \{0\}$ is clearly a continuous homomorphism, hence sends the compact group G on a compact subgroup of the multiplicative group of all reals > 0. But $\{1\}$ is the only such subgroup.

3.17. Exercises. Let G be a locally compact topological group, which, as a topological space, is an open subset of some $\mathbb{R}^n$ with the usual topology. Multiplication in G is thus a mapping

$$((x_1, \ldots, x_n), (y_1, \ldots, y_n)) \to (f_1(x_1, \ldots, y_n), \ldots, f_n(x_1, \ldots, y_n))$$

of $G \times G \subseteq \mathbb{R}^{2n}$ onto $G \subseteq \mathbb{R}^n$. Assume that $f_1, \ldots, f_n$ are at least once continuously differentiable on $G \times G$. Each of the mappings $L_s\colon G \to G$, $R_t\colon G \to G$ $(s, t \in G)$ has thus a continuous *Jacobian* determinant which we denote by $D(L_s)$ and $D(R_t)$, respectively. We assume that for any $s, t \in G$, the functions $D(L_s)$, $D(R_t)$ on G are constants, whose absolute values we denote by $l(s)$, $r(t)$. Write dx for the ordinary Lebesgue integral (its restriction, the Riemann integral, will mostly meet our purposes).

3.17.1. Prove that

$$m(f) = \int_G f(x) \frac{1}{r(x)} dx \qquad (f \in \mathscr{C}^{00}(G, \mathbb{R}))$$

defines a left invariant Haar measure m on G, and that

$$m'(f) = \int_G f(x) \frac{1}{l(x)} dx \qquad (f \in \mathscr{C}^{00}(G, \mathbb{R}))$$

defines a right Haar measure m' on G. (*Hint*: Use Exercise IV.2.7.2.)

3.17.2. Prove that the modular function of G is given by $\Delta(x) = l(x)/r(x)$ $(x \in G)$.

3.17.3. Consider the special case $G = \mathbb{R}\backslash\{0\}$, with the usual multiplication of reals as group multiplication. Prove that $l(x) = |x| = r(x)$ and hence

$$m(f) = \int_{\mathbb{R}\backslash\{0\}} \frac{f(x)}{|x|} dx \qquad (f \in \mathscr{C}^{00}(G, \mathbb{R}))$$

defines a left (= right since G is abelian) invariant Haar measure on G. (Notice that these integrals are always meaningful since every f in question vanishes outside some compact set not containing 0.)

3.17.4. Consider the special case $G = \mathbb{R}^2\backslash\{(0,0)\}$, with complex multiplication $(x, y)(u, v) = (xu - yv, xv - yu)$ as group multiplication. Prove that $l((x, y)) = x^2 + y^2 = r((x, y))$ $((x, y) \in G)$ and hence

$$m(f) = \int \frac{f(x, y)}{x^2 + y^2} dx \qquad (f \in \mathscr{C}^{00}(G, \mathbb{R}))$$

defines a left (and right, of course) invariant Haar measure. (The meaningfulness of the integrals follows as in 3.)

3.17.5. Consider the special case $G = \mathbb{R}^2\backslash(\{0\} \times \mathbb{R})$ with $(x, y)(u, v) = (xu, xv + y)$ as a multiplication. Prove that G is in fact a group with this multiplication. (*Hint*: Use the following identity for 2×2 matrices:

$$\begin{pmatrix} x & y \\ 0 & 1 \end{pmatrix} \begin{pmatrix} u & v \\ 0 & 1 \end{pmatrix} = \begin{pmatrix} xu & xv + y \\ 0 & 1 \end{pmatrix})$$

Prove that $r(x, y) = x^2$ and hence

$$\int \frac{f(x, y)}{x^2} dx\, dy \qquad (f \in \mathscr{C}^{00}(G, \mathbb{R}))$$

defines a left invariant Haar measure on G. Prove that $l(x, y) = |x|$, hence

$$\int \frac{f(x. y)}{|x|} dx\, dy \qquad (f \in \mathscr{C}^{00}(G, \mathbb{R}))$$

defines a right invariant Haar measure on G. Prove that the modular function of G is given by $\Delta(x, y) = 1/|x|$. This exercise provides us with the most popular example of a locally compact group where the left and right Haar measures are not constant multiples of each other; it is said that the example was first calculated by v. NEUMANN.

CHAPTER XIII

SOUSLIN SETS, ANALYTIC SETS, AND CAPACITIES

There are some traditional problems in set theory:

Problem 1. *To find sort of a construction of all members of the σ-field generated by a set system $\mathscr{S}$.*

Problem 2. *To find sufficient conditions under which the forward measurable image of a Borel set is a Borel set again.*

Problem 3. *To prove the continuum hypothesis under the restriction to Borel sets.*

With the help of nonmeasurable sets (Exercise II.6.6.2) it is not hard to show that problem 2 has a positive answer only if we impose some *regularity conditions*. Problem 1 is difficult, as the present chapter will show. Problem 3 arose historically out of difficulties in deducing the general continuum hypothesis from the other axioms of the Zermelo–Fraenkel set theory. As P. COHEN [1, 2] has shown, these difficulties are insuperable: The continuum hypothesis is logically independent of the other Zermelo–Fraenkel axioms. The positive answer given to problem 3 by Theorem 2.8 of this chapter is the more astonishing.

In Section 1 we develop the general "Souslin apparatus," and in Section 2 we apply it to Polish spaces. K. R. PARTHASARATHY [1] and MEYER [2] are our main sources, and are recommended for further study. We make use of abstract σ-compact set systems (Definition V.1.4). Section 3 contains the rudiments of capacity theory, with *Choquet's capacitability theorem* (3.5) as the main result. In Section 4 we apply it in order to obtain another proof of Corollary II.2.4 on the extension of σ-contents and Saks' theorem that *Souslin sets are Borel up to nullsets*, for any σ-content. In Section 5 we derive two versions of the so-called *measurable choice theorem*, one without and one with an exceptional nullset.

Hoffmann-Jørgensen [1], K. R. Parthasarathy [1], and Hildenbrandt [2] are suggested for further reading. The reader should also compare Choquet [5] and Dellacherie [1].

1. SOUSLIN EXTENSIONS

In this section we introduce Souslin's operation (A) and pursue a rather abstract theory related to it, until we have proved the so-called *separation theorem* (1.11). The particular importance of *Polish spaces* in this context will soon become apparent.

Recall that $\mathbb{N}$ denotes the set $\{1, 2, \ldots\}$ of all natural numbers, $\mathbb{N}^n$ the set of all ordered n-tuples of natural numbers $(n = 1, 2, \ldots)$, $\mathbb{N}^*$ the set $\bigcup_{n=1}^{\infty} \mathbb{N}^n$ of all words $=$ n-tuples of any length, of natural numbers, and $\mathbb{N}^{\mathbb{N}}$ the set of all infinite sequences of natural numbers.

1.1. Definition. *Let $\Omega \neq \emptyset$ and $\emptyset \in \mathscr{P} \subseteq \mathscr{P}(\Omega)$.*

1.1.1. *A mapping $F\colon \mathbb{N}^* \to \mathscr{P}$, i.e., an $F \in \mathscr{P}^{\mathbb{N}^*}$ is also called a* **Souslin frame** *over $\mathscr{P}$.*

1.1.2. *If F is a Souslin frame over $\mathscr{P}$, then the subset*

$$A(F) = \bigcup_{(n_1, n_2, \ldots) \in \mathbb{N}^{\mathbb{N}}} \bigcap_{r=1}^{\infty} F(n_1, \ldots, n_r)$$

of Ω is called the **kernel** *of the Souslin frame F.*

1.1.3. *We put $\mathscr{S}(\mathscr{P}) = \{A(F) \mid F \in \mathscr{P}^{\mathbb{N}^*}\}$. The members of $\mathscr{S}(\mathscr{P})$ are called the* **$\mathscr{P}$-Souslin sets** *and $\mathscr{S}(\mathscr{P})$ is called the* **Souslin extension** *of $\mathscr{P}$.*

1.2. Remarks

1.2.1. The passage from a Souslin frame F to its kernel $A(F)$ is also called **Souslin's operation (A)**, and so is sometimes the passage from $\mathscr{P}$ to $\mathscr{S}(\mathscr{P})$.

1.2.2. Notice that Souslin's operation (A) involves uncountable unions, hence we cannot expect σ-fields etc. to be stable under it. Generally we have however

$$\bigcup_{(n_1, n_2, \ldots) \in \mathbb{N}^{\mathbb{N}}} \bigcap_{r=1}^{\infty} F(n_1, \ldots, n_r) \subseteq \bigcap_{r=1}^{\infty} \bigcup_{n_1, \ldots, n_r \in \mathbb{N}} F(n_1, \ldots, n_r).$$

A sufficient condition for equality to hold here is that F be:

1.2.2.1. antitone in the following sense: $r, n_1, \ldots, n_r \in \mathbb{N}$,

$$F(n_1, \ldots, n_r) \supseteq \bigcup_{n \in \mathbb{N}} F(n_1, \ldots, n_r, n);$$

1.2.2.2. disjoint in the following sense: $r, n_1, \ldots, n_r,$

$$j, k \in \mathbb{N},\ j \neq k \quad \Rightarrow \quad F(n_1, \ldots, n_r, j) \cap F(n_1, \ldots, n_r, k) = \varnothing.$$

1.2.3. Remark 1.2.2.2 indicates that disjointness will play an essential role in the theory of Souslin extensions.

1.2.4. The fact that $\mathbb{N}^{\mathbb{N}}$ with the product topology of the discrete topology on $\mathbb{N}$ is a Polish space indicates that *Polish spaces* will play an essential role in the theory of Souslin extensions.

The understanding of the combinatorics in the proof of the next proposition is an acid test for the reader, I believe.

1.3. Proposition. *Let $\Omega \neq \varnothing$ and $\varnothing \in \mathscr{P} \subseteq \mathscr{P}(\Omega)$. Then*

$$\mathscr{P} \subseteq \mathscr{S}(\mathscr{P}) = \mathscr{S}(\mathscr{S}(\mathscr{P})).$$

Proof. If $A \in \mathscr{P}$, then the constant Souslin frame $F \equiv A$ clearly yields $A(F) = A$, hence $\mathscr{P} \subseteq \mathscr{S}(\mathscr{P})$ follows. For the same reason we have $\mathscr{S}(\mathscr{P}) \subseteq \mathscr{S}(\mathscr{S}(\mathscr{P}))$. Let now $A \in \mathscr{S}(\mathscr{S}(\mathscr{P}))$ and $G\colon \mathbb{N}^* \to \mathscr{S}(\mathscr{P})$ be a Souslin frame over $\mathscr{S}(\mathscr{P})$ such that $A(G) = A$:

$$A = \bigcup_{(n_1, n_2, \ldots) \in \mathbb{N}^{\mathbb{N}}} \bigcap_{s=1} G(n_1, \ldots, n_s).$$

Now, since every $G(n_1, \ldots, n_s) \in \mathscr{S}(\mathscr{P})$, there is, for every $(n_1, \ldots, n_s) \in \mathbb{N}^*$, a Souslin frame $F_{n_1, \ldots, n_s}\colon \mathbb{N}^* \to \mathscr{P}$ over $\mathscr{P}$ such that $A(F_{n_1, \ldots, n_s}) = G(n_1, \ldots, n_s)$:

$$G(n_1, \ldots, n_s) = \bigcup_{(m_1, m_2, \ldots) \in \mathbb{N}^{\mathbb{N}}} \bigcap_{r=1}^{\infty} F_{n_1, \ldots, n_s}(m_1, \ldots, m_r).$$

We thus get

$$A = \bigcup_{(n_1, n_2, \ldots) \in \mathbb{N}^{\mathbb{N}}} \bigcap_{s=1}^{\infty} \bigcup_{(m_1, m_2, \ldots) \in \mathbb{N}^{\mathbb{N}}} \bigcap_{r=1}^{\infty} F_{n_1, \ldots, n_s}(m_1, \ldots, m_r).$$

In words: A point $\omega \in \Omega$ belongs to A iff there is some $(n_1, n_2, \ldots) \in \mathbb{N}^{\mathbb{N}}$ such that for every s there is a sequence $(m_1, m_2, \ldots) \in \mathbb{N}^{\mathbb{N}}$ such that $\omega \in \bigcap_{r=1}^{\infty} F_{n_1, \ldots, n_s}(m_1, \ldots, m_r)$. This can be put as follows: $\omega \in A$ iff there is a $(n_1, n_2, \ldots) \in \mathbb{N}^{\mathbb{N}}$ and a $(m_{sr})_{s, r \in \mathbb{N}} \in \mathbb{N}^{\mathbb{N} \times \mathbb{N}}$ such that

$$\omega \in \bigcap_{r=1}^{\infty} F_{n_1, \ldots, n_s}(m_{s1}, \ldots m_{sr}) \qquad (s = 1, 2, \ldots).$$

Hence

$$A = \bigcup_{(n_1, n_2, \ldots) \in \mathbb{N}^{\mathbb{N}}} \bigcup_{(m_{jk})_{j, k \in \mathbb{N}} \in \mathbb{N}^{\mathbb{N} \times \mathbb{N}}} \bigcap_{s, r=1}^{\infty} F_{n_1, \ldots, n_s}(m_{s1}, \ldots, m_{sr}).$$

All we have to do now is to give this the form of Souslin's operation (A) for a suitable Souslin frame over $\mathscr{P}$, by coding $\mathbb{N}^{\mathbb{N}}$ into $\mathbb{N}^{\mathbb{N}} \times \mathbb{N}^{\mathbb{N} \times \mathbb{N}}$ suitably. For this, we proceed as follows: Let $(n(1), m(1))$, $(n(2), m(2))$, ... be any one-to-one enumeration of $\mathbb{N} \times \mathbb{N}$, i.e., a bijection of $\mathbb{N}$ onto $\mathbb{N} \times \mathbb{N}$. This establishes a bijection

$$(k_1, k_2, \ldots) \to ((n(k_1), m(k_1)), (n(k_2), m(k_2)), \ldots)$$

of $\mathbb{N}^{\mathbb{N}} \to \mathbb{N}^{\mathbb{N}} \times \mathbb{N}^{\mathbb{N}}$ $(= (\mathbb{N} \times \mathbb{N})^{\mathbb{N}})$. Now we treat the second factor $\mathbb{N}^{\mathbb{N}}$ here in a special way: Let $(s(1), r(1))$, $(s(2), r(2))$, ... be the diagonal enumeration of $\mathbb{N} \times \mathbb{N}$ after Cauchy. It establishes a bijection

$$(m_1, m_2, \ldots) \to \begin{bmatrix} m_1 & m_2 & m_4 \\ m_3 & m_5 & \cdots \\ m_6 & \cdots & \cdots \\ \cdots & \cdots & \cdots \end{bmatrix}$$

of $\mathbb{N}^{\mathbb{N}} \to \mathbb{N}^{\mathbb{N} \times \mathbb{N}}$. Now we put our bijections together and obtain a bijection

$$(k_1, k_2, \ldots) \to \left[(n(k_1), n(k_2), \ldots), \begin{bmatrix} m(k_1) & m(k_2) & m(k_4) \\ m(k_3) & m(k_5) & \cdots \\ m(k_6) & \cdots & \cdots \\ \cdots & \cdots & \cdots \end{bmatrix} \right]$$

of $\mathbb{N}^{\mathbb{N}} \to \mathbb{N}^{\mathbb{N}} \times \mathbb{N}^{\mathbb{N} \times \mathbb{N}}$. The very nature of the Cauchy diagonal enumeration implies that we know in the matrix that forms the second component here all matrix coefficients standing left of $m(k_t)$ in the same row as $m(k_t)$ as soon as we know $m(k_1), \ldots, m(k_t)$. We should also mention the general inequality $s(t) \leqq t$ $(t = 1, 2, \ldots)$. Clearly, if $(m_{sr})_{s, r \in \mathbb{N}}$ is the matrix in question, then $m_{s(t), r(t)} = m(k_t)$ $(t = 1, 2, \ldots)$. With this notation, we put

$$H(k_1, \ldots, k_t) = F_{n(k_1), \ldots, n(k_{s(t)})}(m_{s(t), 1}, \ldots, m_{s(t), r(t)})$$
$$(t \geqq 1; k_1, \ldots, k_t \in \mathbb{N}),$$

and this is well-defined after our remarks about the Cauchy enumeration. Also, it is clear that

$$\bigcap_{t=1}^{\infty} H(k_1, \ldots, k_t) = \bigcap_{s, r=1} F_{n_1, \ldots, n_s}(m_{s1}, \ldots, m_{sr})$$

if

$$(k_1, k_2, \ldots) \to \left[(n_1, n_2, \ldots), \begin{bmatrix} m_{11} & m_{12} & \cdots \\ m_{21} & m_{22} & \cdots \\ \cdots & \cdots & \cdots \end{bmatrix} \right]$$

in our above bijection. Now we can conclude that

$$A = \bigcup_{(k_1, k_2, \ldots)} \bigcap_{t=1}^{\infty} H(k_1, \ldots, k_t) = A(H),$$

proving $A \in \mathscr{S}(\mathscr{P})$.

As a consequence of the above proposition, we get

1.4. Proposition. *Let $\Omega \neq \varnothing$ and $\varnothing \in \mathscr{P} \subseteq \mathscr{P}(\Omega)$. Then*

$$A_1, A_2, \ldots \in \mathscr{S}(\mathscr{P}) \quad \Rightarrow \quad A_1 \cap A_2 \cap \cdots, \; A_1 \cup A_2 \cup \cdots \in \mathscr{S}(\mathscr{P}),$$

i.e., $\mathscr{S}(\mathscr{P})$ is stable under countable intersections and countable unions.

Proof. Choose the Souslin frame $F(n_1, \ldots, n_r) = A_r$ $(r \geqq 1; n_1, \ldots, n_r \in \mathbb{N})$ in order to get $A_1 \cap A_2 \cap \cdots = A(F) \in \mathscr{S}(\mathscr{S}(\mathscr{P})) = \mathscr{S}(\mathscr{P})$. Choose the Souslin frame $G(n_1, \ldots, n_r) = A_{n_1}$ $(r \geqq 1;\, n_1, \ldots, n_r \in \mathbb{N})$ in order to get $A_1 \cup A_2 \cup \cdots = A(G) \in \mathscr{S}[(\mathscr{S}(\mathscr{P})] = \mathscr{S}(\mathscr{P})$.

Now we can carry out a first step toward the investigation of the relation between the Souslin extension and the generated σ-field of a given set system.

1.5. Proposition. *Let $\Omega \neq \varnothing$ and $\varnothing \in \mathscr{P} \subseteq \mathscr{P}(\Omega)$. Then the following statements are equivalent:*

1.5.1. $\mathscr{B}(\mathscr{P}) \subseteq \mathscr{S}(\mathscr{P})$.

1.5.2. $F \in \mathscr{P} \Rightarrow \Omega \backslash F \in \mathscr{S}(\mathscr{P})$.

Proof. $1 \Rightarrow 2$ is obvious. In order to prove $2 \Rightarrow 1$ consider the set system $\mathscr{F} = \{F \mid F \in \mathscr{S}(\mathscr{P}),\ \Omega\backslash F \in \mathscr{S}(\mathscr{P})\}$. By Proposition 1.4, $\mathscr{F}$ is stable under countable union and countable intersection, and by definition also under passage to complements. Clearly $\mathscr{F}$ contains $\mathscr{P}$. It is now an easy exercise to prove that $\mathscr{F}$ is a σ-field, hence $\mathscr{P} \subseteq \mathscr{B}(\mathscr{P}) \subseteq \mathscr{F} \subseteq \mathscr{S}(\mathscr{P})$.

1.6. Exercises. Let $(\Omega, \mathscr{T})$ be a second countable topological space and $\mathscr{B} = \mathscr{B}(\mathscr{T})$. Prove:

1.6.1. If $(\Omega, \mathscr{T})$ is locally compact Hausdorff, and $\mathscr{P} = \{K \mid K \subseteq \Omega,$ K compact$\}$, then $\mathscr{B} \subseteq \mathscr{S}(\mathscr{P})$. (*Hint*: Every open set is a K_σ here; use Proposition 1.4.)

1.6.2. If $(\Omega, \mathscr{T})$ is Polish and $\mathscr{P} = \{F \mid F \subseteq \Omega$ closed$\}$, then $\mathscr{B} \subseteq \mathscr{S}(\mathscr{P})$. (*Hint*: Every open set is an F_σ here; use Proposition 1.4.)

1.6.3. If Ω is an infinite-dimensional Banach space with norm topology $\mathscr{T}$ and a countable dense subset, and $\mathscr{P} = \{K \mid K \subseteq \Omega,\ K$ compact$\}$, then $\mathscr{B} \subseteq \mathscr{S}(\mathscr{P})$ is false. (*Hint*: Since no ball with positive radius is compact, the complement of every norm compact subset of Ω is open and everywhere dense; by Baire's category theorem and Remark 1.2.2 every set in $\mathscr{S}(\mathscr{P})$ has a dense complement in Ω; in particular, $\Omega \notin \mathscr{S}(\mathscr{P})$.)

1.6.4. Consider $\mathbb{N}$ with the discrete topology and $\mathbb{N}^{\mathbb{N}}$ with the corresponding product topology $\mathscr{T}$. Let $\mathscr{Q}$ be the system of all subsets of $\mathbb{N}^{\mathbb{N}}$ that are either empty or of the form

$$\{n_1\} \times \cdots \times \{n_r\} \times \mathbb{N} \times \mathbb{N} \times \cdots \qquad (r \geqq 0, \quad n_1, \ldots, n_r \in \mathbb{N})$$

and $\mathscr{Q}'$ the system of all finite unions of sets from $\mathscr{Q}$. Prove:

1.6.4.1. $\mathscr{Q}$ and $\mathscr{Q}'$ are σ-compact.

1.6.4.2. $\mathscr{B}(\mathscr{T}) = \mathscr{B}(\mathscr{Q}) = \mathscr{B}(\mathscr{Q}')$.

1.6.4.3. $\mathscr{S}(\mathscr{B}(\mathscr{T})) = \mathscr{S}(\mathscr{Q})$.

1.6.5. Consider the set $\mathbb{N}_0 = \mathbb{N} \cup \{0\}$ with the topology given by the metric $|m, n| = |m^{-1} - n^{-1}|$ (convention: $1/0 = 0$), and $\mathbb{N}_0^{\mathbb{N}}$ with the product topology $\mathscr{T}_0$. Prove:

1.6.5.1. The set systems $\mathscr{Q}$, $\mathscr{Q}'$ of Exercise 1.6.4 can be considered as systems of subsets of $\mathbb{N}_0^{\mathbb{N}}$; as such, they are σ-compact, but $\mathscr{B}(\mathscr{T}_0) \neq \mathscr{B}(\mathscr{Q})$.

1.6.5.2. $(\mathbb{N}_0^{\mathbb{N}}, \mathscr{T}_0)$ is compact metrizable.

1.6.5.3. $\mathbb{N}^{\mathbb{N}}$ is a G_δ subset of $\mathbb{N}_0^{\mathbb{N}}$.

We are now able to derive a first result about cardinalities.

1.7. Proposition. *Let $\Omega \neq \varnothing$ and $\varnothing \in \mathscr{P} \subseteq \mathscr{P}(\Omega)$, $\mathscr{P}$ at most countable. Then $\mathscr{B}(\mathscr{P})$ and $\mathscr{S}(\mathscr{P})$ have at most the power of the continuum.*

Proof. $\mathscr{Q} = \mathscr{P} \cup \{\Omega \backslash F \mid F \in \mathscr{P}\}$ is at most countable, and $F \in \mathscr{Q} \Leftrightarrow \Omega \backslash F \in \mathscr{Q}$, hence $\mathscr{B}(\mathscr{P}) = \mathscr{B}(\mathscr{Q}) \subseteq \mathscr{S}(\mathscr{Q}) \supseteq \mathscr{S}(\mathscr{P})$ by Proposition 1.5. The set of all Souslin frames over $\mathscr{Q}$ is $\mathscr{Q}^{\mathbb{N}^*}$. Here $\mathscr{Q}$ is at most countable and $\mathbb{N}^*$ is countable, hence $\mathscr{Q}^{\mathbb{N}^*}$ has at most the power of the continuum. Thus $\mathscr{S}(\mathscr{Q})$ has at most this power, and thus the same statement holds for $\mathscr{B}(\mathscr{P})$ and $\mathscr{S}(\mathscr{P})$.

We now set out for more characterizations of Souslin sets. In this context it is necessary to recall Definition V.1.4 of σ-compact set systems.

1.8. Proposition. *Let Ω, $X \neq \varnothing$ and $\varnothing \in \mathscr{P} \subseteq \mathscr{P}(\Omega)$, $\varnothing \in \mathscr{Q} \subseteq \mathscr{P}(X)$. Put $[\mathscr{P}, \mathscr{Q}] = \{F \times E \mid F \in \mathscr{P}, E \in \mathscr{Q}\}$ and denote by φ the natural projection $(\omega, x) \to \omega$ of $\Omega \times X$ onto Ω. Assume now that $\mathscr{Q}$ is σ-compact. Then*

$$\varphi(\mathscr{S}([\mathscr{P}, \mathscr{Q}])) \subseteq \mathscr{S}(\mathscr{P}),$$

i.e.,

$$\tilde{F} \in \mathscr{S}([\mathscr{P}, \mathscr{Q}]) \quad \Rightarrow \quad \varphi(\tilde{F}) \in \mathscr{S}(\mathscr{P}).$$

Proof. Choose any $\tilde{F} \in \mathscr{S}([\mathscr{P}, \mathscr{Q}])$ and consider any Souslin frame

$$H: (n_1, \ldots, n_r) \rightarrow F'(n_1, \ldots, n_r) \times G(n_1, \ldots, n_r) \quad (r \geqq 1, \quad n_1, \ldots, n_r \in \mathbb{N})$$

over $[\mathscr{P}, \mathscr{Q}]$ such that $\tilde{F} = A(H)$, i.e.,

$$\tilde{F} = \bigcup_{(n_1, n_2, \ldots) \in \mathbb{N}^{\mathbb{N}}} \bigcap_{r=1}^{\infty} [F'(n_1, \ldots, n_r) \times G(n_1, \ldots, n_r)].$$

Clearly

$$\varphi(\tilde{F}) = \bigcup_{(n_1, n_2, \ldots) \in S} \bigcap_{r=1}^{\infty} F'(n_1, \ldots, n_r),$$

where $S = \{(n_1, n_2, \ldots) | \bigcap_{r=1}^{\infty} G(n_1, \ldots, n_r) \neq \varnothing\}$. Since $\mathscr{Q}$ is σ-compact, we may as well write $S = \{(n_1, n_2, \ldots) | \bigcap_{r=1}^{s} G(n_1, \ldots, n_r) \neq \varnothing \ (s = 1, 2, \ldots)\}$. Put now $W = \{(n_1, \ldots, n_s) | s \geqq 1, n_1, \ldots, n_s \in \mathbb{N}, \bigcap_{r=1}^{s} G(n_1, \ldots, n_r) \neq \varnothing\}$ and

$$F(n_1, \ldots, n_s) = \begin{cases} \varnothing & \text{if } (n_1, \ldots, n_s) \notin W \\ F'(n_1, \ldots, n_s) & \text{if } (n_1, \ldots, n_s) \in W. \end{cases}$$

Then F is a Souslin frame over $\mathscr{P}$ ($\ni \varnothing$!) and $A(F) = \varphi(\tilde{F})$.

For the formulation of the next theorem we use the following notation: If $\mathscr{M}$ is a set system, then $\mathscr{M}_\sigma$ denotes the system

$$\{M_1 \cup M_2 \cup \cdots | M_1, M_2, \ldots \in \mathscr{M}\}$$

of all countable unions of sets from $\mathscr{M}$, and $\mathscr{M}_\delta$ denotes the system $\{M_1 \cap M_2 \cap \cdots | M_1, M_2, \ldots \in \mathscr{M}\}$ of all countable intersections of sets from $\mathscr{M}$. We write $\mathscr{M}_{\sigma\delta}$ for $(\mathscr{M}_\sigma)_\delta$. If $\mathscr{P}$, $\mathscr{Q}$ are two set systems, then we write $[\mathscr{P}, \mathscr{Q}]$ for $\{F \times E | F \in \mathscr{P}, E \in \mathscr{Q}\}$. For any cartesian product $\Omega \times X$, φ denotes the projection $(\omega, x) \rightarrow \omega$ of it onto its first component.

1.9. Theorem. *Let $\Omega \neq \varnothing$ and $\varnothing \in \mathscr{P} \subseteq \mathscr{P}(\Omega)$. Then for every A the following statements are equivalent:*

1.9.1. $A \in \mathscr{S}(\mathscr{P})$.

1.9.2. *There is a set $X \neq \varnothing$ and a σ-compact set system $\varnothing \in \mathscr{Q} \subseteq \mathscr{P}(X)$ and a set $\tilde{F} \in [\mathscr{P}, \mathscr{Q}]_{\sigma\delta}$ such that $A = \varphi(\tilde{F})$.*

1.9.3. *There is a compact Hausdorff space $(Y, \mathscr{T})$ such that, if $\mathscr{K}$ denotes the system of all compact subsets of Y, there is some set $\tilde{G} \in [\mathscr{P}, \mathscr{K}]_{\sigma\delta}$ such that $A = \varphi(\tilde{G})$.*

1.9.4. *There is a compact metric space $(Z, |\cdot, \cdot|)$ such that, if $\mathscr{L}$ denotes* *the system of all compact subsets of* Z, *there is some set* $\tilde{H} \in [\mathscr{P}, \mathscr{L}]_{\sigma\delta}$ *such that* $A = \varphi(\tilde{H})$.

Proof. $1 \Rightarrow 2$. Let $F \in \mathscr{P}^{\mathbb{N}^*}$ be a Souslin frame over $\mathscr{P}$ such that $A = A(F)$, i.e.,

$$A = \bigcup_{(n_1, n_2, \dots) \in \mathbb{N}^{\mathbb{N}}} \bigcap_{r=1}^{\infty} F(n_1, \dots, n_r).$$

Let $X = \mathbb{N}^{\mathbb{N}}$ and let $\mathscr{Q} \subseteq \mathscr{P}(\mathbb{N}^{\mathbb{N}})$ consist of all sets of the form $C_1 \times C_2 \times \cdots$ where $|C_n| < \infty$ for finitely many n and $C_n = \mathbb{N}$ otherwise. $\mathscr{Q}$ contains $\varnothing$ and is σ-compact (Exercise 1.6.4.1). Typical members of $\mathscr{Q}$ are the *special cylinders*

$$[n_1, \dots, n_r] = \{n_1\} \times \cdots \times \{n_r\} \times \mathbb{N} \times \mathbb{N} \times \cdots \; (r \geqq 1, n_1, \dots, n_r \in \mathbb{N}).$$

Put now

$$\tilde{F} = \bigcap_{r=1}^{\infty} \bigcup_{n_1, \dots, n_r \in \mathbb{N}} F(n_1, \dots, n_r) \times [n_1, \dots, n_r].$$

Clearly $\tilde{F} \in [\mathscr{P}, \mathscr{Q}]_{\sigma\delta}$. It is an easy exercise to show that $F = \varphi(\tilde{F})$.

$2 \Rightarrow 1$. By Proposition 1.4 we have $[\mathscr{P}, \mathscr{Q}]_{\sigma\delta} \subseteq \mathscr{S}([\mathscr{P}, \mathscr{Q}])$, and by Proposition 1.8 we get $\varphi([\mathscr{P}, \mathscr{Q}]_{\sigma\delta}) \subseteq \varphi(\mathscr{S}([\mathscr{P}, \mathscr{Q}])) \subseteq \mathscr{S}(\mathscr{P})$.

$4 \Rightarrow 3 \Rightarrow 2$ follow trivially, by specialization.

$2 \Rightarrow 4$. Let $\mathbb{N}_0 = \{0\} \cup \mathbb{N}$ and define a metric on $\mathbb{N}_0$ by $|j, k| = |j^{-1} - k^{-1}|$ $(j, k \in \mathbb{N}_0)$, with the understanding $1/0 = 0$. Clearly $(\mathbb{N}_0, |\cdot, \cdot|)$ is a compact metric space. Thus also $Z = \mathbb{N}_0 \times \mathbb{N}_0 \times \cdots$ is a compact metric space with an obvious metric. It is an easy exercise to prove that all subsets of Z having the form $\{n_1\} \times \cdots \times \{n_r\} \times \mathbb{N}_0 \times \mathbb{N}_0 \times \cdots$, $r \geqq 1$, $n_1, \dots, n_r \in \mathbb{N}$, are compact. The rest of the proof of $2 \Rightarrow 4$ goes literally as the proof of $1 \Rightarrow 2$.

1.10. MEYER [2] took 1.9.2 as his definition of a "$\mathscr{P}$-analytic set."

We now come to the most important result of this section, the so-called *separation theorem*. It needs a short notational preparation. If $\mathscr{M}$ is a set system, then $\mathscr{M}^{\sigma\delta}$ denotes the smallest set system containing $\mathscr{M}$ and stable under countable unions and intersections. It is obtainable, according to the model deductions in Proposition I.4.5, as the intersection of all set systems $\supseteq \mathscr{M}$ that are stable under countable unions and intersections, and are contained in $\mathscr{P}(\Omega)$ for a suitable basic set Ω (e.g., $\Omega = \bigcup_{M \in \mathscr{M}} M$); $\mathscr{P}(\Omega)$ clearly participates in this intersection. Furthermore, we shall say, for two set systems $\mathscr{S}$, $\mathscr{T} \subseteq \mathscr{P}(\Omega)$, that $\mathscr{T}$ *separates* (the members of) $\mathscr{S}$ if A, $B \in \mathscr{S}$, $A \cap B = \varnothing$ implies the existence of U, $V \in \mathscr{T}$ such that $A \subseteq U$, $B \subseteq V$, $U \cap V = \varnothing$.

1.11. Theorem (separation theorem). *Let $\Omega \neq \varnothing$ and $\varnothing \in \mathscr{P} \subseteq \mathscr{P}(\Omega)$, $\mathscr{P}$ σ-compact. Then $(\mathscr{P} \cup \{\Omega\})^{\sigma\delta}$ separates $\mathscr{S}(\mathscr{P})$.*

Proof. We begin with a general observation. Let $E_1, E_2, \ldots \subseteq \Omega \supseteq F_1, F_2, \ldots$ and assume that for any j, k there are U_{jk}, V_{jk} such that $E_j \subseteq U_{jk}$, $F_k \subseteq V_{jk}$, $U_{jk} \cap V_{jk} \neq \varnothing$. Then

$$E_1 \cup E_2 \cup \cdots \subseteq \bigcup_{j=1}^{\infty} \bigcap_{k=1}^{\infty} U_{jk}, \quad F_1 \cup F_2 \cup \cdots \subseteq \bigcup_{k=1}^{\infty} \bigcap_{j=1}^{\infty} V_{jk}$$

and

$$\left[\bigcup_j \bigcap_k U_{jk}\right] \cap \left[\bigcup_k \bigcap_j V_{jk}\right] = \varnothing.$$

In words: If we can "separate" each E_j from each F_k by certain sets U_{jk}, V_{jk}, then we can also "separate" $\bigcup_j E_j$ and $\bigcup_k F_k$ by sets obtainable from the U_{jk}, V_{jk} by countable unions and intersections. Put now $\bar{\mathscr{P}} = (\mathscr{P} \cup \{\Omega\})^{\sigma\delta}$ for brevity. Let $A, B \in \mathscr{S}(\mathscr{P})$, $A \cap B = \varnothing$. By Theorem 1.9.2 we may choose $X \neq \varnothing \neq Y$, a σ-compact $\varnothing \in \mathscr{Q} \subseteq \mathscr{P}(X)$ and a σ-compact $\varnothing \in \mathscr{R} \subseteq \mathscr{P}(Y)$ and sets $\tilde{F} \in [\mathscr{P}, \mathscr{Q}]_{\sigma\delta}$, $\tilde{G} \in [\mathscr{P}, \mathscr{R}]_{\sigma\delta}$ such that $A = \varphi(\tilde{F})$, $B = \psi(\tilde{G})$, where $\varphi\colon (\omega, x) \to \omega$ and $\psi\colon (\omega, y) \to \omega$ denote the projections onto the first components.

If one of the set $\tilde{F}$, $\tilde{G}$ is empty, so is one of the sets A, B, and we are through. If $\tilde{F} \neq \varnothing \neq \tilde{G}$, then clearly there are $\mathscr{R}_{\sigma\delta} \ni F' \neq \varnothing \neq G' \in \mathscr{Q}_{\sigma\delta}$, and it is an easy exercise to prove that $\hat{F} = \tilde{F} \times F' \in [\mathscr{P}, [\mathscr{Q}, \mathscr{R}]]_{\sigma\delta}$, $\hat{G} = \tilde{G} \times G' \in [\mathscr{P}, [\mathscr{Q}, \mathscr{R}]]_{\sigma\delta}$, having A, B as their projections on the Ω-component, respectively. It is another easy exercise to prove that $[\mathscr{Q}, \mathscr{R}]$ is σ-compact. Summing up, we may assume $X = Y$, $\mathscr{Q} = \mathscr{R}$, i.e., $\tilde{F}, \tilde{G} \in [\mathscr{P}, \mathscr{Q}]_{\sigma\delta}$, $\varphi(\tilde{F}) = A$, $\varphi(\tilde{G}) = B$. We may find representations $\tilde{F} = \bigcap_j \bigcup_k C_{jk}$, $\tilde{G} = \bigcap_j \bigcup_k D_{jk}$ with $C_{jk}, D_{jk} \in [\mathscr{P}, \mathscr{Q}]$, i.e., $C_{jk} = E_j \times K_k$, $D_{jk} = F_j \times L_k$, $E_j, F_j \in \mathscr{P}$, $K_k, L_k \in \mathscr{Q}$ ($j, k = 1, 2, \ldots$). Assume now that $\bar{\mathscr{P}}$ does not separate the projections A, B of $\tilde{F}$, $\tilde{G}$. For any $r \geqq 1$, $n_1, \ldots, n_r \in \mathbb{N}$, put

$$C(n_1, \ldots, n_r) = C_{1n_1} \cap \cdots \cap C_{rn_r} \cap \bigcap_{j>r} \bigcup_k C_{jk},$$

$$D(n_1, \ldots, n_r) = D_{1n_1} \cap \cdots \cap D_{rn_r} \cap \bigcap_{j>r} \bigcup_k D_{jk}.$$

Clearly $\tilde{F} = \bigcup_{n_1=1}^{\infty} C(n_1)$, $\tilde{G} = \bigcup_{n_1=1} D(n_1)$ and hence, by projection, $A = \bigcup_j \varphi(C(j))$, $B = \bigcup_k \varphi(D(k))$. Assume now that we cannot "separate" A and B by $\bar{\mathscr{P}}$. Our general observation then implies the existence of j_1,

$k_1 \in \mathbb{N}$ such that $\mathscr{P}$ does not "separate" $\varphi(C(j_1))$ and $\varphi(D(k_1))$, although clearly $\varphi(C(j_1)) \cap \varphi(D(k_1)) \subseteq \varphi(\tilde{F}) \cap \varphi(\tilde{G}) = A \cap B = \varnothing$. Now

$$C(j_1) = \bigcup_j C(j_1, j), \qquad D(k_1) = \bigcup_k D(k_1, k).$$

By the same token, we find $j_2, k_2 \in \mathbb{N}$ such that $\overline{\mathscr{P}}$ does not "separate" $\varphi(C(j_1, j_2))$ and $\varphi(D(k_1, k_2))$, although clearly these sets are disjoint. Going on in this way by induction, we find $(j_1, j_2, \ldots), (k_1, k_2, \ldots) \in \mathbb{N}^{\mathbb{N}}$ such that for every $r = 1, 2, \ldots$, the sets $\varphi(C(j_1, \ldots, j_r)), \varphi(D(k_1, \ldots, k_r))$ are disjoint but not "separated" by $\overline{\mathscr{P}}$. It follows in particular that they are nonempty, hence the sets $K_{j_1} \cap \cdots \cap K_{j_r}, L_{k_1} \cap \cdots \cap L_{k_r}$ are all nonempty, and by σ-compactness of $\mathscr{Q}$ the $K_{j_1} \cap K_{j_2} \cap \cdots$, $L_{k_1} \cap L_{k_2} \cap \cdots$ are nonempty. By the same token the sets $(E_{j_1} \cap \cdots \cap E_{j_r}), (F_{k_1} \cap \cdots \cap F_{k_r})$ have a nonempty intersection since they are in $\mathscr{P}$ which does not "separate" the sets $\varphi(C(j_1, \ldots, j_r)), \varphi(D(k_1, \ldots, k_r))$ contained in them, respectively $(r = 1, 2, \ldots)$. Since the E_j, F_k are in $\mathscr{P}$, which was assumed to be σ-compact, $(E_{j_1} \cap E_{j_2} \cap \cdots) \cap (F_{k_1} \cap F_{k_2} \cap \cdots) \neq \varnothing$. Choose now ω from this set, $x \in K_{j_1} \cap K_{j_2} \cap \cdots$, $x' \in L_{k_1} \cap L_{k_2} \cap \cdots$. Then $(\omega, x) \in \tilde{F}$, $(\omega, x') \in \tilde{G}$, $\omega \in \varphi(\tilde{F}) \cap \varphi(\tilde{G}) = A \cap B$, a contradiction.

1.12. Corollary. *Let $(\Omega, \mathscr{T})$ be a topological space and $\mathscr{F} = \{F \mid F$ closed$\}$. Then $\mathscr{S}(\mathscr{F})$ is separated by $\mathscr{B}(\mathscr{F}) = \mathscr{B}(\mathscr{T})$.*

Proof. $\mathscr{B}(\mathscr{T}) \supseteq [\mathscr{F} \cup \{\Omega\}]^{\sigma\delta}$.

1.13. Exercise. Let $(\Omega, \mathscr{T})$ be a topological space and $\mathscr{F} = \{F \mid \Omega \backslash F \in \mathscr{T}\}$. Prove: If $A_1, A_2, \ldots \in \mathscr{S}(\mathscr{F})$, $A_j \cap A_k = \varnothing$ $(j \neq k)$, then there are $U_1, U_2, \ldots \in \mathscr{B}(\mathscr{T})$ such that $U_j \cap U_k = \varnothing$ $(j \neq k)$ and $A_1 \subseteq U_1$, $A_2 \subseteq U_2, \ldots$.

2. SOUSLIN SETS AND ANALYTIC SETS IN POLISH SPACES

In this section we apply the results of Section 1 in order to carry through a thorough investigation of the structure of *Borel*, *Souslin*, and so-called *analytic sets* in an arbitrary Polish space Ω. A particularly sharp instrument will be the special Polish space $\mathbb{N}^{\mathbb{N}}$ (with the product topology of the discrete topology in $\mathbb{N}$), besides the compact spaces $\mathbb{N}_0^{\mathbb{N}}$ and $\{0, 1\}^{\mathbb{N}}$. We shall settle problems 2 and 3 from the introduction to this chapter to a large extent.

2.1. Definition. *Let $(\Omega, \mathscr{T})$ be a Polish space and $|\cdot, \cdot|$ a complete metric defining $\mathscr{T}$. A Souslin frame $F \in \mathscr{P}(\Omega)^{\mathbb{N}^*}$ is called:*

2.1.1. **pointed** *if (with the general definition* diam $F = \sup\{|\omega, \eta| \,|\, \omega, \eta \in F\}$ $(\varnothing \neq F \subseteq \Omega)$, diam $\varnothing = 0$)

$$(n_1, n_2, \ldots) \in \mathbb{N}^{\mathbb{N}} \quad \Rightarrow \quad \lim\,[\text{diam } F(n_1, \ldots, n_r)] = 0;$$

2.1.2. **antitone** *if* $r \geqq 0, n_1, \ldots, n_{r+1} \in \mathbb{N} \Rightarrow F(n_1, \ldots, n_r, n_{r+1}) \subseteq F(n_1, \ldots, n_r)$ (*compare Remark* 1.2.2);

2.1.3. **standard** *if it is pointed and antitone and if* $r \geqq 1$,

$$n_1, \ldots, n_r \in \mathbb{N} \quad \Rightarrow \quad F(n_1, \ldots, n_r) \neq \varnothing$$

and closed;

2.1.4. **subdivisive** *if* $r \geqq 0$,

$$n_1, \ldots, n_r \in \mathbb{N} \quad \Rightarrow \quad F(n_1, \ldots, n_r) = \bigcup_{n=1}^{\infty} F(n_1, \ldots, n_r, n).$$

2.1.5. **disjoint** *if* $r \geqq 1$, $(m_1, \ldots, m_r)$, $(n_1, \ldots, n_r) \in \mathbb{N}^r$,

$$(m_1, \ldots, m_r) \neq (n_1, \ldots, n_r) \quad \Rightarrow \quad F(m_1, \ldots, m_r) \cap F(n_1, \ldots, n_r) = \varnothing.$$

2.2. Proposition. *Let* $(\Omega, \mathscr{T})$ *be a Polish space,* $\mathscr{F} = \{F \,|\, F \text{ closed}\}$, $|\cdot, \cdot|$ *a complete metric defining* $\mathscr{T}$. *Then for every* $A \subseteq \Omega$ *the following statements are equivalent:*

2.2.1. $A \in \mathscr{S}(\mathscr{F})$.

2.2.2. $A = A(F)$ *for some standard Souslin frame* F.

Proof. $2 \Rightarrow 1$ is obvious.

$1 \Rightarrow 2$. We first settle the case where A is closed. The special case $A = \varnothing$ is trivial. Let now $A \neq \varnothing$. Choose any $\varepsilon > 0$. For every $r = 1, 2, \ldots$, we can find a covering $A \subseteq G_{r1} \cup G_{r2} \cup \cdots$ such that every G_{rn} is closed and satisfies diam $G_{rn} \leqq \varepsilon/r$, $G_{rn} \cap A \neq \varnothing$. Put

$$F(n_1, \ldots, n_r) = G_{1n_1} \cap \cdots \cap G_{rn_r} \qquad (r \geqq 1, \quad n_1, \ldots, n_r \in \mathbb{N}).$$

This defines a standard Souslin frame F. If $\omega \in A$, determine $n_1, n_2, \ldots \in \mathbb{N}$ such that $\omega \in G_{rn_r}$ $(r = 1, 2, \ldots)$. Then $\{\omega\} = \bigcap_r F(n_1, \ldots, n_r)$, i.e., $\omega \in A(F)$. If $\omega \notin A$, let $\delta = \inf\{|\omega, \eta| \,|\, \eta \in A\}$. We have $\delta > 0$ since A is closed. It follows that $\omega \notin G_{rn}$ $(n \in \mathbb{N})$ as soon as $\varepsilon/r < \delta$, and thus $\omega \notin A(F)$. This finishes the proof of $A = A(F)$.

Let now $A \in \mathscr{S}(\mathscr{F})$ be arbitrary and $A = A(F)$ where $F \in \mathscr{F}^{\mathbb{N}^*}$ is a Souslin frame over $\mathscr{F}$. For every $r \geqq 1$, $n_1, \ldots, n_r \in \mathbb{N}$ we may find a standard Souslin frame $F_{n_1, \ldots, n_r} \in \mathscr{F}^{\mathbb{N}^*}$ such that

$$\text{diam } F_{n_1, \ldots, n_r}(n) < \frac{1}{n_1 + \cdots + n_r};$$

we only have to choose $F = F_{n_1, \ldots, n_r}$ $\varepsilon = 1/(n_1 + \cdots + n_r)$ in the above construction. Now we go back to the proof of Proposition 1.3 and see how a Souslin frame H with $A = A(H)$ was composed from the Souslin frames $F_{n_1, \ldots, n_r}$ $(r \geqq 1, n_1, \ldots, n_r \in \mathbb{N})$:

$$H(k_1, \ldots, k_t) = F_{n(k_1), \ldots, n(k_{s(t)})}(m_{s(t), 1}, \ldots, m_{s(t), r(t)})$$

$(t \geqq 1, k_1, \ldots, k_t \in \mathbb{N})$, where $s(1), s(2), \ldots$ runs through the first components of the Cauchy diagonal enumeration of $\mathbb{N} \times \mathbb{N}$, i.e., through 1, 2, 1, 3, 2, 1, 4, 3, 2, 1, ... It follows from our construction that

$$\liminf_t \operatorname{diam} H(k_1, \ldots, k_t) = 0 \qquad ((k_1, k_2, \ldots) \in \mathbb{N}^{\mathbb{N}}).$$

Consequently, if we put

$$G(k_1, \ldots, k_t) = \bigcap_{u=1}^{t} H(k_1, \ldots, k_u) \qquad (t \geqq 0, \quad k_1, \ldots, k_t \in \mathbb{N}),$$

we obtain a standard Souslin frame G over $\mathscr{F}$ which apparently yields $A(G) = A(H) = A$.

2.3. Proposition. *Let $(\Omega, \mathscr{T})$ be a Polish space. Then there is a continuous mapping $f\colon \mathbb{N}^{\mathbb{N}} \to \Omega$ such that $f(\mathbb{N}^{\mathbb{N}}) = \Omega$.*

Proof. Let $|\cdot, \cdot|$ be a complete metric defining $\mathscr{T}$. We have only to slightly modify the construction at the beginning of the proof of Proposition 2.2 in order to obtain a standard frame F such that

$$(n_1, n_2, \ldots) \in \mathbb{N}^{\mathbb{N}} \quad \Rightarrow \quad \bigcap_r F(n_1, \ldots, n_r) \neq \varnothing.$$

In fact we have only to begin with some covering $\Omega = F(1) \cup F(2) \cup \cdots$ of Ω with nonempty closed sets $F(n)$ satisfying $\operatorname{diam} F(n) \leqq 1$ $(n = 1, 2, \ldots)$. Next we find—it is easy—representations $F(n) = (F(n, 1) \cup F(n, 2) \cup \cdots)$ with closed nonempty sets $F(n, j)$ satisfying $\operatorname{diam} F(n, j) \leqq \frac{1}{2}$ $(n, j = 1, 2, \ldots)$. We continue in this way and thus construct $F \in \mathscr{F}^{\mathbb{N}^*}$. It is clear that F is standard. For every $(n_1, n_2, \ldots) \in \mathbb{N}^{\mathbb{N}}$, $F(n_1) \supseteq F(n_1, n_2) \supseteq \cdots$ and $\bigcap_r F(n_1, \ldots, n_r)$ is a one-point set which we denote by $f(n_1, n_2, \ldots)$. This defines a mapping $f\colon \mathbb{N}^{\mathbb{N}} \to \Omega$. It is continuous: $(n_1, n_2, \ldots)$ is close to $(k_1, k_2, \ldots)$ in the topology of $\mathbb{N}^{\mathbb{N}}$ if $n_1 = k_1, \ldots, n_r = k_r$ for a large r. But this implies $|f(n_1, n_2, \ldots), f(k_1, k_2, \ldots)| \leqq \operatorname{diam} F(n_1, \ldots, n_r) \leqq 1/r$. If $\omega \in \Omega$, then we find $n_1, n_2, \ldots \in \mathbb{N}$, not necessarily unique, such that $\omega \in F(n_1, \ldots, n_r)$ $(r = 1, 2, \ldots)$, hence $f(n_1, n_2, \ldots) = \omega$.

2.4. Definition. *Let $(\Omega, \mathscr{T})$ be a Polish space. A subset A of Ω is called* **analytic** *(in Ω) if it is empty or if there is a continuous mapping $f\colon \mathbb{N}^{\mathbb{N}} \to \Omega$ such that $A = f(\mathbb{N}^{\mathbb{N}})$.*

2.5. Remark. Proposition 2.3 implies that every closed subset of a Polish space, being a Polish space again, is analytic in the original space.

2.6. Proposition. *Let $(\Omega, \mathcal{T})$ be a Polish space and $\mathcal{F} = \{F \mid F \text{ closed}\}$. Then for every $A \subseteq \Omega$ the following statements are equivalent:*

2.6.1. $A \in \mathcal{S}(\mathcal{F})$.

2.6.2. *A is analytic in Ω.*

Proof. $1 \Rightarrow 2$. Let $|\cdot, \cdot|$ be a complete metric defining $\mathcal{T}$ and let, by Proposition 2.2, F be a standard Souslin frame such that $A = A(F)$. For every $(n_1, n_2, \ldots) \in \mathbb{N}^{\mathbb{N}}$, there are two cases:

Case I: $F(n_1, \ldots, n_r) = \varnothing$ for some $r \geqq 1$.
Case II: $F(n_1, \ldots, n_r) \neq \varnothing$ $(r = 1, 2, \ldots)$.

Clearly the set of all $(n_1, n_2, \ldots) \in \mathbb{N}^{\mathbb{N}}$ for which case I holds is a union of cylinders $[n_1, \ldots, n_r] = \{(n_1, \ldots, n_r, n_{r+1}, \ldots) \mid n_{r+1}, \ldots \in \mathbb{N}\}$, hence an open subset of the Polish space $\mathbb{N}^{\mathbb{N}}$. The set E of all $(n_1, n_2, \ldots) \in \mathbb{N}^{\mathbb{N}}$ for which case II holds is thus a closed subset of the Polish space $\mathbb{N}^{\mathbb{N}}$. For every $x = (n_1, n_2, \ldots) \in E$ the set $\bigcap_r F(n_1, \ldots, n_r)$ consists of exactly one point; and if we denote this point by $f(x)$, then $f\colon E \to \Omega$ is continuous with $f(E) = A$ (exercise). If $A = \varnothing$, we are through. If $A \neq \varnothing$, then $E \neq \varnothing$, hence E is a Polish space, hence $E = g(\mathbb{N}^{\mathbb{N}})$ for some continuous $g\colon \mathbb{N}^{\mathbb{N}} \to E$ (Proposition 2.3). Now $A = (f \circ g)(\mathbb{N}^{\mathbb{N}})$ shows that A is analytic.

$2 \Rightarrow 1$. Let $f\colon \mathbb{N}^{\mathbb{N}} \to \Omega$ be continuous with $f(\mathbb{N}^{\mathbb{N}}) = A$. For any $r \geqq 1$, $n_1, \ldots, n_r \in \mathbb{N}$, let $F(n_1, \ldots, n_r)$ be the closure of $f([n_1, \ldots, n_r])$ where $[n_1, \ldots, n_r]$ is the cylinder as above. The continuity of f implies that the Souslin frame $F \in \mathcal{F}^{\mathbb{N}}$ thus obtained is pointed. It is antitone by construction and hence standard, given a complete metric $|\cdot, \cdot|$ defining $\mathcal{T}$. If $\omega \in A$, then

$$\omega = f(n_1, n_2, \ldots) \quad \Rightarrow \quad \{\omega\} = \bigcap_r F(n_1, \ldots, n_r),$$

hence $\omega \in A(F)$. If $\omega \notin A$ and $(n_1, n_2, \ldots) \in \mathbb{N}^{\mathbb{N}}$, then we put $\delta = |\omega, f(n_1, n_2, \ldots)|$ and get $\delta > 0$. Find $r \geqq 1$ such that $\operatorname{diam} F(n_1, \ldots, n_r) < \delta$. Then $\omega \notin F(n_1, \ldots, n_r) \ni f(n_1, n_2, \ldots)$ follows and we get $\omega \notin A(F)$. Thus $A = A(F)$ is proved.

2.7. Corollary. *Let $(\Omega, \mathcal{T})$ be a Polish space. Then the system of all analytic sets in Ω is stable under countable unions and intersections. It contains $\mathcal{B}(\mathcal{T})$.*

Proof. Notice that every open set in Ω is an F_σ set and hence in $\mathcal{S}(\mathcal{F})$, by Proposition 1.4, which together with Proposition 1.5 now entails the rest.

The next theorem answers problem 3 from the introduction to this chapter.

2.8. Theorem. *Let $(\Omega, \mathscr{T})$ be a Polish space. Then every uncountable set from $\mathscr{B}(\mathscr{T})$ has the power of the continuum.*

Proof. $\mathbb{N}^{\mathbb{N}}$ has the power of the continuum, hence Ω, being an image of $\mathbb{N}^{\mathbb{N}}$ by Proposition 2.3, has at most the power of the continuum, and hence so has every subset of Ω. The rest of the theorem is now an easy consequence of the next theorem and Corollary 2.7.

2.9. Theorem. *Let $(X, \mathscr{S})$ and $(\Omega, \mathscr{T})$ be Polish spaces and $f\colon X \to \Omega$ be continuous and $f(X)$ an uncountable subset of Ω. Then there is a subset X_0 of X with the following properties:*

2.9.1. *There is a homeomorphism of the Bernoulli space $\{0, 1\}^{\mathbb{N}}$ onto X_0. In particular, X_0 is compact and has the power of the continuum.*

2.9.2. *The restriction of f to X_0 is one-to-one, and hence a homeomorphism of X_0 onto a subset of $f(X)$ which thus has the power of the continuum.*

Proof. We begin with a general preparatory statement. Call a subset X_1 of X dense in itself if every open neighborhood of a point in X_1 contains infinitely many points of X_1. We prove that every uncountable subset M of X contains a subset X_1 that is uncountable and dense in itself. In fact, let $|\cdot, \cdot|$ be a complete metric defining T and $2r(x) = \sup\{r \mid \text{there are at most countably many } y \in M \text{ with } |x, y| < r\}$ $(x \in X)$. Then $X_1' = \{x \mid r(x) > 0\}$ is easily seen to be open. If $X_1' = \varnothing$, then we may clearly put $X_1 \neq M$. If $X_1' \neq \varnothing$, we may find a dense sequence $x_1, x_2, \ldots \in X_1'$; and it is an easy exercise to prove that $X_1' = \bigcup_n U_n$ where $U_n = \{x \mid |x, x_n| < r(x_n)\}$. Since $U_n \cap M$ is countable $(n = 1, 2, \ldots)$, $X_1' \cap M$ is countable too, and hence $X_1 = M \setminus X_1'$ is still uncountable, and obviously dense in itself.

Now we turn to the situation of our theorem. For every $\omega \in f(X)$, we choose, by the axiom of choice, some point $x_\omega \in X$ such that $f(x_\omega) = \omega$. Put $M = \{x_\omega \mid \omega \in f(X)\}$. Since $f(X)$ is uncountable, so is M, and hence M contains an uncountable set X_1 that is dense in itself. What follows now is a construction in countably many steps.

Step 1: Find $x_0, x_1 \in X_1$ such that $f(x_0) \neq f(x_1)$. Find open sets $U(0), U(1) \subseteq X$ with (for some complete metric $|\cdot, \cdot|$ defining $\mathscr{S}$) $\operatorname{diam} U(0) < 1 > \operatorname{diam} U(1)$, and such that $x_0 \in U(0)$, $x_1 \in U(1)$ and their closures have disjoint images under f. This is possible since f is continuous. Clearly $U(0) \cap X_1$, $U(1) \cap X_1$ are uncountable and each is dense in itself.

Step 2: Find $x_{00}, x_{01} \in U(0) \cap X_1$ and $x_{10}, x_{11} \in U(1) \cap X_1$ such that $f(x_{00}), \ldots, f(x_{11})$ are four distinct points. Find open sets $U(0, 0)$, $U(0,1) \subseteq U(0)$, $U(1,0)$, $U(1, 1) \subseteq U(1)$ such that $x_{ij} \in U(i,j)$, diam $U(i,j) < \frac{1}{2}$ $(i, j = 0, 1)$ and the closures of the four sets $f(U(0, 0)), \ldots, f(U(1, 1))$ are pairwise disjoint. This is possible since f is continuous. Clearly each of the four sets $U(i, j) \cap X_1$ is uncountable and dense in itself.

It is clear how to continue. We obtain, for every $n \geqq 1$, and every n-tuple $(a_1, \ldots, a_n)$ an open set $U(a_1, \ldots, a_n)$ of diameter $< 1/n$ such that $U(a_1, \ldots, a_n) \subseteq U(a_1, \ldots, a_{n-1})$ and the closures of these $U(a_1, \ldots, a_n)$ have, for a fixed n, disjoint images under f. Since $|\cdot, \cdot|$ is complete, we find that for every $(a_1, a_2, \ldots) \in \{0, 1\}^{\mathbb{N}}$ the closures of the sets $U(a_1)$, $U(a_1, a_2)$, ... intersect in exactly one point which we call $g(a_1, a_2, \ldots)$. The disjointnesses and the estimates of diameters enforced in our construction imply that the g: $\{0, 1\}^{\mathbb{N}} \to X$ thus defined is continuous and the restriction of f to $X_0 = g(\{0, 1\}^{\mathbb{N}})$ is one-to-one. The rest is obvious.

2.10. Exercises

2.10.1. Let $(X, \mathscr{S})$, $(\Omega, \mathscr{T})$ be Polish spaces. Prove $\mathscr{B}(\mathscr{S} \times \mathscr{T}) = \mathscr{B}(\mathscr{S}) \times \mathscr{B}(\mathscr{T})$, where $\mathscr{S} \times \mathscr{T}$ denotes the product topology in $X \times \Omega$ (see Exercise VI.1.6.3).

2.10.2. Let $(X, \mathscr{S})$ be a Polish space. Prove that the diagonal $\Delta = \{(x, x) | x \in X\}$ is closed in $X \times X$ and belongs to $\mathscr{B}(\mathscr{S}) \times \mathscr{B}(\mathscr{S})$.

2.10.3. Consider the lexicographic ordering $\leqq$ in the Polish space $\mathbb{N}^{\mathbb{N}}$, i.e., write $(m_1, m_2, \ldots) \leqq (n_1, n_2, \ldots)$ iff $m_1 = n_1, \ldots, m_r = n_r$, $m_{r+1} \neq n_{r+1} \Rightarrow m_{r+1} < n_{r+1}$. Write $<$ for $\leqq$ plus inequality. Prove:

2.10.3.1. $\leqq$ is a sharp total ordering in $\mathbb{N}^{\mathbb{N}}$.

2.10.3.2. The sets of the form

$$\{(n_1, n_2, \ldots) | (a_1, a_2, \ldots) \leqq (n_1, n_2, \ldots) < (b_1, b_2, \ldots)\}$$

$((a_1, a_2, \ldots), (b_1, b_2, \ldots) \in \mathbb{N}^{\mathbb{N}})$ form a base of the Polish (= product) topology $\mathscr{U}$ of $\mathbb{N}^{\mathbb{N}}$ and generate $\mathscr{B}(\mathscr{U})$.

2.10.3.3. If $\varnothing \neq F \subseteq \mathbb{N}^{\mathbb{N}}$ is closed, then there is exactly one $(n_1, n_2, \ldots) \in F$ such that $(m_1, m_2, \ldots) \in F \Rightarrow (n_1, n_2, \ldots) \leqq (m_1, m_2, \ldots)$.

In the rest of this section, we are tackling problem 2 from the introduction to this chapter. We begin with a preparatory

2.11. Proposition. *Let $(X, \mathscr{S})$, $(\Omega, \mathscr{T})$ be Polish spaces and f: $X \to \Omega$ be $\mathscr{B}(\mathscr{S})$-$\mathscr{B}(\mathscr{T})$-measurable. Then the graph $\{(x, f(x)) | x \in X\}$ is in $\mathscr{B}(\mathscr{S}) \times \mathscr{B}(\mathscr{T})$ $(= \mathscr{B}(\mathscr{S} \times \mathscr{T}))$.*

Proof. The diagonal $\Delta = \{(\omega, \omega) | \omega \in \Omega\}$ is in $\mathscr{B}(\mathscr{T} \times \mathscr{T}) = \mathscr{B}(\mathscr{T}) \times \mathscr{B}(\mathscr{T})$ by Exercises 2.10.1 and 2.10.2. The mapping

$$\varphi: X \times \Omega \to \Omega \times \Omega$$

defined by $\varphi(x, \omega) = (f(x), \omega)$ is $[\mathscr{B}(\mathscr{S}) \times \mathscr{B}(\mathscr{T})]$-$[\mathscr{B}(\mathscr{T}) \times \mathscr{B}(\mathscr{T})]$-measurable by Proposition VI.1.8.2. Now $\varphi^{-1}\Delta = \{(x, \omega) | f(x) = \omega,\ x \in X\} = \{(x, f(x)) | x \in X\} \in \mathscr{B}(\mathscr{S}) \times \mathscr{B}(\mathscr{T})$ follows.

2.12. Exercise. Let $(X, \mathscr{S})$, $(\Omega, \mathscr{T})$ be Polish spaces, $E \in \mathscr{B}(\mathscr{S})$ and $f: E \to \Omega$ be $[\mathscr{B}(\mathscr{S}) \cap E]$-$\mathscr{B}(\mathscr{T})$-measurable, where

$$\mathscr{B}(\mathscr{S}) \cap E = \{F \cap E | F \in \mathscr{B}(\mathscr{S})\}$$

(see Proposition I.4.7.1). Prove that $\{(x, f(x)) | x \in E\} \in \mathscr{B}(\mathscr{S}) \times \mathscr{B}(\mathscr{T})$.

2.13. Proposition. *Let $(X, \mathscr{S})$, $(\Omega, \mathscr{T})$ be Polish spaces, E an analytic set in X, and $f: E \to \Omega$ be $[\mathscr{B}(\mathscr{S}) \cap E]$-$\mathscr{B}(\mathscr{T})$-measurable. Then $f(E)$ is an analytic set in Ω.*

Proof. By Corollary 2.7 all members of $\mathscr{B}(\mathscr{S})$ are analytic in X. Let us first deal with the case $E \in \mathscr{B}(\mathscr{S})$. By Exercise 2.12 the set $G = \{(x, f(x)) | x \in E\}$ belongs to $\mathscr{B}(\mathscr{S}) \times \mathscr{B}(\mathscr{T}) = \mathscr{B}(\mathscr{S} \times \mathscr{T})$ (see Exercise 2.10.1), hence is an analytic set in the Polish space $(X \times \Omega, \mathscr{S} \times \mathscr{T})$, again by Corollary 2.7. Thus we can find a continuous mapping $g: \mathbb{N}^{\mathbb{N}} \to X \times \Omega$ such that $g(\mathbb{N}^{\mathbb{N}}) = G$. Let $\psi: (x, \omega) \to \omega$ denote the projection of $X \times \Omega$ onto Ω. Clearly ψ is continuous. Hence $f(E) = \psi(G) = \psi(g(\mathbb{N}^{\mathbb{N}})) = (\psi \circ g)(\mathbb{N}^{\mathbb{N}})$ is an analytic set in Ω. Let us now settle the general case. There is a continuous $h: \mathbb{N}^{\mathbb{N}} \to X$ with $h(\mathbb{N}^{\mathbb{N}}) = E$. In particular $f \circ h: \mathbb{N}^{\mathbb{N}} \to \Omega$ is $\mathscr{B}$-$\mathscr{B}(\mathscr{T})$-measurable if $\mathscr{B}$ denotes the σ-field generated by the (Polish) product topology in $\mathbb{N}^{\mathbb{N}}$. By the first part of our proof, $(f \circ h)(\mathbb{N}^{\mathbb{N}})$ is analytic in Ω.

Now we shall see how the separation theorem 1.11 works.

2.14. Theorem. *Let $(X, \mathscr{S})$, $(\Omega, \mathscr{T})$ be Polish spaces and $f: X \to \Omega$ be $\mathscr{B}(\mathscr{S})$-$\mathscr{B}(\mathscr{T})$-measurable, one-to-one, and surjective $(f(X) = \Omega)$. Then $F \in \mathscr{B}(\mathscr{S}) \Rightarrow f(F) \in \mathscr{B}(\mathscr{T})$.*

Proof. $F \in \mathscr{B}(\mathscr{S}) \Rightarrow X \backslash F \in \mathscr{B}(\mathscr{S})$. Clearly $f(X \backslash F) = \Omega \backslash f(F)$. By Proposition 2.13, $f(F)$, $f(X \backslash F) \in \mathscr{S}(\mathscr{F})$, where $\mathscr{F} = \{F | F \subseteq \Omega,\ \Omega \backslash F \in \mathscr{T}\}$. By Corollary 1.12 to the separation theorem 1.11 we may find $U, V \in \mathscr{B}(\mathscr{T})$ such that $U \cap V = \varnothing$, $f(F) \subseteq U$, $\Omega \backslash f(F) \subseteq V$. It follows that $f(F) = U$, $\Omega \backslash f(F) = V$. In particular, $f(F) = U \in \mathscr{B}(\mathscr{T})$.

In order to prepare for subsequent results, we need a new tool: *totally disconnected Polish spaces.* Recall that spaces with their discrete topology are always totally disconnected and that a product of totally disconnected spaces is totally disconnected again. Consequently, countable product spaces of totally disconnected Polish spaces are totally disconnected Polish spaces again. $\mathbb{N}^{\mathbb{N}}$ and the Bernoulli space $\{0, 1\}^{\mathbb{N}}$ are good examples.

2.15. Exercises

2.15.1. Let $(\Omega, \mathscr{T})$ be an uncountable totally disconnected Polish space and $|\cdot, \cdot|$ a metric defining $\mathscr{T}$. Prove that there is a disjoint subdivisive standard Souslin frame F such that $r \geqq 1, n_1, \ldots, n_r \in \mathbb{N} \Rightarrow F(n_1, \ldots, n_r)$ is clopen and has diam $F(n_1, \ldots, n_r) \leqq 1/r$, and $\bigcup_n F(n) = \Omega$.

2.15.2. Let $(X_1, \mathscr{S}), (X_2, \mathscr{S}), \ldots$ be Polish spaces such that $X_j \cap X_k = \varnothing$ $(j \neq k)$. Put $X = X_1 + X_2 + \cdots$ and define

$$\mathscr{S} = \{G_1 + G_2 + \cdots | G_1 \in \mathscr{S}_1, G_2 \in \mathscr{S}_2, \ldots).$$

Prove:

2.15.2.1. $(X, \mathscr{S})$ is a Polish space.

2.15.2.2. If $(X_1, \mathscr{S}_1), (X_2, \mathscr{S}_2), \ldots$ are totally disconnected, then $(X, \mathscr{S})$ is totally disconnected.

2.16. Proposition. *Let $(X, \mathscr{S})$, $(\Omega, \mathscr{T})$ be Polish spaces, $f\colon X \to \Omega$ be continuous and injective. Assume that $(X, \mathscr{S})$ is totally disconnected. Then $f(X) \in \mathscr{B}(\mathscr{T})$.*

Proof. Let $|\cdot, \cdot|$ denote a complete metric defining the topology, both in X and in Ω. Only the case where X is uncountable poses a problem. Let $F\colon \mathbb{N}^* \to \mathscr{T}$ be a disjoint subdivisive standard Souslin frame such that $r \geqq 1, n_1, \ldots, n_r \in \mathbb{N} \Rightarrow F(n_1, \ldots, n_r)$ is clopen with diam $F(n_1, \ldots, n_r) \leqq 1/r$, and $\bigcup_n F(n) = X$ (Exercise 2.15). Put $G(n_1, \ldots, n_r) = f(F(n_1, \ldots, n_r))$. Clearly this defines a disjoint subdivisive standard Souslin frame $G \in \mathscr{P}(\Omega)^{\mathbb{N}^*}$ such that every $G(n_1, \ldots, n_r)$ is analytic in Ω and $f(X) = A(G)$. We shall now, by systematic use of Exercise 1.13 (hence of Corollary 1.12), replace the $G(n_1, \ldots, n_r)$ by sets $H(n_1, \ldots, n_r) \in \mathscr{B}(\mathscr{T})$ such that the resulting Souslin frame $H \in \mathscr{B}(\mathscr{T})^{\mathbb{N}^*}$ still satisfies $A(H) = f(X)$ and is disjoint and antitone. Once we have that, we finish our proof with the help of Remark 1.2.2:

$$f(X) = A(H) = \bigcap_{r=1}^{\infty} \bigcup_{(n_1, \ldots, n_r) \in \mathbb{N}^r} H(n_1, \ldots, n_r),$$

and this is in $\mathscr{B}(\mathscr{T})$ since only countably many operations on sets from $\mathscr{B}(\mathscr{T})$ are involved here any more.

We now go into the details of the step-by-step construction of H. In the first step we use Exercise 1.13 in order to determine $H'(1)$,

$H'(2), \ldots \in \mathscr{B}(\mathscr{T})$ such that $H'(j) \cap H'(k) = \varnothing$ $(j \neq k)$ and $G(1) \subseteq H'(1)$, $G(2) \subseteq H'(2), \ldots$. In the second step we determine for every $n \in \mathbb{N}$, sets $H'(n, 1), H'(n, 2), \ldots \subseteq H'(n)$ such that $H'(n, j) \cap H'(n, k) = \varnothing$ $(j \neq k)$ and $G(n, 1) \subseteq H'(n, 1)$, $G(n, 2) \subseteq H'(n, 2), \ldots$. It is clear how to go on to infinity. Let generally $\overline{M}$ denote the closure of a set $M \subseteq \Omega$ and put $H(n_1, \ldots, n_r) = H'(n_1, \ldots, n_r) \cap \overline{G(n_1, \ldots, n_r)}$ $(r \geqq 1; n_1, \ldots, n_r \in \mathbb{N})$. Clearly this defines a disjoint standard Souslin frame $H \in \mathscr{B}(\mathscr{T})^{\mathbb{N}}$. From $G(n_1, \ldots, n_r) \subseteq H'(n_1, \ldots, n_r)$ $(r \geqq 1;\ n_1, \ldots, n_r \in \mathbb{N})$ we deduce $f(X) = A(G) \subseteq A(H)$. We have to show that also $\supseteq$ holds here. Let thus $\omega \in A(H)$ and choose $(n_1, n_2, \ldots) \in \mathbb{N}^{\mathbb{N}}$ such that $\omega \in H(n_1) \cap H(n_1, n_2) \cap \cdots$, hence $\omega \in \overline{G(n_1)} \cap \overline{G(n_1, n_2)} \cap \cdots$. Consequently, $G(n_1)$, $G(n_1, n_2), \ldots$ are all nonempty; and since $G(n_1, \ldots, n_r) = f(F(n_1, \ldots, n_r))$ $(r = 1, 2, \ldots)$, we see that $F(n_1), F(n_1, n_2), \ldots$ are all nonempty. But here we have a descending sequence of clopen sets with diameter tending to 0, hence $F(n_1) \cap F(n_1, n_2) \cap \cdots = \{x\}$ for some $x \in X$ and the continuity of f implies $\{f(x)\} = \overline{G(n_1)} \cap \overline{G(n_1, n_2)} \cap \cdots$. Thus $\omega = f(x) \in F(X) = A(G)$. This does it.

It is clear what we have to prove next:

2.17. Proposition. *Let $(\Omega, \mathscr{T})$ be a Polish space and $\varnothing \neq F \in \mathscr{B}(\mathscr{T})$. Then there is a totally disconnected Polish space $(X, \mathscr{S})$ and a continuous injective $f\colon X \to \Omega$ such that $f(X) = F$.*

Proof. Let $\mathscr{P}$ be the system of all sets $\subseteq \Omega$ that are either empty or have the property stated in our proposition. We first prove that $\Omega \in \mathscr{P}$. This is trüe if $\Omega = [0, 1]$; in fact, dyadic expansion of reals represents $[0, 1]$ as a continuous image of the totally disconnected Polish space $\{0, 1\}^{\mathbb{N}}$ minus some countable subset of the latter; the subtraction of such a subset leads to a G_δ subset $X \neq \varnothing$ of $\{0, 1\}^{\mathbb{N}}$, hence to a totally disconnected Polish space again. $\Omega \in \mathscr{P}$ is equally true if $\Omega = [0, 1]^{\mathbb{N}}$; a continuous one-to-one $\psi\colon X \to [0, 1]$ immediately leads to a continuous one-to-one $\varphi = \psi^{\mathbb{N}}\colon X^{\mathbb{N}} \to [0, 1]^{\mathbb{N}}$, and $X^{\mathbb{N}}$ is Polish and totally disconnected again. Finally, the Urysohn–Alexandrov theorem represents an arbitrary Polish Ω as a homeomorphic image of some G_δ subset of $[0, 1]^{\mathbb{N}}$, which immediately leads to Ω as a one-to-one continuous image of some G_δ set $X^{\mathbb{N}}$, which is a totally disconnected Polish space again. (By the same token every closed subset $\neq \varnothing$ of Ω, being a Polish space again, is in $\mathscr{P}$.) Next we see that $\mathscr{P}$ is stable under countable intersections. In fact let $X_1, X_2, \ldots$ be totally disconnected Polish spaces and $f_n\colon X_n \to \Omega$ continuous and injective. In the totally disconnected Polish space $X_1 \times X_2 \times \cdots$ consider the subset $X = \{(x_1, x_2, \ldots) | x_1 \in X_1, x_2 \in X_2, \ldots, f(x_1) = f(x_2) = \cdots\}$. Clearly X is closed, hence a totally disconnected Polish space, and $f = f_1 \times f_2 \times \cdots$ sends X injectively onto $f_1(X_1) \cap f_2(X_2) \cap \cdots$. Finally, we prove that $\mathscr{P}$ is stable under countable unions. For this, let again $X_1, X_2, \ldots$ be totally disconnected Polish spaces and $f_1: X_1 \to \Omega, f_2: X_2 \to \Omega, \ldots$ continuous one-to-one mappings. We may assume $X_j \cap X_k = \emptyset$ $(j \neq k)$. Apply Exercise 2.15.2 in order to make $X = X_1 + X_2 + \cdots$ into a totally disconnected Polish space again. Define $f: X \to \Omega$ by $f(x) = f_n(x)$ $(x \in X_n, n = 1, 2, \ldots)$. Then f is one-to-one and continuous, and $f(X) = f_1(X_1) \cup f_2(X_2) \cup \cdots$. From what we have shown so far we may now deduce $\mathscr{P} \supseteq \mathscr{B}(\mathscr{T})$. In fact, $\mathscr{P}$ contains every closed subset of Ω, and by countable union also every open subset of Ω. $\mathscr{Q} = \{M \mid M \in \mathscr{P}, \Omega \setminus M \in \mathscr{P}\}$ thus contains $\mathscr{T}$, is stable under countable unions and intersections, and under passage to complements, hence it is a σ-field and $\mathscr{P} \supseteq \mathscr{Q} \supseteq \mathscr{B}(\mathscr{T})$ follows.

We are now in the position to prove the famous

2.18. Theorem (Kuratowski). *Let $(X, \mathscr{S})$, $(\Omega, \mathscr{T})$ be Polish spaces and $f: X \to \Omega$ injective and $\mathscr{B}(\mathscr{S})$-$\mathscr{B}(\mathscr{T})$-measurable. Then $f(\mathscr{B}(\mathscr{S})) \subseteq \mathscr{B}(\mathscr{T})$,* i.e., *$F \in \mathscr{B}(\mathscr{S}) \Rightarrow f(F) \in \mathscr{B}(\mathscr{T})$.*

Proof. Let $F \in \mathscr{B}(\mathscr{S})$. Only the case where F is uncountable is of interest. By Exercise 2.12 the set $G = \{(x, f(x)) \mid x \in F\}$ is in $\mathscr{B}(\mathscr{S}) \times \mathscr{B}(\mathscr{T}) = \mathscr{B}(\mathscr{S} \times \mathscr{T})$ (Exercise 2.10.1) and hence, by Proposition 2.17, there is a totally disconnected Polish space $(Y, \mathscr{R})$ and a continuous injective mapping $h: Y \to X \times \Omega$ such that $h(Y) = G$. The projection $\psi: (x, \omega) \to \omega$ of $X \times \Omega \to \Omega$ is continuous; and since f is injective, the restriction of ψ to G is one-to-one, hence $\psi \circ h: Y \to \Omega$ is continuous and injective. By Proposition 2.16 we now have $f(F) = (\psi \circ h)(Y) \in \mathscr{B}(\mathscr{T})$.

2.19. Exercises

2.19.1. Scan the text of this Section 2 and try to weaken assumptions here and there without losing conclusions. (*Hint*: In many cases the assumption that a certain topological space be Polish can be replaced by the assumption that it be metrizable and first countable.)

2.19.2. Let $(\Omega, \mathscr{T})$ be a compact Polish space. Prove that there is closed subset X of $[0, 1]$ and a continuous mapping $f: X \to \Omega$ such that $f(X) = \Omega$.

2.19.3. Let $(X, \mathscr{S})$, $(\Omega, \mathscr{T})$ be compact Polish spaces and $f: X \to \Omega$ continuous with $f(X) = \Omega$. Prove that there is a set $X_0 \in \mathscr{B}(\mathscr{S})$ such that $f(X_0) = \Omega$ and the restriction of f to X_0 is one-to-one.

2.19.4. Let $\Omega = X = \mathbb{R}$ and $\mathscr{B}$ be the σ-field of all Borel sets in $\mathbb{R}$. Call $x, y \in \mathbb{R}$ equivalent if $x - y \in \mathbb{Q}$ and let $\mathscr{B}_0$ denote the σ-field of all those Borel sets in $\mathbb{R}$ consisting of full equivalence classes, i.e., which are invariant under rational translations. Prove:

2.19.4.1. Every $\mathscr{B}_n$-$\mathscr{B}$-measurable $f: \mathbb{R} \to \mathbb{R}$ is constant on every equivalence class.

2.19.4.2. If $f: \mathbb{R} \to \mathbb{R}$ attains different values on different equivalence classes, then $f(\mathbb{R}) \notin \mathscr{B}$. (*Hint*: See Exercise II.6.6.2.)

3. CAPACITIES

While Sections 1 and 2 were purely set theoretical, we are now stepping into the theory of *set functions* again and present the rudiments of *capacity theory* here. Capacities are in general nonadditive; and where some concept like additivity appears in their theory, it is in a fashion that makes it advisable to treat them not as in Chapter VII but rather in connection with *Souslin sets* whose theory is needed for an understanding of capacities.

The basic result of this section is *Choquet's theorem on the "capacitability" of Souslin sets.* For a systematic approach to capacity theory see CHOQUET [1], HOFFMAN–JØRGENSEN [1].

In Section 4 we use capacity theory in order to present after Meyer an alternative approach to the σ-content extension theory of Chapter II. In Section 5 we use Choquet's capacitability theorem in order to prove the so-called measurable choice theorem of V. NEUMANN [1, 4] (see also T. PARTHASARATHY [1] and HILDENBRAND [2]). It has important applications in economic theory (HILDENBRAND [2]). The basic field of applications for capacity theory is modern potential theory (see MEYER [2]). For applications in information theory, see STRASSEN [1]. For applications in statistics, see HUBER [1] and HUBER and STRASSEN [1, 2]. Set functions closely related to capacities appear in economic theory; see ROSENMÜLLER and WEIDNER [1].

3.1. Definition. *Let $\Omega \neq \varnothing$ and $\varnothing \in \mathscr{C} \subseteq \mathscr{P}(\Omega)$, and $\mathscr{C}$ be stable under finite unions and finite intersections.*

3.1.1. *A function $m: \mathscr{P}(\Omega) \to \overline{\mathbb{R}}$ $(= \mathbb{R} \cup \{-\infty, \infty\})$ is called a* **(Choquet) $\mathscr{C}$-capacity** *if:*

3.1.1.1. *m is isotone, i.e., $E \subseteq F \subseteq \Omega \Rightarrow m(E) \leq m(F)$;*

3.1.1.2. *m is lower σ-continuous, i.e., $F_1 \subseteq F_2 \subseteq \cdots \subseteq F \subseteq \Omega$, $F_1 \cup F_2 \cup \cdots = F \Rightarrow \lim_n m(F_n) = m(F)$;*

3.1.1.3. *m is upper σ-continuous under approximation from $\mathscr{C}$, i.e., C_1, $C_2, \ldots \in \mathscr{C}$, $C_1 \supseteq C_2 \supseteq \cdots$, $C_1 \cap C_2 \cap \cdots = F \Rightarrow \lim_n m(C_n) = m(F)$.*

3.1.2. *Let* $m \in \overline{\mathbb{R}}^{\mathscr{P}(\Omega)}$ *be a* $\mathscr{C}$*-capacity. A set* $F \subseteq \Omega$ *is said to be* ($\mathscr{C}$*-m-*)**capacitable** *if*

$$\text{(1)} \qquad m(F) = \sup\{m(E) \mid \mathscr{C}_\delta \ni E \subseteq F\},$$

where $\mathscr{C}_\delta = \{C_1 \cap C_2 \cap \cdots \mid C_1, C_2, \ldots \in \mathscr{C}\}$ *as usual.*

We emphasize that a capacity is always defined on *all* subsets of the given basic set. Since the set system $\mathscr{C}$ is assumed to be stable under finite intersections, we may always represent a countable intersection of sets from $\mathscr{C}$ as the intersection of a decreasing sequence of sets from $\mathscr{C}$.

3.2. Proposition. *Let* $\Omega \neq \varnothing$ *and* $\varnothing \in \mathscr{C} \subseteq \mathscr{P}(\Omega)$, $\mathscr{C}$ *be stable under finite unions and finite intersections,* $m \in \overline{\mathbb{R}}^{\mathscr{P}(\Omega)}$ *a* $\mathscr{C}$*-capacity. Then every* $E \in \mathscr{C}_{\sigma\delta}$ (= *the system of all countable intersections of countable unions of sets from* $\mathscr{C}$) *is* $\mathscr{C}$*-m-capacitable.*

Proof. Let $E \in \mathscr{C}_{\sigma\delta}$. If $m(E) = -\infty$, (1) holds trivially, by isotony of m. Assume now $m(E) > -\infty$. Choose any $\mathbb{R} \ni \alpha < m(E)$. We have to find some $D \in \mathscr{C}_\delta$ such that $D \subseteq E$ and $m(D) \geqq \alpha$. Choose $U_1, U_2, \ldots \in \mathscr{C}_\sigma$ such that $U_1 \cap U_2 \cap \cdots = E$. Since $\mathscr{C}$ is stable under finite intersections, so is $\mathscr{C}_\sigma$ (exercise); hence we may assume $U_1 \supseteq U_2 \supseteq \cdots$. For every $k = 1, 2, \ldots$, we may choose $C_{k1}, C_{k2}, \ldots \in \mathscr{C}$ with $\bigcup_n C_{kn} = U_k$; and since $\mathscr{C}$ is stable under finite unions, we may assume $C_{k1} \subseteq C_{k2} \subseteq \cdots$. We shall now construct successively $\mathscr{C} \ni C_1 \subseteq U_1, \mathscr{C} \ni C_2 \subseteq U_2, \ldots$ such that

$$m(E \cap C_1 \cap \cdots \cap C_n) > \alpha \qquad (n = 1, 2, \ldots).$$

For the construction of C_1 observe that

$$E = E \cap U_1 = E \cap (C_{11} \cup C_{12} \cup \cdots) = \bigcup_n (E \cap C_{1n})$$

since $C_{11} \subseteq C_{12} \subseteq \cdots$. Lower σ-continuity of m implies the existence of some C_{1n}, which we choose as C_1, such that $m(E \cap C_1) > \alpha$. For the construction of C_2, observe that

$$E \cap C_1 = E \cap C_1 \cap U_2 = \bigcup_n E \cap C_1 \cap C_{1n}.$$

Again lower σ-continuity of m implies the existence of some C_{2n}, which we choose as C_2, such that $m(E \cap C_1 \cap C_2) > \alpha$. It is clear how to continue this procedure. Put now $D_n = C_1 \cap \cdots \cap C_n$. Clearly $D_n \in \mathscr{C}$ since $\mathscr{C}$ is stable under finite intersections ($n = 1, 2, \ldots$). Put $D = D_1 \cap D_2 \cap \cdots$. Then $D \in C_\delta$. Upper σ-continuity of m under approximation from $\mathscr{C}$ implies

$$m(D) = \lim_n m(D_n) = \lim_n m(C_1 \cap \cdots \cap C_n) \geqq \lim_n m(E \cap C_1 \cap \cdots \cap C_n) \geqq \alpha.$$

But $D \subseteq U_1 \cap U_2 \cap \cdots = E$. Hence we are through.

In order to prepare for *Choquet's capacitability theorem,* we have to elaborate a bit on Souslin sets. We shall use Theorem 1.9.2 which makes use of σ-compact set systems (Definition V.1.4). We have thus to deal with such set systems in some detail.

3.3. Proposition. *Let $X \neq \emptyset$ and $\emptyset \in \mathcal{Q} \subseteq \mathcal{P}(X)$ be σ-compact. Then:*

3.3.1. *The system $\mathcal{Q}' = \{Q_1 \cup \cdots \cup Q_n | n \geqq 1, Q_1, \ldots, Q_n \in \mathcal{Q}\}$ of all finite unions of sets in $\mathcal{Q}$ is σ-compact again.*

3.3.2. *Let $\Omega \neq \emptyset$ and $\varphi: (\omega, x) \to \omega$ be the projection of $\Omega \times X$ onto Ω. Let $\tilde{C}_1, \tilde{C}_2, \ldots$ be finite unions of sets from $[\mathcal{P}(\Omega), \mathcal{Q}]$, $\tilde{C}_1 \supseteq \tilde{C}_2 \supseteq \cdots$. Then*

$$\bigcap_n \varphi(\tilde{C}_n) = \varphi\left(\bigcap_n \tilde{C}_n\right)$$

Proof. 1. Let $Q_1', Q_2', \ldots \in \mathcal{Q}'$ be such that $Q_1' \cap \cdots \cap Q_n' \neq \emptyset$ $(n = 1, 2, \ldots)$. Then $\{Q_1' \cap \cdots \cap Q_n' | n = 1, 2, \ldots\}$ is a filter base in X. Let $\mathcal{F}$ be an ultrafilter containing it. Clearly every Q_n', containing $Q_1' \cap \cdots \cap Q_n' \in \mathcal{F}$, belongs to $\mathcal{F}$, Let $Q_n' = Q_{n1} \cup \cdots \cup Q_{nr_n}$ with $Q_{n1}, \ldots, Q_{nr_n} \in \mathcal{Q}$. Since $\mathcal{F}$ is an ultrafilter, $Q_n' \in \mathcal{F} \Rightarrow Q_{n\rho_n} \in \mathcal{F}$ for some $1 \leqq \rho_n \leqq r_n$. From $Q_{1\rho_1}, Q_{2\rho_2}, \ldots \in \mathcal{F}$ we deduce $Q_{1\rho_1} \cap \cdots \cap Q_{n\rho_n} \neq \emptyset$ $(n = 1, 2, \ldots)$, hence $\bigcap_n Q_{n\rho_n} \neq \emptyset$ since $\mathcal{Q}$ is σ-compact. We conclude now $\bigcap_n Q_n' \supseteq \bigcap_n Q_{n\rho_n} \neq \emptyset$, and the σ-compactness of $\mathcal{Q}'$ follows.

2. $\bigcap_n \varphi(\tilde{C}_n) \supseteq \varphi(\bigcap_n \tilde{C}_n)$ is trivial. Let now $\omega \in \bigcap_n \varphi(\tilde{C}_n)$. Clearly every set $\varphi^{-1}(\{\omega\}) \cap \tilde{C}_n$ has the form $\{\omega\} \times Q_n'$ where $Q_n' \in \mathcal{Q}'$. From $\omega \in \varphi(\tilde{C}_n)$ $(n = 1, 2, \ldots)$ we deduce $\{\omega\} \times Q_n' \neq \emptyset$ $(n = 1, 2, \ldots)$, hence

$$\begin{aligned}\{\omega\} \times (Q_1' \cap \cdots \cap Q_n') &= (\{\omega\} \times Q_1') \cap \cdots \cap (\{\omega\} \times Q_n') \\ &= \varphi^{-1}(\{\omega\}) \cap \tilde{C}_1 \cap \cdots \cap \tilde{C}_n = \varphi^{-1}(\{\omega\}) \cap \tilde{C}_n \\ &= \{\omega\} \times Q_n' \neq \emptyset,\end{aligned}$$

hence $Q_1' \cap \cdots \cap Q_n' \neq \emptyset$ $(n = 1, 2, \ldots)$, hence $Q_1' \cap Q_2' \cap \cdots \neq \emptyset$ after 1. Choose $x \in Q_1' \cap Q_2' \cap \cdots$. Then

$$\begin{aligned}(\omega, x) \in \{\omega\} \times (Q_1' \cap Q_2' \cap \cdots) &= \varphi^{-1}(\{\omega\}) \cap (\tilde{C}_1 \cap \tilde{C}_2 \cap \cdots) \\ &= \tilde{C}_1 \cap \tilde{C}_2 \cap \cdots\end{aligned}$$

and $\omega \in \varphi(\bigcap_n \tilde{C}_n)$ follows.

It seems to be feasible to unfold the abstract ideas of the above proposition in the form of some

3.4. Exercises

3.4.1. Let $X \neq \emptyset$ and $\emptyset \in \mathcal{Q} \subseteq \mathcal{P}(X)$ be σ-compact. Prove: $\mathcal{Q}_\delta$ and the system of all countable intersections of finite unions of sets from $\mathcal{Q}$ are σ-compact.

3.4.2. Let $I \neq \varnothing$ and for every $\iota \in I$, let $X_\iota \neq \varnothing$, $\mathcal{Q}_\iota \subseteq \mathcal{P}(X_\iota)$ σ-compact. Prove:

3.4.2.1. $\{\prod_{\iota \in I} Q_\iota | Q_\iota \in \mathcal{Q}_\iota$ for finitely many ι and $Q_\iota = X_\iota$ else$\}$ is σ-compact.

3.4.2.2. If $X_\iota \cap X_\kappa = \varnothing$ $(\iota \neq \kappa)$, then $\{\sum_{\iota \in I} Q_\iota | Q_\iota \in \mathcal{Q}_\iota$ for finitely many ι and $Q_\iota = \varnothing$ else$\}$ is σ-compact.

3.4.3. Let Ω, $Y \neq \varnothing$, $\varnothing \in \mathcal{Q} \subseteq \mathcal{P}(Y)$ and $\varphi\colon Y \to \Omega$ a mapping such that for every $\omega \in \Omega$ the set system $\{\varphi^{-1}(\{\omega\}) \cap Q | Q \in \mathcal{Q}\}$ is σ-compact. Prove that for every sequence $Q_1, Q_2, \ldots \in \mathcal{Q}$,

$$Q_1 \supseteq Q_2 \supseteq \cdots \quad \Rightarrow \quad \varphi\left(\bigcap_n Q_n\right) = \bigcap_n \varphi(Q_n).$$

The basic result of this section is

3.5. Theorem (Choquet's capacitability theorem). *Let $\Omega \neq \varnothing$ and $\varnothing \in \mathcal{C} \subseteq \mathcal{P}(\Omega)$, and $\mathcal{C}$ be stable under finite unions and finite intersections, $m \in \overline{\mathbb{R}}^{\mathcal{P}(\Omega)}$ a $\mathcal{C}$-capacity. Then every $\mathcal{C}$-Souslin set is $\mathcal{C}$-m-capacitable.*

Proof. Let $A \in \mathcal{S}(\mathcal{C})$. By Theorem 1.9 there is a set $X \neq \varnothing$ and a σ-compact set system $\varnothing \in \mathcal{Q} \subseteq \mathcal{P}(X)$ and a set $\tilde{F} \in [\mathcal{C}, \mathcal{Q}]_{\sigma\delta} = \{C \times Q | C \in \mathcal{C}, Q \in \mathcal{Q}\}_{\sigma\delta}$ such that $A = \varphi(\tilde{F})$, where φ is the projection $(\omega, x) \to \omega$ of $\Omega \times X$ onto Ω. Let $\tilde{\mathcal{C}}$ denote the system of all finite unions of sets from $[\mathcal{C}, \mathcal{Q}]$. Clearly $\tilde{\mathcal{C}}$ is stable under finite unions. Since $[\mathcal{C}, \mathcal{P}]$ is stable under finite intersections, so is $\tilde{\mathcal{C}}$ (exercise). It is thus meaningful to speak of $\tilde{\mathcal{C}}$-capacities. We define $\tilde{m} \in \overline{\mathbb{R}}^{\mathcal{P}(\Omega \times X)}$ by $\tilde{m}(\tilde{F}) = m(\varphi(\tilde{F}))$ $(\tilde{F} \subseteq \Omega \times X)$. Let us show that $\tilde{m}$ is a $\tilde{C}$-capacity. Clearly $\tilde{m}$ is isotone. Next let us show that $\tilde{m}$ is lower σ-continuous:

$$\begin{aligned}
&\tilde{F}_1 \subseteq \tilde{F}_2 \subseteq \cdots \subseteq \tilde{F} \subseteq \Omega \times X,\ \tilde{F}_1 \cup \tilde{F}_2 \cup \cdots = \tilde{F} \\
&\quad\Rightarrow \quad \varphi(\tilde{F}_1) \cup \varphi(\tilde{F}_2) \cup \cdots = \varphi(\tilde{F}) \\
&\quad\Rightarrow \quad \lim_n \tilde{m}(\tilde{F}_n) = \lim_n m(\varphi(F_n)) \\
&\qquad\qquad = m(\varphi(\tilde{F})) = \tilde{m}(\tilde{F}).
\end{aligned}$$

Let us now show that $\tilde{m}$ is upper σ-continuous under approximation from $\tilde{\mathcal{C}}$. For this, let $\tilde{C}_1, \tilde{C}_2, \ldots \in \tilde{\mathcal{C}}$, $\tilde{C}_1 \supseteq \tilde{C}_2 \supseteq \cdots$, $\tilde{C}_1 \cap \tilde{C}_2 \cap \cdots = \tilde{F}$. By Proposition 3.3.2 we have $\bigcap_n \varphi(\tilde{C}_n) = \varphi(\bigcap_n \tilde{C}_n) = \varphi(\tilde{F})$. Moreover, every $\tilde{C}_n$ is of the form $(D_1 \times Q_1) \cup \cdots \cup (D_r \times Q_r)$, $D_1, \ldots, D_r \in \mathcal{C}$, $Q_1, \ldots, Q_r \in \mathcal{Q}$, hence $\varphi(\tilde{C}_n)$ is the union of some of these D_ρ, namely those for which

$Q_\rho \neq \varnothing$, hence $\varphi(\tilde{C}_n) \in \mathscr{C}$ since $\mathscr{C}$ is stable under finite unions. Now we obtain

$$\lim_n \tilde{m}(\tilde{C}_n) = \lim_n m(\varphi(\tilde{C}_n))$$
$$= m\left(\bigcap_n \varphi(\tilde{C}_n)\right) = m\left(\varphi\left(\bigcap_n \tilde{C}_n\right)\right) = m(\varphi(\tilde{F})) = \tilde{m}(\tilde{F}).$$

Thus $\tilde{m}$ is a $\tilde{\mathscr{C}}$-capacity. By Proposition 3.2, $\tilde{F}$ is $\tilde{\mathscr{C}}$-$\tilde{m}$-capacitable. Excluding the trivial case $\tilde{m}(\tilde{F}) = m(A) = -\infty$, we may, for any $\mathbb{R} \ni \alpha < m(A) = \tilde{m}(\tilde{F})$, find some $\tilde{C} \in \tilde{\mathscr{C}}$ such that $\tilde{m}(\tilde{C}) > \alpha$. Put $C = \varphi(\tilde{C})$. By our above argument we get $C \in \mathscr{C}$, and moreover $m(C) = m(\varphi(\tilde{C})) = \tilde{m}(\tilde{C}) > \alpha$. This does it.

The rest of this section is devoted to the construction of $\mathscr{C}$-capacities from set functions defined only on $\mathscr{C}$. Certain additivitylike properties will play an essential role. Only nonnegative set functions will be considered.

3.6. Definition. *Let $\Omega \neq \varnothing$ and $\varnothing \in \mathscr{C} \subseteq \mathscr{P}(\Omega)$ be stable under finite unions and finite intersections. A function $m \in \overline{\mathbb{R}}_+^{\mathscr{C}}$ is called:*

3.6.1. **strongly subadditive** *if it is isotone and satisfies*

$$C, D \in \mathscr{C} \quad \Rightarrow \quad m(C \cup D) + m(C \cap D) \leqq m(C) + m(D);$$

3.6.2. **strongly additive** *if it is isotone and satisfies*

$$C, D \in \mathscr{C} \quad \Rightarrow \quad m(C \cup D) + m(C \cap D) = m(C) + m(D).$$

3.7. Remark. If $\Omega \neq \varnothing$ and $\varnothing \neq \mathscr{R} \subseteq \mathscr{P}(\Omega)$ is a set ring, then $m \in \overline{\mathbb{R}}_+^{\mathscr{R}}$, m additive $\Rightarrow$ m strongly additive since then

$$C, D \in \mathscr{R} \quad \Rightarrow \quad m(C \cup D) + m(C \cap D)$$
$$= m(C \backslash D) + m(C \cap D) + m(D \backslash C) + m(C \cap D)$$
$$= m(C) + m(D).$$

3.8. Proposition. *Let $\Omega \neq \varnothing$ and $\varnothing \in \mathscr{C} \subseteq \mathscr{P}(\Omega)$ be stable under finite unions and finite intersections. Then for every isotone $m \in \overline{\mathbb{R}}_+^{\mathscr{C}}$ the following statements are equivalent:*

3.8.1. *m is strongly subadditive.*

3.8.2. *$C, D, E \in \mathscr{C} \Rightarrow m(C \cup D \cup E) + m(E) \leqq m(C \cup E) + m(D \cup E)$.*

3.8.3. *$C_1, C_2, D_1, D_2 \in \mathscr{C}$, $C_1 \subseteq D_1$, $C_2 \subseteq D_2$ imply*

$$m(D_1 \cup D_2) + m(C_1) + m(C_2) \leqq m(C_1 \cup C_2) + m(D_1) + m(D_2).$$

Proof. $1 \Rightarrow 2$. Observe that $E \subseteq (C \cup E) \cap (D \cup E)$ and apply the definition of strong subadditivity with $C \cup E$, $D \cup E$ in place of C, D, plus the isotony of m.

$2 \Rightarrow 3$. Apply 2 with $C = D_1$, $D = D_2$, $E = C_1$ and get

$$m(D_1 \cup D_2 \cup C_1) + m(C_1) \leqq m(D_1 \cup C_1) + m(D_2 \cup C_1).$$

But the hypotheses of 3 imply also

$$D_1 \cup D_2 \cup C_1 = D_1 \cup D_2, D_1 \cup C_1 = D_1, D_2 \cup C_1 = D_2 \cup C_1 \cup C_2,$$

and thus we get

$$m(D_1 \cup D_2) + m(C_1) \leqq m(D_1) + m(D_2 \cup C_1 \cup C_2).$$

Adding $m(C_2)$ we get

$$m(D_1 \cup D_2) + m(C_1) + m(C_2) \leqq m(D_1) + m(C_2) + m(D_2 \cup C_1 \cup C_2)$$

which is, by a second application of 2,

$$\leqq m(D_1) + m(D_2 \cup C_2) + m(C_1 \cup C_2) = m(C_1 \cup C_2) + m(D_1) + m(D_2).$$

$3 \Rightarrow 1$. Apply 3 with $C_1 = C \cap D$, $D_1 = D$, $C_2 = D_2 = C$ and get $m(C \cup D) + m(C \cap D) + m(C) \leqq m(C) + m(D) + m(C)$. If $m(C) = \infty$, then $m(C \cup D) + m(C \cap D) \leqq m(C) + m(D)$ is trivial. If $m(C) < \infty$, we may subtract $m(C)$ from both members to get the desired inequality.

3.9. Remark. Let $\Omega \neq \varnothing$ and $\varnothing \in \mathscr{C} \subseteq \mathscr{P}(\Omega)$ be stable under finite unions and intersections. Let $m \in \overline{\mathbb{R}}_+^{\mathscr{C}}$ be strongly subadditive. Proposition 3.8 and induction yield

$$\begin{aligned} &C_1, \ldots, C_n, D_1, \ldots, D_n \in \mathscr{C}, \ C_1 \subseteq D_1, \ldots, C_n \subseteq D_n \\ &\quad \Rightarrow m(D_1 \cup \cdots \cup D_n) + m(C_1) + \cdots + m(C_n) \\ &\qquad \leqq m(C_1 \cup \cdots \cup C_n) + m(D_1) + \cdots + m(D_n). \end{aligned}$$

If $m(C_1)$, $\ldots$, $m(C_n)$, $m(C_1 \cup \cdots \cup C_n) < \infty$, then we may rewrite this conclusion as

$$\begin{aligned} &m(D_1 \cup \cdots \cup D_n) - m(C_1 \cup \cdots \cup C_n) \\ &\quad \leqq [m(D_1) - m(C_1)] + \cdots + [m(D_n) - m(C_n)]. \end{aligned}$$

The ideas proferred in the next theorem and its proof resemble Daniell's extension theory (Chapter III, Sections 1–4) to a certain extent and are not alien to Chapter II, Sections 1–2 either.

3.10. Theorem. *Let $\Omega \neq \varnothing$ and $\varnothing \in \mathscr{C} \subseteq \mathscr{P}(\Omega)$ be stable under finite* *unions and finite intersections. Let* $m \in \overline{\mathbb{R}}_+^{\mathscr{C}}$ *be strongly subadditive and lower* σ*-continuous.*

3.10.1. *Define* $\underline{m}: \mathscr{C}_\sigma \to \overline{\mathbb{R}}_+$ *by* $E \in \mathscr{C}_\sigma \Rightarrow \underline{m}(E) = \sup\{m(C) \mid \mathscr{C} \ni C \subseteq E\}$. *(Recall that* $\mathscr{C}_\sigma$ *is the system of all countable unions of sets from* $\mathscr{C}$.*)*

3.10.2. *Define* $\tilde{m} \in \overline{\mathbb{R}}_+^{\mathscr{P}(\Omega)}$ *by* $M \subseteq \Omega \Rightarrow \tilde{m}(M) = \inf\{\underline{m}(E) \mid M \subseteq E \in \mathscr{C}_\sigma\}$ *(with the usual convention* $\inf \varnothing = \infty$*). Then:*

3.10.3. *The restriction of* $\underline{m}$ *to* $\mathscr{C}$ *is* m.

3.10.4. *The restriction of* $\tilde{m}$ *to* $\mathscr{C}_\sigma$ *is* $\underline{m}$*, hence in particular* $\tilde{m}$ *is an extension of* m.

3.10.5. $\tilde{m}$ *is isotone and lower* σ*-continuous.*

3.10.6. $M_1, N_1, M_2, N_2, \ldots \subseteq \Omega$, $M_1 \subseteq N_1$, $M_2 \subseteq N_2, \ldots$ *imply*

$$\tilde{m}(N_1 \cup N_2 \cup \cdots) + \tilde{m}(M_1) + \tilde{m}(M_2) + \cdots \\ \leqq \tilde{m}(M_1 \cup M_2 \cup \cdots) + \tilde{m}(N_1) + \tilde{m}(N_2) + \cdots.$$

3.10.7. $\tilde{m}$ *is a* $\mathscr{C}$*-capacity iff it is upper* σ*-continuous under approximation from* $\mathscr{C}$*, i.e., iff* $C_1, C_2, \ldots \in \mathscr{C}$, $C_1 \supseteq C_2 \supseteq \cdots$,

$$C_1 \cap C_2 \cap \cdots = F \quad \Rightarrow \quad \lim_n m(C_n) = \tilde{m}(F).$$

In this case, $\tilde{m}$ *is called the* **outer (**$\mathscr{C}$**-)capacity derived from** m.

Proof. 3 is obvious by isotony of m. Clearly $\underline{m}$ is isotone. This in turn implies 4 plus the isotony of $\tilde{m}$. Next we show that $\underline{m}$ is lower σ-continuous. We first show $E \in \mathscr{C}_\sigma$, $C_1, C_2, \ldots \in \mathscr{C}$, $C_1 \subseteq C_2 \subseteq \cdots$,

$$C_1 \cup C_2 \cup \cdots = E \quad \Rightarrow \quad \lim_n m(C_n) = \underline{m}(E).$$

In fact, $\leqq$ is obvious here. Let now $\mathscr{C} \ni C \subseteq E$ be arbitrary. Then the σ-continuity and the isotony of m imply

$$\lim_n m(C_n) \geqq \lim_n m(C \cap C_n) = m(C).$$

Varying C, we get the desired $\geqq$ above.

Let now $E_1, E_2, \ldots \in \mathscr{C}_\sigma$, $E_1 \subseteq E_2 \subseteq \cdots$. Put $E = E_1 \cup E_2 \cup \cdots$. Clearly $E \in \mathscr{C}_\sigma$. Find $C_{jk} \in \mathscr{C}$ $(j, k = 1, 2, \ldots)$ with $\bigcup_k C_{jk} = E_j$ $(j = 1, 2, \ldots)$. Since $\mathscr{C}$ is stable under finite unions, we may replace C_{jk} by $\bigcup_{n \leqq j;\, i \leqq k} C_{ni}$ and thus assume $C_{ni} \subseteq C_{jk}$ $(n \leqq j;\ i \leqq k)$. Clearly $E = C_{11} \cup C_{22} \cup \cdots$. Now we get

$$\lim_n \underline{m}(E_n) \geqq \lim_n m(C_{nn}) = \underline{m}(E).$$

But $\leqq$ is obvious here and thus $\lim_n m(E_n) = m(E)$ follows.

By a nearly obvious passage to suprema we see that $\underline{m}$ is strongly subadditive. Let us notice a sketchy proof: $\mathscr{C} \ni C_n \nearrow E$, $\mathscr{C} \ni D_n \nearrow F$ imply

$$\begin{aligned} \underline{m}(E \cup F) + \underline{m}(E \cap F) &= \lim_n m(C_n \cup D_n) + \lim_n m(C_n \cap D_n) \\ &= \lim_n \, [m(C_n \cup D_n) + m(C_n \cap D_n)] \\ &\leqq \lim_n \, [m(C_n) + m(D_n)] \\ &= \lim_n m(C_n) + \lim_n m(D_n) = m(E) + m(F). \end{aligned}$$

Now we show that $\tilde{m}$ is strongly subadditive. Let $M, N \subseteq \Omega$. If $\tilde{m}(M) + \tilde{m}(N) = \infty$, we are through. Assume now $m(M) + m(N) < \infty$ and choose any $\mathbb{R} \ni \alpha > \tilde{m}(M)$, $\mathbb{R} \ni \beta > \tilde{m}(N)$. Find $E, F \in \mathscr{C}_\sigma$ with $E \supseteq M$, $F \supseteq N$, $\underline{m}(E) < \alpha$, $\underline{m}(F) < \beta$. Clearly $E \cup F \supseteq M \cup N$, $E \cap F \supseteq M \cap N$. Thus we get

$$\tilde{m}(M \cup N) + \tilde{m}(M \cap N) \leqq \underline{m}(E \cup F) + \underline{m}(E \cap F) \leqq \underline{m}(E) + \underline{m}(F) < \alpha + \beta.$$

Varying α, β, we get $m(M \cup N) + m(M \cap N) \leqq m(M) + m(N)$ as desired.

Now we prove that m is lower σ-continuous. Let $M_1 \subseteq M_2 \subseteq \cdots \subseteq \Omega$, $M = M_1 \cup M_2 \cup \cdots$. If $\lim_n \tilde{m}(M_n) = \infty$, then $\lim_n \tilde{m}(M_n) = \tilde{m}(M)$ follows. Assume now $\lim_n \tilde{m}(M_n) < \infty$ and choose any $\varepsilon > 0$. It will be sufficient to find $E_1, E_2, \ldots \in \mathscr{C}_\sigma$ such that $E_1 \subseteq E_2 \subseteq \cdots$, $M_n \subseteq E_n$, $\underline{m}(E_n) < \tilde{m}(M_n) + \varepsilon$ $(n = 1, 2, \ldots)$, since this entails $M \subseteq E_1 \cup E_2 \cup \cdots$,

$$\underline{m}(E_1 \cup E_2 \cup \cdots) = \lim_n \underline{m}(E_n) \leqq \lim \tilde{m}(M_n) + \varepsilon;$$

we would then get $\lim_n \tilde{m}(M_n) \geqq \tilde{m}(M)$; and since $\leqq$ is obvious, we would be through.

Now we construct $E_1, E_2, \ldots$ as follows. Find $\mathscr{C}_\sigma \ni F_n \supseteq M_n$ with

$$\underline{m}(F_n) < \tilde{m}(M_n) + \frac{\varepsilon}{2^n} \qquad (n = 1, 2, \ldots).$$

Put $E_n = F_1 \cup \cdots \cup F_n$. Clearly $E_1, E_2, \ldots \in \mathscr{C}_\sigma$, $E_1 \subseteq E_2 \subseteq \cdots$. Now we show

$$\underline{m}(E_n) < \tilde{m}(M_n) + \varepsilon\left(1 - \frac{1}{2^n}\right) \qquad (n = 1, 2, \ldots)$$

by induction. For $n = 1$, it is obvious. If we have it for n, we get, using the strong subadditivity of $\underline{m}$, $E_{n+1} = E_n \cup F_{n+1}$, $M_n \subseteq E_n \cap F_{n+1}$ and thus

$$\tilde{m}(M_n) \leqq \underline{m}(E_n \cap F_{n+1}) \leqq \tilde{m}(M_n) + \varepsilon\left(1 - \frac{1}{2^n}\right),$$

hence

$$\underline{m}(E_n) - \underline{m}(E_n \cap F_{n+1}) < \varepsilon\left(1 - \frac{1}{2^n}\right),$$

$$\begin{aligned}\tilde{m}(M_{n+1}) &\leqq \underline{m}(E_n \cup F_{n+1}) \\ &\leqq \underline{m}(E_n) + \underline{m}(F_{n+1}) - \underline{m}(E_n \cap F_{n+1}) \\ &< \tilde{m}(M_{n+1}) + \frac{\varepsilon}{2^{n+1}} + \varepsilon\left(1 - \frac{1}{2^n}\right) \\ &= \tilde{m}(M_{n+1}) + \varepsilon\left(1 - \frac{1}{2^{n+1}}\right).\end{aligned}$$

This does it.

6 follows from Remark 3.9 for $n \to \infty$. The rest of the theorem is now trivial.

3.11. Theorem. *Let $\Omega \neq \varnothing$ and $\varnothing \in \mathscr{C} \subseteq \mathscr{P}(\Omega)$ be stable under finite unions and finite intersections. Let $m \in \mathbb{R}_+^{\mathscr{C}}$ be strongly additive and lower σ-continuous. Define $\underline{m} \in \overline{\mathbb{R}}_+^{\mathscr{C}_\sigma}$ and $\tilde{m} \in \overline{\mathbb{R}}_+^{\mathscr{P}(\Omega)}$ as in Theorem 3.10 and assume that $\tilde{m}$ is upper σ-continuous under approximation from $\mathscr{C}$. Then the outer capacity $\tilde{m}$ derived from m is strongly additive on the system $\mathscr{S}(\mathscr{C})$ and the strong additivity formula $\tilde{m}(A \cup B) + \tilde{m}(A \cap B) = \tilde{m}(A) + \tilde{m}(B)$ holds even for all $\mathscr{C}$-$\tilde{m}$-capacitable sets A, B.*

Proof. We first prove that m is strongly additive when restricted to $\mathscr{C}_\delta$. This is a meaningful statement since $\mathscr{C}_\delta$, the system of all countable intersections of sets from $\mathscr{C}$, is actually stable under finite unions and intersections (exercise). Now let $E, F \in \mathscr{C}_\delta$. We may then find $C_1, D_1, C_2, D_2, \ldots \in \mathscr{C}$ such that $C_1 \supseteq C_2 \supseteq \cdots$, $C_1 \cap C_2 \cap \cdots = E$, $D_1 \supseteq D_2 \supseteq \cdots$, $D_1 \cap D_2 \cap \cdots = F$. Now we get, by upper σ-continuity under approximation from $\mathscr{C}$,

$$\begin{aligned}\tilde{m}(E \cup F) + \tilde{m}(E \cap F) &= \lim_n m(C_n \cup D_n) + \lim_n m(C_n \cap D_n) \\ &= \lim_n m(C_n) + \lim_n m(D_n) = \tilde{m}(E) + \tilde{m}(F).\end{aligned}$$

That every $A \in \mathscr{S}(\mathscr{C})$ is m-capacitable follows from Choquet's theorem 3.5. Let now $A, B \subseteq \Omega$ be $\mathscr{C}$-$\tilde{m}$-capacitable. For any $\mathbb{R} \ni \alpha < \tilde{m}(A)$, $\mathbb{R} \ni \beta < \tilde{m}(B)$, choose $E, F \in \mathscr{C}_\delta$ such that $E \subseteq A$, $F \subseteq B$ and $\tilde{m}(E) > \alpha$, $\tilde{m}(F) > \beta$. From the strong additivity of m on $\tilde{\mathscr{C}}_\delta$ we conclude now

$$\tilde{m}(A \cup B) + \tilde{m}(A \cap B) \geqq \tilde{m}(E \cup F) + \tilde{m}(E \cap F)$$
$$= \tilde{m}(E) + \tilde{m}(F) > \alpha + \beta.$$

Varying α, β, we obtain $\tilde{m}(A \cup B) + \tilde{m}(A \cap B) \geqq \tilde{m}(A) + \tilde{m}(B)$. Since $\leqq$ is obvious here, the strong additivity formula

$$\tilde{m}(A \cup B) + \tilde{m}(A \cap B) = \tilde{m}(A) + \tilde{m}(B)$$

follows.

3.12. Exercises

3.12.1. Let $\Omega = \mathbb{R}$, $\mathscr{C} = \mathscr{D} =$ the system of all finite unions of half-open dyadic intervals (Example I.1.4), m the geometric content on $\mathscr{D}$ (Example I.1.10). Prove:

3.12.1.1. $\mathscr{C}$ and m satisfy all hypotheses of Theorem 3.11.

3.12.1.2. Use Example IV.4.20 in order to obtain a set $M \subseteq [0, 1]$ such that M and $[0, 1] \backslash M$ have outer Lebesgue σ-content 1.

3.12.1.3. Prove that $A =]-\infty, 0[\cup M$ and $B = M \cup]1, \infty[$ are $\mathscr{C}$-$\tilde{m}$-capacitable with $\tilde{m}(A) = \tilde{m}(B) = \infty$.

3.12.1.4. Prove that $A \cap B = M$ is not $\mathscr{C}$-$\tilde{m}$-capacitable.

3.12.2. Let Ω, $\mathscr{C}$, m, $\tilde{m}$ be as in Theorem 3.11. Prove that for any $\mathscr{C}$-$\tilde{m}$-capacitable sets $A, B \subseteq \Omega$ with $\tilde{m}(A), \tilde{m}(B) < \infty$ also $A \cup B$ and $A \cap B$ are $\mathscr{C}$-$\tilde{m}$-capacitable.

4. THE CAPACITY APPROACH TO σ-CONTENT EXTENSION

In this section we apply, following MEYER [2], Choquet's capacitability theorem to capacities derived from σ-contents. We obtain a proof of Corollary II.2.4 on the extension of σ-contents once more, bringing the total number of proofs of that corollary in this book to three (number two was Remark III.6.2.2). It should be observed that there are ideas which are common to all three proofs. Our present approach brings however one new result: The Souslin sets considered here are all in the completion of the extension of the given σ-content (SAKS [3, 4]). This might be stated briefly as: Souslin sets are Borel up to nullsets.

4.1. Theorem. *Let $(\Omega, \mathscr{R}, m)$ be a σ-content prespace (Definition* I.1.7.2). *Then there is a unique σ-content (again denoted by m) on the local σ-ring*

$\mathscr{B}^{00} = \mathscr{B}^{00}(\mathscr{R})$ whose restriction to $\mathscr{R}$ is the given m. For every $\mathscr{R}$-Souslin set A contained in some set Ω_0 from $\mathscr{B}^{00}$ there are a set $E \in \mathscr{R}_{\delta\sigma}$ and a set $F \in \mathscr{R}_{\sigma\delta}$ such that $E \subseteq A \subseteq F \subseteq \Omega_0$ (which implies E, $F \in \mathscr{B}^{00}$) and $m(E) = m(F)$. In particular, A belongs to the completion of $\mathscr{B}^{00}$ for m (defined in Theorem II.6.2.2).

Proof. Every member of $\mathscr{B}^{00}$ is, by Proposition I.4.7.1, contained in some $\Omega_0 \in \mathscr{R}$. Fix an $\Omega_0 \in \mathscr{R}$ and put $\mathscr{R}_0 = \mathscr{R} \cap \Omega_0 = \{F \mid F \in \mathscr{R}, F \subseteq \Omega_0\}$. It is an easy exercise to show that every $A \in \mathscr{S}(\mathscr{R})$ with $A \subseteq \Omega_0$ belongs to $\mathscr{S}(\mathscr{R}_0)$. We may thus forget $\Omega \backslash \Omega_0$, i.e., assume $\Omega = \Omega_0 \in \mathscr{R}$ for a while. Doing this, we observe that $m \in \mathbb{R}_+^{\mathscr{R}}$ is strongly additive and lower σ-continuous. Now we apply Theorem 3.10 and construct $\tilde{m} \in \mathbb{R}_+^{\mathscr{P}(\Omega)}$ (clearly no $\tilde{m}$-values $= \infty$ occur). Theorem 3.10.5 implies that $\tilde{m}$ is lower σ-continuous. We now prove, using the definition of $\tilde{m}$ and the σ-additivity of $m \in \mathbb{R}_+^{\mathscr{R}}$, that $\tilde{m}$ is upper σ-continuous under approximation from $\mathscr{R}$. For this it is clearly sufficient to prove the following: If C_1, D_1, C_2, $D_2, \ldots \in \mathscr{R}$, $C_1 \supseteq C_2 \supseteq \cdots$, $D_1 \subseteq D_2 \subseteq \cdots$ and $\bigcap_n C_n \subseteq \bigcup_n D_n$, then

$$\lim_n m(C_n) \leqq \lim_n m(D_n).$$

But clearly $C_1 \backslash D_1 \supseteq C_2 \backslash D_2 \supseteq \cdots$, $\bigcup_n (C_n \backslash D_n) = \varnothing$ and thus

$$\begin{aligned}\lim_n m(C_n) - \lim_n m(D_n) &= \lim_n [m(C_n) - m(D_n)] \\ &= \lim_n [m(C_n \backslash D_n) + m(C_n \cap D_n) - m(D_n)] \\ &\leqq \lim_n m(C_n \backslash D_n) = 0;\end{aligned}$$

here we have used the additivity, isotony, and σ-additivity of $m \in \mathbb{R}_+^{\mathscr{R}}$. Now we can apply Choquet's capacitability theorem 3.5 and see that every $A \in \mathscr{S}(\mathscr{R})$ is $\mathscr{R}$-$\tilde{m}$-capacitable. Proposition 1.5 applies to the effect that $\mathscr{B}^{00} \subseteq \mathscr{S}(\mathscr{R})$. Theorem 3.11 shows that the restriction of $\tilde{m}$ to $\mathscr{B}^{00}$ is strongly additive. Since clearly $\tilde{m}(\varnothing) = 0$, we see that $\tilde{m}$ is additive on $\mathscr{B}^{00}$; and since $\tilde{m}$ is lower σ-continuous, it is σ-additive on $\mathscr{B}^{00}$ and hence provides us with the, by Proposition I.5.4 unique, σ-additive extension of m to $\mathscr{B}^{00}$. We shall write m for this extension again. The very definition of m shows that for any $A \in \mathscr{S}(\mathscr{R})$, there is a sequence F_1, $F_2, \ldots \in \mathscr{R}_\sigma \subseteq \mathscr{B}^{00}$ such that $F_1, F_2, \ldots \supseteq A$, $\inf m(F_n) = \tilde{m}(A)$. Since $\mathscr{R}_\sigma$ is stable under finite intersections (exercise), we may choose $F_1 \supseteq F_2 \supseteq \cdots$. If we now put $F = F_1 \cap F_2 \cap \cdots$, then $F \in \mathscr{R}_{\sigma\delta} \subseteq \mathscr{B}^{00}$ and $m(F) = \tilde{m}(A)$. The very definition of $\mathscr{R}$-$\tilde{m}$-capacitability shows that there are E_1, $E_2, \ldots \in \mathscr{R}_\delta \subseteq \mathscr{B}^{00}$ such that $E_1, E_2, \ldots \subseteq A$, $\tilde{m}(A) = \sup_n m(E_n)$. Since $\mathscr{R}_\delta$ is stable under finite unions (exercise), we may assume $E_1 \subseteq E_2 \subseteq \cdots$.

If we now put $E = E_1 \cup E_2 \cup \cdots$, then $E \in \mathscr{R}_{\sigma\delta} \subseteq \mathscr{B}^{00}$ and $m(E) = \lim_n m(E_n) = \tilde{m}(A)$. This proves the theorem in the case $\Omega \in \mathscr{R}$. The extension to the general case is nearly obvious: It suffices to show that it does not matter which $\Omega_0 \in \mathscr{R}$ we choose in order to let everything happen within Ω_0. The details are left as an exercise to the reader.

4.2. Exercises

4.2.1. Let $(\Omega, \mathscr{E}, m)$ be a σ-measure space (Definition I.8.1.7). Put $\overline{\Omega} = \Omega \times \overline{\mathbb{R}}_+$ and $\overline{\mathscr{C}} = \{\{(\omega, \alpha) | \omega \in \Omega, \ 0 \leqq \alpha \leqq f(\omega)\} | f \in \mathscr{E}_+\}$. Define $\overline{m}: \overline{\mathscr{C}} \to \mathbb{R}_+$ by $\overline{m}(\{(\omega, \alpha) | \omega \in \Omega, 0 \leqq \alpha \leqq f(\omega)\}) = m(f)$ $(f \in \mathscr{E}_+)$. Prove:

4.2.1.1. $\overline{m}$ is well defined on $\overline{\mathscr{C}}$.

4.2.1.2. $\overline{\mathscr{C}}$ is stable under finite unions and intersections.

4.2.1.3. $\overline{m}$ is strongly additive and lower σ-continuous on $\overline{\mathscr{C}}$.

4.2.1.4. If $\tilde{m}$ is defined on $\mathscr{P}(\overline{\Omega})$ according to Theorem 3.10, then $\tilde{m}$ is a $\overline{\mathscr{C}}$-capacity.

4.2.1.5. $\overline{A} \subseteq \overline{\Omega}$ is $\tilde{\mathscr{C}}$-$\tilde{m}$-capacitable iff there is an m-σ-integrable $h \in \overline{\mathbb{R}}_+^{\Omega}$ such that $\overline{A} = \{(\omega, \alpha) | \omega \in \Omega, 0 \leqq \alpha \leqq h(\omega)\}$. In this case $\tilde{m}(\overline{A}) = \int h \, dm$.

4.2.2. Let $(\Omega, \mathscr{T})$ be a topological space and $\mathscr{C}$ the system of all compact sets $\subseteq \Omega$. Observe that $\mathscr{C}$ is stable under finite unions and arbitrary (e.g., countable) intersections. Let $m: \mathscr{C} \to \mathbb{R}_+$ be strongly additive and lower σ-continuous, $m(\varnothing) = 0$. Show that there is a unique σ-content on $\mathscr{B}(\mathscr{C})$ that is inner $\mathscr{C}$-regular (and hence extends m).

5. THE MEASURABLE CHOICE THEOREM

Let $\Omega \neq \varnothing \neq X$ and $M \subseteq \Omega \times X$ be such that for the projection $\varphi: (\omega, x) \to \omega$ of $\Omega \times X$ onto Ω, the relation $\varphi(M) = \Omega$ holds. The classical axiom of choice implies that then there is a mapping $f: \Omega \to X$ such that $(\omega, f(\omega)) \in M$ $(\omega \in \Omega)$. Actually, this is nothing but an equivalent formulation of that axiom.

The *measurable choice problem* can be stated as follows. Let $\Omega \neq \varnothing \neq X$ and $\mathscr{B} \subseteq \mathscr{P}(\Omega)$, $\mathscr{S} \subseteq \mathscr{P}(X)$ be σ-fields. Assume that M is a set from the product σ-field $\mathscr{B} \times \mathscr{S}$ such that $\varphi(M) = \Omega$. Does there exist a $\mathscr{B}$-$\mathscr{S}$-measurable $f: \Omega \to X$ such that $(\omega, f(\omega)) \in M$ $(\omega \in \Omega)$?

The known positive answers to the measurable choice problem involve analytic resp. Souslin sets, partly in Polish spaces. They fall moreover into two classes. One class of answers makes assumptions of a *topological* nature about Ω. The other class avoids this, operates with a σ-content on

Ω instead, and yields $(\omega, f(\omega)) \in M$ *only* a.e. We shall give one example of each class here and ask the reader to consult HOFFMANN-JØRGENSEN [1] and T. PARTHASARATHY [1] for systematic studies.

5.1. Theorem. *Let $(\Omega, \mathscr{T})$ and $(X, \mathscr{S})$ be Polish spaces and $A \subseteq \Omega \times X$ analytic (for $\mathscr{T} \times \mathscr{S}$) such that $\varphi(A) = \Omega$, where φ denotes the projection $(\omega, x) \to \omega$ of $\Omega \times X$ onto Ω. Let $\mathscr{A}$ denote the system of all analytic subsets of Ω. Then there is a $\mathscr{B}(\mathscr{A})$-$\mathscr{B}(\mathscr{T})$-measurable mapping $f\colon \Omega \to X$ such that*

$$\omega \in \Omega \quad \Rightarrow \quad (\omega, f(\omega)) \in A.$$

Proof. By Definition 2.4 of an analytic set in $\Omega \times X$ (with $\mathscr{T} \times \mathscr{S}$) there are continuous mappings $g\colon \mathbb{N}^{\mathbb{N}} \to \Omega$, $h\colon \mathbb{N}^{\mathbb{N}} \to X$ such that $(g \times h)(\mathbb{N}^{\mathbb{N}}) = \{(g(n_1, n_2, \ldots), h(n_1, n_2, \ldots)) \mid (n_1, n_2, \ldots) \in \mathbb{N}^{\mathbb{N}}\} = A$. For every $\omega \in \Omega$, the set $g^{-1}(\{\omega\}) = \{(n_1, n_2, \ldots) \mid g(n_1, n_2, \ldots) = \omega\} \subseteq \mathbb{N}^{\mathbb{N}}$ is nonempty (since $\varphi(A) = \Omega$) and closed. By Exercise 2.10.3.3 there is a unique $\psi(\omega) \in g^{-1}(\{\omega\})$ that is lexicographically $\leqq$ every element of $g^{-1}(\{\omega\})$. It is obvious that $(b_1, b_2, \ldots) \in \mathbb{N}^{\mathbb{N}} \Rightarrow$

$$\begin{aligned}\psi^{-1}&(\{(n_1, n_2, \ldots) \mid (n_1, n_2, \ldots) < (b_1, b_2, \ldots)\})\\ &= g(\{(n_1, n_2, \ldots) \mid (n_1, n_2, \ldots) < (b_1, b_2, \ldots)\}) \in \mathscr{A}.\end{aligned}$$

It follows that ψ is $\mathscr{B}(\mathscr{A})$-$\mathscr{B}(\mathscr{R})$-measurable, where $\mathscr{R}$ denotes the Polish (= product) topology in $\mathbb{N}^{\mathbb{N}}$. It follows that $f = h \circ \psi$ is $\mathscr{B}(\mathscr{A})$-$\mathscr{B}(\mathscr{S})$-measurable. Moreover

$$\omega \in \Omega \quad \Rightarrow \quad (\omega, f(\omega)) = ((g \circ \psi)(\omega), (h \circ \psi)(\omega)) \in (g \times h)(\mathbb{N}^{\mathbb{N}}) = A.$$

As a σ-content theoretical sample, we choose a theorem which covers an old result of v. NEUMANN [1, 4] and is of use in economic theory (see HILDENBRAND [2] whose proof we take over to a large extent).

5.2. Theorem. *Let $(\Omega, \mathscr{B}, m)$ be a σ-content space such that $\Omega \in \mathscr{B}$, and let $(X, \mathscr{S})$ be a Polish space. Let $A \in \mathscr{A}([\mathscr{B}, \mathscr{B}(\mathscr{S})])$ and assume $\varphi(A) = \Omega$, where φ is the projection $(\omega, x) \to \omega$ of $\Omega \times X$ onto Ω. Then there is a $\mathscr{B}$-$\mathscr{B}(\mathscr{S})$-measurable $f\colon \Omega \to X$ such that $(\omega, f(\omega)) \in A$ for m-almost all $\omega \in \Omega$.*

Proof. We begin with a reduction to the case where $X = \mathbb{N}^{\mathbb{N}}$. By Proposition 2.3 there is a continuous $h\colon \mathbb{N}^{\mathbb{N}} \to X$ such that $h(\mathbb{N}^{\mathbb{N}}) = X$. Let $\mathscr{R}$ be the topology in $\mathbb{N}^{\mathbb{N}}$. The mapping

$$\psi\colon (\omega, (n_1, n_2, \ldots)) \to (\omega, h(n_1, n_2, \ldots))$$

of $\Omega \times \mathbb{N}^{\mathbb{N}} \to \Omega \times X$ is $[\mathscr{B} \times \mathscr{B}(\mathscr{R})]$-$[\mathscr{B} \times \mathscr{B}(\mathscr{S})]$-measurable. It follows that $\psi^{-1}A \in \mathscr{A}([\mathscr{B}, \mathscr{B}(\mathscr{R})])$ (exercise). If $f'\colon \Omega \to \mathbb{N}^{\mathbb{N}}$ is $\mathscr{B}$-$\mathscr{B}(\mathscr{R})$-measurable, then $h \circ f'\colon \Omega \to X$ is $\mathscr{B}$-$\mathscr{B}(\mathscr{S})$-measurable and $(\omega, f'(\omega) \in \psi^{-1}A$ implies

$(\omega, (h \circ f')(\omega)) = \psi(\omega, f'(\omega)) \in A$. Thus we may assume $(X, \mathscr{S}) = (\mathbb{N}^{\mathbb{N}}, \mathscr{R})$. Let $\mathscr{Q}$ denote the system of all sets that are either empty or cylinders

$$[n_1, \ldots, n_r] = \{(n_1, \ldots, n_r, n_{r+1}, \ldots) | n_{r+1}, \ldots \in \mathbb{N}\}.$$

It is easy to see that $\mathscr{Q}$ is σ-compact and $\mathscr{B}(\mathscr{R}) = \mathscr{B}(\mathscr{Q})$ (Exercise 1.6.4). In particular $A \in A([\mathscr{B}, \mathscr{Q}])$ (exercise). Let now $\mathscr{C}$ be the system of all finite unions of sets from $[\mathscr{B}, \mathscr{Q}]$. Define $m^*: \mathscr{P}(\Omega) \to \mathbb{R}_+$ as the outer content derived from m (Definition II.1.1). Since m is σ-additive, m^* is lower σ-continuous (Proposition II.1.3.3). Define now $\tilde{m}: \mathscr{P}(\Omega \times \mathbb{N}^{\mathbb{N}}) \to \mathbb{R}_+$ by $\tilde{m}(M) = m^*(\varphi\tilde{M})$ $(\tilde{M} \subseteq \Omega \times \mathbb{N}^{\mathbb{N}})$. Clearly $\tilde{m}$ is isotone. $\tilde{m}$ is lower σ-continuous since

$$\begin{aligned}
\tilde{M}_1 \subseteq \tilde{M}_2 \subseteq \cdots \subseteq \Omega \times \mathbb{N}^{\mathbb{N}} \quad &\Rightarrow \quad \varphi\tilde{M}_1 \subseteq \varphi\tilde{M}_2 \subseteq \cdots \subseteq \Omega \\
&\Rightarrow \quad \lim_n \tilde{m}(\tilde{M}_n) = \lim_n m^*(\varphi\tilde{M}_n) = m^*\left(\bigcup_n \varphi\tilde{M}_n\right) \\
&\qquad\qquad = m^*\left(\varphi\left(\bigcup_n \tilde{M}_n\right)\right) \\
&\qquad\qquad = \tilde{m}\left(\bigcup_n \tilde{M}_n\right).
\end{aligned}$$

Finally, $\tilde{m}$ is upper σ-continuous under approximation from $\tilde{\mathscr{C}}$. In fact, we can apply Proposition 3.3.2 and get, for any $\tilde{C}_1, \tilde{C}_2, \ldots \in \tilde{\mathscr{C}}$ with $\tilde{C}_1 \supseteq \tilde{C}_2 \supseteq \cdots$,

$$\begin{aligned}
\tilde{m}\left(\bigcap_n \tilde{C}_n\right) = m^*\left(\varphi\left(\bigcap_n \tilde{C}_n\right)\right) = m^*\left(\bigcap_n \varphi\tilde{C}_n\right) \\
= \lim_n m^*(\varphi\tilde{C}_n) = \lim_n \tilde{m}(\tilde{C}_n).
\end{aligned}$$

Now we see that we can apply Choquet's capacitability theorem 3.5 and find, for any $\varepsilon > 0$, some $\tilde{C} \in \tilde{C}_\delta$ such that $\tilde{m}(\tilde{C}) > \tilde{m}(\Omega \times \mathbb{N}^{\mathbb{N}}) - \varepsilon = m(\Omega) - \varepsilon$. Now $\varphi\tilde{C} \subseteq \mathscr{B}$ and Proposition 1.8 imply $\varphi(\tilde{C}) \in \mathscr{S}(\mathscr{B})$. By Theorem 4.1 we find some $E \in \mathscr{B}$ with $E \subseteq \varphi(\tilde{C})$, $m(E) = m^*(\varphi(C^*)) = \tilde{m}(\tilde{C}) > m(\Omega) - \varepsilon$. For every $\omega \in E$, we have $\varphi^{-1}(\{\omega\}) \cap \tilde{C} \neq \varnothing$, and this is a countable intersection of sets from the system of all finite unions of sets from $\mathscr{Q}$, hence a closed subset of $\mathbb{N}^{\mathbb{N}}$ (exercise). Let $f(\omega)$ be the unique lexicographically minimal element in $\varphi^{-1}(\{\omega\}) \cap \tilde{C}$ (Exercise 2.10.3.3). For every $(b_1, b_2, \ldots) \in \mathbb{N}^{\mathbb{N}}$, we have

$$\begin{aligned}
&f^{-1}\{(n_1, n_2, \ldots) | (n_1, n_2, \ldots) < (b_1, b_2, \ldots)\} \\
&\quad = \{\omega | \varphi^{-1}(\{\omega\}) \cap \tilde{C} \cap \{(n_1, n_2, \ldots) | (n_1, n_2, \ldots) < (b_1, b_2, \ldots)\} \neq \varnothing\} \\
&\quad = \varphi[\tilde{C} \cap (\Omega \times \{(n_1, n_2, \ldots) | (n_1, n_2, \ldots) < (b_1, b_2, \ldots)\})],
\end{aligned}$$

and this is a countable union of sets of the form $\varphi[\tilde{C} \cap (\Omega \times [n_1, \ldots, n_r])]$ with $r \geqq 0$, $n_1, \ldots, n_r \in \mathbb{N}$. But these sets are obviously in $\mathscr{B}$. Working on $\Omega \backslash E$ in the same way as we just worked on Ω, but with some $\varepsilon' < \varepsilon$, and iterating this procedure, we get some Ω_0 that is a countable disjoint union of sets from $\mathscr{B}$ and hence in $\mathscr{B}$ again, and some $f\colon \Omega_0 \to \mathbb{N}^{\mathbb{N}}$ such that

$$f^{-1}\{(n_1, n_2, \ldots) \,|\, (n_1, n_2, \ldots) < (b_1, b_2, \ldots)\} \in \mathscr{B} \qquad ((b_1, b_2, \ldots) \in \mathbb{N}^{\mathbb{N}}).$$

Giving f any constant value $\in \mathbb{N}^{\mathbb{N}}$ in $\Omega \backslash \Omega_0$, we get some f with the desired properties (use Exercise 2.10.3.2).

CHAPTER XIV

ATOMS, CONDITIONAL ATOMS, AND ENTROPY

In this chapter we treat a topic which is comparatively isolated in σ-content theory: HANEN and NEVEU's [1] (see also NEVEU [3]) theory of *atoms* and *conditional atoms* on the one hand, and *entropy theory* on the other hand. Part of the latter is an application of the former. Entropy theory, for its part, has important application in the mathematical branches of *information theory* (see SHANNON and WEAVER [1], KHINTCHINE [1], WOLFOWITZ [1], AUGUSTIN [1]), *statistical mechanics* (see RUELLE [1]), and notably *ergodic theory* (see KATOK *et al.* [1], SINAI [1], JACOBS [2], ORNSTEIN [1], PARRY [1]).

Section 1 presents the theory of atoms and conditional atoms. Section 2 contains the rudiments of entropy theory. Throughout this chapter we deal with σ-probability contents.

1. ATOMS AND CONDITIONAL ATOMS

In this section we present, following HANEN and NEVEU [1] and NEVEU [3], a theory of atoms and conditional atoms of normalized σ-content spaces. Nearly all relations among sets considered here are relations modulo nullsets.

1.1. Definition. *Let* $(\Omega, \mathscr{B}, m)$ *be a* σ*-probability space, i.e., a* σ*-content space with* $\Omega \in \mathscr{B}$, $m(\Omega) = 1$.

1.1.1. *A set* $A \in \mathscr{B}$ *is called an* **atom** *of* $(\Omega, \mathscr{B}, m)$ *if*

$$\mathscr{B} \ni F \subseteq A \quad \Rightarrow \quad m(F)m(A \backslash F) = 0,$$

i.e., $F = A$ *or* $= \varnothing$ mod m.

1.1.2. $(\Omega, \mathscr{B}, m)$ *is called* **atomless** *if every atom of* $(\Omega, \mathscr{B}, m)$ *is an m-nullset.*

1.1.3. *Let* $\mathscr{B}_0 \subseteq \mathscr{B}$ *be a* σ*-field. A set* $A \in \mathscr{B}$ *is called a* **(conditional)** $\mathscr{B}_0$**-atom** (*of* $(\Omega, \mathscr{B}, m)$) *if* $\mathscr{B} \cap A = \mathscr{B}_0 \cap A \bmod m$, *i.e., if for every* $F \in \mathscr{B}$ *there is an* $F_0 \in \mathscr{B}_0$ *such that* $F \cap A = F_0 \cap A \bmod m$.

1.2. Remarks. Let $(\Omega, \mathscr{B}, m)$ be a σ-probability space and $\mathscr{B}_0 \subseteq \mathscr{B}$ a σ-field.

1.2.1. For $\mathscr{B}_0 = \{\varnothing, \Omega\}$, the concepts of an atom and a $\mathscr{B}_0$-atom coincide.

1.2.2. $A \in \mathscr{B}$ is a $\mathscr{B}_0$-atom iff $\mathscr{B} \ni F \subseteq A \Rightarrow F = A \cap F_0 \bmod m$ for a suitable $F_0 \in \mathscr{B}_0$.

1.2.3. It should be emphasized that m is essential in determining whether a set $A \in \mathscr{B}$ is a $\mathscr{B}_0$-atom, and not $\mathscr{B}_0$ alone since we use the m-nullsets. We may briefly state that $A \in \mathscr{B}$ is a $\mathscr{B}_0$-atom iff "$\mathscr{B}$ looks like $\mathscr{B}_0$ inside A, mod m."

1.2.4. It is obvious from the definition that a subset (from $\mathscr{B}$) of a $\mathscr{B}_0$-atom is a $\mathscr{B}_0$-atom again.

1.3. Exercises

1.3.1. Let $(\Omega, \mathscr{B}, m)$ be a σ-probability space. Prove:

1.3.1.1. If $A, A' \in \mathscr{B}$ are atoms of $(\Omega, \mathscr{B}, m)$, then either $A = A' \bmod m$ or $A \cap A' = \varnothing \bmod m$.

1.3.1.2. There are mod m at most countably many atoms of $(\Omega, \mathscr{B}, m)$. There is a mod m unique $\Omega_0 \in \mathscr{B}$ such that Ω_0 is a disjoint union of at most countably many atoms of $(\Omega, \mathscr{B}, m)$, and every atom $\subseteq \Omega \backslash \Omega_0$ is an m-nullset.

1.3.2. Let $(\Omega_0, \mathscr{B}_0, m_0)$, $(\Omega_1, \mathscr{B}_1, m_1)$ be σ-probability spaces and $(\tilde{\Omega}, \tilde{\mathscr{B}}, \tilde{m}) = (\Omega_0, \mathscr{B}_0, m_0) \times (\Omega_1, \mathscr{B}_1, m_1)$ (Theorem VI.2.1). Let $\tilde{\mathscr{B}}_0 = \mathscr{B}_0 \times \{\varnothing, \Omega_1\} = \{F_0 \times \Omega_1 | F_0 \in \mathscr{B}_0\}$, $\tilde{\mathscr{B}}_1 = \{\varnothing, \Omega_0\} \times \mathscr{B}_1 = \{\Omega_0 \times F_1 | F_1 \in \mathscr{B}_1\}$. Prove:

1.3.2.1. $A_1 \in \mathscr{B}_1$ is an atom of $(\Omega_1, \mathscr{B}_1, m_1)$ iff $\Omega_0 \times A_1$ is a $\tilde{\mathscr{B}}_0$-atom. (*Hint*: For the "if" part, choose any $\mathscr{B}_1 \ni F_1 \subseteq A_1$ and then an $\tilde{F}_0 = F_0 \times \Omega_1 \in \tilde{\mathscr{B}}_0$ with $F_0 \cap (\Omega_0 \times A_1) = \Omega_0 \times F_1 \bmod \tilde{m}$; exploit

$$m((\Omega_0 \times F_1) \cap [(\Omega_0 \backslash F_0) \times A_1]) = 0;$$

for the "only if" part, choose any $\tilde{\mathscr{B}} \ni \tilde{F} \subseteq \Omega_0 \times A_1$ and put $\tilde{F}_0 = \{\omega_0 | m_1(A_1 \backslash \tilde{F}_{\omega_0}) = 0\} \times \Omega_1$; here $\tilde{F}_{\omega_0} = \{\omega_1 | (\omega_0, \omega_1) \in \tilde{F}\}$ as usual.)

1.3.2.2. If $F_{01}, F_{02}, \ldots \in \mathscr{B}_0$, $F_{0j} \cap F_{0k} = \varnothing$ $(j \neq k)$, $A_{11}, A_{12}, \ldots \in \mathscr{B}_1$ are atoms of $(\Omega_1, \mathscr{B}_1, m_1)$, then $\sum_n (F_{0n} \times A_{1n})$ is a $\tilde{\mathscr{B}}_0$-atom.

1.3.3. Let $(\Omega, \mathscr{B}, m)$ be a σ-probability space and $\mathscr{B}_0 \subseteq \mathscr{B}$ a σ-field. Let $E_0\colon L_m^1 \to L_m^1$ denote the conditional expectation for $\mathscr{B}_0$ (Proposition VIII.6.1). Assume that there is a stochastic $\mathscr{B}$-measurable σ-content kernel P from Ω to Ω (Definitions IV.4.1.2 and IV.4.12.2) which represents E_0, i.e., $P\colon \Omega \times \mathscr{B} \to \mathbb{R}_+$ is such that $P(\cdot, F)$ is $\mathscr{B}_0$-measurable for every $F \in \mathscr{B}$, and $P(\omega, \cdot)$ is a σ-probability content on $\mathscr{B}$ for every $\omega \in \Omega$, moreover $P(\omega, F) = E_0 1_F \bmod m$ $(F \in \mathscr{B})$. Prove:

1.3.3.1. If $A \in \mathscr{B}$ is an atom of $(\Omega, \mathscr{B}, P(\omega, \cdot))$ for m-a.e., $\omega \in \Omega$, then A is a $\mathscr{B}_0$-atom. (*Hint*: If $\mathscr{B} \ni F \subseteq A$, put $F_0 = \{\omega \mid P(\omega, F) = P(\omega, A)\}$.)

1.3.3.2. If A is a $\mathscr{B}_0$-atom and $\mathscr{B}$ is generated by some countable set system, then A is an atom of $(\Omega, \mathscr{B}, P(\omega, \cdot))$ for m-a.e., $\omega \in \Omega$. (*Hint*: Choose $\mathscr{B} \ni F \subseteq A$ and some $F_0 \in \mathscr{B}_0$ with $F = A \cap F_0$, and prove $P(\omega, F) = 1_{F_0}(\omega)P(\omega, A) \bmod m$.)

1.4. Remark. In Exercise 1.3.3 we have assumed a conditional expectation to be represented by a kernel. Such a *kernel representation* does not always exist (see the kernel representation problem IV.4.18 and Example IV.4.20). Exercise 1.3.3 represents a leading idea for our present section. The theory presented here was designed in order to circumvent difficulties arising from the possible nonexistence of kernel representations.

1.5. Proposition. *Let $(\Omega, \mathscr{B}, m)$ be a σ-probability space, $\mathscr{B}_0 \subseteq \mathscr{B}$ a σ-field, $E_0\colon L_m^1 \to L_m^1$ the conditional expectation for $\mathscr{B}_0$ (Proposition* VIII.6.1). *Then the following statements hold for every $A \in \mathscr{B}$:*

1.5.1. $\{E_0 1_A > 0\}$ *is* mod m *the smallest set in $\mathscr{B}_0$ containing A, i.e.,*

1.5.2. $\{E_0 1_A > 0\} \in \mathscr{B}_0$.

1.5.3. $E_0 1_A > 0$ m-a.e. *on* A.

1.5.4. $A \subseteq F_0 \in \mathscr{B}_0 \Rightarrow E_0 1_A = 0$ m-a.e. *on* $\Omega \setminus F_0$.

We call $\{E_0 1_A > 0\}$ *the* $\mathscr{B}_0$-**hull** *of A and denote it by* $A^{\mathscr{B}_0}$.

Proof. 2 is obvious. For a proof of 3 and 4, choose any $F_0 \in \mathscr{B}_0$. Theorem VIII.6.4 implies $0 \leq E_0(1_{F_0 \cap A}) = E_0(1_{F_0} 1_A) = 1_{F_0} E_0 1_A \bmod m$, and $\int E_0(1_{F_0 \cap A})\, dm = m(F_0 \cap A)$. In order to get 3 specify F_0 such that $E_0 1_A = 0 \bmod m$ on F_0, i.e., $1_{F_0} E_0 1_A = 0 \bmod m$. We conclude $m(F_0 \cap A) = 0$, i.e., $F_0 \cap A = \varnothing \bmod m$; if we choose in particular $F_0 = \{E_0 1_A = 0\}$, we obtain $A \subseteq \{E_0 1_A > 0\} \bmod m$. In order to get 4 we observe that

$$\begin{aligned} F_0 \supseteq A \bmod m \quad &\Rightarrow \quad 1_{F_0 \cap A} = 1_A \bmod m \quad \Rightarrow \quad 1_{F_0} E_0 1_A = E_0 1_A \bmod m \\ &\Rightarrow \quad F_0 \supseteq \{E_0 1_A > 0\} \bmod m. \end{aligned}$$

1.6. Proposition. *Let $(\Omega, \mathscr{B}, m)$ be a σ-probability space, $\mathscr{B}_0 \subseteq \mathscr{B}$ a σ-field, and $E_0: L^1_m \to L^1_m$ the conditional expectation for $\mathscr{B}_0$. Let $A, B \in \mathscr{B}$. Then:*

1.6.1. $m(A) = 0 \Rightarrow m(A^{\mathscr{B}_0}) = 0.$

1.6.2. $A \subseteq B \bmod m \Rightarrow A^{\mathscr{B}_0} \subseteq B^{\mathscr{B}_0} \bmod m.$

1.6.3. $(A \cap B)^{\mathscr{B}_0} \subseteq A^{\mathscr{B}_0} \cap B^{\mathscr{B}_0}.$

1.6.4. *Let* $f_0, g_0 \in \mathscr{L}^1_m \cap \text{mble}(\Omega, \mathscr{B}_0, \overline{\mathbb{R}})$. *Then:*

1.6.4.1. $f_0 \leqq g_0$ *m*-a.e. *on* $A \Rightarrow f_0 \leqq g_0$ *m*-a.e. *on* $A^{\mathscr{B}_0}$;

1.6.4.2. $f_0 < g_0$ *m*-a.e. *on* $A \Rightarrow f_0 < g_0$ *m*-a.e. *on* $A^{\mathscr{B}_0}$.

1.6.5. *A is a $\mathscr{B}_0$-atom iff* $f \in L^1_m \Rightarrow E_0(f 1_A) = f E_0 1_A$ *m*-a.e. *on* $A^{\mathscr{B}_0}$.

1.6.6. *Let A, B be $\mathscr{B}_0$-atoms. Then* $E_0 1_A = E_0 1_B$ *m*-a.e. *on* $A \cap B$.

Proof. 1 and 2 are obvious, and 3 easily follows from 2. In order to prove 4.1 consider $F_0 = \{f_0 \leq g_0\}$. We have $A \subseteq F_0 \in \mathscr{B}_0 \bmod m$, hence $A^{\mathscr{B}_0} \subseteq F_0 \bmod m$. The same idea clearly proves 4.2. For the "only if" part of 5, we first observe that obvious approximation techniques allow us to restrict attention to the case $f = 1_F$, $F \in \mathscr{B}$ (see Theorem III.6.4). Find $F_0 \in \mathscr{B}_0$ with $A \cap F = A \cap F_0 \bmod m$. We conclude

$$\begin{aligned} 1_A[1_F E_0 1_A] &= 1_{A \cap F} E_0 1_A = 1_{A \cap F_0} E_0 1_A \\ &= 1_A 1_{F_0} E_0 1_A = 1_A[E_0(1_{F_0} 1_A)] \\ &= 1_A E_0(1_{F_0 \cap A}) = 1_A E_0(1_{F \cap A}) = 1_A[E_0(1_F 1_A)] \end{aligned}$$

(using Theorem VIII.6.4.7). This does it. In order to get the "if" part, choose any $F \in \mathscr{B}$. Since $E_0 1_A >$ mod m on A (Proposition 1.5), we get $1_F = E_0(1_F 1_A)/E_0(1_A)$ m-a.e. on A. Put $F_0 = \{E_0 1_A > 0, E_0(1_F 1_A)/E_0 1_A = 1\}$. Then $F_0 \in \mathscr{B}_0$, $F \cap A = F_0 \cap A$, as requested. 6 follows from 5 and

$$\begin{aligned} 1_{A \cap B} E_0 1_A &= 1_{A \cap B} 1_B E_0 1_A = 1_{A \cap B} E_0(1_B 1_A) \\ &= 1_{A \cap B} E_0(1_{A \cap B}) = 1_{A \cap B} E_0 1_B \end{aligned}$$

(by symmetry).

1.7. Exercises. Let $(\Omega, \mathscr{B}, m)$ be a σ-probability space, $\mathscr{B}_0 \subseteq \mathscr{B}$ a σ-field, and $E_0: L^1_m \to L^1_m$ the conditional expectation for $\mathscr{B}_0$. Prove:

1.7.1. If $A \in \mathscr{B}$ is a $\mathscr{B}_0$-atom and $\mathscr{B} \ni F \subseteq A$, then $F^{\mathscr{B}_0}$ is the mod m unique set in $\mathscr{B}_0$ contained in $A^{\mathscr{B}_0}$ and intersecting A in F. (*Hint*: Apply Proposition 1.6.4.1 to indicator functions.)

1.7.2. Use Exercise 1.3.2 in order to construct $(\Omega, \mathscr{B}, m)$, $\mathscr{B}_0$, and $\mathscr{B}_0$-atoms A, B such that $A \cap B = \varnothing$, $A^{\mathscr{B}_0} = B^{\mathscr{B}_0} \neq \varnothing \bmod m$.

For the next definition we need Exercise III.6.5.9, which implies that the set of all equivalence classes mod m of measurable sets in a σ-probability space $(\Omega, \mathscr{B}, m)$ is a *complete lattice* with the order relation derived for classes from the relation $\subseteq$ for sets. This implies that for any subsystem $\mathscr{S}$ of $\mathscr{B}$ there is a set $S \in \mathscr{B}$ such that $A \in \mathscr{S} \Rightarrow A \subseteq S \bmod m$ and $[E \in \mathscr{B}, A \subseteq E \bmod m\ (A \in \mathscr{S})] \Rightarrow S \subseteq E \bmod m$. In particular, S is uniquely determined mod m. We call it the *supremum* mod m of $\mathscr{S}$. Similarly, the infimum of $\mathscr{S}$ is defined.

1.8. Definition. *Let $(\Omega, \mathscr{B}, m)$ be a σ-probability space and $\mathscr{B}_0 \subseteq \mathscr{B}$ a σ-field. The supremum* mod m *of the system of all $\mathscr{B}_0$-atoms is called the* **$\mathscr{B}_0$-atomary part** *of $(\Omega, \mathscr{B}, m)$.*

1.9. Proposition. *Let $(\Omega, \mathscr{B}, m)$ be a σ-probability space and $\mathscr{B}_0 \subseteq \mathscr{B}$ a σ-field. Then the $\mathscr{B}_0$-atomary part of Ω can be written as an at most countable disjoint union of $\mathscr{B}_0$-atoms.*

Proof. Let $\mathscr{A}$ be the system of all $\mathscr{B}_0$-atoms and Ω_0 the $\mathscr{B}_0$-atomary part of $(\Omega, \mathscr{B}, m)$. Clearly

$$m(\Omega_0) = \sup\{m(A_1 \cup \cdots \cup A_r) \mid r \geqq 1, A_1, \ldots, A_r \in \mathscr{A}\},$$

hence Ω_0 can be written as a countable unions of $\mathscr{B}_0$-atoms, up to an m-nullset, but m-nullsets are also $\mathscr{B}_0$-atoms. Since every subset ($\in \mathscr{B}$) of a $\mathscr{B}_0$-atom is a $\mathscr{B}_0$-atom, the first entrance decomposition does the job.

1.10. Proposition. *Let $(\Omega, \mathscr{B}, m)$ be a σ-probability space, $\mathscr{B}_0 \subseteq \mathscr{B}$ a σ-field E_0: $L_m^1 \to L_m^1$ the conditional expectation for $\mathscr{B}_0$. Let $\zeta = \{A_1, A_2, \ldots\} \subseteq \mathscr{B}$, $A_j \cap A_k = \varnothing$ $(j \neq k)$, $\Omega = \sum_n A_n$. Then the following statements hold:*

1.10.1. *$F \in \mathscr{B}(\mathscr{B}_0 \cup \zeta)$ iff there are $F_1, F_2, \ldots$ such that*

(1) $F_1, F_2, \ldots \in \mathscr{B}_0$.

(2) $F = \sum_n (A_n \cap F_n)$.

(3) $F_n \subseteq A_n^{\mathscr{B}_0}$ $(n = 1, 2, \ldots)$.

In this case, $F_1, F_2, \ldots$ are uniquely determined by F and (1)–(3).

1.10.2. *If $\mathscr{B} = \mathscr{B}(\mathscr{B}_0 \cup \zeta)$ and $A_1, A_2, \ldots$ are $\mathscr{B}_0$-atoms, then an F satisfying (1)–(3) is a $\mathscr{B}_0$-atom iff $m(F_j \cap F_k) = 0$ $(j \neq k)$.*

Proof. 1. $\mathscr{B}(\mathscr{B}_0 \cup \zeta) = \{F \mid F$ is of the form (2) with (1)$\}$ is an easy exercise. Since A_n is contained in $A_n^{\mathscr{B}_0} \in \mathscr{B}_0$ for every $n = 1, 2, \ldots$, the set $A_n \cap F_n$ remains unchanged if we replace $F_n \in \mathscr{B}_0$ by $F_n \cap A_n^{\mathscr{B}_0} \in \mathscr{B}_0$. Thus we may always assume (3) whenever we have (1), (2). Now clearly F

determines $A_n \cap F_n = A_n \cap F$ ($n = 1, 2, \ldots$) uniquely. But $A_n \cap F$, (2), and (3) determine F_n uniquely, by Exercise 1.7.1.

2. Assume that F is of the form (2), with (1), (3), and a $\mathscr{B}_0$-atom. Fix any $n \in \mathbb{N}$. From $A_n \cap F_n \subseteq F$ we deduce the existence of a $F_{0n} \in \mathscr{B}_0$ with $A_n \cap F_n = F \cap F_{0n}$, and from $A_n \cap F_n \subseteq A_n$ we deduce that we may assume $F_{0n} \subseteq A_n^{\mathscr{B}_0}$ (Exercise 1.7.1). Now

$$A_n \cap F_n = A_n \cap (A_n \cap F_n) = A_n \cap (F \cap F_{0n}) \subseteq A_n \cap F_{0n}$$

follows. Since $F_n, F_{0n} \in \mathscr{B}_0$, $F_n, F_{0n} \subseteq A_n^{\mathscr{B}_0}$, we may apply Exercise 1.7.1 and see that $F_n \subseteq F_{0n}$. Now for $j \neq k$, we obtain $A_k \cap F_k \cap F_j \subseteq A_k \cap F_k \cap F_{0j} = A_k \cap F_k \cap F \cap F_{0j} = (A_k \cap F_k) \cap (A_j \cap F_j) = \varnothing$. Exercise 1.7.1 now implies $F_k \cap F_j = \varnothing$. Let, conversely, $F_j \cap F_k = \varnothing$ $(j \neq k)$. Choose any $\mathscr{B} \ni E \subseteq F$. Represent E in the form (2), with (1) and (3):

$$E = \sum_n (A_n \cap E_n), \qquad E_j \in \mathscr{B}_0, \qquad E_j \subseteq A_j^{\mathscr{B}_0} \qquad (j = 1, 2, \ldots).$$

Clearly $E \subseteq F$ implies $E_1 \subseteq F_1$, $E_2 \subseteq F_2, \ldots$; and since the F_j are disjoint, we get $E_j \cap E_k = \varnothing$ $(j \neq k)$, $E = F \cap (E_1 + E_2 + \cdots)$, and here $E_1 + E_2 + \cdots \in \mathscr{B}_0$, hence F is a $\mathscr{B}_0$-atom.

Applying the above result to the situation prevailing in the $\mathscr{B}_0$-atomary part, we get

1.11. Theorem. *Let $(\Omega, \mathscr{B}, m)$ be a σ-probability space, $\mathscr{B}_0 \subseteq \mathscr{B}$ a σ-field, $E_0\colon L_m^1 \to L_m^1$ the conditional expectation for $\mathscr{B}_0$, and Ω_0 the $\mathscr{B}_0$-atomary part of $(\Omega, \mathscr{B}, m)$, $A_1, A_2, \ldots$ $\mathscr{B}_0$-atoms with $A_j \cap A_k = \varnothing$ $(j \neq k)$ and $\Omega_0 = A_1 + A_2 + \cdots$ (Proposition 1.10). Then every set $\mathscr{B} \ni F \subseteq \Omega_0$ has a unique representation*

$$F = \sum_{n=1} (A_n \cap F_n)$$

such that $F_1, F_2, \ldots \in \mathscr{B}_0$ and

$$F_n \subseteq A_n^{\mathscr{B}_0} \qquad (n = 1, 2, \ldots).$$

F is a $\mathscr{B}_0$-atom iff $F_j \cap F_k = \varnothing$ $(j \neq k)$ in this representation.

Proof. Forget about $\Omega \backslash \Omega_0$ and observe that $\mathscr{B} \cap \Omega_0$ is generated by $\mathscr{B}_0 \cup \{A_1, A_2, \ldots\}$, according to the very definition of $\mathscr{B}_0$-atoms, and of Ω_0. Now apply Proposition 1.10.

1.12. Theorem. *Let $(\Omega, \mathscr{B}, m)$ be a σ-probability space, $\mathscr{B}_0 \subseteq \mathscr{B}$ a σ-field, $E_0\colon L_m^1 \to L_m^1$ the conditional expectation for $\mathscr{B}_0$. Then the $\mathscr{B}_0$-atomary part Ω_0 of $(\Omega, \mathscr{B}, m)$ has a representation*

$$\Omega_0 = A_1 + A_2 + \cdots$$

where $A_1, A_2, \ldots \in \mathscr{B}$ *are* $\mathscr{B}_0$*-atoms such that* $A_j \cap A_k = \varnothing$ $(j \neq k)$ *and* $E_0 1_{A_n} \geqq E_0 1_{A_1} \geqq \cdots$ *m*-a.e.

Proof. By Proposition 1.10 there are $\mathscr{B}_0$-atoms $C_1, C_2, \ldots \in \mathscr{B}$ such that $C_j \cap C_k = \varnothing$ $(j \neq k)$, $\Omega_0 = C_1 + C_2 + \cdots$. Now $1 \geqq 1_{\Omega_0} = 1_{C_1} + 1_{C_2} + \cdots$ implies $1 \geqq E_0 1_{\Omega_0} = E_0 1_{C_1} + E_0 1_{C_2} + \cdots$ *m*-a.e. Thus for *m*-almost every $\omega \in \Omega$, the real numbers $(E_0 1_{C_1})(\omega)$, $(E_0 1_{C_2})(\omega)$, ... are $\geqq 0$ and form a convergent series, hence tend to 0, hence in particular every subset of this set of at most countably many reals $\geqq 0$ contains a largest element. It makes sense therefore, for such an ω, to form

$$a_1(\omega) = \min\{k \mid (E_0 1_{C_k})(\omega) \geqq (E_0 1_{C_j})(\omega)\ (j \in \mathbb{N})\},$$
$$a_2(\omega) = \min\{k \mid k \neq a_1(\omega), (E_0 1_{C_k})(\omega) \geqq (E_0 1_{C_j})(\omega)\ (j \in \mathbb{N} \setminus \{a_1(\omega)\})\}$$

and so on. Clearly

$$\begin{aligned}\{\omega \mid a_1(\omega) = k\} &= \{\omega \mid (E_0 1_{C_k})(\omega) \geqq (E_0 1_{C_j})(\omega)\ (j \in \mathbb{N}),\\ &\qquad (E_0 1_{C_k})(\omega) > (E_0 1_{C_j})(\omega)\ (j < k)\},\\ \{a_2 = k\} &= \sum_{i \neq k} \{a_1 = i,\ E_0 1_{C_k} \geqq E_0 1_{C_j}\ (j \neq i),\\ &\qquad E_0 1_{C_k} > E_0 1_{C_j}\ (i \neq j < k)\}\end{aligned}$$

and so on; hence $\{a_n = k\} \in \mathscr{B}_0$ $(n, k = 1, 2, \ldots)$. Put now

$$A_n = \sum_k \{a_n = k\} \cap C_k \qquad (n = 1, 2, \ldots).$$

For every $n = 1, 2, \ldots$, the sets $\{a_n = 1\}, \{a_n = 2\}, \ldots$ are in $\mathscr{B}_0$ and pairwise disjoint, hence A_n is a $\mathscr{B}_0$-atom (Theorem 1.11). Clearly $A_j \cap A_k = \varnothing$ $(j \neq k)$ and $\Omega_0 = A_1 + A_2 + \cdots$. Moreover we always have

$$1_{\Omega_0} = \sum_k 1_{\{a_n = k\}} = \sum_j 1_{\{a_r = j\}},$$

hence $n \leqq r$ implies

$$\begin{aligned}E_0 1_{A_n} &= \sum_k 1_{\{a_n = k\}} E_0 1_{C_k} = \sum_{k, j} 1_{\{a_n = k,\, a_r = j\}} E_0 1_{C_k}\\ &\geqq \sum_{k, j} 1_{\{a_n = k,\, a_r = j\}} E_0 1_{C_j} = \sum_j 1_{\{a_r = j\}} E_0 1_{C_j} = E_0 1_{A_r};\end{aligned}$$

here the $\geqq$ comes in by the very definition of the a_n, a_r.

1.13. Remark. It is advisable that the reader reconsider Exercises 1.3.2 and 1.3.3 under the viewpoint suggested by the preceding propositions and theorems.

1.14. Theorem. *Let $(\Omega, \mathscr{B}, m)$ be a σ-probability space, $\mathscr{B}_0 \subseteq \mathscr{B}$ a σ-field and $E_0: L_m^1 \to L_m^1$ the conditional expectation for $\mathscr{B}_0$. Let $f_0 \in \text{mble}(\Omega, \mathscr{B}_0, \mathbb{R})$, $0 \leqq f_0 \leqq 1$. Then there is an $F \in \mathscr{B}$ and a $\mathscr{B}_0$-atom A such that $E_0 1_F \leqq f_0 \leqq E_0 1_F + E_0 1_A$. In particular, if every $\mathscr{B}_0$-atom is an m-nullset, then there is an $F \in \mathscr{B}$ such that $f_0 = E_0 1_F$.*

Proof. Let $\Sigma^0 = \{F \mid F \in \mathscr{B},\ E_0 1_F \leqq f_0 \ m\text{-a.e.}\}$. Clearly $\varnothing \in \Sigma^0$. It is nearly obvious that the supremum of mod m (in the sense of Exercises II.2.6.7, III.12.15) of a totally ordered (by $\subseteq$ mod m, of course) subset of Σ^0 is in Σ^0 again (exercise). Thus we may apply Zorn's lemma and obtain a maximal F in Σ^0: $F \subseteq E \in \Sigma^0 \Rightarrow F = E$ mod m. Let now

$$\Sigma_0 = \{C \mid C \in \mathscr{B}, m(F \cap C) = 0, f_0 \leqq E_0(1_F + 1_C)\ m\text{-a.e.}\}.$$

Clearly $\Omega \backslash F \in \Sigma_0$. It is nearly obvious that the infimum mod m (in the sense of Exercises II.2.6.7, III.12.15) of a totally ordered subset of Σ_0 is in Σ_0 again (exercise). Thus we may apply Zorn's lemma and obtain a minimal $A \in \Sigma_0: \Sigma_0 \ni E \subseteq A \Rightarrow E = A$ mod m. We have to show that A is a $\mathscr{B}_0$-atom. Let $\mathscr{B} \ni G \subseteq A$. Put $G_0 = \{E_0(1_F + 1_G) > f_0\}$. Clearly $G_0 \in \mathscr{B}_0$. It suffices to show $G = A \cap G_0$ mod m. For this, let $G_0' = \Omega \backslash G_0 = \{E_0(1_F + 1_G) \leqq f_0\}$. We observe that G is $\subseteq A$, hence disjoint mod 0 from F, and prove that $F + (G \cap G_0') \in \Sigma^0$. In fact,

$$\begin{aligned} 1_{G_0'} E_0(1_{F+(G \cap G_0')}) &= 1_{G_0'} E_0(1_F + 1_G 1_{G_0'}) \\ &\leqq 1_{G_0'} E_0(1_F + 1_G) \leqq 1_{G_0'} f_0 \end{aligned}$$

and

$$\begin{aligned} 1_{G_0} E_0(1_{F+(G \cap G_0')}) &= E_0(1_{G_0} 1_F + 1_{G_0} 1_{G_0'}) = E_0(1_{G_0} 1_F) \\ &= 1_{G_0} E_0 1_F \leqq 1_{G_0} f_0 \quad \text{mod } m. \end{aligned}$$

Putting these two results together, we obtain $E_0(1_{F+(G \cap G_0')}) \leqq f_0$ mod m, i.e., $F + (G \cap G_0') \in \Sigma^0$. The maximality of F now yields $m(G \cap G_0') = \varnothing$, hence $E_0(1_F + 1_G) > f_0$ m-a.e. on G. On the other hand, $G + [(A \backslash G) \cap G_0']$ belongs to Σ_0:

$$\begin{aligned} 1_{G_0'} E_0(1_F + 1_{G+[(A\backslash G) \cap G_0']}) &= 1_{G_0'} E_0(1_F + 1_G + 1_{G_0'}(1_A - 1_G)) \\ &= E_0(1_{G_0'}(1_F + 1_G) + 1_{G_0'}(1_A - 1_G)) \\ &= 1_{G_0'} E_0(1_F + 1_A) \geqq 1_{G_0'} f_0 \end{aligned}$$

and

$$\begin{aligned} 1_{G_0} E_0(1_F + 1_{G+[(A\backslash G) \cap G_0']}) &= E_0(1_{G_0}(1_F + 1_G) + 1_{G_0} 1_{G_0'}(1_A - 1_G)) \\ &= E_0(1_{G_0}(1_F + 1_G)) = 1_{G_0} E_0(1_F + 1_G) \\ &\geqq 1_{G_0} f_0 \quad \text{mod } m. \end{aligned}$$

Putting these two results together we obtain

$$E_0(1_F + 1_{G+[(A\backslash G)\cap G_0']}) \geqq f_0 \quad \text{mod } m,$$

hence indeed $G + [(A\backslash G) \cap G_0'] \in \Sigma_0$. The minimality of A now implies $(A\backslash G) \cap G_0' = A\backslash G$ mod m, i.e., $A\backslash G \subseteq G_0'$ mod m. This and $m(G \cap G_0') = 0$ yields $G = A \cap G_0$ mod m, and the theorem is proved.

As a consequence of this theorem, we obtain

1.15. Theorem. *Let $(\Omega, \mathscr{B}, m)$ be a σ-probability space, $\mathscr{B}_0 \subseteq \mathscr{B}$ a σ-field, and $E_0\colon L_m^1 \to L_m^1$ the conditional expectation for $\mathscr{B}_0$. Let $F \in \mathscr{B}$ and assume that every $\mathscr{B}_0$-atom $\subseteq F$ is an m-nullset. Then for every $f_0 \in \text{mble}(\Omega, \mathscr{B}_0, \mathbb{R})$ with $0 \leqq f_0 \leqq 1$ there is a $\mathscr{B} \ni E \subseteq F$ such that*

$$E_0\, 1_E = f_0\, E_0\, 1_F \quad \text{mod } m.$$

Proof. Forget about $\Omega\backslash F$ and apply Theorem 1.14. It follows that there is a $\mathscr{B} \ni E \subseteq F$ with $E_0^F\, 1_E = f_0$, where E_0^F denotes the conditional expectation with respect to $\mathscr{B}_0 \cap F$ on F. Within F we have $E_0\, 1_F > 0$ (Proposition 1.5) and $E_0^F\, 1_E = E_0\, 1_E / E_0\, 1_F$. In fact, if $G_0 \in \mathscr{B}_0$, then

$$\begin{aligned}
\int 1_{G_0 \cap F} \frac{E_0\, 1_E}{E_0\, 1_F}\, dm &= \int 1_{G_0}\, 1_F \frac{E_0\, 1_E}{E_0\, 1_F}\, dm \\
&= \int 1_{G_0}\, E_0\, 1_F \frac{E_0\, 1_E}{E_0\, 1_F}\, dm = \int 1_{G_0}\, E_0\, 1_E\, dm \\
&= \int 1_{G_0}\, 1_E\, dm = \int 1_{G_0 \cap F}\, 1_E\, dm.
\end{aligned}$$

(This is exercise VIII.6.5.4.) It follows that $f_0 = E_0\, 1_E / E_0\, 1_F$ on F, hence (remember $\Omega\backslash F$ again) on $F^{\mathscr{B}_0}$. But outside $F^{\mathscr{B}_0}$ both $E_0 1_E$ and $E_0 1_F$ vanish, hence $E_0 1_E = f_0\, E_0 1_E$ remains valid all over Ω (mod m, of course).

As a further consequence, we obtain

1.16. Theorem. *Let $(\Omega, \mathscr{B}, m)$ be a probability space, $\mathscr{B}_0 \subseteq \mathscr{B}$ a σ-field and $E_0\colon L_m^1 \to L_m^1$ the conditional expectation for $\mathscr{B}_0$. Let $p_1, p_2, \ldots$ be reals $\geqq 0$, with $p_1 + p_2 + \cdots = 1$. Then every $F \in \mathscr{B}$ such that every $\mathscr{B}_0$-atom $\subseteq F$ is an m-nullset can be split in the form $F = F_1 + F_2 + \cdots$ with $F_1, F_2, \ldots \in \mathscr{B}$, $F_j \cap F_k = \varnothing$ $(j \neq k)$, such that*

$$E_0\, 1_{F_k} = p_k\, E_0\, 1_F \qquad (k = 1, 2, \ldots)$$

holds mod m.

Proof. Use Theorem 1.15 and find $\mathscr{B} \ni F_1 \subseteq F$ with $E_0 1_{F_1} = p_1 E_0 1_F$. Use Theorem 1.15 again and find $\mathscr{B} \ni F_2 \subseteq F \backslash F_1$ with

$$E_0 1_{F_2} = \frac{p_2}{p_2 + p_3 + \cdots} E_0 1_{F \backslash F_1} \qquad \text{(assuming that } p_2 + p_3 + \cdots > 0).$$

Clearly

$$E_0 1_{F_2} = \frac{p_2}{1 - p_1}(E_0 1_F - E_0 1_{F_1}) = p_2 E_0 1_F.$$

We continue in this fashion until $p_n + p_{n+1} + \cdots = 0$, or ad infinitum, getting $F_1, F_2, \ldots \in \mathscr{B}$ pairwise disjoint with $E_0 1_F = p_n E_0 1_F$ $(n = 1, 2, \ldots)$. But now

$$\begin{aligned} m(F_1 + F_2 + \cdots) &= \int E_0 1_{F_1}\, dm + \int E_0 1_{F_2}\, dm + \cdots \\ &= p_1 \int E_0 1_F\, dm + p_2 \int E_0 1_F\, dm + \cdots \\ &= (p_1 + p_2 + \cdots) \int E_0 1_F\, dm = m(F), \end{aligned}$$

and hence $F = F_1 + F_2 + \cdots \bmod m$ follows.

The last theorem of this section compares conditional atoms for two sub-σ-fields.

1.17. Theorem. *Let $(\Omega, \mathscr{B}, m)$ be a σ-probability space and $\mathscr{B}_0, \mathscr{B}_1 \subseteq \mathscr{B}$ σ-fields with $\mathscr{B}_0 \subseteq \mathscr{B}_1$. Let $E_k\colon L_m^1 \to L_m^1$ denote the conditional expectation for $\mathscr{B}_k$ and Ω_k the $\mathscr{B}_k$-atomary part of $(\Omega, \mathscr{B}, m)$ $(k = 0, 1)$. Let Ω_{10} denote the $\mathscr{B}_0$-atomary part of $(\Omega, \mathscr{B}_1, m)$. Then:*

1.17.1. $\Omega_0 = \Omega_1 \cap \Omega_{10} \bmod m$.

1.17.2. *If A_1 is a $\mathscr{B}_1$-atom of $(\Omega, \mathscr{B}, m)$ and A_{10} is a $\mathscr{B}_0$-atom of $(\Omega, \mathscr{B}_1, m)$, then $A_1 \cap A_{10}$ is a $\mathscr{B}_0$-atom of $(\Omega, \mathscr{B}, m)$ and*

$$E_0(1_{A_1 \cap A_{10}}) = (E_1 1_{A_1})(E_0 1_{A_{10}}) \qquad (\text{mod } m, \quad \text{on } A_1 \cap A_{10})$$

holds.

Proof. We begin with 2. By the very definition of conditional atoms, $\mathscr{B}$ coincides mod m with $\mathscr{B}_1$ on A_1, and $\mathscr{B}_1$ coincides mod m with $\mathscr{B}_0$ on A_{10}, hence $\mathscr{B}$ coincides mod m with $\mathscr{B}_0$ on $A_1 \cap A_{10}$, hence $A_1 \cap A_{10}$ is a $\mathscr{B}_0$-atom for $(\Omega, \mathscr{B}, m)$. By Proposition 1.6.5, applied twice, we get

$$\begin{aligned} 1_{A_{10}} E_0 1_{A_1 \cap A_{10}} &= 1_{A_{10}} E_0(E_1(1_{A_1} 1_{A_{10}})) \\ &= 1_{A_{10}} E_0(1_{A_{10}} E_1 1_{A_1}) \\ &= 1_{A_{10}}(E_1 1_{A_1})(E_0 1_{A_{10}}) \quad \text{mod } m. \end{aligned}$$

Here we have used the fact that $A_{10} \in \mathscr{B}_1$ and A_{10} is a $\mathscr{B}_0$-atom of $(\Omega, \mathscr{B}_1, m)$. We thus get 2, and hence immediately $\Omega_0 \supseteq \Omega_1 \cap \Omega_{10}$ mod m for 1. In order to complete the proof of 1 we have to show $\Omega_0 \subseteq \Omega_1 \cap \Omega_{10}$ mod m. For this it is sufficient to show that any $\mathscr{B}_0$-atom A_0 of $(\Omega, \mathscr{B}, m)$ can be represented in the form $A_0 = A_1 \cap A_{10}$ where A_1 is a $\mathscr{B}_1$-atom of $(\Omega, \mathscr{B}, m)$ and A_{10} is a $\mathscr{B}_0$-atom of $(\Omega, \mathscr{B}_1, m)$. Now $\mathscr{B}$ coincides mod m with $\mathscr{B}_0$ on A_0. The more $\mathscr{B}_1$ coincides mod m with $\mathscr{B}_0$ on A_0, and hence $\mathscr{B}$ coincides mod m with $\mathscr{B}_1$ on A_0. We thus may put $A_1 = A_0$. Put now $A_{10} = \{E_1 1_{A_0} > 0\}$. Then we have $\mathscr{B}_1 \ni A_{10} \supseteq A_0$ mod m and hence $A_0 = A_{10} \cap A_0$ mod m. It now suffices to show that A_{10} is a $\mathscr{B}_0$-atom of $(\Omega, \mathscr{B}_1, m)$. Choose any $\mathscr{B}_1 \ni F_1 \subseteq A_{10}$. Since A_0 is a $\mathscr{B}_0$-atom of $(\Omega, \mathscr{B}, m)$, we may find $F_0 \in \mathscr{B}_0$ with $A_0 \cap F_1 = A_0 \cap F_0$ mod m. Now we get

$$\begin{aligned} 1_{F_1} E_1 1_{A_0} &= E_1 1_{F_1 \cap A_0} = E_1 1_{F_0 \cap A_0} \\ &= E_1(1_{F_0} 1_{A_0}) = 1_{F_0} E_1 1_{A_0}, \end{aligned}$$

which implies $A_{10} \cap F_1 = A_{10} \cap F_0$, hence A_{10} is a $\mathscr{B}_0$-atom for $(\Omega, \mathscr{B}_1, m)$.

1.18. Exercises

1.18.1. Let $(\Omega, \mathscr{B}, m)$ be a σ-probability space that is atomless in the sense that every $\{\varnothing, \Omega\}$-atom of it is an m-nullset.

1.18.1.1. Prove that for every $F \in \mathscr{B}$ and every real $0 \leqq \alpha \leqq m(F)$, there is an $F \supseteq E \in \mathscr{B}$ with $m(E) = \alpha$. (*Hint*: Use Theorem 1.15.)

1.18.1.2. Prove that there is, for every $n = 1, 2, \ldots$, a partition $\zeta_n = \{A_{n1}, \ldots, A_{n2^n}\} \subseteq \mathscr{B}$ of Ω such that

1.18.1.2.1. ζ_{n+1} is finer than ζ_n.

1.18.1.2.2. $A \in \zeta_n \Rightarrow m(A) = 2^{-n}$ $(n = 1, 2, \ldots)$.

1.18.1.3. Prove that there is an $f \in \text{mble}(\Omega, \mathscr{B}, \mathbb{R})$ such that $0 \leqq f \leqq 1$ and f transports m into the Lebesgue σ-content in $[0, 1]$. (*Hint*: Use 1.2 in order to construct, for any finite string $(x_1, \ldots, x_n)$ of symbols $x_k = 0, 1$, a set $F_{x_1, \ldots, x_n}$ such that: (a) for every fixed n the lexicographic relation $(x_1, \ldots, x_n) < (y_1, \ldots, y_n)$ (excluding equality) implies $F_{x_1, \ldots, x_n} \subseteq F_{y_1, \ldots, y_n}$

and $m(F_{y_1,\ldots,y_n} \backslash F_{x_1,\ldots,x_n}) \geqq 2^{-n}$, and $F_{0,\ldots,0} = \varnothing$, $F_{1,\ldots,1} = \Omega$; (b) $F_{x_1,\ldots,x_n,0} = F_{x_1,\ldots,x_n}$ $(n = 1, 2, \ldots; x_1, \ldots, x_n = 0, 1)$; put $f(\omega) = \inf\{2^{-1}x_1 + \cdots + 2^{-n}x_n \mid n \geqq 1, \omega \in F_{x_1,\ldots,x_n}\}$.)

1.18.1.4. Refine 1.2 in the following way: Given $F_1, F_2, \ldots \in \mathscr{B}$, we can construct the A_{nk} with the properties mentioned in 1.2 in such a fashion that for every N and every $\varepsilon > 0$, there is a union E of finitely many sets A_{nk} such that $m(E \,\Delta\, F_N) < \varepsilon$. (*Hint*: Adapt the A_{nk} suitably to the partitions $\{F_1, \Omega\backslash F_1\} \vee \cdots \vee \{F_N, \Omega\backslash F_N\}$.)

1.18.2. Let $\mathscr{B}$ be a σ-field in Ω, $n \geqq 1$, and $m_1, \ldots, m_n$ σ-probability contents such that $(\Omega, \mathscr{B}, m_k)$ is atomless $(k = 1, \ldots, n)$.

1.18.2.1. Prove that for every $F \in \mathscr{B}$ with $m_k(F) > 0$ $(k = 1, \ldots, n)$, there is an $F \supseteq E \in \mathscr{B}$ with $0 < m_k(E) < m_k(F)$ $(k = 1, \ldots, n)$.

1.18.2.2. Prove that for every $F \in \mathscr{B}$ and any real $0 \leqq \alpha \leqq 1$, there is an $F \supseteq E \in \mathscr{B}$ such that $m_k(E) = \alpha m_k(F)$ $(k = 1, \ldots, n)$. (*Hint*: Form $\Sigma^0 = \{C \mid F \supseteq C \in \mathscr{B}, m_k(C) \leqq \alpha m_k(F)\ (k = 1, \ldots, n)\}$ and construct E as a maximal element of Σ^0 for the partial ordering $\subseteq$ mod m; use Theorem 1.15, induction over n, and 1 in order to show that E has the required properties.)

1.18.2.3. Prove that for $F, G \in \mathscr{B}$, $F \cap G = \varnothing$, $0 \leqq \alpha \leqq 1$, there is a $\mathscr{B} \ni E \subseteq F + G$ with $m_k(E) = \alpha m_k(F) + (1 - \alpha) m_k(G)$ $(k = 1, \ldots, n)$.

1.18.2.4. Prove that $\{(m_1(F), \ldots, m_n(F)) \mid F \in \mathscr{B}\}$ is a convex subset of $\mathbb{R}^n$.

2. ENTROPY

In this section we present the rudiments of the theory of entropy and conditional entropy, based on the theory of atoms and conditional atoms given in Section 1. We prove all limit theorems needed for the application of entropy theory in *ergodic theory* (see, e.g., PARRY [1], JACOBS [2], SMORODINSKI [1], WALTERS [1], BROWN [1], DENKER *et al.* [1]).

2.1. Lemma. *Let $(\Omega, \mathscr{B}, m)$ be a σ-probability space and $\mathscr{B}_0, \mathscr{B}_1$ σ-fields $\subseteq \mathscr{B}$ such that $\mathscr{B}_0 \subseteq \mathscr{B}_1$. Let $E_k\colon L^1_m \to L^1_m$ denote the conditional expectation for $\mathscr{B}_k$ $(k = 0, 1)$ and Ω_0 the $\mathscr{B}_0$-atomary part of $(\Omega, \mathscr{B}_1, m)$. Then there is a* mod m *unique function $\rho_{\mathscr{B}_1|\mathscr{B}_0} \in \mathrm{mble}(\Omega, \mathscr{B}_1, \mathbb{R})$ such that:*

2.1.1. $\rho_{\mathscr{B}_1|\mathscr{B}_0} = E_0 1_A$ *m-a.e. on A, for every $\mathscr{B}_0$-atom A of $(\Omega, \mathscr{B}_1, m)$.*

2.1.2. $\rho_{\mathscr{B}_1|\mathscr{B}_0} = 0$ *m-a.e. on $\Omega\backslash\Omega_0$.*

We call $\rho_{\mathscr{B}_1|\mathscr{B}_0}$ the **conditional density** *of $(\Omega, \mathscr{B}_1, m)$ for $\mathscr{B}_0$.*

2.1.3. $\rho_{\mathscr{B}_1|\mathscr{B}_0}$ *is* m-a.e. *strictly positive on* Ω_0 *and it satisfies*

$$\rho_{\mathscr{B}_1|\mathscr{B}_0}(\omega) = \begin{cases} \sum_{k=1}^{\infty} 1_{A_k}(\omega)(E_0 1_{A_k})(\omega) & (\omega \in \Omega_0) \\ 0 & (\textit{otherwise}) \end{cases}$$

for any representation $\Omega_0 = A_1 + A_2 + \cdots$ *with pairwise disjoint* $\mathscr{B}_0$*-atoms* $A_1, A_2, \ldots$ *of* $(\Omega, \mathscr{B}_1, m)$ *for* $\mathscr{B}_0$ (*Proposition* 1.9).

Proof. By Proposition 1.9 we may write $\Omega_0 = A_1 + A_2 + \cdots$ with conditional $\mathscr{B}_0$-atoms $A_1, A_2, \ldots$ of $(\Omega, \mathscr{B}_1, m)$, $A_j \cap A_k = \varnothing$ $(j \neq k)$. Define

$$\rho_{\mathscr{B}_1|\mathscr{B}_0}(\omega) = \begin{cases} (E_0 1_{A_k})(\omega) & (\omega \in A_k, \quad k = 1, 2, \ldots) \\ 0 & (\text{otherwise}) \end{cases}$$

having fixed representatives of the $E_0 1_{A_k}$, of course. Let now A be any $\mathscr{B}_0$-atom for $(\Omega, \mathscr{B}_1, m)$. By Proposition 1.6.6 we have $E_0 1_A = E_0 1_{A_k}$ m-a.e. on A_k, for $k = 1, 2, \ldots,$ and thus $E_0 1_A = \rho_{\mathscr{B}_1|\mathscr{B}_0}$ m-a.e. on A $(\subseteq A_1 + A_2 + \cdots \bmod m)$. The rest is obvious.

2.2. Definition. *Let* $(\Omega, \mathscr{B}, m)$ *be a* σ*-probability space,* $\mathscr{B}_0 \subseteq \mathscr{B}_1 \subseteq \mathscr{B}$ σ*-fields, and* $\rho_{\mathscr{B}_1|\mathscr{B}_0}$ *the conditional density of* $(\Omega, \mathscr{B}_1, m)$ *for* $\mathscr{B}_0$.

2.2.1. *The function* $\in \overline{\mathbb{R}}_+^{\Omega}$ *defined uniquely* mod m *by*

$$h_{\mathscr{B}_1|\mathscr{B}_0}(\omega) = \begin{cases} -\log \rho_{\mathscr{B}_1|\mathscr{B}_0}(\omega) & (\rho_{\mathscr{B}_1|\mathscr{B}_0}(\omega) > 0) \\ \infty & (\textit{otherwise}) \end{cases}$$

is called the **conditional entropy density** *of* $(\Omega, \mathscr{B}_1, m)$ *for* $\mathscr{B}_0$. *By*

$$H(\mathscr{B}_1|\mathscr{B}_0) = \begin{cases} \int h_{\mathscr{B}_1|\mathscr{B}_0}\, dm & \textit{if} \quad h_{\mathscr{B}_1|\mathscr{B}_0} \in \mathscr{L}_m^1 \\ \infty & \textit{otherwise,} \end{cases}$$

we define the **conditional entropy** *of* $(\Omega, \mathscr{B}_1, m)$ *for* $\mathscr{B}_0$.

2.2.2. *If* $\mathscr{B}_0 = \{\varnothing, \Omega\}$, *then we write* $\rho_{\mathscr{B}_1}$ *for* $\rho_{\mathscr{B}_1|\mathscr{B}_0}$ *and* $h_{\mathscr{B}_1}$ *for* $h_{\mathscr{B}_1|\mathscr{B}_0}$ *and call this latter function the* **entropy density** *of* $\mathscr{B}_1$, *and we write* $H(\mathscr{B}_1)$ *for* $H(\mathscr{B}_1|\mathscr{B}_0)$ *and call this the* **entropy** *of* $\mathscr{B}_1$.

2.3. Remarks

2.3.1. Since $-\log x$ is a convex function, Jensen's inequality yields

$$H(\mathscr{B}_1|\mathscr{B}_0) = \int [-\log \rho_{\mathscr{B}_1|\mathscr{B}_0}]\, dm \geqq -\log \int \rho_{\mathscr{B}_1|\mathscr{B}_0}\, dm$$

$$= -\log\left[\sum_{k=1}^{\infty} \int_{A_k} E_0 1_{A_k}\, dm\right] = -\log m(\Omega_0) \geqq 0$$

if $h_{\mathscr{B}_1|\mathscr{B}_0} \in \mathscr{L}_m^1$ and we use the representation of $\rho_{\mathscr{B}_1|\mathscr{B}_0}$ given in Lemma 2.1.3.

2.3.2. If $\mathscr{B}_0 = \{\varnothing, \Omega\}$, then conditional $\mathscr{B}_0$-atoms of $(\Omega, \mathscr{B}_1, m)$ are just atoms of $(\Omega, \mathscr{B}_1, m)$, conditional expectations for $\mathscr{B}_0$ are just expectations, i.e., integrals, and we get $\Omega_0 = A_1 + A_2 + \cdots$ in a mod m unique fashion, where $A_1, A_2, \ldots$ runs over a maximal choice of pairwise disjoint atoms of $(\Omega, \mathscr{B}_1, m)$ and Ω_0 is the atomary part of $(\Omega, \mathscr{B}_1, m)$. Thus we have, in this case,

$$h_{\mathscr{B}_1}(\omega) = \begin{cases} -\sum_{k=1}^{\infty} 1_{A_k}(\omega) \log m(A_k) & (\omega \in \Omega_0) \\ \infty & (\text{otherwise}) \end{cases}$$

$$H(\mathscr{B}_1) = \begin{cases} -\sum_{k=1}^{\infty} m(A_k) \log m(A_k) & \begin{array}{l}(\text{if this sum is finite} \\ \text{and } \Omega_0 = \Omega \\ \bmod m)\end{array} \\ \infty & (\text{otherwise}) \end{cases}$$

2.3.3. For some purposes it is convenient to define $\rho_{\mathscr{B}_1|\mathscr{B}_0}$, $h_{\mathscr{B}_1|\mathscr{B}_0}$, and $H(\mathscr{B}_1|\mathscr{B}_0)$ even if $\mathscr{B}_0 \subseteq \mathscr{B}_1$ does not hold. We understand these symbols as synonyms of $\rho_{\mathscr{B}_1 \vee \mathscr{B}_0|\mathscr{B}_0}$, $h_{\mathscr{B}_1 \vee \mathscr{B}_0|\mathscr{B}_0}$, $H(\mathscr{B}_1 \vee \mathscr{B}_0|\mathscr{B}_0)$, respectively. Here $\mathscr{B}_1 \vee \mathscr{B}_0$ stands for $\mathscr{B}(\mathscr{B}_0 \cup \mathscr{B}_1)$.

2.4. Exercises

2.4.1. Let $(\Omega_0, \mathscr{B}_0, m_0)$, $(\Omega_1, \mathscr{B}_1, m_1)$ be σ-probability spaces and $(\tilde{\Omega}, \tilde{\mathscr{B}}, \tilde{m}) = (\Omega_0, \mathscr{B}_0, m_0) \times (\Omega_1, \mathscr{B}_1, m_1)$, $\tilde{\mathscr{B}}_1 = \{\Omega_0 \times F_1 | F_1 \in \mathscr{B}_1\}$, $\tilde{\mathscr{B}}_0 = \{F_0 \times \Omega_1 | F_0 \in \mathscr{B}_0\}$ (compare Exercise 1.3.2). Let $A_{11}, A_{12}, \ldots$ be pairwise disjoint atoms of $(\Omega_1, \mathscr{B}_1, m_1)$ such that $m_1(A_{1k}) > 0$ $(k = 1, 2, \ldots)$ and every atom of $(\Omega_1, \mathscr{B}_1, m_1)$ contained in $\Omega_1 \backslash \bigcup_k A_{1k}$ is an m_1-nullset. Put $\Omega_{10} = A_{11} + A_{12} + \cdots$.

2.4.1.1. Prove that $\tilde{\Omega}_1 = \Omega_0 \times \Omega_{10}$ is the $\tilde{\mathscr{B}}_0$-atomary part of $(\tilde{\Omega}, \tilde{\mathscr{B}}, m)$.

2.4.1.2. Prove that (using Remark 2.3.3)

$$h_{\tilde{\mathscr{B}}_1|\tilde{\mathscr{B}}_0}(\omega_0, \omega_1) = \begin{cases} -\sum_k 1_{A_{1k}}(\omega_1) \log m_1(A_{1k}) & (\omega_1 \in \Omega_{10}) \\ \infty & (\text{otherwise}) \end{cases}$$

$$H(\tilde{\mathscr{B}}_1|\tilde{\mathscr{B}}_0) = \begin{cases} -\sum_k m(A_{1k}) \log m_1(A_{1k}) & (m_1(\Omega_1 \backslash \Omega_{10}) = 0) \\ \infty & (\text{otherwise}) \end{cases}$$

$$= H(\mathscr{B}_1).$$

2.4.2. Let $(\Omega, \mathscr{B}, m)$ be a σ-probability space and $\mathscr{B}_0 \subseteq \mathscr{B}$ a σ-field. Let $E_0 \colon L_m^1 \to L_m^1$ denote the conditional expectation for $\mathscr{B}_0$ (Proposition

VIII.6.1). Assume that there is a stochastic $\mathscr{B}_0$-measurable σ-content kernel P from Ω to Ω (Definitions IV.4.1.2, IV.4.12.2) which represents E_0 (see Exercise 1.3.3). Let $A_1, A_2, \ldots \in \mathscr{B}$ be $\mathscr{B}_0$-atoms such that $A_j \cap A_k = \varnothing$ $(j \neq k)$ and $\Omega_0 = A_1 + A_2 + \cdots$ is the $\mathscr{B}_0$-atomary part of $(\Omega, \mathscr{B}, m)$ (Proposition 1.9).

2.4.2.1. Prove that

$$h_{\mathscr{B}|\mathscr{B}_0}(\omega) = \begin{cases} -\sum_k 1_{A_k}(\omega) \log P(\omega, A_k) & (\omega \in \Omega_0) \\ \infty & (\text{otherwise}) \end{cases}$$

$$H(\mathscr{B}|\mathscr{B}_0) = \begin{cases} -\sum_k \int m(d\omega)P(\omega, A_k) \log P(\omega, A_k) & (m(\Omega \backslash \Omega_0) = 0) \\ \infty & (\text{otherwise}) \end{cases}$$

2.4.2.2. Prove that $H(\mathscr{B}|\mathscr{B}_0) < \infty$ implies the following: For m-a.e. $\omega \in \Omega$, the $A_1, A_2, \ldots$ are atoms of $(\Omega, \mathscr{B}, P(\omega, \cdot))$, $P(\omega, \Omega\backslash\Omega_0) = 0$ holds and $H(\mathscr{B})$, computed for $P(\omega, \cdot)$, is finite, the integral (for m) of this $\mathscr{B}_0$-measurable, m-a.e. finite function being $H(\mathscr{B}|\mathscr{B}_0)$ (for m).

2.4.2.3. Assume that $H(\mathscr{B})$, computed for $P(\omega, \cdot)$, is finite for m-a.e. ω and that the $\mathscr{B}_0$-measurable m-a.e. defined function thus obtained is in $\mathscr{L}_m^1$. Prove that $H(\mathscr{B}|\mathscr{B}_0)$ (for m) is finite and coincides with the m-integral of the mentioned function.

2.5. Theorem. *Let $(\Omega, \mathscr{B}, m)$ be a σ-probability space and $\mathscr{B}_0, \mathscr{B}_1, \mathscr{B}_2 \subseteq \mathscr{B}$ σ-fields. Then (we are writing $\mathscr{B}_j \vee \mathscr{B}_k$ for $\mathscr{B}(\mathscr{B}_j \cup \mathscr{B}_k)$):*

2.5.1. $h_{\mathscr{B}_2 \vee \mathscr{B}_1|\mathscr{B}_0} = h_{\mathscr{B}_2|\mathscr{B}_1 \vee \mathscr{B}_0} + h_{\mathscr{B}_1|\mathscr{B}_0}$ *holds* m-a.e.; *and*

2.5.2. $H(\mathscr{B}_2 \vee \mathscr{B}_1|\mathscr{B}_0) = H(\mathscr{B}_2|\mathscr{B}_1 \vee \mathscr{B}_2) + H(\mathscr{B}_1|\mathscr{B}_2)$.

In particular, we have:

2.5.3. $h_{\mathscr{B}_2 \vee \mathscr{B}_1} = h_{\mathscr{B}_2|\mathscr{B}_1} + h_{\mathscr{B}_1}$;

2.5.4. $H(\mathscr{B}_2 \vee \mathscr{B}_1) = H(\mathscr{B}_2|\mathscr{B}_1) + H(\mathscr{B}_1)$.

Proof. We begin by considering the special case $\mathscr{B}_0 \subseteq \mathscr{B}_1 \subseteq \mathscr{B}_2$. Let Ω_k denote the $\mathscr{B}_k$-atomary part of $(\Omega, \mathscr{B}_2, m)$ and $\Omega_{10} \in \mathscr{B}_1$ the $\mathscr{B}_0$-atomary part of $(\Omega, \mathscr{B}_1, m)$. By Theorem 1.17 we have $\Omega_0 = \Omega_1 \cap \Omega_{10}$, and Lemma 2.1 plus Theorem 1.17.2 imply $\rho_{\mathscr{B}_2|\mathscr{B}_0} = \rho_{\mathscr{B}_2|\mathscr{B}_1} \; \rho_{\mathscr{B}_1|\mathscr{B}_0}$. Now $-\log(xy) = -\log x - \log y$ holds even for $x, y \geqq 0$ if we put $-\log 0 = \infty$. Thus we get $h_{\mathscr{B}_2|\mathscr{B}_0} = h_{\mathscr{B}_2|\mathscr{B}_1} + h_{\mathscr{B}_1|\mathscr{B}_0}$. Integration yields 2, and $\mathscr{B}_0 = \{\varnothing, \Omega\}$ leads to 3 and 4 in our present special case. In the general case we make use of Remark 2.3.3 and write out, using what we have already proved,

$$\begin{aligned} h_{\mathscr{B}_2 \vee \mathscr{B}_1 | \mathscr{B}_0} &= h_{\mathscr{B}_2 \vee \mathscr{B}_1 \vee \mathscr{B}_0 | \mathscr{B}_0} \\ &= h_{\mathscr{B}_2 \vee \mathscr{B}_1 \vee \mathscr{B}_0 | \mathscr{B}_1 \vee \mathscr{B}_0} + h_{\mathscr{B}_1 \vee \mathscr{B}_0 | \mathscr{B}_0} \\ &= h_{\mathscr{B}_2 | \mathscr{B}_1 \vee \mathscr{B}_0} + h_{\mathscr{B}_1 | \mathscr{B}_0}. \end{aligned}$$

The theorem follows now also in the general case by integration resp. passage to the special case.

Our next proposition collects some important monotonicity properties of entropy densities and entropies.

2.6. Proposition. *Let* $(\Omega, \mathscr{B}, m)$ *be a* σ*-probability space and* $\mathscr{B}_0$, $\mathscr{B}_1$, $\mathscr{B}_2 \subseteq \mathscr{B}$ σ*-fields. Then (we write* $\mathscr{B}_j \vee \mathscr{B}_k$ *for* $\mathscr{B}(\mathscr{B}_j \cup \mathscr{B}_k)$*):*

2.6.1. $\mathscr{B}_1 \subseteq \mathscr{B}_2$ *implies:*

2.6.1.1. $h_{\mathscr{B}_2 | \mathscr{B}_0} \geqq h_{\mathscr{B}_1 | \mathscr{B}_0}$ *m-a.e.;*

2.6.1.2. $H(\mathscr{B}_2 | \mathscr{B}_0) \geqq H(\mathscr{B}_1 | \mathscr{B}_0)$;

2.6.1.3. $h_{\mathscr{B}_2} \geqq h_{\mathscr{B}_1}$ *m-a.e.;*

2.6.1.4. $H(\mathscr{B}_2) \geqq H(\mathscr{B}_1)$.

2.6.2. $\mathscr{B}_0 \subseteq \mathscr{B}_1$ *implies*

$$H(\mathscr{B}_2 | \mathscr{B}_1) \leqq H(\mathscr{B}_2 | \mathscr{B}_0) \leqq H(\mathscr{B}_2).$$

Proof. 1. Read Theorem 2.5.1 for the special case $\mathscr{B}_1 \subseteq \mathscr{B}_2$. We obtain $h_{\mathscr{B}_2 | \mathscr{B}_0} = h_{\mathscr{B}_2 | \mathscr{B}_1 \vee \mathscr{B}_0} + h_{\mathscr{B}_1 | \mathscr{B}_0}$, and this is $\geqq h_{\mathscr{B}_1 | \mathscr{B}_0}$ since conditional entropy densities are nonnegative. Integration resp. passage to the special case $\mathscr{B}_0 = \{\varnothing, \Omega\}$ yields the rest of 1.

2. The last inequality is a special case of the first, and the first inequality is nontrivial only for $H(\mathscr{B}_2 | \mathscr{B}_0) < \infty$. Let us therefore consider this case. But here we have a disjoint decomposition $\Omega = A_1 + A_2 + \cdots$ where $A_1, A_2, \ldots$ are $\mathscr{B}_0$-atoms of $\mathscr{B}_2 \vee \mathscr{B}_0$. It is easy to see that the A_k are also $\mathscr{B}_1$-atoms of $\mathscr{B}_2 \vee \mathscr{B}_1$. In fact they belong to $\mathscr{B}_2 \vee \mathscr{B}_0 \subseteq \mathscr{B}_2 \vee \mathscr{B}_1$. Furthermore, the system of all sets of the form $\sum_k A_k \cap F_k$ with $F_1, F_2, \ldots \in \mathscr{B}_1$ is a σ-field which is clearly contained in $\mathscr{B}_1 \vee \mathscr{B}_2$ and contains the σ-fields $\mathscr{B}_2 \vee \mathscr{B}_0$ (let the F_k vary over $\mathscr{B}_0$ and use the fact that the A_k are $\mathscr{B}_0$-atoms of $\mathscr{B}_2 \vee \mathscr{B}_0$) and hence $\mathscr{B}_2$, but also $\mathscr{B}_1$. In particular, for every $A_{k_0} \supseteq F \in \mathscr{B}_2$ there are $F_1, F_2, \ldots \in \mathscr{B}_1$ with $F = \sum_k A_k \cap F_k$, hence $A_{k_0} \cap F = A_{k_0} \cap F_{k_0}$, and thus A_k is a $\mathscr{B}_1$-atom of $\mathscr{B}_1 \vee \mathscr{B}_2$. Now we use the concavity of the function $-x \log x$ and Jensen's

inequality (Theorem III.5.13) in order to obtain (E_i denotes the conditional expectation for $\mathscr{B}_i$)

$$
\begin{aligned}
H(\mathscr{B}_2|\mathscr{B}_0) &= -\sum_k \int 1_{A_k} \log(E_0\, 1_{A_k})\, dm \\
&= -\sum_k \int E_0(1_{A_k} \log(E_0\, 1_{A_k}))\, dm = -\sum_k \int (E_0\, 1_{A_k}) \log(E_0\, 1_{A_k})\, dm \\
&= -\sum_k \int (E_0(E_1 1_{A_k})) \log(E_0(E_1 1_{A_k}))\, dm \\
&\geqq -\sum_k \int E_0[(E_1 1_{A_k}) \log(E_1 1_{A_k})]\, dm \\
&= -\sum_k \int (E_1 1_{A_k}) \log(E_1 1_{A_k})\, dm = H(\mathscr{B}_2|\mathscr{B}_1).
\end{aligned}
$$

2.7. Remark. The analogue of Proposition 2.6.2 for conditional entropy densities does not hold in general. We sketch an example. Let us distribute mass 1 over n^2 points according to a matrix $p = (p_{jk})_{j,\, k=1,\, \ldots,\, n}$:

$$
\begin{bmatrix} p_{11} & \cdots & p_{1n} \\ \vdots & & \vdots \\ p_{n1} & \cdots & p_{nn} \end{bmatrix}
$$

Assume that the first column consists of equal positive numbers: $p_{11} = \cdots = p_{n1} > 0$. Let $\mathscr{B}_0$ be the trivial σ-field and $\mathscr{B}_1$ correspond to the columns, $\mathscr{B}_2$ to the rows of the matrix. Then $h_{\mathscr{B}_2|\mathscr{B}_1} = \text{const} = \log n$ on the first column while $h_{\mathscr{B}_2}$, being constant on every row, attains values bigger and smaller than $\log n$ if the row sums are not all equal. The next proposition shows that the analogue of Proposition 2.6.2 holds for conditional entropy densities in a special case.

2.8. Proposition. *Let* $(\Omega, \mathscr{B}, m)$ *be a* σ*-probability space and* $\mathscr{B}_0 \subseteq \mathscr{B}_1 \subseteq \mathscr{B}_2 \subseteq \mathscr{B}$ σ*-fields. Then*

$$
h_{\mathscr{B}_2|\mathscr{B}_1} \leqq h_{\mathscr{B}_2|\mathscr{B}_0} \quad m\text{-a.e.}
$$

Proof. Here we have $h_{\mathscr{B}_2|\mathscr{B}_0} = h_{\mathscr{B}_2|\mathscr{B}_1} + h_{\mathscr{B}_1|\mathscr{B}_0} \geqq h_{\mathscr{B}_2|\mathscr{B}_1}$.

The objective of our next proposition is a minimal property of conditional entropy densities; this will be the basis of limiting theorems later.

2.9. Proposition. *Let* $(\Omega, \mathscr{B}, m)$ *be a* σ*-probability space and* $\mathscr{B}_0 \subseteq \mathscr{B}_1 \subseteq \mathscr{B}$ σ*-fields. Then:*

2.9.1. $E_0 f \geqq f e^{-h_{\mathscr{B}_1|\mathscr{B}_0}}$ *m-a.e.* $(0 \leqq f \in \mathscr{L}_m^1 \cap \text{mble}(\Omega, \mathscr{B}_1, \overline{\mathbb{R}}))$.

2.9.2. *If* $0 \leqq h \in \mathrm{mble}(\Omega, \mathscr{B}_1, \overline{\mathbb{R}})$ *and*

$$(1) \qquad E_0 f \geqq fe^{-h} \ m\text{-a.e.} \qquad (0 \leqq f \in \mathscr{L}_m^1 \cap \mathrm{mble}(\Omega, \mathscr{B}, \overline{\mathbb{R}}))$$

then $h \geqq h_{\mathscr{B}_1|\mathscr{B}_0}$ *m-a.e.*

In other words, $h_{\mathscr{B}_1|\mathscr{B}_0}$ *is the smallest among all measurable functions* $h \geqq 0$ *that satisfy* (1).

Proof. 1. Let A be any $\mathscr{B}_0$-atom of $(\Omega, \mathscr{B}_1, m)$. Then

$$0 \leqq f \in \mathscr{L}_m^1 \quad \Rightarrow \quad 1_A E_0 f \geqq 1_A E_0(f 1_A) = 1_A f E_0 1_A = 1_A f e^{-h_{\mathscr{B}_1|\mathscr{B}_0}} \ m\text{-a.e.}$$

It follows that the desired inequality holds m-a.e. on the $\mathscr{B}_0$-atomic part of $(\Omega, \mathscr{B}_1, m)$. But else it holds trivially because of $h_{\mathscr{B}_1|\mathscr{B}_0} = \infty$.

2. Let $0 \leqq h \in \mathrm{mble}(\Omega, \mathscr{B}_1, \overline{\mathbb{R}})$ satisfy (1). Then we choose any $\mathscr{B}_0$-atom A of $(\Omega, \mathscr{B}_1, m)$ and choose $f = 1_A$. Now we get $1_A E_0 f = 1_A E_0 1_A = e^{-h_{\mathscr{B}_1|\mathscr{B}_0}} 1_A$, and this is $\geqq e^{-h} 1_A$ m-a.e. by assumption, hence $h \geqq h_{\mathscr{B}_1|\mathscr{B}_0}$ m-a.e. on A follows, and now we get this inequality m-a.e. on the $\mathscr{B}_0$-atomary part Ω_0 of $(\Omega, \mathscr{B}_1, m)$. We have still to prove $h = \infty$ m-a.e. on $\Omega \backslash \Omega_0$. But $\Omega \backslash \Omega_0 \in \mathscr{B}_1$ does not contain any $\mathscr{B}_0$-atom for $(\Omega, \mathscr{B}_1, m)$ except m-nullsets; hence by Theorem 1.16 we may split $\Omega \backslash \Omega_0$ for every $n = 1, 2, \ldots,$ into n disjoint sets $F_1, \ldots, F_n \in \mathscr{B}_1$ such that $E_0 1_{F_k} = n^{-1} E_0(1_{\Omega \backslash \Omega_0})$. Choosing $f = 1_{F_k}$ in (1), we get

$$E_0 f = \frac{1}{n} E_0(1_{\Omega \backslash \Omega_0}) \geqq fe^{-h} = 1_{F_k} e^{-h},$$

and hence $h \geqq \log n$ m-a.e. on F_k, hence on $\Omega \backslash \Omega_0$. Since n is arbitrary, $h = \infty$ m-a.e. on $\Omega \backslash \Omega_0$ follows.

We now obtain the basic limit theorems.

2.10. Theorem. *Let* $(\Omega, \mathscr{B}, m)$ *be a* σ*-probability space and* $(\mathscr{B}_\iota)_{\iota \in I}$ *an increasingly filtered family of* σ*-fields* $\subseteq \mathscr{B}$. *Put* $\mathscr{B}_\infty = \mathscr{B}(\bigcup_{\iota \in I} \mathscr{B}_\iota)$ *and let* $\mathscr{B}_0$ *be any* σ*-field contained in all* $\mathscr{B}_\iota$ $(\iota \in I)$. *Then*

$$h_{\mathscr{B}|\mathscr{B}_0} = \sup_{\iota \in I} h_{\mathscr{B}_\iota|\mathscr{B}_0} \quad \mathrm{mod}\ m.$$

Proof. The right-hand member of this equality has, of course, to be understood in the sense of Exercise III.12.15. Now $h_{\mathscr{B}|\mathscr{B}_0} \geqq \sup_{\iota \in I} h_{\mathscr{B}_\iota|\mathscr{B}_0}$ is obvious from Proposition 2.6.1.1. In order to prove $\leqq$, we use Proposition 2.9 with $\mathscr{B}_\infty$ in the position of $\mathscr{B}_1$. We show that $h = \sup_{\iota \in I} h_{\mathscr{B}_\iota|\mathscr{B}_0}$ satisfies

$$0 \leqq f \in \mathscr{L}_m^1 \cap \mathrm{mble}(\Omega, \mathscr{B}_\infty, \overline{\mathbb{R}}) \quad \Rightarrow \quad E_0 f \geqq fe^{-h}.$$

We begin with indicator functions $f = 1_F$, $F \in \mathscr{B}_\infty$. Now it is an easy exercise to show that the set of all $F \in \mathscr{B}_\infty$ for which $f = 1_F$ satisfies the above conclusion is a σ-field containing $\bigcup_{\iota \in I} \mathscr{B}_\iota$. Thus it is $\mathscr{B}_\infty$ itself. Standard approximation techniques show that the above conclusion holds for arbitrary $0 \leqq f \in \mathscr{L}_m^1 \cap \text{mble}(\Omega, \mathscr{B}_\infty, \overline{\mathbb{R}})$. The desired inequality $\leqq$ now follows from Proposition 2.9.

2.11. Theorem. *Let $(\Omega, \mathscr{B}, m)$ be a σ-probability space and $(\mathscr{B}_\iota)_{\iota \in I}$ a decreasingly filtered family of σ-fields $\subseteq \mathscr{B}$ such that $\mathscr{B}_\iota$ is complete for m ($\iota \in I$; see Chapter* II, *Section* 6). *Let $\mathscr{B}_\infty = \bigcap_{\iota \in I} \mathscr{B}_\iota$. Let $\mathscr{B}_0 \subseteq \mathscr{B}$ be a σ-field with $\mathscr{B}_0 \subseteq \mathscr{B}_\iota$ $(\iota \in I)$. Then*

$$(2) \qquad h_{\mathscr{B}_\infty|\mathscr{B}_0} = \inf_{\iota \in I} h_{\mathscr{B}_\iota|\mathscr{B}_0} \ m\text{-a.e.} \qquad \text{on} \qquad \sup_{\iota \in I} \{h_{\mathscr{B}_\iota|\mathscr{B}_0} < \infty\} \quad \text{mod } m$$

Proof. The supremum of sets considered here has of course to be taken in the spirit of Exercise II.2.6.7 resp. III.12.15. Let denote E_ι generally the conditional expectation for $\mathscr{B}_\iota$. Fix any $\kappa \in I$ and a $\mathscr{B}_0$-atom A of $(\Omega, \mathscr{B}_\kappa, m)$. For every $\mathscr{B}_\iota \subseteq \mathscr{B}_\kappa$, put $A_\iota = \{E_\iota 1_A > 0\}$. The proof of Theorem 1.17 shows that A_ι is a $\mathscr{B}_0$-atom of $(\Omega, \mathscr{B}_\iota, m)$. By Proposition 1.5.1, A_ι is the mod m smallest set in $\mathscr{B}_\iota$ containing A, hence it is the bigger the smaller $\mathscr{B}_\iota$ is. In the spirit of Exercise III.12.15 the supremum $\overline{A}$ mod m of all those A_ι is well defined mod m and clearly contained in $\mathscr{B}_\infty$. Since it contains A mod m, we find $\overline{A} \supseteq \{E_\infty 1_A > 0\}$ mod m. But here equality holds since $\{E_\infty 1_A > 0\}$ is in every $\mathscr{B}_\iota$ in question and contains A, hence A_ι, mod m. Now obviously $\leqq$ holds in (2) by Proposition 2.6.1.1. On the other hand, the very definition of $h_{\mathscr{B}_\iota|\mathscr{B}_0}$ shows that $e^{-h_{\mathscr{B}_\iota|\mathscr{B}_0}} = E_0 1_{A_\iota}$ m-a.e. on A_ι, hence m-a.e. on A. Applying Exercise VIII.6.5.5 to the 1_{A_ι}, we find

$$\begin{aligned}\exp(-\inf_\iota h_{\mathscr{B}_\iota|\mathscr{B}_0}) &= \sup_\iota \exp(-h_{\mathscr{B}_\iota|\mathscr{B}_0}) = \sup_\iota E_0 1_{A_\iota} \\ &= E_0(\sup 1_{A_\iota}) = E_0 1_{\overline{A}} \geqq \exp(-h_{\mathscr{B}_\infty|\mathscr{B}_0}) E_0 1_{\overline{A}}\end{aligned}$$

m-a.e. on $\overline{A}$ by Proposition 2.9, hence $\geqq$ holds m-a.e. on $\overline{A} = \{E_\infty 1_A > 0\}$, hence in particular m-a.e. on A, and thus m-a.e. on the $\mathscr{B}_0$-atomary part of $(\Omega, \mathscr{B}_\kappa, m)$. Since $\kappa \in I$ was arbitrary, (2) follows in full.

2.12. Theorem. *Let $(\Omega, \mathscr{B}, m)$ be a σ-probability space and $(\mathscr{B}_\iota)_{\iota \in I}$ an increasingly filtered family of σ-fields $\subseteq \mathscr{B}$. Put $\mathscr{B}_\infty = \mathscr{B}(\bigcup_{\iota \in I} \mathscr{B}_\iota)$ and let $\mathscr{B}_1$ be any σ-field $\supseteq \mathscr{B}_\infty$. Then*

$$(3) \qquad h_{\mathscr{B}_1|\mathscr{B}_\infty} = \inf_{\iota \in I} h_{\mathscr{B}_1|\mathscr{B}_\iota} \quad m\text{-a.e.} \qquad \textit{on} \qquad \sup_{\iota \in I} \{h_{\mathscr{B}_1|\mathscr{B}_\iota} < \infty\} \quad \text{mod } m.$$

Proof. The last sup has to be taken in the spirit of Exercise II.2.6.7 resp. III.12.15, of course. Fix some $\kappa \in I$. It suffices to prove that the equality (3) holds m-a.e. on $\{h_{\mathscr{B}_1|\mathscr{B}_\kappa} < \infty\}$, and for this it is, by Proposition 2.8, clearly sufficient to treat the case $\mathscr{B}_\iota \supseteq \mathscr{B}_\kappa$ $(\iota \in I)$. But here we have $h_{\mathscr{B}_1|\mathscr{B}_\kappa} = h_{\mathscr{B}_1|\mathscr{B}_\iota} + h_{\mathscr{B}_\iota|\mathscr{B}_\kappa}$. Since all three functions in this equation are non-negative, they are all finite m-a.e. on $\{h_{\mathscr{B}_1|\mathscr{B}_\kappa} < \infty\}$. Now we get (3), from Theorem 2.10 (with $\mathscr{B}_\kappa$ in place of $\mathscr{B}_0$).

2.13. Theorem. *Let $(\Omega, \mathscr{B}, m)$ be a σ-probability space and $(\mathscr{B}_\iota)_{\iota \in I}$ a decreasingly filtered family of σ-fields $\subseteq \mathscr{B}$ such that $\mathscr{B}_\iota$ is complete for m $(\iota \in I$; see Chapter* II, *Section* 6). *Let $\mathscr{B}_\infty = \bigcap_{\iota \in I} \mathscr{B}_\iota$ and $\mathscr{B}_1 \subseteq \mathscr{B}$ be any σ-field containing all $\mathscr{B}_\iota$ $(\iota \in I)$, Then*

$$h_{\mathscr{B}_1|\mathscr{B}_\infty} = \sup_{\iota \in I} h_{\mathscr{B}_1|\mathscr{B}_\iota} \quad m\text{-a.e.} \quad \textit{on} \quad \sup_{\iota \in I} \{h_{\mathscr{B}_1|\mathscr{B}_\iota} < \infty\} \mod m.$$

Proof. Proceed as in the proof of Theorem 2.12, making use of Theorem 2.11.

2.14. Remark. Analyzing the proof of Theorem 2.11, we find that the assumption of completeness of all $\mathscr{B}_\iota$ can be replaced by the assumption that $\mathscr{B}_\iota$ contains all m-nullsets from $\mathscr{B}$ $(\iota \in I)$.

2.15. Exercises. Let $(\Omega, \mathscr{B}, m)$ be a σ-probability space.

2.15.1. Prove that

$$H(\mathscr{B}) = \sup\left\{-\sum_{k=1}^{n} m(A_k) \log m(A_k) \,\middle|\, n \geqq 1,\right.$$

$$\left. A_1, \ldots, A_n \in \mathscr{B}, A_j \cap A_k = \varnothing \; (j \neq k), \Omega = A_1 + \cdots + A_n\right\}$$

(where we put $0 \log 0 = 0$, of course).

2.15.2. Let $\mathscr{B}_0, \mathscr{B}_1 \subseteq \mathscr{B}$ be finite and $A_1, \ldots, A_r, C_1, \ldots, C_s \in \mathscr{B}$ be such that $\mathscr{B}_0$ is the system of all unions of some A_ρ, and $\mathscr{B}_1$ is the system of all unions of some C_σ. Show that

$$H(\mathscr{B}_1|\mathscr{B}_0) = -\sum_{m(A_\rho)>0} m(A_\rho) \sum_{\sigma=1}^{s} \frac{m(C_\sigma \cap A_\rho)}{m(A_\rho)} \log \frac{m(C_\sigma \cap A_\rho)}{m(A_\rho)},$$

(where we put $0 \log 0 = 0$ again, of course).

2.15.3. Let $\mathscr{B}_0, \mathscr{B}_1 \subseteq \mathscr{B}$ be arbitrary. Prove that

$$H(\mathscr{B}_1|\mathscr{B}_0) = \sup_{\mathscr{C} \subseteq \mathscr{B}_1 \text{ finite}} \left[\inf_{\mathscr{A} \subseteq \mathscr{B}_0 \text{ finite}} H(\mathscr{C}|\mathscr{A})\right].$$

2.15.4. Let $\mathscr{B}_0, \mathscr{B}_1 \subseteq \mathscr{B}$ be arbitrary and assume $H(\mathscr{B}_1|\mathscr{B}_0) < \infty$.

Prove that there are $A_1, A_2, \ldots \in \mathscr{B}$ such that $A_j \cap A_k = \varnothing$ $(j \neq k)$, $A_1 + A_2 + \cdots = \Omega$ and

$$H(\mathscr{B}_1|\mathscr{B}_0) = \int \left[-\sum_{k=1}^{\infty} E_0 1_{A_k} \log E_0 1_{A_k} \right] dm.$$

(*Hint*: Choose the A_k as $\mathscr{B}_0$-atoms of $\mathscr{B}_1 \vee \mathscr{B}_0$ and observe that

$$\int \rho_{\mathscr{B}_1|\mathscr{B}_0}\, dm = \int (E_0\, \rho_{\mathscr{B}_1|\mathscr{B}_0})\, dm.)$$

2.15.5. Let $\mathscr{B}_0 \subseteq \mathscr{B}$ be a σ-field and let E_0 denote the conditional expectation for $\mathscr{B}_0$. Assume that Ω is the $\mathscr{B}_0$-atomary part of $(\Omega, \mathscr{B}, m)$ and that $\zeta = \{A_1, A_2, \ldots\}$ is a partition of Ω into $\mathscr{B}_0$-atoms such that $E_0 1_{A_1} \geqq E_0 1_{A_2} \geqq \cdots$ m-a.e. (Theorem 1.12). Let $\mathscr{B}_1 = \mathscr{B}(\zeta)$. Prove:

$$H(\mathscr{B}|\mathscr{B}_0) < 1 \quad \Rightarrow \quad H(\mathscr{B}_1) \leqq 6\sqrt{H(\mathscr{B}|\mathscr{B}_0)}$$

(*Hints*: (1) Apply the general observation $p_1, p_2, \ldots \in \mathbb{R}_+$, $p_1 + p_2 + \cdots = 1$, $p_1 \geqq p_2 \geqq \cdots \Rightarrow p_n \leqq n^{-1}$ $(n = 1, 2, \ldots)$ to $E 1_{A_1} + E 1_{A_2} + \cdots = 1$ m-a.e. (2) Put $H(\mathscr{B}|\mathscr{B}_0) = \delta^2$ and deduce

$$\delta^2 \geqq \sum m(A_n) \log n, \qquad \delta^2 \geqq -m(A_1) \log m(A_1).$$

(3) Choose any $s \geqq 2$ and put

$$\delta_2 = -\sum_{n \geqq 2,\, m(A_n) \geqq 1/n^s} m(A_n) \log m(A_n).$$

Deduce $\delta_2 \leqq s\delta^2$. (4) Put

$$\delta_3 = -\sum_{n \geqq 2,\, m(A_n) < 1/n^s} m(A_n) \log m(A_n)$$

and deduce

$$\delta^3 \leqq -\sum_n \frac{1}{n^s} \log \frac{1}{n^s} \leqq s \int_1^{\infty} t^{-s} \log t\, dt + \frac{1}{s} \leqq \frac{5}{s}.$$

(5) Assume $\delta < 1$, specify $s = \sqrt{5}/\delta$ $(> 2!)$ and deduce

$$H(\mathscr{B}_1) \leqq \delta(\sqrt{5} + \sqrt{5} + 1) \leqq 6\delta.)$$

CHAPTER XV

CONVEX COMPACT SETS AND THEIR EXTREMAL POINTS

In this chapter we treat the theory of convex compact subsets of locally convex topological vector spaces. Our main goals are the theorems of Krein–Milman and Choquet on extremal points and barycentric representation. Although part of the theory could be carried out for complex vector spaces, we stick to real vector spaces throughout.

Section 1 presents what I think to be the shortest way from the definition of a topological vector space to the Krein–Milman theorem. The swiftness of exposition entails that arguments are packed together into one proof, and easy-to-prove facts are treated as exercises, which usually fill whole little theories. A reader who wishes to have a broader exposition is referred to the books of BOURBAKI [1], KÖTHE [1, 2], ROBERTSON and ROBERTSON [1, 2], and SCHAEFER [1].

In Section 2 Hervé's proof of Choquet's theorem on barycentric representation is presented, plus the theory of simplexes and uniqueness of barycentric representation. We restrict ourselves essentially to the case of a metrizable compact convex set. The theory has enormous ramifications today, including the nonmetrizable case. The reader is referred to CHOQUET [5], PHELPS [1], and BAUER [3].

The Exercises in Sections 1 and 2 are essential for understanding the whole theory.

In Section 3 two applications are given. The KREIN–MILMAN [1] theorem proves LIAPOUNOV's [1] convexity theorem (after LINDENSTRAUSS [1]), and Choquet's theorem yields ergodic decomposition. It would also yield Bochner's theorem on positive-definite functions, but this application requires too many techniques to be included here. The reader should enjoy reading CHOQUET [5].

1. LOCALLY CONVEX TOPOLOGICAL VECTOR SPACES

We shall deal with real vector spaces H in this section. Linear forms on H will frequently be used. The value of a linear form $x^*: H \to \mathbb{R}$ at $x \in H$ is denoted by $x^*(x)$ or $\langle x, x^* \rangle$. Our main purpose are to present the rudiments of the theory of topological vector spaces and to prove the *Krein–Milman theorem.*

1.1. Definition. *Let H be a real vector space and $\mathcal{T}$ a topology in H. The couple $(H, \mathcal{T})$ is called a* **topological vector space** *if:*

1.1.1. *the mapping $(x, y) \to x - y$ of $H \times H \to H$ is continuous (with the product topology $\mathcal{T} \times \mathcal{T}$ in $H \times H$);*

1.1.2. *the mapping $(\alpha, x) \to \alpha x$ of $\mathbb{R} \times H \to H$ is continuous (with the product topology in $\mathbb{R} \times H$);*

A topological vector space $(H, \mathcal{T})$ is called **separated** *if it is Hausdorff, and* **locally convex** *if there is a neighborhood base for $\mathcal{T}$ at every $x \in H$ that consists of convex sets.*

1.2. Remarks. Let $(H, \mathcal{T})$ be a topological vector space.

1.2.1. Clearly $(H, \mathcal{T})$ with addition is an abelian topological group. Thus every translation $x \to x + x_0$ of H by a vector $x_0 \in H$ is a homeomorphism of H onto itself; it sends convex sets into convex sets. It follows that translation by x_0 sends every neighborhood base at some $x \in H$ into a neighborhood base at $x + x_0$. In particular, every base $\mathcal{U}$ of neighborhoods of the zero vector $0 \in H$ leads to a base $x_0 + \mathcal{U} = \{x_0 + U \mid U \in \mathcal{U}\}$ of neighborhoods at x_0, and $(H, \mathcal{T})$ is locally convex iff there is a base of convex neighborhoods of $0 \in H$.

1.2.2. Let $I \neq \varnothing$ and $(H_\iota, \mathcal{T}_\iota)$ be a topological vector space ($\iota \in I$). Then $(\prod_{\iota \in I} H_\iota, \prod_{\iota \in I} \mathcal{T}_\iota)$ is a topological vector space (linear operations in $\prod_{\iota \in I} H$ being defined componentwise), separated if $(H_\iota, \mathcal{T}_\iota)$ is separated for every $\iota \in I$, and locally convex if $(H_\iota, \mathcal{T}_\iota)$ is locally convex for every $\iota \in I$.

1.3. Examples

1.3.1. Let $(H, \|\cdot\|)$ be a real normed vector space. Let $\mathcal{T} \subseteq \mathcal{P}(H)$ be the norm topology, i.e., $U \in \mathcal{T}$ iff for every $x \in U$, there is an $\varepsilon > 0$ such that $y \in H$, $\|x - y\| < \varepsilon \Rightarrow y \in U$. Then $(H, \mathcal{T})$ is a separated locally convex topological vector space. In particular, every Banach space falls under these conditions.

1.3.2. Let H be a real vector space and $I \neq \varnothing$, $(H_\iota, \mathcal{T}_\iota)$ a separated topological vector space ($\iota \in I$). For every $\iota \in I$, let $\varphi_\iota: H \to H_\iota$ be a linear mapping. Assume that the family $(\varphi_\iota)_{\iota \in I}$ separates the points of H; this is tantamount to the existence, for every $0 \neq x \in H$, of an $\iota \in I$ such that $\varphi_\iota(x) \neq 0 \in H_\iota$. Then the topology $\mathscr{T}$ generated by $(\varphi_\iota)_{\iota \in I}$, i.e., the coarsest (= smallest) topology $\mathscr{T}$ in H such that φ_ι is $\mathscr{T}$-$\mathscr{T}_\iota$-continuous (= $\mathscr{T}$-$\mathscr{T}_\iota$-measurable) turns H into a separated topological vector space $(H, \mathscr{T})$, and $(H, \mathscr{T})$ is locally convex if $(H_\iota, \mathscr{T}_\iota)$ $(\iota \in I)$ are all locally convex. We call $\mathscr{T}$ the *tôpology generated by the family* $(\varphi_\iota)_{\iota \in I}$. Since every set becomes a family when indexed by (the identity mapping of it onto) itself, we may as well speak of the *topology generated by a set of linear mappings* that separate the points of H, and get a separated topological vector space in this way.

1.3.3. Example 2 applies in particular if $H_\iota = \mathbb{R}$ $(\iota \in I)$, with the usual topology in $\mathbb{R}$, of course. The φ_ι are then linear forms on H. If $(H, \|\cdot\|)$ is a real normed vector space, then the norm continuous linear forms on H separate the points of H (Example I.6.2). The topology generated by them is called the weak or $\sigma(H, H^*)$ topology in H, and turns H into a separated locally convex topological vector space. Let, on the other hand, H^* be the set of all norm continuous linear forms on H. Recall that $\langle x, x^* \rangle$ denotes the value of the linear form $x^* \in H^*$ at the point $x \in H$. Then

$$\|x^*\| = \sup\{|\langle x, x^* \rangle| \,|\, x \in H, \|x\| \leqq 1\}$$

defines a norm in H^*, the dual norm of the original one; and it is an easy exercise in uniform convergence of linear forms on norm bounded sets to prove that $(H^*, \|\cdot\|)$ is a Banach space, hence a locally convex separated topological vector space with the norm topology. Every $x \in H$ induces a linear form $x^* \to \langle x, x^* \rangle$ on H. Trivially these linear forms separate the points of H^*. The topology generated in H^* by these linear forms is called the $\sigma(H^*, H)$ *or the* w^*-*topology* in H^*. With this topology, H^* becomes a separated locally convex topological vector space again. It is an easy exercise in ultrafilters (say) to prove that

$$E^* = \{x^* \,|\, x^* \in H^*, \|x\| \leqq 1\}$$

is a w^*-*compact* set (see Chapter XI, introduction).

1.4. Definition. *Let H be a real vector space.*

1.4.1. *A linear subspace $H_0 \neq H$ of H is called a* **hyperspace** *of H if $x \in H \backslash H_0 \Rightarrow \operatorname{lin}(H_0 \cup \{x\}) = H$.*

1.4.2. *Translates of hyperspaces of H are called* **hyperplanes.**

1.5. Proposition. *Let $(H, \mathscr{T})$ be a separated topological vector space. Then for a set $H_0 \subseteq H$, the following statements are equivalent:*

1.5.1. *H_0 is a closed hyperplane of H.*

1.5.2. *There is a continuous linear form $x^*: H \to \mathbb{R}$ and a real number α such that $x^* \neq 0$ and $H_0 = \{x | x \in H, \langle x, x^* \rangle = \alpha\}$.*

Proof. $2 \Rightarrow 1$ is nearly trivial.

$1 \Rightarrow 2$. It is clearly sufficient to settle the case $0 \in H_0$, i.e., H_0 is a hyperspace of H. Choose any $x_1 \in H \backslash H_0$. Since $H = \text{lin}(H_0 \cup \{x_1\})$, every $x \in H$ has a representation $x = \beta x_1 + x_0$ with $\beta \in \mathbb{R}$ and $x_0 \in H_0$. This representation is obviously unique, with β depending linearly on x. We may thus write $x = \beta(x)x_1 + x_0$, $x_0 \in H_0$, where β: $x \to \beta(x)$ defines a linear form $\neq 0$ on H. Clearly $H_0 = \{x | \beta(x) = 0\}$. We have still to show that β: $H \to \mathbb{R}$ is continuous. For this, a few preparations are in order. Multiplication with a constant $\gamma \neq 0$ is clearly a homeomorphism of H since it is continuous with the continuous inverse $x \to \gamma^{-1}x$. Thus V open, $\gamma \neq 0 \Rightarrow \gamma V$ open. Call a set $U \subseteq H$ *circled* if $|\lambda| \leqq 1 \Rightarrow \lambda U \subseteq U$. We show that every neighborhood W of $0 \in H$ contains a circled neighborhood. In fact we may, by the continuity of $(\beta, x) \to \beta x$, find an open neighborhood V of $0 \in H$ and a $\beta_0 > 0$ such that $|\beta| < \beta_0$, $x \in V \Rightarrow \beta x \in W$. The open set $U = \bigcup_{0 < |\beta| < \beta_0} \beta V$ is thus a neighborhood of $0 \in H$ contained in W, and clearly circled. Apparently β sends every circled $U \in H$ into an interval with center $0 \in \mathbb{R}$. Let us show that it has positive length if U is open. But this is clear because $\gamma\beta(x_1) = \beta(\gamma x_1)$ is $\neq 0$ for $\gamma \neq 0$, and $\gamma x_1 \in U$ for $|\gamma|$ sufficiently small. Let $\varepsilon > 0$ be arbitrary. Since H_0 is closed and $\varepsilon x_1 \notin H_0$, we may find a circled neighborhood U of $0 \in H$ such that $(\varepsilon x_1 + U) \cap H_0 = \varnothing$, or equivalently, $\varepsilon x_1 \notin H_0 + U$. It follows that $\varepsilon\beta(x_1) = \beta(\varepsilon x_1) \notin \beta(H_0 + U) = \beta(U)$, i.e., the symmetric neighborhood $\beta(U)$ of $0 \in \mathbb{R}$ does not contain $\varepsilon\beta(x_1)$, hence has a length $\leqq \varepsilon|\beta(x_1)|$. Since $\varepsilon > 0$ was arbitrary, the continuity of β at $0 \in H$, and thus on all of H, follows. Rewriting $\beta = x^*$, $\beta(x) = \langle x, x^* \rangle$ $(x \in H)$, our proposition follows.

1.6. Exercises. Let $(H, \mathscr{T})$ be a separated topological vector space. Prove:

1.6.1. The closure

1.6.1.1. of a linear subspace of H is a linear subspace of H;

1.6.1.2. of a convex subset of H is a convex subset of H;

1.6.1.3. of a linear manifold $\subseteq H$ is a linear manifold $\subseteq H$.

1.6.2. The set of all interior points of a convex set $\subseteq H$ is convex.

1.6.3. If $K \subseteq H$ is convex, x is an inner point of K, and y a point of the closure $\overline{K}$ of K, then $0 < \alpha \leqq 1 \Rightarrow \alpha x + (1 - \alpha)y$ is an inner point of K.

1.6.4. A hyperplane of H is either closed or dense in H. (*Hint*: Its closure is either a hyperplane or H.)

1.6.5. If two linear forms $x_1^*, x_2^* \neq 0$ on H and two constants α_1, α_2 are such that $\{x \mid \langle x, x_1^* \rangle = \alpha_1\} = \{x \mid \langle x, x_2^* \rangle = \alpha_2\}$, then there is a real $\gamma \neq 0$ such that $x_1^* = \gamma x_2^*$.

1.6.6. For every closed hyperplane $H_0 \subseteq H$, there are four sets H_1, H_2, $\overline{H}_1$, $\overline{H}_2 \subseteq H$ such that for every continuous linear form $x^*: H \to \mathbb{R}$ and every $\alpha \in \mathbb{R}$ such that $H_0 = \{x \mid \langle x, x^* \rangle = \alpha\}$, we have either

$$H_1 = \{x \mid \langle x, x^* \rangle < \alpha\}, \qquad H_2 = \{x \mid \langle x, x^* \rangle > \alpha\},$$
$$\overline{H}_1 = \{x \mid \langle x, x^* \rangle \leqq \alpha\}, \qquad \overline{H}_2 = \{x \mid \langle x, x^* \rangle \geqq \alpha\};$$

or

$$H_1 = \{x \mid \langle x, x^* \rangle > \alpha\}, \qquad H_2 = \{x \mid \langle x, x^* \rangle < \alpha\},$$
$$\overline{H}_1 = \{x \mid \langle x, x^* \rangle \geqq \alpha\}, \qquad \overline{H}_2 = \{x \mid \langle x, x^* \rangle \leqq \alpha\}.$$

H_1 and H_2 are called the open half-spaces, and $\overline{H}_1$, $\overline{H}_2$ are called the closed half-spaces associated with H_0. Prove that $\overline{H}_i$ is in fact the closure of H_i $(i = 1, 2)$.

1.7. Definition. *Let $(H, \mathscr{T})$ be a separated topological vector space. Two sets A, $B \subseteq H$ are said to be:*

1.7.1. **separated** *by the closed hyperplane H_0 of H if, for the closed half-spaces $\overline{H}_1$, $\overline{H}_2$ associated with H_0, either $A \subseteq \overline{H}_1$, $B \subseteq \overline{H}_2$ or $A \subseteq \overline{H}_2$, $B \subseteq \overline{H}_1$;*

1.7.2. **strictly separated** *by the closed hyperplane H_0 of H if, for the open half-spaces H_1, H_2 associated with H_0, either $A \subseteq H_1$, $B \subseteq H_2$ or $A \subseteq H_2$, $B \subseteq H_1$.*

1.8. Remarks. Let $(H, \mathscr{T})$ be a separated topological vector space.

1.8.1. Clearly Definition 1.7 rests heavily on Exercise 1.6.6.

1.8.2. Let A, $B \subseteq H$. Clearly A and B are

1.8.2.1. separated by some closed hyperplane of H iff there is a continuous linear form $x^*: H \to \mathbb{R}$ and a constant $\alpha \in \mathbb{R}$ such that $x \in A \Rightarrow \langle x, x^* \rangle \leqq \alpha$, and $x \in B \Rightarrow \langle x, x^* \rangle \geqq \alpha$, or, equivalently,

$$\sup_{x \in A} \langle x, x^* \rangle \leqq \inf_{x \in B} \langle x, x^* \rangle$$

for some continuous linear form x^* on H;

1.8.2.2. strictly separated by some closed hyperplane of H iff there exists a continuous linear form x^* on H and a real α such that $x \in A \Rightarrow \langle x, x^* \rangle < \alpha$, and $x \in B \Rightarrow \langle x, x^* \rangle > \alpha$.

1.9. Theorem. *Let $(H, \mathscr{T})$ be a separated topological vector space. Let A, $B \subseteq H$ be nonempty convex sets such that A has at least one interior* *point and B does not contain any interior point of A. Then A and B can be separated by a closed hyperplane of H.*

Proof. We proceed in several steps.

1. Let A_0 denote the set of all interior points of A. By Exercise 1.6.2, A_0 is convex. By assumption we have $A_0 \neq 0$, $A_0 \cap B = \varnothing$. Let $\Sigma = \{(A', B') | A_0 \subseteq A', B \subseteq B', A', B' \text{ convex}, A' \cap B' = \varnothing\}$. For (A', B'), (A'', B'') define $(A', B') \prec (A'', B'')$ by $A' \subseteq A''$, $B' \subseteq B''$. It is nearly obvious that $(\Sigma, \prec)$ satisfies the hypothesis of Zorn's lemma, hence there is some maximal (A', B') in Σ. From the maximality of (A', B') we deduce $A' \cup B' = H$. In fact, assume that there is some $x_0 \in H \backslash (A' \cup B')$; we show that either $\text{conv}(A' \cup \{x_0\}) \cap B' = \varnothing$ or $A' \cap \text{conv}(B' \cup \{x_0\}) = \varnothing$, which leads to a contradiction to the maximality of (A', B'); now if, say, $\text{conv}(A' \cup \{x_0\}) \cap B' \ni x'$ and $A' \cap \text{conv}(B' \cup \{x_0\}) \ni y'$, then there are $a \in A, b \in B, 0 < \alpha, \beta < 1$ such that $x' = \alpha x_0 + (1 - \alpha)a$, $y' = \beta x_0 + (1 - \beta)b$. We calculate

$$\begin{aligned} \beta x' &= \alpha\beta x_0 + (1-\alpha)\beta a, \quad \alpha y' = \alpha\beta x_0 + \alpha(1-\beta)b \\ &\Rightarrow \quad \beta x' - \alpha y' = (1-\alpha)\beta a - \alpha(1-\beta)b \\ &\Rightarrow \quad \beta x' + \alpha(1-\beta)b = \alpha y' + (1-\alpha)\beta a. \end{aligned}$$

Clearly $\beta + \alpha(1 - \beta) = \alpha + (1 - \alpha)\beta > 0$. Find $\gamma > 0$ with

$$\gamma\beta + \gamma\alpha(1-\beta) = \gamma\alpha + \gamma(1-\alpha)\beta = 1.$$

Then $\gamma\beta x' + \gamma\alpha(1 - \beta)b = \gamma\alpha y' + \gamma(1 - \alpha)\beta a$ is both a convex combination of x', $b \in B'$ and of y', $a \in A'$, hence in $A' \cap B'$, a contradiction. Thus we have $A' \cup B' = H$.

2. We now want to construct a closed hyperplane H_0 separating A and B as the intersection $H_0 = \bar{A}' \cap \bar{B}'$. By Exercise 1.6.1.2 this H_0 is a closed convex set. Actually, it is the set of all boundary points of $\bar{A}'$ as well as the set of all boundary points of $H \backslash A' = B'$. From this we deduce $H_0 \neq 0$: Choose any $a \in A'$, $b \in B'$; then some $\alpha a + (1 - \alpha)b$ with $0 \leqq \alpha \leqq 1$ is a boundary point of A' and B', hence in H_0 (we shall apply this argument once again below).

Now $H_0 \neq \varnothing$ implies that we may (by translation) assume $0 \in H_0$. We still have to prove that H_0 is a hyperspace of H. We begin by proving that H_0 is a linear subspace of H. Since H_0 is convex and $0 \in H_0$, it is certainly sufficient to prove $x \in H_0 \Rightarrow \lambda x \in H_0$ for $\lambda < 0$ and $\lambda > 1$ (the case $0 \leqq \lambda \leqq 1$ being settled by convexity). Now if $\lambda > 1$, $\lambda x \notin H_0$, then λx is, say, in the interior of A'. But now Exercise 1.6.3 shows that $x = (1/\lambda)(\lambda x)$ is in the interior of A', a contradiction. If $\lambda < 0$ and λx is in the interior of, say, A', then

$$0 = \frac{1}{1-\lambda}\lambda x - \frac{\lambda}{1-\lambda}x$$

and Exercise 1.6.3 show that 0 is in the interior of A', again a contradiction. Thus we have shown that H_0 is a linear subspace of H. Choose now any $x_1 \in H\backslash H_0$. In order to show $H = \text{lin}(H_0 \cup \{x_1\})$ we first observe that x_1 is in the interior of, say, A'. Now it follows that $-x_1$ is in the interior of B' because otherwise $0 = \frac{1}{2}(x_1 + (-x_1))$ would be in the interior of A', and not in H_0. Let now $x \in H\backslash H_0$ be arbitrary. If $x - x_1 \in H_0$, we are through. If $x - x_1 \notin H_0$, then $x - x_1$ is in the interior of, say, A'. By a previous argument, we may find $0 \leqq \alpha \leqq 1$ with

$$\alpha(x - x_1) + (1 - \alpha)(-x_1) = x_0 \in H_0.$$

Actually, $x - x_1, -x_1 \notin H_0$ shows $0 < \alpha < 1$. But now we get

$$x = \left(1 + \frac{1-\alpha}{\alpha}\right)x_1 + \frac{1}{\alpha}x_0 \in \text{lin}(H_0 \cup \{x_1\}).$$

Thus $H_0 = \bar{A}' \cap \bar{B}'$ is a closed hyperplane of H.

3. Now we apply Proposition 1.5 and get a continuous linear form $x^* \neq 0$ on H such that $H_0 = \{x \mid \langle x, x^*\rangle = 0\}$ (recall that we have, by a translation, passed to the assumption $0 \in H_0$). If x is an interior point of A, $\langle x, x^*\rangle \neq 0$ follows. Multiplying x^* by -1 if necessary, we may assume $\langle x, x^*\rangle < 0$. It follows that $y \in A' \Rightarrow \langle y, x^*\rangle \leqq 0$ because $\langle y, x^*\rangle > 0$ would allow us to find some $0 < \alpha < 1$ with $\langle \alpha x + (1-\alpha)y, x^*\rangle = \alpha\langle x, x^*\rangle + (1-\alpha)\langle y, x^*\rangle = 0$, i.e., $\alpha x + (1-\alpha)y \in H_0$; on the other hand, Exercise 1.6.3 shows that $\alpha x + (1-\alpha)y$ is an interior point of A', hence not in H_0, and we get a contradiction. By a similar argument we see that $x \in B' \Rightarrow \langle x, x^*\rangle \geqq 0$. The theorem is proved.

1.10. Exercises

1.10.1. Let H be a real vector space.

1.10.1.1. Let $A, B \subseteq H$ be nonempty convex sets, $A \cap B = \varnothing$. Prove that there is a linear form x^* on H and a real α such that $x \in A \Rightarrow \langle x, x^*\rangle \leqq \alpha$, and $x \in B \Rightarrow \langle x, x^*\rangle \geqq \alpha$. (*Hint*: Copy the proof of Theorem 1.9 in a purely algebraic fashion.)

1.10.1.2. Determine x^*, α in the special case $H = \mathbb{R}^2$,

$$A = \{(\xi, \eta) \mid \xi, \eta \in \mathbb{R}, \xi < 0\},$$
$$B = \{(\xi, \eta) \mid 0 < \xi, \eta \in \mathbb{R}, \xi\eta > 1\}.$$

1.10.1.3. Prove Theorem I.6.1 (Hahn–Banach) once again, using 1.1.

1.10.2. Let $(H, \mathscr{T})$ be a separated topological vector space.

1.10.2.1. Let $p\colon H \to \mathbb{R}$ be a continuous majorant function (Theorem I.6.1) and H_0 a linear subspace, $\beta\colon H_0 \to \mathbb{R}$ a linear mapping such that $x \in H_0 \Rightarrow \beta(x) \leqq p(x)$. Prove that there is a continuous linear form x^* on H such that

$$x \in H_0 \quad \Rightarrow \quad \langle x, x^* \rangle = \beta(x) \qquad \text{and} \qquad x \in H \quad \Rightarrow \quad \langle x, x^* \rangle \leqq p(x).$$

(*Hint*: Apply Theorem 1.9 to the open convex set $\{x \mid p(x) < 1\}$ and a suitable linear manifold on which β equals 1 (unless $\beta \equiv 0$, and $x^* = 0$ is a solution).)

1.10.2.2. Let $A, B \subseteq H$ be open, nonempty, convex, and disjoint. Prove that A and B are strictly separated by some closed hyperplane. (*Hint*: Apply Theorem 1.9 to the closures $\bar{A}$, $\bar{B}$.)

1.10.2.3. Let $A, B \subseteq H$ be nonempty, convex, and compact. Prove that $\operatorname{conv}(A \cup B) = \{\alpha x + (1 - \alpha)y \mid 0 \leqq \alpha \leqq 1, x \in A, y \in B\}$ and that this set is convex and compact.

1.10.3. Let $(H, \mathscr{T})$ be a separated locally convex topological vector space.

1.10.3.1. Let $\varnothing \neq A \subseteq H$ be closed and convex, and $x_0 \in H \backslash A$. Prove that there is a continuous linear form x^* on H such that

$$\langle x_0, x^* \rangle < \inf_{x \in A} \langle x, x^* \rangle.$$

In particular, A and $\{x_0\}$ can be strictly separated by a closed hyperplane. (*Hint*: Apply Theorem 1.9 to A and an open convex neighborhood U of x_0 such that $A \cap U = \varnothing$.)

1.10.3.2. Let $\varnothing \neq A, B \subseteq H$ be closed, convex, and disjoint. Let A be compact. Prove that there is a continuous linear form x^* on H such that $\sup_{x \in A} \langle x, x^* \rangle < \inf_{x \in B} \langle x, x^* \rangle$. In particular, A and B can be strictly separated by a closed hyperplane. (*Hint*: Apply Exercise 3.1 to the closed convex set $A - B = \{a - b \mid a \in A, b \in B\}$ and the point $x_0 = 0$.)

We now proceed to the study of extremal points.

1.11. Definition. *Let H be a real vector space and $\varnothing \neq K \subseteq H$ be a convex set.*

1.11.1. *A point $x_0 \in K$ is called an* **extremal point** *of K if $K \backslash \{x_0\}$ is still convex, or equivalently, if a representation*

$$x_0 = \alpha x + (1 - \alpha)y, \qquad 0 < \alpha < 1, \quad x, y \in K, \quad x \neq y$$

is impossible. The set of all extremal points of K is denoted by $\operatorname{ext}(K)$.

1.11.2. *A convex subset $W \neq \emptyset$ of K is called a* **face** *of K if*

$$x_0 \in W,\ x, y \in K,\ 0 < \alpha < 1,\ x_0 = \alpha x + (1 - \alpha)y \quad \Rightarrow \quad x, y \in W.$$

1.12. Remark. Clearly extremal points are nothing but one-point faces. Every convex set is a face of itself.

1.13. Theorem (Krein–Milman). *Let $(H, \mathscr{T})$ be a separated locally convex topological vector space and $\emptyset \neq K \subseteq H$ a convex compact set. Then $K = \overline{\mathrm{conv}(\mathrm{ext}(K))}$, i.e., K is the closure of the convex hull of the set of all extremal points of K. In particular* $\mathrm{ext}(K)$ *is nonempty.*

Proof. Let W be a face of K (e.g., $W = K$) and $\Sigma(W)$ be the system of all closed (and hence compact) faces $\subseteq W$ of K. It is obvious that the intersection of the members of a totally ordered (by $\subseteq$) subsystem of $\Sigma(W)$ is a face of K again. Thus Zorn's lemma applies, and we obtain a face $W_0 \subseteq W$ of K not containing a proper subset that is still a face of K. Let us show that W_0 consists of a single point. Assume W_0 contains two different points x_1, x_2. Apply Exercise 1.10.3.1 and obtain a continuous linear form x^* on H such that $\langle x_1, x^* \rangle < \langle x_2, x^* \rangle$. Let $\alpha = \inf_{x \in W_0} \langle x, x^* \rangle$. Since W_0 is compact, the set

$$W_0' = \{x \mid x \in W_0, \langle x, x^* \rangle = \alpha\}$$

is nonempty. It is clearly a compact convex subset of W_0, hence of W and hence of K. Since $x_2 \notin W_0'$, we get $W_0' \neq W_0$. If we can prove that W_0' is a face of K, then we get a contradiction, and W_0 consists of one point only. Now let $y_0 \in W_0'$, $y_1, y_2 \in K$, $0 < \beta < 1$, $y_0 = \beta y_1 + (1 - \beta)y_2$. Since W_0 is a face of K, $y_1, y_2 \in W_0$ follows. Clearly

$$\alpha = \langle y_0, x^* \rangle = \beta\langle y_1, x^* \rangle + (1 - \beta)\langle y_2, x^* \rangle, \qquad \alpha \leqq \langle y_1, x^* \rangle, \langle y_2, x^* \rangle.$$

We conclude $\alpha = \langle y_1, x^* \rangle = \langle y_2, x^* \rangle$, hence $y_1, y_2 \in W_0'$. We have proved that every face of K contains at least one extremal point of K, and we have seen how to get a face of a compact convex set by intersection with a suitable hyperplane. Form now the convex compact set $K_0 = \overline{\mathrm{conv}(\mathrm{ext}(K))}$. It is clearly nonempty and contained in K. Assume, it does not equal K. Then there is a $x_0 \in K \backslash K_0$. Apply Exercise 1.10.3.1 and get a continuous linear form x^* on H such that $\langle x_0, x^* \rangle < \inf_{x \in K_0} \langle x, x^* \rangle$. Put $\alpha = \inf_{x \in K} \langle x, x^* \rangle$. Clearly

$$W = \{x \mid x \in K, \langle x, x^* \rangle = \alpha\}$$

is a face of K (use our previous reasoning) and disjoint from K_0. On the other hand, W contains an extremal point of K, hence cannot be disjoint from K_0. We are at a contradiction, and $K_0 = K$ follows.

1.14. Exercises. Determine all extremal points of each of the following sets:

1.14.1. an open disk in the plane;

1.14.2. a closed disk in the plane;

1.14.3. a convex polygon in the plane;

1.14.4. the convex hull of the union of the following sets in $\mathbb{R}^3$: $\{(0, 0, 1)\}$, $\{(0, 0, -1)\}$, $\{(\zeta, \eta, 0) | \zeta, \eta \in \mathbb{R}, (\zeta - 1)^2 + \eta^2 = 1\}$.

2. BARYCENTERS

The purpose of this section is to present the rudiments of *Choquet's theory of barycentric representation.* His main theorem will be presented for the *metric case* only. The definition of a *simplex* and the theorem about *uniqueness* of barycentric representation are included as well.

We remind the reader that for a compact space X every positive measure $m: \mathscr{C}(X, \mathbb{R}) \to \mathbb{R}$ is σ- and even τ-continuous (Theorem I.10.1). Recall that weak convergence for measures m means the convergence of all integrals $m(f) = \int f \, dm$ $(f \in \mathscr{C}(X, \mathbb{R}))$. The set of all probability measures $m: \mathscr{C}(X, \mathbb{R}) \to \mathbb{R}$ is convex and weakly compact (Chapter XI, introduction, Section 1, Exercise XI.2.8.6).

2.1. Definition. *Let $(H, \mathscr{T})$ be a locally convex separated topological vector space.*

2.1.1. *A function $f \in \mathbb{R}^H$ is said to be* **affine** *if there is a linear form x^* on H and a real α such that $f(x) = \langle x, x^* \rangle + \alpha$ $(x \in H)$.*

2.1.2. *Let $\varnothing \neq X \subseteq H$. Restrictions to X of affine functions on H are called* **affine functions on** X. *The set of all affine functions on X is denoted by* aff$(X, \mathbb{R})$ *or* aff(X). *The set of all continuous affine functions on X is denoted by $\mathscr{C}^{\mathrm{aff}}(X, \mathbb{R})$ or $\mathscr{C}^{\mathrm{aff}}(X)$.*

2.2. Proposition. *Let $(H, \mathscr{T})$ be a locally convex separated topological vector space and $\varnothing \neq X \subseteq H$ be convex and compact. Then for every probability measure $m: \mathscr{C}(X, \mathbb{R}) \to \mathbb{R}$, there is a unique $x = x(m) \in X$ such that*

$$f \in \mathscr{C}^{\mathrm{aff}}(X, \mathbb{R}) \quad \Rightarrow \quad f(x(m)) = \int f(x) \, m(dx).$$

We call $x(m)$ the **barycenter** *of m. For any probability measures m, m' on X and any $0 \leqq \alpha \leqq 1$,*

$$x(\alpha m + (1 - \alpha)m') = \alpha x(m) + (1 - \alpha)x(m').$$

Proof. Let $\mathscr{P}$ denote the set of all probability measures on X, $\mathscr{P}_0$ the set of all $m \in P$ for which the existence statement of our proposition is true, and

$$\mathscr{P}_1 = \left\{ \sum_{k=1}^{n} \alpha_k \delta_{x_k} \middle| n \geqq 1, \alpha_1, \ldots, \alpha_n \geqq 0, \alpha_1 + \cdots + \alpha_n = 1, x_1, \ldots, x_n \in X \right\}.$$

It is an easy exercise to show that $\mathscr{P}_0$ is convex and weakly closed. (*Hint*: Work with ultrafilters and recall that weak convergence of measures m means convergence of all

$$m(f) = \int f\, dm \qquad (f \in \mathscr{C}^{\mathrm{b}}(X, \mathbb{R}) = \mathscr{C}(X, \mathbb{R}) \supseteq \mathscr{C}^{\mathrm{aff}}(X, \mathbb{R})).$$

Clearly $\mathscr{P}_1 \subseteq \mathscr{P}_0$:

$$x\left(\sum_{k=1}^{n} \alpha_k \delta_{x_k} \right) = \sum_{k=1}^{n} \alpha_k x_k .$$

By Proposition XI.1.5, $\mathscr{P}_1$ is weakly dense in $\mathscr{P}$, hence so is $\mathscr{P}_0$, and $\mathscr{P}_0 = \mathscr{P}$ follows, proving the existence statement of our proposition. We are obviously through with the uniqueness statement as soon as we know that $\mathscr{C}^{\mathrm{aff}}(X, \mathbb{R})$ separates the points of X. But Exercise 1.10.3.1 shows that the continuous linear forms separate the points of H. This does it for the uniqueness. The last statement of our proposition is an easy exercise which we leave to the reader.

2.3. Definition. *Let $(H, \mathscr{T})$ be a real vector space and $\varnothing \neq X \subseteq H$ convex. A function $f \in \mathbb{R}^X$ is said to be:*

2.3.1. **convex** *if*

$$x, y \in X, 0 \leqq \alpha \leqq 1 \quad \Rightarrow \quad f(\alpha x + (1 - \alpha)y) \leqq \alpha f(x) + (1 - \alpha) f(y);$$

2.3.2. **strictly convex** *if $x, y \in X$, $x \neq y$, $0 < \alpha < 1$ imply*

$$f(\alpha x + (1 - \alpha)y) < \alpha f(x) + (1 - \alpha) f(y);$$

2.3.3. **concave** *if the function $-f$ is convex.*

2.4. Proposition. *Let $(H, \mathscr{T})$ be a locally convex separated topological vector space, $\varnothing \neq X \subseteq H$ convex and compact, and $f \in \mathscr{C}(X, \mathbb{R})$. For every $x \in X$, let $\mathscr{P}_x$ denote the set of all probability measures m: $\mathscr{C}(X, \mathbb{R}) \rightarrow \mathbb{R}$ such that the barycenter $x(m)$ of m is x. Then:*

2.4.1. *$\mathscr{P}_x$ is convex and compact $(x \in X)$.*

2.4.2. *For every $x \in X$, there is some $m_x \in \mathscr{P}_x$ (in general not unique) such that*

$$\int f\,dm = \sup\left\{\int f\,dm \,\middle|\, m \in \mathscr{P}_x\right\}.$$

The function $\bar{f} \in \mathbb{R}^X$ defined by $\bar{f}(x) = \int f\,dm_x$ is bounded, concave, and upper semicontinuous. We call f the **concave upper envelope** *of f.*

Proof. 1. The set $\mathscr{P}$ of all probability measures $m\colon \mathscr{C}(X, \mathbb{R}) \to \mathbb{R}$ is convex and weakly compact.

$$\mathscr{P}_x = \bigcap_{f \in \mathscr{C}^{\mathrm{aff}}(X)} \left\{m \,\middle|\, m \in \mathscr{P},\ f(x) = \int f\,dm\right\}$$

immediately shows that $\mathscr{P}_x$ is convex and compact.

2. For every $x \in X$, the mapping $m \to \int f\,dm$ of $\mathscr{P}_x$ into $\mathbb{R}$ is continuous for the weak topology in $\mathscr{P}_x$. Thus $\left\{\int f\,dm \,\middle|\, m \in \mathscr{P}_x\right\}$ is a compact interval contained between any constant bounds for f, proving the existence of an m_x with the desired property, and the boundedness of $\bar{f}$. In order to show that $\bar{f}$ is concave, choose any x_0, x, $y \in X$, $0 \leqq \alpha \leqq 1$ such that $x_0 = \alpha x + (1 - \alpha)y$. Then

$$\begin{aligned}\alpha\bar{f}(x) + (1 - \alpha)\bar{f}(y) &= \alpha \int f\,dm_x + (1 - \alpha)\int f\,dm_y \\ &= \int f\,d(\alpha m_x + (1 - \alpha)m_y) \leq \bar{f}(x_0)\end{aligned}$$

since the barycenter of $\alpha m_x + (1 - \alpha)m_y$ is $\alpha x + (1 - \alpha)y = x_0$ by Proposition 2.2. In order to show that $\bar{f}$ is upper semicontinuous, choose any ultrafilter $\mathscr{F}$ in X converging to a certain $x_0 \in X$. Let

$$m(g) = \lim_{\mathscr{F}} \int g\,dm_x \qquad (g \in \mathscr{C}(X, \mathbb{R})).$$

Clearly $m\colon \mathscr{C}(X, \mathbb{R}) \to \mathbb{R}$ is a probability measure. For every $g \in \mathscr{C}^{\mathrm{aff}}(X, \mathbb{R})$, we have

$$\int g\,dm = \lim_{\mathscr{F}} \int g\,dm_x = \lim_{\mathscr{F}} g(x) = g(x_0),$$

hence $m \in \mathscr{P}_{x_0}$ follows. We find now

$$\lim_{\mathscr{F}} f(x) = \lim_{\mathscr{F}} \int f \, dm_x = \int f \, dm \leqq \bar{f}(x_0),$$

proving the upper semicontinuity of f.

2.5. Lemma. *Let $(H, \mathscr{T})$ be a locally convex separated topological vector space, $\varnothing \neq X \in H$ convex and compact, $g \in \mathbb{R}^X$ bounded, concave, and upper semicontinuous. Let $x_0 \in X$ be arbitrary, and m: $\mathscr{C}(X, \mathbb{R}) \to \mathbb{R}$ be a probability measure with barycenter x_0. Then*

$$g(x_0) \geqq \int g \, dm.$$

Proof. We mention beforehand that $\int g \, dm$ is meaningful since m is a τ-measure and g a bounded τ-lower function. In the locally convex separated topological vector space $H \times \mathbb{R}$ (with the obvious product topology, see Example 1.2.2) we consider the set $A = \{(x, \xi) | x \in X, \xi \leqq g(x)\}$ which is clearly convex since X is convex and g is concave, and closed since X is closed and g is bounded and upper semicontinuous (exercise). Fix any $\xi_0 > g(x_0)$ and separate the point $(x_0, \xi_0) \notin A$ from the convex compact set A by a closed hyperplane. More precisely, apply Exercise 1.10.3.2 and find a continuous linear form β: $H \times \mathbb{R} \to \mathbb{R}$ such that $\sup\{\beta(x, \xi) | (x, \xi) \in A\} < \beta(x_0, \xi_0)$. Clearly $\beta \neq 0$. It is an easy exercise to find a continuous linear form x^* on H (namely, the restriction of β to $H \times \{0\}$, up to a trivial identification) and a real γ such that

$$\beta(x, \xi) = \langle x, x^* \rangle + \gamma\xi \qquad (x \in H, \quad \xi \in \mathbb{R}).$$

Since $\langle x_0, x^* \rangle + \gamma\xi < \langle x_0, x^* \rangle + \gamma\xi_0$ for all $\xi \leqq g(x_0) < \xi_0$, we find $\gamma > 0$, and clearly may assume $\gamma = 1$ henceforth. The function $h \in \mathbb{R}^X$ defined by $h(x) = -\langle x, x^* \rangle + \langle x_0, x^* \rangle + \xi_0$ clearly belongs to $\mathscr{C}^{\mathrm{aff}}(X)$ and satisfies $g(x) \leqq h(x)$ since this inequality means nothing but $\beta(x, g(x)) \leqq \beta(x_0, \xi_0)$ $(x \in X)$. We now obtain, for every probability measure m with barycenter x_0,

$$\xi_0 = h(x_0) = \int h \, dm \geqq \int g \, dm.$$

Since $\xi_0 > g(x_0)$ was arbitrary, $g(x_0) \geqq \int g \, dm$ follows.

2.6. Lemma. *Let $(H, \mathscr{T})$ be a locally convex separated topological vector space and $\varnothing \neq X \subseteq H$ convex and compact. Assume that the topology $\mathscr{T} \cap X = \{U \cap X | U \in \mathscr{T}\}$ in X is metrizable. Then there is a continuous strictly convex real function on X.*

Proof. Since $(X, \mathcal{T} \cap X)$ is compact metrizable, there is a countable subset in $\mathscr{C}(X, \mathbb{R})$ that is dense for uniform approximation. Let $\mathscr{A}$ be the algebra generated by $\mathscr{C}^{\text{aff}}(X)$. Since $\mathscr{C}^{\text{aff}}(X)$ separates the points of X (Exercise 1.10.3.2), $\mathscr{A}$ is dense in $\mathscr{C}(X, \mathbb{R})$. Consequently, there is a countable subset $\mathscr{C}$ of $\mathscr{A}$ that is still dense in $\mathscr{C}(X, \mathbb{R})$. Taking the affine functions used for representations of members of $\mathscr{C}$ in polynomial form, we end up with the sequence $f_1, f_2, \ldots \in \mathscr{C}^{\text{aff}}(X)$ separating the points of X. Clearly we may assume $0 \leqq f_n \leqq 1/2^n$ $(n = 1, 2, \ldots)$. The function $h = f_1^2 + f_2^2 + \cdots$ belongs to $\mathscr{C}(X, \mathbb{R})$ by uniform convergence. Let $x, y \in X$, $x \neq y$, $0 < \alpha < 1$. Find n with $f_n(x) \neq f_n(y)$. Then

$$\begin{aligned} \alpha f_k^2(x) + (1-\alpha) f_k^2(y) &\geqq (\alpha f_k(x) + (1-\alpha) f_k(y))^2 \\ &= f_k(\alpha x + (1-\alpha) y)^2 \qquad (k = 1, 2, \ldots), \end{aligned}$$

and $>$ holds here if $f_k(x) \neq f_k(y)$, in particular for $k = n$. Summing up, we find $\alpha h(x) + (1-\alpha) h(y) > h(\alpha x + (1-\alpha) y)$, i.e., h is strictly convex.

2.7. Theorem (Choquet). *Let $(H, \mathcal{T})$ be a locally convex separated topological vector space and $\varnothing \neq X \subseteq H$ a compact convex set such that the topology $\mathcal{T} \cap X$ in X is* **metrizable.** *Then the set $\text{ext}(X)$ of all extremal points of X is a G_δ set for $\mathcal{T} \cap X$, hence belongs to $\mathscr{B}(\mathcal{T} \cap X)$. For every $x_0 \in X$, there is at least one probability measure $m\colon \mathscr{C}(X, \mathbb{R}) \to \mathbb{R}$ such that the σ-content m derived from m satisfies $m(X \backslash \text{ext}(X)) = 0$ and x_0 is the barycenter of m.*

Proof. By Lemma 2.6 there is a strictly convex $h \in \mathscr{C}(X, \mathbb{R})$. Let $\bar{h}$ be the concave upper envelope of h. Then $\text{ext}(X) = \{x \mid x \in X, \bar{h}(x) = h(x)\}$. In fact $x \in X \backslash \text{ext}(X)$ implies the existence of $y, z \in X$, $y \neq z$, and $0 < \alpha < 1$ such that $x = \alpha y + (1-\alpha) z$. But then $m = \alpha \delta_y + (1-\alpha)\delta_z$ has barycenter x and satisfies $\bar{h}(x) \geqq \int h\, dm = \alpha h(y) + (1-\alpha) h(z) > h(x)$ by strict convexity of h. Let, conversely, $\bar{h}(x) > h(x)$. Then there is some probability measure m with barycenter x and $\int h\, dm > h(x)$. The set S of all $y \in X$ such that $m(U) > 0$ (we pass to the derived σ-content freely) for every open neighborhood U of y (in $\mathcal{T} \cap X$) is easily found to be a closed subset of X (exercise). If it consisted of only one point, we would have $m = \delta_x$, contradicting $\int h\, dm > h(x)$. Thus there are at least two points y, z in S, and we may find convex neighborhoods $y \in U$, $z \in V$ such that $\bar{U} \cap \bar{V} = \varnothing$. Let H_1 and H_2 the two open half-spaces belonging to a closed hyperplane H_0 separating $\bar{U}$ and $\bar{V}$ (Exercise 1.10.3.2), and assume $y \in H_1$, $z \in H_2$ (Exercise 1.6.6). Since $U \subseteq H_1$, $V \subseteq H_2$, we have

$$m(H_1 \cap X) > 0 < m(H_2 \cap X).$$

Observe that, for a continuous linear form x^* on H and a real α such that $H_0 = \{x \mid \langle x, x^* \rangle = \alpha\}$, neighboring values α' of α still yield separation of $\bar{U}$ and $\bar{V}$ by $\{x \mid \langle x, x^* \rangle = \alpha'\}$. These sets, when intersected with X, are all in $\mathscr{B}(\mathscr{T} \cap X)$, and disjoint for different values of α'. Since there are uncountably many, there are m-nullsets among them, hence we may assume $m(H_0 \cap X) = 0$. Define now $\alpha = m(H_1 \cap X)$, get $0 < \alpha < 1$ and $1 - \alpha = m(H_2 \cap X)$, and define two σ-probability contents m_1, m_2 on $\mathscr{B}(\mathscr{T} \cap X)$ by

$$m_1(E) = \frac{1}{\alpha} m(E \cap H_1), \qquad m_2(E) = \frac{1}{1 - \alpha} m(E \cap H_2).$$

Let $x_1 = x(m_1)$, $x_2 = x(m_2)$. For the affine function $f\colon x \to \langle x, x^* \rangle$ on X, we have $f < \alpha$ m_1-a.e., hence $f(x_1) = \int f \, dm_1 < \alpha$. Similarly we get $\alpha < f(x_2)$. We conclude $x_1 \neq x_2$. On the other hand, $g \in \mathscr{C}^{\text{aff}}(X) \Rightarrow$

$$g(\alpha x_1 + (1 - \alpha)x_2) = \alpha g(x_1) + (1 - \alpha)g(x_2) = \alpha \int g \, dm_1 + (1 - \alpha) \int g \, dm_2$$

$$= \int g \, d(\alpha m_1 + (1 - \alpha)m_2) = \int g \, dm,$$

and thus $\alpha x_1 + (1 - \alpha)x_2 = x(m) = x$. Thus we have proved $x \in X \backslash \text{ext}(X)$ and finished the proof of $\text{ext}(X) = \{x \mid x \in X, \bar{h}(x) = h(x)\}$. Now $\bar{h} - h$ is $\geqq 0$ and upper semicontinuous (Proposition 2.4). $\text{ext}(X) = \bigcap_{n=1}^{\infty} \{x \mid x \in X, \bar{h}(x) - h(x) < 1/n\}$ thus shows that $\text{ext}(X)$ is a G_δ set in X.

Fix now any $x_0 \in X$. Apply Proposition 2.4 and find a probability measure $m\colon \mathscr{C}(X, \mathbb{R}) \to \mathbb{R}$ with $\bar{h}(x_0) = \int h \, dm$. Lemma 2.5 yields $\bar{h}(x_0) \geqq \int \bar{h} \, dm$, and this is $\geqq \int h \, dm$ since $\bar{h} \geqq h$. We conclude $\int (\bar{h} - h) \, dm = 0$. The set $X \backslash \text{ext}(X) = \{x \mid x \in X, \bar{h}(x) > h(x)\}$ is therefore an m-nullset. This finishes the proof.

2.8. Exercises

2.8.1. Determine all probability measures m with barycenter x_0 and $m(X \backslash \text{ext}(X)) = 0$ in the following cases:

2.8.1.1. $H = \mathbb{R}^2$ (usual topology), $x_0 = (0, 0)$,

$$X = \{(\xi, \eta) \mid \xi, \eta \in \mathbb{R}, \xi^2 + \eta^2 \leqq 1\} \qquad \text{or} \qquad X = \{(\xi, \eta) \mid -1 \leqq \xi, \eta \leqq 1\}.$$

2.8.1.2. $H = \mathbb{R}^3$ (usual topology), X as in Exercise 1.14.4, $x_0 = (0, 0, 0)$.

2.8.2. Let $H = \mathbb{R}^2$ (usual topology) and $X = \{(\xi, \eta) \mid \xi, \eta \in \mathbb{R}, |\xi| + \eta^2 \leqq 1\}$. Prove that X is convex and compact. Prove that $f\colon (\xi, \eta) \to \eta^2$ defines a convex function $f \in \mathscr{C}(X, \mathbb{R})$ such that the concave upper envelope $\bar{f}$ of f is not affine.

2.8.3. Let $(H, \mathscr{T})$ be a locally convex separated topological vector space and $\varnothing \neq X \subseteq H$ be convex and compact. Analyze the proof of Theorem 2.7 and prove:

2.8.3.1. For any $x, y \in X$, $x \neq y$, there is an $f \in \mathscr{C}^{\text{aff}}(X)$ with $f(x) \neq f(y)$.

2.8.3.2. For any $x \in X \backslash \text{ext}(X)$, there is a convex $f \in \mathscr{C}(X, \mathbb{R})$ with $\bar{f}(x) > f(x)$ (again $\bar{f}$ denotes the concave upper envelope of f).

2.8.3.3. For every $f \in \mathscr{C}(X, \mathbb{R})$, put $E_f = \{x \mid x \in X, f(x) = \bar{f}(x)\}$. Prove that $f, g \in \mathscr{C}(X, \mathbb{R})$ convex $\Rightarrow E_{f+g} \subseteq E_f \cap E_g$. (*Hint*: Use Lemma 2.5.)

2.8.3.4. With the E_f of 3.3, $\text{ext}(X) = \bigcap_{f \in \mathscr{C}(X, \mathbb{R}) \text{ convex}} E_f$.

2.8.3.5. For every $x \in X$ and every $f \in \mathscr{C}(X, \mathbb{R})$, there are $y, z \in X$, $0 \leqq \alpha \leqq 1$ such that $\alpha y + (1 - \alpha)z = x$, $\alpha f(y) + (1 - \alpha) f(z) = \bar{f}(x)$.

2.8.3.6. $f, g \in \mathscr{C}(X, \mathbb{R}) \Rightarrow \overline{f + g} \leqq \bar{f} + \bar{g}$.

2.8.3.7. For probability measures m, m': $\mathscr{C}(X, \mathbb{R}) \to \mathbb{R}$ define $m \prec m'$ by $\int f\, dm \leqq \int f'\, dm'$ ($f \in \mathscr{C}(X, \mathbb{R})$ convex). Then $\prec$ is a partial ordering and $m \prec m'$ implies that m and m' have the same barycenter.

2.8.3.8. Call a probability measure m: $\mathscr{C}(X, \mathbb{R}) \to \mathbb{R}$ *maximal* if it is maximal for the partial ordering $\prec$. For every m, there exists a maximal $m' \succ m$. (*Hint*: Apply Zorn's lemma to the weakly compact set $\{m' \mid m \prec m'\}$.)

2.8.3.9. A probability measure m: $\mathscr{C}(X, \mathbb{R}) \to \mathbb{R}$ is maximal iff $m(E_f) = 1$ ($f \in \mathscr{C}(X, \mathbb{R})$ convex) iff $\int f\, dm = \int \bar{f}\, dm$ ($f \in \mathscr{C}(X, \mathbb{R})$ convex). (*Hint*: For the "only if" part take a maximal m and use Theorem I.6.1 (Hahn–Banach) with the majorant function $\int \bar{f}\, dm$ ($f \in \mathscr{C}(X, \mathbb{R})$) in order to get an $m' \succ m$ with $\int f\, dm = \int \bar{f}\, dm$ for any specified convex $f \in \mathscr{C}(X, \mathbb{R})$; for the "if" part take an m with $\int f\, dm = \int \bar{f}\, dm$ ($f \in \mathscr{C}(X, \mathbb{R})$ convex) and majorize it (for $\succ$) by a maximal m' (3.8); represent every $\bar{f}$ as a pointwise infimum of continuous concave functions (namely, finite infima of continuous affine functions; see the proof of Lemma 3.5) and get $\int \bar{f}\, dm' \leqq \int \bar{f}\, dm$ ($f \in \mathscr{C}(X, \mathbb{R})$); use this for a proof of $m' = m$.)

2.8.3.10. For every $x \in X$, there is a probability measure m: $\mathscr{C}(X, \mathbb{R}) \to \mathbb{R}$ with barycenter x and $\int f\, dm = \int \bar{f}\, dm$ ($f \in \mathscr{C}(X, \mathbb{R})$ convex), namely a maximal majorant of δ_x.

We have seen in the exercises that there can be many σ-probability measures living on the external points of a compact convex set and having the same barycenter. We now tackle the problem of uniqueness. In finite dimension the answer is clear: We certainly have uniqueness if our compact convex set is a simplex. Everyone who is familiar with barycentric calculus in linear algebra knows how to define the notion of a simplex and to prove the mentioned statement.

We thus aim at a suitable generalization of the notion of a simplex to our in general infinite-dimensional case.

2.9. Definition. *Let H be a real vector space. A convex set $C \subseteq H$ is called a* **convex cone with tip x_9** $\in H$ *if* $x \in C, 0 \leqq \alpha \in \mathbb{R} \Rightarrow x_0 + \alpha(x - x_0) \in C$.

2.10. Exercises Let H be a real vector space.

2.10.1. Prove that for every convex set $0 \neq X \subseteq H$ and every $x_0 \in H$ the set $\{x_0 + \alpha(x - x_0) | x \in X, 0 \leqq \alpha \in R\}$ is a convex cone with tip x_0. It is called the cone with tip x_0, *generated* by X.

2.10.2. For every convex cone $C \subseteq H$ with tip x_0, and for any $x, y \in H$, write $x \leqq_C y$ iff $x_0 + (y - x) \in C$. Prove that $(H, \leqq_C)$ is a partially ordered vector space.

2.10.3. Call two convex cones $C_1, C_2 \subseteq H$, with tips x_1, x_2, *isomorphic* if there is a bijection $\varphi: C_1 \to C_2$ such that

$$\varphi(x_1 + \alpha(x - x_1) + \beta(y - x_1)) = x_2 + \alpha(\varphi(x) - x_2) + \beta(\varphi(y) - x_2)$$

$(x, y \in C_1, 0 \leqq \alpha, \beta \in \mathbb{R})$. Prove that for isomorphic convex cones C_1, $C_2 \subseteq H$ the partial orderings $\leqq_{C_1}$ and $\leqq_{C_2}$ are related by

$$x, y \in C_1 \quad \Rightarrow \quad [x \leqq_{C_1} y \Leftrightarrow \varphi(x) \leqq_{C_2} \varphi(y)].$$

2.10.4. Let $0 \neq X \subseteq H$ be convex and $x_1, x_2 \in H$, $H_1, H_2 \subseteq H$ hyperplanes such that $x_1 \notin H_1$, $x_2 \notin H_2$, $X \subseteq H_1 \cap H_2$. Prove that the cones with tips x_1 resp. x_2, generated by X, are isomorphic.

2.10.5. Call the convex cone $C \subseteq H$ a *lattice cone* if $(C, \leqq_C)$ is a lattice. Prove that of two isomorphic cones either both or none is a lattice cone.

2.10.6. Prove that a convex cone $C \subseteq H$ with tip x_0 is a lattice cone iff

$$x, y \in C \quad \Rightarrow \quad (C + x - x_0) \cap (C + y - x_0)$$

is isomorphic to C.

2.10.7. Let in particular $H = \mathbb{R}^2$. Prove that every convex cone $C \subseteq H$ is either $= H$ or a lattice cone.

2.10.8. Let in particular $H = \mathbb{R}^3$. Prove that $C_1 = \{(\xi, \eta, \zeta) | 0 \leqq \xi, \eta, \zeta \in \mathbb{R}\}$ is a lattice cone; $C_2 = \{(\xi, \eta, \zeta) | \xi, \eta, \zeta \in \mathbb{R}, \xi \geqq 0, \eta^2 + \zeta^2 \leqq \xi^2\}$ is a convex cone but not a lattice cone.

2.11. Definition. *Let H be a real vector space. A convex nonempty subset X of H is called a* **simplex** *if there is a real vector space H', a hyperplane H'_0 of H', and an $x'_0 \in H'$ such that $x'_0 \notin H'_0$ and:*

2.11.1. *H is a linear subspace of H'.*

2.11.2. $X \subseteq H'_0$.

2.11.3. *The convex cone with tip x'_0 generated by X is a lattice cone.*

2.12. Remarks. Let H be a real vector space.

2.12.1. Exercises 2.10.4–6 show that a convex $0 \neq X \subseteq H$ is a simplex iff for any choice of a real vector space H', a hyperplane H'_0 in H', and an $x'_0 \notin H'_0$ such H is a linear subspace of H' and $X \subseteq H'_0$, the convex cone with tip x'_0 generated by X is a lattice cone.

2.12.2. It is nearly obvious that the usual simplexes in $\mathbb{R}^n$ satisfy the above definition of a simplex. It is not so easy to prove that no other subsets of $\mathbb{R}^n$ are simplexes in the sense of the above definition.

2.13. Exercises

2.13.1. Let H be a real vector space and $\varnothing \neq C \subseteq H$ a convex lattice cone with tip $0 \in H$. Prove: If $x_1, \ldots, x_r, y_1, \ldots, y_s \in C$, $x_1 + \cdots + x_r = y_1 + \cdots + y_s$, then there are z_{jk} $(j = 1, \ldots, r;\ k = 1, \ldots, s)$ in C such that

$$x_j = \sum_{k=1}^{s} z_{jk}, \qquad y_k = \sum_{j=1}^{r} z_{jk} \qquad (j = 1, \ldots, r; \quad k = 1, \ldots, s).$$

(*Hint*: For $r = s = 2$, choose, exploiting the lattice property of C, $z_{11} = x_1 \wedge y_1$, $z_{12} = x_1 - z_{11}$, $z_{21} = y_1 - z_{11}$, $z_{22} = y_2 - z_{11}$; pass to the general case by induction, first over r, then over s.)

2.13.2. Prove that every polynomial $\subseteq \mathbb{R}^{\mathbb{R}}$ can be written as a difference of two convex functions. (*Hint*: Even powers are convex, odd powers are differences of two convex functions.)

2.13.3. Let $(H, \mathcal{T})$ be a locally convex separated topological vector space, $\varnothing \neq K \subseteq H$ convex and compact. Prove:

2.13.3.1. $\mathscr{C}^{\text{aff}}(X)$ separates the points of X.

2.13.3.2. If $h \in \mathbb{R}^{\mathbb{R}}$ is convex, then $f \in \mathscr{C}^{\text{aff}}(X) \Rightarrow h(f)$ is a continuous convex function on X.

2.13.3.3. The set $\{f - g \mid f, g \in \mathscr{C}(X, \mathbb{R}), f, g \text{ convex}\}$ is dense in $\mathscr{C}(X, \mathbb{R})$ for uniform approximation. (*Hint*: Use Exercises 2.13.2 and 2.13.3.1,2 in order to show that the set in question is a subalgebra of $\mathscr{C}(X, \mathbb{R})$ separating the points of X; apply the Weierstrass–Stone theorem.)

2.14. Proposition. *Let $(H, \mathcal{T})$ be a locally convex separated topological vector space and $\varnothing \neq X \subseteq H$ a convex compact simplex. Then:*

2.14.1. *For every convex $f \in \mathscr{C}(X, \mathbb{R})$ the concave upper hull $\bar{f}$ of f is* *affine. Actually, $\bar{f}$ is the pointwise infimum of a decreasingly filtered system of affine continuous functions on X.*

2.14.2. *If f, $g \in \mathscr{C}(X, \mathbb{R})$ are convex, and $0 \leqq \alpha, \beta \in \mathbb{R}$, then $\overline{\alpha f + \beta g} = \alpha\bar{f} + \beta\bar{g}$.*

Proof. Embedding H into $H \times \mathbb{R}$ and imposing a translation, if necessary, we may assume that there is a closed hyperplane containing X but not $0 \in H$, and that the cone $\{\lambda x | 0 \leqq \lambda \in \mathbb{R},\ x \in X\}$ with tip $0 \in H$ generated by X is a lattice cone. For every $0 \neq y \in C$, there is a unique $x \in X$ and a unique $\lambda > 0$ with $y = \lambda x$. Every $f \in \mathbb{R}^X$ has a unique extension—again denoted by f—to C such that $f(\beta y) = \beta f(y)$ $(0 \leqq \beta \in \mathbb{R},$ $y \in C)$. We shall work freely with these notations. They enable us to abstain from writing out multiplicative constants.

1. We know already that $\bar{f}$ is concave (Proposition 2.4.2). In order to show that $\bar{f}$ is affine, it now clearly suffices to prove

$$x_1, x_2 \in X \quad \Rightarrow \quad \bar{f}(x_1 + x_2) \leqq \bar{f}(x_1) + \bar{f}(x_2).$$

Upon applying Exercise 2.8.3.5 we may find, for any given $\varepsilon > 0$, y_1, $y_2 \in C$ with $y_1 + y_2 = x_1 + x_2$ and $f(y_1) + f(y_2) \geqq \bar{f}(x_1 + x_2) - \varepsilon$. Exercise 2.13.1 (we need only the case $r = s = 2$ there) yields $z_{11}, z_{12}, z_{21}, z_{22} \in C$ such that $x_1 = z_{11} + z_{12}$, $x_2 = z_{21} + z_{22}$, $y_1 = z_{11} + z_{21}$, $y_2 = z_{12} + z_{22}$. But then we have, exploiting the convexity of f,

$$f(y_1) + f(y_2) \leqq f(z_{11}) + f(z_{12}) + f(z_{21}) + f(z_{22}) \leqq \bar{f}(x_1) + \bar{f}(x_2).$$

Let $\mathscr{F} = \{g | g \in \mathscr{C}^{\text{aff}}(X), \inf_{x \in X} [g(x) - \bar{f}(x)] > 0\}$. The proof of Lemma 2.5 shows that $\bar{f}(x) = \inf\{g(x) | g \in \mathscr{F}\}$ (add small positive constants to the affine functions constructed there in order to bring them to $\mathscr{F}$). We prove now that $\mathscr{F}$ is decreasingly filtered. In the locally convex separated topological vector space $H \times \mathbb{R}$ (with product topology) consider the set $A = \{(x, \xi) | x \in X,\ \xi \leqq \bar{f}(x)\}$. Since X is convex and compact, and $\bar{f}$ is concave and upper semicontinuous (Proposition 2.4.2), A is closed. Since $\bar{f}$ is bounded (as is f), $\mathscr{F}$ contains some constants at least. Let now g, $h \in \mathscr{F}$ and $\alpha \in \mathbb{R}$ be such that $g, h \leqq \alpha$ on X. Consider the sets

$$B_g = \{(x, \xi) | x \in X, g(x) \leqq \xi \leqq \alpha\}, \qquad B_h = \{(x, \xi) | x \in X, h(x) \leqq \xi \leqq \alpha\}.$$

They are both convex and compact. Let $\varepsilon > 0$ be such that $\bar{f} + \varepsilon \leqq g, h$. Such an ε exists by the definition of $\mathscr{F}$. Since $\bar{f}$ is affine, we have x, $y \in X$, $0 \leqq \alpha \leqq 1$

$$\begin{aligned} \Rightarrow \bar{f}(\alpha x + (1 - \alpha)y) + \varepsilon &= \alpha[\bar{f}(x) + \varepsilon] + (1 - \alpha)[\bar{f}(y) + \varepsilon] \\ &\leqq \alpha g(x) + (1 - \alpha)h(y), \end{aligned}$$

and this immediately translates itself into $A \cap [\text{conv}(B_g \cup B_h)] = \varnothing$. But

Exercise 1.10.2.3 shows that $\operatorname{conv}(B_g \cup B_h)$ is compact. Clearly $\operatorname{conv}(B_g \cup B_h)$ is convex and compact and disjoint from A. Apply Exercise 1.10.3.2 to find a continuous linear form β: $H \times \mathbb{R} \to \mathbb{R}$ such that

$$\sup\{\beta(x, \xi) \mid (x, \xi) \in A\} < \inf\{\beta(x, \xi) \mid (x, \xi) \in \operatorname{conv}(B_g \cup B_h).$$

Let $\delta > 0$ be the difference of these two reals. Find a continuous linear form x^*: $H \to \mathbb{R}$ and a constant $\gamma \in \mathbb{R}$ with $\beta(x, \xi) = \langle x, x^* \rangle + \gamma\xi$ $(x \in H, \xi \in \mathbb{R})$. Since $x_0 \in X$, $\xi \leqq \bar{f}(x_0)$

$$\Rightarrow \quad \xi < g(x_0), \langle x_0, x^* \rangle + \gamma\xi < \langle x_0, x^* \rangle + \gamma g(x_0),$$

we conclude $\gamma > 0$, hence we may and shall assume $\gamma = 1$. Define now l: $X \to \mathbb{R}$ by

$$l(x) = -\langle x, x^* \rangle + \inf\{\beta(x, \xi) \mid (x, \xi) \in \operatorname{conv}(B_g \cup B_h)\}.$$

Clearly $l \in \mathscr{C}^{\mathrm{aff}}(X)$. In addition we have $x \in X \Rightarrow$

$$l(x) \leqq -\langle x, x^* \rangle + \beta(x, g(x)) = -\langle x, x^* \rangle + \langle x, x^* \rangle + g(x) = g(x),$$

and similarly $l \leqq h$ follows. On the other hand

$$\begin{aligned} x \in X \quad \Rightarrow \quad \bar{f}(x) &= -\langle x, x^* \rangle + \beta(x, \bar{f}(x)) \\ &\leqq -\langle x, x^* \rangle + \inf\{\beta(x, \xi) \mid (x, \xi) \in \operatorname{conv}(B_g \cup B_h)\} - \delta \\ &= l(x) - \delta, \end{aligned}$$

and thus $\inf_{x \in X} [l(x) - \bar{f}(x) \geqq \delta > 0$ follows, i.e., $l \in \mathscr{F}$.

2. Let $x \in X$ and m: $\mathscr{C}(X, \mathbb{R}) \to \mathbb{R}$ be a probability measure with barycenter x. We conclude

$$\int (\alpha f + \beta g)\, dm = \alpha \int f\, dm + \beta \int g\, dm \leqq \alpha\bar{f}(x) + \beta\bar{g}(x).$$

Varying m, we find $(\overline{\alpha f + \beta g})(x) \leqq \alpha\bar{f}(x) + \beta\bar{g}(x)$, hence $\overline{\alpha f + \beta g} \leqq \alpha\bar{f} + \beta\bar{g}$ follows (this is essentially Exercise 2.8.3.6). Let us now prove the reverse inequality. Since $\alpha\bar{f} = \overline{\alpha f}$ $(\alpha \geqq 0)$ is obvious, it suffices to prove $\overline{f + g} \geqq \bar{f} + \bar{g}$. Choose some $\varepsilon > 0$ and apply Exercise 2.8.3.5 in order to get, for a fixed $x \in X$, points $x_1, x_2, y_1, y_2 \in C$ such that (we make use of extensions to C as mentioned at the beginning of the proof) $x_1 + x_2 = x = y_1 + y_2, f(x_1) + f(x_2) > \bar{f}(x) - \varepsilon, g(y_1) + g(y_2) > \bar{g}(x) - \varepsilon$. Apply Exercise 2.13.1 and get $z_{11}, z_{12}, z_{21}, z_{22} \in C$ with $z_{11} + z_{12} = x_1$, $z_{21} + z_{22} = x_2$, $z_{11} + z_{21} = y_1$, $z_{12} + z_{22} = y_2$. From the convexity of f and g we deduce

$$\begin{aligned} \bar{f}(x) + \bar{g}(x) - 2\varepsilon &< f(x_1) + f(x_2) + g(y_1) + g(y_2) \\ &\leqq f(z_{11}) + f(z_{12}) + f(z_{21}) + f(z_{22}) \\ &\quad + g(z_{11}) + g(z_{21}) + g(z_{12}) + g(z_{22}) \\ &\leqq (\overline{f + g})(z_{11} + \cdots + z_{22}) = (\overline{f + g})(x). \end{aligned}$$

Since $\varepsilon > 0$ was arbitrary, we get $\bar{f}(x) + \bar{g}(x) \leqq (\overline{f + g})(x)$, and $\overline{f + g} \geqq \bar{f} + \bar{g}$ follows.

2.15. Theorem. *Let $(H, \mathcal{T})$ be a locally convex separated topological vector space and $\varnothing \neq X \subseteq H$ a convex compact simplex. Then for every $x \in X$, there is at most one probability measure m: $\mathscr{C}(X, \mathbb{R}) \to \mathbb{R}$ with barycenter x such that $\int f\, dm = \int \bar{f}\, dm$ for every convex $f \in \mathscr{C}(X, \mathbb{R})$ (whose concave upper envelope is denoted by $\bar{f}$).*

Proof. Let $x \in X$ and m, m': $\mathscr{C}(X, \mathbb{R}) \to \mathbb{R}$ be two probability measures with the stated properties. By Theorem I.10.1, m and m' are τ-measures. For a given convex $f \in \mathscr{C}(X, \mathbb{R})$, find a decreasingly filtered system $\mathscr{F} \subseteq \mathscr{C}^{\mathrm{aff}}(X)$ such that $\bar{f}(y) = \inf_{u \in \mathscr{F}} u(y)$ $(y \in X)$.

We conclude that $\bar{f}$ is a τ-lower function for the elementary domain $\mathscr{C}(X, \mathbb{R})$, hence τ-integrable for m and m', with

$$\int f\, dm = \int \bar{f}\, dm = \inf_{u \in \mathscr{F}} \int u\, dm = \inf_{u \in \mathscr{F}} u(x) = \bar{f}(x) = \int f\, dm'.$$

It follows that m and m' coincide on all continuous convex functions on X, hence on all difference of such functions, hence, by Exercise 2.13.3.3, on $\mathscr{C}(X, \mathbb{R})$, which means $m = m'$.

2.16. Corollary. *Let $(H, \mathcal{T})$ be a locally convex separated topological vector space and $\varnothing \neq X \subseteq H$ be a convex compact simplex such that the topology $\mathcal{T} \cap X$ on X is* **metrizable.** *Then for every $x \in X$, there is exactly one probability measure m: $C(X, \mathbb{R}) \to \mathbb{R}$ with barycenter x such that $m(X \backslash \mathrm{ext}(X)) = 0$.*

Proof. Let $f \in \mathscr{C}(X, \mathbb{R})$ be convex. By Exercise 2.8.3.4 the set $\{f < \bar{f}\}$ is contained in the m-nullset $X \backslash \mathrm{ext}(X)$, hence $\int f dm = \int \bar{f}\, dm$ follows, and Theorem 2.15 applies.

2.17. Remark. Theorem 2.15 and Corollary 2.16 are due to CHOQUET and MEYER [1]. Actually, compact simplexes are characterized by the uniqueness property stated in Theorem 2.15.

2.18. Exercises

2.18.1. Let $(H, \mathcal{T})$ be a locally convex separated topological vector space and $\varnothing \neq X \subseteq H$ be convex and compact. For every $f \in \mathscr{C}(X, R)$, let $\bar{f}$ denote the concave upper envelope of f, and let $\underline{f} = -(\overline{-f})$, which we call (obviously justifiably) the convex lower envelope of f. Recall Exercise 2.8.3 and the notion of a maximal probability measure m introduced there (an equivalent definition is $\int \bar{f}\, dm = \int f\, dm$ $(f \in \mathscr{C}(X, \mathbb{R})$ convex).

Call $f \in \mathscr{C}(X, \mathbb{R})$ *resolutive* if m maximal $\Rightarrow \int \underline{f}\, dm = \int \bar{f}\, dm \left(= \int f\, dm\right)$.

2.18.1.1. Prove that every convex $f \in \mathscr{C}(X, \mathbb{R})$ is resolutive. (*Hint*: $\underline{f} = f$.)

2.18.1.2. Prove that the sum and the difference of two resolutive functions is resolutive. (*Hint*: Use $\overline{f + g} \leqq \bar{f} + \bar{g}$, as proved in the proof of Proposition 2.14 (Exercise 2.8.3.6).)

2.18.1.3. Prove that all $f \in \mathscr{C}(X, \mathbb{R})$ are resolutive. (*Hint*: Use Exercise 2.13.3.3.)

2.18.1.4. Prove that for every maximal measure $m\colon \mathscr{C}(X, \mathbb{R}) \to \mathbb{R}$ and every compact G_δ set $K \subseteq X \backslash \operatorname{ext}(X)$, $m(K) = 0$ holds.

2.18.2. Let $H = \operatorname{meas}^\tau(C([0, 1], \mathbb{R}))$ and $\mathscr{T}$ be the weak topology in H (i.e., weak convergence of m means convergence of all integrals $\int f\, dm$ $(f \in \mathscr{C}([0, 1], \mathbb{R})))$. Put $X =$ the set of all probability measures on $\mathscr{C}([0, 1], \mathbb{R})$.

2.18.2.1. Prove that X is a simplex. (*Hint*: Use Theorem IX.2.3.)

2.18.2.2. Prove that X is compact for $\mathscr{T}$ and the topology $\mathscr{T} \cap X$ in X is metrizable. (*Hint*: Use the fact that there is a countable set $\mathscr{C}_0$ that is dense in $\mathscr{C}([0, 1], \mathbb{R})$ for uniform approximation.)

2.18.2.3. Let $m_0 \in X$ be Lebesgue measure on $[0, 1]$. For every $m' \in X$, let $m' = m_0' + m_1'$, $m_0' \ll m_0$, $m_1' \perp m_0$ be the Lebesgue decomposition with respect to m_0 (Theorem IX.3.8) and $F(m') = m_1'(1)$. Prove that this $F \in \mathbb{R}^X$ is affine.

2.18.2.4. Prove that the $F \in \mathbb{R}^X$ defined in 2.3 is $\mathscr{B}(\mathscr{T} \cap X)$-measurable. (*Hint*: $0 \leqq f \in \mathscr{C}([0, 1], \mathbb{R}) \Rightarrow (m' \wedge (nm_0))(f) = \inf\{\min[m'(g), nm_0(g)] + \min[m'(h), nm_0(h)] \mid g, h \in \mathscr{C}_0, g + h \leqq f\}$, where $\mathscr{C}_0$ is as in 2.2 shows that $m' \to (m' \wedge (nm_0))(1)$ is measurable; $F(m') = \lim_n (m' \wedge (nm_0))(1)$ by Theorem IX.3.8.)

2.18.2.5. Show that the mapping $\varphi\colon [0, 1] \to X$ defined by $\varphi(\xi) = \delta_\xi$ $(0 \leqq \xi \leqq 1)$ is measurable (for the obvious σ-fields).

2.18.2.6. Show that φm_0 has barycenter m_0.

2.18.2.7. Show that $F(m_0) = 1$ but $\int F\, d(\varphi m_0) = 0$.

3. APPLICATIONS

In this section we prove one application of Theorem 1.13 (Krein–Milman), and one of Theorem 2.7 (Choquet) and Corollary 2.16. The first is LIAPUNOV's [1] convexity theorem, and the second is ergodic decomposition. I would have liked to present Bochner's theorem on positive definite functions as an application of Choquet's theorem, but I found the proofs too technical to be included in this book and thus refer the reader to CHOQUET [5].

3.1. Theorem (LIAPOUNOV [1]). *Let* $n \geqq 1$ *and* $(\Omega, \mathscr{B}, m_1), \ldots, (\Omega, \mathscr{B}, m_n)$ *be atomless σ-probability content spaces. Then the set* $M = \{(m_1(F), \ldots, m_n(F)) | F \in \mathscr{B}\}$ *is a convex compact subset of* $\mathbb{R}^n$.

Proof. (after LINDENSTRAUSS [1]). Consider the σ-content $m_0 = m_1 + \cdots + m_n$. Clearly $m_1, \ldots, m_n \leqq m_0$, hence $m_1, \ldots, m_n \ll m_0$ and $m_k = f_k m_0$ $(k = 1, \ldots, n)$ where $0 \leqq f_1, \ldots, f_n \in \text{mble}\,^b(\Omega, \mathscr{B}, \mathbb{R})$, $f_1 + \cdots + f_n = 1$. In particular, we have $f_1, \ldots, f_n \in \mathscr{L}^1_{m_0}$. In the dual Banach space $L^\infty_{m_0}$ of $L^1_{m_0}$ (Theorem X.2.4, Remark X.2.2.2) we consider the w*-topology introduced by $L^1_{m_0}$ and the set

$$K = \{h | h \in L^\infty_{m_0}, 0 \leqq h \leqq 1 \ (\text{mod } m_0)\}.$$

Clearly K is w*-compact and convex (exercise; see Example 1.3.3). For every $k = 1, \ldots, n$, the mapping $h \to \int hf_k \, dm_0$ of $L^\infty_m \to \mathbb{R}$ is linear and w*-continuous. Thus the mapping $\varphi\colon h \to \left(\int hf_1 \, dm_0, \ldots, \int hf_n \, dm_0\right)$ of $L^\infty_m \to \mathbb{R}^n$ sends the convex w*-compact set K onto some convex compact set $\overline{M} \subseteq \mathbb{R}^n$. Clearly $M \subseteq \overline{M}$. We prove $M = \overline{M}$, and this will do it. For this, we define $K(x) = \varphi^{-1}\{x\} \cap K$ for every $x \in \overline{M}$. It is clearly sufficient to prove that every $K(x)$ contains an indicator function. Fix some $x \in \overline{M}$. Since $\{x\}$ is convex and compact, and φ is linear and w*-continuous, $K(x)$ is a convex w*-compact subset of K. Apply Theorem 1.13 (Krein–Milman) in order to get an extremal point h of $K(x)$. We show that h is, mod m_0, an indicator function. We pass to a representative of the equivalence class h mod m_0 in mble$(\Omega, \mathscr{B}, \mathbb{R})$ and denote it by h again. If h is not an indicator function, we may find some $0 < \varepsilon < \frac{1}{2}$ such that $F = \{\varepsilon \leqq h \leqq 1 - \varepsilon\} \in \mathscr{B}$ is not an m_0-nullset. Now we have to insert an induction over n. If $n = 1$, we exploit the fact that $m_0 = m_1$ and F is not an m_1-atom. There is thus some splitting $F = A + B$, $A \cap B = \varnothing$, $A, B \in \mathscr{B}$, $m_1(A)m_1(B) > 0$. Find $0 < s, t < \varepsilon$ such that $sm_1(A) - tm_1(B) = 0$ and put $h_1 = h + s1_A - t1_B$, $h_2 = h - s1_A + t1_B$. Clearly $0 \leqq h_1, h_2 \leqq 1$, $h_1, h_2 \in \text{mble}(\Omega, \mathscr{B}, \mathbb{R})$,

$$\int h_1 \, dm_1 = \int h \, dm_1 + sm_1(A) - tm_1(B)$$

$$= \int h \, dm_1 = \int h \, dm_1 - sm_1(A) + tm_1(B) = \int h_2 \, dm_1,$$

hence $h_1, h_2 \in K(x)$. On the other hand, $h_1 \neq h \neq h_2 \bmod m_0$, $h = \frac{1}{2}(h_1 + h_2)$, contradicting the extremality of h in $K(x)$. It follows that h is, mod m_0, an indicator function, and our theorem is proved for $n = 1$. Assume now $n > 1$ and our theorem to be true for $n - 1$. There is some $k \leqq n$ with $m_k(F) > 0$; and since m_k is atomless, we may find a splitting $F = A + B$, $A \cap B = \varnothing$, $A, B \in \mathscr{B}$, $m_k(A)m_k(B) > 0$, hence $m_0(A)m_0(B) > 0$. By induction hypothesis, there is some $A_0 \subseteq A$, $A_0 \in \mathscr{B}$ with $m_k(A_0) = \frac{1}{2}m_k(A)$, and some $B_0 \subseteq B$, $B_0 \in \mathscr{B}$ with $m_k(B_0) = \frac{1}{2}m_k(B)$ $(k = 2, \ldots, n)$. It is easy to find $\alpha, \beta \in \mathbb{R}$ with $0 < |\alpha| + |\beta| < \varepsilon$ such that

$$\alpha(m_1(A) - 2m_1(A_0)) - \beta(m_1(B) - 2m_1(B_0)) = 0.$$

Put now

$$h_1 = h + \alpha(1_A - 2 \cdot 1_{A_0}) - \beta(1_B - 2 \cdot 1_{B_0}),$$
$$h_2 = h - \alpha(1_A - 2 \cdot 1_{A_0}) + \beta(1_B - 2 \cdot 1_{B_0}).$$

Obviously, $0 \leqq h_1, h_2 \leqq 1$, $h_1, h_2 \in \text{mble}(\Omega, \mathscr{B}, \mathbb{R})$,

$$\int h_1\, dm_1 = \int h\, dm_1 + \alpha(m_1(A) - 2m_1(A_0)) - \beta(m_1(B) - 2m_1(B_0)) = \int h\, dm_1$$
$$= \int h\, dm_1 - \alpha(m_1(A) - 2m_1(A_0)) + \beta(m_1(B) - 2m_1(B_0)) = \int h_2\, dm_1,$$

and

$$\int h_1\, dm_k = \int h\, dm_k + \alpha(m_k(A) - 2m_k(A_0)) - \beta(m_k(B) - 2m_k(B_0))$$
$$= \int h\, dm_k = \int h_2\, dm_k \qquad (k = 2, \ldots, n).$$

Thus $h_1, h_2 \in K(x)$. On the other hand, $h_1 \neq h \neq h_2 \bmod m_0$, $h = \frac{1}{2}(h_1 + h_2)$, contradicting the extremality of h in $K(x)$. It follows that h is, mod m_0, an indicator function, and our theorem is proved.

3.2. Theorem. *Let* $(\Omega, \mathscr{S})$ *be a compact metrizable space and* $T: \Omega \to \Omega$ *a homeomorphism. Then:*

3.2.1. *The weak topology* $\mathscr{T}$ *in* $\text{cont}^\sigma(\mathscr{B}(\mathscr{S}))$ *(Definition* XI.1.2*) is metrizable.*

3.2.2. *The set* $\mathscr{P}$ *of all* σ*-probability contents on* $\mathscr{B}(\mathscr{S})$ *is a convex weakly compact subset of* $\text{cont}^\sigma(\mathscr{B}(\mathscr{S}))$ *and* $m \to Tm$ $(m \in \mathscr{P})$ *defines a weakly continuous homeomorphism (again denoted by)* T *of* $\mathscr{P}$ *onto* $\mathscr{P}$.

3.2.3. *The set* $X = \{m | m \in \mathscr{P}, Tm = m\}$ *is a nonempty, convex, and weakly compact simplex in* $\text{cont}^\sigma(\mathscr{B}(\mathscr{S}))$.

3.2.4. $m, m' \in \text{ext}(X)$, $m \neq m' \Rightarrow m \perp m'$.

3.2.5. *$m \in \mathrm{ext}(X) \Leftrightarrow (\Omega, \mathscr{B}(\mathscr{S}), m, T)$ is an ergodic dynamical system (Definitions IV.3.1, IV.3.19).*

3.2.6. *For every $m \in X$, there is a unique σ-probability content M on $\mathscr{B}(\mathscr{T} \cap X)$ such that m is the barycenter of M and $M(\mathrm{ext}(X)) = 1$. M is called the ergodic decomposition of m.*

Proof. 1. This is Exercise XI.2.8.6. Solution: Let $f_1, f_2, \ldots$ be dense in $\mathscr{C}(\Omega, X)$ for uniform approximation, and $|f_k| \leqq \alpha_k \in \mathbb{R}$, say. Then

$$|m, m'| = \sum_{k=1}^{\infty} \frac{1}{2^k \alpha_k} |m(f_k) - m'(f_k)| \qquad (m, m' \in \mathrm{cont}^\sigma(\mathscr{B}(\mathscr{S})))$$

defines a metric in $\mathrm{cont}^\sigma(\mathscr{B}(\mathscr{S}))$ which clearly describes the weak topology.

2. $\mathscr{P}$ is the intersection of the weakly closed subsets

$$\{m \mid 0 \leqq m \in \mathrm{cont}^\sigma(\mathscr{B}(\mathscr{S}))\}, \{m \mid m(1) = 1\},$$

hence weakly closed. Consider the Banach space $(\mathscr{C}(\Omega, \mathbb{R}), \|\cdot\|)$, where $\|\cdot\|$ is the supremum norm. Every norm-continuous linear form m on $\mathscr{C}(\Omega, \mathbb{R})$ is bounded on order bounded sets, hence a difference of two positive linear forms (Theorem IX.2.3) which are automatically τ-continuous (Theorem I.10.1), hence $m \in \mathrm{meas}^\sigma(\mathscr{C}(\Omega, \mathbb{R}))$ which can be identified with $\mathrm{cont}^\sigma(\mathscr{B}(\mathscr{S}))$ (Theorem IX.4.1.3). Thus $\mathrm{meas}^\sigma(\mathscr{B}(\mathscr{S}))$ can be considered as the dual Banach space of $\mathscr{C}(\Omega, \mathbb{R})$, and $\mathscr{P}$ is trivially contained in the unit ball of that dual, whose w*-topology is nothing but the weak topology. Thus $\mathscr{P}$ is weakly compact. Most of this is essentially a repetition of the introduction to Chapter XI. By Proposition XI.1.7, $T: \Omega \to \Omega$ induces a weakly continuous mapping T of $\mathrm{cont}^\sigma(\mathscr{B}(\mathscr{S}))$ into itself, and T^{-1}, likewise continuous, does the same. $T\mathscr{P} = \mathscr{P}$ is now obvious.

3. In order to show that X is nonempty, take any $m \in \mathscr{P}$ (e.g., $m = \delta_\omega$ for some $\omega \in \Omega$) and find a weak limiting point $\overline{m} \in \mathscr{P}$ of the sequence $m_1, m_2, \ldots \in \mathscr{P}$, where $m_n = n^{-1} \sum_{k=1}^{n} T^k m$. Since $\mathscr{P}$ is weakly compact, such an m exists. Now if, e.g., $\lim_k m_{n_k} = \overline{m}$, then $\lim_k Tm_{n_k} = T\overline{m}$ since T is weakly continuous. On the other hand, a simple calculation shows $Tm_{n_k} - m_{n_k} = 1/n_k\,(T^{n_k}m - m)$, and this obviously tends weakly to 0. We conclude $T\overline{m} = \overline{m}$, i.e., $\overline{m} \in X$, and $X \neq \varnothing$ follows. Since X is the intersection of $\mathscr{P}$ with the set $\{m \mid m \in \mathrm{cont}^\sigma(\mathscr{B}(\mathscr{S})), Tm = m\}$, which is weakly closed since T is weakly continuous, the weak compactness of X follows. Convexity of X is obvious. Let us now show that X is a simplex. The hyperplane $\{m \mid m(1) = 1\}$ contains X but not $0 \in \mathrm{cont}^\sigma(\mathscr{B}(\mathscr{S}))$. It will thus be sufficient to show that the cone $C = \{m \mid 0 \leqq m \in \mathrm{cont}^\sigma(\mathscr{B}(\mathscr{S})), Tm = m\}$ with tip $0 \in \mathrm{cont}^\sigma(\mathscr{B}(\mathscr{S}))$ generated by X is a lattice cone. For this it clearly suffices to prove that it is stable under the lattice operations $\wedge$ and $\vee$ in $\mathrm{cont}^\sigma(\mathscr{B}(\mathscr{S}))$. Let thus $0 \leqq m, m' \in \mathrm{cont}^\sigma(\mathscr{B}(\mathscr{S}))$, and put

$m_0 = m \wedge m'$. Since T is obviously order preserving, $Tm_0 \leqq Tm = m$, $Tm' = m'$, i.e., $Tm_0 \leqq m \wedge m' = m_0$ follows. But $(Tm_0)(\Omega) = m_0(\Omega)$ shows that $Tm_0 \neq m_0$ is impossible. Thus C is stable under $\wedge$. Stability under $\vee$ can be deduced either herefrom or by analogous reasoning.

4. $m, m' \in \operatorname{ext}(X)$, $m \neq m' \Rightarrow m \wedge m' \neq m$. If $m \wedge m' \neq 0$, then the real $\alpha = (m \wedge m')(\Omega)$ satisfies $0 < \alpha < 1$, $1 - \alpha = (m - (m \wedge m'))(\Omega)$. We conclude that

$$m_1 = \frac{1}{\alpha}(m \wedge m'), \qquad m_2 = \frac{1}{1-\alpha}(m - (m \wedge m')) \in X,$$

and

$$\alpha m_1 + (1 - \alpha)m_2 = m.$$

Since m is an extremal point of X, $m = m_1$, i.e., $m \wedge m' = \alpha m$ follows. We get $m \wedge m' = \beta m$, $\beta > 0$ symmetrically. But $m(1) = m'(1)$ now implies $m = m'$, a contradiction.

5. Assume that $(\Omega, \mathscr{B}(\mathscr{S}), m, T)$ is not ergodic. Then there is a decomposition $\Omega = \Omega_1 + \Omega_2$, $\Omega_1, \Omega_2 \in \mathscr{B}(\mathscr{S})$, $\Omega_1 \cap \Omega_2 = \varnothing$, $T\Omega_1 = \Omega_1 \bmod m$, $T\Omega_2 = \Omega_2 \bmod m$, $m(\Omega_1) > 0 < m(\Omega_2)$. Define

$$m_1(F) = \frac{m(F \cap \Omega_1)}{m(\Omega_1)}, \qquad m_2(F) = \frac{m(F \cap \Omega_2)}{m(\Omega_2)} \qquad (F \in \mathscr{B}(\mathscr{S})).$$

Clearly

$$m_1(T^{-1}F) = \frac{m(T^{-1}F \cap \Omega_1)}{m(\Omega_1)} = \frac{m(T^{-1}(F \cap \Omega_1))}{m(\Omega_1)} = m_1(F) \qquad (F \in \mathscr{B}(\mathscr{S})),$$

i.e., $Tm_1 = m_1$, and likewise $Tm_2 = m_2$ follows. With $\alpha = m(\Omega_1)$, we have $1 - \alpha = m(\Omega_2)$ and $m = \alpha m_1 + (1 - \alpha)m_2$. Since clearly $m_1 \neq m_2$, m cannot be an extremal point of X. Let now conversely $m \in X \backslash \operatorname{ext}(X)$, and m_1, $m_2 \in X$, $m_1 \neq m_2$, $0 < \alpha < 1$, $m = \alpha m_1 + (1 - \alpha)m_2$. Put

$$m_1' = m_1 - (m_1 \wedge m_2), \qquad m_2' = m_2 - (m_1 \wedge m_2).$$

Then $m_1', m_2' \neq 0$ and $m_1' \wedge m_2' = 0$. Clearly $Tm_1' = m_1'$, $Tm_2' = m_2'$. By Proposition VIII.2.5 there is a decomposition $\Omega = \Omega_1' + \Omega_2'$, $\Omega_1', \Omega_2' \in \mathscr{B}(\mathscr{S})$, $\Omega_1' \cap \Omega_2' = \varnothing$ such that m_1' lives on Ω_1' and m_2' lives on Ω_2'. Put

$$\Omega_1 = \bigcap_{n \in \mathbb{Z}} T^n\Omega_1', \qquad \Omega_2 = \bigcup_{n \in \mathbb{Z}} T^n\Omega_2'.$$

Then $\Omega = \Omega_1 + \Omega_2$, $\Omega_1, \Omega_2 \in \mathscr{B}(\mathscr{S})$, $\Omega_1 \cap \Omega_2 = \varnothing$, $T\Omega_1 = \Omega_1$, $T\Omega_2 = \Omega_2$. From $m_1(T^n\Omega_1') = m_1(\Omega_1')$ $(n \in \mathbb{Z})$ we deduce that m_1' still lives on Ω_1, and likewise m_2' lives on Ω_2. This implies $0 < m_1'(\Omega_1) \leqq m_1(\Omega_1)$, and hence

$m(\Omega_1) > 0$. Symmetrically, we get $m(\Omega_2) > 0$. This shows that m is not ergodic.

6 follows from Theorem 2.7 (Choquet) and Corollary 2.16.

3.3. Exercises

3.3.1. Let $(\Omega, \mathscr{S})$ be a compact Hausdorff space.

3.3.1.1. Let $(T_\iota)_{\iota \in I}$ be a family of homeomorphisms of Ω such that $T_\iota \circ T_\kappa = T_\kappa \circ T_\iota$ $(\iota, \kappa \in I)$. Prove that $X = \{m | m: \mathscr{C}(\Omega, \mathbb{R}) \to \mathbb{R}$ is a probability measure, $T_\iota m = m$ $(\iota \in I)\}$ is a nonempty convex compact simplex in $\text{meas}(\mathscr{C}(\Omega, \mathbb{R}))$ $(= \text{meas}^\tau(\mathscr{C}(\Omega, \mathbb{R})))$. (*Hint*: Define $X_\iota = \{m | m$ is a probability measure, $T_\iota m = m\}$ $(\iota \in I)$ and show $X_{\iota_1} \cap \cdots \cap X_{\iota_n} \neq \varnothing$ $(\iota_1, \ldots, \iota_n \in I)$ by repeating the argument in the proof of Theorem 3.2.3; imitate the proof of that theorem.)

3.3.1.2. Let X be as in 1.1. Prove that $m, m' \in \text{ext}(X)$, $m \neq m' \Rightarrow m \perp m'$.

3.3.1.3. Let X be as in 1.1. Assume that I is at most countable. Prove that $m \in \text{ext}(X) \Leftrightarrow [E \in \mathscr{B}(\mathscr{S}), T_\iota E = E \bmod m$ $(\iota \in I) \Rightarrow m(E)m(\Omega \backslash E) = 0]$.

3.3.1.4. Let X be as in 1.1. Assume that $(\Omega, \mathscr{S})$ is metrizable. Prove that every $m \in X$ is the barycenter of some probability σ-content living on $\text{ext}(X)$.

3.3.1.5. Let $P: \Omega \times \mathscr{C}(\Omega, \mathbb{R}) \to \mathbb{R}$ be a stochastic (τ-) measure kernel such that $f \in \mathscr{C}(\Omega, \mathbb{R}) \Rightarrow P(\cdot, f) \in \mathscr{C}(\Omega, \mathbb{R})$. Put $X = \{m | m: \mathscr{C}(\Omega, \mathbb{R}) \to \mathbb{R}$ is a probability measure, $mP = m\}$. Formulate and prove the analogue of Theorem 3.2.

3.3.1.6. ¨Carry 1.5 over to commuting families of stochastic kernels (see 1.1–1.4).

3.4. Remark. It is possible to make further specifications about the ergodic decomposition M in the situation of Theorem 3.2 if one employs the tools of ergodic theory skillfully. What we are alluding to is the so-called KRYLOV–BOGOLIOUBOV [1] theory, classically presented by OXTOBY [1] (see also JACOBS [1, 2]).

CHAPTER XVI

LIFTING

The *lifting problem* can, in a standard situation, be described as follows. Let $(\Omega, \mathscr{B}, m)$ be a σ-probability (content) space, i.e., $\Omega \in \mathscr{B}$ and m a σ-content on $\mathscr{B}$ with $m(\Omega) = 1$. Equality mod m defines an equivalence relation in the space of all bounded real-valued $\mathscr{B}$-measurable functions; we denote that space by $\text{mble}^{\text{b}}(\Omega, \mathscr{B}, \mathbb{R})$ and the set of all equivalence classes mod m in it by $\text{mble}^{\text{b}}_m(\Omega, \mathscr{B}, \mathbb{R})$. The lifting problem is, roughly speaking, the following: *Can we pick from every equivalence class in* $\text{mble}^{\text{b}}_m(\Omega, \mathscr{B}, \mathbb{R})$ *exactly one representative in such a fashion that finite linear and lattice operations plus multiplication under which we know* $\text{mble}^{\text{b}}(\Omega, \mathscr{B}, \mathbb{R})$ *to be stable lead from picked representatives to picked representatives again*?

The purpose of the present chapter is an introduction into the rudiments of lifting theory. We shall prove the *existence* of lifting, not only in the case of σ-probability spaces, but also for σ-content spaces with a *disjoint base* (Definition VIII.8.4). One basic feature of the theory is the necessity of assuming the involved σ-content space to be *complete* (i.e., the σ-field resp. local σ-ring has to contain all m-nullsets resp. all local m-nullsets). The basic idea of proof is to begin (let us stick to the case of a complete σ-probability space $(\Omega, \mathscr{B}, m)$ for the moment) with the trivial σ-field of all m-nullsets and their complements. Since the measurable functions are constants mod m in this case, the existence of a lifting is a triviality here. We then work up our way to $\mathscr{B}$ by *extension procedures* and *Zorn's lemma.* The extension procedures involve the *martingale convergence theorem* VIII.7.8 and the use of *ultrafilters.* At one point we loose multiplicativity in a limiting procedure. This necessitates the introduction of the notion of a *linear lifting,* plus an *extremal point argument* (to be found, e.g., in PHELPS [1]) for the regaining of multiplicativity. We mention that there is another approach to the existence of liftings, via associated topologies. See A. and C. IONESCU-TULCEA [1], BICHTELER [1].

The theory, as it stands today has the form acquired in A. and C. IONESCU-TULCEA [1] and forerunning papers (see also DINCULEANU [1]). The history of lifting goes back to a question by Alfred Haar and an early paper of v. Neumann who proved the existence of a lifting in the case of Lebesgue σ-content. DOROTHY MAHARAM [1] was the first to prove existence for the general case of a σ-finite m. One might ask whether the lifting problem could as well be solved for L_m^1, but it can be shown (see A. and C. IONESCU-TULCEA [1]) that there is no solution except in the trivial "atomic" case.

Our presentation essentially follows A. and C. IONESCU-TULCEA [1] and DINCULEANU [1]. We deviate from Ionescu-Tulcea by avoiding function algebras and working strictly with σ-fields resp. local σ-rings.

Section 1 is devoted to the solution of the lifting problem first in the case of a complete σ-probability space, and then in the case of a complete σ-content space with a *disjoint base* (Definition VIII.8.4).

In Section 2 we define the concept of a *strong lifting* and prove the existence of strong liftings under a separability assumption. This covers the case of a locally compact space with a countable base for the topology.

A variety of applications is presented in Section 3.

1. THE NOTION AND EXISTENCE OF A LIFTING

In this section we give an exact definition of the notion of *lifting* and prove the *existence of liftings*, first for complete σ-probability content spaces and then for σ-content spaces with a *disjoint base* (Definition VIII.8.4) and an appropriate *completeness property*. Extensive use of Zorn's lemma (i.e., the axiom of choice) is made: We shall use *ultrafilters*, the *Krein–Milman theorem* (XV.1.13) plus a direct application of *Zorn's lemma*. Another important tool is Theorem VIII.7.8 on the a.e. convergence of *submartingales* (actually we use only its specialization to the case of generated *martingales* (Exercise VIII.7.3)).

1.1. Definition. *Let* $(\Omega, \mathscr{B}, m)$ *be a* σ*-probability (content) space and* $\mathscr{B}_0 \subseteq \mathscr{B}$ a σ-field.

1.1.1. $\text{mble}^b(\Omega, \mathscr{B}_0, \mathbb{R})$ *denotes the set of all bounded* $\mathscr{B}_0$*-measurable real functions.*

1.1.2. *A mapping* ρ_0 *of* $\text{mble}^b(\Omega, \mathscr{B}_0, \mathbb{R})$ *into itself is called a* **linear lifting** *for* $(\Omega, \mathscr{B}_0, m)$ *if it has the following properties:*

1.1.2.1. $f \in \text{mble}(\Omega, \mathscr{B}_0, \mathbb{R}) \Rightarrow \rho_0(f) = f \bmod m$.

1.1.2.2. $f, g \in \text{mble}(\Omega, \mathscr{B}_0, \mathbb{R}), f = g \bmod m \Rightarrow \rho_0(f) = \rho_0(g)$.

1.1.2.3. *ρ_0 is linear.*

1.1.2.4. *ρ_0 is positive, i.e., $0 \leqq f \in \text{mble}(\Omega, \mathscr{B}_0, \mathbb{R}) \Rightarrow \rho_0(f) \geqq 0$.*

1.1.2.5. *$\rho_0(1)$ is the constant 1.*

1.1.3. *A linear lifting ρ_0 for $(\Omega, \mathscr{B}_0, m)$ is called a* **lifting** *if it is* **multiplicative,** *i.e., if $f, g \in \text{mble}^b(\Omega, \mathscr{B}_0, \mathbb{R}) \Rightarrow \rho_0(fg) = \rho_0(f)\rho_0(g)$.*

1.2. Remarks. Let $(\Omega, \mathscr{B}, m)$ be a σ-probability space and $\mathscr{B}_0$ a σ-field $\subseteq \mathscr{B}$ containing all m-nullsets from $\mathscr{B}$.

1.2.1. We emphasize that the restriction of m to $\mathscr{B}_0$ is again denoted by m for brevity. $\text{mble}^b(\Omega, \mathscr{B}_0, \mathbb{R})$ is a vector lattice that is stable under bounded convergence m-a.e. (use Proposition III.8.4 and the fact that $\mathscr{B}_0$ contains all m-nullsets). All equalities and inequalities of functions that appear in Definition 1.1 and are not specified as m-a.e. are understood to hold *strictly* (= *everywhere* = *pointwise*).

1.2.2. It is clear that Definition 1.1 makes precise the intuitive explanations given in the introduction to this chapter. Properties 1.1.2.1 and 2 can be worded as: ρ_0 picks exactly one representative out of every equivalence class mod m in $\text{mble}^b(\Omega, \mathscr{B}, \mathbb{R})$. The existence of a lifting for $(\Omega, \mathscr{B}_0, m)$ with $\mathscr{B}_0 = \{F \mid F \in \mathscr{B}, m(F) = 0 \text{ or } 1\}$ is obvious: $\text{mble}^b(\Omega, \mathscr{B}_0, \mathbb{R})$ consists of all m-a.e. constant functions, and we have only to send every such function into the exact constant to which it is m-a.e. equal in order to establish a lifting for $(\Omega, \mathscr{B}_0, m)$. Our aim is, of course, the existence of a lifting for $(\Omega, \mathscr{B}, m)$.

1.2.3. We have chosen the space $\text{mble}(\Omega, \mathscr{B}, \mathbb{R})$ consisting of strictly bounded real functions as a domain for a lifting in order to make an argument occurring in the proof for the existence of a lifting a little bit more straightforward. It should however be clear that every equivalence class mod m in the space $\mathscr{L}_m^\infty$ (Exercise III.12.14.6, Definition X.2.1) intersects $\text{mble}^b(\Omega, \mathscr{B}, \mathbb{R})$ and that this enables us trivially to extend liftings from $\text{mble}^b(\Omega, \mathscr{B}, \mathbb{R})$ to $\mathscr{L}_m^\infty$.

1.3. Exercises. Let $(\Omega, \mathscr{B}, m)$ be a σ-probability space and $\mathscr{B}_0 \subseteq \mathscr{B}$ a σ-field that contains all m-nullsets from $\mathscr{B}$.

1.3.1. Let ρ_0 be a linear lifting for $(\Omega, \mathscr{B}_0, m)$. Prove that it commutes with uniform convergence of functions in $\text{mble}^b(\Omega, \mathscr{B}_0, \mathbb{R})$.

1.3.2. Let ρ_0 be a lifting for $(\Omega, \mathscr{B}_0, m)$. Prove:

1.3.2.1. $f \in \text{mble}^b(\Omega, \mathscr{B}_0, \mathbb{R})$, $h \in \mathscr{C}(\mathbb{R}, \mathbb{R}) \Rightarrow h(f) \in \text{mble}^b(\Omega, \mathscr{B}_0, \mathbb{R})$ and $\rho_0(h(f)) = h(\rho_0(f))$. (*Hint*: Use the Weierstrass–Stone theorem.)

1.3.2.2. ρ_0 commutes with all finite vector lattice operations.

1.3.2.3. ρ_0 sends indicator functions into indicator functions. This, by the way, would enable us to define the notion of a lifting for sets alone; we refrain from doing so since we have need of linear methods, in particular linear liftings, for technical reasons; the reader will, however, observe that we have to resort to sets now and then.

1.3.3. Let $\mathscr{B}_0 \neq \mathscr{B}$ and $E_1 \in \mathscr{B} \backslash \mathscr{B}_0$. Put $\mathscr{B}_1 = \mathscr{B}(\mathscr{B}_0 \cup \{E_1\})$. Prove that $\mathscr{B}_1 = \{(E \cap E_1) + (F \cap (\Omega \backslash E_1)) | E, F \in \mathscr{B}_0\}$ and that $\mathrm{mble}^b(\Omega, \mathscr{B}_1, \mathbb{R})$ consists of exactly those functions f_1 that have a representation $f_1 = f 1_{E_1} + g 1_{\Omega \backslash E_1}$ with $f, g \in \mathrm{mble}^b(\Omega, \mathscr{B}_0, \mathbb{R})$. Prove that we may always choose the f and g of such a representation in such a fashion that $\inf f_1 \leqq f, g \leqq \sup f_1$.

We begin by showing *how to extend a lifting "one step."* From now on we shall always make the assumption that $(\Omega, \mathscr{B}, m)$ is complete.

1.4. Proposition. *Let $(\Omega, \mathscr{B}, m)$ be a complete σ-probability space and $\mathscr{B}_0 \subseteq \mathscr{B}$ a σ-field that contains all m-nullsets, $\mathscr{B}_0 \neq \mathscr{B}$. Let $E_1 \in \mathscr{B} \backslash \mathscr{B}_0$ and $\mathscr{B}_1 = \mathscr{B}(\mathscr{B}_0 \cup \{E_1\})$. Then $\mathscr{B}_1 = \{(A \cap E) + (\Omega \backslash A) \cap F | E, F \in \mathscr{B}_0\}$ and $\mathscr{B}_1$ is a σ-field that contains all m-nullsets. Assume that we are given a lifting ρ_0 for $(\Omega, \mathscr{B}_0, m)$. Then there is a lifting ρ_1 for $(\Omega, \mathscr{B}_1, m)$ that extends ρ_0, i.e., coincides with ρ_0 on* $\mathrm{mble}^b(\Omega, \mathscr{B}_0, \mathbb{R})$.

Proof. We proceed in several steps.

1. Consider $\mathscr{E}_0 = \{g | g \in \mathrm{mble}^b(\Omega, \mathscr{B}_0, \mathbb{R})\ 0 \leqq g \leqq 1_{E_1}\ m\text{-a.e.}\}$. Clearly $\mathscr{E}_0$ is stable under all finite lattice operations, and $f \in \mathscr{E}_0 \Rightarrow ng \wedge 1 \in \mathscr{E}_0$ $(n = 1, 2, \ldots)$. Since ρ_0 sends all constants to themselves, and commutes with all finite vector lattice operations, the same statements hold about $\rho_0 \mathscr{E}_0 = \{\rho_0(g) | g \in \mathscr{E}_0\}$. Moreover, all functions in $\rho_0 \mathscr{E}_0$ are pointwise between 0 and 1. Define $e_0 \in \mathbb{R}^\Omega$ as the pointwise supremum of $\rho_0 \mathscr{E}_0$, i.e., $e_0(\omega) = \sup\{\rho_0(g)(\omega) | g \in \mathscr{E}_0\}$ $(\omega \in \Omega)$. Since $e_0(\omega) > 0$ implies the existence of some $g \in \mathscr{E}_0$ with $\rho_0(g)(\omega) > 0 \Rightarrow [(n\rho_0(g)) \wedge 1](\omega) = \rho_0((ng) \wedge 1)(\omega) = 1$ for some $n > 0 \Rightarrow e_0(\omega) = 1$, we find that $e_0 = 1_{E_0}$ for some set $E_0 \subseteq \Omega$. Let us show that $E_0 \in \mathscr{B}_0$. For this, we choose $g_1, g_2, \ldots \in \mathscr{E}_0$ such that $g_1 \leqq g_2 \leqq \cdots$ and even $g_{n+1} \geqq (ng_n) \wedge 1$ $(n = 1, 2, \ldots)$ and $\lim_n \int g_n\, dm = \sup\left\{\int g\, dm | g \in \mathscr{E}_0\right\}$. It is obvious that this can be done and that the pointwise supremum e of $g_1, g_2, \ldots$ is (a) in $\mathscr{E}_0$ again, (b) the indicator function 1_E of some set $E \in \mathscr{B}_0$ (use the above reasoning about e_0), and (c) majorizes mod m every $g \in \mathscr{E}_0$ (see Theorem III.5.8.5, 6 and its proof). We conclude that $\rho_0(e) \geqq \rho_0(g)$ pointwise for every $g \in \mathscr{E}_0$, hence $e_0 \leqq \rho_0(e) \leqq e_0$ and thus $\rho_0(e) = e_0$ everywhere. Since $\rho_0(e) = e$ mod m, we conclude $1_{E_0} = e_0 = e = 1_E$ mod m, i.e., $E_0 = E$ mod m; since $\mathscr{B}_0$ contains all m-nullsets, we get $E_0 \in \mathscr{B}_0$. Since $1_E \in \mathscr{E}_0$, we conclude that $E_0 \subseteq E_1$ up

to an m-nullset. From $\rho_0(e) = e_0$ we conclude $\rho_0(e_0) = \rho_0(\rho_0(e)) = \rho_0(e) = e_0$ everywhere.

2. Put $F_1 = \Omega \backslash E_1$ and deal with F_1 in the same way as we did with E_1 before: We define $\mathscr{F}_0 = \{f \mid f \in \text{mble}^b(\Omega, \mathscr{B}_0, \mathbb{R}),\ 0 \leqq f \leqq 1_{F_1}\ m\text{-a.e.}\}$ and find that the pointwise supremum f_0 of $\rho_0 \mathscr{F}_0$ is the indicator function of some set $F_0 \in \mathscr{B}_0$ and satisfies $\rho_0(f_0) = f_0$ everywhere.

3. It is easy to see that $E_0 \cap F_0 = \varnothing$. In fact, $g \in \mathscr{E}_0$, $g' \in \mathscr{F}_0$

$$\begin{aligned}
&\Rightarrow\quad g \wedge g' \leqq 1_{E_1} \wedge 1_{E_1} = 0\ m\text{-a.e.}\\
&\Rightarrow\quad \rho_0(g) \wedge \rho_0(g') = \rho_0(g \wedge g') = 0 \text{ everywhere}\\
&\Rightarrow\quad e_0 \wedge f_0 = 1_{E_0} \wedge 1_{F_0} = 1_{E_0 \cap F_0} = 0 \text{ everywhere.}
\end{aligned}$$

4. We show that $f \in \text{mble}^b(\Omega, \mathscr{B}_0, \mathbb{R})$, $f = 0$ m-a.e. on $E_1 \Rightarrow \rho_0(f) = 0$ everywhere on $\Omega \backslash F_0$. Since ρ_0 commutes with all vector lattice operations, it suffices to consider the case $0 \leqq f \leqq 1_{F_1}$ m-a.e. But then we have $f \in \mathscr{F}_0$ and $\rho_0(f) \leqq f_0 = 1_{F_0}$ follows, which does it. We have, of course, a symmetric statement about F_1.

5. Put $P = (E_1 \cup E_0) \backslash F_0$. Clearly $P \in \mathscr{B}_1$. Since $E_0 \subseteq E_1$ up to an m-nullset, and, symmetrically, $F_0 \subseteq F_1 = \Omega \backslash E_1$ up to an m-nullset, we find $P = E_1 \bmod m$. Clearly $E_0 \subseteq P$, $F_0 \subseteq \Omega \backslash P$.

6. Now we define $\rho_1 : \text{mble}^b(\Omega, \mathscr{B}_1, \mathbb{R}) \to \text{mble}^b(\Omega, \mathscr{B}_1, \mathbb{R})$ in the following way. By Exercise 1.3.3 we have

$$\text{mble}^b(\Omega, \mathscr{B}_1, \mathbb{R}) = \{f 1_{E_1} + g 1_{F_1} \mid f, g \in \text{mble}^b(\Omega, \mathscr{B}_0, \mathbb{R})\}.$$

If $f, f', g, g' \in \text{mble}^b(\Omega, \mathscr{B}_0, \mathbb{R})$, $f 1_{E_1} + g 1_{F_1} = f' 1_{E_1} + g' 1_{F_1} \bmod m$, then $f - f'$ vanishes m-a.e. on E_1, and $g - g'$ vanishes m-a.e. on F_1. From 4 above we conclude that $\rho_0(f) = \rho_0(f')$ everywhere on $\Omega \backslash F_0$, hence on P, and $\rho_0(g) = \rho_0(g')$ everywhere on $\Omega \backslash E_0$, hence on $\Omega \backslash P$. Thus we find $\rho_0(f) 1_P + \rho_0(g) 1_{\Omega \backslash P} = \rho_0(f') 1_P + \rho_0(g') 1_{\Omega \backslash P}$ everywhere on Ω. Thus

$$\rho_1(f 1_{E_1} + g 1_{F_1}) = \rho_0(f) 1_P + \rho_0(g) 1_{\Omega \backslash P}$$

well defines ρ_1 as a mapping of $\text{mble}^b(\Omega, \mathscr{B}_1, \mathbb{R})$ into itself. Clearly it extends ρ_0 (put $f = g$).

7. We verify now easily that this ρ_1 is a lifting for $(\Omega, \mathscr{B}_1, m)$. We have already seen in 6 that $\rho_1(f) = f \bmod m$ and $f = f' \bmod m \Rightarrow \rho_1(f) = \rho_1(f')$ everywhere ($f, f' \in \text{mble}^b(\Omega, \mathscr{B}_1, \mathbb{R})$). Obviously $\rho_1(1) = 1$. It is an easy exercise to derive the positivity, linearity, and multiplicativity of ρ_1 from the corresponding properties of ρ_0; we leave it to the reader.

One of our main technical problems is the passage from a linear lifting to a lifting that "essentially does the same." In order to make the latter expression precise, we set up the

1.5. Definition. *Let* $(\Omega, \mathscr{B}, m)$ *be a complete* σ*-probability space and* ρ *a linear lifting for it.*

1.5.1. *For every* $E \in \mathscr{B}$, *we define* $\underline{\rho}(E) = \{\omega \mid \rho(1_E)(\omega) = 1\}$ *and* $\overline{\rho}(E) = \{\omega \mid \rho(1_E)(\omega) > 0\}$.

1.5.2. *The mapping* $\underline{\rho}\colon \mathscr{B} \to \mathscr{B}$ *is called the* **lower density** *and the mapping* $\overline{\rho}\colon \mathscr{B} \to \mathscr{B}$ *is called the* **upper density** *of the linear lifting* ρ.

1.6. Remarks. Let $(\Omega, \mathscr{B}, m)$ be a complete σ-probability space and ρ a linear lifting for it, $\underline{\rho}$ the lower and $\overline{\rho}$ the upper density for ρ.

1.6.1. We emphasize that $\underline{\rho}$ and $\overline{\rho}$ actually send $\mathscr{B}$ into $\mathscr{B}$ since $\rho(1_E) \in \text{mble}(\Omega, \mathscr{B}, \mathbb{R})$ for every $E \in \mathscr{B}$.

1.6.2. It is obvious that $1_{\underline{\rho}(E)} \leqq \rho(1_E) \leqq 1_{\overline{\rho}(E)}$ everywhere in Ω $(E \in \mathscr{B})$. Both inequalities are actually equalities mod m since $\rho(1_E) = 1_E$ mod m. They turn into strict, i.e., pointwise equalities, if $\rho(1_E)$ is an indicator function. Since indicator functions can be characterized as being their own squares, we see that $\rho(1_E)$ is certainly an indicator function if ρ is a lifting: $\rho(1_E)^2 = \rho(1_E)\rho(1_E) = \rho(1_E\, 1_E) = \rho(1_E)$. A full exploitation of this argument can be worded as follows: The above two inequalities turn into equalities wherever ρ behaves like a lifting. This motivates part of the next

1.7. Proposition. *Let* $(\Omega, \mathscr{B}, m)$ *be a* σ*-probability space and* ρ *a linear lifting for it,* $\underline{\rho}\colon \mathscr{B} \to \mathscr{B}$ *the lower and* $\overline{\rho}\colon \mathscr{B} \to \mathscr{B}$ *the upper density* for ρ. *Then there is at least one lifting* ρ_0 *for* $(\Omega, \mathscr{B}, m)$ *such that*

$$1_{\underline{\rho}(E)} \leqq \rho_0(1_E) \leqq 1_{\overline{\rho}(E)} \qquad (E \in \mathscr{B}).$$

Proof. Let K denote the set of all linear liftings ρ' for $(\Omega, \mathscr{B}, m)$ such that $1_{\underline{\rho}(E)} \leqq \rho'(1_E) \leqq 1_{\overline{\rho}(E)}$ $(E \in \mathscr{B})$ holds. K is a convex set in an obvious sense. We intend to apply the Krein–Milman theorem (XV.1.13) and obtain ρ_0 as an extremal point of K. To this end, we have to identify K with some convex compact set in some separated locally convex topological vector space H. We construct $H = \mathbb{R}^I$ (Exercise XV.1.3.2) with the index set $I = \text{mble}^b(\Omega, \mathscr{B}, \mathbb{R}) \times \Omega = \{(f, \omega) \mid f \in \text{mble}^b(\Omega, \mathscr{B}, \mathbb{R}), \omega \in \Omega\}$ and identify K, of course, with the subset $\{(\rho'(f)(\omega))_{(f,\omega) \in I} \mid \rho' \in K\}$. For conciseness, we employ the same letter K for this set again. We know already that it is convex. It is clearly contained in the product of the compact intervals $J_{(f,\omega)} = [\inf f, \sup f]$, which is compact by Tychonov's theorem. Thus it suffices to prove that K is closed. For this, we have to show that the limit ρ'' (componentwise in $H = \mathbb{R}^I$) of linear liftings satisfying the inequalities defining K is a linear lifting again. The limit would have to be made precise by the use of some filter or net, but we content ourselves with a rather verbal argument. Now it is clear that ρ'' is linear, positive, and sends every constant into itself since all this was the case for the approximating linear liftings. The crucial point is to show that ρ'' again picks one representative out of every equivalence class mod m in $\mathrm{mble}^b(\Omega, \mathcal{B}, \mathbb{R})$. Now for indicator functions this is a consequence of $1_{\underline{\rho}(E)} \leqq \rho''(1_E) \leqq 1_{\bar{\rho}(E)}$ and the mod m equalities $1_{\underline{\rho}(E)} = 1_E = 1_{\bar{\rho}(E)}$ (Remark 1.6.2). By linearity it follows for linear combinations of indicator functions. Now it is obvious that ρ'' is continuous with respect to uniform convergence of functions. But the linear combinations of indicator functions are dense in $\mathrm{mble}^b(\Omega, \mathcal{B}, \mathbb{R})$ with respect to uniform approximation (Proposition III.8.13). Thus the desired statement follows, and the compactness of K is established. By Theorem XV.1.13 (Krein–Milman) K has an extremal point ρ_0. We prove that ρ_0 is multiplicative, i.e., a lifting. Choose any $g \in \mathrm{mble}^b(\Omega, \mathcal{B}, \mathbb{R})$. We have to prove $\rho_0(fg) = \rho_0(f)\rho_0(g)$ $(f \in \mathrm{mble}^b(\Omega, \mathcal{B}, \mathbb{R}))$. By decomposition of g into positive and negative parts, and multiplication with constants, it is sufficient to settle the case $0 \leqq g \leqq 1$. Form $\rho^g(f) = \rho_0(fg) - \rho_0(f)\rho_0(g)$. Clearly ρ^g is a linear mapping of $\mathrm{mble}^b(\Omega, \mathcal{B}, \mathbb{R})$ into itself sending every constant into 0. We show now that $\rho_0 + \rho^g$, $\rho_0 - \rho^g \in K$. First, $\rho_0 + \rho^g$ and $\rho_0 - \rho^g$ are linear mappings of $\mathrm{mble}^b(\Omega, \mathcal{B}, \mathbb{R})$ into itself sending every constant function into itself. It is obvious that $\rho_0 + \rho^g$ and $\rho_0 - \rho^g$ send all members of one equivalence class into one

$$\begin{aligned} \rho_0(f) + \rho^g(f) &= \rho_0(f) + \rho_0(fg) - \rho_0(f)\rho_0(g) \\ &= \rho_0(f)(1 - \rho_0(g)) + \rho_0(fg) \geqq 0 \end{aligned}$$

everywhere. $\rho_0(f) - \rho^g(f) \geqq 0$ is proved in a similar way (exercise). Let now $E \in \mathcal{B}$. Clearly $0 \leqq \rho_0(1_E) + \rho^g(1_E) \leqq 1$ and $0 \leqq \rho_0(1_E) - \rho^g(1_E) \leqq 1$. Since

$$\rho_0(1_E) = \tfrac{1}{2}(\rho_0(1_E) + \rho^g(1_E)) + (\rho_0(1_E) - \rho^g(1_E)),$$

we see that $\rho_0(1_E) + \rho^g(1_E)$ and $\rho_0(1_E) - \rho^g(1_E)$ are $= 0$ resp. $= 1$ where $\rho_0(1_E)$ is. We conclude that

$$1_{\underline{\rho}(E)} \leqq \rho_0(1_E) \pm \rho^g(1_E) \leqq 1_{\bar{\rho}(E)}$$

everywhere. Our above approximation argument now shows that $\rho_0 + \rho^g$ and $\rho_0 - \rho^g$ are linear liftings, actually they belong to K. But $\rho_0 = \frac{1}{2}(\rho_0 + \rho^g) + (\rho_0 - \rho^g)$ and the extremal property of ρ_0 now imply $\rho^g = 0$, i.e., $\rho_0(fg) = \rho_0(f)\rho_0(g)$.

Our next aim is to pass to the limit of a countable sequence of successive extensions of liftings.

1.8. Proposition. *Let $(\Omega, \mathcal{B}, m)$ be a complete σ-probability field and $\mathcal{B}_1 \subseteq \mathcal{B}_2 \subseteq \cdots$ be σ-fields such that $\mathcal{B}_1$ contains all m-nullsets. Assume that* *for every* $n = 1, 2, \ldots$, *we are given a lifting* ρ_n *of* $(\Omega, \mathscr{B}_n, m)$ *such that* ρ_{n+1} *coincides with* ρ_n *on* $\mathrm{mble}^b(\Omega, \mathscr{B}_n, \mathbb{R})$ $(n = 1, 2, \ldots)$. *Let* $\mathscr{B}_\infty = \mathscr{B}(\bigcup_{n=1}^\infty \mathscr{B}_n)$. *Then there is a lifting* ρ_∞ *for* $(\Omega, \mathscr{B}_\infty, m)$ *such that* ρ_n *is the restriction of* ρ_∞ *to* $\mathrm{mble}^b(\Omega, \mathscr{B}_n, \mathbb{R})$ $(n = 1, 2, \ldots)$.

Proof. We proceed in two steps. In the first step we use a passage to limits in order to obtain a linear lifting for $(\Omega, \mathscr{B}_\infty, m)$ extending ρ_1, $\rho_2, \ldots$. In the second step we apply the Krein–Milman theorem to all linear liftings that extend $\rho_1, \rho_2, \ldots$ in order to find a lifting, i.e., a multiplicative one among them. For every $n = 1, 2, \ldots$, the conditional expectation $E_n\colon L_m^1 \to L_m^1$ for $\mathscr{B}_n$ (Proposition VIII.6.1) sends every equivalence class mod m in $\mathrm{mble}^b(\Omega, \mathscr{B}_\infty, \mathbb{R})$ into exactly one equivalence class mod m in $\mathrm{mble}^b(\Omega, \mathscr{B}_n, \mathbb{R})$ (Theorem VIII.6.4). Thus

$$\varepsilon_n\colon \mathrm{mble}(\Omega, \mathscr{B}_\infty, \mathbb{R}) \quad\to\quad \mathrm{mble}(\Omega, \mathscr{B}_n, \mathbb{R})$$

is well defined by $\varepsilon_n(f) = \rho_n(E_n f)$ $(f \in \mathrm{mble}^b(\Omega, \mathscr{B}_\infty, \mathbb{R}))$. For every fixed $\omega \in \Omega$ and every fixed $f \in \mathrm{mble}^b(\Omega, \mathscr{B}_\infty, \mathbb{R})$, clearly $f \to E_n(f)(\omega)$ $(n = 1, 2, \ldots)$ defines a bounded sequence of real numbers; if $f \in \mathrm{mble}^b(\Omega, \mathscr{B}_{n_0}, \mathbb{R})$, then we have $E_n f = f$ $(n \geqq n_0)$ and our sequence is $= E_{n_0}(f)(\omega)$ for $n \geqq n_0$. Consider now an ultrafilter $\mathscr{F}$ refining the Fréchet filter in $\{1, 2, \ldots\}$. Then $\rho(f)(\omega) = \lim_{\mathscr{F}} \varepsilon_n(f)(\omega)$ is defined, and obvious approximation arguments show that the mapping

$$\rho\colon \mathrm{mble}^b(\Omega, \mathscr{B}_\infty, \mathbb{R}) \to \mathbb{R}^\Omega$$

thus defined is linear, positive, sends every constant function onto itself, and coincides with ε_n on $\mathrm{mble}^b(\Omega, \mathscr{B}_n, \mathbb{R})$ $(n = 1, 2, \ldots)$. In order to show that ρ is a linear lifting for $(\Omega, \mathscr{B}_\infty, m)$, we use the submartingale convergence theorem VIII.7.8. In fact, Exercise VIII.7.3.1 shows that for every $f \in \mathrm{mble}^b(\Omega, \mathscr{B}_\infty, \mathbb{R})$ any choice of representatives from the equivalence classes $E_1 f, E_2 f, \ldots \in L_m^1$ yields a martingale. Let us make such a choice via $\rho_1, \rho_2, \ldots$. Thus we see that $\varepsilon_1(f) = \rho_1(E_1 f)$, $\varepsilon_2(f) = \rho_2(E_2 f), \ldots$ is a martingale, hence a submartingale, and obviously an L^1-norm bounded one. It follows that it is m-a.e. convergent. Clearly $\rho(f)$ is an m-a.e. limit of it. On the other hand, it is, by Exercise VIII.7.3.1.2, L^1-norm convergent to f. Hence $\rho(f) = f \bmod m$. Also, it is obvious that $f = f' \bmod m \Rightarrow \varepsilon_n(f) = \varepsilon_n(f')$ $(n = 1, 2, \ldots) \Rightarrow \rho(f) = \rho(f')$. We have thus in fact established a linear lifting ρ for $(\Omega, \mathscr{B}_\infty, m)$.

Now we apply Proposition 1.7 with $\mathscr{B}_\infty$ in place of $\mathscr{B}$ and find, with $\underline{\rho}\colon \mathscr{B}_\infty \to \mathscr{B}_\infty$ as the lower and $\bar{\rho}\colon \mathscr{B}_\infty \to \mathscr{B}_\infty$ as the upper density for ρ, a lifting ρ_0 with $1_{\underline{\rho}(E)} \leqq \rho_0(1_E) \leqq 1_{\bar{\rho}(E)}$ $(E \in \mathscr{B}_\infty)$. Let us now prove that ρ_0 coincides with ρ_n on $\mathrm{mble}^b(\Omega, \mathscr{B}_n, \mathbb{R})$. By uniform approximation by linear combinations of indicator functions we see that it suffices to prove

$\rho_0(1_E) = \rho_n(1_E)$ $(E \in \mathscr{B}_n)$. But ρ_n is a lifting, hence $\rho_n(1_E)$ is, for every $E \in \mathscr{B}_n$, an indicator function again, hence $\rho(1_E) = \rho_n(1_E)$ is an indicator function, and $\underline{\rho}(E) = \bar{\rho}(E)$ follows, whence $\rho_0(1_E) = \rho(1_E)$ since both of these functions are everywhere between $1_{\underline{\rho}(E)}$ and $1_{\bar{\rho}(E)}$.

We are now in a position to prove

1.9. Theorem. *Let* $(\Omega, \mathscr{B}, m)$ *be a complete* σ*-probability space. Then there is a lifting for* $(\Omega, \mathscr{B}, m)$.

Proof. Let $\Sigma = \{(\mathscr{B}_0, \rho_0) | \mathscr{B}_0 \subseteq \mathscr{B}$ is a σ-field containing all m-nullsets, ρ_0 is a lifting for $(\Omega, \mathscr{B}_0, m)\}$. Clearly Σ is nonempty: Choose $\mathscr{B}_0 = \{F | F \in \mathscr{B}, m(F) = 0 \text{ or } = 1\}$; then $\text{mble}^b(\Omega, \mathscr{B}_0, \mathbb{R})$ consists of all m-a.e. constant functions, and we need only put $\rho_0(f) =$ the constant that equals f m-a.e. in order to obtain a lifting ρ_0 for $(\Omega, \mathscr{B}_0, m)$. We introduce the following obvious partial ordering $\prec$ in

$$\Sigma: (\mathscr{B}_0, \rho_0) \prec (\mathscr{B}_1, \rho_1) \quad \Leftrightarrow \quad \mathscr{B}_0 \subseteq \mathscr{B}_1$$

and ρ_1 coincides with ρ_0 on $\text{mble}^b(\Omega, \mathscr{B}_0, \mathbb{R})$. We show now that $(\Sigma, \prec)$ satisfies the hypotheses of Zorn. Let thus Σ_0 be a totally ordered subset of Σ. We have to find a majorant for Σ_0 (and $\prec$) in Σ. *Case I*: There is no countable $\Sigma_1 \subseteq \Sigma_0$ such that every $(B_0, \rho_0) \in \Sigma_0$ has a majorant in Σ_1. Put $\mathscr{B}_\infty = \bigcup_{(B_0, \rho_0) \in \Sigma_0} \mathscr{B}_0$. Clearly $\mathscr{B}_\infty$ is a set field containing all m-nullsets. In order to show that $\mathscr{B}_\infty$ is a σ-field, we choose $E_1, E_2, \ldots \in \mathscr{B}_\infty$, say $E_1 \in \mathscr{B}_1$, $E_2 \in \mathscr{B}_2, \ldots, (B_1, \rho_1), (B_2, \rho_2), \ldots \in \Sigma_0$. In our case I there is some $(B', \rho') \in \Sigma_0$ that has no majorant among $(B_1, \rho_1), \ldots,$ hence $(B_n, \rho_n) \prec (B', \rho')$ $(n = 1, 2, \ldots)$. We conclude $E_1, E_2, \ldots \in \mathscr{B}'$ and thus $E_1 \cup E_2 \cup \cdots \in \mathscr{B}' \subseteq \mathscr{B}_\infty$. A similar argument shows that

$$\text{mble}^b(\Omega, \mathscr{B}_\infty, \mathbb{R}) = \bigcup_{(B_0, \rho_0) \in \Sigma_0} (\Omega, \mathscr{B}_0, \mathbb{R}),$$

that there is a unique mapping ρ_∞ of $\text{mble}^b(\Omega, \mathscr{B}_\infty, \mathbb{R})$ into itself whose restriction to $\text{mble}^b(\Omega, \mathscr{B}_0, \mathbb{R})$ is ρ_0 for every $(B_0, \rho_0) \in \Sigma_0$, and that ρ_0 is a lifting (exercise). We conclude that $(\mathscr{B}_\infty, \rho_\infty) \in \Sigma$ and that it is a majorant of Σ_0. *Case II*: There is a countable $\Sigma_1 \subseteq \Sigma_0$ such that every $(B_0, \rho_0) \in \Sigma_0$ has a majorant in Σ_1. Clearly we may assume that $\Sigma_1 = \{(B_1, \rho_1), (B_2, \rho_2), \ldots\}$ with $(B_1, \rho_1) \prec (B_2, \rho_2) \prec \ldots$. Now Proposition 1.8 yields a majorant $(\mathscr{B}_\infty, \rho_\infty)$ for Σ_1, hence for Σ_0. We now apply Zorn's lemma and obtain a maximal (B', ρ') in Σ. It is sufficient to prove $\mathscr{B}' = \mathscr{B}$. But if $\mathscr{B}' \neq \mathscr{B}$, then Proposition 1.4 yields a $(\mathscr{B}'', \rho'') \neq (\mathscr{B}', \rho')$ that majorizes $(\mathscr{B}', \rho')$, contradicting the maximality of the latter. Thus $\mathscr{B}' = \mathscr{B}$ follows and ρ' is a lifting for $(\Omega, \mathscr{B}, m)$.

1.10. Remark. Looking back over the steps that led to the proof of

Theorem 1.9, it is clear that several ideas have been combined into the proofs of one proposition. A reader who prefers to have it in the "one idea, one theorem" way should consult Ionescu-Tulcea.

We are now going to extend the existence of lifting to more general σ-content spaces. Recall Definition VIII.8.1 of locally measurable and integrable functions and local nullsets.

1.11. Definition. *Let* $(\Omega, \mathscr{B}^{\infty}, m)$ *be a* σ*-content space.*

1.11.1. *By* loc $\text{mble}^b(\Omega, \mathscr{B}^{00}, \mathbb{R})$ *we denote the set of all bounded locally* $\mathscr{B}^{00}$*-measurable real-valued functions on* Ω. *Note that* loc $\text{mble}^b(\Omega, \mathscr{B}^{00}, \mathbb{R})$ *is a vector lattice containing all constants (see also Exercise* VIII.8.3.3). *It is also stable under multiplication.*

1.11.2. *A mapping* ρ *of* loc $\text{mble}^b(\Omega, \mathscr{B}^{00}, \mathbb{R})$ *into itself is called a* **lifting for** $(\Omega, \mathscr{B}^{00}, m)$ *if it has the following properties:*

1.11.2.1. $f \in \text{loc mble}^b(\Omega, \mathscr{B}^{00}, \mathbb{R}) \to \rho(f) = f$ loc mod m.

1.11.2.2. $f, g \in \text{loc mble}^b(\Omega, \mathscr{B}^{00}, \mathbb{R}), f = g$ loc mod $m \Rightarrow \rho(f) = \rho(g)$.

1.11.2.3. ρ *is linear.*

1.11.2.4. ρ *is positive.*

1.11.2.5. $\rho(1)$ *is the constant* 1.

1.11.2.6. ρ *is multiplicative.*

1.12. Remarks. Let $(\Omega, \mathscr{B}^{00}, m)$ be a σ-content space.

1.12.1. Exercise VIII.8.3.3 shows that

$$\text{loc mble}^b(\Omega, \mathscr{B}^{00}, \mathbb{R}) = \text{mble}^b(\Omega, \mathscr{B}, \mathbb{R})$$

for the σ-field $\mathscr{B} = \{F \mid E \in \mathscr{B}^{00} \Rightarrow E \cap F \in \mathscr{B}^{00}\}$. We adhere however to the "loc" notation, in order to emphasize the idea of "locality" underlying the whole construction.

1.12.2. Remark 1.2.3 applies *mutatis mutandis*: It is the same thing to define liftings on loc $\text{mble}^b(\Omega, \mathscr{B}^{00}, \mathbb{R})$ and on loc $\mathscr{L}_m^{\infty}$.

1.12.3. It is our next aim to prove the existence of a lifting for every σ-content space $(\Omega, \mathscr{B}^{00}, m)$ with all local m-nullsets in $\mathscr{B}^{00}$ and with a disjoint base (Definition VIII.8.4). One is tempted to do it in the following way. Let $\mathscr{D} \subseteq \mathscr{B}^{00}$ be a disjoint base for $(\Omega, \mathscr{B}^{00}, m)$. For every fixed $D \in \mathscr{D}$, forget for a moment what happens in $\Omega \backslash D$, apply Theorem 1.9 (after an obvious renorming of the restriction of m to D) and get a lifting ρ_D "within D." For any $f \in \text{loc mble}^b(\Omega, \mathscr{B}^{00}, \mathbb{R})$, define $\rho(f)(\omega)$ to be $= \rho_D(f_D)(\omega)$ $(\omega \in D \in \mathscr{D})$, where f_D denotes the restriction of f to D. The problem that remains open when this procedure is applied is how to define $\rho(f)(\omega)$ for $\omega \in \Omega \backslash \bigcup_{D \in \mathscr{D}} D$. The policy of putting $\rho(f)(\omega) = 0$ there for all

f is not feasible since the constants have to remain constants. We thus find ourselves forced to prove a preparatory

1.13. Proposition. *Let $(\Omega, \mathscr{B}^{00}, m)$ be a σ-content space. Then there is a linear form λ^* on* loc L_m^∞ *that is multiplicative and sends every equivalence class containing a constant onto that same constant.*

Proof. It is in fact meaningful to speak of multiplication in loc L_m^∞ (Exercise X.2.3.2). We essentially copy the extremal point argument used in the proof of Proposition 1.7. Let H^* be the dual Banach space of loc L_m^∞ and E^* its unit ball. Apply the Hahn–Banach theorem (I.6.1) to the subspace H_0 of loc L_m^∞ consisting of all constants (loc mod m), and the majorant function p: loc $L_m^\infty \to \mathbb{R}$ defined by $p(f) =$ loc ess sup f in order to obtain some $\lambda^* \in E^*$ which carries every equivalence class in loc L_m^∞ that contains a constant into that same constant, and is a positive linear form. (*Hint*: Look at Exercise I.8.10.2.)

The set $K^* \subseteq E^*$ of all $\bar{\lambda}^*$ with these properties is thus nonempty. It is clearly convex and closed in the w*-topology of H^*, hence, being contained in the w*-compact set E^*, a convex compact set $\subseteq H^*$ (see Chapter XI, introduction).

By Theorem XV.1.13 (Krein–Milman) there is an extremal point λ^* of K. We have now only to prove that λ^* is multiplicative, i.e., $\lambda^*(fg) = \lambda^*(f)\lambda^*(g)$ ($f, g \in$ loc L_m^∞). It clearly suffices to prove this for all $f \in$ loc L_m^∞ and a fixed $g \in$ loc L_m^∞ with $0 \leqq g \leqq 1$ (loc mod m). Define $\lambda_g^* \in H^*$ by $\lambda_g^*(f) = \lambda^*(fg) - \lambda^*(f)\lambda^*(g)$ ($f \in$ loc L_m^∞). Clearly λ_g^* i is a positive linear form that sends every constant (loc mod m) into 0. Thus $\lambda^* + \lambda_g^*$ and $\lambda^* - \lambda_g^*$ are elements of H^* that send constants into themselves. Let us prove that $\lambda^* + \lambda_g^*, \lambda^* - \lambda_g^* \in K^*$. In fact $0 \leqq f \in$ loc L_m^∞ implies

$$\lambda^*(f) - \lambda_g^*(f) = \lambda^*(f) - \lambda^*(fg) + \lambda^*(f)\lambda^*(g) \geqq \lambda^*(f)\lambda^*(g) \geqq 0$$

(note that $fg \leqq f$), and similarly the positivity of $\lambda^* + \lambda_g^*$ is proved. Since $\lambda^* = \frac{1}{2}(\lambda^* + \lambda_g^*) + (\lambda^* - \lambda_g^*)$ and λ^* is an extremal point of K^*, we find $\lambda^* + \lambda_g^* = \lambda^*$, i.e., $\lambda_g^* = 0$ which means $\lambda^*(fg) = \lambda^*(f)\lambda^*(g)$.

We are now in the position to prove

1.14. Theorem. *Let $(\Omega, \mathscr{B}^{00}, m)$ be a σ-content space with a disjoint base $\mathscr{D}$ (Definition* VIII.8.4). *Assume that $\mathscr{B}^{00}$ contains all local m-nullsets. Then there is a lifting for $(\Omega, \mathscr{B}^{00}, m)$.*

Proof. Use the ideas and notations of Remark 1.12.3 plus some λ^*: loc $L_m^\infty \to \mathbb{R}$ according to Proposition 1.13. Clearly every equivalence class loc mod m in loc mble$^b(\Omega, \mathscr{B}^{00}, \mathbb{R})$ is contained in some equivalence class $\in$ loc L_m^∞. Hence λ^* defines a λ: loc mble$^b(\Omega, \mathscr{B}^{00}, \mathbb{R}) \to \mathbb{R}$ that is constant on every equivalence class in loc mble$^b(\Omega, \mathscr{B}^{00}, \mathbb{R})$ and attains the value prescribed by λ^* there. Clearly λ is a multiplicative, positive linear form on loc mble$^b(\Omega, \mathscr{B}^{00}, \mathbb{R})$ sending every constant onto itself. Clearly

$$\rho(f)(\omega) = \begin{cases} \rho_D(f_D)(\omega) & (\omega \in D \in \mathscr{D}) \\ \lambda(f) & (\text{otherwise}) \end{cases}$$

is a lifting for $(\Omega, \mathscr{B}^{00}, m)$.

1.15. Exercise. Let $(\Omega, \mathscr{B}^{00}, m)$ be a σ-content space such that $\mathscr{B}^{00}$ contains all local m-nullsets. Assume that there is a lifting ρ for $(\Omega, \mathscr{B}^{00}, m)$ and prove that $(\Omega, \mathscr{B}^{00}, m)$ has a disjoint base. (*Hint*: Choose $\mathscr{D}' \subseteq \mathscr{B}^{00}$ such that $E \in \mathscr{D}' \Rightarrow m(E) > 0$, and $E, F \in \mathscr{D}'$, $E \neq F \Rightarrow m(E \cap F) = 0$ and $\mathscr{D}'$ is maximal with these properties; then $\mathscr{D} = \{D \mid 1_D = \rho(1_E) \text{ for some } E \in \mathscr{D}'\}$ is a disjoint base.)

2. STRONG LIFTING

Every lifting sends a huge space of measurable functions into a comparatively small one which: (1) is a system of representatives for the equivalence classes (for the given σ-content) of the huge space; (2) is stable under all linear operations, multiplication, and uniform approximation, and (3) consists of exactly those functions in the huge space that are mapped into themselves by the lifting. Let us concentrate upon the third property for the moment. What are the functions for which we can say a priori that they are sent into themselves by the lifting? Our Definition 1.11 requires nothing but that the constants be among them. Now if a topology is given in the basic space and $\mathscr{B}^{00}$ is related to it in a natural fashion, then it is natural to make the *stronger requirement* that the lifting send, e.g., every bounded continuous function into itself; or, in other words: *From those equivalence classes that contain a continuous function, such a function is picked.*

One faces severe technical difficulties when trying to prove the existence of liftings with such strong additional properties without making further assumptions. A very obvious difficulty arises if it can happen that an equivalence class contains more than one continuous function; this is prevented by assuming that every open set $\neq \varnothing$ has a subset with strictly positive content.

In this section we set up an abstract definition of strong lifting in accordance with the above discussion and prove their existence under certain separability assumptions. This includes the case of locally compact Hausdorff spaces with a countable base and a σ-content that is >0 on every conditionally compact nonempty open set. At present the existence of a strong lifting without separability assumptions is an open problem even in the compact case. The case of a Polish space is unsettled as well.

2.1. Definition. *Let* $(\Omega, \mathscr{B}^{00}, m)$ *be a* σ*-content space.*

2.1.1. *A set* $\mathscr{A} \subseteq \mathbb{R}^{\Omega}$ *is called a* **lattice algebra** (*of real functions on* Ω) *if it is stable under linear operations and multiplication and if it is a vector lattice.*

2.1.2. *A lattice algebra* $\mathscr{A} \subseteq \mathbb{R}^{\Omega}$ *is called* **feasible** *for* $(\Omega, \mathscr{B}^{00}, m)$ *if:*

2.1.2.1. $\mathscr{A} \subseteq \text{loc mble}^{\text{b}}(\Omega, \mathscr{B}^{00}, \mathbb{R})$;

2.1.2.2. $f, g \in \mathscr{A}, f = g$ loc mod $m \Rightarrow f = g$.

2.1.2.3. $f \in \mathscr{A}$, $f \leqq 1$ loc mod $m \Rightarrow f \leqq 1$ everywhere.

2.1.3. *Let* $\mathscr{A}$ *be a lattice algebra feasible for* $(\Omega, \mathscr{B}^{00}, m)$. *A lifting* ρ *for* $(\Omega, \mathscr{B}^{00}, m)$ *is said to be* **strong** *for* $\mathscr{A}$ *iff* $f \in \mathscr{A} \Rightarrow \rho(f) = f$.

2.2. Remarks. Let $(\Omega, \mathscr{B}^{00}, m)$ be a σ-content space.

2.2.1. Every lifting for $(\Omega, \mathscr{B}^{00}, m)$ is strong for the feasible lattice algebra of all constants, by the very definition of a lifting.

2.2.2. A standard situation in which a feasible lattice algebra presents itself in a most natural fashion is the following. Let $(\Omega, \mathscr{T})$ be a topological space such that every set $\varnothing \neq U \in \mathscr{T}$ contains a set $E \in \mathscr{B}^{00}$ with $m(E) > 0$. Assume further that $\mathscr{C}^{\text{b}}(\Omega, \mathbb{R}) \subseteq \text{loc mble}^{\text{b}}(\Omega, \mathscr{B}^{00}, \mathbb{R})$. Then every lattice algebra $\mathscr{A} \subseteq \mathscr{C}^{\text{b}}(\Omega, \mathbb{R})$ is feasible for $(\Omega, \mathscr{B}^{00}, m)$. In fact, let $f, g \in \mathscr{A}$, $f = g$ loc mod m. Assume that $f \neq g$. Then by continuity there is a $\varnothing \neq U \in \mathscr{T}$ with $f \neq g$ on U. Find $\mathscr{B}^{00} \ni E \subseteq U$ with $m(E) > 0$ in order to lead $f = g$ loc mod m to a contradiction.

2.2.3. Let us list two popular special cases of 2.2.2.

2.2.3.1. $(\Omega, \mathscr{T})$ is locally compact. $(\Omega, \mathscr{B}^{00}, m)$ is obtained from a positive τ-measure $m\colon \mathscr{C}^{00}(\Omega, \mathbb{R}) \to \mathbb{R}$ (i.e., by Theorem I.10.2, an arbitrary positive linear form on $\mathscr{C}^{00}(\Omega, \mathbb{R})$) by passing to the τ-derived σ-content (Chapter III, Section 7) and adjoining all local nullsets (Remark VIII.8.2.3). The lattice algebra $\mathscr{C}^{00}(\Omega, \mathbb{R})$ is then feasible for $(\Omega, \mathscr{B}^{00}, m)$, provided the measure m from which we started has the property $0 \leqq f \in \mathscr{C}^{00}(\Omega, \mathbb{R})$, $f \neq 0 \Rightarrow m(f) > 0$. This is, e.g., the case for Haar measures on a locally compact topological group (see Theorem XII.3.10.3). It is unknown whether a strong lifting for $\mathscr{C}^{00}(\Omega, \mathbb{R})$ exists in this case. If the topology $\mathscr{T}$ has a countable base, then the existence follows from Theorem 2.4 below.

2.2.3.2. $(\Omega, \mathscr{T})$ is a Polish space and $(\Omega, \mathscr{B}(\mathscr{T}), m)$ is a σ-content space such that $\varnothing \neq U \in \mathscr{T} \Rightarrow m(U) > 0$. From $(\Omega, \mathscr{B}, m)$ by completion of

$(\Omega, \mathscr{B}(\mathscr{T}), m)$ (Theorem II.6.2.2). Then $\mathscr{C}^{\mathrm{b}}(\Omega, \mathbb{R})$ is a feasible lattice algebra for $(\Omega, \mathscr{B}, m)$.

We intend to prove the existence of strong liftings under some separability assumption about the feasible algebra. For technical reasons, we first prove the

2.3. Proposition. *Let $(\Omega, \mathscr{B}^{00}, m)$ be a σ-content space such that $\mathscr{B}^{00}$ contains all local m-nullsets. Let $\mathscr{A}$ be a feasible lattice algebra for $(\Omega, \mathscr{B}^{00}, m)$. Then for every $\omega \in \Omega$, there is a mapping λ_ω:* loc mble$^{\mathrm{b}}(\Omega, \mathscr{B}^{00}, \mathbb{R}) \to \mathbb{R}$ *with the following properties:*

2.3.1. $f, g \in$ loc mble$^{\mathrm{b}}(\Omega, \mathscr{B}^{00}, \mathbb{R}), f = g$ loc mod $m \Rightarrow \lambda_\omega(f) = \lambda_\omega(g)$.

2.3.2. $f \in \mathscr{A} \Rightarrow \lambda_\omega(f) = f(\omega)$.

Proof. Every $f \in \mathscr{A}$ is in some equivalence class from loc L_m^∞, and different f's from $\mathscr{A}$ are in different classes (Definition 2.1.2.2). We may thus consider $\mathscr{A}$ as a subset of loc L_m^∞. As such, it is a subalgebra of the algebra loc L_m^∞. Fix some $\omega \in \Omega$ and define λ_ω^*: $\mathscr{A} \to \mathbb{R}$ by $\lambda_\omega^*(f) = f(\omega)$. Clearly λ_ω^* is a linear form on $\mathscr{A}$; and if we adopt the majorant function $p(h) =$ loc ess sup h on loc L_m^∞, we certainly have $\lambda_\omega^*(f) \leqq p(f)$ $(f \in \mathscr{A})$. The Hahn–Banach theorem (I.6.1) provides us with an extension λ^* of λ_ω^* from $\mathscr{A}$ to loc L_m^∞, such that $\lambda^* \leqq p$. This shows that the set K^* of all linear forms λ^* on loc L_m^∞ that extend λ_ω^* and satisfy $|\lambda^*(f)| \leqq 1$ $(f \in$ loc L_m^∞, $\|f\|_{\mathrm{loc}\ \infty} \leqq 1)$ and $\lambda^*(f) \geqq 0$ $(0 \leqq f \in$ loc $L_m^\infty)$ is nonempty. It is clearly a convex closed subset of the unit ball in the dual Banach space of the Banach space loc L_m^∞, hence is compact in the w*-topology of that space. By Theorem XV.1.13 (Krein–Milman), K^* has an extremal point which we denote by λ_ω. Let us prove that λ_ω is multiplicative. This is now achieved in the same way as in the proof of Proposition 1.13, with only a little complication. For any $g \in$ loc L_m^∞, let $\lambda_g(h) = \lambda(gh) - \lambda(g)\lambda(h)$. We begin with the special case $g \in \mathscr{A}$, $0 \leqq g \leqq 1$ and see that λ_g is a linear form on loc L_m^∞ such that $h \in \mathscr{A} \Rightarrow gh \in \mathscr{A} \Rightarrow$

$$\lambda_g(h) = \lambda(gh) - \lambda(g)\lambda(h) = g(\omega)h(\omega) - g(\omega)h(\omega) = 0.$$

Thus $\lambda_\omega + \lambda_g$ and $\lambda_\omega - \lambda_g$ coincide with λ_ω^* on $\mathscr{A}$. It is an easy exercise to show that they actually belong to K^*. Since

$$\lambda_\omega = \tfrac{1}{2}[(\lambda_\omega + \lambda_g) + (\lambda_\omega - \lambda_g)]$$

and λ_ω is an extremal point of K^*, we conclude

$$\lambda_\omega + \lambda_g = \lambda_\omega \quad \Rightarrow \quad \lambda_g = 0 \quad \Rightarrow \quad \lambda_\omega(gh) = \lambda_\omega(g)\lambda_\omega(h) \qquad (h \in \text{loc } L_m^\infty).$$

Since $\mathscr{A}$ is a vector lattice, we conclude that $\lambda_\omega(gh) = \lambda_\omega(g)\lambda_\omega(h)$ $(g \in \mathscr{A},\ h \in$ loc $L_m^\infty)$. Now we repeat the above argument, but with

$g \in \operatorname{loc} L_m^\infty$, $0 \leqq g \leqq 1$. By what we know already we get $\lambda_g(h) = 0$ $(h \in \mathscr{A})$. This does it.

We are now in a position to prove

2.4. Theorem. *Let $(\Omega, \mathscr{B}^{00}, m)$ be a σ-content space such that $\mathscr{B}^{00}$ contains all local m-nullsets. Let $\mathscr{A}$ be a feasible lattice algebra for $(\Omega, \mathscr{B}^{00}, m)$ and assume that there is a countable set $\mathscr{A}_0 \subseteq \mathscr{A}$ such that every $f \in \mathscr{A}$ is a uniform limit of a sequence from $\mathscr{A}_0$. Then there is a strong lifting for $(\Omega, \mathscr{B}^{00}, m)$ and $\mathscr{A}$.*

Proof. Let ρ_0 be a lifting for $(\Omega, \mathscr{B}^{00}, m)$ (Theorem 1.14). Let $\mathscr{A}_0 = \{f_1, f_2, \ldots\}$. For every $n = 1, 2, \ldots$, let $N_n = \{\omega \mid \rho_0(f_n)(\omega) \neq f_n(\omega)\}$. Put $N = N_1 \cup N_2 \cup \cdots$. Clearly N is a local m-nullset and we have $\omega \in \Omega \backslash N \Rightarrow \rho_0(f)(\omega) = f(\omega)$ $(f \in \mathscr{A}_0)$. Since ρ_0 commutes with uniform approximation (compare Exercise 1.3.1), we get even

$$\omega \in \Omega \backslash N \Rightarrow \rho_0(f)(\omega) = f(\omega) \qquad (f \in \mathscr{A}).$$

For every $\omega \in N$, we choose some λ_ω according to Proposition 2.3. For every $f \in \operatorname{loc} \operatorname{mble}^b(\Omega, \mathscr{B}^{00}, \mathbb{R})$, we put

$$\rho(f)(\omega) = \begin{cases} \rho_0(f)(\omega) & (\omega \in \Omega \backslash N) \\ \lambda_\omega(f) & (\omega \in N). \end{cases}$$

It is now an easy exercise to verify that ρ is a strong lifting for $(\Omega, \mathscr{B}^{00}, m)$ and $\mathscr{A}$.

2.5. Exercises. Let $(\Omega, \mathscr{T})$ be a locally compact Hausdorff space and $\mathscr{B}^{00}(\mathscr{T})$ the local σ-ring of all conditionally compact Borel sets in Ω. Let m be a positive inner compact regular σ-content on $\mathscr{B}^{00}(\mathscr{T})$ such that $m(U) > 0$ for every open nonempty $U \in \mathscr{B}^{00}(\mathscr{T})$. Let m also denote the unique σ-content that extends m to the smallest local σ-ring $\mathscr{B}^{00}$ containing $\mathscr{B}^{00}(\mathscr{T})$ and all local m-nullsets.

2.5.1. Assume the existence of a countable base for $\mathscr{T}$. Prove that there is a strong lifting for $(\Omega, \mathscr{B}^{00}, m)$ and $\mathscr{C}^{00}(\Omega, \mathbb{R})$. (*Hint*: Use Remark 2.2.3.1 and the fact that there is a countable subset of $\mathscr{C}^{00}(\Omega, \mathbb{R})$ that is dense in $\mathscr{C}^{00}(\Omega, \mathbb{R})$ for uniform approximation.)

2.5.2. Assume that for every compact $K \subseteq \Omega$, there is a countable base for the relative topology $\mathscr{T} \cap K = \{U \cap K \mid U \in \mathscr{T}\}$ in K. Prove that there is a strong lifting for $(\Omega, \mathscr{B}^{00}, m)$ and $\mathscr{C}^b(\Omega, \mathbb{R})$. (*Hint*: Use Theorem VIII.8.7 in order to get a disjoint base $\mathscr{D}$ for $(\Omega, \mathscr{B}^{00}, m)$ consisting of compact sets; apply Exercise 2.5.1 "locally" on every $D \in \mathscr{D}$ and use Proposition 2.3 as well as the ideas of the proofs for Theorems 1.14 and 2.4.)

3. APPLICATIONS

In this section we present several applications of lifting: *kernel representations* of mappings, conditional probabilities, *disintegration of measures, Strassen's theorem,* and a result on so-called *separability* of stochastic processes. Many of these topics have been treated in the literature before the existence of lifting was known. Lifting supplants elegantly all the cumbersome older devices which were designed in their time in order to ensure some additivity, linearity, etc.

It should be noted that the scope of applications of lifting is considerably enlarged if one can dispose of a theory of integration of Banach-space-valued functions, which is not the case in the present book. The non-specialist might, though, be grateful to find here important special cases of famous theorems whose more general versions do not show at a glance how much ground they really cover. Readers with specialized interests are referred to A. and C. IONESCU-TULCEA [1] and BICHTELER [1].

1. Kernel Representations of Positive Linear Mappings

The reader is reminded of the discussion of Problem IV.4.18 concerning kernel representation given in subsection 5 of Chapter IV, Section 4. We present here a few answers to Problem IV.4.18.

3.1. Theorem. *Let* $(\Omega', \mathcal{T}')$ *be a compact Hausdorff space and* $(\Omega, \mathcal{E}, m)$, $(\Omega', \mathcal{C}(\Omega', \mathbb{R}), m')$ *σ-probability measure spaces. Let* $P{:}\, P{:}\, L^1_m \leftarrow L^1_{m'}$ *be a positive linear mapping that sends the constant* 1 (mod m) *into the constant* 1 (mod m'), *and preserves integrals, i.e., satisfies*

$$\int (Pf')\, dm = \int f'\, dm' \qquad (f' \in L^1_m).$$

Then there is a stochastic σ-measure kernel P *from* Ω *to* $(\Omega', \mathcal{C}(\Omega', \mathbb{R}))$ *such that*

3.1.1. *The kernel* P *is* m*-integrable and* $mP = m'$ (*Theorem* IV.4.8.1).

3.1.2. *The kernel* P *induces the mapping* $P{:}\, L^1_m \leftarrow L^1_m$ (*Theorem* IV.4.10).

Proof. Let $(\Omega, \mathcal{B}_m, m)$ be the σ-content space derived from the σ-measure space $(\Omega, \mathcal{E}, m)$. Clearly it is a σ-probability space and complete (Theorem III.6.1). Let ρ: $\text{mble}^b(\Omega, \mathcal{B}_m, \mathbb{R}) \to \text{mble}^b(\Omega, \mathcal{B}_m, \mathbb{R})$ be a lifting for $(\Omega, \mathcal{B}_m, m)$ (Theorem 1.9; actually, a linear lifting would do it). Define $P(\omega, f') = \rho(Pf')(\omega)$ $(\omega \in \Omega,\ f' \in \mathcal{C}(\Omega', \mathbb{R}))$. This is meaningful since the equivalence class Pf' intersects mble$^b(\Omega, \mathscr{B}_m, \mathbb{R})$ in exactly one equivalence class. For every $\omega \in \Omega$, this defines a positive linear form, i.e., a τ-measure $P(\omega, \cdot)$ on $\mathscr{C}(\Omega', \mathbb{R})$ (use the linearity properties (everywhere!) of a lifting and Theorem I.10.1). Thus we have a σ-measure kernel (even a τ-measure kernel) from Ω to $(\Omega', \mathscr{C}(\Omega', \mathbb{R}))$. Moreover, it is a $\mathscr{B}_m$-measurable stochastic kernel (use more properties of a lifting). In order to show that this kernel represents the mapping $P\colon L^1_m \leftarrow L^1_{m'}$ of functions, we essentially copy the proof of Theorem IV.4.10, i.e., we go through the σ-extension theory (Chapter III.), both for the σ-measure m' and the σ-measures $P(\omega, \cdot)$ on $\mathscr{C}(\Omega', \mathbb{R})$. We present a general argument about monotone convergence which we shall apply repeatedly later. Let

$$0 \leqq f'_1 \leqq f'_2 \leqq \cdots \in \overline{\mathbb{R}}^{\Omega}, \ f'(\omega') = \lim_n f'_n(\omega') \qquad (\omega' \in \Omega')$$

and assume that we know already the following: For every $n = 1, 2, \ldots,$ we have $f'_n \in \mathscr{L}^1_\sigma(m')$ and there is an m-nullset N_n such that

$$\omega \in \Omega \backslash N_n \quad \Rightarrow \quad f'_n \in L^1_\sigma(P(\omega, \cdot)),$$

and the m-a.e. defined function $\int P(\cdot, d\omega') f'_n(\omega')$ belongs to the equivalence class Pf'_n mod m. Assume further $\lim_n \int f'_n \, dm' < \infty$. We shall prove the analogous statement about f'. In fact, the monotone convergence theorem (III.5.3), when applied to the σ-measure m', implies $f' \in \mathscr{L}^1_\sigma(m')$ and $\int f' \, dm' = \lim_n \int f'_n \, dm'$, hence $\int (Pf') \, dm = \lim_n \int (Pf'_n) \, dm$. The monotone convergence theorem, when applied to the σ-measure m, shows $Pf' = \lim_n Pf'_n$ mod m, and this proves

$$(Pf')(\omega) = \lim_n \int P(\omega, d\omega') f'_n(\omega') \quad \text{mod } m.$$

In particular, there is an m-nullset N such that $\lim_n \int P(\omega, d\omega') f'_n(\omega') < \infty$ $(\omega \in \Omega \backslash N)$. The monotone convergence theorem for $P(\omega, \cdot)$ $(\omega \in \Omega \backslash N)$ shows $\omega \in \Omega \backslash N \Rightarrow f' \in \mathscr{L}^1_\sigma(P(\omega, \cdot))$, $\int P(\omega, d\omega') f'(\omega') = \lim_n \int P(\omega, d\omega') f'_n(\omega')$, hence the m-a.e. defined function $\int P(\cdot, d\omega') f'(\omega')$ represents Pf'. It is obvious how to turn this "upward" argument into a "downward" argument.

Going through the σ-extension process with this argument, we see: If $u' \in \overline{\mathbb{R}}^{\Omega'}$ is a σ-upper or σ-lower function for $\mathscr{C}(\Omega', \mathbb{R})$ which is in $\mathscr{L}^1_\sigma(m')$, then there is an m-nullset N such that $u' \in \mathscr{L}^1_\sigma(P(\omega, \cdot))$ for $\omega \in \Omega \backslash N$, and the m-a.e. defined function $\int P(\cdot, d\omega') u'(\omega')$ represents Pu'. The same follows for countable suprema of σ-lower and countable infima of σ-upper functions. Let now $h' \in \mathscr{L}^1_\sigma(m')$. Find a countable supremum u' of σ-lower functions and a countable infimum v' of σ-upper functions such that $u' \leqq h' \leqq v'$ everywhere on Ω' and $\int u'\, dm' = \int v'\, dm'$, i.e., $u' = h' = v' \mod m'$. We conclude $Pu' = Ph' = Pv' \mod m$ and find an m-nullset N such that

$$\omega \in \Omega\backslash N \quad \Rightarrow \quad \int P(\omega, d\omega')u'(\omega') = \int P(\omega, d\omega')v'(\omega').$$

Thus $\omega \in \Omega\backslash N \Rightarrow h' \in \mathscr{L}^1_\sigma(P(\omega, \cdot))$ and $\int P(\omega, d\omega')u'(\omega') = \int P(\omega, d\omega')h'(\omega')$. The m-a.e. defined function $\int P(\cdot, d\omega')h'(\omega')$ therefore represents $Pu' = Ph'$. This does it.

3.2. Theorem. *Let $(\Omega', \mathscr{T}')$ be a Polish space and $(\Omega, \mathscr{B}, m)$, $(\Omega', \mathscr{B}(\mathscr{T}'), m')$ be σ-probability (content) spaces. Assume that $(\Omega, \mathscr{B}, m)$ is complete. Let $P\colon L^1_m \leftarrow L^1_{m'}$ be a positive linear mapping that sends the constant* 1 (mod m') *into the constant* 1 (mod m) *and preserves integrals, i.e., satisfies*

$$\int (Pf')\, dm = \int f'\, dm' \qquad (f' \in L^1_{m'}).$$

Then there is a stochastic σ-content kernel P from Ω to $(\Omega', \mathscr{B}(\mathscr{T}'))$ that is $\mathscr{B}$-measurable and satisfies:

3.2.1. *The kernel P is m-integrable and satisfies $mP = m'$ (Theorem* IV.4.8.3).

3.2.2. *The kernel P induces the mapping $P\colon L^1_m \leftarrow L^1_{m'}$ (Exercise* IV.4.11.1).

Proof. There is a countable set field $\mathscr{F}' \subseteq \mathscr{B}(\mathscr{T}')$ satisfying $\mathscr{B}(\mathscr{F}') = \mathscr{B}(\mathscr{T}')$ (take, e.g., the set field generated by some countable base of the Polish topology $\mathscr{T}'$). By Theorem V.5.3, m' is inner compact regular. Thus for every $E' \in \mathscr{F}'$ there is a sequence $K'_1 \subseteq K'_2 \subseteq \cdots$ of compact sets $\subseteq E'$ such that $\lim_n m'(K'_n) = m'(E')$. Let now

$$\rho\colon \mathrm{mble}^{\mathrm{b}}(\Omega, \mathscr{B}, \mathbb{R}) \to \mathrm{mble}^{\mathrm{b}}(\Omega, \mathscr{B}, \mathbb{R})$$

be a lifting for $(\Omega, \mathscr{B}, m)$ (Theorem 1.9; actually, a linear lifting would do it). For every $E' \in \mathscr{B}(\mathscr{T}')$, we put $P_0(\omega, E') = \rho(P1_{E'})(\omega)$ $(\omega \in \Omega)$. This is meaningful since the equivalence class $P1_{E'}$ intersects $\mathrm{mble}^{\mathrm{b}}(\Omega, \mathscr{B}, \mathbb{R})$ in exactly one equivalence class mod m. Clearly $P_0(\omega, \cdot)$ is, for every $\omega \in \Omega$, a probability content on $\mathscr{B}(\mathscr{T}')$. Fix some $E' \in \mathscr{F}'$ and choose $K'_1 \subseteq K'_2 \subseteq \cdots \subseteq E'$ as above. Clearly $\lim_n m'(K'_n) = m'(E')$ means

$\lim_n \int (1_{E'} - 1_{K'_n})\, dm' = 0$, and this implies $\lim_n \int m(d\omega)P_0(\omega, E'\backslash K'_n) = 0$. There is thus a m-nullset $N(E')$ such that

$$\omega \in \Omega\backslash N(E') \quad \Rightarrow \quad \lim_n P_0(\omega, K'_n) = P_0(E').$$

Let $N = \bigcup_{E' \in \mathscr{F}'} N(E')$. Clearly N is an m-nullset and $\omega \in \Omega\backslash N \Rightarrow$ the content $P_0(\omega, \cdot)$, when restricted to $\mathscr{F}'$, is inner regular for the σ-compact system of all compact subsets of $\Omega' \Rightarrow P_0(\omega, \cdot)$ is σ-additive on $\mathscr{F}'$ (Proposition V.1.6). Choose any $\omega'_0 \in \Omega'$ and put

$$P(\omega, E') = \begin{cases} P_0(\omega, E') & (\omega \in \Omega\backslash N) \\ 1_{E'}(\omega'_0) & (\omega \in N) \end{cases}$$

for $E' \in \mathscr{F}'$. This defines a stochastic σ-content kernel P from Ω to Ω' with $\mathscr{F}'$, which is $\mathscr{B}$-measurable. By Remarks IV.4.2.3 and IV.4.13.3 we obtain a $\mathscr{B}$-measurable stochastic σ-content kernel P from Ω to Ω' with $\mathscr{B}(\mathscr{T}') = \mathscr{B}(\mathscr{F}')$ if we take, for every $\omega \in \Omega$, the unique σ-extension of the σ-content $P(\omega, \cdot)$ from $\mathscr{F}'$ to $\mathscr{B}(\mathscr{T}')$ and denote it by $P(\omega, \cdot)$ again. The stochastic σ-content kernel P thus obtained clearly satisfies $P(\cdot, E') = P1_{E'}$ mod m. An obvious monotone extension procedure (see the proof of Theorem 3.1 resp. of Theorem IV.4.10) shows that this kernel induces the mapping $P: L^1_m \leftarrow L^1_{m'}$.

3.3. Exercises

3.3.1. Carry Theorem 3.2 over to the case where $(\Omega', \mathscr{B}', m')$ is a σ-probability space such that $\mathscr{B}$ contains a σ-compact class $\mathscr{K}'$ of sets for which m' is inner regular.

3.3.2. Try to carry Theorem 3.1 over to cases where $(\Omega', \mathscr{T}')$ is locally compact.

It is of particular interest to know when a mapping $P: L^1_m \leftarrow L^1_{m'}$ is induced by a point mapping rather than a kernel. We shall treat this problem explicitly in the Polish case and leave its treatment in other, related cases to exercises.

We begin with a preparatory

3.4. Theorem. *Let $(\Omega, \mathscr{T})$ be a Polish space and $(\Omega, \mathscr{B}(\mathscr{T}), m)$ a σ-probability content space. Then the following two statements are equivalent:*

3.4.1. *m is concentrated on one point ω_0, i.e., $m(E) = 1_E(\omega_0)$ $(E \in \mathscr{B}(\mathscr{T}))$.*

3.4.2. *m is multiplicative, more precisely*

$$m(E \cap F) = m(E)m(F) \qquad (E, F \in \mathscr{B}(\mathscr{T})).$$

Proof. $1 \Rightarrow 2$ is obvious.

$2 \Rightarrow 1$. Let $\mathscr{K} = \{K \mid K \subseteq \Omega,\ K \text{ compact}\}$. Clearly $\mathscr{K} \subseteq \mathscr{B}(\mathscr{T})$ and m is inner $\mathscr{K}$-regular (Theorem V.5.3). On the other hand, $E \in \mathscr{B}(\mathscr{T}) \Rightarrow m(E) = m(E \cap E) = m(E)m(E) \Rightarrow m(E) = 0$ or $= 1$. Let $\mathscr{K}_1 = \{K \mid K \in \mathscr{K}, m(K) = 1\}$. Clearly $\mathscr{K}_1$ is stable under finite intersections and consists of nonempty compact sets. Thus $K_1 = \bigcap_{K \in \mathscr{K}_1} K$ is nonempty. Choose $\omega_0 \in K_1$. Then $m(\{\omega_0\}) = 1$ because otherwise $K \in \mathscr{K}_1 \Rightarrow m(K \setminus \{\omega_0\}) > 0 \Rightarrow m(K') = 1$ for some compact $K' \subseteq K \setminus \{\omega_0\}$ by inner compact regularity and the fact that all m-values are 0 or 1. But then $\omega_0 \notin K_1$ follows, a contradiction. This does it.

3.5. Theorem. *Let $(\Omega', \mathscr{T}')$ be a Polish space. Let $(\Omega, \mathscr{B}, m)$, $(\Omega', \mathscr{B}(\mathscr{T}'), m')$ be σ-probability content spaces and assume that $(\Omega, \mathscr{B}, m)$ is complete. Let $P\colon L_m^1 \leftarrow L_{m'}^1$ be linear, positive, and integral preserving: $\int (Pf')\, dm = \int f'\, dm$ $(f' \in L_m^1)$. Then the following statements are equivalent:*

3.5.1. *There is a $\mathscr{B}$-$\mathscr{B}(\mathscr{T}')$-measurable mapping $T\colon \Omega \to \Omega'$ such that $Tm = m'$ and $Pf' = f' \circ T \bmod m$ $(f' \in L_m^1)$.*

3.5.2. *P is multiplicative on indicator functions, more precisely*

$$E', F' \in \mathscr{B}(\mathscr{T}') \quad \Rightarrow \quad P1_{E' \cap F'} = (P1_{E'})(P1_{F'}) \mod m.$$

Proof. $1 \Rightarrow 2$ is obvious.

$2 \Rightarrow 1$. Let ρ be a lifting (now a linear lifting would not do any more) for $(\Omega, \mathscr{B}, m)$ and construct $\mathscr{F}'$ and the $\mathscr{B}$-measurable stochastic σ-content kernel P from Ω to Ω' (with $\mathscr{B}(\mathscr{T}')$) as in the proof of Theorem 3.2. $E', F' \in \mathscr{F}', \omega \in \Omega \Rightarrow P(\omega, E' \cap F') = P(\omega, E')P(\omega, F')$ is now obvious from the multiplicativity of a lifting (plus the construction of $P(\omega, \cdot)$ for ω in the exceptional set N). For every $\omega \in \Omega$ and $E' \in \mathscr{F}'$, the set system $\{F' \mid F' \in \mathscr{B}(\mathscr{T}'),\ P(\omega, E' \cap F') = P(\omega, E')P(\omega, F')$ is clearly a monotone class containing $\mathscr{F}'$, and hence equal to $\mathscr{B}(\mathscr{T}')$ (Exercise I.5.5). Thus $\omega \in \Omega$, $E' \in \mathscr{F}'$, $F' \in \mathscr{B}(\mathscr{T}') \Rightarrow P(\omega, E' \cap F') = P(\omega, E')P(\omega, F')$. Repeating the same argument, now with $E' \in \mathscr{B}(\mathscr{T}')$, we get $\omega \in \Omega$, E', $F' \in \mathscr{B}(\mathscr{T}') \Rightarrow P(\omega, E' \cap F') = P(\omega, E')P(\omega, F')$. For every $\omega \in \Omega$, we now apply Theorem 3.4 to $P(\omega, \cdot)$ in order to obtain a point $T\omega \in \Omega'$ such that $P(\omega, E') = 1_{E'}(T\omega)$ $(E' \in \mathscr{B}(\mathscr{T}'))$. It is obvious from the $\mathscr{B}$-measurability of the kernel P that the mapping T is $\mathscr{B}$-$\mathscr{B}(\mathscr{T}')$-measurable. Moreover, we get $Pf' = f' \circ T \bmod m$, first for $f' = 1_{E'}$, $E' \in \mathscr{B}(\mathscr{T}')$, and then, by obvious extension procedures, for $f' \in \mathscr{L}_{m'}^1$.

3.6. Exercises

3.6.1. Carry Theorem 3.5 over to the case of a compact Hausdorff space $(\Omega', \mathscr{T}')$, a complete σ-probability space $(\Omega, \mathscr{B}, m)$, and a σ-measure space $(\Omega', \mathscr{C}^{00}(\Omega', \mathbb{R}), m')$.

3.6.2. Call two σ-probability content spaces $(\Omega', \mathscr{B}', m')$, $(\Omega'', \mathscr{B}'', m'')$ *isomorphic* if there are an m'-nullset $N' \in \mathscr{B}'$, an m''-nullset $N'' \in \mathscr{B}''$, and a bijection φ: $\Omega' \backslash N' \to \Omega'' \backslash N'$ that sends $\mathscr{B}' \cap (\Omega' \backslash N')$ into $\mathscr{B}'' \cap (\Omega'' \backslash N'')$ forward and vice versa backward and satisfies $m'(F') = m''(\varphi(F' \backslash N'))$ $(F' \in \mathscr{B}')$. φ is called an *isomorphism* between the two σ-probability spaces, of course.

3.6.2.1. Carry Theorem 3.5 over to the case where $(\Omega', \mathscr{B}(\mathscr{T}'), m')$ is replaced by a σ-probability space that is isomorphic to a σ-probability space whose σ-field is generated by a Polish topology. (*Hint*: Apply Theorem 3.5 to the Polish space and "transfer" the result via isomorphism.)

3.6.2.2. Let $(\Omega, \mathscr{B}, m)$, $(\Omega', \mathscr{B}', m')$ be σ-probability spaces which are each isomorphic to a σ-probability space whose σ-field is generated by a Polish topology. Let P: $L^1_m \leftarrow L^1_{m'}$ be a bijective linear positive mapping such that both P and P^{-1} commute with multiplication for indicator functions of sets from $\mathscr{B}'$ resp. $\mathscr{B}$ (mod m). Prove that $(\Omega, \mathscr{B}, m)$ and $(\Omega', \mathscr{B}', m')$ are isomorphic by finding an isomorphism inducing P.

3.6.2.3. Weaken the hypothesis of Exercise 3.6.2.2 to the following: There are set systems $\mathscr{M} \subseteq \mathscr{B}$, $\mathscr{M}' \subseteq \mathscr{B}'$ that are stable under finite intersections such that $(\Omega, \mathscr{B}, m)$ has the same completion as $(\Omega, \mathscr{B}(\mathscr{M}), m)$ and $(\Omega', \mathscr{B}', m')$ has the same completion as $(\Omega', \mathscr{B}'(\mathscr{M}'), m')$, and P sends the set of indicator functions of sets from $\mathscr{M}'$ bijectively on indicator functions of sets from $\mathscr{M}$ (mod m' resp. mod m, of course) commuting with multiplication of such functions as well as P^{-1}.

3.6.3. Let $(\Omega, \mathscr{B}, m)$ be a σ-probability content space and $(\Omega, \mathscr{B}_m, m)$ its completion.

3.6.3.1. Prove that every $f \in \text{mble}\ (\Omega, \mathscr{B}_m, \mathbb{R})$ equals mod m some $f \in \text{mble}(\Omega, \mathscr{B}, \mathbb{R})$. (*Hint*: Begin with indicator functions and use approximation by step functions with rational values.)

3.6.3.2. Prove that the completeness assumptions in Theorems 3.2 and 3.5 can be dropped. (*Hint*: Enlarge the exceptional sets used in the proofs of those theorems by countable unions of nullsets coming from an application of Exercise 3.6.3.1.)

2. Conditional Probabilities

We now specialize the theory of subsection 1 to a particular class of linear positive mappings: *conditional expectations* (Chapter VIII, Section 6). In view of Exercise VI.2.7.5 the following results settle also the problem of representing a σ-content in a product space in the form $m_0 \times P$ (Theorem VI.2.3, Example VI.2.6).

3.7. Theorem. *Let $(\Omega, \mathscr{T})$ be a Polish space and $(\Omega, \mathscr{B}(\mathscr{T}), m)$ a σ-probability content space. Let $\mathscr{B}_0 \subseteq \mathscr{B}(\mathscr{T})$ be a σ-field and E_0: $\mathscr{L}_m^1 \cap \text{mble}(\Omega, \mathscr{B}_0, \mathbb{R}) \leftarrow \mathscr{L}_m^1$ the conditional expectation for $\mathscr{B}_0$. For every $\omega \in \Omega$, let $[\omega]_0 = \bigcap_{\omega \in E \in \mathscr{B}_0} E$. Then there is a countably generated σ-field $\mathscr{B}_{00} \subseteq \mathscr{B}_0$ such that $(\Omega, \mathscr{B}_{00}, m)$ and $(\Omega, \mathscr{B}_0, m)$ have the same completion,* and *a $\mathscr{B}_{00}$-measurable σ-probability content kernel P from Ω to Ω (with $\mathscr{B}(\mathscr{T})$) such that the following hold:*

3.7.1. *P induces E_0: for every $h \in \mathscr{L}_m^1 \cap \text{mble}(\Omega, \mathscr{B}(\mathscr{T}), \mathbb{R})$, there is an m-nullset $N_0 \in \mathscr{B}_{00}$ such that $\omega \in \Omega \backslash N_0 \Rightarrow h \in \mathscr{L}_\sigma^1(P(\omega, \cdot))$ and the m-a.e. defined function $h_0(\omega) = \int P(\omega, d\eta) h(\eta)$ equals $E_0 h$ mod m.*

3.7.2. *Define, for every $\omega \in \Omega$, $[\omega]_{00} = \bigcap_{\omega \in F \in \mathscr{B}_{00}} F$. Then*

$$\omega \in \Omega \quad \Rightarrow \quad [\omega]_{00} \in \mathscr{B}_{00},$$

and there is an m-nullset $N_{00} \in \mathscr{B}_{00}$ such that

$$\omega \in \Omega \backslash N_{00} \quad \Rightarrow \quad P(\omega, [\omega]_{00}) = 1.$$

Proof. Consider the pseudometric defined in $\mathscr{B}$ by $|E, F| = m(E \,\Delta\, F)$ $(E, F \in \mathscr{B})$. Clearly there is a countable dense subset of $\mathscr{B}$ for this pseudometric (use a countable field generating $\mathscr{B}$, and Exercise II.2.6.1). Consequently, there is also a countable dense subset $\mathscr{F}_0$ in $\mathscr{B}_0 \subseteq \mathscr{B}$, for the same pseudometric; clearly we may assume that $\mathscr{F}_0$ is a field. Let $\mathscr{B}_{00}$ be the σ-field generated by $\mathscr{F}_0$. It is an easy exercise to prove that $(\Omega, \mathscr{B}_{00}, m)$ and $(\Omega, \mathscr{B}_0, m)$ have the same completion. Notice that, for the restriction m_{00} of m to $\mathscr{B}_{00}$, $L_m^1 \cap \text{mble}(\Omega, \mathscr{B}_0, \mathbb{R}) = L_{m_{00}}^1$. Apply Theorem 3.5 and Exercise 3.6.3 in order to obtain the $\mathscr{B}_{00}$-measurable σ-probability content kernel P that induces E_0: $L_{m_{00}}^1 \leftarrow L_m^1$. For every $F \in \mathscr{F}_0$, we have $1_F \in \mathscr{L}_{m_{00}}^1$, hence $P(\cdot, F) = 1_F$ mod m; let $N(F) \in \mathscr{B}_{00}$ be an m-nullset such that $\omega \in \Omega \backslash N(F) \Rightarrow P(\omega, F) = 1_F(\omega)$. Put $N_{00} = \bigcup_{F \in \mathscr{F}_0} N(F)$. Since $\mathscr{F}_0$ is countable, N_{00} is an m-nullset from $\mathscr{B}_{00}$. For $\omega \in \Omega \backslash N_{00}$, we have $P(\omega, F) = 1_F(\omega)$ first for $F \in \mathscr{F}_0$, and then, by unique σ-additive extension, also for $F \in \mathscr{B}_{00}$. It is clear that, for every $\omega \in \Omega$, the set $[\omega]_{00}$ consists of exactly those $\eta \in \Omega$ that cannot be separated from ω by sets from $\mathscr{B}_{00}$. But the system of all sets that do not separate a certain η from ω form a σ-field (exercise). It follows that $[\omega]_{00}$ is as well the set of all those $\eta \in \Omega$ that cannot be separated from ω by sets from $\mathscr{F}_0$. Thus $[\omega]_{00} = \bigcap_{\omega \in F \in \mathscr{F}_0} F$, and this is in $\mathscr{B}_{00}$ since $\mathscr{F}_0$ is countable. Now $\omega \in \Omega \backslash N_{00}$, $\omega \in F \in \mathscr{F}_0 \Rightarrow P(\omega, F) = 1_F(\omega) = 1$, hence $\omega \in \Omega \backslash N_{00} \Rightarrow P(\omega, [\omega]_{00}) = 1$.

3. Disintegration of Measures

Disintegration of measures is nothing but another variant of Theorem 3.7.

3.8. Theorem. *Let $(\Omega, \mathcal{T})$ and $(\Omega', \mathcal{T}')$ be compact Hausdorff spaces and $\varphi: \Omega \to \Omega'$ a continuous mapping with $\varphi\Omega = \Omega'$. Let $(\Omega, \mathscr{C}(\Omega, \mathbb{R}), m)$ and $(\Omega', \mathscr{C}(\Omega', \mathbb{R}), m')$ be (τ-)probability measure spaces such that $m' = \varphi m$ and assume that there exists a strong lifting ρ' for the complete σ-content space τ-derived from $(\Omega', \mathscr{C}(\Omega', \mathbb{R}), m')$, and for $\mathscr{C}(\Omega', \mathbb{R})$. Then there is a stochastic τ-measure kernel P from Ω' to Ω such that:*

3.8.1. $m'P = m$;

3.8.2. *for every $\omega' \in \Omega'$, $\Omega \backslash \varphi^{-1}\{\omega'\}$ is a $P(\omega', \cdot)$-nullset.*

Proof. Clearly $\Omega = \sum_{\omega' \in \Omega'} \varphi^{-1}\{\omega'\}$ is a disjoint decomposition of Ω into compact subsets and $f' \to f' \circ \varphi$ sends $\mathbb{R}^{\Omega'}$ bijectively into the set of all $f \in \mathbb{R}^{\Omega}$ that are constant on every $\varphi^{-1}\{\omega\}$. Let ψ denote the inverse of this mapping, i.e., $\psi(f' \circ \varphi) = f'$ $(f' \in \mathbb{R}^{\Omega'})$. Clearly f and $\psi(f)$ have the same range, hence f is bounded iff $\psi(f)$ is bounded, and bounds are the same in that case. By Exercise IV.1.7.3, φ is $\mathscr{B}(\mathscr{C}(\Omega, \mathbb{R}))$-$\mathscr{B}(\mathscr{C}(\Omega', \mathbb{R}))$-measurable. Every member of the σ-field $\mathscr{B}_0 = \varphi^{-1}\mathscr{B}(\mathscr{C}(\Omega', \mathbb{R}))$ is a union of some of the sets $\varphi^{-1}\{\omega'\}$, hence every $\mathscr{B}_0$-measurable function is constant on every $\varphi^{-1}\{\omega'\}$ and thus in the domain of ψ. By Proposition IV.1.6, ψ sends mble$(\Omega, \mathscr{B}_0, \mathbb{R})$ onto mble$(\Omega', \mathscr{B}(\mathscr{C}(\Omega', \mathbb{R})), \mathbb{R})$. It is practically obvious that ψ sends mod m equivalent functions into mod m' equivalent ones. Let now $E_0: \mathscr{L}_m^1 \to \mathscr{L}_m^1 \cap$ mble$(\Omega, \mathscr{B}_0, \mathbb{R})$ denote the conditional expectation for $\mathscr{B}_0$. Define the stochastic (τ-)measure kernel P from Ω' to Ω by $P(\omega', f) = \rho'(\psi(E_0 f))(\omega)$ $(\omega' \in \Omega', f \in \mathscr{C}(\Omega, \mathbb{R}))$. If $f' \in \mathscr{C}(\Omega', \mathbb{R}), f = f' \circ \varphi$, then $f \in$ mble$(\Omega, \mathscr{B}_0, \mathbb{R})$, hence $E_0 f = f$, hence $\psi(E_0 f) = f' = \rho'(f')$. Thus we conclude $P(\omega', f' \circ \varphi) = f'(\omega')$ $(\omega' \in \Omega', f' \in \mathscr{C}(\omega', \mathbb{R}))$. The set of all $f' \in$ mble$^b(\Omega', \mathscr{B}(\mathscr{C}(\Omega', \mathbb{R})), \mathbb{R})$ with $0 \leqq f' \leqq 1$ and $\int P(\omega', d\omega) f'(\varphi(\omega)) = f'(\omega')$ is obviously invariant under pointwise convergence of increasing or decreasing sequences of functions. Thus Exercise VI.1.3.7.4 applies and $P(\omega', \varphi^{-1}F') = 1_{F'}(\omega')$ $(\omega' \in \Omega', F' \in \mathscr{B}(\mathscr{C}(\Omega', \mathbb{R})))$ follows. Fix now any $\omega' \in \Omega'$. Urysohn's theorem implies

$$1_{\{\omega'\}} = \inf\{f' \mid f' \in \mathscr{C}(\Omega', \mathbb{R}), 0 \leqq f' \leqq 1, f'(\omega') = 1\}.$$

From this we deduce $1_{\varphi^{-1}\{\omega'\}} = \inf_{f \in \mathscr{F}} \mathscr{F}$, where

$$\mathscr{F} = \{f' \circ \varphi \mid f' \in \mathscr{C}(\mathscr{T}', \mathbb{R}), 0 \leqq f' \leqq 1, f'(\omega') = 1\}.$$

Clearly $\mathscr{F} \in \mathscr{C}(\Omega, \mathbb{R})$, $\mathscr{F}$ is decreasingly filtered, hence (compare Chapter III, Section 7), $1_{\varphi^{-1}\{\omega'\}}$ is a τ-lower function and

$$\int P(\omega', d\omega) 1_{\varphi^{-1}\{\omega'\}}(\omega) = \inf_{f \in \mathscr{F}} P(\omega'. f) = 1.$$

We conclude $P(\omega', \varphi^{-1}\{\omega'\}) = 1$ for the σ-probability content $P(\omega', \cdot)$ τ-derived from the τ-measure $P(\omega', \cdot)$, i.e., $\Omega \setminus \varphi^{-1}\{\omega'\}$ is a $P(\omega, \cdot)$-nullset.

4. Strassen's Theorem

In this subsection we prove a famous theorem of STRASSEN [2] in a form rather close to Strassen's original version; an application in the theory of barycentric representation (Chapter XV, Section 2) is given as Exercise 3.10.3. For more general versions of Strassen's theorem, see A. and C. IONESCU-TULCEA [1] and BICHTELER [1].

3.9. Theorem (STRASSEN [2]). *Let $(\Omega, \mathscr{B}, m)$ be a complete σ-probability content space and ρ:* $\text{mble}^b(\Omega, \mathscr{B}, \mathbb{R}) \to \text{mble}^b(\Omega, \mathscr{B}, \mathbb{R})$ *a lifting for it. Let $(H, \|\cdot\|)$ be a Banach space and $(H^*, \|\cdot\|)$ its dual Banach space. For every $\omega \in \Omega$, let p_ω: $H \to \mathbb{R}$ be a majorant function (see Theorem* I.6.1 *(Hahn–Banach)) such that, for a certain m-nullset N, the following holds:*

3.9.1. *For every $x \in H$,*

3.9.1.1. *the function $p_\bullet(x)$: $\omega \to p_\omega(x)$ is bounded and $\mathscr{B}$-measurable*:

3.9.1.2. $\omega \in \Omega \setminus N \Rightarrow p_\omega(x) \leqq \rho(p_\bullet(x))(\omega)$;

3.9.2. $\sup\{|p_\omega(x)| \mid \omega \in \Omega, x \in H, \|x\| \leqq 1\} = \alpha < \infty$.

Then the following holds:

3.9.3. $p(x) = \int p_\omega(x) m(d\omega)$ $(x \in H)$ defines a majorant function p on H.

3.9.4. *If $x^* \in H^*$ satisfies*

$$\langle x, x^* \rangle \leqq p(x) \qquad (x \in H), \tag{1}$$

then there is a mapping P: $\Omega \times H \to \mathbb{R}$ such that the following hold:

3.9.4.1. *For every $\omega \in \Omega$, the mapping $P(\omega, \cdot)$: $x \to P(\omega, x)$ of $H \to \mathbb{R}$ belongs to H^* (i.e., is a norm continuous linear form on H) and satisfies* $\omega \in \Omega \setminus N$, $x \in H \Rightarrow P(\omega, x) \leqq p_\omega(x)$, and $x \in H$, $\|x\| \leqq 1 \Rightarrow P(\omega, x) \leqq \alpha$.

3.9.4.2. $x \in H \Rightarrow \langle x, x^* \rangle = \int P(\omega, x)\, m(d\omega)$.

Proof. 3 is practically obvious.

4. Let $\tilde{\tilde{H}}$ be the system of all mappings $\Omega \to H$ of the form

$$\omega \to \sum_{k=1}^{n} x_k 1_{F_k}(\omega), \qquad \text{where} \quad n \geqq 1, \quad x_1, \ldots, x_n \in H,$$

$F_1, \ldots, F_n \in \mathscr{B}$. Clearly $\tilde{\tilde{H}}$ is a real vector space. Call two mappings in $\tilde{\tilde{H}}$ equivalent mod m if they coincide except on an m-nullset. It is clear that this defines an equivalence relation in $\tilde{\tilde{H}}$ and that the linear operations in $\tilde{\tilde{H}}$ are class operations for this equivalence. Denote by $\tilde{H}$ the set of all equivalence classes in $\tilde{\tilde{H}}$. In an obvious way, $\tilde{H}$ is a real vector space. Upon identifying $x \in H$ with the equivalence class of $x1_\Omega$, we may consider H as a linear subspace of $\tilde{H}$. Clearly

$$\sum_{k=1}^{n} x_k 1_{F_k} \to \sum_{k=1}^{n} \int p_\omega(x_k) 1_{F_k}(\omega)\, m(d\omega)$$

uniquely defines a mapping $\tilde{\tilde{p}}: \tilde{\tilde{H}} \to \mathbb{R}$ that is constant on every equivalence class in $\tilde{\tilde{H}}$ and hence defines a mapping $\tilde{p}: \tilde{H} \to \mathbb{R}$. It is obvious that $\tilde{p}$ is a majorant function and that the restriction of $\tilde{p}$ to H is p. If $x^* \in H^*$ satisfies (1), then x^* is a linear form on the linear subspace H of $\tilde{H}$ that is majorized by $\tilde{p}$ there. By Theorem I.6.1 (Hahn–Banach) there is a linear form $\tilde{x}^*: x \to \langle \tilde{x}, \tilde{x}^* \rangle$ on $\tilde{H}$ such that $\tilde{x} \Rightarrow \tilde{H}\langle \tilde{x}, \tilde{x}^* \rangle \leqq \tilde{p}(\tilde{x})$. For every $x \in H$, consider the set function defined on $\mathscr{B}$ by $F \to \langle x1_F, \tilde{x}^* \rangle$. It is clearly additive. Since

$$|\langle x1_F, \tilde{x}^* \rangle| \leqq |\tilde{p}(x1_F)| = \left| \int p_\omega(x) 1_F(\omega)\, m(d\omega) \right|$$

$$\leqq \int 1_F(\omega) |p_\omega(x)|\, m(d\omega) \qquad (F \in \mathscr{B}),$$

our set function is a signed σ-content that is absolutely continuous with respect to the positive σ-content m, dominated by the signed σ-content $p_\bullet(x)m$, and dominated in absolute value by the positive σ-content $|p_\bullet(x)|m$. By Theorem VIII.5.2 (Radon–Nikodym) and Proposition VIII.5.1 there is a function $g(\cdot, x) \in \text{mble}^b(\Omega, \mathscr{B}, \mathbb{R})$ such that $g(\omega, x) \leqq p_\omega(x)$, $|g(\omega, x)| \leqq |p_\omega(x)|$ for m-a.e. $\omega \in \Omega$, and

$$\langle x1_F, \tilde{x}^* \rangle = \int g(\omega, x) 1_F(\omega)\, m(d\omega).$$

Employ now the lifting ρ and put $P(\omega, x) = \rho(g(\cdot, x))(\omega)$ $(\omega \in \Omega, x \in H)$. By tedious but obvious arguments we see: For every $\omega \in \Omega$, $P(\omega, \cdot)$: $H \to \mathbb{R}$ is a linear form majorized by $\rho(p_\bullet(\cdot))(\omega)$, hence by α if $\|x\| \leqq 1$, and by p_ω in case $\omega \in \Omega \backslash N$. In particular, we have $P(\omega, \cdot) \in H^*$. The rest is obvious.

3.10. Exercises

3.10.1. Carry Theorem 3.9 over to the following situation: $(\Omega, \mathscr{B}^{00}, m)$ is a σ-content space with a disjoint base, such that $\mathscr{B}^{00}$ contains all local m-nullsets; ρ is a lifting for $(\Omega, \mathscr{B}^{00}, m)$. (*Hint*: Recall Definition VIII.8.4, Theorem I.14, Exercise 1.15, and Theorem VIII.8.8.)

3.10.2. Prove the following modification of Theorem 3.9: $(H, \|\cdot\|)$ is separable, 3.9.1.2 is dropped. (*Hint*: Prove that 3.9.1.2 can be assumed.)

3.10.3. Let $(H, \mathscr{S})$ be a locally convex separated topological vector space and $\varnothing \neq X \subseteq H$ convex and compact. For two probability measure $m, m'\colon \mathscr{C}(X, \mathbb{R}) \to \mathbb{R}$ define $m \prec m'$ by $\int f\, dm \leqq \int f\, dm'$ ($f \in \mathscr{C}(X, \mathbb{R})$ convex). Employ the definition of Chapter XV, Section 2, in particular the concave upper envelope $\bar{f}$ of a $f \in \mathscr{C}(X, \mathbb{R})$.

3.10.3.1. Prove that $m \prec m' \Rightarrow m$ and m' have the same barycenter.

3.10.3.2. Let $P\colon X \in \mathscr{C}(X, \mathbb{R}) \to \mathbb{R}$ be a stochastic (τ-)measure kernel such that for every $x \in X$, the probability measure $P(x, \cdot)$ has barycenter x. Prove that $mP \succ m$ for every probability measure m. (*Hint*: Use Lemma XV.2.5.)

3.10.3.3. Let $m, m'\colon \mathscr{C}(\Omega, \mathbb{R}) \to \mathbb{R}$ be two probability measures such that $m \prec m'$. Prove that there is a stochastic kernel $P\colon X \times \mathscr{C}(X, \mathbb{R}) \to \mathbb{R}$ such such that for every $x \in X$, the probability measure $P(x, \cdot)$ has barycenter x, and $mP = m'$. (*Hint*: Apply Theorem 3.9 with the majorant functions p_x ($x \in X$) defined on the Banach space $\mathscr{C}(X, \mathbb{R})$ (with the sup norm) by $p_x(f) = \bar{f}(x) \in (f \in \mathscr{C}(X, \mathbb{R}))$.)

3.10.4. Use Theorem 3.9 in order to prove Theorem 3.8 (disintegration of measures) once again. (*Hint*: On the Banach space $\mathscr{C}(\Omega, \mathbb{R})$ use the majorant functions p_ω ($\omega \in \Omega$) defined by

$$p_\omega(f) = \inf\{g(\varphi(\omega)) \mid g \in \mathscr{C}(\Omega', \mathbb{R}),\, g \circ \varphi \geqq f\} \qquad (f \in \mathscr{C}(\Omega, \mathbb{R})).)$$

5. Lifting of Functions with Values in a Compact Metrizable Space. Separability of Stochastic Processes

In this subsection we transfer *liftings* from bounded measurable real functions to *measurable mappings into a completely regular topological space*. An application to the so-called *separability* of stochastic processes (Doob [2]) is presented as an application.

Remember that a topological space $(X, \mathscr{T})$ is called *completely regular* if for every $x \in U \in \mathscr{T}$, there is an $f \in \mathscr{C}(X, \mathbb{R})$ with $0 \leqq f \leqq f(x) = 1$, $f \leqq 1_U$. This implies that for every compact $K \subseteq X$ the decreasingly filtered system $\mathscr{C}_K = \{f \mid f \in \mathscr{C}(X, \mathbb{R}),\ 1_K \leqq f \leqq 1\}$ has pointwise infimum 1_K.

For the next theorem we need a preparatory

3.11. Lemma. *Let $(X, \mathscr{T})$ be a compact Hausdorff space and $m: \mathscr{C}(X, \mathbb{R}) \to \mathbb{R}$ a probability (τ-)measure that is multiplicative, i.e., satisfies $m(fg) = m(f)m(g)$ $(f, g \in \mathscr{C}(X, \mathbb{R}))$. Then there is a unique $x \in X$ such that $m(f) = f(x)$ $(f \in \mathscr{C}(X, \mathbb{R}))$.*

Proof. Multiplicativity immediately extends, by τ-continuity of m, to bounded τ-lower functions, in particular to indicator functions of compact sets (exercise). For any compact $K \subseteq X$, we have thus (m denotes the σ-content τ-derived from m as well) $m(K) = m(1_K) = m(1_K \cdot 1_K) = m(1_K)m(1_K) = m(K)^2$, hence $m(K) = 0$ or $= 1$. Let $\mathscr{K}_1 = \{K \mid K \subseteq X,\ K$ compact, $m(K) = 1\}$. Clearly $X \in \mathscr{K}_1$ and $\mathscr{K}_1$ is stable under finite intersections. Since every $K \in \mathscr{K}_1$ is nonempty, the existence of some $x \in \bigcap_{K \in \mathscr{K}_1} K$ follows. We claim $\{x\} \in \mathscr{K}_1$, i.e., $m(\{x\}) = 1$. If $m(\{x\}) = 0$, then $m(X \backslash \{x\}) = 1$. By inner compact regularity (Theorem V.2.1.2.3) there is a $\mathscr{K}_1 \ni K \subseteq X \backslash \{x\}$, hence $x \notin \bigcap_{K \in \mathscr{K}_1} K$, a contradiction. Thus $m(\{x\}) = 1$ follows. This implies $m(f) = \int f\, dm = f(x)$ $(f \in \mathscr{C}(X, \mathbb{R}))$. Since $\mathscr{C}(X, \mathbb{R})$ separates the points of X, the uniqueness of x follows.

3.12. Theorem. *Let $(\Omega, \mathscr{B}, m)$ be a complete σ-probability content space and ρ: $\text{mble}^b(\Omega, \mathscr{B}, \mathbb{R}) \to \text{mble}^b(\Omega, \mathscr{B}, \mathbb{R})$ a lifting for it. Let $(X, \mathscr{T})$ be a completely regular Hausdorff space and $\mathscr{M}$ the set of all $\mathscr{B}$-$\mathscr{B}(\mathscr{C}(X, \mathbb{R}))$-measurable mappings $\varphi: \Omega \to X$ with conditionally compact $\varphi\Omega$. Then there is a unique mapping ρ_X: $\mathscr{M} \to \mathscr{M}$ such that:*

3.12.1. $f(\rho_X(\varphi)) = f(\varphi) \bmod m$ $(\varphi \in \mathscr{M}, f \in \mathscr{C}(X, \mathbb{R}))$.

3.12.2. $\varphi, \psi \in \mathscr{M}$, $\varphi = \psi \bmod m \Rightarrow \rho_X(\varphi) = \rho_X(\psi)$ everywhere on Ω.

3.12.3. $\varphi \in \mathscr{M}, f \in \mathscr{C}(X, \mathbb{R}) \Rightarrow f(\rho_X(\varphi)) = \rho(f(\varphi))$ everywhere on Ω.

ρ_X is called the **lifting of $\mathscr{M}$ associated with ρ.**

Proof. We emphasize that 3 makes sense since $f(\varphi) \in \text{mble}^b(\Omega, \mathscr{B}, \mathbb{R})$ for any $\varphi \in \mathscr{M}$ and $f \in \mathscr{C}(X, \mathbb{R})$, by the very definition of the $\mathscr{B}$-$\mathscr{B}(\mathscr{C}(X, \mathbb{R}))$-measurability of φ; the compactness of $\overline{\varphi(\Omega)}$ in fact ensures the boundedness of $f(\varphi)$. We now define ρ_X. Fix some $\varphi \in \mathscr{M}$ and some $\omega \in \Omega$. Choose any compact $\varphi\Omega \subseteq K \subseteq X$. Define m_ω: $\mathscr{C}(K, \mathbb{R}) \to \mathbb{R}$ by $m_\omega(g) = \rho(g(\varphi(\omega)))$. Clearly m_ω is a probability (τ-)measure with the property $m_\omega(g \cdot h) = m_\omega(g)m_\omega(h)$ $(g, h \in \mathscr{C}(K, \mathbb{R}))$ following from the multiplicativity of lifting. By Lemma 3.11 we get a unique $x \in K$ with $m_\omega(g) = g(x)$ $(g \in \mathscr{C}(K, \mathbb{R}))$. We put now $\rho_X(\varphi)(\omega) = x$. Then 3 holds by definition. 3 implies that ρ_X is uniquely determined because $\mathscr{C}(X, \mathbb{R})$ separates the points of X.

2 and 1 are obvious consequences of 3.

3.13. Remark. Theorem 3.12.1 does not mean that in general $\{\omega \mid \rho_X(\varphi)(\omega) \neq \varphi(\omega)\}$ is an m-nullset (see Exercise 3.17.2).

3.14. Lemma. *Let $(\Omega, \mathscr{B}, m)$ be a complete σ-probability content space and ρ: $\text{mble}^b(\Omega, \mathscr{B}, \mathbb{R}) \to \text{mble}^b(\Omega, \mathscr{B}, \mathbb{R})$ a lifting for it. Let $(X, \mathscr{T})$ be a completely regular Hausdorff space, $\mathscr{M}$ the set of all $\mathscr{B}$-$\mathscr{B}(\mathscr{C}(X, \mathbb{R})$-measurable mappings of Ω into X, and ρ_X: $\mathscr{M} \to \mathscr{M}$ the lifting of $\mathscr{M}$ associated with ρ. Then for every $\varphi \in \mathscr{M}$ with $\rho_X(\varphi) = \varphi$, and for every compact $K \subseteq X$, we have $\rho(1_K(\varphi)) \leqq 1_K(\varphi)$ everywhere.*

Proof. Put $\mathscr{C}_K = \{f \mid 1_K \leqq f \leqq 1,\ f \in \mathscr{C}(X, \mathbb{R})\}$. Clearly $\mathscr{C}_K$ is decreasingly filtered with pointwise infimum 1_K. By definition of ρ_X we have $f \in \mathscr{C}_K \Rightarrow \rho(1_K(\varphi)) \leqq \rho(f(\varphi)) = f(\rho_X(\varphi)) = f(\varphi)$ everywhere $(f \in \mathscr{C}_K)$. It follows that

$$\rho(1_K(\varphi)) \leqq \inf_{f \in \mathscr{C}_K} f(\varphi) = 1_K(\varphi)$$

everywhere.

3.15. Lemma. *Let $(\Omega, \mathscr{B}, m)$ be a complete σ-probability content space and ρ a lifting for it. Let $\mathscr{F} \subseteq \text{mble}^b(\Omega, \mathscr{B}, \mathbb{R})$ be decreasingly filtered, and assume $0 \leqq \rho(f) \leqq f \leqq 1$ $(f \in \mathscr{F})$. Then*

$$f_0(\omega) = \inf_{f \in \mathscr{F}} f(\omega) \qquad (\omega \in \Omega)$$

defines an $f_0 \in \text{mble}^b(\Omega, \mathscr{B}, \mathbb{R})$ such that

$$\int f_0\, dm = \inf_{f \in \mathscr{F}} \int f\, dm \qquad \text{and} \qquad \rho(f_0) \leqq f_0 .$$

Proof. Put $\alpha = \inf_{f \in \mathscr{F}} \int f\, dm$. It is easy to find $f_1, f_1, \ldots \in \mathscr{F}$ such that $f_1 \geqq f_2 \geqq \cdots$ and $\lim_n \int f_n\, dm = \alpha$. Put $f'_0(\omega) = \lim_n f_n(\omega)$ $(\omega \in \Omega)$. Clearly $f'_0 \in \text{mble}(\Omega, \mathscr{B}, \mathbb{R})$, $\int f'_0\, dm = \alpha$. Moreover, $f \in \mathscr{F} \Rightarrow f'_0 \leqq f m$-a.e. compare Theorem III.5.8.5,6 and its proof). We conclude $\rho(f'_0) \leqq \rho(f) \leqq f$ everywhere $(f \in \mathscr{F})$, and hence $\rho(f'_0) \leqq f_0 \leqq f'_0$. Now $\rho(f'_0) = f'_0 \bmod m$ and the completeness of $(\Omega, \mathscr{B}, m)$ imply $f_0 \in \text{mble}^b(\Omega, \mathscr{B}, \mathbb{R})$, $\rho(f) = \rho(f'_0)$, $\int f_0\, dm = \int f'_0\, dm = \alpha$.

3.16. Theorem. *Let $(\Omega, \mathscr{B}, m)$ be a complete σ-probability content space and ρ: $\mathrm{mble}^b(\Omega, \mathscr{B}, \mathbb{R}) \to \mathrm{mble}^b(\Omega, \mathscr{B}, \mathbb{R})$ a lifting for it. Let $(X, \mathscr{T})$ be a completely regular Hausdorff space and $\mathscr{M}$ the set of all $\mathscr{B}$-$\mathscr{B}(\mathscr{C}(X, \mathbb{R}))$-measurable mappings of $\Omega \to X$. Let ρ_X: $\mathscr{M} \to \mathscr{M}$ be the lifting of $\mathscr{M}$ associated with ρ. Let $I \neq \varnothing$ and $\varphi_\iota \in \mathscr{M}$ $(\iota \in I)$. Then for every compact $K \subseteq X$ and every $\varnothing \neq I_0 \subseteq I$, the set $\{\omega \mid \rho_X(\varphi_\iota)(\omega) \in K \ (\iota \in I_0)\}$ belongs to $\mathscr{B}$ and satisfies*

$$m(\{\omega \mid (\rho_X(\varphi_\iota))(\omega) \in K \ (\iota \in I_0)\}) = \inf_{J \subseteq I_0 \text{ finite}} m(\{\omega \mid (\rho_X(\varphi_\iota))(\omega) \in K \ (\iota \in J)\}).$$

Proof. Put $\psi_\iota = \rho_X(\varphi_\iota)$ $(\iota \in I)$ for brevity. Let $\mathscr{F}$ be the system of all indicator functions of sets of the form $\{\omega \mid \psi_\iota(\omega) \in K \ (\iota \in J)\}$ with finite $\varnothing \neq J \subseteq I_0$. Clearly $\mathscr{F}$ is decreasingly filtered, and its pointwise infimum is the indicator function of $\{\omega \mid \psi_\iota(\omega) \in K \ (\iota \in I_0)\}$. By Lemma 3.15 we are through as soon as we know $\rho(f) \leqq f$ for every $f \in \mathscr{F}$. But if $\varnothing \neq J \subseteq I_0$ is finite, then Lemma 3.14 shows

$$\begin{aligned} 1_{\{\omega \mid \psi_\iota(\omega) \in K(\iota \in J)\}} &= \inf_{\iota \in J} 1_K(\psi_\iota) \geqq \inf_{\iota \in J} \rho(1_K(\psi_\iota)) \\ &= \rho\Big(\inf_{\iota \in J} 1_K(\psi_\iota)\Big) \\ &= \rho(1_{\{\omega \mid \psi_\iota(\omega) \in K \ (\iota \in J)\}}), \end{aligned}$$

as desired.

3.17. Exercises

3.17.1. Carry Theorems 3.12 and 3.16 over to the case of an arbitrary σ-content space $(\Omega, \mathscr{B}^{00}, m)$ with a disjoint base and all local m-nullsets in $\mathscr{B}^{00}$.

3.17.2. Let $I = \Omega_\iota = [0, 1]$, $\mathscr{B}_\iota$ be the Borel σ-field in $[0, 1]$, and m_ι the Lebesgue σ-content on $\mathscr{B}_\iota (\iota \in I)$. Let $(\Omega, \mathscr{B}, m)$ be the completion of the σ-probability content space $\prod_{\iota \in I} (\Omega_\iota, \mathscr{B}_\iota, m_\iota)$.

3.17.2.1. Let φ_ι: $\Omega \to [0, 1]$ denote the ιth component mapping $(\iota \in I)$. Prove that $\bigcup_{\iota \in I} \varphi_\iota^{-1}\{\frac{1}{2}\}$ is not an m-nullset.

3.17.2.2. Define φ, ψ: $\Omega \to \Omega$ by $\varphi_\iota(\varphi(\omega)) = 0$ and

$$\varphi_\iota(\psi(\omega)) = \begin{cases} 0 & \text{if} \quad \varphi_\iota(\omega) \neq \frac{1}{2}, \\ 1 & \text{if} \quad \varphi_\iota(\omega) = \frac{1}{2}. \end{cases}$$

Let $\mathscr{T}$ be the product topology in Ω. Prove that φ and ψ are $\mathscr{B}$-$\mathscr{B}(\mathscr{C}(\Omega, \mathbb{R}))$-measurable, $\{\omega \mid f(\varphi(\omega)) \neq f(\psi(\omega))\}$ is an m-nullset for every $f \in \mathscr{C}(\Omega, \mathbb{R})$, but $\{\omega \mid \varphi(\omega) \neq \psi(\omega)$ is not an m-nullset.

3.17.3. Let $(X, \mathcal{T})$ be a compact metric space (such as $\overline{\mathbb{R}} = \mathbb{R} \cup \{-\infty, \infty\}$ with the obvious topology). Prove that Theorem 3.16.3 implies that $\{\omega \mid \rho_X(\varphi)(\omega) \neq \varphi(\omega)\}$ is an m-nullset in this case.

3.18. Remark. Exercise 3.17.3 makes Theorem 3.16 comprise the main result of the theory of so-called separability of stochastic processes, as treated in DOOB [2].

APPENDIX A

THE PERRON–WARD INTEGRAL AND RELATED CONCEPTS

JAROSLAW KURZWEIL, Prague

In this appendix we present some concepts of integrability which go essentially back to PERRON [1] and WARD [1]. See also KAMKE [2] for a concise treatment of the Perron integral. Perron's basic aim was to establish integration as the inverse of differentiation also in cases usually not accessible to ordinary Riemann integration. The leading technical idea, as presented here, is to imitate the classical Riemann procedure, but to measure the fineness of a partition by a positive function rather than a positive constant. To be precise, we do this for "pointed" partitions, i.e., partitions with a specified point in every atom. The theory employs the geometry of the basic space to an extent that forces us to consider essentially only $\mathbb{R}$ as a possible basic domain. An extension to $\mathbb{R}^n$ is not difficult, but more or less of technical interest only, and will not be described here. Another typical feature is the non-appearance of function values $\pm\infty$.

Basic notions and notations are given in Section 1. Section 2 presents what we call the S-integral, Section 3 the V-integral, and Section 4 the Perron–Ward integral. The equivalence of all three concepts is established. We work essentially with S-integrability for the rest of the appendix. A monotone convergence theorem analogous to Theorem III.5.3 is proved in Section 5. In Section 6 we compare S-integrability with Daniell integrability. It turns out that for real-valued functions, S-integrability leads at least as far as Daniell integrability. Example 4.4 yields an $\mathbb{R}$-valued f' that is S-integrable while $|f'|$ is not. Thus there are S-integrable functions that are not in L_m^1.

1. NOTATIONS

Throughout this appendix we deal with subsets M of $\mathbb{R}$. The interior of M is denoted by int M. An interval $K \subseteq \mathbb{R}$ is always tacitly understood to be compact and of length > 0. A finite union of such intervals is called a figure. A couple (K, t), where K is an interval and $t \in K$, is called a pointed interval. For every $M \subseteq \mathbb{R}$ we denote by:

$\mathscr{I}(M)$ the set of all intervals M;
$\mathscr{F}(M)$ the set of all figures M;
$\mathscr{P}\mathscr{I}(M)$ the set of all pointed intervals (K, t) with $K \subseteq M$.

Let $F \in \mathscr{F}(M)$ and $Z \subseteq \mathscr{P}\mathscr{I}(M)$, say $Z = \{(K_1, t_1), \ldots, (K_n, t_n)\}$. We call Z a pointed decomposition of F if $j \neq k \Rightarrow \text{int } K_j \cap \text{int } K_k = \varnothing$ and $K_1 \cup \cdots \cup K_n = F$. Let $0 < \delta \in \mathbb{R}^M$. For a $(K, t) \in \mathscr{P}\mathscr{I}(M)$, we write (K, t) sub δ if $K \subseteq [t - \delta(t), t + \delta(t)]$. For a Z as above, we write Z sub δ if (K_j, t_j) sub δ $(j = 1, \ldots, n)$. By an easy compactness argument we see that for every figure $F \in \mathscr{F}(M)$ and every $0 < \delta \in \mathbb{R}^M$, there is a decomposition Z sub δ of F. In the case of a constant $0 < \delta \in \mathbb{R}^M$ we come back to a concept of fineness of a decomposition which is customary in the classical theory of the Riemann integral.

A real function m on $\mathscr{I}(M)$ is called:

∂-subadditive if $J, K, J \cup K \in \mathscr{I}(M)$, int $J \cap$ int $K = \varnothing$ imply

$$m(J \cup K) \leqq m(J) + m(K);$$

∂-superadditive if $J, K, J \cup K \in \mathscr{I}(M)$, int $J \cap$ int $K = \varnothing$ imply

$$m(J \cup K) \geqq m(J) + m(K);$$

∂-additive if it is subadditive and superadditive.

∂-additivity is defined also for functions on $\mathscr{F}(M)$ in an obvious way.

It is clear that every additive function on $\mathscr{I}(M)$ has a unique additive extension to $\mathscr{F}(M)$.

The prefix ∂ is to remind one of the special role played by boundary points here.

A real function U on $\mathscr{P}\mathscr{I}(M)$ is called ∂-additive if (J, t), $(K, t) \in \mathscr{P}\mathscr{I}(M)$, $\text{int } J \cap \text{int } K = \varnothing \Rightarrow U(J \cup K, t) = U(J, t) + U(K, t)$. Observe here that $t \in J \cap K$ implies $J \cup K \in \mathscr{I}(M)$. Two real functions U, V on $\mathscr{P}\mathscr{I}(M)$ are called equivalent on $J \in \mathscr{I}(M)$ if for any $0 < \varepsilon \in \mathbb{R}$ there is a ∂-superadditive $m \geqq 0$ on $\mathscr{I}(J)$ and a $0 < \delta \in \mathbb{R}^J$ such that $m(J) < \varepsilon$, and $(K, t) \in \mathscr{P}\mathscr{I}(J)$, (K, t) sub $\delta \Rightarrow |U(K, t) - V(K, t)| \leqq m(K)$.

For a real function U on $\mathscr{P}\mathscr{J}(M)$ and a pointed decomposition $Z = \{(K_1, t_1), \ldots, (K_n, t_n)\}$ of a figure $F \in \mathscr{F}(M)$, we put

$$S_F(U; Z) = \sum_{j=1}^{n} U(K_j, t_j).$$

2. THE S-INTEGRAL

We introduce a first type of integrals by

2.1. Definition. *Let $M \subseteq \mathbb{R}$ and U be a real function on $\mathscr{P}\mathscr{J}(M)$.*

2.1.1. *For any $0 < \delta \in \mathbb{R}^M$ and $F \in \mathscr{F}(M)$, we define*

$$\Delta S_F(U; \delta) = [\inf\{S_F(U; Z) | Z \text{ sub } \delta\}, \sup\{S_F(U; Z) | Z \text{ sub } \delta\}].$$

2.1.2. *U is called* S*-integrable over $F \in \mathscr{F}(M)$ with* S*-integral $\alpha = (\mathrm{S}) \int_F U \in \mathbb{R}$ if*

$$\bigcap_{0 < \delta \in \mathbb{R}^M} \Delta S_F(U; \delta) = \{\alpha\}.$$

2.1.3. *Let $f \in \mathbb{R}^M$ and m be a real function on $\mathscr{J}(M)$. If the function U defined on $\mathscr{P}\mathscr{J}(M)$ by $U(K, t) = f(t)m(K)$ $((K, t) \in \mathscr{P}\mathscr{J}(M))$ is* S*-integrable over $F \in \mathscr{F}(M)$ with* S*-integral α, we say that f is* S*-integrable for m over F and write $(\mathrm{S}) \int_F f\, dm$ instead of $(\mathrm{S}) \int_F U$.*

2.1.4. *U is said to be* S*-integrable if it is* S*-integrable over every $F \in \mathscr{F}(M)$.*

2.2. Remarks. Let M, F, U be as above.

2.2.1. Clearly $0 < \delta \leqq \eta \in \mathbb{R}^M \Rightarrow \Delta S_F(U; \delta) \subseteq \Delta S_F(U; \eta)$, i.e., the $\Delta S_F(U; \delta)$ are filtering downward as the δ filter down to 0.

2.2.2. Clearly U is S-integrable over F with S-integral $\alpha \in \mathbb{R}$ iff for every $\varepsilon > 0$ there is a $0 < \delta \in \mathbb{R}^M$ such that Z sub $\delta \Rightarrow |S_F(U; Z) - \alpha| < \varepsilon$.

It is obvious that the S-integral is unique if it is defined and that it is a positive linear functional on the real vector space of all S-integrable functions for a given $F \in \mathscr{F}(M)$.

2.2.3. It should be clear that the usual Riemann integral is an extreme special case of the S-integral.

2.3. Proposition. *Let $M \subseteq \mathbb{R}$, $U \in \mathbb{R}^{\mathscr{P}\mathscr{J}(M)}$.*

2.3.1. *$F \in \mathscr{F}(M)$, U* S*-integrable over F, $E \in \mathscr{F}(M)$, $E \subseteq F \Rightarrow U$* S*-integrable over E.*

2.3.2. $E, F, G \in \mathscr{F}(M)$, $\text{int } E \cap \text{int } F = \varnothing$, $E \cup F = G$, *U S-integrable over $G \Rightarrow U$ S-integrable over E and F, and*

$$(\mathrm{S}) \int_G U = (\mathrm{S}) \int_E U + (\mathrm{S}) \int_F U.$$

The proof is an easy exercise. *Hint for* 2.3.1: assume $E \neq F$ and find $G \in \mathscr{F}(M)$ such that $\text{int } E \cap \text{int } G = \varnothing$, $E \cup G = F$. For any $0 < \delta \in \mathbb{R}^M$, prove that $\Delta_E(U; \delta)$ is not longer than $\Delta_F(U; \delta)$.

Hint for 2.4.2: It is easy to show $\Delta_E(U; \delta) + \Delta_F(U; \delta) \subseteq \Delta_G(U; \delta)$ for any $0 < \delta \in \mathbb{R}^M$.

Recall the concept of ∂-additivity for real functions on $\mathscr{P}\mathscr{J}(M)$, as given in Section 1. It follows from 2.3.2 that $(\mathrm{S}) \int_J U$ depends ∂-additively on J provided that U is S-integrable. Of course, ∂-additivity extends from $\mathscr{J}(M)$ to $\mathscr{F}(M)$.

2.4. Example. Let M, U be as above $[-1, 1] \subset M$. It can happen that both integrals $(\mathrm{S}) \int_{[-1,0]} U$, $(\mathrm{S}) \int_{[0,1]} U$ exist, but $(\mathrm{S}) \int_{[-1,1]} U$ does not exist. As an example, define

$$U(K, t) = \begin{cases} 0 & \text{if } 0 \notin K, \\ 1 & \text{if } 0 \in K. \end{cases}$$

It is easy to prove directly from the definition of the S-integral that U is S-integrable over $[-1, 0]$ and $[0, 1]$ with $(\mathrm{S}) \int_{[-1,0]} U = 1 = (\mathrm{S}) \int_{[0,1]} U$ and that

$$\Delta S_{[-1,1]}(U, \delta) = [1, 2] \qquad \text{for any} \quad 0 < \delta \in \mathbb{R}^F.$$

Observe that:

1. U does not depend on t, so that we may write $U(K, t) = m(K)$ $\left(\text{we proved that } (\mathrm{S}) \int_{[-1,1]} dm = (\mathrm{S}) \int_{[-1,1]} 1\, dm \text{ does not exist}\right)$.
2. U is not ∂-additive.

2.5. Proposition. *Let $M \subseteq \mathbb{R}$ and assume that $U \in \mathbb{R}^{\mathscr{P}\mathscr{J}(M)}$ is ∂-additive. Let U be S-integrable over $E \in \mathscr{F}(M)$ as well as over $F \in \mathscr{F}(M)$. Then U is S-integrable over $E \cup F$.*

Proof. It is easy to see that it is sufficient to assume $E = [a, b]$, $F = [b, c]$ with $a < b < c$, and to restrict attention to functions $0 < \delta \in \mathbb{R}^M$ with $\delta(t) < |t - b|$ $(t \neq b)$. But the latter restriction implies for every pointed decomposition $Z = \{(K_1, t_1), \ldots, (K_n, t_m)\}$ of $E \cup F$ that:

Either $b \in \operatorname{int} K_j$ for some j; in this case we have $t_j = b$, and we form Z' by replacing (K_j, b) by $(K_j \cap E, b)$ plus $(K_j \cap F, b)$; this yields $\Delta_{E \cup F}(U; Z) = \Delta_{E \cup F}(U, Z')$ by the ∂-additivity of U, and $\Delta_{E' \cup F}(U, Z') = \Delta_E(U, Z'') + \Delta_F(U, Z''')$ for two obvious pointed decompositions Z'' sub δ, Z''' sub δ of E, F.

Or b is the common endpoint of K_j and K_{j+1} for some j, and $t_j = b = t_{j+1}$. The rest is as in the preceding case.

This yields enough material to prove the proposition.

2.6. Remarks

2.6.1. The point in this argument is that $\delta(t) < |t - b|$ $(t \neq b)$, $b \in K$, (K, t) sub δ forces $t = b$.

2.6.2. The result of Proposition 2.3 can be stated as follows: If U is S-integrable, then $\mathscr{F}(M) \ni F \to (\mathrm{S}) \int_F U$ defines a ∂-additive real function on $\mathscr{I}(M)$, and hence on $\mathscr{F}(M)$. Proposition 2.5 is a refined statement covering some points of subtle interest.

3. THE V-INTEGRAL

Let $M \subseteq \mathbb{R}$ be given. In this section we present another type of integral for real functions on $\mathscr{P}\mathscr{I}(M)$, the so-called variational or V-integral. It is strongly based on the concept of equivalence for real functions on $\mathscr{P}\mathscr{I}(M)$, as defined in Section 1. Every real function I on $\mathscr{I}(M)$ gives rise to a real function, again called I, on $\mathscr{P}\mathscr{I}(M)$ in a trivial way: $I(K, t) = I(K)$. We can thus speak of equivalence between a function on $\mathscr{P}\mathscr{I}(M)$ and a function on $\mathscr{I}(M)$, as well as of equivalence of two functions of the latter type. It is clear, further, that every ∂-additive function on $\mathscr{I}(M)$ can be extended to a ∂-additive function on $\mathscr{F}(M)$ in exactly one way.

3.1. Definition. *Let $M \subseteq \mathbb{R}$ and U be a real function on $\mathscr{P}\mathscr{I}(M)$. Let I be a ∂-additive real function on $\mathscr{I}(M)$ (or, equivalently on $\mathscr{F}(M)$). U is said to be* V-*integrable with* V-*integral I if U is equivalent to I on every $K \in \mathscr{I}(M)$. I is then called the* V-*integral of U, and we write*

$$I = (\mathrm{V}) \int U, \qquad I(F) = (\mathrm{V}) \int_F U.$$

In the special case where U comes from an m: $\mathscr{I}(M) \to \mathbb{R}$ and an f: $M \to \mathbb{R}$ via $U(K, t) = f(t)m(K)$, we write $(\mathrm{V}) \int_F f\, dm$ in place of $(\mathrm{V}) \int U$, in case the V-*integral exists.*

The question to whether the V-integral is uniquely determined if it exists is settled by an easy

3.2. Lemma. *Two ∂-additive functions I, I' on $\mathscr{J}(M)$ are equivalent iff they coincide.*

For the proof, we have only to recall the definition of equivalence and to read, for a pointed decomposition $\{(K_1, t_1), \ldots, (K_n, t_n)\}$ of some $J \in \mathscr{J}(M)$, the estimates

$$\begin{aligned} |I(J) - I'(J)| &= |(I(K_1) + \cdots + I(K_n)) - (I'(K_1) + \cdots + I'(K_n))| \\ &\leqq |I(K_1) - I'(K_1)| + \cdots + |I(K_n) - I'(K_n)| \\ &\leqq m(K_1) + \cdots + m(K_n) \leqq m(J) < \varepsilon. \end{aligned}$$

We have already seen that S-integrability of some $U \in \mathbb{R}^{\mathscr{P}\mathscr{J}(M)}$ leads to a ∂-additive function on $\mathscr{F}(M)$. The validity of our next proposition is therefore to be expected.

3.3. Proposition. *Let $M \subseteq \mathbb{R}$ and $U \in \mathbb{R}^{\mathscr{P}\mathscr{J}(M)}$. Then the following statements are equivalent:*

3.3.1. *U is S-integrable.*

3.3.2. *U is V-integrable.*

If both of these hold, and $I \in \mathbb{R}^{\mathscr{F}(M)}$ denotes the V-integral of U, then

$$I(F) = (\mathrm{S}) \int_F U \qquad (F \in \mathscr{F}(M)).$$

Proof. $1 \Rightarrow 2$. Let U be S-integrable. Define $I \in \mathbb{R}^{\mathscr{F}(M)}$ by $I(F) = (\mathrm{S}) \int_F U$ $(F \in \mathscr{F}(M))$. We shall show that U is V-integrable with V-integral I. Choose any $J \in \mathscr{J}(M)$ and any $\varepsilon > 0$ and determine some $0 < \delta \in \mathbb{R}^M$ such that the interval $\Delta S_J(U;\ \delta)$ is shorter than ε. For any $K \in \mathscr{J}(J)$, denote by $m(K)$ the length of the interval $\Delta S_K(U;\ \delta)$. The proof that this m is ∂-superadditive is an easy exercise which we leave to the reader. Observing (X, t) sub $\delta \Rightarrow U(K, t) \in \Delta S_K(U;\ \delta)$, we find $\{(K, t)\}$ sub $\delta \Rightarrow |U(K, t) - I(K)| \leqq m(K)$. This proves the equivalence of U and I, i.e., the desired statement on V-integrability of U.

$2 \Rightarrow 1$. Let U be V-integrable with V-integral I. We shall show that U is S-integrable with $(\mathrm{S}) \int_H U = I(J)$ $(J \in \mathscr{J}(M))$. To show this, we choose some $J \in \mathscr{J}(M)$, any $\varepsilon > 0$, and a ∂-superadditive $0 \leqq m \in \mathbb{R}^{\mathscr{J}(M)}$ with $m(J) < \varepsilon$ according to the definition of equivalence. For any pointed decomposition

$Z = \{(K_1, t_1), \ldots, (K_n, t_n)\}$ of J, we obtain, in case Z sub δ,

$$\begin{aligned}|S_J(U; Z) - I(J)| &= |(U(K_1, t_1) + \cdots + U(K_n, t_n)) - (I(K_1) + \cdots + I(K_n))| \\ &\leqq |U(K_1, t_1) - I(K_1)| + \cdots + |U(K_n, t_n) - I(K_n)| \\ &\leqq m(K_1) + \cdots + m(K_n) \leqq m(J) < \varepsilon.\end{aligned}$$

This does it.

4. THE PERRON–WARD INTEGRAL

In this section we present a concept of an integral over intervals in $\mathbb{R}$ which goes back to ideas of PERRON [1] and WARD [1]. The basic idea is to establish the integral as the inverse of a derivative also under circumstances where nobody would dare to think of differentiation. For a clear and concise exposition, see KAMKE [1], SAKS [4].

4.1. Definition. *Let $M \subseteq \mathbb{R}$ and $U \in \mathbb{R}^{\mathscr{P}\mathscr{J}(M)}$, $J \in \mathscr{J}(M)$. A real function u on $\mathscr{J}(J)$ is called:*

4.1.1. *a lower function for U on J if:*

4.1.1.1. *u is ∂-subadditive, and*

4.1.1.2. *there is a $0 < \delta \in \mathbb{R}^M$ with*

$$\mathscr{P}\mathscr{J}(J) \ni (K, t) \text{ sub } \delta \quad \Rightarrow \quad u(K) \leqq U(K, t);^J$$

4.1.2. *an upper function for U on J if:*

4.1.2.1. *u is ∂-superadditive, and*

4.1.2.2. *there is a $0 < \delta \in \mathbb{R}^M$ with*

$$\mathscr{P}\mathscr{J}(J) \ni (K, t) \text{ sub } \delta \quad \Rightarrow \quad u(K) \geqq U(K, t).$$

4.1.3. *We denote by*

$\mathscr{L}(U{:}\ J)$ the set of all lower functions for U on J.
$\mathscr{U}(U; J)$ the set of all upper functions for U on J.

4.1.4. *If*

$$\sup\{u(J) \mid u \in \mathscr{L}(U; J)\} = \inf\{v(J) \mid v \in \mathscr{U}(U; J)\},$$

and if this is a finite real number α, then U is said to be P-W-*integrable over J, with the* P-*integral*

$$(\mathrm{P}) \int_J U = \alpha$$

4.1.5. *If $f \in \mathbb{R}^M$, $m \in \mathbb{R}^{\mathscr{J}(M)}$ and U is of the form $U(K, t) = f(t)m(K)$ $((K, t) \in \mathscr{P}\mathscr{J}(M))$, and if U is* P-W-*integrable over some $J \in \mathscr{J}(M)$, then we say that f is* P-W-*integrable for m over J and write* (P) $\int_J f\,dm$ *in place of* (P) $\int_J U$.

4.2. Remarks

4.2.1. It is obvious that the $0 < \delta \in \mathbb{R}^M$ appearing in the definitions of lower and upper functions can be replaced by any minorant of the same type, without danger for the conclusions. In particular, we may deal with one and the same $0 < \delta \in \mathbb{R}^M$ for any finite number of lower and upper functions.

4.2.2. Let $U \in \mathbb{R}^{\mathscr{P}\mathscr{J}(M)}$ and u be a lower with v an upper function for U, $0 < \delta \in \mathbb{R}^M$ a function working for both of them. Let

$$Z = \{(K_1, t_1), \ldots, (K_n, t_n)\} \text{ sub } \delta$$

be a pointed decomposition of J. Then the ∂-subadditivity of u and the ∂-superadditivity of v imply

$$\begin{aligned} u(J) &\leqq u(K_1) + \cdots + u(K_n) \leqq U(K_1, t_1) + \cdots + U(K_n, t_n) \\ &= S_J(U; Z) \leqq v(K_1) + \cdots + v(K_n) \leqq v(J). \end{aligned}$$

This shows that the interval $[u(J), v(J)]$ contains the interval $\Delta_J(U; \delta)$.

4.3. Proposition. *Let $M \subseteq \mathbb{R}$, $J \in \mathscr{J}(M)$, $U \in \mathbb{R}^{\mathscr{P}\mathscr{J}(M)}$. Then the following statements are equivalent:*

4.3.1. *U is* S-*integrable over J.*

4.3.2. *U is* P-W-*integrable over J.*

If both statements hold, then

$$\text{(S)} \int_J U = \text{(P)} \int_J U.$$

Proof. $2 \Rightarrow 1$ is an immediate consequence of Remark 4.2.2.

$1 \Rightarrow 2$. Assume that U is S-integrable over J. Choose any $\varepsilon > 0$ and determine $0 < \delta \in \mathbb{R}^M$ such that $\Delta_J(U; \delta)$ has length $< \varepsilon$. Define u, v on $\mathscr{J}(J)$ by $\Delta_K(U; \delta) = [u(K), v(K)]$ $(J \supseteq K \in \mathscr{J}(J))$. It is an easy exercise to show that u is a lower and v is an upper function for U on J, with δ serving for both of them. In particular $0 \leqq v(J) - u(J) < \varepsilon$. This does it (see Remark 4.2.2).

We devote the rest of this section to an informal discussion of the relationship between differentiation and the P-W-integral.

Let $f \in \mathbb{R}^{\mathbb{R}}$ be everywhere differentiable in the usual sense and f' its derivative. Choose an interval $J = [a, b]$ with $a < b$ and put

$$U(K, t) = f'(t)|K| \qquad ((K, t) \in \mathcal{P}\mathcal{J}(J))$$

where $|K|$ denotes the length of the interval K.

For any $\varepsilon > 0$ and $K = [c, d] \in \mathcal{J}(J)$, we put

$$u_\varepsilon(K) = f(d) - f(c) - \varepsilon|K|, \qquad v_\varepsilon(K) = f(d) - f(c) + \varepsilon|K|.$$

Define $0 < \delta \in \mathbb{R}^{\mathbb{R}}$ such that

$$([c, d], t) \text{ sub } \delta \quad \Rightarrow \quad \left| \frac{f(d) - f(c)}{d - c} - f'(t) \right| < \varepsilon.$$

This is possible since f is differentiable. The argument is obvious in case $t = c$ or $t = d$. In case $c < t < d$ we have to consider

$$\left| \frac{f(d) - f(c)}{d - c} - f'(t) \right|$$

$$\leqq \frac{t - c}{d - c} \left| \frac{f(t) - f(c)}{t - c} - f'(t) \right| + \frac{d - t}{d - c} \left| \frac{f(d) - f(t)}{d - t} - f'(t) \right|.$$

This convex combination becomes $< \varepsilon$ by the differentiability of f at t, provided $|t - c|$, $|d - t| < \delta(t)$ is sufficiently small.

Clearly

$$u_\varepsilon(K) < U(K, t) < v_\varepsilon(K) \qquad (\mathcal{P}\mathcal{J}(J)) \ni (K, t) \text{ sub } \delta).$$

The subadditivity of u_ε and the superadditivity of v_ε are immediate, hence $u_\varepsilon \in \mathcal{L}(U; J)$, $v_\varepsilon \in \mathcal{U}(U, J)$. Moreover, $|v_\varepsilon(J) - u_\varepsilon(J)| < 2\varepsilon|J|$. Hence U is P-W-integrable, i.e., f' is P-W-integrable for the usual interval length function $m(K) = |K|$, over J, and

$$(\mathrm{P}) \int_J U = (\mathrm{P}) \int_J f \, dm = f(b) - f(a);$$

i.e., the P-W-integral yields precisely the inverse of differentiation, without any assumptions beyond pointwise differentiability.

4.4. Example. Let $\alpha > 1$, $\beta > 0$ be reals. Define $f \in \mathbb{R}^{\mathbb{R}}$ by

$$f(t) = \begin{cases} t^\alpha \sin t^{-\beta} & (t \neq 0) \\ 0 & (t = 0). \end{cases}$$

Then f is everywhere differentiable and

$$f'(t) = \begin{cases} \alpha t^{\alpha - 1} \sin t^{-\beta} - \beta t^{\alpha - \beta - 1} \cos t^{-\beta} & (t \neq 0) \\ 0 & (t = 0). \end{cases}$$

We have, by our previous argument,

$$(\mathrm{P})\int_{[a,\,b]} f'\,dm = f(b) - f(a).$$

Thus f' is S-integrable by Proposition 4.3. The function $|f'|$, however, is, for $\beta > \alpha$, no longer S-integrable since we have for the ordinary Riemann integral, the relation

$$\lim_{s\to 0+0}\int_{-1}^{-s} |f'(t)|\,dt = \infty.$$

5. MONOTONE CONVERGENCE

Sections 2–4 have provided us with three equivalent concepts of integrability for real functions defined on pointed intervals. We may thus pass freely from one of these three concepts to any other.

In this section we state and prove a monotone convergence theorem which is, in a way, an analogue of Theorem III.5.3.

5.1. Definition. *Let* $M \subseteq \mathbb{R}$, $F \in \mathscr{F}(M)$, *and* $W, U_1, U_2, \ldots \in \mathbb{R}^{\mathscr{P}\mathscr{I}(F)}$ *be such that* $U_1 \leqq U_2 \leqq \cdots$ *everywhere on* $\mathscr{P}\mathscr{I}(F)$.

5.1.1. *Assume that for every* $\varepsilon > 0$, *there is a* ∂*-superadditive* $0 \leqq m \in \mathbb{R}^{\mathscr{I}(F)}$, *and* $0 < N, \delta \in \mathbb{R}^M$ *such that*

$$\begin{aligned} m(F) < \varepsilon,\ t \in F,\ n \geqq N(t),\ \mathscr{P}\mathscr{I}(F) \ni (K, t) \text{ sub } \delta \\ \Rightarrow\quad U_n(K, t) \geqq W(K, t) - m(K). \end{aligned}$$

Then we write

$$\operatorname*{S\text{-}lim}_{n\to\infty} U_n \geqq W.$$

5.1.2. *Assume* $\text{S-lim}_{n\to\infty} U_n \geqq W$ *and* $U_1 \leqq U_2 \leqq \cdots \leqq W$. *Then we write*

$$\operatorname*{S\text{-}lim}_{n\to\infty} U_n = W.$$

5.1.3. *Let* $U \in \mathbb{R}^{\mathscr{P}\mathscr{I}(F)}$ *and assume that* $U_1 = U_2 = \cdots = U$ *satisfies* $\text{S-lim}_{n\to\infty} U_n \geqq W$. *Then we write* $U \geqq_{\mathrm{S}} W$.

5.1.4. *We carry over the preceding notations to functions on* $\mathscr{I}(F)$ *via the functions on* $\mathscr{P}\mathscr{I}(F)$ *trivially corresponding to them* $(U(K, t) = U(K))$.

5.2. Theorem (monotone convergence). *Let* $M \subseteq \mathbb{R}$, $F \in \mathscr{F}(M)$, U_1, $U_2, \ldots \in \mathbb{R}^{\mathscr{P}\mathscr{J}(F)}$ *be such that* $U_1 \leqq U_2 \leqq \cdots$ *and*

$$\sup\left\{(\mathrm{S})\int_F U_n \,\middle|\, n = 1, 2, \ldots\right\} < \infty.$$

Then

5.2.1. *The formula*

$$I(J) = \lim_{n\to\infty} (\mathrm{S})\int_J U_n \qquad (J \in \mathscr{J}(F))$$

defines a ∂*-additive function* $I \in \mathbb{R}^{\mathscr{J}(F)}$.

5.2.2. *Let* $W \in \mathbb{R}^{\mathscr{P}\mathscr{J}(F)}$ *be such that* $\text{S-lim}_{n\to\infty} U_n \geqq W$. *Then* $I \geqq_{\mathrm{S}} W$.

5.2.3. *Let* $W \in \mathbb{R}^{\mathscr{P}\mathscr{J}(F)}$ *be such that* $\text{S-lim}_{n\to\infty} U_n = W$. *Then* W *is S-integrable over* F *and*

$$I(J) = (\mathrm{S})\int_J W \qquad (J \in \mathscr{J}(F)).$$

Proof. 1 is an easy exercise, on the basis of Proposition 2.4.

2. Choose any $\varepsilon > 0$. We have to find a ∂-superadditive $0 \leqq m \leqq \mathbb{R}^{\mathscr{J}(F)}$ and a $0 < \delta \in \mathbb{R}^F$ such that $m(F) < \varepsilon$ and $\mathscr{P}\mathscr{J}(F) \ni (K, t)$ sub $\delta \Rightarrow I(K) \geqq W(K, t) - m(K)$. From the V-integrability of $U_1, U_2, \ldots$ we get ∂-superadditive $m_1, m_2, \ldots \mathbb{R}^{\mathscr{J}(F)}$ and functions $0 < \delta_1, \delta_2, \ldots \in \mathbb{R}^F$ such that the functions $I_n(J) = (\mathrm{S})\int_J U_n$ $(J \in \mathscr{J}(F), n = 1, 2, \ldots)$ satisfy

$$\mathscr{P}\mathscr{J}(F) \ni (K, t) \text{ sub } \delta_n \quad \Rightarrow \quad |U_n(K, t) - I_n(K)| < m_n(K),$$

while

$$m_n(F) < \varepsilon/2^{n+1} \qquad (n = 1, 2, \ldots).$$

Define $m' = m_1 + m_2 + \cdots$. Clearly m' is ∂-superadditive and $m'(F) < \varepsilon/2$. Moreover

$$\mathscr{P}\mathscr{J}(F) \ni (K, t) \text{ sub } \delta_n \quad \Rightarrow \quad I(K) \geqq I_n(K) \geqq U_n(K) - m'(K) \qquad (n = 1, 2, \ldots).$$

Now $\text{S-lim}_n U_n \geqq W$ works to the effect that we get a ∂-superadditive $0 \leqq m'' \in \mathbb{R}^{\mathscr{J}(F)}$ and functions $0 < \delta_0, N \in \mathbb{R}^F$ such that $m''(F) < \varepsilon/2$

$$\mathscr{P}\mathscr{J}(F) \ni (K, t) \text{ sub } \delta_0,\ n \geqq N(t) \quad \Rightarrow \quad U_n(K, t) \geqq W(K, t) - m'(K).$$

Put now

$$m = m' + m''$$
$$\delta(t) = \min\{\delta_0(t), \delta_{N(t)}(t)\} \qquad (t \in F).$$

Then

$$m(F) = m'(F) + m''(F) < \tfrac{1}{2}\varepsilon + \tfrac{1}{2}\varepsilon = \varepsilon$$
$$\mathscr{P}\mathscr{J}(F) \ni (K, t) \text{ sub } \delta \quad \Rightarrow \quad I(K) \geqq U_{N(t)}(Kt) - m'(K) \geqq W(K, t) - m(K).$$

This does it.

3. We can exploit what we did for the proof of 2. It is then clearly sufficient to find a ∂-superadditive $0 \leqq \overline{m} \in \mathbb{R}^{\mathscr{J}(F)}$ and a $0 < \overline{\delta} \in \mathbb{R}^F$ such that $\overline{m}(F) < \varepsilon$ and

$$\mathscr{P}\mathscr{J}(F) \ni (K, t) \text{ sub } \overline{\delta} \quad \Rightarrow \quad W(K, t) \leqq I(K) + \overline{m}(K).$$

For this, we determine n such that $I_n(F) > I(F) - \varepsilon/2$ and put $\overline{m}' = I - I_n$. Then $\overline{m}'$ is $\geqq 0$, ∂-superadditive (even ∂-additive) and satisfies $\overline{m}'(F) < \frac{1}{2}\varepsilon < \varepsilon$. Put $\overline{\delta} = \delta_n$ for the chosen n. Then

$$\begin{aligned} \mathscr{P}\mathscr{J}(F) \ni (K, t) \text{ sub } \overline{\delta} \quad \Rightarrow \quad I(K) &= I_n(K) + \overline{m}'(K) \\ &\leqq U_n(K, t) + m_n(K) + \overline{m}'(K) \\ &\leqq W(K, t) + \overline{m}(K) \end{aligned}$$

if we put $\overline{m} = \overline{m}' + m_n$. Now clearly $\overline{m}$ is ∂-superadditive and satisfies $\overline{m}(F) < \frac{1}{2}\varepsilon + \frac{1}{2}\varepsilon = \varepsilon$. This does it.

6. RELATION TO THE DANIELL INTEGRAL

We have presented three equivalent integration theories (S, V, and P-W) in Sections 2–4 of this appendix. On the other hand, we have the monotone integration theory (after Daniell), as presented in Chapter III. In the present section we are going to find out whether (S) $\int_J f\, dm = \int_J f\, dm$ provided that both integrals exist. For this it is feasible to fix an $F \in \mathscr{F}(R)$, consider the natural σ-field $B \cap F = \{E \mid E \text{ borel}, E \subset F\}$ in F, and to fix a positive σ-content m on $B \cap F$. Having chosen $f \in \mathbb{R}^F$ we define $U\colon \mathscr{P}\mathscr{J}(F) \to \mathbb{R}$ by $U(K, t) = f(t)m(K)$ and put (S) $\int_J f\, dm = $ (S) $\int_J U$.

We are going to answer the following question: Given m, does (S) $\int_J f\, dm = \int_J f\, dm$ hold for all $f \in L^1_m \cap \mathbb{R}^F$ and all $J \in \mathscr{J}(F)$? If so, then putting $f = 1$, we obtain that $m(J) = \int_J 1\,dm = (\mathrm{S})\int_J 1\,dm$. By 2.3 the integral $(\mathrm{S})\int_J 1\,dm$ depends ∂-additively on $J \in \mathcal{J}(F)$, so that m must be ∂-additive. It is obvious that m is ∂-additive iff

$$(1) \qquad m(\{t\}) = 0 \qquad \text{for} \quad t \in \operatorname{int} F.$$

Thus (1) is a necessary condition for the affirmative answer to the above question. We are going to prove that it is sufficient also.

Our most effective tool will be

6.1. Theorem (monotone integration). *Let F and m be as above (m need not satisfy (1). Let $0 \leqq f_1 \leqq f_2 \leqq \cdots \in \mathbb{R}^F \ni f$ be such that*

$$\lim_{n\to\infty} f_n(t) = f(t) \qquad (t \in F).$$

Assume that $f_1, f_2, \ldots$ are S-*integrable for m over F and*

$$\lim_{n\to\infty} (\mathrm{S})\int_F f_n\,dm < \infty.$$

Then f is S-*integrable for m over F and*

$$(\mathrm{S})\int_F f\,dm = \lim_{n\to\infty} (\mathrm{S})\int_F f_n\,dm.$$

Proof. Put $U_n(K, t) = f_n(t)m(K)$, $U(K, t) = f(t)m(K)$ $((K, t) \in \mathcal{P}\mathcal{J}(F))$. By Theorem 5.2 we have only to show $\text{S-lim}_{n\to\infty} U_n \geqq U$. Define, for a given $\varepsilon > 0$, $\overline{m} = (\varepsilon/2m(F))m$. Then $\overline{m}$ is ∂-additive, hence ∂-superadditive, and $\overline{m}(F) = \varepsilon/2 < \varepsilon$. For every $t \in F$ define $N(t)$ such that $n \geqq N(t)$ implies

$$f_n(t) \geqq f(t) - \frac{\varepsilon}{2m(F)}.$$

Then $(K, t) \in \mathcal{P}\mathcal{J}(F)$, $n \geqq N(t)$ imply

$$U_n(K, t) = f_n(t)m(K) \geqq f(t)m(K) - m(K)$$

and $\text{S-lim}_{n\to\infty} U_n \geqq U$ follows; we did not even have to bother about some $0 < \delta \in \mathbb{R}^F$.

There is evidently a "downward" analogue to this "upward" theorem.

What we need next is the stability of S-integrability with respect to finite lattice operations.

6.1. Proposition. *Let F and m be as above, m satisfies (1). Let $f, g, h \in \mathbb{R}^F$ be* S-*integrable for m over F. Assume $h \geqq 0$, $|f|, |g| \leqq h$. Then $f \vee g$, $f \wedge g$ are* S-*integrable for m over F.*

Proof. It is sufficient to deal with $u = f \wedge g$. For $(K, t) \in \mathscr{P}\mathscr{J}(F)$, we define

$$U(K, t) = f(t)m(K), \qquad W(K, t) = u(t)m(K),$$
$$V(K, t) = g(t)m(K), \qquad X(K, t) = h(t)m(K).$$

Determine $0 < \delta_1 \in \mathbb{R}^F$ such that $\Delta_F(X, \delta_1)$ is bounded. Let b be the upper endpoint of $\Delta_F(X, \delta_1)$. Then for any pointed decomposition Z sub δ_1 of F, the reals $S_F(U, Z)$, $S_F(V, Z)$, $S_F(W, Z)$, $S_F(X, Z)$ are in $[-b, b]$. For $\eta \in \mathbb{R}^F$, $0 < \eta \leqq \delta_1$, denote by $a(\eta)$ the lower end of $\Delta_F\,S(W, \eta)$ and define

$$\alpha = \sup\{a(\eta) \mid \eta \in \mathbb{R}^F, 0 < \eta \leqq \delta_1\}.$$

Obviously we have $a(\eta), \alpha \in [-b, b]$; we are going to prove that

$$\alpha = (\mathrm{S}) \int_F u\, dm.$$

Let $0 < \varepsilon \in \mathbb{R}$; find $\delta_2 \in \mathbb{R}^F$, $0 < \delta_2 \leqq \delta_1$ and a nonnegative superadditive m' on $J(F)$ such that $m'(F) < \varepsilon/3$ and that

$$\{(K, t)\} \text{ sub } \delta_2 \quad \Rightarrow \quad \left\{ \begin{aligned} &\left| (\mathrm{S}) \int_K f\, dm - f(t)m(K) \right| \leqq m'(K), \\ &\left| (\mathrm{S}) \int_K g\, dm - g(t)m(K) \right| \leqq m'(K). \end{aligned} \right.$$

Find $\delta_3 \in \mathbb{R}^F, 0 < \delta_3 \leqq \delta_2$ such that $\alpha < a(\delta_3) + \varepsilon$. Let Z sub δ_3 be a pointed decomposition of F, $Z = \{(K_1, t_1), (K_2, t_2), \ldots, (K_n, t_n)\}$ such that

$$S_F(W, Z) = \sum_{j=1}^{n} u(t_j)m(K_j) < a(\delta_3) + \frac{\varepsilon}{3}.$$

Find $\delta_4 \in \mathbb{R}^F, 0 < \delta_4 \leqq \delta_3$, such that $\delta_4(t) <$ the distance of t to any endpoint of any K_j, provided this endpoint is different from t.

Then $\mathscr{P}\mathscr{J}(F) \ni (K, t)$ sub δ_4 implies: Either K is in the interior of some K_j, or t is the endpoint of some $K_j \supseteq K$, or t is the inner point of K and the common point of two adjacent K_j (the union of which contains K). For every pointed decomposition X sub δ_4, we replace every $(K, t) \in X$, which falls under the third case by (K', t), (K'', t), where K', K'' are in two different ones among the K_j from Z and $K' \cup K'' = K$. This leads to a pointed decomposition $Y = \{(L_1, s_1), (L_2, s_2), \ldots, (L_k, s_k)\}$; Y is still sub δ_4 and $S_F(W, Z) = S_F(W, Y)$ (since m is additive by (1)). For every

$j = 1, 2, \ldots, n$, let Y_j be the restriction of Y to K_j, a pointed decomposition of K_j. Obviously we have Y_j sub δ_4. Now $\delta_4 \leqq \delta_2$ yields

$$u(s_i)m(L_i) \leqq f(s_i)m(L_i) \leqq (\mathrm{S}) \int_{L_i} f\, dm + m'(L_i),$$

$$u(s_i)m(L_i) \leqq g(s_i)m(L_i) \leqq (\mathrm{S}) \int_{L_i} g\, dm + m'(L_i) \qquad (i = 1, 2, \ldots, k).$$

Since $K_j = \bigcup \{L_i \,|\, L_i \subseteq K_j\}$, we obtain (summing the above inequalities over such i that $L_i \subseteq K_j$)

$$S_{K_j}(W, Y_j) \leqq (\mathrm{S}) \int_{K_j} f\, dm + m'(K_j),$$

$$S_{K_j}(W, Y_j) \leqq (\mathrm{S}) \int_{K_j} g\, dm + m'(K_j) \qquad (j = 1, 2, \ldots, n).$$

But (K_j, t_j) sub δ_2 implies that

$$(\mathrm{S}) \int_{K_j} f\, dm \leqq f(t_j)m(K_j) + m'(K_j),$$

$$(\mathrm{S}) \int_{K_j} g\, dm \leqq g(t_j)m(K_j) + m'(K_j).$$

Thus

$$S_{K_j}(W, Y_j) \leqq f(t_j)m(K_j) + 2m'(K_j)$$

and simultaneously

$$S_{K_j}(W, Y_j) \leqq g(t_j)m(K_j) + 2m'(K_j),$$

so that

$$S_{K_j}(W, Y_j) \leqq u(t_j)m(K_j) + 2m'(K_j).$$

Summing over j we obtain

$$\begin{aligned}
\alpha - \varepsilon < a(\delta_3) \leqq S_F(W, X) = S_F(W, Y) &= \sum_{j=1}^{n} S_{K_j}(W, Y_j) \\
&\leqq \sum_{j=1}^{n} (u(t_j)m(K_j) + 2m'(K_j)) \\
&\leqq S_F(W, Z) + 2\varepsilon/3 \leqq a(\delta_3) + \varepsilon \leqq \alpha + \varepsilon.
\end{aligned}$$

This does it.

6.3. Proposition. *Let F and m be as before (m need not satisfy (1)). Then $\mathscr{S}(m, F) = \{f \mid f \in \mathbb{R}^F,\ f$ and $|f|$ S-integrable for m over $F\}$ is a vector lattice.*

It is easy to prove that the constant function 1 is S-integrable over F provided that (1) holds. We can now imitate a device called "blowing up the hat" which was employed in the proof of Theorem III.6.3. It yields (exercise)

6.4. Proposition. *Let F and m be as before, let (1) be satisfied, $f \geqq 0$ S-integrable for m over F, and $0 < \alpha \in \mathbb{R}$. Then $1_{\{f > \alpha\}}$ is S-integrable for m over F.*

6.5. Corollary. *Let F and m be as before, m satisfying (1), $f \geqq 0$ S-integrable. Then for every $n = 1, 2, \ldots$, the function $f_n \in \mathbb{R}^F$ defined by*

$$f_n = \sum_{k=1}^{2^{2n}} \frac{k}{2^n} [1_{\{f > (k-1)/2^n\}} - 1_{\{f > k/2^n\}}] + 2^n 1_{\{f > 2^n\}}$$

is S-integrable for m over F. We have $0 \leqq f_1 \leqq f_2 \leqq \cdots \leqq f$ and $\lim_n f_n = f$ pointwise on F and

$$\lim_n \text{(S)} \int_F f_n \, dm = \text{(S)} \int f \, dm.$$

6.6. Theorem. *Let F and m be as before, m satisfying (1). Then every $f \in \mathscr{L}_m^1 \cap \mathbb{R}^F$ is S-integrable for m over F and*

$$\int_F f \, dm = \text{(S)} \int_F f \, dm.$$

(Of course, if $J \in \mathscr{J}(F)$, then the restriction f_J of f to J satisfies $f_J \in \mathscr{L}_m^1 \cap \mathbb{R}^J$ so that $\int_J f \, dm = \text{(S)} \int_J f \, dm$.)

Proof. We may assume $f \geqq 0$. We intend to pursue Daniell's extension procedure with the help of Theorems 6.1 and III.5.3. We have, however, to be careful to never deal with functions attaining the value ∞. This forces us to proceed as follows. By ordinary Riemann integrability, our theorem is true for continuous f. Applying Theorems 6.1 and III.5.3 with a constant bound, we arrive at the truth of our theorem for bounded σ-upper and σ-lower functions $0 \leqq f \in \mathscr{L}_m^1 \cap \mathbb{R}^F$. Iterating the same argument, we establish it for bounded countable infima of σ-upper functions and bounded suprema of σ-lower functions. If $g \leqq f \leqq h$, g and h are S-integrable for m over F and $\text{(S)} \int g \, dm = \text{(S)} \int h \, dm$, then f is S-integrable for m over F. This establishes our theorem in the case of a bounded $0 \leqq f \in \mathscr{L}_m^1 \cap \mathbb{R}^F$. For any $0 \leqq f \in \mathscr{L}_m^1 \cap \mathbb{R}^F$, we see that $f_n = f \wedge n$ is S-integrable for m over F. Now $f_n \nearrow f$ and Theorem 6.1 yields the desired result for this f, and the theorem is proved.

Let us briefly describe the situation if (1) is not satisfied. Put

$$T = \{t \in \operatorname{int} F \mid m(\{t\}) > 0\}$$

Obviously T is a countable set and $\sum_{t \in T} m(t) < \infty$. It can be proved from the definition of the S-integral that if (S) $\int_F f\, dm$ exists, then $f(t) = 0$ for $t \in T$ (cf. Example 2.4). Nevertheless, the following result is close to Theorem 6.6. Let χ be the characteristic function of int F. Define

$$\hat{m}([a, b]) = m([a, b]) - \tfrac{1}{2}m(\{a\})\chi(a) - \tfrac{1}{2}m(\{b\})\chi(b).$$

6.7. Theorem. *Let F, m, $\hat{m}$ be as above. Then every $f \in L_m^1 \cap \mathbb{R}^F$ is S-integrable for m over F and*

$$\int_F f\, dm = (\mathrm{S}) \int_F f\, d\hat{m}.$$

The proof can be obtained if we decompose $m = m_c + m_d$, putting $m_d(E) = \sum_{t \in E \cap T} m(\{t\})$ and apply Theorem 6.6 to $\int_F f\, dm_c$.

7. SOME RESULTS ON THE S-INTEGRAL

7.1. Dependence of (S) $\int_J U$ on J. Let $M = [\alpha, \beta]$, $U \in \mathbb{R}^{\mathscr{P}\mathscr{I}(M)}$, and assume that (S) $\int_M U$ exists. Then (S) $\int_{[\alpha, y]} U$ exists for every $y \in]\alpha, \beta[$ by Proposition 2.3.1, and it is not difficult to prove that

7.1.1.
$$\lim_{y \to \beta - 0} \left((\mathrm{S}) \int_{[\alpha, y]} U + U([y, \beta], \beta) \right) = (\mathrm{S}) \int_M U.$$

The following proposition can be proved directly from the definition of the S-integral.

7.1.2. Proposition. *Let $M = [\alpha, \beta]$, $U \in \mathbb{R}^{\mathscr{P}\mathscr{I}(M)}$, and assume that:*

(1) *the integral* (S) $\int_{[\alpha, y]} U$ *exists for every* $y \in]\alpha, \beta[$;

(2) $\lim_{y \to \beta - 0} \left((\mathrm{S}) \int_{[\alpha, y]} U + U([y, \beta], \beta) \right) = c \in \mathbb{R}$.

Then (S) $\int_M U$ *exists and is equal to* c.

Especially, let $\theta \in \mathbb{R}^M$ *be continuous,* $m([\gamma, \delta]) = \theta(\delta) - \theta(\gamma)$ *for* $[\gamma, \delta] \in \mathcal{J}(M)$, $f \in \mathbb{R}^M$. *If* (S) $\int_{[\alpha, y]} f\, dm$ *exists for every* $y \in]\alpha, \beta[$ *and if* $\lim_{y \to \beta - 0}$ (S) $\int_{[\alpha, y]} f\, dm = c \in \mathbb{R}$, then (S) $\int_M f\, dm$ *exists and is equal to c.*

7.2. Integration by parts. Let $M = [\alpha, \beta], f, g \in \mathbb{R}^M$, $K = [\gamma, \delta] \in \mathcal{J}(M)$, $t \in K$. Then

7.2.1.
$$f(t)(g(\delta) - g(\gamma)) + g(t)(f(\delta) - f(\gamma)) = W(K, t) + f(\delta)g(\delta) - f(\gamma)g(\gamma)$$

where

$$W(K, t) = -(f(\delta) - f(t))(g(\delta) - g(t)) + (f(t) - f(\gamma))(g(t) - g(\gamma)).$$

Put $\tilde{g}(K) = g(\delta) - g(\gamma)$, $\tilde{f}(K) = f(\delta) - f(\gamma)$. Let

$$Z = \{(K_1, t_1), (K_2, t_2), \ldots, (K_n, t_n)\} \in \mathcal{P}\mathcal{J}(M).$$

From 7.2.1 we obtain that

7.2.2.
$$\sum_{i=1}^{n} f(t_i)\tilde{g}(K_i) + \sum_{i=1}^{n} g(t_i)\tilde{f}(K_i) = f(\beta)g(\beta) - f(\alpha)g(\alpha) + \sum_{i=1}^{n} W(K_i, t_i).$$

Since 7.2.2 holds for any $Z \in \mathcal{P}\mathcal{J}(M)$, we have

7.2.3. Proposition. *If any two from the three integrals in*

7.2.4.
$$\text{(S)} \int_M f\, d\tilde{g} + \text{(S)} \int_M g\, d\tilde{f} = f(\beta)g(\beta) - f(\alpha)g(\alpha) + \text{(S)} \int_M W$$

exist, then the third one exists too and 7.2.4 holds.

If (S) $\int_M W = 0$, *then 7.2.4 is the usual formula for integration by parts. For the analogue of 7.2.4 for functions* $U \in \mathbb{R}^{\mathcal{P}\mathcal{J}(M)}$, *see* KURZWEIL [1].

7.3. Multipliers of S-integrable functions. Let $M \in J(\mathbb{R})$, q be a σ-content on $B \cap M = \{E | E \text{ borel}, E \subset M\}$, $h \in L_q^1$. Then $fh \in L_q^1$ provided that $f \in \mathbb{R}^M$ is bounded and measurable. For $f = \operatorname{sgn} h$, this implies that $|h| \in L_q^1$. Therefore an analogous result for the S-integral cannot hold (cf. Example 4.4) and the conditions on f must be more restrictive.

7.3.1. Proposition. *Let* $f \in \mathbb{R}^M$ *be of bounded variation,* $U \in \mathbb{R}^{\mathcal{P}\mathcal{J}(M)}$, $V(J, t) = f(t)U(J, t)$. *Assume that* (S) $\int_M U$ *exists and that*

7.3.2. *to every* $t \in M$ *and* $\varepsilon > 0$ *there exists such an* $\eta > 0$ *that* $|U(J, t)| < \varepsilon$ *if* $t \in J \subset [t - \eta, t + \eta] \cap M$.

Then (S) $\int_M V$ *exists.*

Proof. Let $M = [\alpha, \beta]$. Put $g(y) = \text{(S)} \int_{[\alpha, y]} U$ for $y \in]\alpha, \beta[$, $g(\alpha) = 0$,

$\tilde{g}([\gamma, \delta]) = g(\delta) - g(\gamma)$ for $[\gamma, \delta] \in \mathscr{J}(M)$. g is continuous by 7.3.2 and 7.1.1. Put $Y(J, t) = f(t)\tilde{g}(J)$. U is variationally equivalent to $\tilde{g}$; and since f is bounded, V is variationally equivalent to Y. Thus it is sufficient to prove that (S) $\int_M f\, d\tilde{g}$ exists.

Put $\tilde{f}([\gamma, \delta]) = f(\delta) - f(\gamma)$ for $[\gamma, \delta] \in \mathscr{J}(M)$ and define W as in 7.2.1. Since g is continuous and f is of bounded variation, both integrals (S) $\int_M g\, d\tilde{f}$ and (S) $\int_M W$ exist (moreover, (S) $\int_J W = 0$ for $J \in \mathscr{J}(M)$). by Proposition 7.2.1, (S) $\int_M f\, d\tilde{g}$ exists. This does it. For an extension to the multidimensional case see KURZWEIL [3].

Let $\theta \in \mathbb{R}^M$ be continuous, $m([\gamma, \delta]) = \theta(\delta) - \theta(\gamma)$ for $[\gamma, \delta] \in \mathscr{J}(M)$. The function $f \in \mathbb{R}^M$ is called a multiplier of S-integrable functions (for m) if (S) $\int_M fh\, dm$ exists whenever (S) $\int_M h\, dm$ exists. We proved that every function f of bounded variation is a multiplier of S-integrable functions (for m).

7.4. Transformation of the S-integral. The following proposition follows immediately from the definition of the S-integral.

7.4.1. Proposition. *Let $K, M \in \mathscr{J}(R)$, let $\varphi \in M^K$ be continuous and strictly increasing. $\varphi(K) = M$. Let $m \in \mathbb{R}^{\mathscr{J}(M)}$, $f \in \mathbb{R}^M$, and let* (S) $\int_M f\, dm$ *exist. Put $\hat{f}(t) = f(\varphi(t))$ for $t \in K$ and $\hat{m}([\gamma, \delta]) = m([\varphi(\gamma), \varphi(\delta)])$ for $[\gamma, \delta] \in \mathscr{J}(K)$. Then* (S) $\int_K \hat{f}\, d\hat{m}$ *exists and is equal to* (S) $\int_K f\, dm$.

7.5. The multidimensional and abstract S-integrals. The main results that were obtained in Sections 1–7 can be extended to the r-dimensional case, i.e., $M, K, J, \ldots$ are r-dimensional intervals of the form

7.5.1. $$M = [\alpha_1, \beta_1] \times [\alpha_2, \beta_2] \times \cdots \times [\alpha_r, \beta_r]$$

with $\alpha_j, \beta_j \in \mathbb{R}$, $\alpha_j < \beta_j$. The following two results have no analogue in the one-dimensional case:

7.5.2. for the S-integral a good analogue of the Fubini theorem is valid (see KURZWEIL [2]),

7.5.3. orthonormal transformations cannot be applied in general to the S-integrals (this is not surprising since the set of intervals of form 7.5.1 is not invariant with respect to orthonormal transformations).

The abstract approach to the S-integral will be presented in a forthcoming paper. In this approach $M, J, K, \ldots$ are no longer intervals, but elements of a family A of subsets of an abstract set; the values of U are elements of an abelian semigroup with a limiting process.

APPENDIX B

CONTENTS WITH GIVEN MARGINALS

In this appendix we present the results of HANSEL-TROALLIC [1], Mesures marginales et théorème de Ford–Fulkerson, *Z. Wahrscheinlichkeitstheorie und Verw. Gebiete* **43** (1978), 245–251, on contents and σ-contents with given marginals. These results comprise previous results of Strassen, Kellerer, and Furstenberg as special cases. Moreover, their method of approach is much simpler than their predecessor's: An ingenious application of the combinatorial max-flow–min-cut theorem of Ford and Fulkerson. The problem to be solved is the following: Given a product space $\Omega = \Omega_0 \times \Omega_1$, and probabilities p_0 in Ω_0, p_1 in Ω_1; does there exist a probability m in Ω such that the natural projections of Ω onto Ω_0, Ω_1 send m into p_0 resp. p_1? The answer is trivially positive, namely $m = p_0 \times p_1$ if no further requirements are made. The problem becomes interesting if we require m to be majorized by some given σ-content q in Ω. Since $\varphi_0 m$ and $\varphi_1 m$ are usually called the marginals (Chapter VI, Section 5) of m, our problem is also called the problem of given marginals.

We begin our discussion with Ford–Fulkerson's theorem (Section 1) and then proceed to the given marginal problem for matrices (Section 2). Extension and approximation techniques of more or less routine character lead us (in Section 3) to existence results in the domain of contents. Additional considerations lead to corresponding results involving σ-additivity. Results involving topologies in Ω_0, Ω_1 are presented in Section 4.

1. THE FORD–FULKERSON THEOREM

The Ford–Fulkerson theorem deals with flows in networks. The edges of the network (of pipelines, e.g.) are supposed to bear capacity restrictions which must not be violated by the flow. The question is how strong a flow can be.

We begin by the "bipartite" special case of the Ford–Fulkerson theorem.

A complete *bipartite* network consists of two finite sets X, Y, and two additional points 0, 1 such that the four sets $\{0\}$, X, Y, $\{1\}$ are pairwise disjoint. 0 is linked to any $j \in X$ by an edge e_j, every $j \in X$ is linked to

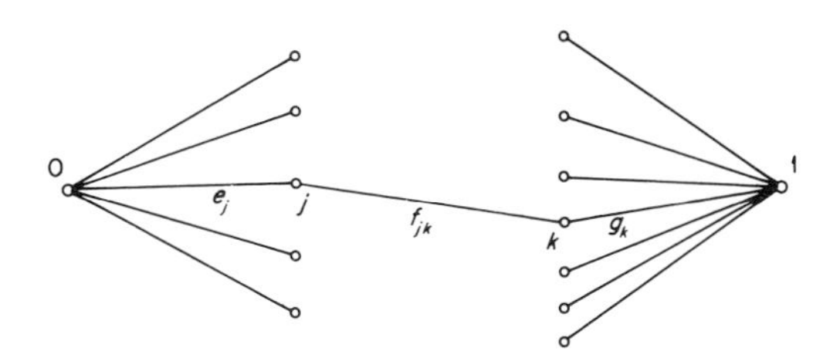

every $k \in Y$ by an edge f_{jk}, and every $k \in Y$ is linked to 1 by an edge g_k. By no means all f_{jk} have been drawn in the figure. To every e_j we attach an integer $p_j \geqq 0$, to every f_{jk} an integer $q_{jk} \geqq 0$, and to every g_k an integer $r_k \geqq 0$. The p_j, q_{jk}, r_k are understood as "capacities." A triple (e_j, f_{jk}, g_k) is called a path from 0 to 1. A subset C of the set of all edges e_j, f_{jk}, g_k is called a **cut** if every path from 0 to 1 contains an edge from C. The sum of the capacities of all edges in a cut C is called the **capacity** $c(C)$ of the cut.

A function F defined on the set of all edges, with nonnegative integer values, is called a **flow** if

$$F(e_j) = \sum_k F(f_{jk}) \qquad (j \in X)$$

$$\sum_j F(j, k) = F(g_k) \qquad (k \in Y),$$

which means that whatever flows into j flows out from j again, and the same for k. F is said to **conform** to the given capacities if

$$F(e_j) \leqq p_j, \qquad F(f_{jk}) \leqq q_{jk}, \qquad F(g_k) \leqq r_k \qquad (j \in X, \quad k \in Y)$$

Clearly we have

$$\sum_j F(e_j) = \sum_k F(g_k)$$

for every flow. This integer $\geqq 0$ is called the **strength** $\|F\|$ of the flow. It is obvious that

$$\|F\| \leqq c(C)$$

for any cut C if F conforms to the given capacities.

1.1. Theorem (Ford–Fulkerson, bipartite case). *Under the above assumptions,*

$$\max \|F\| = \min c(C)$$

where F runs through all flows conforming with the given capacities, and C runs through all cuts. There is a flow F_0 and a cut C_0 such that

$$\max \|F\| = \|F_0\| = c(C_0) = \min c(C).$$

Proof. It is obvious that there is a flow F_0 of maximal strength among those conforming with the capacities. We shall construct a cut C_0 with $c(C_0) = \|F_0\|$ by means of a **marking algorithm** which runs as follows:

First, mark 0. Secondly, mark any $j \in X$ such that $F_0(e_j) < p_j$. Thirdly, mark every $k \in Y$ for which there is a $j \in X$ already marked such that $F_0(f_{jk}) < q_{jk}$. Fourthly, mark 1 if there is a $k \in Y$ already marked such that $F_0(g_k) < r_k$. Stop if 1 was marked and find $j_0 \in X$, $k_0 \in Y$ such that $F_0(e_{j_0}) < p_{j_0}$, $F_0(f_{j_0k_0}) < q_{j_0k_0}$, $F_0(g_{k_0}) < r_{k_0}$. Define a flow F_1 by

$$F_1(e_{j_0}) = F_0(e_{j_0}) + 1, \qquad F_1(f_{j_0k_0}) = F_0(f_{j_0k_0}) + 1,$$
$$F_1(g_{k_0}) = F_0(g_{k_0}) + 1,$$

and $F_1 = F_0$ otherwise. Clearly F_1 still respects the capacities, but $\|F_1\| = \|F_0\| + 1$, contradicting the maximality of $\|F_0\|$. If 1 could not be marked as above, go back to X and mark every $j \in X$ not previously marked such that there is a $k \in Y$ previously marked such that $F_0(f_{jk}) > 0$. Next, mark every $k \in Y$ not previously marked for which there is some $j \in X$ already marked such that $F_0(f_{jk}) < q_{jk}$. Attempt anew to mark 1 according to the above rule. Go back and forth with marking until you can mark 1. Then there is a zigzag sequence $j_0, k_0, j_1, k_1, \ldots, j_n, k_n$ such that

$$F_0(e_{j_0}) < p_{j_0}, \qquad F_0(f_{j_0}) < q_{j_0k_0}, \qquad F_0(f_{j_1k_0}) > 0,$$
$$F_0(f_{j_1k_1}) < q_{j_1k_1}, \ldots, F_0(g_{k_n}) < r_{k_n}.$$

Going up by 1 on e_{j_0}, $f_{j_0k_0}$, $f_{j_1,k_1}, \ldots, g_{k_n}$, and down by 1 on $f_{j_1k_0}, \ldots, f_{j_n,k_{n-1}}$, we modify F_0 into a flow F_{n-1} that still conforms with the capacities, but has $\|F_{n-1}\| = \|F_0\| + 1$, a contradiction. We conclude that marking stops after a finite number of steps since there is no possibility of further marking within the given rules, and yet 1 has not been marked. Let C consist of all edges with left end marked, right end unmarked. Since 1 is unmarked, every path from 0 to 1 must contain an edge in C, hence C is a cut. For every edge e in C, $F_0(e)$ equals the capacity of e because otherwise the right end of e could be marked. An easy exercise shows that $\|F_0\| = c(C)$. The theorem is proved.

A slightly more formal proof would proceed as follows: Define $A_0 \subseteq X$ by

$$A_0 = \{j \,|\, F(e_j) < p_j\}.$$

Define, for any $M \subseteq X$, $P(M) = \{k \mid k \in Y, F_0(f_{jk}) < q_{jk}$ for some $j \in M\}$. Put

$$B_0 = P(A_0).$$

For any $N \subseteq Y$, define $Q(N) = \{j \mid j \in X, F_0(f_{jk}) > 0$ for some $k \in N\}$. Put

$$A_1 = Q(B_0) \backslash A_0,$$

and, inductively,

$$B_m = P(A_m) \backslash (B_1 \cup \cdots \cup B_{m-1})$$
$$A_{m+1} = Q(B_m) \backslash (A_1 \cup \cdots \cup A_m)$$

for $m = 1, 2, \ldots$. Put

$$A = A_0 \cup A_1 \cup \cdots, \qquad B = B_0 \cup B_1 \cup \cdots.$$

Clearly there is an n such that $A = A_0 \cup \cdots \cup A_n$, $B = B_0 \cup \cdots \cup B_n$. Put $X_0 = X \backslash A$, $Y_0 = Y \backslash B$. It is obvious that $k \in B$ implies $F_0(g_k) = r_k$ because otherwise we could, in case $k \in B_m$, find $j_0 \in A_0$, $k_0 \in B_0$, $j_1 \in A_1$, $k_1 \in B_1$, $\ldots$, $j_m \in A_m$, $k_m = k \in B_m$ such that the zigzag argument for a strengthening of F_0 by 1 given in the above proof applies.

We are thus led to the conclusion that

$$F_0(g_k) = r_k \qquad (k \in B).$$

The construction of A, B implies that

$$F_0(f_{jk}) = q_{jk} \qquad (j \in A, \quad k \in Y_0)$$
$$F_0(f_{jk}) = 0 \qquad (j \in X_0, \quad k \in B)$$
$$F_0(e_j) = p_j \qquad (j \in X_0).$$

We conclude

$$\begin{aligned}
\|F_0\| &= \sum_{j \in X} F(e_j) = \sum_{j \in A} F_0(e_j) + \sum_{j \in X_0} F_0(e_j) \\
&= \sum_{j \in A} \sum_{k \in Y} F_0(f_{jk}) + \sum_{j \in X_0} p_j \\
&= \sum_{j \in A} \sum_{k \in Y_0} q_{jk} + \sum_{j \in X_0} p_j + \sum\sum_{j \in A,\, k \in B} F_0(f_{jk}) \\
&= \sum_{j \in A} \sum_{k \in Y_0} q_{jk} + \sum_{j \in X_0} p_j + \sum\sum_{j \in X,\, k \in B} F_0(f_{jk}) \\
&= \sum_{j \in A} \sum_{k \in Y_0} q_{jk} + \sum_{j \in X_0} p_j + \sum_{k \in B} F(g_k) \\
&= \sum_{j \in A} \sum_{k \in Y_0} q_{jk} + \sum_{j \in X_0} p_j + \sum_{k \in B} r_k \\
&= c(C)
\end{aligned}$$

with $C = \{e_j \mid j \in X_0\} \cup \{f_{jk} \mid j \in A, k \in Y_0\} \cup \{g_k \mid k \in B\}$, which apparently is a cut. The Ford–Fulkerson theorem in its full generality has the same max-flow–min-cut conclusion as its above "bipartite" special version. Since the latter one is sufficient for our purposes, we shall neither formulate nor prove the general theorem. It can be found in VOGEL [1], JACOBS [3]. The reader might however try to achieve it on his own, guided by the following hints. One considers a network which is a directed graph with one source and one sink and without circuits. Nonnegative real capacities are assigned to the edges. A flow arises by assigning nonnegative reals to the edges in such a fashion that there is "no storage." It is obvious what it means that a flow conforms to the capacities, and it is obvious by compactness that there is a flow of maximal strength. A marking algorithm is carried out in a helter-skelter zigzag fashion until no marking is possible anymore. If the sink has been marked, the flow can be strengthened by reconstructing how the marking of the sink was achieved. If the sink has not been marked, the edges with a marked tail and an unmarked tip form a cut whose capacity equals the strength of the flow.

2. MATRICES WITH GIVEN MARGINALS

Let X, Y be two nonempty finite sets. For any real X-Y-matrix $m = (m_{jk})_{j \in X, k \in Y}$, the row sums $m_j^X = \sum_{k \in Y} m_{jk}$ form the X-marginal X-vector $m^X = (m_j^X)_{j \in X}$ of m, and the column sums $m_k^Y = \sum_{j \in X} m_{jk}$ form the Y-marginal Y-vector $m^Y = (m_k^Y)_{k \in Y}$ of m. Let

$$
\begin{aligned}
p &= (p_j)_{j \in X} \text{ be any } X\text{-vector,} \\
q &= (q_{jk})_{j \in X, k \in Y} \text{ be any } X\text{-}Y\text{-matrix,} \\
r &= (r_k)_{k \in Y} \text{ be any } Y\text{-vector}
\end{aligned}
$$

with real coefficients. The real X-Y-matrix m is called *p-q-r-admissible* if the inequalities

$$m^X \leqq p, \qquad m \leqq q, \qquad m^Y \leqq r$$

hold componentwise.

Throughout this section we restrict attention to vectors and matrices p, q, r, m with nonnegative components. The sum of all components of p is denoted by $\|p\|$, and $\|q\|$, $\|r\|$, $\|m\|$ are likewise defined as component sums. Clearly we always have

$$\|m\| = \|m^X\| = \|m^Y\|.$$

We are aiming at $m^X = p$, $m^Y = r$, and thus are essentially only interested in cases with $\|p\| = \|r\|$.

For every X-vector $p = (p_j)_{j \in X}$ and any $E \subseteq X$, we define $p(E) = \sum_{j \in E} p_j$, and analogous definitions are made for Y-vectors and X-Y-matrices.

2.1. Theorem. *Let p, q, r be as above, but with integer components; put*

$$(1) \qquad c(p, q, r) = \inf\{p(X \backslash E) + q(E \times F) + r(Y \backslash F) \,|\, E \subseteq X, F \subseteq Y\}.$$

Then:

2.1.1. *If m is p-q-r-admissible, then $\|m\| \leqq c(p, q, r)$.*

2.1.2. *There is at least one p-q-r-admissible m with $m = c(p, q, r)$.*

Proof. 1. Let m be p-q-r-admissible. Since $E \subseteq X$, $F \subseteq Y$

$$\Rightarrow \quad [(X \backslash E) \times Y] \cup [E \times F] \cup [X \times (Y \backslash F)] = X \times Y,$$

we get

$$\begin{aligned} \|m\| &= m(X \times Y) \leqq m((X \backslash E) \times Y) + m(E \times F) + m(X \times (Y \backslash F)) \\ &= m^X(X \backslash E) + m(E \times F) + m^Y(Y \backslash F) \leqq p(X \backslash E) + q(E \times F) + r(Y \backslash F). \end{aligned}$$

Varying E, F, we get $\|m\| \leqq c(p, q, r)$.

2. We may assume that the four sets $\{0\}$, X, Y, $\{1\}$ are pairwise disjoint. Construct the bipartite network with source 0 and sink 1 and understand the p_j, q_{jk}, r_k ($j \in X$, $k \in Y$) as capacities, as described in Section 1. An X-Y-matrix m together with its marginal vectors m^X, m^Y is apparently the same thing as a flow in that network, and p-q-r-admissibility of m means nothing than the conformity of the flow with the capacities. Any cut in our bipartite network can be described by telling:

(a) what edges starting from 0 are in it—this amounts to a subset E' of X;
(b) what edges ending at 1 are in it—this amounts to some subset F' of Y;
(c) what edges from X and Y are in it—this amounts to some subset of $X \times Y$; but this subset has (for $E = X \backslash E'$, $F = Y \backslash F'$) to contain $E \times F$, otherwise we could devise a path from 0 to E, from E to F, from F to 1 avoiding the cut, which would be no cut then; on the other hand, we can omit from any cut all edges not in $E \times F$ without destroying its property of being a cut.

Thus $c(p, q, r)$ is nothing but the minimum capacity of a cut. Hence the Ford–Fulkerson theorem (bipartite version) 1.1 proves our present theorem.

2.2. Remark. The above theorem does not need the assumption $\|p\| = \|r\|$.

2.3. Theorem. *Let p, q, r be as above but with integer coefficients, $\|p\| = \|r\|$. Then the following statements are equivalent:*

2.3.1. *There is an X-Y-matrix $m \leqq q$ such that $m^X = p$ and $m^Y = r$.*

2.3.2. $E \subseteq X,\ F \subseteq Y \Rightarrow q(E \times F) + \|p\| \geqq p(E) + r(F)$.

Proof. $1 \Rightarrow 2$. $E \subseteq X$, $F \subseteq Y$ imply

$$\begin{aligned} p(E) + q(F) &= m^X(E) + m^Y(F) = m(E \times Y) + m(X \times F) \\ &= m(E \times F) + m([E \times (Y \backslash F)] \cup [X \times F]) \\ &\leqq m(E \times F) + \|m\| \leqq q(E \times F) + \|p\|. \end{aligned}$$

$2 \Rightarrow 1$. $E \subseteq X$, $F \subseteq Y$ imply

$$\begin{aligned} p(X \backslash E) + q(E \times F) + r(Y \backslash F) &= \|p\| - p(E) + q(E \times F) + \|r\| - r(F) \\ &\geqq p(E) + r(F) - p(E) + \|r\| - r(F) \\ &= \|r\| = \|p\| \end{aligned}$$

with equality, e.g., for $E = X$, $F = \varnothing$. Thus $c(p, q, r) = \|r\|$. Theorem 2.1 yields the existence of a p-q-r-admissible m with $\|m\| = \|r\|$. Now, $\|m^X\| = \|p\|, m^X \leqq p$ implies $m^X = p$, and likewise we get $m^Y = q$. This does it.

2.4. Remark. The above theory carries over to vectors and matrices with rational entries $\geqq 0$ immediately. An obvious approximation argument yields the same result for real vectors and matrices, too. The most important special case is where the components of p and r sum up to 1 (probability vectors).

3. CONTENTS AND σ-CONTENTS WITH GIVEN MARGINALS

We are dealing with positive contents and σ-contents throughout this section, except for the discussion at the end.

Let $\Omega_0 \neq \varnothing \neq \Omega_1$ and $\Omega = \Omega_0 \times \Omega_1$, and let $\varphi_i \colon \Omega \to \Omega_i$ be the natural projection ($i = 0, 1$). In Ω_i a set field $\mathscr{F}_i$ is assumed to be given ($i = 0, 1$). Let $\mathscr{F}$ be the field generated by all rectangles $A_0 \times A_1$ ($A_0 \in \mathscr{F}_0$, $A_1 \in \mathscr{F}_1$). Clearly φ_i is $\mathscr{F}$-$\mathscr{F}_i$-measurable and hence sends contents on $\mathscr{F}$ into contents on $\mathscr{F}_i$, preserving the total mass.

3.1. Definition. *Let*

p_i be a content on $\mathscr{F}_i$ ($i = 0, 1$),

q be a content on $\mathscr{F}$.

A content m on $\mathscr{F}$ is called **p_0-q-p_1-admissible** *if*

$$\varphi_0 m \leqq p_0, \qquad m \leqq q, \qquad \varphi_1 m \leqq p_1$$

hold. We define

(1)

$$c(p_0, q, p_1) = \inf\{p_0(\Omega_0 \backslash A_0) + q(A_0 \times A_1) + p_1(\Omega_1 \backslash A_1) \mid A_0 \in \mathscr{F}_0, A_1 \in \mathscr{F}_1\}.$$

3.2. Theorem. *With the above notations, we have:*

3.2.1. *If a content m on $\mathscr{F}$ is p_0-q-p_1-admissible, then $\|m\| \leqq c(p_0, q, p_1)$.*

3.2.2. *There is at least one p_0-q-p_1-admissible m with $\|m\| = c(p, q, r)$. If q is a σ-content, so is m.*

Proof. 1. For any finite disjoint decomposition $\mathscr{D}_i \subseteq \mathscr{F}_i$ of Ω_i $(i = 0, 1)$, let $\mathscr{D} = \{D_0 \times D_1 \mid D_0 \in \mathscr{D}_0, D_1 \in \mathscr{D}_1\}$ and $\mathscr{F}(\mathscr{D}_0)$, $\mathscr{F}(\mathscr{D}_1)$, $\mathscr{F}(\mathscr{D})$ be the corresponding set fields. Clearly φ_i is $\mathscr{F}(\mathscr{D})$-$\mathscr{F}(\mathscr{D}_i)$-measurable $(i = 0, 1)$. Apply Theorem 2.1 (and Remark 2.4) with $X = \mathscr{D}_0$, $Y = \mathscr{D}_1$, $p_j = p_0(j)$ $q_{jk} = q(j \times k)$, $r_k = p_1(k)$ $(j \in \mathscr{D}_1, k \in \mathscr{D}_2)$. Omitting some obvious details, we find

$$\begin{aligned}(2) \quad m(\Omega) &\leqq c(\mathscr{D}_0, \mathscr{D}_1) \\ &= \inf\{p_0(\Omega_0 \backslash A_0) + q(A_0 \times A_1) + p_1(\Omega_1 \backslash A_1) \mid A_0 \in \mathscr{F}(\mathscr{D}_0), \\ &\qquad A_1 \in \mathscr{F}(\mathscr{D}_1)\}.\end{aligned}$$

If $\mathscr{D}_i$ filters up to finer and finer partitions such that $\mathscr{F}(\mathscr{D}_i)$ filters up to $\mathscr{F}_i$ $(i = 0, 1)$, then clearly $\mathscr{F}(\mathscr{D})$ filters up to $\mathscr{F}$, and $c(\mathscr{D}_0, \mathscr{D}_1)$ filters down to $c(p_0, q, p_1)$.

2. Applying Theorem 2.1 (and Remark 2.4), we find, for any partition $\mathscr{D}_i \subseteq \mathscr{F}_i$ of Ω_i $(i = 0, 1)$, a content $m_{\mathscr{D}_0\mathscr{D}_1}$ on $\mathscr{F}(\mathscr{D})$ that is admissible with respect to the restriction of p_0 resp. q resp. p_1 to $\mathscr{F}(\mathscr{D}_0)$ resp. $\mathscr{F}(\mathscr{D})$ resp. $\mathscr{F}(\mathscr{D}_1)$ and satisfies $m_{\mathscr{D}_0\mathscr{D}_1}(\Omega) = c(\mathscr{D}_0, \mathscr{D}_1)$. Let $\mathscr{D}_i$ filter up such that $\mathscr{F}(\mathscr{D}_i)$ filters up to $\mathscr{F}_i$ $(i = 0, 1)$, and hence $\mathscr{F}(\mathscr{D})$ filters up to $\mathscr{F}$. An obvious usage of, e.g., ultrafilters lead to a limiting m on $\mathscr{F}$ such that $m = m(\Omega) = c(p_0, q, p_1)$. Clearly m is p_0-q-p_1-admissible. If q is a σ-content, $0 \leqq m \leqq q$ implies that m is a σ-content as well (exercise, see Proposition VII.3.1).

3.3. Theorem (Strassen [2]). *With the above notations assume that p_0, p_1 are probability σ-contents and q is a σ-content. Then the following are equivalent:*

3.3.1. *There is a probability σ-content $m \leqq q$ on $\mathscr{F}$ such that $\varphi_i m = p_i$ $(i = 0, 1)$.*

3.3.2. $A_0 \in \mathscr{F}_0$, $A_1 \in \mathscr{F}_1 \Rightarrow q(A_0 \times A_1) + 1 \geqq p_0(A_0) + p_1(A_1)$.

Proof. It is obvious how to combine the ideas employed in the proof of Theorem 3.2 with Theorem 2.3. Note that in the proof of $2 \Rightarrow 1$ we get an additive $m \leqq q$, which, by the σ-additivity of q, turns out to be σ-additive (exercise, see Proposition VII.3.1).

3.4. Discussion. The existence results of the above theorem can be obtained under weaker assumptions. We list the most important possibilities.

3.4.1. The limiting procedure in the proof of Theorem 3.2 can easily be modified so as to yield the existence of an additive $\overline{\mathbb{R}}_+$-valued m with $\|m\| = c(p_0, q, p_1)$ even in the case of additive $\overline{\mathbb{R}}_+$-valued set functions p_0, q, p_1.

3.4.2. In that case the question arises how to ensure σ-additivity for m. We can certainly form $(\varphi_0 m)(A_0) = m(\varphi_0^{-1} A_0) = m(A_0 \times \Omega_1)$ and get an additive $\overline{\mathbb{R}}_+$-valued $\varphi_0 m$, and likewise we may form $\varphi_0 q$. Now, the σ-additivity follows certainly if we make the following two assumptions:

(a) $\varphi_0 m \leqq p_0$ and p_0 is $\mathbb{R}$-valued and σ-additive;
(b) $\varphi_0 q$ is σ-finite and q is σ-additive.

Namely, let $E, E_1, E_2, \ldots \in \mathscr{F}$ be such that $E_j \cap E_k = \varnothing$ $(j \neq k)$, $E_1 \cup E_2 \cap \cdots = E$. We want to prove $m(E) = m(E_1) + m(E_2) + \cdots$. This is interesting only if $m(E_1), m(E_2), \ldots < \infty$. Let $B_0, B_1, B_2, \ldots \in \mathscr{F}_0$ be such that $B_0 \times \Omega_1 \supseteq E$ and $B_j \cap B_k = \varnothing$ $(j \neq k)$, $B_1 \cup B_2 \cup \cdots = B_0$, $\varphi_0 q(B_j) = q(B_j \times \Omega_1) < \infty$ $(j = 1, 2, \ldots)$. Then

$$m(E) = \sum_{n=1}^{\infty} m(E \cap (B_n \times \Omega_1))$$

follows from the finite additivity of m plus the estimates

$$\sum_{n=r}^{\infty} m(E \cap (B_n \times \Omega_1)) \leqq \sum_{n=r}^{\infty} m(B_n \times \Omega_1)$$

$$= \sum_{n=r}^{\infty} \varphi_0 m(B_n) \leqq \sum_{n=r}^{\infty} p_0(B_n),$$

which tends to 0 as $r \to \infty$ since p_0 is $\mathbb{R}$-valued and σ-additive. We can now continue, by the σ-additivity of $q \geqq m$ and by $q(B_j \times \Omega_1) < \infty$, writing

$$m(E) = \sum_{n=1}^{\infty} \sum_{r=1}^{\infty} m(E_r \cap (B_n \times \Omega_1))$$

$$= \sum_{r=1}^{\infty} \sum_{n=1}^{\infty} m(E_r \cap (B_n \times \Omega_1))$$

$$= \sum_{r=1}^{\infty} m(E_r)$$

by an argument involving the σ-additivity of $q \geqq m$ in a similar fashion as the σ-additivity of p_0 was involved before. Accordingly, Theorem 3.3 generalizes to the case of a possibly ∞-valued but σ-additive q with a σ-finite $\varphi_0 q$.

3.4.3. Under assumptions of σ-additivity, one can use well-known extension procedures, assume $\mathscr{F}_0, \mathscr{F}_1$ to be σ-fields, and take for $\mathscr{F}$ the product σ-field $\mathscr{F}_0 \times \mathscr{F}_1$.

We sketch a generalization of Theorem 3.3 to the case where upper and lower bounds, and not upper bounds only, are prescribed.

3.5. Theorem. *Let Ω_i, $\mathscr{F}_i$ $(i = 0, 1)$ and Ω, $\mathscr{F}$ be as in the beginning of this section. Let p_i, r_i be contents $\geqq 0$ on $\mathscr{F}_i$ such that $r_i \leqq p_i$ $(i = 0, 1)$. Let $q \geqq 0$ be a content on $\mathscr{F}$. Then the following statements are equivalent:*

3.5.1. *There is a content m on $\mathscr{F}$ such that*

$$r_0 \leqq \varphi_0 m \leqq p_0, \qquad m \leqq q, \qquad r_1 \leqq \varphi_1 m \leqq p_1.$$

3.5.2. *$A_0 \in \mathscr{F}_0$, $A_1 \in \mathscr{F}_1$ imply*

$$q(A_0 \times A_1) + p_0(\Omega_0 \backslash A_0) \geqq r_1(A_1), \qquad q(A_0 \times A_1) + p_1(\Omega_1 \backslash A_1) \geqq r_0(A_0).$$

Proof. $1 \Rightarrow 2$ is nearly obvious. We have, e.g.,

$$\begin{aligned} r_1(A_1) &\leqq \varphi_1 m(A_1) = m(\Omega_0 \times A_1) \\ &\leqq m(A_0 \times A_1) + m((\Omega_0 \backslash A_0) \times \Omega_1) \\ &= m(A_0 \times A_1) + \varphi_0 m(\Omega_0 \backslash A_0) \\ &\leqq q(A_0 \times A_1) + p_0(\Omega_0 \backslash A_0). \end{aligned}$$

$2 \Rightarrow 1$. By obvious approximation procedures we may restrict attention to the case where Ω_0, Ω_1 are finite and $\mathscr{F}_i = \mathscr{P}(\Omega_i)$ $(i = 0, 1)$. We then come down to a problem with vectors and matrices, and we may clearly assume that all components are nonnegative integers. We could transpose our problem into a problem of flows, but prefer to deal with contents instead. Consider all p_0-q-p_1-admissible contents on $\Omega_0 \times \Omega_1$. The constant

0 is one of them, but let us take any one, say m. Assume that $r_0 < \varphi_0 m$ or $r_1 \leqq \varphi_1 m$ is false. We may assume, e.g., one $j \in \Omega_0$ with $\varphi_0 m(\{j\}) < r_0(\{j\})$. For any $A \subseteq \Omega_0$, define

$$P(A) = \bigcup_{i \in A} \{k \mid k \in \Omega_1, m(\{(i, k)\}) < q(\{(i, k)\})\}.$$

For any $B \subseteq \Omega_1$ define

$$Q(B) = \bigcup_{k \in B} \{i \mid i \in \Omega_0, m(\{(i, k)\}) > 0\}.$$

Put $A_0 = \{i \mid i \in \Omega_0,\ \varphi_0 m(\{i\}) < r_0(\{i\}),\ B_0 = P(A_0),\ A_1 = Q(B_0)\backslash A_0,\ B_1 = P(A_1)\backslash B_0, \ldots,$

$$A_n = Q(B_{n-1})\backslash(A_0 \cup \cdots \cup A_{n-1}),\ B_n = P(A_n)\backslash(B_0 \cup \cdots \cup B_{n-1})$$

and assume $Q(B_n) \subseteq A_0 \cup \cdots \cup A_n$. Put $A = A_0 \cup \cdots \cup A_n$, $B = B_0 \cup \cdots \cup B_n$. By construction we have

$$m(A \times (\Omega_1 \backslash B)) = q(A \times (\Omega_1 \backslash B)), \qquad m((\Omega_0 \backslash A) \times B) = 0,$$

$\varphi_0 m(\Omega_0 \backslash A) \geqq r_0(\Omega_0 \backslash A)$. Now, making use of 2, we get

$$\begin{aligned}
\varphi_0 m(A) + p_1(B) &= m(A \times \Omega_1) + p_1(B) \\
&= m(A \times (\Omega_1 \backslash B)) + m(A \times B) + p_1(B) \\
&= q(A \times (\Omega_1 \backslash B)) + p_1(B) + m(A \times B) \\
&\geqq r_0(A) + m(A \times B) \\
&= r_0(A) + m(A \times B) + m((\Omega_0 \backslash A) \times B) \\
&= r_0(A) + m(\Omega_0 \times B) = r_0(A) + \varphi_1 m(B),
\end{aligned}$$

hence $p_1(B) - \varphi_1 m(B) \geqq r_0(A_0) - \varphi_0 m(A_0) + r_0(A \backslash A_0) - \varphi_0 m(A \backslash A_0)$. Now $r_0(A_0) - \varphi_0 m(A_0) > 0$ by construction of A_0. We conclude that either there is an $i \in A \backslash A_0$ with $r_0(\{i\}) < \varphi_0 m(i)$ or there is a $k \in B$ with $p_1(\{k\}) > \varphi_1 m(\{k\})$ (or both). In the first case we find $1 \leqq \nu \leqq n$ with $i \in A_\nu$. We determine $k_{\nu-1} \in B_{\nu-1}$ such that $m(\{(i, k)\}) > 0$, by the construction of A_ν; zigzaging in an obvious manner, we see that we can increase certain $m(\{(h, l)\})$ by one, and decrease others by one, in such a fashion that a new $\overline{m}$ arises on Ω such that $\varphi_0 \overline{m}(\{j\}) = \varphi_0 m(\{j\}) + 1$, for one $j \in A_0$, and

$$\varphi_0 \overline{m}(\{i\}) = \varphi_0 m(\{i\}) - 1 \geqq r_0(\{i\}),$$

while $\varphi_0 \overline{m}(\{h\}) = \varphi_0 m(\{h\})$, $\varphi_1 \overline{m}(\{l\}) = \varphi_1 m(\{l\})$ for $h \neq i, j$ and all l. We have, in particular, $\varphi_0 \overline{m} \leqq p_0$, $\overline{m} \leqq q$, $\varphi_1 \overline{m} \leqq p_1$, but the "defect" from the desired relation $r_0 < \varphi_0 m$ has decreased by one. In the second case an analogous technique yields an $\overline{\overline{m}}$ with $\varphi_0 \overline{\overline{m}} \leqq p_0$, $\overline{\overline{m}} \leqq q$, $\varphi_1 \overline{\overline{m}} \leqq p_1$, and the "defect" from the desired relation $r_1 \leqq \varphi_1 m$ decreased by one. Iterating that procedure we end up with an m with no more "defects" at all, i.e., all properties required for the conclusion of our theorem.

3.6. Discussion. We list some remarks concerning essentially the question of σ-additivity of the m obtained in the above theorem.

3.6.1. It is practically obvious that m is σ-additive, i.e., a σ-content, if q is a σ-content. Obvious extension procedures allow one to prove the analogous theorem for the case of σ-contents on σ-fields.

3.6.2. Theorem 3.5 can clearly be proved as well under the assumption that $\mathscr{F}_0$, $\mathscr{F}_1$ are σ-fields, $\mathscr{F} = \mathscr{F}_0 \times \mathscr{F}_1$ their product σ-field, r_0, p_0, q, r_1, p_1 σ-additive $\overline{\mathbb{R}}_+$-valued functions. We obtain an additive $\overline{\mathbb{R}}_+$-valued m. In each of the following cases we get the σ-additivity of m as well:

3.6.2.1. Assume that $\varphi_0 q$ and $\varphi_1 q$ are σ-finite. Then $\Omega = \Omega_0 \times \Omega_1$ splits into a countable number of "rectangles" each of which represents the situation of 3.6.1. The σ-additivity of m now easily follows (exercise).

3.6.2.2. p_0 is a (finite) σ-content and $\varphi_1 q$ is σ-finite (exercise, see 3.4.2).

3.6.3. Theorem 3.5 and the above discussion yield the following result of Kellerer [1]: Let p_0, q, p_1 be defined on the σ-algebras $\mathscr{F}_0$ in Ω_0, $\mathscr{F}_1$ in Ω_1, and the product σ-algebra $\mathscr{F}$ on $\Omega_0 \times \Omega_1$. Let $\varphi_0 q$, $\varphi_1 q$ be σ-finite. Then the following statements are equivalent:

3.6.3.1. There is a σ-additive $\overline{\mathbb{R}}_+$-valued m on $\mathscr{F}$ such that $\varphi_0 m = p_0$, $m \leqq q$, $\varphi_1 m = p_1$.

3.6.3.2. $A_0 \in \mathscr{F}_0$, $A_1 \in \mathscr{F}_1$ imply

$$q(A_0 \times A_1) + p_0(\Omega_0 \backslash A_0) \geqq p_1(A_1), \qquad q(A_0 \times A_1) + p_1(\Omega_1 \backslash A_1) \geqq p_0(A_0).$$

4. RESULTS INVOLVING TOPOLOGY

Some further results on contents with given marginals can be obtained by methods involving topologies. Thus we shall throughout this section assume Ω_0 and Ω_1 to be **Polish spaces** and $\mathscr{B}_i$ to be the natural (Baire–Borel) σ-field in Ω_i $(i = 0, 1)$. The product σ-field $\mathscr{B}$ in $\Omega = \Omega_0 \times \Omega_1$ is then the natural σ-field of the Polish space Ω. The natural projection $\varphi_i\colon \Omega \to \Omega_i$ is continuous and sends sets from $\mathscr{B}$ into analytic sets in Ω_i (Proposition XIII.2.13). Since analytic sets are the same as Souslin sets (over the closed sets, say) by Proposition XIII.2.6 and, whatever σ-content we consider, Souslin sets are Borel up to nullsets (Theorem XIII.4.1), we may and shall form the values of σ-contents for sets $\varphi_i E$, $E \subseteq \Omega$ Borel, without hesitation $(i = 0, 1)$.

4.1. Theorem. *Let* $\Omega_0, \Omega_1, \Omega, \mathscr{B}_0, \mathscr{B}_1, \mathscr{B}, \varphi_0, \varphi_1$ *be as above. Let* p_i *be a* σ*-content on* $\mathscr{B}_i (i = 0, 1)$, q *a* σ*-additive* $\overline{\mathbb{R}}_+$*-valued function on* $\mathscr{B}$. *Assume that*

$$q(E) = \inf\{q(G) | E \subseteq G \subseteq \Omega, G \text{ open}\}$$

for every $E \in \mathscr{B}$. *Define*

$$c(p_0, q, p_1) = \inf\{p_0(\Omega_0 \setminus A_0) + q(A_0 \times A_1) + p_1(\Omega_1 \setminus A_1) | A_0 \in \mathscr{B}_0, A_1 \in \mathscr{B}_1\}.$$

Then there is a σ*-content* $m \geqq 0$ *on* $\mathscr{B}$ *such that*

$$\varphi_0 m \leqq p_0, \qquad m \leqq q, \qquad \varphi_1 m \leqq p_1$$

and

$$m(\Omega) = c(p_0, q, p_1).$$

Proof. Let $\mathscr{F}$ be the set field generated by $(\varphi_0^{-1}\mathscr{B}_0) \cup (\varphi_1^{-1}\mathscr{B}_1)$, i.e., the system of all finite unions of "rectangles" $E_0 \times E_1$ with $E_i \in \mathscr{B}_i$ $(i = 0, 1)$. Theorem 3.2 ensures the existence of an additive $m: \mathscr{F} \to \overline{\mathbb{R}}_+$ with $\varphi_0 m \leqq p_0$, $m \leqq q$ on $\mathscr{F}$, $\varphi_1 m \leqq p_1$. Since $p_0(\Omega_0) < \infty > p_1(\Omega_1)$, we find $m(\Omega) < \infty$. The only thing still to be achieved is the extension of m to $\mathscr{B}$, preserving $m \leqq q$. To this end we first show that m is σ-additive on $\mathscr{F}$. This is easily done by inner compact regularity and Proposition V.1.6. Inner compact regularity of m is easy: Take any $E_i \in \mathscr{B}_i$ $(i = 0, 1)$ and consider $m_0(F_0) = m(F_0 \times E_1)$ $(F_0 \in \mathscr{B}_0)$. Clearly this is additive on $\mathscr{B}_0$, and even σ-additive since $m_0 \leqq p_0$ obviously holds. Thus m_0 is tight (Theorem V.5.4) and, given any $\varepsilon > 0$, the existence of a compact $K_0 \subseteq E_0$ with $m_0(K_0) > m_0(E_0) - \frac{1}{2}\varepsilon$, i.e., $m(K_0 \times E_1) > m(E_0 \times E_1) - \frac{1}{2}\varepsilon$, follows. Repeating this argument, we find a compact $K_1 \subseteq E_1$ such that $m(K_0 \times K_1) > m(K_0 \times E_1) - \frac{1}{2}\varepsilon > m(E_0 \times E_1) - \varepsilon$. But $K_0 \times K_1$ is compact and the inner compact regularity, hence the σ-additivity, of m on $\mathscr{F}$ is established. Let m also denote the unique σ-content on $\mathscr{B}$ extending our given m on $\mathscr{F}$. Since every open $G \subseteq \Omega$ is a countable union of rectangles, we find $m \leqq q$ on all open sets. Let now $E \in \mathscr{B}$ arbitrary. Then

$$\begin{aligned} m(E) &\leq \inf\{m(G) | E \subseteq G \subseteq \Omega, G \text{ open}\} \\ &\leq \inf\{q(G) | E \subseteq G \subseteq \Omega, G \text{ open}\} = q(E) \end{aligned}$$

follows from our hypothesis concerning q.

The following theorem is due to STRASSEN [2].

4.2. Theorem. *Let* $\Omega_0, \Omega_1, \Omega, \mathscr{B}_0, \mathscr{B}_1, \mathscr{B}, \varphi_0, \varphi_1$ *be as above. Let* p_i *be a probability* σ*-content on* $\mathscr{B}_i$ $(i = 0, 1)$, *and let* $F \subseteq \Omega$ *be closed. Then for any* $\varepsilon \geqq 0$, *the following statements are equivalent:*

4.2.1. *There is a probability σ-content m on $\mathscr{B}$ such that $\varphi_0 m = p_0$, $\varphi_1 m = p_1$, and $m(F) \geqq 1 - \varepsilon$.*

4.2.2. $E_0 \in \mathscr{B}_0$, $E_1 \in \mathscr{B}_1$, $(E_0 \times E_1) \cap F = \varnothing$ *imply*

$$p_0(E_0) + p_1(E_1) \leqq 1 + \varepsilon.$$

4.2.3. $F_1 \subseteq \Omega_1$ *closed implies*

$$p_1(F_1) \leqq p_0(\varphi_0[(\Omega_0 \times F_1) \cap F]) + \varepsilon$$

Proof. $1 \Rightarrow 3$.

$$\begin{aligned} p_1(F_1) = \varphi_1 m(F_1) = m(\Omega_0 \times F_1) &= m((\Omega_0 \times F_1) \cap F) + m((\Omega_0 \times F_1)\backslash F) \\ &\leqq m((\Omega_0 \times F_1) \cap F) + \varepsilon \\ &\leqq m(\varphi_0^{-1}[\varphi_0((\Omega_0 \times F_1) \cap F)]) + \varepsilon \\ &\leqq \varphi_0 m(\varphi_0[(\Omega_0 \times F_1) \cap F]) + \varepsilon \\ &= p_0(\varphi_0[(\Omega_0 \times F_1) \cap F]) + \varepsilon. \end{aligned}$$

$3 \Rightarrow 2$. $(E_0 \times E_1) \cap F = \varnothing$ implies

$$[\varphi_0((\Omega_0 \times E_1) \cap F)] \cap E_0 = \varnothing$$

$$\Rightarrow \quad p_0(E_0) + p_1(E_1) \leqq p_0(E_0) + p_0(\varphi_0[(\Omega_0 \times E_1) \cap F]) + \varepsilon \leqq 1 + \varepsilon.$$

$2 \Rightarrow 1$. Define q on $\mathscr{B}$ by

$$q(E) = \begin{cases} \infty & \text{if } E \cap F \neq \varnothing \\ 0 & \text{otherwise.} \end{cases}$$

Clearly q is $\overline{\mathbb{R}}_+$-valued, σ-additive, and satisfies

$$q(E) = \inf\{q(G) \mid E \subseteq G \subseteq \Omega, G \text{ open}\}$$

since every set disjoint from F is contained in the open set $\Omega\backslash F$ disjoint from F. We apply now Theorem 4.1 and obtain some σ-content $\tilde{m}$ on $\mathscr{B}$ such that $\varphi_0 \tilde{m} \leqq p_0$, $\tilde{m} \leqq q$, $\varphi_1 \tilde{m} \leqq p_1$, and

$$\begin{aligned} \tilde{m}(\Omega) &= c(p_0, q, p_1) \\ &= \inf\{p_0(\Omega_0\backslash E_0) + q(E_0 \times E_1) + p_1(\Omega_1\backslash E_1) \mid E_0 \in \mathscr{B}_0, \\ &\qquad\qquad E_1 \in \mathscr{B}_1\}. \end{aligned}$$

$\tilde{m} \leqq q$ says that $m(\Omega\backslash F) = 0$. If $c(p_0, q, p_1) = 1$, then $\tilde{m}$ is a probability, hence so are $\varphi_0 m$ and $\varphi_1 m$, and $\varphi_0 m = p_0$, $\varphi_1 m = p_1$ follows. We shall then put $m = \tilde{m}$. Assume now $\tilde{m}(\Omega) = c(p_0, q, p_1) < 1$. But $E_0 \in \mathscr{B}_0$, $E_1 \in \mathscr{B}_1$ implies

$$p_0(\Omega_0\backslash E_0) + q(E_0 \times E_1) + p_1(\Omega_1\backslash E_1) = \infty$$

in case $(E_0 \times E_1) \cap F \neq \varnothing$, and equals

$$1 - p_0(E_0) + 1 - p_1(E_1) = 2 - (p_0(E_0) + p_1(E_1)) \geqq 2 - (1 + \varepsilon) = 1 - \varepsilon$$

in case $(E_0 \times E_1) \cap F = \varnothing$, hence $c(p_0, q, p_1) \geqq 1 - \varepsilon$. We may now take

$$m = \tilde{m} + \frac{1}{1 - c(p_0, q, p_1)} [(p_0 - \varphi_0 m) \times (p_1 - \varphi_1 m)]$$

and obtain $m(\Omega) = 1$, $\varphi_0 m = p_0$, $\varphi_1 m = p_1$, and clearly $m(F) \geqq 1 - \varepsilon$.

The following theorem is essentially due to FURSTENBERG [1].

4.3. Theorem. *Let Ω_0, Ω_1, Ω, $\mathscr{B}_0$, $\mathscr{B}_1$, $\mathscr{B}$, φ_0, φ_1 be as above. Let X be a Polish space with natural σ-field $\mathscr{S}$, and $\psi_i\colon \Omega_i \to X$, be continuous $(i = 0, 1)$. Let p_i be a probability σ-content on $\mathscr{B}_i$ $(i = 0, 1)$ such that*

$$\psi_0 p_0 = \psi_1 p_1.$$

Then there is a probability σ-content m on Ω such that

$$\varphi_0 m = p_0, \qquad \varphi_1 m = p_1,$$

and m lives on the closed set $F = \{(\omega_0, \omega_1) | \psi_0(\omega_0) = \psi_1(\omega_1)\}$.

Proof. Find compact sets $K_i \subseteq \Omega_i$ $(i = 0, 1)$ such that

$$p_0(K_0) + p_1(K_1) > 1.$$

Put $p = \psi_0 p_0 = \psi_1 p_1$. We obtain

$$p(\psi_0 K_0) + p(\psi_1 K_1) > 1.$$

It follows that $p((\psi_0 K_0) \cap (\psi_1 K_1)) > 0$, hence $(\psi_0 K_0) \cap (\psi_1 K_1) \neq \varnothing$, hence $(K_0 \times K_1) \cap F \neq \varnothing$. By inner compact regularity in Polish spaces (Theorem V.5.4) we find that

$$E_0 \in B_0,\ E_1 \in B_1,\ p_0(E_0) + p_1(E_1) > 1 \quad \Rightarrow \quad (E_0 \times E_1) \cap F \neq \varnothing.$$

Thus we get $(E_0 \times E_1) \cap F = \varnothing$ which implies

$$p_0(E_0) + p_1(E_1) \leqq 1.$$

Applying Theorem 4.2 with $\varepsilon = 0$, our result follows.

SELECTED BIBLIOGRAPHY

AKCOGLU, L.
[1] A pointwise ergodic theorem in L_p-spaces, *Can. J. Math.* **27** (1975), 1075–1082.
ALAOGLU, L., and G. BIRKHOFF
[1] General ergodic theorems, *Ann. of Math.* **41** (1940), 293–309.
ALEKSANDROV, A. D.
[1] Additive functions in abstract spaces I–III, *Mat. Sb.* **8(50)** (1940), 307–348, **9(51)** (1941), 563–628.
ALEXANDROFF, P. S.
[1] Über die Äquivalenz des Perron'schen und des Denjoy'schen Integralbegriffs, *Math. Z.* **20** (1924), 213–222.
ALEXANDROFF, P. S., and P. URYSON
[1] Mémoire sur les espaces topologiques compacts, *Verh. Konink. Akad. Wetensch* **14** (1929), 1–96.
ANDERSEN, E. S., and B. JESSEN
[1] On the introduction of measures in an abstract set, *Danske Vid. Selsk. Mat–Fys. Medd.* **25** (1948), no. 4.
ANGER, B., and H. BAUER
[1] "Mehrdimensionale Integration." de Gruyter, Berlin and New York, 1976.
AUGUSTIN U.
[1] Noisy channels, Lecture Notes (to appear).
AUMANN, G.
[1] "Reelle Funktionen," Springer–Verlag, Berlin, 1954.
BAIRE, R.
[1] Sur les fonctions des variables reelles, *Ann. Mat. Pura Appl.* **3(3)** (1899), 1–122.
[2] "Leçons sur les fonctions discontinues." Gauthier-Villars, Paris, 1905.
BANACH, S.
[1] Sur un théorème de M. Vitali, *Fund. Math.* **5** (1924), 130–136.
[2] Sur les lignes rectifiables et les surfaces dont l'aire est finie, *Fund. Math.* **7** (1925), 225–237.
[3] Sur les fonctionnelles linéaires, *Studia Math.* **1** (1929), 211–216.
[4] "Théorie des opérations linéaires." Warszawa, 1932.
BAUER, H.,
[1] Eine Riesz'sche Banderzerlegung im Raum der Bewertungen eines Verbandes, *Sb. Bay. Akad. Wiss* (1953), 89–117.

[2] Sur l'équivalence des théories de l'intégration selon N. Bourbaki et selon M. H. Stone, *Bull. Soc. Math. France* **85** (1957), 51–75.
[3] "Konvexität in topologischen Vektorräumen." Skriptum, Hamburg, 1964.
[4] "Wahrscheinlichkeitstheorie und Grundzüge der Maßtheorie." de Gruyter, Berlin and New York, 1974.
[5] "Probability Theory and Elements of Measure Theory." Holt, New York, 1972.

BECK, A.
[1] Probability in Banach spaces, *Int. Conf. Prob. Banach Spaces, 1st, July 1975, Oberwolfach*, Lect. Notes in Math. 526. Springer–Verlag, Berlin and New York, 1976.

BELLOW, A., and K. KÖLZOW (eds.)
[1] Measure theory, *Proc. Conf. Oberwolfach, June 15–21, 1975*, Lecture Notes in Math. 541. Springer–Verlag, Berlin and New York, 1976.

BERBERIAN, S.
[1] "Measure and integration." Chelsea, New York, 1965.

BICHTELER, K.
[1] "Integration Theory" (with special attention to vector measures), Lect. Notes in Math. 315. Springer–Verlag, Berlin and New York, 1973.

BILLINGSLEY, P.
[1] "Convergence of Probability Measures." Wiley, New York, 1968.

BIRKHOFF, G. D.
[1] Proof of the ergodic theorem, *Proc. Nat. Acad. Sci. U.S.* **17** (1932), 656–660.

BISHOP, E., and K. DE LEEUW
[1] The representation of linear functionals by measures on sets of extreme points, *Ann. Inst. Fourier* **9** (1959), 305–331.

BLACKWELL, D.
[1] On a class of probability spaces, *Proc. Berkeley Symp. Prob. Stat., 3rd, 1954/55*, Vol. II, pp. 1–6, Univ. California Press, California, 1956.

BLUMENTHAL, R., and R. GETOOR.
[1] "Markov Processes and Potential Theory." Academic Press, New York, 1968.

BOCHNER, S.
[1] "Vorlesungen über Fouriersche Integrale." Akad. Verlagsges., Leipzig, 1932.

BOREL, E.
[1] "Leçons sur la théorie des fonctions." Paris, 1898.
[2] Sur l'intégration des fonctions non bornées et sur les définitions constructives, *Ann. El. Norm.* **36** (1919), 71–91.

BOURBAKI, N.
[1] "Espaces Vectoriels Topologiques," Chapters I–IV. Hermann, Paris, 1953, 1964.
[2] "Integration," Chapters I–VIII. Hermann, Paris, 1952, 1956, 1959, 1963.
[3] "Topologie générale" (fascicule des résultats). Hermann, Paris, 1964.

BREIMAN, L.
[1] "Probability." Addison-Wesley, Reading, Massachusetts, 1968.

BROWN, J. R.
[1] "Ergodic Theory and Topological Dynamics." Academic Press, New York, 1976.

BUCY, R. S., and G. MALTESE
[1] Extreme positive definite functions and Choquet's representation theorem, *J. Math. Anal. Appl.* **12** (1966), 371–377.

BURKILL, J. C.
[1] Functions of intervals, *Proc. London Math. Soc.* **22(2)** (1924), 275–310.
[2] The expression of area as an integral, *Proc. London Math. Soc.* **22(2)** (1924), 311–336.
[3] The derivatives of functions of intervals, *Fund. Math.* **5** (1924), 321–327.

CARATHÉODORY, C.

[1] "Vorlesungen über reelle Funktionen." Leipzig and Berlin, 1918.

[2] "Maß und Integral und ihre Algebraisierung." Birkhäuser, Basel and Stuttgart, 1956.

[3] "Algebraic theory of measure and integration." Chelsea, New York, 1963.

CARTAN, H.

[1] Sur la mesure de Haar, *C. R. Acad. Sci. Paris* **211** (1940), 759–762.

CECH, E.

[1] On bicompact space, *Ann. of Math.* **38** (1937), 823–844.

[2] "Topological Spaces." Wiley (Interscience), New York, 1966.

CHANG, C. C., and H. J. KEISLER

[1] "Model Theory." North Holland Publ., Amsterdam, 1973.

CHOQUET, G.

[1] Theory of capacities, *Ann. Inst. Fourier Grenoble* **5** (1955), 131–295.

[2] Existence et unicité des représentations intégrales au moyen des points extrémaux dans les cônes convexes, *Sém. Bourbaki Exposé* **139** (1956).

[3] Forme abstraite du théorème de capacitabilité, *Ann. Inst. Fourier Grenoble* **9** (1959), 83–89.

[4] Le théorème de représentation intégrale dans les ensembles convexes compacts, *Ann. Inst. Fourier Grenoble* **10** (1960), 333–344.

[5] "Lectures on analysis I–III." Benjamin, New York, 1969.

CHOQUET, G., and P. A. MEYER

[1] Existence et unicité des représentations intégrales dans les ensembles convexes compacts quelconques, *Ann. Inst. Fourier Grenoble* **13** (1963), 139–154.

CLARKSON, J. A.

[1] Uniformly convex spaces, *Trans. Amer. North. Soc.* **40** (1936), 396–414.

COHEN, P.

[1] The independence of the continuum hypothesis, *Proc. Nat. Acad. Sci. U.S.* **50** (1963), 1143–1148; **51** (1964), 105–110.

[2] Independence results in set theory. The theory of models, *Proc. Int. Symp. Berkeley, 1963*, pp. 39–54 (1965).

DANIELL, P. J.

[1] A general form of integral, *Ann. of Math.* **19(2)** (1917/18), 279–294.

[2] Further properties of the general integral, *Ann. of Math.* **21(2)** (1920), 203–220.

DE LEEUW, and I. GLICKSBERG

[1] Applications of almost periodic compactifications, *Acta Math.* **105** (1961), 63–97.

[2] Almost periodic functions on semigroups, *Acta Math.* **105** (1961), 99–140.

DELLACHERIE, C.

[1] "Capacités et processus stochastiques." Springer-Verlag, Berlin and New York, 1972.

DENKER, M., Chr. GRILLENBERGER, and K. SIGMUND

[1] "Ergodic Theory on Compact Spaces," Lect. Notes in Math. 527. Springer-Verlag, Berlin and New York, 1976.

DIEUDONNÉ, J.

[1] Sur le théorème de Lebesgue-Nikodym III, *Ann. Inst. Fourier Grenoble* **23** (1948), 25–53.

[2] Sur le théorème de Lebesgue-Nikodym IV, *J. Indian Math. Soc. N.S.* **15** (1951), 77–86.

DINCULEANU, N.

[1] "Vector Measures." Oxford Univ. Press (Pergamon), London and New York, 1967.

DINI, U.

[1] "Fondamenti per la teoria delle funzioni di variabili reali." Pisa ca. 1877.

DOOB, J. L.

[1] Regularity properties of certain families of chance variables, *Trans. Amer. Math. Soc.* **47** (1940), 455–486.

[2] "Stochastic Processes," 5th printing. Wiley, New York, 1964.
DOUBROWSKI, V.
[1] On some properties of completely additive set functions, *Izv. Akad. Nauk SSSR, Ser. Mat.* **9** (1945), 311–320.
DOWKER, Y. N.
[1] Finite and σ-finite invariant measures, *Ann. of Math.* **54** (1951), 595–608.
DUNFORD, N., and J. T. SCHWARTZ
[1] "Linear operators I," 4th printing. Wiley (Interscience), New York, 1967.
DYNKIN, E. B.
[1] "Die Grundlagen der Theorie der Markoffschen Prozesse." Springer-Verlag, Berlin and New York, 1961.
[2] "Markov Processes I, II." Springer-Verlag, Berlin and New York, 1965.
EGOROV, D.Th.
[1] Sur les suites des fonctions mésurables, *C. R. Acad. Sci. Paris* **152** (1911), 244–246.
ELLIS, R.
[1] Locally compact transformation groups, *Duke Math. J.* **24** (1957), 119–126.
FATOU, P.
[1] Series trigonométriques et séries de Taylor, *Acta Math.* **30** (1906), 335–400.
FELLER, W.
[1] Diffusion processes in one dimension, *Trans. Amer. Math. Soc.* **77** (1954), 1–31.
[2] The general diffusion operator and positivity preserving semigroups in one dimension, *Ann. of Math.* **60(2)** (1954), 417–436.
[3] "An Introduction to Probability Theory and Its Applications," Vol. II. Wiley, New York, 1966.
FISCHER, E.
[1] Sur la convergence en moyenne, *C. R. Acad. Sci. Paris* **144** (1907), 1022–1024.
[2] Applications d'un théorème sur la convergence en moyenne, *C. R. Acad. Sci. Paris* **144** (1907), 1148–1151.
FREMLIN, D. H.
[1] "Topological Riesz Spaces and Measure Theory." Cambridge Univ. Press, London and New York, 1974.
FUBINI, G.
[1] Sugli integrali multipli, *Rend. Accad. Lincei Roma* **16** (1907), 608–614.
[2] Sulla derivazione per serie, *Rend. Accad. Lincei Roma* **24** (1915), 204–206.
FURSTENBERS, H.
[1] Disjointness in ergodic theory, minimal sets and a problem in Diophantine approximations, *Math. Syst. Th.* **1** (1967), 1–49.
GARSIA, A.
[1] A simple proof of Hopf's maximal ergodic theorem, *J. Math. Mech.* **14** (1965), 381–382.
GÜNZLER, H.
[1] Integral representations with prescribed lattices, *Rend. Sem. Mat. Fis. d. Milano* **XLV** (1975), 107–168.
[2] Stonean Lattices, Measures and Completeness, *ISNM* **25** (1974), 113–126.
[3] Linear functionals which are integrals, *Rend. Sem. Mat. Fis. Milano* **43** (1973), 167–176.
HAAR, A.
[1] Zur Theorie der orthogonalen Funktionensysteme (erste Mitteilung), *Math. Ann.* **69** (1910), 331–371.
[2] Zur Theorie der orthogonalen Funktionensysteme (zweite Mitteilung), *Math. Ann.* **71** (1911), 38–53.
[3] Der Maßbegriff in der Theorie der unendlichen Gruppen, *Ann. of Math.* **34(2)** (1933), 147–169.

HAHN, H.

[1] Über Folgen linearer Operationen, *Monatsh. Math. Phys.* **32** (1922), 3–88.

[2] Über lineare Gleichungen in linearen Räumen, *J. Math.* **157** (1927), 214–229.

[3] "Theorie der reellen Funktionen." Springer, New York, 1921.

HAHN, H., and A. ROSENTHAL

[1] "Set Functions." Univ. of New Mexico Press, Albuquerque, New Mexico, 1948.

HALL, Ph.

[1] On representatives of subsets, *J. London Math. Soc.* **10** (1935), 26–30.

HALMOS, P. R.

[1] "Introduction to Hilbert Space and the Theory of Spectral Multiplicity." Chelsea, New York, 1951.

[2] "Naive Set Theory," 5th printing. Van Nostrand–Reinhold, Princeton, New Jersey, 1965 and Springer–Verlag, Berlin and New York, 1974.

[3] "Naive Mengenlehre." Vandenhoeck and Ruprecht, Göttingen, 1972.

[4] "Measure theory." Springer–Verlag, Berlin and New York, 1974.

HALMOS, P. R. and J. L. SAVAGE

[1] Applications of the Radon-Nikodym theorem to the theory of sufficient statistics, *Ann. Math. Stat.* **20** (1949), 225–241.

HALMOS, P. R., and H. VAUGHAN

[1] The marriage problem, *Amer. J. Math.* **72** (1950), 214–215.

HANEN, A., and J. NEVEU

[1] Atomes conditionnels d'un espace de probabilité, *Acta Math. Acad. Sci. Hung.* **17** (1966), 443–449.

HANSEL, G., and J. P. TROALLIC

[1] Mesures marginales et théorème de Ford–Fulkerson, *Z. Wahrsch.* **43** (1978), 245–251.

HAUPT, O. G., AUMANN, and Chr. PAUC

[1] "Differential- und Integralrechnung, III." de Gruyter, Berlin, 1955.

HAUSDORFF, F.

[1] "Grundzüge der Mengenlehre." Berlin, 1914.

HENSTOCK, R.

[1] "Theory of Integration." Butterworths, Washington, D.C., 1963.

HERMES, H.

[1] "Einführung in die mathematische Logik." Teubner, Stuttgart, 1963.

[2] "Introduction to Mathematical Logic." Springer–Verlag, Berlin and New York, 1972.

HERVÉ, M.

[1] Sur les représentations intégrales à l'aide des points extrémaux dans un ensemble compact convexe métrisable, *C. R. Acad. Sci. Paris* **253** (1961), 366–368.

HEWITT, E., and K. A. ROSS

[1] "Abstract Harmonic Analysis I." Springer–Verlag, Berlin and New York, 1963.

[2] "Abstract Harmonic Analysis II." Springer–Verlag, Berlin and New York, 1970.

HILDEBRANDT, T. H.

[1] "Introduction to the Theory of Integration." Academic Press, New York, 1963.

HILDENBRAND, W.

[1] Über straffe Funktionale, *Z. Wahrsch.* **4** (1966), 269–292.

[2] "Core and Equilibria of Large Economics." Princeton Univ. Press, Princeton, New Jersey, 1974.

HILL, D. G. B.

[1] σ-finite invariant measures on infinite product spaces, Thesis, Yale Univ. (1968).

HÖLDER, O.

[1] Über einen Mittelwertsatz, *Gött. Nachr.* (1889), 38–47.

HOFFMANN-JØRGENSEN, J.
[1] "The theory of analytic spaces," Lecture Notes Aarhus, 1970.
HUBER, P.
[1] The use of Choquet capacities in statistics, *Bull. Inst. Int. Stat.* **45** (1973), book 4, 181–191.
HUBER, P., and V. STRASSEN
[1] Minimax Tests and the Neyman-Pearson Lemma for Capacities, *Ann. of Stat.* **1** (1973), 251–263.
[2] Correction: Minimax tests and the Neyman-Pearson lemma for capacities, *Ann. Statist.* **1** (1973), 251–263; **2** (1974), 223–224.
IONESCU-TULCEA, C.
[1] Mesures dans les espaces produits, *Atti Acad. Naz. Lincei, Ser. 8 Rend. Cl. Sci. Fis. Mat. Nat.* **7** (1949), 208–211.
IONESCU-TULCEA, A., and C.
[1] "Topics in the Theory of Lifting." Springer–Verlag, Berlin and New York, 1969.
JACOBS, K.
[1] "Neuere Methoden und Ergebnisse der Ergodentheorie." Springer–Verlag, Berlin and New York, 1960.
[2] "Ergodic Theory I, II." Lecture Notes Aarhus, 1962/63.
[3] Der Heiratssatz, *Selecta Math.* **1** (1969), 103–141.
JECH, T.
[1] "Lectures in Set Theory with Particular Emphasis on the Method of Forcing," Lect. Notes in Math. 217. Springer–Verlag, Berlin and New York, 1971.
JENSEN, J. L. W. V.
[1] Sur les fonctions convexes et les inégalités entre les valeurs moyennes, *Acta Math.* **30** (1906), 175–193.
JESSEN, B.
[1] The theory of integration in a space of an infinite number of dimensions, *Acta Math.* **63** (1934), 249–323.
KAKUTANI, S.
[1] On equivalence of infinite product measures, *Ann. of Math.* **49(2)** (1948), 214–224.
KALLIANPUR, G.
[1] The topology of weak convergence of probability measures, *J. Math. Mech.* **10** (1961), 947–969.
KAMKE, E.
[1] "Mengenlehre." de Gruyter, Berlin, 1969.
[2] "Das Lebesgue-Stieltjes'sche Integral." Teubner, Leipzig, 1956.
KAPPOS, D. A.
[1] "Strukturtheorie der Wahrscheinlichkeitsfelder und -räume." Springer–Verlag, Berlin and New York, 1960.
[2] "Probability Algebras and Stochastic Spaces." Academic Press, New York, 1969.
KATOK, A. B., Ja. G. SINAI, and A. M. STEPIN
[1] Theory of dynamical systems and general groups of mappings with invariant measure, *Mat. Anal.* **13** (1975), 129–264 (Russian).
KELLERER, H.
[1] Funktionen auf Produkträumen mit vorgegebenen Marginal-Funktionen, *Math. Ann.* **144** (1961), 323–344.
KELLEY, J. L.
[1] "General Topology." Springer–Verlag, Berlin and New York, 1975.
KHINTCHINE, A. I.
[1] "Mathematical Foundations of Information Theory." Dover, New York, 1957.

KINNEY, J. R.
[1] Continuity properties of sample functions of Markov processes, *Trans. Amer. Math. Soc.* **74** (1953), 280–302.
KÖLZOW, D.
[1] Charakterisierung der Maße, welche zu einem Integral im Sinne von Stone oder von Bourbaki gehören, *Arch. Math.* **16** (1965), 200–207.
[2] Adaptations-und Zerlegungseigenschaften von Maßen, *Math. Z.* **94** (1966), 309–321.
[3] Topologische Eigenschaften des abstrakten Integrals im Sinne von Bourbaki, *Arch. Math.* **17** (1966), 244–252.
[4] "Differentiation von Maßen," Lect. Notes in Math. 65. Springer-Verlag, Berlin and New York, 1968.
KÖNIG, D.
[1] Über Graphen und ihre Anwendungen, *Math. Ann.* **77** (1916), 453–465.
[2] "Theorie der endlichen und unendlichen Graphen." Akad. Verlagsges., Leipzig, 1936.
KÖTHE, G.
[1] "Topologische lineare Räume, 2." Springer-Verlag, Berlin and New York, 1966.
[2] "Topological Vector Spaces." Springer-Verlag, Berlin and New York, 1969.
KOLMOGOROFF, A. N.
[1] "Grundbegriffe der Wahrscheinlichkeitsrechnung." Springer, New York, 1933 and Springer-Verlag, Berlin and New York, 1973.
[2] "Foundations of the Theory of Probability." Chelsea, New York, 1956.
KREIN, M., and D. MILMAN
[1] On extreme points of regular convex sets, *Studia Math.* **9** (1940), 133–138.
KRENGEL, U.
[1] Über perfekte Maße, 57 S., Staatsexamensarbeit Göttingen 1961.
KRICKEBERG, K.
[1] Convergence of Martingales with a directed index set, *Trans. Amer. Math. Soc.*, **83** (1956), 313–337.
KRICKEBERG, K., and CHR. PAUC
[1] Martingales et dérivation, *Bull. Soc. Math. France* **91** (1963), 455–543.
KRYLOFF, N., and N. BOGOLIOUBOFF
[1] La théorie générale de la mesure dans son application à l'étude des systèmes dynamiques de la mécanique non-linéaire, *Ann. of Math.* **38** (1937), 65–113.
KURATOWSKI, K.
[1] Une méthode d'elimination des nombres transfinis des raisonnements mathématiques, *Fund. Math.* **3** (1922), 76–108.
[2] "Topologie I." Warszawa, 1948.
[3] "Topologie II." Warszawa-Wroclaw 1950.
KURZWEIL, J.
[1] On integration by parts, *Czechoslovak Math. J.* **8(83)** (1958), 356–359.
[2] On Fubini theorem for general Perron Integral, *Czechoslovak Math. J.* **23(98)** (1973), 286–279.
[3] On multiplication of Perron-integrable functions, *Czechoslovak Math. J.* **23(98)** (1973), 542–566.
LEBESGUE, H.
[1] Sur une généralisation de l'intégrale définie, *C. R. Acad. Sci. Paris* **132** (1901), 1025.
[2] Intégrale, longeur, aire, Thèse. Bernardoni-Rebeschini, Milano, 1902; *Ann. Mat. Pura Appl.* **7** (1902), 231–359.
[3] "Leçons sur l'intégration et la recherche des fonctions primitives." Gautheir Villars, Paris, 1904.
[4] Sur l'intégration des fonctions discontinues, *Ann. Sci. Ecole Norm. Sup.* **27** (1910), 361–450.

LEVI, B.
[1] Sopra l'integrazione delle serie, *Rend. del. R. Inst. Lomb. Sci. Lett.* **39(2)** (1906), 775–780.
LIAPUNOV, A.
[1] Sur les fonctions-vecteurs complètement additives, *Izv. Akad. Nauk SSSR* **4** (1940), 465–478.
LINDENSTRAUSS, J.
[1] A short proof of Liapounoff's convexity theorem., *J. Math. Mech.* **15** (1966), 971–972.
LOÈVE, M.
[1] "Probability Theory I." Springer–Verlag, Berlin and New York, 1977.
LOOMIS, L. H.
[1] Haar measure in uniform structures, *Duke Math. J.* **16** (1949), 193–208.
[2] "An Introduction to Abstract Harmonic Analysis." Von Nostrand–Reinhold, Princeton, New Jersey, 1953.
[3] Unique direct integral decompositions on convex sets, *Amer. J. Math.* **84** (1962), 509–526.
LUSIN, N.
[1] Sur les propriétés des fonctions mesurables, *C. R. Acad. Sci. Paris* **154** (1912), 1688–1690.
[2] Sur les ensembles analytiques, *Fund. Math.* **10** (1927), 1–95.
[3] "Leçons sur les ensembles analytiques et leurs applications." Gauthier-Villars, Paris, 1930.
LUXEMBURG, W. A. J., and A. C. ZAANEN
[1] "Riesz Spaces I." North Holland Publ., Amsterdam, 1971.
MAAK, W.
[1] Eine neue Definition der fastperiodischen Funktionen, *Abh. Math. Sem. Univ. Hamb.* **11** (1936), 240–244.
[2] "Fastperiodische Funktionen." Springer–Verlag, Berlin and New York, 1950, 2. unveränd. Aufl. 1967.
MAHARAM, D.
[1] On a theorem of von Neumann, *Proc. Amer. Math. Soc.* **9** (1958), 987–994.
MARCZEWSKI, E.
[1] On compact measures, *Fund. Math.* **40** (1953), 113–124.
MARKOFF, A. A.
[1] Extension of the law of large numbers to dependent events, *Bull. Soc. Phys. Math. Kazan* **15(2)** (1906), 135–156 (Russian).
[2] "Calculus of Probability." Moscow, 1924 (Russian).
MCSHANE, E. J.
[1] "Integration." Princeton Univ. Press, Princeton, New Jersey, 1947.
[2] A Riemann-type integral that includes Lebesgue-Stieltjes, Bochner and stochastic integrals, "Memoirs of the American Mathematical Society," Vol. 88, American Mathematical Society, Providence, Rhode Island, 1969.
MEYER, P. A.
[1] Sur les démonstrations nouvelles du théorème de Choquet, Sém. Brelot-Choquet-Deny, 6ième année, No. 7, 1962.
[2] "Probability and Potentials." Ginn (Blaisdell), Boston, Massachusetts, 1966.
[3] "Processus de Markov," Lect. Notes in Math. 26. Springer–Verlag, Berlin and New York, 1967.
MINKOWSKI, H.
[1] "Geometrie der Zahlen I." Teubner, Leipzig, 1896; Vervollst. Ausgabe 1910.
MÜLLER, D. W.
[1] Verteilungs-Invarianzprinzipien für das starke Gesetz der großen Zahl, *Z. Wahrsch.* **10** (1968), 173–192.

[2] Nonstandard proofs of invariance principles in probability theory, "Appl. of Model Th. to Alg., Anal. and Prob." (W. A. J. Luxemburg, ed.), pp. 186–194. Holt, New York, 1969.

NACHBIN, L.

[1] "The Haar Integral." Van Nostrand–Reinhold, Princeton, New Jersey, 1965.

NELSON, E.

[1] Representation of a Markovian semigroup and its infinitesimal generator, *J. Math. Mech.* **7** (1958), 977–987.

NEUMANN, J. v.

[1] Algebraische Repräsentanten der Funktionen "bis auf eine Menge von Maße Null". *Crelle J.* **165** (1931), 109–115.

[2] Proof of the quasi-ergodic hypothesis, *Proc. Nat. Acad. Sci. U.S.* **18** (1932), 263–266.

[3] "Invariant Measures," notes by P. R. Halmos. Princeton Univ. Press, Princeton, New Jersey, 1940–1941.

[4] On rings of operators, reduction theory. *Ann. of Math.* **50** (1949), 401–485.

NEVEU, J.

[1] "Bases mathématiques du calcul des probabilités." Masson, Paris, 1964.

[2] "Mathematical Foundations of the Calculus of Probability." Holden-Day, San Francisco, California, 1965.

[3] Atomes conditionnels d'espaces de probabilité et théorie de l'information, *Symp. Prob. Meth. Anal., Loutraki* (1966) (Kappos, ed.) pp. 256–271, Lect. Notes in Math. 31 Springer–Verlag, Berlin and New York, 1967.

[4] "Martingales à temps discret." Masson, Paris, 1972.

NIKODYM, O. M.

[1] Sur les fonctions d'ensembles, *C.R.I. Congr. Math. Pays Slaves Warszawa* (1929), 304–313.

[2] Sur une géneralization des intégrales de M. J. Radon, *Fund. Math.* **15** (1930), 131–179.

ORNSTEIN, D.

[1] "Ergodic Theory, Randomness and Dynamical Systems." Yale Univ. Press, New Haven, Connecticut, 1974.

OXTOBY, J. C.

[1] Ergodic sets, *Bull. Amer. Math. Soc.* **58** (1952), 116–136.

[2] "Measure and category." Springer–Verlag, Berlin and New York, 1971.

[3] "Maß und Kategorie." Springer–Verlag, Berlin and New York, 1971.

PARRY, W.

[1] "Entropy and Generators in Ergodic Theory." Benjamin, New York, 1969.

PARTHASARATHY, K. R.

[1] "Probability Measures on Metric Spaces." Academic Press, New York, 1967.

PARTHASARATHY, T.

[1] "Selection Theorems and Their Applications," Lecture Notes in Math. 263. Springer–Verlag, Berlin and New York, 1972.

PERRON, O.

[1] Über den Integralbegriff., *S. B. Heidelberger Akad. Wiss. Abt. A* (1914), 14. Abh.

PFANZAGL, J., and W. PIERLO

[1] "Compact Systems of Sets," Lect. Notes in Math. 16. Springer–Verlag, Berlin and New York, 1966.

PHELPS, R. R.

[1] "Lectures on Choquet's Theorem." Van Nostrand–Reinhold, Princeton, New Jersey, 1966.

PRINGSHEIM, A.

[1] Zur Theorie der ganzen transzendenten Funktionen (Nachträge), *Sber. Bay. Akad. Wiss.* **32** (1902), 295–304.

PROHOROV, Yu. V.

[1] Convergence of random processes and limit theorems in probability theory, *Theor. Probability Appl.* **1** (1956), 157–214 (English transl.).

RADO, T.

[1] Sur un problème rélatif à un théorème de Vitali, *Fund. Math.* **11** (1928), 228–229.

RADON, J.

[1] Theorie und Anwendungen der absolut additiven Mengenfunktionen., *S. B. Akad. Wiss. Wien* **122** (1913), 1295–1438.

RAY, D.

[1] Stationary Markov processes with continuous paths, *Trans. Amer. Math. Soc.* **82** (1956), 452–493.

RÉNYI, A.

[1] "Foundations of Probability." Holden-Day, San Francisco, California, 1970.

RICHTER, H.

[1] "Wahrscheinlichkeitstheorie." Springer–Verlag, Berlin and New York, 1966.

RIESZ, F.

[1] Über orthogonale Funktionensysteme, *Nachr. Gött.* (1907), 116–122.

[2] Sur quelques notions fondamentales dans la théorie générale des opérations linéaires, *Ann. of Math.* **41** (1940), 174–206.

RIESZ, F., and B. SZ-NAGY

[1] "Leçons d'analyse fonctionnelle." Akadémiai Kiadó, Budapest, 1952.

ROBERTSON, A. P., and W. ROBERTSON

[1] "Topological Vector Spaces," 2nd ed. Cambridge Univ. Press, London and New York, 1973.

[2] "Topologische Vektorräume." Bibl. Inst., Mannheim, 1967.

ROKHLIN, V. A.

[1] On the fundamental ideas of measure theory, *Mat. Sb.* **25** (67) (1949), 107–150 [English transl.: *Trans. Amer. Math. Soc.* **71** (1952)].

ROSENMÜLLER, J., and H. G. WEIDNER

[1] A class of extreme convex set functions with finite carrier, *Advances in Math.* **10** (1973), 1–38.

[2] Extreme convex set functions with finite carrier: general theory, *Discr. Math.* **10** (1974), 343–382.

ROTA, G.-C.

[1] Une théorie unifiée des martingales et des moyennes ergodiques, *C. R. Acad. Sci. Paris* **252** (1961), 2064–2066.

RUELLE, D.

[1] "Statistical Mechanics. Rigorous Results." Benjamin, New York, 1969.

RYLL-NARDZEWSKI, C.

[1] On quasi-compact measures, *Fund. Math.* **40** (1953), 125–130.

SAKS, S.

[1] On some functionals, *Trans. Amer. Math. Soc.* **35** (1933), 549–556, 965–966.

[2] Addition to the note on some functionals, *Trans. Amer. Math. Soc.* **35** (1933), 967–974.

[3] "Théorie de l'intégrale." Warszawa, 1933.

[4] "Theory of the Integral." Stechert, Warszawa and New York, 1937.

SCHAEFER, H.

[1] "Topological Vector Space." p. 3. Springer–Verlag, New York and Berlin, 1971.

SCHUR, I.

[1] Über lineare Transformationen in der Theorie der unendlichen Reihen, *Crelles J.* **151** (1921), 79–111.

SCHWARTZ, L.
[1] "Radon Measures on Arbitrary Topological Spaces and Cylindrical Measures. Oxford Univ. Press, London and New York, 1973.
SHANNON, C., and W. WEAVER
[1] "The Mathematical Theory of Communication," Univ. of Illinois Press, Urbana, Illinois, 1959.
SINAI, Ja. G.
[1] "Theory of Dynamical Systems," Part I, Ergodic Theory, Lecture Notes, Aarhus, 1970.
SION, M.
[1] "A Theory of Semigroup Valued Measures," Lect. Notes in Math. 355. Springer-Verlag, Berlin and New York, 1973.
SKOROKHOD, A. V.
[1] "Studies in the Theory of Random Processes." Addison-Wesley, Reading, Massachusetts, 1965.
SMORODINSKI, M.
[1] "Ergodic Theory, Entropy," Lect. Notes in Math. 214, Springer-Verlag, Berlin and New York, 1971.
SOUSLIN, M.
[1] Sur une définition des ensembles mesurables B sans nombres transfinis, *C. R. Acad. Sci. Paris* **164** (1917), 88–91.
STEEN, L. A., and J. A. SEEBACH Jr.
[1] "Counterexamples in Topology." Holt, New York, 1970.
STEINHAUS, H.
[1] Some remarks on the generalization of limit, *Prace Mat. Fiz.* **22** (1911), 121–134 (Poln.)
STONE, M. H.
[1] Applications of the theory of Boolean rings to general topology, *Trans. Amer. Math. Soc.* **41** (1937), 375–481.
[2] The generalized Weierstrass approximation theorem, *Math. Mag.* **21** (1948), 167–184.
[3] Notes on integration I-IV, *Proc. Nat. Acad. Sci. U.S.* **34** (1948), 336–342, 447–455, 483–490; **35** (1949), 50–58.
STRASSEN, V.
[1] Meßfehler und Information, *Z. Wahrsch.* **2** (1964), 273–305.
[2] The existence of probability measures with given marginals, *Ann. Math. Statist.* **36** (1965), 423–439.
TAKEUTI, G., and W. M. ZARING
[1] "Introduction to Axiomatic Set Theory." Springer-Verlag, Berlin and New York, 1971.
[2] "Axiomatic Set Theory." Springer-Verlag, Berlin and New York, 1973.
THORIN, G.
[1] Convexity theorems, *Comm. Math. Sem. Univ. Lund* **9** (1948), 1–58.
TOPSØE, F.
[1] Preservation of weak convergence under mappings, *Ann. Math. Statist.* **38** (1967), 1661–1665.
[2] On the connection between P-continuity and P-uniformity in weak convergence, *Teor. Ver. Prim.* [*English transl.*: *Theor. Probability Appl.* **12** (1967), 281–290].
[3] A criterion for weak convergence of measures with an application to convergence of measures on $D[0, 1]$, *Math. Scand.*, to appear.
[4] On the Glivenko-Cantelli theorem, *Z. Wahrsch.* **14** (1970), 239–250.
[5] Compactness in spaces of measures, *Studia Math.*
[6] "Topology and Measure," Lecture Notes in Math. 133. Springer-Verlag, Berlin and New York, 1970.

TYCHONOV, A.

[1] Über die topologische Erweiterung von Räumen, *Math. Ann.* **102** (1929), 544–561.

[2] Über einen Funktionenraum, *Math. Ann.* **111** (1935), 762–766.

URYSOHN, P.

[1] Über die Mächtigkeit des zusammenhängenden Raumes, *Math. Ann.* **94** (1925), 262–295.

[2] Über Metrization des kompakten topologischen Raumes, *Math. Ann.* **92** (1924), 275–293.

VARADARAJAN, V. S.

[1] Weak convergence of measures on separable metric spaces, *Sankhyā* **19** (1958), 15–22.

[2] Measures on topological spaces, *Mat. Sb.* **55** (1961), 35–100 (Russian).

VITALI, G.

[1] Sull' integrazione per serie, *Rend. Circ. Mat. Palermo* **23** (1907), 137–155.

[2] Sui gruppi di punti e sulle funzioni di variabili reali, *Atti. Acad. Sci. Torino Cl. Sci. Fis. Rend.* **43** (1908), 75–92.

[3] Sulle funzioni integrali, *Atti. Accad. Sci. Torino Cl. Sci. Fis. Rend.* **40** (1905), 753–766.

VOGEL, W.

[1] "Lineares Optimieren." Akad. Verlagsges., Leipzig, 1967.

WALTERS, P.

[1] "Ergodic Theory—Introductory Lectures. Lecture Notes in Math. 458. Springer–Verlag, Berlin and New York, 1975.

WARD, A.

[1] The Perron-Stieltjes integral, *Math. Z.* **41** (1936), 578–604.

WEIERSTRASS, K.

[1] Über die analytische Darstellbarkeit sogenannter willkürlicher Functionen reeller Argumente, *Sb. Kgl. Akad. Wiss. Berlin* (1885), 633–639, 789–805.

WEIL, A.

[1] "L'intégration dans les groupes topologiques et ses applications," 2nd éd. Hermann, Paris, 1951.

WEYL, H.

[1] Über die Gleichverteilung von Zahlen mod 1, *Math. Ann.* **77** (1916), 313–352.

[2] Almost periodic invariant vector sets in a metric vector space, *Amer. J. Math.* **71** (1949), 178–205.

WOLFOWITZ, J.

[1] "Coding Theorems of Information Theory," 2nd ed. Springer–Verlag, Berlin and New York, 1964.

YOSIDA, K.

[1] Vector lattices and additive set functions, *Proc. Imp. Acad. Tokyo* **17** (1941), 228–232.

YOSIDA, K., and E. HEWITT

[1] Finitely additive measures, *Trans. Amer. Math. Soc.* **72** (1952), 46–66.

ZORN, M.

[1] A remark on method in transfinite algebra, *Bull. Amer. Math. Soc.* **41** (1935), 667–670.

INDEX

C

D

E

F

J

K

L

Q

R

S

T

U

Probability and Mathematical Statistics

A Series of Monographs and Textbooks

Thomas Ferguson. Mathematical Statistics: A Decision Theoretic Approach. 1967

Howard Tucker. A Graduate Course in Probability. 1967

K. R. Parthasarathy. Probability Measures on Metric Spaces. 1967

P. Révész. The Laws of Large Numbers. 1968

H. P. McKean, Jr. Stochastic Integrals. 1969

B. V. Gnedenko, Yu. K. Belyayev, and A. D. Solovyev. Mathematical Methods of Reliability Theory. 1969

Demetrios A. Kappos. Probability Algebras and Stochastic Spaces. 1969

Ivan N. Pesin. Classical and Modern Integration Theories. 1970

S. Vajda. Probabilistic Programming. 1972

Sheldon M. Ross. Introduction to Probability Models. 1972

Robert B. Ash. Real Analysis and Probability. 1972

V. V. Fedorov. Theory of Optimal Experiments. 1972

K. V. Mardia. Statistics of Directional Data. 1972

H. Dym and H. P. McKean. Fourier Series and Integrals. 1972

Tatsuo Kawata. Fourier Analysis in Probability Theory. 1972

Fritz Oberhettinger. Fourier Transforms of Distributions and Their Inverses: A Collection of Tables. 1973

Paul Erdös and Joel Spencer. Probabilistic Methods in Combinatorics. 1973

K. Sarkadi and I. Vincze. Mathematical Methods of Statistical Quality Control. 1973

Michael R. Anderberg. Cluster Analysis for Applications. 1973

W. Hengartner and R. Theodorescu. Concentration Functions. 1973

Kai Lai Chung. A Course in Probability Theory, Second Edition. 1974

L. H. Koopmans. The Spectral Analysis of Time Series. 1974

L. E. Maistrov. Probability Theory: A Historical Sketch. 1974

William F. Stout. Almost Sure Convergence. 1974

E. J. McShane. Stochastic Calculus and Stochastic Models. 1974

Robert B. Ash and Melvin F. Gardner. Topics in Stochastic Processes. 1975

Avner Friedman, Stochastic Differential Equations and Applications, Volume 1, 1975; Volume 2. 1975

Roger Cuppens. Decomposition of Multivariate Probabilities. 1975

Eugene Lukacs. Stochastic Convergence, Second Edition. 1975

H. Dym and H. P. McKean. Gaussian Processes, Function Theory, and the Inverse Spectral Problem. 1976

N. C. Giri. Multivariate Statistical Inference. 1977

Lloyd Fisher and John McDonald. Fixed Effects Analysis of Variance. 1978

Sidney C. Port and Charles J. Stone. Brownian Motion and Classical Potential Theory. 1978

Konrad Jacobs. Measure and Integral. 1978